Principles of Wood Science and Technology

II

Wood Based Materials

Franz F. P. Kollmann   Edward W. Kuenzi
Alfred J. Stamm

Springer-Verlag New York Heidelberg Berlin 1975

Franz F. P. Kollmann

o. ö. Professor (em.) an der Universität München
Institut für Holzforschung und Holztechnik

Edward W. Kuenzi

Professor at the U.S. Forest Products Laboratory
Madison, Wis.

Alfred J. Stamm

Professor (em.) at the North Carolina State University
School of Forest Resources
Department of Wood and Paper
Science, Raleigh, N.C.

ISBN 0-387-06467-2 Springer-Verlag New York Heidelberg Berlin
ISBN 3-540-06467-2 Springer-Verlag Berlin Heidelberg New York

PREFACE

The principal author published the first edition of a book entitled in German "Technologie des Holzes" (Springer-Verlag) in the year 1936. This book was available only in photostat copy form in the United States during World War II. A translation into the English language was later made in Australia. Due to many hampering factors caused by World War II it was not until 1951 that the second edition of the first volume of this book with the enlarged title (translated) "Technology of Wood and Wood Based Materials" was published. The second volume followed in 1955.

The enormous expansion of wood science and technology within less than two decades is evident from a comparison of the number of pages of these two editions. The first edition comprises 764 pages, the two volumes of the second edition 2231 pages.

More recently, it became evident that an English version in a condensed up-to-date form was needed. Therefore, the principal author decided to publish a book on "Principles of Wood Science and Technology" in English. The first volume dealt with solid wood, this second volume deals with wood based materials. Using the word "Principles" in the title of the condensed book this led directly to abandoning all detailed historical reviews, explanations, and discussions of machines and processes.

The principal author asked for direct and indirect assistance by outstanding experts in compiling both volumes. He was happy that Prof. *W. A. Côté*, Jr. wrote chapters on structure, fine structure, chemistry, and pathology of solid wood for the first volume. In this second volume he obtained valuable contributions of complete chapters by Prof. *A. J. Stamm* on "Solid Modified Wood" and Mr. *Edward W. Kuenzi*, P. E., on "Sandwich Composites". In writing four chapters of this second volume, the principal author is extremely grateful that colleagues abroad reviewed and corrected, as far as necessary, his original drafts. He wishes to express his sincere gratitude to

Prof. *A. P. Schniewind* (Chapters "Adhesion and Adhesives" and "Fiberboard"),
Mr. *F. C. Lynam* (Chapters "Veneer, Plywood, and Laminates" and "Particleboard"),
Prof. *Raymond A. Moore* (Chapter "Veneer, Plywood, and Laminates"),
Prof. *Alan A. Marra* (Chapter "Particleboard"), and
Prof. *A. J. Stamm* (Chapter "Fiberboard").

Mr. *A. Kaila* kindly gave permission to make use of the content of the chapter "Boardindustrien" (p. 821 to 1109), contributed by him to "Träindustriell Handbok" (Ed.: *O. Heikinheimo*; Stockholm, 1968: AB Svenks Trävarutidning). Mr. *S. Åke Lundgren* also kindly gave permission to use his article "Wood-based sheet as a structural material", published by the Swedish Wallboard Manufacturers' Association in the booklet "Swedish Fibreboard Information", Stockholm 1969.

The author is furthermore indebted to colleagues in Germany: Dr. *Dietger Grosser*, Dr.-Ing. *Max Kufner*, Prof. Dr. *Erich Plath*, Dr. *Eberhard Schmidt*, and

Dr. *Reinwald Teichgräber*. He also received printed material and letter communications from the industry concerned either with production of wood based materials or with manufacture of machinery and equipment for these industries.

Miss *R. Preissler* wrote the numerous letters necessary in conjunction with the whole work and wrote the drafts and final manuscripts of the chapters. Drafts of pictures and diagrams were made with skill and care by Ing. *H. Sanzi* and the many necessary mimeographed copies of sketches and diagrams and glossy prints by Mrs. *H. Bauer*. The authors of this book are sincerely indebted to Springer-Verlag publishers of the volumes in the high standard of printing and illustrating admired throughout the world.

München, Germany   **Franz F. P. Kollmann**
Madison, Wis., U.S.A.   **Edward W. Kuenzi**
Raleigh, N.C., U.S.A., 1974   **Alfred J. Stamm**

CONTENTS

Contents   VII

1. ADHESION AND ADHESIVES FOR WOOD

1.0 General Considerations about Gluing

Adhesion, adhesives and adherents are elements of a general adhesion technology which is of great importance to many fields of engineering, including the gluing of wood. Glue joints play and important role in economical woodworking since they allow the improvement of blockboards and of lowgrade plywood cores by veneering, the manufacture of rather complicated constructions, the production of plywood and laminated board, beams and arches, the utilization of smaller wood sections and the binding in particleboards, Practically no wood based material — except densified solid wood and flexible wood — is manufactured without gluing of sheets or particles.

Stumbo (1965) compiled an interesting historical table. In the introduction to this table he wrote: "The art of gluing was developed long before men began to record history." He pointed out that mud and clay were probably the first adhesive substances, waxes and resins followed, and later blood, eggs, casein and boiled hides and bones were used. By 4000 B. C. resins and bituminous cements were already in use. At the time of the Pharaohs the application of casein adhesives, probably also fish or animal glues and starches were known. The Goths bonded coins to wooden boxes with adhesives based on egg white and lime. Pliny the Elder reports in the Natural History on the veneering of citrus wood and the use of beech veneers. After 400 A. D. until about 1500 A. D. the art of gluing fell into disuse. Between 1500 and 1700, veneering, probably with animal glues, and inlaying flourished again. The first patent for manufacturing fish glue was issued in Britain 1754. Then various glues were rapidly developed: about 1800 commercial manufacture of casein glue in Switzerland and Germany; in 1814 a U.S.A. patent was granted for an adhesive from animal bones. In the following hundred years rubber cements, water resistant casein cements, starch adhesives, phenolic resins, blood albumin glues followed. After World War II there was increased interest in the U.S.A. in soybean adhesives. In the year 1929 the firm Th. Goldschmidt at Essen obtained patents on phenolic films. The TEGO-films were introduced in Europe in the year 1934. *Jarvi* (1967) dealt with exterior glues for plywood. He mentioned that the TEGO-film used in the 1930'ies in Germany was too expensive and not convenient to use by West Coast Plywood Manufacturers. In the early 1930'ies *Nevins* developed a liquid phenolic resin for plywood. His resins were obtained by the sodium hydroxide catalyzed reaction between formaldehyde and cresols and xylenols. Press temperature exceeded 149 °C (300 °F).

Urea-formaldehyde adhesives came into use in 1937. Resorcinol-formaldehyde adhesives and epoxy adhesives as well as polyurethanes were developed during World War II. The formerly difficult problem of gluing metal sheets to wood has been solved. *Halligan* (1969) discussed recent glues and gluing research applied to particleboard.

Adhesives can be classified according to several systems. One such system is given below with the chemical composition of their main components:

Natural glues and allied products:

1. Starches, dextrins and vegetable gums;

2. Protein glues;
2.1 Animal glues, made from hides, sinews, bones, hoofs, horns, fish-skins;
2.2 Blood albumin adhesives made from the soluble blood albumin, a by-product in slaughter-houses;
2.3 Casein adhesives made from the curd of soured milk, mixed with lime and other chemical ingredients;
2.4 Vegetable protein glues on the basis of soybean flour, peanuts a. s. o.;

3. Shellac;

4. Rubber, synthetic rubber;

5. Asphalt, mastix;

6. Sodium silicate, magnesium oxychloride etc.

Synthetic glues:

1. Thermosetting resins (on the basis of urea, melamine, phenols, resorcinol, furan, epoxy, unsaturated polyesters);

2. Thermoplastic resins (cellulose esters and ethers, alkyd and acryl esters, polyamide, polystyren, polyvinyl alcohols, and derivates).

This list ist complied with many alterations according to *Blomquist* (1963). *Blomquist* remarks that the system given above works well for the simpler adhesives, but does not cover many new formulations in which several different components are combined. An example are metal bonding adhesives consisting of a phenolic resin with a synthetic rubber and/or a thermoplastic resin.

1.1 Physical-Chemical Principles of Gluing

1.1.1 Cohesion and Adhesion

An understanding of the fundamentals of gluing developed only slowly. Now we know the close relation between cohesion and adhesion. Cohesion is caused by the forces of attraction between atoms or molecules. In the process of separation energy is consumed. Only thus can the forces of attraction be overcome and simultaneously new surfaces be created. This is the reason why we can estimate the amount of cohesion forces based on the tearing strenght.

Fundamentally the forces of cohesion and of adhesion are identical. Generally adhesion is the adhesive power of adjacent molecules. This adhesive power can only be developed if the molecules are sufficiently close together (distance below 3×10^{-8} cm). In this proximity molecules can interact. For solid bodies such an extremely small distance between the molecules of two parts is nearly impossible to achieve since even surfaces after lapping are too irregular and are contaminated by oxidation, humidity and dust.

In practice adhesion is effected by the use of liquid glues which can adapt to the profile of the surface due to their rheological behavior. At present all technical procedures of gluing solid bodies use liquid glues. They wet the surfaces to be glued and they form glue joints after having been set or hardened. Even glue films and contact stickers during the gluing process are liquid or have a quasi-liquid phase.

According to *McBain* (1922, 1926, 1932), a liquid which is able to wet the surface of a solid body and which afterwards can be hardened may be considered a glue. The same author distinguishes:

a) *Mechanical adhesion* caused by the interlocking network of hardened glue in the joint and in the pores of the wood;

b) *Specific adhesion*, based on molecular forces.

With respect to mechanical adhesion the entrance of liquid glue into wood pores is facilitated by capillary forces. Microscopic investigations show that the glue is penetrating vessels and other hollow cells of the wood before it is hardened. From there follows a three-dimensional mechanical branching of the hardened threads or fingers of glue. If one tries to separate the glued parts then one has to overcome the shearing resistance of this network. Fig. 1.1 shows a proper glue joint in basswood plywood made with an animal glue of a correct consistency ("slightly chilled") and pressure. If one applies glue with too low a concentration and/or one chooses the pressure too high, then the glue oozes away and the result are "starved joints" (Fig. 1.2).

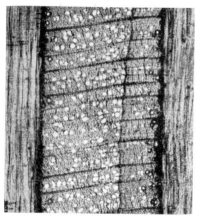

Fig. 1.1. Cross section through basswood plywood glued with animal glue of correct concentration and appropriate pressure; good joint. From *Truax* (1929)

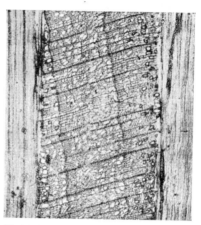

Fig. 1.2. Cross section though basswood plywood glued with animal glue of too low concentration and with too high pressure; typical "starved joint". From *Truax* (1929)

Mechanical adhesion is produced by penetration into open pores on the surfaces of wood. *Voss* (1947) and *Browne* and *Truax* (1926, p. 258—269) pointed out that the mechanical effect contributes only a small part to the total joint strenght in the case of wood. "If the penetrating tendrils of the adhesive were mainly responsible for bond, it would have to follow that the adhesive itself was highly resistant to tensile and shear forces. This we know is not true, in most cases. It is quite likely that large ions, especially those which have incomplete or unsaturated outer electronic orbits, may change their shape and dimensions under the influence of the electrons of neighboring ions. This may cause concentrations of repulsion and subsequent attraction of considerable extent." (*Voss*, 1947, p. 22/23). A diagrammatic sketch of bonding two different materials A and B having different surface structures with an adhesive C is given in Fig. 1.3.

Condensed in part from *Rinker* and *Kline* (1946, p. 12), Table 1.1 gives a survey of the nature of bonding.

Koch (1967) examined the minimizing and predicting delamination of southern plywood in exterior exposure. He cleared the interrelationship of seven variables (eight including time exposure) affecting glue bond durability, the merit of percentage of wood failure and several other parameters as predictor of durability.

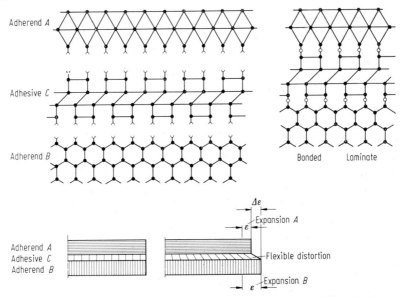

Fig. 1.3. Diagrammatic sketch of bonding two different materials *A* and *B* having different surface structures with an adhesive *C*. From *Voss* (1947)

Table 1.1. Nature of Bonding

Bond	Adhesive	Type of Bond
Wood	Phenolic resin, urea-formaldehyde, casein, animal glue	Phenolic, hydrogen bridging hydroxyl groups; water molecules induce hydroxyl groups for others
Metal	Rubber and phenolic resin	Hydrogen bridging between hydroxyl groups (wood and resin); resin and rubber, complex structure
Wood to rubber	Rubber and phenolic resin	Hydrogen bridging (wood surface); interdiffusion (rubber surface)
Plastics	Phenolic resin or compatible resin	Hydrogen bridging between hydroxyl groups. Interdiffusion at plastic face if thermoplastic. Hydrogen bridging at plastic face if thermoset.

Hse (1968) investigated the gluability of southern pine, earlywood and latewood. Glue-bond quality as tested wet and dry in tension, was best with earlywood and poorest with latewood to latewood; earlywood to latewood was intermediate. Earlywood cells in the vicinity of the glue line were compressed and impregnated with resin. These cells form the transition layer between glue

line and the undeformed wood substrate. The dense, thick-walled latewood showed
no such cell deformation and resin impregnation was confined to the cells immediately adjacent to the glue line.

Originally there was the opinion that the mechanical adhesion is the most
important and effective factor causing strong glue joints. Recently the specific
adhesion gained more importance.

Dibuz and *Shelton* (1967) discussed the question of glue line identification.
Through the techniques of infrared spectrophotometry and microchemical
analysis in most cases positive identification of the adhesives can be made.

1.1.2 Intermolecular Forces, Polarity

Between atoms and molecules electrical forces (London-, Debye- and Keeson-
forces) (*de Bruyne* and *Houwink*, 1951, p. 18) are acting. As a rule one distinguishes
(*de Bruyne* 1951, p. 1) two classes of attracting forces:

1. Primary valences;
2. Secondary valences or partial valences.

The primary valences or bonding forces, ranging from 10 to 100 or more
kcal/mole are strong forces of attraction between atoms. They determine the
structure of molecules and appear in some chemical processes. They are without
any importance for the phenomenon of adhesion.

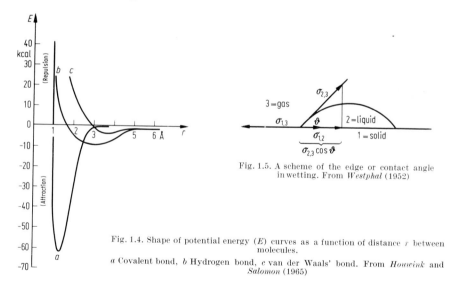

Fig. 1.5. A scheme of the edge or contact angle in wetting. From *Westphal* (1952)

Fig. 1.4. Shape of potential energy (*E*) curves as a function of distance *r* between molecules.

a Covalent bond, *b* Hydrogen bond, *c* van der Waals' bond. From *Houwink* and *Salomon* (1965)

The weaker secondary valences (2 to 4 kcal/mole) are much less directional
than covalent forces (*Salomon*, in *Houwink* and *Salomon*, 1965, p. 13); they
decide the amount of adhesion (Fig. 1.4). They cause the main resistance against
sliding of molecules or micells. Without the existence of secondary valences
neither the solid nor the liquid state of material would be possible.

Secondary valences cause surface tension in liquids. Surface tension can formally be described as a tangential force which tries to reduce the surface. Due to
surface tension a small amount of falling liquid forms a drop. If such a drop is
put on a surface of a solid body it becomes more or less flat. A measure for the

wetting is the edge or contact angle ϑ. Fig. 1.5 illustrates the phenomenon and the interrelationships between the various angles. Between the three surface tensions there exists an equilibrium.

According to *Young* (1805) and *Dupré* (1869) the following equation holds (*Westphal* 1952, p. 700/701)

$$\sigma_{1.3} - \sigma_{1.2} = \sigma_{2.3} \cos \vartheta = \sigma_{\text{Adh}}, \tag{1.1}$$

where $\sigma_{1.3}$ is the surface tension between the two phases solid/gaseous, $\sigma_{1.2}$ between the phases solid/liquid, $\sigma_{2.3}$ between the two phases liquid/gaseous and σ_{Adh} is the tension of adhesion.

Following *Dupré*'s thermodynamic consideration the energy of adhesion A_σ may be expressed by the term

$$A_\sigma = \sigma_{1.3} + \sigma_{2.3} - \sigma_{1.2}. \tag{1.2}$$

Though it is not possible to measure exact values of $\sigma_{1.3}$ and $\sigma_{1.2}$, one can combine Eq. (1.1) with Eq. (1.2). The result is an equation which makes it possible to determine the energy of adhesion A_0 by experiments:

$$A_\sigma = \sigma_{2.3}(1 + \cos \vartheta). \tag{1.3}$$

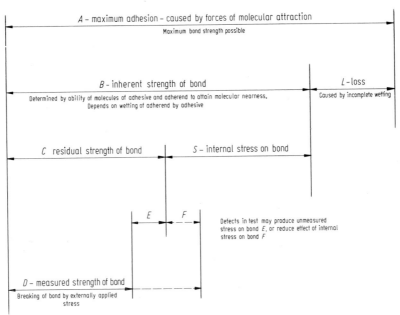

Fig. 1.6. Relations between the factors involved in specific adhesion. This chart is qualitative, not quantitative, no significance is attached to the relative lengths of the lines. A is always greater than B, and B greater than C. From *Reinhart* (1954)

A qualitative scheme of the relations between the factors involved in specific adhesion is shown in Fig. 1.6 (*Reinhart*, 1954, p. 128).

Kitazawa (1946) found that the various strata of the secondary cell wall, the primary wall and the middle lamella very probably cause different adhesion forces. Fig. 1.7, designed hypothetically, illustrates the possible adhesive attractions which are in force in the adjoining excised two cell walls. The secondary wall consists of: a) an inner, b) a middle and c) an outer stratum. The primary wall is schematically represented by d) and the middle lamella by e). The total thickness of all strata is t. The height of the cell wall is assumed as $h = 1$. A_2 represents the unexcised cell wall area. Dotted lines between upper and lower diagrams relate positions of strata. The lower diagram is a hypothetical graph of adhesion stress values which represent relative magnitudes, not real values. Nevertheless, the total force of adhesion acting on an area of unit height across

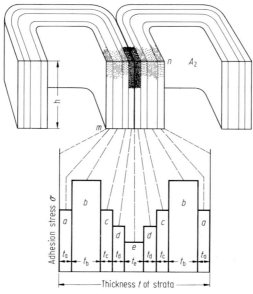

Fig. 1.7. Hypothetical graph of the possible adhesive attractions in force in the several layers of two contiguous cells and in secondary layer. Upper diagram showing two adjoining excised cell walls. Secondary wall consisting of a inner stratum, b central stratum and, c outer stratum. Compound middle lamella consisting of d primary wall and e middle lamella; h is the height of the excised cell wall area; A_2 the unexcised cell wall area. The lower diagram showing hypothetical graph of adhesion stress values (σ) existing on the parts of two wall layers according to the upper diagram. From *Kitazawa* (1946)

the compound cell wall can be calculated approximately, or better estimated, on this hypothesis. Reliable values of the shear stress in the different strata, as shown in Fig. 1.7, cannot be found in the literature. Only the strength of adhesion between glue and excised wood surfaces has been measured. Van der Waals' forces and chemical bonds between the adhesive and the wood are cooperating. This is a summary explanation of specific adhesion. Evaluating the lower diagram in Fig. 1.7, one can assume that the stress acting on the central stratum of the secondary wall is of greater magnitude than the other stresses. The conclusion is simple: the adhesive strength exerted by an excised wall is more or less controlled by the central stratum. To make the considerations complete for wood, account must be taken of the two following factors: 1. The force of adhesion between the adhesive and the surface of the unexcised inner stratum of the secondary wall presented in the cell cavity. 2. The forces of resistance in shear of the glue projecting into cavities whose wall area is represented by A_2. This shear of glue (mechanical adhesion) may be induced by the shape and alignment of the elements forming the wood tissues.

The total adhesion in wood is therefore dependent on specific adhesion and mechanical adhesion. *Marian* (1953) also has investigated the importance of the wetting conditions between the different layers in cell walls and the strenght of adhesion.

The measurement of the wetting angle or edge angle is doubtful because this angle changes immediately after the falling of the first drop of a liquid adhesive on any porous surface. One has to consider the different strenght of adhesion between excised cell walls and unexcised or natural cell walls, the former generally being greater than the latter.

A defective glue line usually is caused by poor contact and air pockets. Crazing is present in urea-formaldehyde glue lines and is absent in resorcinol-formaldehyde glue lines. Penetration of glue occurs mainly throughout cell ends. The depth of penetration is a function of the grain angle. The total adhesion is a function of the nature of the porous surface. The type of adhesives used determines via the necessary gluing pressure the plasticity (deformation) of the wood. Solid wood shear strength is related linearly to specific gravity. The shear behavior allows to compute to some extent the comparative glue strenght. Glue efficiency may be estimated under certain conditions by the coefficient of correlation between shear strenght and fiber failure.

The forces of molecular attraction (van der Waals' forces) play an important role in the adhesion between wood an glue. Many scientists tried to investigate the nature of this bond. *Pauling* and *Corey* (1951) have dealt with this problem and the covalent links in chemistry were the objects of various scientists. The results are not very encouraging since many transitions and intermediate states are combined, so we can only by abstraction and idealization approach the difficult problem. *Mark* (1943) has calculated the stresses to overcome the van der Waals' forces.

As a simplification one can say that secondary forces between atoms and molecules are either "polar forces" or "nonpolar forces". These forces are electrical. The decisive factor is the unequal distribution of residual attractive energies which are relatively great for some molecules. Such molecules have a permanent "dipole moment". Polar molecules have an asymmetrical structure. Polar solutions are characterised by a high dielectric constant and a big power factor. Table 1.2 gives a condensed survey of various polar and nonpolar chemicals.

Table 1.2. Polar and Nonpolar Chemicals

Polar chemicals	Nonpolar chemical
Cellulose (wood, cotton, paper)	Caoutchouc
Phenol-formaldehyde- and	Polystyrene
Urea-formaldehyde-Resins	
prior to hardening	Polyethylene
Materials with hydroxyl groups in the molecule	Teflon (polymers of tetrafluorethylene)
Water	Benzene
Alcohol	Mineral oils
Oxides of metals	

de Bruyne (1951), one of the pioneers in the researches of adhesion, has stated "An Adhesion Rule". Strong joints can never be made to polar adherents with nonpolar adhesives or to nonpolar adherents with polar adhesives. Some adhesives like fish glue will stick to almost anything, but fish glue is a mixture of proteins and other materials of uncertain composition. The most important synthetic glues which are based on urea-formaldehyde or phenol-formaldehyde condensation products are strongly polar adhesives which adhere well to polar materials but poorly to metals and fully hardened bakelites.

1.1.3 Influence of Temperature, Concentration and Amount of Polymerization

For most liquids the rate of change of surface tension σ is a constant, equal to

$$\frac{d\sigma}{dT} = -K. \tag{1.4}$$

Delmonte (1947, p. 325) has explained that the use of heat in curing the adhesive and causing it to flow before setting contributes to lessening its surface tension. *Gibbs* (1876, p. 439) has stated a relationship between the surface tension σ and the concentration c for a solution:

$$\frac{d\sigma}{dc} = -\frac{\mu RT}{c}, \tag{1.5}$$

where R = gas constant (8.314 · 10^7 erg/°C mole), T = absolute temperature and c = concentration, μ = fluid on the surface in g mole per cm². Most chemical glues are not pure but mixtures.

Under this point of view the surface tension is of high importance since it can be lowered remarkably by the addition of some materials. Adhesion is improved if fractions with low molecular weights are used since they are chemically more active and have a better penetration capacity.

In this respect it may be mentioned that in a limited range the transition from high to low molecular weights facilitates the penetration. The relationships are complicated and reference is made to *Delmonte* (1947, p. 326) who gave a survey of the problems, knowing and expressing that it is incomplete. Even 20 years later many details of adhesion are not fully investigated, perhaps for the reason that some adhesives are combined materials. Nevertheless, for thermoplastic and high polymer materials, e.g. polyvinyl esters and derivatives of cellulose, the best are obtained with medium molecular weights. Even glues from animal hides and bones produce the best joints if their viscosity is average.

Armonas (1970) described gel chromatography. Polymer materials are molecular mixtures, containing molecules of varying size and weight, hence the term "average molecular weight" must be used in describing their structural composition. Molecular weight distribution affects such mechanical properties as tensile strenght, elongation, impact strenght and elasticity. Gel permeation chromatography is a rapid procedure for separating polymer molecules according to their molecular size.

1.1.4 Influence of pH Value on the Hardening of Glue Joints

The influence of the pH value on hardening glue joints is remarkable. This problem has been investigated very early. The result of the investigation is that strong alkalies or acids reduce the strenght of the joint. *Delmonte* (1947, p. 349) mentioned that cellulosic materials like wood are especially affected. Fig. 1.8 shows the interrelationship between pH value and pressing time.

Machined wood surfaces of 8 tropical wood species were treated by *Chen* (1970) with a 10% solution of sodium hydroxide, acetone, and alcohol benzene, respectively. After reconditioning, wood blocks were glued with urea-formaldehyde and resorcinol-formaldehyde resin adhesives. Untreated wood blocks were glued for comparison, and joint strength was determined by the ASTM standard glue block shear test. Extractive content and pH determinations were made and wettability was measured by advancing contact angle with distilled water. Adhesive joint strength was improved by all treatments in all but one species. Extractive removal treatment improves wettability and increases pH of the wood. A positive linear correlation exists between wettability and joint strength of blocks glued with urea-formaldehyde, not for blocks glued with resorcinol.

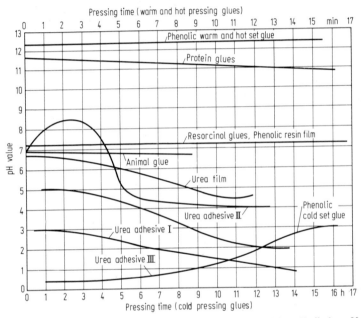

Fig. 1.8. Dependence of pH values of various glues for wood on pressing time. Schematically from *Marian* (1953)

1.1.5 Colloidal State of Wood Glues, Sols and Gels

Practically all glues for wood have relatively high molecular weights. We have to distinguish and differentiate between the following conditions:

1. Colloidal solutions pass through very fine filters (ultrafilters).
2. Colloidal parts will not diffund and dialyze or only very slowly.
3. Colloidal solutions exert an extraordinarily low osmotic pressure and therefore lead to a reduction of the freezing point.
4. It is impossible to detect such colloidal parts by an ordinary light microscope.
5. Under the influence of electrical fields the colloidal parts move in the solvent.

We also have to distinguish between sols and gels. Most wood glues are dissolved in water, but occasionally such adhesives as phenolic resins are dissolved in alcohol.

Some glues are viscous like syrup and they have the tendency to convert into more or less gel state, that means strength building or slimy state.

The hardening of synthetic resin glues depends on the catalysts. It is easy to influence this process by relatively low pH values, but now we know that very low pH values have a deleterious effect on wood.'The wood fibers on both sides of the glue joints at some time and under certain conditions may be affected; the wood joints fail.

1.1.6 Rheology of Glues

As mentioned already glues used for woodworking have rather high molecular weights. This affects the rheological behavior of the binding agents. Certainly one must know that gluing is a process which takes place in several stages. The liquid binding agent is spread on the surfaces to be glued. It is necessary to create a coherent layer. The thickness of this layer should be as thin as possible according to recent investigations. The contact angle can be neglected because it is a changing and doubtful property.

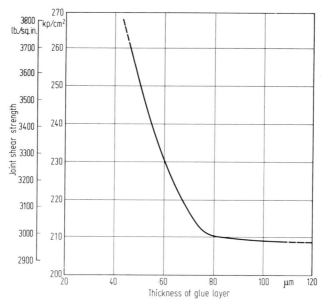

Fig. 1.9. Dependence of the shear strength of glue joints on the thickness of the glue line. From *Maxwell* (1945)

The surfaces of the wood to be glued, even after planing or sanding, to some extent are rough. Therefore the tracheids or vessels, cut on, allow the penetration of glue. This is the so-called mechanical adhesion. The glue penetrating into such holes, after having hardened, forms fingers or tentacles. American investigations (*Browne* and *Truax*, 1929, p. 74, *Browne*, 1931, p. 290) proved that no direct inter-relationship exists between the strength of a joint and glue line thickness. Of course, the glue joint must have an adequate thickness. In contrast *McBain* and *Lee* (1932) have formulated the law that the heigth of bonding strength of glue joints increases with decreasing thickness of glue lines. Later on *Maxwell* (1945) (Fig. 1.9) and *Farrow, Hamley* and *Smith* (1946) and *Plath* (1951) came to similar results. *Poletika* (1943) found that the shear strength in the joints of plywood is approximately inversely proportional to the thickness of the glue line.

The theory is not complete insofar as

1. The stresses of the shrinkage are higher in a thicker glue line (*Hoekstra* and *Fritzius*, 1951);

2. The probability of faults in very thin glue lines is lower than in thicker ones (*Bikerman*, 1941, 1947);

3. Creeping or plasticity is remarkably reduced in thin glue lines (*Zschokke* and *Montandon*, 1944).

Too deep a penetration of the glue into the wood is unfavorable by two rea-sons: It produces a waste of glue and starved joints are the result. In this respect it is necessary to adjust the pressure during the gluing operation to the viscosity of the liquid glue. One has to consider that the viscosity of the glue during the pressing process is remarkably varying. Fig. 1.10 (Aero-Research Techn. Notes, 1943) shows that the viscosity at first during heating period decreases rapidly up to point A, the chemical reaction between glue and hardener then commences, the viscosity remains nearly constant during the following period, but as point B is reached hardening proceeds rapidly and the glue becomes a solid body. Under each particular set of conditions, the kind of glue, consistency, temperature and pressure, and hardener or other chemicals added, must be thoroughly studied.

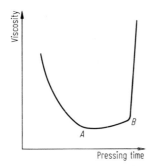

Fig.1.10. Interdependence between viscosity of an urea-formaldehyde resin glue and pressing time in hot gluing plywood. From Aero Research Ltd (1943)

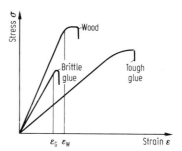

Fig. 1.11. Stress-strain diagram for wood, brittle and tough glue. From *Plath* (1951)

The change of the liquid glue with low viscosity to a near-gel is always present and essential for the gluing process. The transitions are complex and complicated. The dispersion agent continously is diminished by diffusion and evaporation and the effect — often surprising — is a rapid hardening.

Thermoplastic glues are not to be compared with thermosetting ones. They deliver rather strong joints, even able to sustain shock influences, but in the case of long duration loading and/or warming there is danger of creep and the joints may lose their strength. From this point of view such thermoplastic glues as polyvinyl acetates and polyvinyl chlorides are applicable only in a very limited range of temperatures.

It is interesting to study the mechanical interrelationships between wood and hardened glues. Some of the glues are rather brittle, others are tough. Fig. 1.11 elucidates the facts. There is a simple mathematical relationship which must be understood as an introduction

$$\left(\frac{\sigma}{E}\right)_W = \left(\frac{\sigma}{E}\right)_G, \qquad (1.6\,\mathrm{a})$$

where σ is the stress applied (kp/cm²) and E the modulus of elasticity or Young's modulus (kp/cm²). The indices have the following meaning: W = wood, G = glue.

In the case that glue joints should be obtained which when overloaded induce failures in the wood and not in the glue lines, then the breaking stresses in wood (σ_{W_b}) and in the glue line (σ_{G_b}) must be introduced into Eq. (1.6a), so

$$\left(\frac{\sigma}{E}\right)_{W_b} \leqq \left(\frac{\sigma}{E}\right)_{G_b}. \qquad (1.6\,\text{b})$$

Generally glues should be applied producing joints with high strains at the breaking stresses, i.e. "tough glues". The modulus of elasticity of the fully hardened glue always should be lower than that of the material glued. Other factors are the rheological properties of the glues influenced by the loss of the dispersion agents (in unfavorable cases this may last a few months), and the reaching of moisture content equilibrium with the surrounding atmosphere. The conditions are complicated since at first the moving of the dispersion agent is accompanied by a wetting and swelling of the wood layers near to the glue lines. Then, of course, drying and shrinkage of the glue joints are unavoidable and unfavorable in their effects. The elastic-plastic properties of the glue are changing. The glue becomes more brittle. This is the reason why in glue lines thicker than 0.3 mm ($^1/_8$ in.) dangerous stresses may be caused. The attraction between glue and the surfaces of wood to be glued is known under the term "bond" (*Marian* and *Fickler*, 1953, p. 18 and 37). Hardening means the chemical reactions which occur in the glue substances separately from the wood surfaces. For improving the bond strength especially against the influences of moisture some chemicals, known as strengthening agents, may be added.

1.1.7 Phenomena of Hardening

There exist hardening and non-hardening glue types. The first type includes the animal glues (hide glues). During the bonding of these glues their chemical composition is not changed, in the latter ones there are always chemical reactions, e.g. under the influence of aldehydes or metal oxides, etc. According to *Browne* and *Brouse* (1929, p. 80) it is interesting to observe for different glue types the transition from the sol to the gel state. *Plath* (1951) and *Marian* and *Fickler* (1953) have explained that the gel state is influenced by a great variety of major and minor factors in a various extent. It is nearly impossible to classify the glues according to their geling properties but the following principal factors may influence the bond:

1. Loss of the dispersion agent by evaporation and diffusion. Most reversible glues such as hide glues, emulsions of polyvinyl, and also glues dissolved in organic solvents like polyvinyl acetate, belong to this group.

2. Loss of the dispersion agent and cooling, e.g. most of the hide glues used in the wood working industry. The process is reversible and the application of elevated temperatures is dangerous insofar as it induces hydrolytic processes which destroy the glue.

3. Loss of the dispersion agent and coagulation by the application of heat. A typical example for this type of glues are the blood albumin glues.

4. Loss of the dispersion agent and chemical reaction:

a) The following glues may be hardened without significant increase in temperatures. Some glues on the basis of casein and soybean and special types of synthetic resin glues like carbamide-, melamine-, phenol-, cresol- and resorcinol-glues.

b) By application of heat for hot-setting protein glues as well as for hot-setting synthetic resin glues on the basis of carbamide, melamine, phenol, cresol and resorcinol.

5. Reaching a gel state without any remarkable loss of dispersion agent. To this type belong all glue films and the redux powder applications.

Bonding is decisive for the success of the whole gluing process. After spreading the glue or overlaying a glue film an "open time" is unavoidable. Occasionally this time is called "waiting time". Immediately after this time the wood parts or veneers to be glued are assembled and the desired pressure is applied by suitable means. Normally the pressure in combination with heat causes a flow of the glue; the effect is a wetting of the surfaces and the glue penetrates partly into the wood. The bonding must be seen as a conversion from the sol state into the gel state (cf. Section 1.1.5). This is the prerequisite to the proper hardening of the glue substance. The influence of the pH value on hardening has been mentioned already and can be controlled (cf. Fig. 1.8). Other factors are temperature and pressure. Of high importance is the migration of the dispersion agents. Usual glues for wood contain between 30 and 60% of water or any organic solvent. For assembly gluing the amount of glue to be spread ranges between 80 and 350 g/m² (16 and 70 lb./1,000 sq. ft.). In the last decade an excess glue has been avoided for two reasons: First, of course, too much glue is very expensive and second, and perhaps the more important, the bonding strenght is increased if not too much glue is applied. Considering the figures mentioned above, in industrial wood working generally 50 to 175 g/m² (10 to 35 lb./1,000 sq. ft.) of dispersion agents must be removed. This is practically impossible as far as solid wood parts are concerned. Therefore one has applied at first glue films. In the meantime very concentrated glues on the basis of urea-formaldehyde, phenol-formaldehyde or resorcinol-formaldehyde have altered the situation.

Glue fillers are inavoidable if via diffusion of the dispersion agent the wood moisture becomes too high. Another problem occurs if the diffusion is retarded and thus the bonding process is disturbed. Fig. 1.12 shows how different glues behave in this respect. Evaluating the curves it has to be taken into consideration that the thermosetting Kaurit-glue contains 30% water, thermoplastic glues (Mowi-coll) about 35%. If one calculates the amount of evaporated water in g/m² for both glues as a function of open time the amount of evaporated water is higher for the Mo-wicoll than for the Kaurit-glue. The reason for this difference is that both glues retain the dispersion agent to a varying extent due to a different effect between the colloidal parts on the one hand and the molecules of the dispersion agent on the other hand. The evaporation of water or any dispersion agent depends on the surrounding climate. Higher temperatures and lower relative humidities of the air are favorable. The open time, of course, must be adjusted to the gluing conditions and the process of heating and pressing. One has to distinguish between two terms: "hardening" and "strengthening". Most synthetic resin glues need hardeners that control the gluing. It is well known that such glues as phenol-formaldehydes normally harden at rather high temperatures, approximately at 165 °C (329 °F), but they can be hardened at room temperature with very strong mineral acid (e.g. H_2SO_4). At first a thin glue joint is created within a short time, but under the influence of the acids the wood fibers adjacent to the glue line are destroyed sooner or later and bond failure occurs. Therefore, these hardeners are not applied any more (cf. Section 1.1.5).

The hardeners can be delivered as powders or, occasionally, in a liquid state, e.g. for carbamide glues. They are a part of the glue mixture and become active after the mixing. The hardeners may be mixed with strengthening agents and with extending agents in order to alter prior to use important properties of the glue mixture, e.g. to influence the viscosity or the penetration capacity and to reduce the cost of the glue mixture.

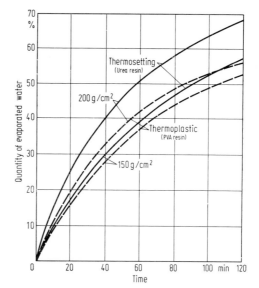

Fig. 1.12. Dependence of the evaporation of water of a thermosetting urea-formaldehyde resin glue (Kaurit) and a thermoplastic (Mowicoll) on the basis of polyvinyl acetate on the heating time. From *Plath* (1951)

1.1.8 Fortifying (Upgrading), Filling, Extending of Glues for Wood

It is possible to improve the water resistance of urea-formaldehyde resins by addition of melamine resins. Fig. 1.13 shows how important the addition of melamine-formaldehyde to urea-formaldehyde glue is. The effect depends on wood species, amount of melamine added and the boiling time.

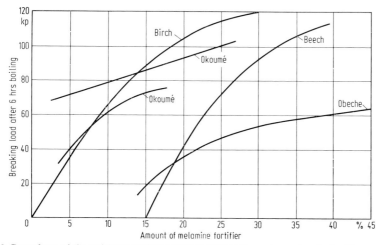

Fig. 1.13. Dependence of the resistance against boiling of various plywood samples glued with upgraded urea-formaldehyde resin glues on different amount of melamine-formaldehyde fortifier. From *Scales* (1951)

The relatively high price of synthetic resin glues for many years has led to the addition of so-called filling agents. One point is that these filling agents are inexpensive. Under this point of view various filling agents have been used all over the world. Flours made of beans, nuts (especially peanuts), nut shells, roots and vetches, but also powders from blood albumin and phenol resins and even anorganic materials like chalk and plaster have been used. Doubtless those filling

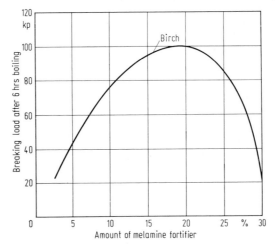

Fig. 1.14. Dependence of the resistance against boiling for joints glued with urea-formaldehyde resins mixed with melamine. From *Scales* (1951)

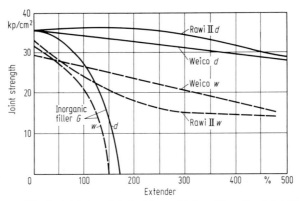

Fig. 1.15. Dependence of the glue joint strength in dry (*d*) or wet (*w*) condition after gluing with urea-formaldehyde resin Melocol H on the ratio of extenders. From *Frey* (1947)

materials are the best which either have adherent properties or which develop adhesive forces with the glues. An example is the glue "Kaurit WHK" which consists of liquid urea-formaldehyde resin and phenol-formaldehyde resin powder (*Klemm*, 1938). With this glue it was possible to produce rather thick gap-filling glue joints without any loss of strength.

There is a maximum for the addition of strengthening material (Fig. 1.14, *Scales*, 1951, p. 89). A successful way is that of addition of a self-gluing filling material, e.g. ground vetches (*Frey*, 1947, p. 83) (Fig. 1.15).

1.2 Strength Improvement and Stresses in Glue Joints

There are still some problems in achieving sufficient strength and with stresses in glue joints. They are not yet completely clarified. The change of state from sol to gel within the glue line is a process of varying speed; the joint increases slowly. It is not completely explained how glue joint strength is developed. It is not a continuous process and after a maximum strength is reached there follows a decrease until finally the glues reach an equilibrium strength. Fig. 1.16 elucidates details of the process as it normally occurs. After the initial increase (in Fig. 1.16 about 3 to 4 h[1] clamping time) the glue strength rapidly decreases and this may be dangerous if the glued parts, not clamped any more, are set under external stresses.

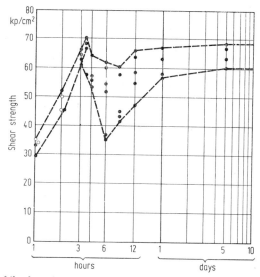

Fig. 1.16. Dependence of the shear strength of glued joints (pine wood glued with animal glues) upon time. From *Plath* (1951)

Plath (1951, p. 32) called the phenomenon the "feverish phase". In this state failures in the glue joints are normal. The reason for this decrease of strength may be the moistening of the wood near the glue lines. This assumption is confirmed by the fact that the decrease in strength is especially great for laminated wood.

In the hardened glue joints the stress conditions are very complicated. In the literature normally the breaking strength of glue joints is expressed as the ratio of maximum load to the area of the joints. This is an error because the stresses in the glue joints are not all equally distributed. Depending on the geometry of the glued parts, peaks of stresses occur. They induce the failure and are responsible for the strength of the whole glue joint.

McNamara and *Waters* (1970) compared the rate of glue-line strength development for oak and maple. They found that glued with a vinyl type adhesive the ring porous red oak required approximately twice the clamping time to develop

---

[1] h is identical with hour (hr) or hours (hrs) as customary in the U. S. A.

50% of its 8 h strength than did the diffuse porous Sugar maple. Density gradients between springwood and summerwood in oak are suggested as a reason for the onger clamping time required by the ring porous wood.

1.3 Testing of Glue Joints

Mylonas and *de Bruyne* (1951, p. 91) gave a condensed summary on the (static) mechanics of glue joints. In the meantime the non-destructive testing of glue joints has been rapidly developed. *Salomon* in the second edition of "Adhesion and Adhesives" by *Houwink* and *Salomon* (1965, p. 110) dealt with this question. The following possibilities may be mentioned:

1. Absorption of sound;
2. Ultrasonic technique;
3. Photoelastic analysis;
4. Technique of stress coating.

Matting and *Ulmer* (1963, p. 334 and 387) have calculated the stress distribution in a lap shear joint based on the stress-strain behavior of the adhesives. Since the stress-strain curves are non-linear, the method is only approximate but it allows an estimate of the influence of plastic yielding of the adhesives when low tensile loads are applied. The results show that the classical theories of *Volkerson* (1938, p. 41) and *Goland* and *Reissner* (1944, p. 417) led to shear stress maxima never reached in adhesive joints due to flow or creep. The thickness and rigidity of the adherent contributes to the behavior of the adhesives (*Salomon*, 1965, p. 113). *Hahn* (1959, 1961) reported on lap shear and creep testing of metal-to-metal adhesive bonds.

Freeman and *Kreibich* (1968) made measurements of glue line viscosity utilizing a rotary motion in order to subject the glue line in shear stresses during the cure cycle.

The distribution of tensile and compressive stresses, the maximum stresses at the edges of the overlap and the influence of the thickness of the strips can be investigated by coating the surfaces with photoelastic materials. The method apparently is more versatile and efficient than the classical photoelastic analysis with isotropic models applied by *Mylonas* (1948, 1951). *Sneddon* (1961, p. 46) and *de Bruyne* (1944, 1947 a, b, 1951) tried to reconcile theoretical considerations with the results of laboratory tests.

The goal is always to make a joint strong enough and from this point of view the length of the overlap is important. *de Bruyne* came to the conclusion that when the shear modulus of the adhesive, the Young's modulus of the adherent and the thickness of the adhesive are kept constant, a joint factor is determined. This joint factor is expressed by the ratio $\sqrt{t/l}$ where t is the thickness of the adherent and l the length of the overlap. Fig. 1.17 shows the dependence of the average shear strength on the joint factor. Fig. 1.18 gives an impression of the influence of position in the glue joint on the one hand and the joint factor on the other hand. It must be stated that the methods are highly empirical, some experimental results are not reproducible, and some phenomena are not yet clearly understood. This is also valid for the destructive testing. The strength of any glue joint between two wooden parts can be tested parallel to the grain, perpendicular to the grain or at an angle to the grain. Various methods and geometrical conditions are used in static tests. Fig. 1.19 shows the test piece (accord-

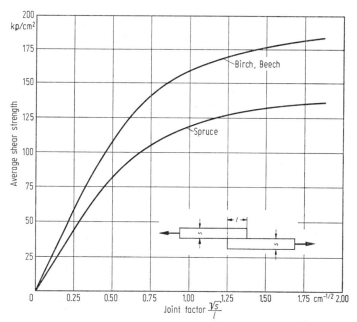

Fig. 1.17. Dependence of the ratio of effective shear stress to average shear stress on the position in the glue joint and on the joint factor. From *de Bruyne* (1947 b)

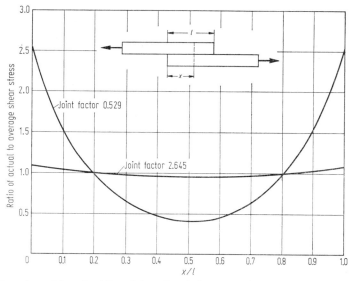

Fig. 1.18. Dependence of the ratio of the real shear stress to the average shear stress from the position in the overlapped glue joint.

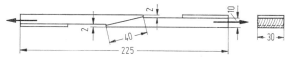

Fig. 1.19. Test sample for shear tension tests of glue joints. According to Deutsche Versuchsanstalt für Luftfahrt and later German Standard Specification DIN 53253

ing to the German Standard Specification DIN 53 253) for obtaining the shear tensile strength and Fig. 1.20 specimens for the same purpose according to English and American recommendations.

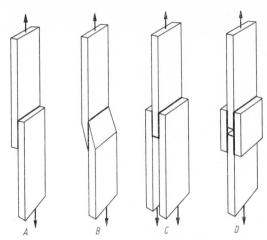

Fig. 1.20. Shear tensile specimens according to English and American recommendations. *A* Single lap (good, practical), *B* Bevelled lap (very good, but difficult to be machined), *C* Double lap (good, difficult to balance load), *D* Double strap (good, sometimes desirable; if bevelled double strap lap joints are used, the results become very good, but the production of the best samples is difficult)

Kreibich and *Freeman* (1970) reported on the effect of specimen stressing of eight wood adhesives. Lap joints prepared from Douglas-fir veneer, using eight adhesives, were exposed in Washington and tested periodically over four years. One half of the joints was unstressed while the other half was stressed in bending. The stress specimens lost shear strength more rapidly while the wood failure response showed a stressing to accelerate a degradation of the less durable adhesives. For the testing of thick glue joints, British Standard Specification BS 1204 has been developed (Fig. 1.21).

Thick glue layer

Fig. 1.21. Test samples for determination of strength of thick glue layers. According to British Standard Specifications BS 1204

It is outside the scope of this book to discuss in detail the testing of glue joints in plywood and laminated wood, the peeling tests and the possible dynamic tests. Nevertheless, it may be mentioned that *Johnson* (1970) dealt with flexural tests of large glue-laminated beams made of nondestructively tested lumber. The results were encouraging and indicated advantages for NDT-methods in construction of glued-laminated beams. The lumber available can be orientated in the beam to gain maximum stiffness (modulus of elasticity) and good breaking strength (modulus of rupture). The modulus of elasticity of the beams was well predictable, less the modulus of rupture. Strength and durability of glue joints depend on the moisture content of the wood, the relative humidity of the surrounding air and the temperature. *Freeman* and *Kreibich* (1968) estimated the durability of wood adhesives in vitro. Already in 1964 *Northcott* published a report based on 94 reference sources, dealing with durability of wood adhesives.

Accelerated tests and long-term exposures on glue joints are possible. The U. S. Forest Products Laboratory (1946) during a period of more than 40 years has developed a list of exposure conditions for glued parts:

1. *Continuous soaking in water at room temperature.* Primarily the water resistance is tested, although bacterial action may be involved, particularly with protein glues.

2. *Continuous exposure to 97% rel. humidity and 27 °C (80 °F);* tested is the ability to withstand the effects of moisture content approaching fiber saturation of the wood as well as the resistance to a combination of mold and other micro-organisms.

3. *Soaking-drying cycle.* This test consists of repeated cycles of alternate soaking in water at room temperature for 2 days and drying at 27 °C (80 °F) and 30% rel. humidity for 12 days; tested is resistance to mechanical stresses to water, and possibly to biological deterioration.

4. *High and low cycle.* In this test two periods of two weeks a) at 27 °C (80 °F) and 97% relative humidity and b) at 27 °C (80 °F) and 30% r.h. are subsequent. Mechanical strength, resistance to molds and other micro-organisms and ability to withstand the effects of moisture content approaching fiber saturation is measured.

5. *Continuous exposure at 27 °C (80 °F) and 65% relative humidity.* The effect of aging of the glue joint under mild conditions is measured.

6. *Continuous exposure at 70 °C (158 °F) and 60% relative humidity.* The effect of high temperature and moderate moisture content is indicated.

7. *Continuous exposure at 70 °C (158 °F) and 60% relative humidity.* In this test the combined effect of high temperature and moderate moisture content is indicated.

8. *Room-temperature-high temperature cycle.* This cycle, consisting of repeated alternate exposures for 16 hours at 27 °C (80 °F) and 65% relative humidity and 8 hours at 70 °C (158 °C) and 20% relative humidity, ascertains the combined effect of high temperature and moderate to low moisture content, together with resistance to light swelling and shrinking stresses.

9. *Room-temperature—low temperature cycle.* This repeated cycle of alternate exposures for 16 hours at 27 °C (80 °F) and 65% relative humidity and 8 hours at −29 °C (−20 °F) measures the resistance of moisture-containing glue joints to extremely low temperatures. Early results obtained in this test were based upon a low temperature of −55 °C (−67 °F).

10. *Continuous exposure at 93 °C (200 °F) and 21% relative humidity.* This is an accelerated test method of the comparative resistance of glue and wood to high temperature.

Kreibich and *Freeman* (1968) discussed development and design of an accelerated boiling test machine. Three adhesives were tested

a) an epoxy resin;

b) a hot press phenolic plywood gluing formulation;

c) a commercial aminoplast formulation.

Strickler (1968) studied specimen designs for accelerated tests. The block shear test more nearly approached a state of pure shear than any other test method studied.

1.4 Summarized Results of Tests for Various Glues

The many results of various tests carried out in laboratories as well as in the open air in various countries of different climates allow the following summarized conclusions:

1. Animal glues (hide glues, bone glues) have a very high geling power and the initial strength of the glue joint develops rapidly. The joints generally are not resistant against moisture and microorganisms.

2. Bloodalbumin glues, after heat-treating, give moderately strong and very water resistant bonds, but are sensitive to bacterial and fungal attack.

3. Casein glues produce joints with a strength which decreases under wet conditions for extended periods of time (*Brouse*, 1938, p. 306). The stability of joints kept dry lasts indefinitely. Various facto rsaffect the joint strength (*Knight*, 1952). Failure may be induced by mechanical stresses set up when the wood swells or shrinks, by alkaline hydrolysis and by attack of microorganisms.

4. Soybean glue joints are similar to casein glue joints in their physical properties and their water resistance.

5. Most types of urea-formaldehyderesin glues offer moderate to low resistance under the influence of elevated temperatures, especially when combined with high relativ humidity. Urea-formaldehyde glues can be very much improved ("upgraded") by adding 10 to 20% melamine. The stability then becomes nearly as great as that of phenol-formaldehyde resin glues.

6. Melamine-formaldehyde resins are used to a limited extent in the plywood industry. The wet strenght ranges between that of plywood manufactured with phenol and with urea glues. Under severe conditions of exposure to heat and moisture the durability of melamine adhesives is not lower than that of most alkaline phenol and resorcinol glues. Melamine-formaldehyde resin glue joints are resistant to boiling water, if the immersion period is not too long.

7. Phenol resin glues are the most resistant under all conditions of exposure. Phenol resin glues which are hardened with alkaline substances are superior to such which are hardened with acids. The reason is that acid residues weaken the wood fibers (*Blomquist*, 1949).

8. Resorcinol-resin glues are resistant against the influences of heat and humidity and deliver durable glue joints as phenol resin glues.

1.5 Effect of Wood Species and Moisture Content on the Strength of Glue Joints

The properties of the wood adherent always have a distinct effect on the properties of the glue joints. Generally hardwoods offer more difficulties than softwoods. The higher the moisture content of the wood is, the lower becomes the strength of the glue joint and delamination is possible. For the various types of glues the moisture content must be observed within certain limits in order to avoid failures, e.g. for phenolic resin glue films the moisture content of the wood never should decrease below 6%. The influence of wood preservatives on the strenght of glue joints depends mainly on the type of the preservatives. In some cases the strength is practically not influenced.

There are some endeavors to improve the testing procedures of glue joints. Fig. 1.22 shows how the bending shear strength is investigated at the Federal Institute for Testing of Material (EMPA) at Zürich. This test is suited for the static as well as for fatigue experiments. The maximum horizontal stress τ_{max} in the glue joint (neutral axis) can be calculated as follows:

$$\tau_{max} = \frac{Q_{max} \cdot S_x}{I_x \cdot w_2},$$

$$(1.7\,a)$$

where Q_{max} is the maximum vertical shear force $(= P_{max}/2)$, $S_x =$ the static moment of one half of the cross section in regard to the neutral axis $x - x$, $I_x =$ the moment of inertia of the total cross section in the relation to the x-axis, and w_2 is the width of the glue joint $(= b_2$ in Fig. 1.22).

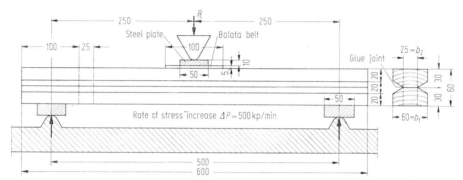

Fig. 1.22. Bending shear specimen according to Federal Institute for Testing Material EMPA, Zürich

For the cross section according to Fig. 1.22 the formula is as follows:

$$\tau_{max} = 0.05 P_{max}. \qquad (1.7\,b)$$

The results of such bending shear tests on specimens glued with urea-formaldehyde glue (Melocol H) are shown in Table 1.3.

Table 1.3. Results of Bending Shear Tests on Glued Specimens of Spruce and Beech Wood.
(From *Roš*, 1945)

	Moisture content H	Spruce	Beech
	at gluing $12 \cdots 15\%$	Bending shear strength kp/cm² (lb./sq. in.)	
Static tests			
Dry storing		60 (853)	125 (1,778)
Wet storing		55 (782)	100 (1,422)
	<22	55 (782)	110 (1,565)
High moisture content in use	>25	50 (711)	80 (1,138)
Dynamic tests			
Dry storing		25 (356)	40 (569)
Wet storing		22 (313)	33 (469)
	30	20 (285)	—
High moisture content	60	—	30 (427)

Tests on the bending strength of glue joints are of doubtful value. Fig. 1.23 shows how results may be obtained. It is certain that the strength of joints produced by different glues cannot be compared in such a way. The same problems apply to biological tests insofar as the influence of microorganisms, bacteria and molds on the glue joints varies and is not easily reproduced. It just may be mentioned that prior to application glues are tested as to viscosity, solid content, pH value and to chemical properties and composition (content of fat, protein and other substances).

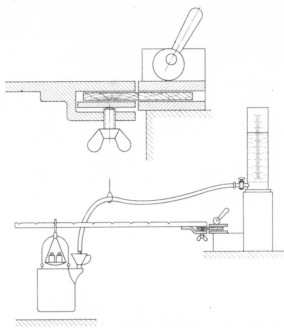

Fig. 1.23. Arrangement for testing the beinding strength of glued joints across the grain. According to German Committee for Economic Production AWF

1.6 Pretreatment of Wood Prior to Gluing

It is not the function of a book dealing with the principles of wood science and technology to explain the technique of the application of glue and pressure to glue joints. Nevertheless, it may be mentioned that the following points should be strictly observed:

1. A proper conditioning of the wood prior to gluing. Table 1.4 gives some guides according to *Perry* (1948, p. 27).

Table 1.4. Proper Moisture Contents for Gluing
(From *Perry*, 1948)

Type of glue	Type of glued material	Moisture content %
Casein and animal glues	Plywood	3··· 5
Starch and soybean glues	Plywood	3··· 5
Casein and animal glues	Gluing of wooden parts	5···10
Urea-formaldehyde glues, cold pressed	Plywood	7··· 9
Urea-formaldehyde glues, hot pressed	Plysood	5··· 7
Urea-formaldehyde glues, cold pressed	Gluing of wooden parts	7[1]···9
Phenol-formaldehyde liquid, hot pressed	Plywood	4··· 6
Phenol-formaldehyde glue films,	Plywood, Veneers	8[1]···10
hot pressed	Core	5··· 6

[1] The low moisture content may be dangerous.

2. Internal stresses in gluing can be avoided if the pieces to be glued have about the same alignment of annual rings in the cross section.

3. If the wood surfaces contain a high amount of resin, it should be removed by suitable solvents.

4. The wood surfaces must be clean and free from dust.

5. The curing is dependent on temperature and therefore during the winter season wood to be glued must be stored in heated rooms prior to gluing.

1.7 Glue Spreading

Proper spreading, especially uniform distribution of the liquid glue on the surfaces is essential. The surfaces to be glued should be flat and should mate properly. Rough, wavy and uneven surfaces do not allow an even and economical spread of glue.

If the liquid glue contains a water content between 40 and 60%, the amount of glue will be relatively high, and the thicker glue joints cause a weaker adhesion to the adjacent wood layers. A higher amount of water must be absorbed by the wood or it must evaporate for complete curing. Longer pressing times are necessary and the danger of steam blisters exists. On the other hand, if the spread of glue is insufficient the water diffuses rapidly into the wood and adhesion is reduced.

Fig. 1.24. Example for glue spreading on two wood surfaces. From *Truax* (1929)

Higher spread, e.g. by the addition of fillers, can equalize some defects between the surfaces, but the result is not always encouraging.

Glue can be applied to one or both surfaces of veneer (Fig. 1.24). In the plywood industry single surface glue spreading is rare.

Spreading of the liquid glue onto the surface can be done either by hand or mechanically. For plywood and laminated coreboard mechanical glue spreaders are common. Fig. 1.25 shows how the various rolls can be arranged. In

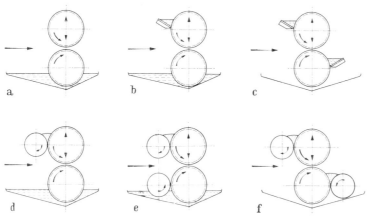

Fig. 1.25 a—f. Schematic view of various possibilities for dosing the glue with 2 or 3 and 4 rolls. From *Truax* (1929)

Fig. 1.25 (d, e, f) a roll is shown (in the literature called "doctor roll") which is smaller in diameter and revolves more slowly than the spreading rolls. It allows a more uniform distribution of spread. Nevertheless, in industry machines like those shown in Fig. 1.25a, b, c are used. Fig. 1.26 shows the dimensions of a glue spreader with 4 rolls. Figure 1.27 schematically explains how the glue is dosed. In a plywood factory the economical consumption of adhesive is of great importance. It is a major factor in the cost of production and with the increased use of rather expensive resins, the necessity for avoiding waste has become more and more important (*Wood*, 1963).

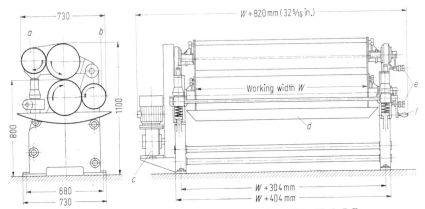

Fig. 1.26. Dimensions of a four roll glue spreader. From *Ritter, Fleck, Roller*

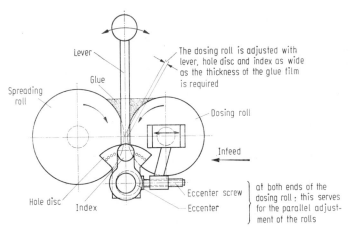

Fig. 1.27. Scheme of glue dosing

Mechanical spreaders have been considerably improved during the last decades. They can apply an even spread of adhesive to one or both sides of veneer.

The rolls are made of steel, uniform in diameter and rigid to prevent any deflection. The steel rolls are grooved in various ways; the corrugation may run longitudinally or spirally, but must be evenly cut.

For resin glues steel cores are usually covered with rubber. The most modern rollers are grooved every 0.7 mm (1/32″) in a regular pattern.

Automatic feeding to and from the spreader is predominant. In all spreading operations the thinnest practical glue line should be obtained. It is, of course, of highest importance that every glue spreader is kept clean and corrugations from chips or hardened lumps of adhesives must be avoided.

The spreader roll application of phenolic resin adhesives to plywood core veneer was studied by *Barnes* (1970). Variables included resin source, adhesive formulation, roll-groove pattern, the degree of wear of the roll, and variations in veneer thickness. Wear of the roll grooving caused a marked reduction in spreading efficiency and variations in veneer thickness had a remarkable effect on glue spread levels.

Webb (1970) wrote:

"Of prime importance in the wood laminating industry is the application of adhesive of the board surfaces. Conventional spreading equipment in most laminating plants is a roll spreader. Other methods that have been used to apply adhesive to board surfaces have included curtain coating, spraying and brushing. The idea of ribbon spreading is not necessarily new to the industry. Hot melt adhesives are applied in a bead form."

Conventional phenol-resorcinol-formaldehyde adhesives used in lumber laminating were found not to be too thin to perform satisfactorily in ribbon spreading. The ribbons of the adhesive "sagged" when vertically positioned. The use of organic fillers and thixotropic agents reduced the sag tendency.

1.8 Pressing of Glued Parts and of Laminated Wood

Storing, blending, foaming of glues are technical questions which cannot be described here in detail. The parts to be glued must be pressed to get a joint of high strength. The following conditions are to be observed:

1. The surfaces to be glued must come into complete contact on all positions.

2. Air and superfluous glue should be expressed at the edge of the joints.

3. A thin glue line of equal thickness over the whole surface of the joint must be produced.

4. Tentacles of the hardened glue in wood cells are necessary for increasing the mechanical adhesion.

5. It is necessary to keep the parts to be glued under some pressure during the major part of the curing.

There are two possible methods: cold pressing and hot pressing.

According to *Redfern* (1950) the disadvantages of cold pressing are as follows:

1. Limitation to plywood for interior use, except resorcinol glues are used;

2. Higher gluing costs since the amount of spread is bigger than for hot pressing;

3. To some extent great weights per unit of glued area and therefore higher freight costs;

4. Longer duration of press cycles;

5. Longer periods prior to sanding and transporting of boards.

Hot pressing offers the following advantages:

1. Wide spread application on the production of plywood for interior as well as for exterior use;

2. Relatively low gluing costs since the amount of spread is lower than for cold pressing;

3. Lower board weight due to drier veneers;

4. Remarkably shorter pressing time;

5. Possibility to sand and transport the boards quickly.

The disadvantages are as follows:

1. High investment costs.

2. The veneer must be dried to 3 to 5% moisture content, since otherwise the development of blisters is unavoidable; an exception is the gluing with phenol-formaldehyde glue films (cf. Table 1.4).

3. The low moisture content of the veneer is to some extent dangerous (increased brittleness) and conditioning (re-moistening) may be necessary.

4. Very careful grading and preparation of the veneer is necessary.

There are many possibilities to apply pressure during the gluing process ranging from wedges and screws to pneumatic and hydraulic pressure elements.

In many cases pneumatic pressure elements are useful and very versatile. It may be mentioned that the specific pressure of a pneumatic element is limited to about 2 to 3 kp/cm² (28 to 43 lb./sq. in.). If high pressures are necessary, e.g. for the manufacture of laminated wood with hardwoods or in densifying woods, hydraulic elements are necessary. The following values may be taken as normal: For ordinary laminated products from softwoods up to 15 kp/cm² (213 lb./sq. in.) and from hardwoods up to 25 kp/cm² (356 lb./sq. in.), for surface treatment of laminated products, e.g. by plastic films, up to 120 kp/cm² (1,450 lb./sq. in.), for densified laminated wood up to 200 kp/cm² (2,844 lb./sq. in.) and for manufacture of densified solid woods up to 350 kp/cm² (4,980 lb./sq. in.).

1.9 Difficulties and Defects in Gluing

As has already been mentioned, the gluing process is very complex and influenced by many factors. There are various types of glues, more and more mixtures of different glues, many of them are filled, extended or upgraded. The properties of the components of any mixture may be quite different but the goal is to produce the best possible joint (of long duration and low sensitivity against the influences of humidity and higher temperatures). Another question is the composition and the content of solvents in the blended glues, for wood glues mostly water. The history of the glue: storage-life, pot-life, open and closed waiting time of the glued assembled material prior to pressing is important. Concentration and viscosity of the glue mixture has to be controlled. The pre-treatment of the wood to be glued should be taken into consideration as well as species and moisture content. Impurities can reduce the strength of glue joints if they appear in the glue line or on the surface of the wood. The thickness of the glue line dependent not only on the amount of spread per unit area of surface but also on the structure of the wood is important (cf. Fig. 1.9). Other factors are duration of cold or hot pressing and conditioning after pressing. This is not a complete survey but it may elucidate the difficulties in a theoretically based knowledge and a practically applied technology of gluing.

It may be briefly repeated that the nature of adhesion is still not completely understood and that it is nearly impossible to draw clearly the limits between mechanical and specific adhesion. Nevertheless adhesion is closely related to adsorption (*Blomquist*, 1963). Another problem is the geometry of the joints. *Blomquist* (1963) stated some general principles for good joint design:

1. The bonded area should be as large as possible.

2. The maximum proportion of the bonded area should contribute to the joint strength.

3. The adhesive should be stressed in the direction of its maximum strength (as in shear rather than in tension or peel), and

4. Stresses in the weakest direction of the adhesive line should be minimized.

Lap joints can be improved (cf. Fig. 1.20), but not always in an economical manner. Scarf joints (cf. Fig. 1.19) are good, but usually (except in testing) impractical since machining is difficult. Butt joints (cf. Fig. 1.23) are unsatisfactory.

Open glue joints with very low or even scarcely not existent strength or other defects can be caused by the following reasons (cf. *Truax* 1929):

A. As far as *wood* is concerned:

1. Incorrect drying of the wood prior to the gluing to a moisture content which is either too high or too low (cf. Table 1.4). The different behavior of the various types of glue must be observed in this respect. Not uniform distribution of the moisture content also is detrimental. Under most circumstances casehardened wood can not be glued easily. Rigorous drying of wood and veneer leads to case-hardening. It is necessary to condition such wood or at least to sand it in the fiber direction prior to gluing.

2. Careless machining and improper tools, especially sawing producing rough surfaces and/or uneven planes and in the case of laminated products veneers or laminates of unequal thickness.

3. Insufficient cleaning of the wood surfaces so that dust, dirt, oil, resins, waxes or other impurities remain.

4. Warped woods due to unbalanced construction, cross-grained or decayed wood, improper drying, poorly fit joints.

5. Surface defects, e.g. uneven spots due to knots, limb markings, open joints, blisters (cf. C5), checks, sunken joints, corrugated appearance.

B. As far as the *glues* are concerned:

1. Bad condition of the glue, dried or jellied, partly bonded, too low concentration (mixed too thin), too much extended, not properly blended prior to the spreading.

2. Too low glue spread. Spreading is of great importance for the gluing of wood because the liquid glue spread has the tendency to fill capillaries and other voids.

3. Unequal glue spread. The glue must have a viscosity high enough to be able to form a film of appropriate thickness after spreading. If glue spread is insufficient or the viscosity is too low and/or the setting rate is too slow, the heating of wood is excessive, then the glue will penetrate too far into the wood tissues

adjacent to the glue line. In this case a continuous film of glue cannot be produced and the result is a "starved joint".

4. Application of interior glues for exterior exposure.

C. Finally mistakes may happen during the *gluing process*, e.g. the following:

1. Inadequate pressure (due to warped stock, jellied glue, uneven application of pressure, uneven press platerns or cauls.

2. Wrong temperature during the pressing cycle and heating up the press prior to the application of pressure.

3. Too short a pressing cycle and excessively fast closing of the press.

4. So-called "chilled joints" result if the wood is too cold or the glue is too viscous to allow the necessary assembly time.

5. In veneering blisters may appear as gluing defects. A blister is a spot or area where the veneer does not adhere and bulges like a blister (*Perry*, 1948, p. 4/5). In veneering the following causes for the development of blisters may be named: Unequal thickness of the wood, knots, resin in the core, high amount in volatile substances in the veneers, uneven distribution of glue, too much extension of the glue, bad wetting of the wood, improper pressing with inadequate equipment, too low pressure, unequal distribution of pressure, too short pressing periods or heating of the press before the pressure is applied, excessively hot cauls, too early cooling in the press etc.

Another point which must be taken into consideration is the dependence of the boiling temperature of water on pressure as by the following example may be clarified: for a pressure of 10 kp/cm² (142 lb./sq. in.) the boiling temperature amounts to 183 °C (361 °F). If the temperature applied in the press is only 135 °C (275 °F) then the moisture present in the veneer cannot be converted into steam, but if the press is quickly opened the pressure (2.2 kp/cm²) (31 lb/sq. in.) corresponding to the boiling point will be reached. A very quick conversion of water in the wood into steam follows. Since the steam cannot escape through the wood tissues steam blisters develop and the veneer bulges. Therefore, in using press temperatures much over 100 °C (212 °F) the pressure should be reduced slowly allowing a harmless diffusion of steam. The allowable rate depends on various factors such as pressing temperature, pressure, moisture content of the wood, wood species and thickness of veneer. This portion of the cycle may vary from several seconds to several minutes.

D. Staining and the development of spots in gluing of decorative veneer mostly is due to chemical influences. The faults may be detected immediately after unloading from the press, but it is possible that they appear later or even after some months. The content of tannin generally is responsible for the staining. The content of tannin is especially high in oak, mahogany, chestnut and walnut. Woods with a high content of tannin are spoiled by dark spots if they are glued with glues containing a relatively high amount of alkali, e.g. casein and starch glues. Light German oak wood, for instance, is already stained in contact with glues of a very low alkali content (pH = 7.5 to 8). The reactions for commercial casein glues with pH = 12 are significantly stronger. The following review shows the tendency of various woods to stain under the influence of alkaline glues according to *Blankenstein* (1937, p. 109).

The development of stains is less pronounced if the wood has been dried carefully and if a proper assembly time of the veneer has been observed.

Stains due to room temperature curing glues scarcely can be removed. For bleaching the following recipe may be recommended: solution of 7% oxalic acid in water with addition of per liter (= 1.06 U. S. liquid quarts) 10 to 20 grams (1 gram is equal to 15.4 troy grains) diluted hydrochloric-acid or acetic acid. For neutralizing the acids afterwards the wood should be washed with a solution of 5 grams sodium carbonate in 1 liter of water.

Very sensitive	Moderately sensitive	Slightly sensitive or not sensitive
Maple	Spruce	Ash
Birch	Lime-tree	Pine
Beech	Plane-tree	Poplar
Cedar	White-(yellow)	Pitch-pine
Oak	pine	Elm
Chestnut		Whitewood
Cherry		
Mahogany		
Walnut		
Satinwood		
Zebrano		

The tannin in wood reacts also with acids in the presence of water which contains some ferriferous substances. The result are chemical compounds of the same type as used in ink. It is well known that during drying of oak wood, especially in some older kilns, dark blue-black stains may occur if condensed water from iron equipment drips on the load. Urea-formaldehyde resin glues and hardeners with an acid character have the same effect. Glue of high acidity solves less iron. It is impossible to remove stains produced by the reaction of iron and tannin, but normally the stains do not penetrate into deeper wood layers.

Stained spots also may be produced if colored hardeners or very dark glues, e.g. blood albumin glues, are used. They can penetrate light and thin veneers. Veneers with thickness in the range of 0.5 to 0.8 mm (1/20 to 1/12.5 in.) often tend toward glue penetration. The problem arises when the glue is not absorbed by the core, but preferentially by porous veneers with rather high moisture content. Especially dangerous are glues with low viscosity, e.g. bone glues, and some synthetic resin glues. Filling is a remedy. Penetrated hide glues may be removed by means of hot water and soap, using a brass wire brush. It is possible to bleach blood albumin glue spots with the aid of solutions of hydrogen oxide. Spots caused by synthetic resins cannot be removed. If dark, highly porous veneers are used, then it is necessary to stain the glue, otherwise the pores become too bright. In these cases — e.g. for walnut or sapeli — brown dye-stuff or soot should be added.

E. Checks and waves (cf. A 5) often appear during and after gluing. There are several reasons for this:

1. Bad slicing or peeling of the veneer.

2. Incorrect drying or too high moisture content in the veneer.

3. Open assembly times, which are too short cause the same defects.

4. Improper constructions can induce internal stresses which cause checks. The more valuable veneer is, veneer sliced from stumps or roots, the more expressed is the tendency for checking.

5. Waves occur if the structure in the different layers (veneers) varies significantly. This is especially so with core stock when the direction of the annual rings is not the same in all elements.

6. The outer veneer should not be too thin.

7. The veneers should be produced in a carefull manner free of small groves.

8. Glue spread should be equal.

F. Gluing of *impregnated* or *treated* woods is an old problem. The result of laboratory experiments do not always correspond with experience in practice. Nevertheless, a few remarks and recommendations are possible:

1. Some glues and preservatives are compatible.

2. Important fundamental rules of the gluing technique for untreated wood are valid also for treated wood, e.g. the fact that difficulties in gluing increase with higher wood densities.

3. It is well known that untreated wood which contains oil, fat or waxes is difficult to be glued, therefore wood, impregnated with coal tar which has a tendency to bleeding can scarcely be glued. It may be possible to clean the surfaces of such coal tar treated wood by means of vapors.

4. The bonding in gluing coal tar treated wood is improved according to the U. S. Forest Products Laboratory with increasing pressing temperature.

5. The success in gluing wood treated with oily solutions of pentachlorophenol mainly depends on the type of solvent or carrier. Naphthalene used as solvent did not affect the gluability of maple, red oak, Douglas fir, and southern pine while petroleum gave worse results. In any case the surfaces of the wood must be cleaned prior to gluing.

6. Highly volatile solvents lead to very good gluing results.

7. Copper naphthanate in the same solvents as used for pentachlorophenol does not interfere with the gluing of the treated wood when phenol or resorcinol glues are used. Melamine resins are not satisfactory in the case of treated red oak.

8. If water soluble preservative are applied, drying after treatment is necessary, since otherwise the moisture content is too high. Excellent results were obtained in the laboratory when wood treated with preservatives of the Wolman salt type were glued with phenol or resorcinol glues. The temperature of gluing was about 94 °C (201 °F). Melamine glues gave high joint strength for treated maple and red oak for the same temperature. Some resorcinol glues gave rather good results for treated Douglas fir and yellow pine even for only 27 °C (81 °F) curing temperature.

9. Celcure, zinc meta arsenite, chromated zinc cloride and similar salts generally are compatible with glues based on melamine, resorcinol, and phenol resorcinol. A proper curing temperature (e.g. 94 °C (201 °F)) must be applied.

10. The effect of fire retardant chemicals on glues used in plywood manufacture has been studied by *Black* (1943).

The following may be stated in a very condensed manner referring to *Black*'s (1943, p. 46) report for more detailed information:

Type of fire retardant chemical	Type of glue	Results in gluing
Monoammonium phosphate	Blood albumin (hot pressed) Phenolic film glues	good
	Soybean Starch Casein Phenolic, liquid	failure
	Urea (cold pressed) Urea (hot pressed)	moderate bad
Diammonium phosphate	Phenolic film glue	good
	Phenolic, liquid (hot pressed) Urea (cold pressed)	acceptable
	Casein	bad
Ammonium sulphate	Phenolic film glues Urea (cold or hot pressed)	good good
Boric acid	Urea (cold pressed)	good
	Urea (hot pressed) Phenolic, liquid (cold or hot pressed) Phenolic film glues	bad
Borax (Sodium tetraborate)	Many types	failure
Borax in mixture with Boric acid (3:2)	Urea (cold pressed) Urea (upgraded), Phenol-resorcinol Resorcinol Melamine	good

The short survey shows that even fire retardants may vary in their effect on glues ranging from "good" to "failure". General rules cannot be established since too many variables are involved such as:

Type (including mixtures) of fire retardants, type (including mixtures) of glue; kind and amount of upgrading, filling or extending agents; type of catalyst or hardener; amount of glue spread, pH value during the whole reaction; species of wood, smoothness and cleanliness of the surfaces of the wood; amount and distribution of moisture in the treated wood prior to gluing and temperature and duration of pressing.

These are perhaps not all the variables influencing the gluing of treated wood but they show how complex and complicated this problem is, probably one of growing importance for the practice in the future. Scientifically based principles are not known. The formulations of the glues as well as the fire retardants are not always well known and the substrates are very inhomogeneous. Therefore, the gluing of treated wood needs close cooperation of experts from the chemical and woodworking industries. They have to find the proper way by consultation and experiments. Fire tests should be carried out as well. The glue joints should at least withstand elevated temperatures or even fire to the same extent as the adjacent wood layers. Delamination of plywood and other laminated products in the case of fire is very dangerous and should be avoided under all circumstances (*Kollmann* and *Teichgrüber*, 1961, p. 173, 186).

1.10 Natural Glues

1.10.1 Starches, Dextrins and Natural Gums

Starch glues for the production of wood joints were developed in the U.S.A. between 1905 and 1910 (*Perry*, 1948, p. 96—99, *Truax*, 1932). Starch glues may be extracted commercially from many vegetable sources throughout the world, such as cereals, roots, tubers, pith of plants, corn, rice, potatoes, and sweet potatoes. The principal raw material used in manufacturing starch glue is tapioca from cassava or manioc roots.

Vegetable adhesives are produced either from raw starch (*Jones*, 1927, p. 21 to 27) or from processed starch (*Kirby*, 1965, p. 167—185).

Dextrins are degradation products of starch, obtained by heating, usually in the presence of hydrolyzing agents (acids or acid-producing chemicals) or other catalysts. Dextrins are not suitable for the gluing of solid wood.

Starch glues are, as a rule, delivered as white powders or granules, and in Germany occasionally in liquid form. The dry glue is first mixed with cold water and then mechanically stirred to a mixture of a uniform consistency and of a rather high viscosity. The proportions of water and dry glue vary according to the methods of manufacture of starch glues from 1.5 parts of water to 1 part by weight of dry glue, to 4.25 parts of water to 1 part by weight of dry glue. The type of work to be done must also be taken into consideration in most cases a proportion of 2.25 parts of water to 1 part by weight of dry glue is adequate.

After having mixed the dry starch with cold water there are three different methods for preparing the glues (*Truax*, 1929, p. 8).

1. Caustic soda (about 3% by weight of the dry glue) is dissolved in a small amount of cold water and then slowly stirred into the starch water mixture. The resulting mixture is subsequently under constant stirring heated in a steam-jacketed heater to the temperature recommended by the starch manufacturer. The result is a viscous and translucent mass.

2. Caustic soda (6% or more) is thoroughly stirred into the glue. The caustic itself heats and agglutinates the starch to the wanted viscosity.

3. The starch and the water are mixed without caustic soda. The starch is converted into glue by the application of external heat.

After being prepared by one of the methods described by *Truax*, the glue is allowed to cool. It is then ready for use by spreading or storing in a vat. Starch glues have a pot life of several days, they are thick, stringy mixtures ("tacky in character"). *Perry* (1948, p. 97) published two typical formulas given below:

Ingredients	General standard	For short clamping periods
	Parts by weight	
Cassava flour, blended	75	100
Water	100	130
Caustic soda	2.25	3
Water	10	20
Solid content	34%	40%
Cooking period required	30 min	45 min

Starch glues have been used as adhesives in the American plywood industry since the beginning of this century. Especially the adhesives from alkali conversion have found an increasing application as lowcost veneer glues. They offer the following advantages:

1. Low cost.

2. Long pot-life (free from decomposition of the glue, though due to evaporation of water the viscosity will be slowly increased.

3. Suitable for cold or hot pressing.

4. Relatively low gluing pressures (about 1 kp/cm² or 14.2 lb./sq. in.) are sometimes sufficient but for well glued joints the relationship between consistency of the glue at the time of pressing and the pressure must be observed (Fig. 1.28). For plywood made with starch glues customary pressure is between about 5 and 7 kp/cm² (about 70 to 100 lb./sq. in.). Too little pressure may prevent sufficient surface contact while too high pressure may cause starved joints.

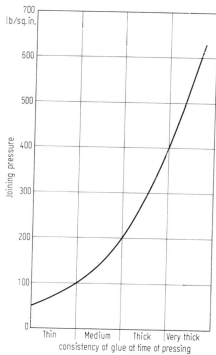

Fig. 1.28. Pressures recommended for starch glues of different consistency at the time of pressing

5. Pressing time for hot-gluing 0.75 to 2 h, for cold-gluing 2 to 24 h.

6. Strong joints (shear strength 97 to 250 kp/cm², corresponding 1,380 to 3,560 lb./sq. in., wood failure, that is proportion of the joint area of the test samples where 35 to 100% wood fibers were torn away in testing (*Truax*, 1929, p. 43).

Disadvantages of starch glues are:

1. Spreading is to some extent difficult since the starch glues are viscous. *Truax* (1929, p. 41) writes: "As excessive glue spread wastes glue, adds more water to the wood than necessary and may cause the wood to slip out of position

while being pressed." The general relation of quantity of glue spread to joint strength is illustrated in Fig. 1.29. It is evident that about 365 g (mixed glue) per m² (73 lb./1,000 sq. ft.) for plywood gives the highest strength.

2. Usually starch glues contain caustic soda reducing the viscosity and prolonging the pot-life, but causing staining. The following wood species have the tendency to be stained by strongly alkaline glues:

Beech, birch, black cherry, southern cypress, Douglas fir, red gum, mahogany, red and white oak, redwood, and black walnut (*Truax*, 1929, p. 49).

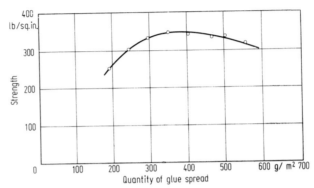

Fig. 1.29. Effect of quantity of glue spread on joint strength of vegetable glue; three-ply cross-banded panels of 1/16 in. (= 1.6 mm) birch. From *Truax* (1929)

Caustic free glues have a shorter working life, but they should be used where staining due to the effect of caustic must be avoided.

3. Assembly periods for vegetable glues may vary in the wide range between 0.5 and 20 min. In gluing thick pieces of wood and in edge gluing about 1 min is enough. In all other cases 6 to 8 minutes assembly time create maximum joint strength.

4. The moisture content of veneer and lumber, on which starch glues will be spread, should be between 3 to 6%. More than 8% moisture content is hazardous. All starch glues introduce (as glue solvent) surplus water into the wood which must be removed, therefore redrying is necessary. Kilns for redrying should be operated at approximately 47 °C (117 °F) and 35% relative humidity. Redrying is necessary and should be continued until the plywood (proved by tests) reaches 6 to 8% moisture content. The time required for this process depends on the thickness of veneer, the absorption capacity of the wood species, and the number of glue lines per panel. A period of 12 h is a minimum requirement but 48 h are safer. Low water-content mixtures are less sensitive, especially with respect to redrying.

Summarizing it may be said that starch glues especially in the U.S.A. are still used. The advantages and disadvantages were explained. In Germany starch glues are used only on special occasions.

Natural gums have been studied from the point of view of their adhesive capacity since long time (*Whistler*, 1959), Adhesiveness depends on many factors and most of the natural gums are not good adhesives. Hydrophilic gums found some application in a mixture with thermosetting synthetic resins, for creating a bond between wood and metal or rubber.

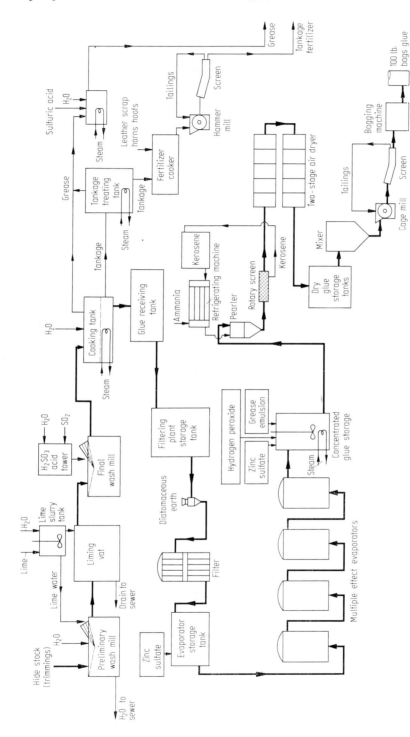

Fig. 1.30. Flow sheet for the production of animal glue. From *Hull* and *Bangert* (1952)

1.10.2 Protein Glues

1.10.2.1 Animal Glues, Manufactured from Hide, Sinews, Bones. The term "animal glue" generally refers to glues manufactured principally from hide or skin, sinews, bones and fish-wastes. Casein and blood albumin glues are also animal glues, but are treated separately here.

The process of manufacturing animal glues, as defined above, is not simple. Fig. 1.30 shows the flow sheet in a modern plant for the production of hide glues. Hide glues offer some distinct advantages: a rather high setting rate and no danger that veneers are penetrated. Unfortunately, they also have disadvantages such as: complicated handling and no resistance of the joints against high relative humidity, moisture and microorganisms. Nevertheless, the hide glues are very suitable for the furniture industry, for veneering, but not for exterior woodwork. There are various kinds of animal glues containing about 10 to 20% solids based on the solution.

Most of the skin converted into glue is the waste from tanneries. The whole process of glue action is not completely understood. The term "animal glue" is normally confined to glues on the basis of mammalian collagen. The collagens are fiberproteins which have a helix-structure (*Pauling* and *Corey*, 1951) Collagen belongs to the group of protein macromolecules which have about the following schematic structure (*Baumann*, 1967, p. 119):

$$H \left[\begin{array}{c} R_1 \quad\quad R_2 \quad\quad\quad R \\ | \quad\quad\quad | \quad\quad\quad\quad | \\ -N-C-C-N-C-C....N-C-C- \\ | \ | \ \| \ \ | \ | \ \| \quad\quad | \ | \ \| \\ H \ H \ O \ H \ H \ O \quad\quad H \ H \ O \end{array} \right] OH$$

I

Collagenes from skin and bones differ externally from each other: collagen from skin is tough-elastic and contains water, collagen from bones is hard and poor of water (*Sauer*, 1958, 0. 22). Nevertheless, there is no difference between glues manufactured from both kinds of collagen under the assumption that the degradation is the same.

Hide glues should be colorless, transparent, amorphous and chemically neutral without taste or odor. The technically purest form of the glue is gelatine. Formerly animal glues were offered in the form of small slabs or cakes, but now, for improving the swelling properties they can be delivered in the form of powder, granules, pearls, or small cubes.

Animal glues for use in smaller quantities can be obtained as jelly glues, which are less expensive than dry glues and ready for heating. For industries with a large consumption of animal glues the delivery of warm concentrated solutions by tank-trucks is possible.

Skin or bone glues are practically free of acids. For bone glues the lowest permissible pH is 4. Dry animal glue should not contain more than 17% moisture content, based on the dry weight. There is a very strong interrelationship between viscosity and concentration. According to *Plath* (1951) increasing temperature lowers the viscosity (Fig. 1.31).

Before closing this section it may be mentioned that warm animal glue, spread on a wood surface, rapidly cools down and passes through a tacky phase. It is very important that the joint is formed before the glue has passed through this phase. In this way the original strength of the wood may be reached. Two different influences are effective:

1. The progression of geling (maturing), and
2. The loss of water from the glue joint by diffusion.

A very important aspect is proper usage of animal glues in the woodworking industries or workshops. Overheating is dangerous as shown for example in Fig. 1.32.

Blending of animal glues with fillers or extenders is common. The influence on the bending strength of the glue joints is remarkable if the content of dry

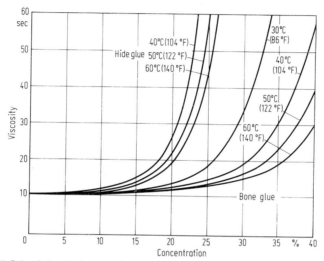

Fig. 1.31. Interrelationship between viscosity, concentration and temperature for hide glues and bone glues. From *Plath* (1951), p. 68

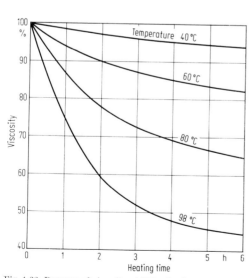

Fig. 1.32. Decrease of viscosity of hide glue by overheating. From *Meess* (1930)

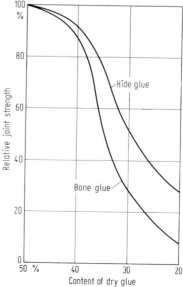

Fig. 1.33. Decrease of relative glue joint strength for hide glues and bone glues with decrease of concentration. From *Meess* (1930)

glue decreases below 40% (Fig. 1.33). It is possible to make hide glues suitable for cold pressing by addition of acids or salts. In this case the "feverish phase" is less pronounced than for normal animal glues, but the gluing pressures must be increased to 6 kp/cm² (85 lb./sq. in.). All animal glues normally are not resistant against the influence of elevated moisture content (Fig. 1.34). On the other hand, according to the experiences and results of the U.S. Forest Products Laboratory in Madison (*Truax*, 1929) it is possible to "temper or harden" animal glues by addition of tanning agents for improving the resistance against humidity. Fig. 1.35 shows to which extent the working life of tanned animal glues is improved for several temperatures with addition of oxalic acids.

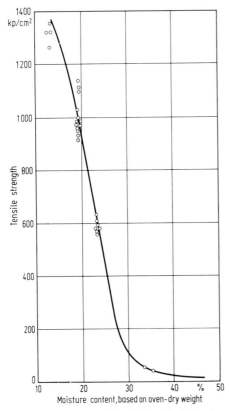

Fig. 1.34. Dependence of tensile strength of hide glue on moisture content. From *Bateman* and *Town* (1923)

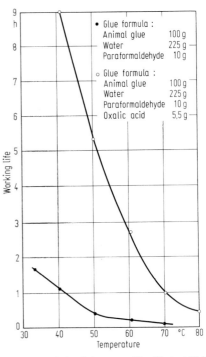

Fig. 1.35. Relation between working life (pot-life) and temperature of animal glues containing paraformaldehyde and oxalic acid. From *Browne* and *Hrubesky* (1927)

The pressure to be applied and the temperature of the room where the wood is stored have a remarkable influence on the shear strength of the glue joints. *Truax* (1929) has published Fig. 1.36.

Important are the temperatures. Fig. 1.36 shows that at room temperature (21 °C = 70 °F) wood gives the highest possible shear strength if a pressure of at least 30 kp/cm² (about 420 lb./sq. in.) is applied. For a temperature of 32 °C (90 °F) a pressure of only about 5 kp/cm² (70 lb./sq. in.) delivers the maximum joint strength. In this connection the problem of starved joints has to be observed.

Many conditions may affect the efficiency of glue joints, e.g. not only the temperature and pressure, but also the open and closed assembly time. The lower the temperature of room and wood the shorter the so-called open time may be kept (< 3 min.)

Staining of veneer is less for hide glues than for bone glues; it is influenced by the free acidity of the glues.

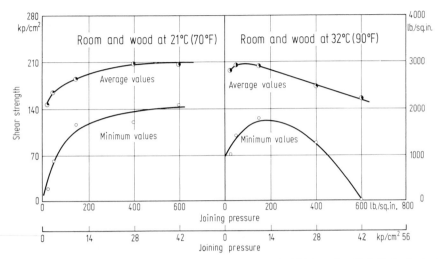

Fig. 1.36. Relation between gluing (joining) pressure and shear strength of joints; yellow birch blocks glued with an animal glue. From *Truax* (1929)

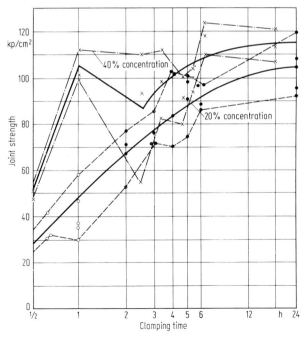

Fig. 1.37. Joint strength as function of clamping time. From *Plath* (1951)

The duration of pressing (dependent on the rate of drying of the glue, the viscosity of the glue, the moisture content and thickness of the wood) for hide glues normally ranges between 30 and 60 min. In veneering several hours are necessary, less for hot pressing. In this connection once more the "feverish phase" may be mentioned (cf. Fig. 1.37).

Animal glues are still used in the whole world because they are to some extent low in cost and they form rather strong joints for interior wood work. Of course, they are sensitive to the influence of moisture content and microorganisms. There is an influence of the wood species on the shear strength on the one hand and on the relationship to wood failure on the other hand. The shear strength is not so instructive as the ratio of failure is. *Truax* (1929) has shown that the shear strength varies irregularly between about 86 and 254 kp/cm² (1,220 to 3,620 lb./sq. in.) for 40 species but that the ratio of wood failures decreases in a nearly parabolic function from 100% to 55%.

1.10.2.2 Casein Glues. Casein glues have a long history. They were used in ancient China, Hellas, Rome and Egypt. The monk *Theophilus* described in the 11th or 12th century how to produce water resistant casein glues. Until about 1935 casein glues were the only adhesives suitable for high stresses, e.g. for aircraft constructions.

Casein glues are made by dissolving casein, obtained from milk, in an aqueous alkaline solvent.

Casein glues are available on the market in powder form. Casein glues have the following advantages:

1. Relatively high resistance of the glue joints against the influence of moisture. Therefore the structural parts glued with casein glues do not tend to delaminate.

2. A low sensitivity against elevated temperatures. According to investigations in the U.S. Forest Products Laboratory (1956) casein glues may be superior in their behavior to some synthetic resin glues up to 33 °C (91 °F) temperature and relative humidity as low as 20%.

3. Casein is a better gap-filling glue than many others. Therefore, it is possible to glue even relatively rough wood surfaces with favorable results but one has to consider that the rate of bonding decreases with joint thickness.

4. Casein glues are easy to be handled.

5. The pressure to be applied is low.

6. The alkalinity of the casein glues offers two aspects. On the one hand for valuable veneer containing tannin it may be dangerous with respect to staining, on the other hand it can facilitate the gluing of wood species containing resin or oils.

7. Casein glues are compatible with blood albumin glues.

Of course, the advantages are in balance with disadvantages as follows:

1. Due to the high amount of alkali staining of wood especially of highly popular species like oak, walnut, and mahogany is nearly unavoidable.

2. Since casein glues contain normally a high amount of mineral substances they contribute to rapid wearing of woodworking tools.

Practically all the casein for gluing purposes is produced from skimmed milk by precipitation with suitable acids (*Spellacy*, 1953). The composition of commercial casein may vary between the following approximate limits: protein 78 to 90%, water 7 to 12%, ash 1 to 4%, butter fat 0.1 to 3%, and lactose 0 to 4%. The density of chemically pure casein is 1.259 g/cm³, according to *Sutermeister* and *Brühl* (1932). The glues are prepared by dissolving casein in an alkaline water solution. The alkalinity of casein glues may vary between pH 9 to 13. There exists a dependence between alkalinity and time on the one hand and viscosity on the other hand (*Browne* and *Brouse*, in *Delmonte*, 1947, p. 260). Curves obtained by *Zoller* (1920, p. 635) are instructive. With the exception of Ammonia ($NH_3 \cdot H_2O$) for all tested alkalies the viscosity of the casein-mixtures rapidly decreased when the pH value 9.2 was reached (Fig. 1.38). This phenomenon is due to the quick splitting of the casein molecules. The first split products still have a high adhesive power but with proceeding hydrolysis or under the influence of bacteria, the glue becomes insufficient. It is possible to prevent biological decomposition by the addition of phenol (C_5H_5OH), thymol ($CH_3(C_3H_7)C_6H_3OH$), salol or phenyl salicylate ($C_6H_4(OH)COOC_6H_5$), phenyl ester ($HOC_6H_4COOC_6H_5$), β-naphthol ($C_{10}H_7OH$) or — according to American experiments — by the addition of copper salts. The U.S. Forest Products Laboratory recommends dissolving 2 to 3 parts per weight of copper chloride ($CuCl_2$) or copper sulfate ($CuSO_4$) in 30 parts of water. This solution is sufficient for 300 parts by weight of glue mixture. On the other hand there is no practical means to retard hydrolysis. This is the main reason that cold-setting casein-glues in the liquid state have only a limited pot-life, seldom more than three to four hours.

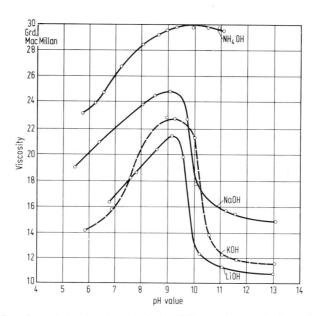

Fig. 1.38. Dependence of viscosity of casein solutions (9%) in various alkalies. From *Zoller* (1920)

After the evaporation of water, casein glue joints are strong but not very resistant against moisture. If irreversible gels are to be obtained the following possibilities do exist:

1. Hardening by means of formaldehyde as for animal glues which is not really satisfying;

2. Conversion of casein-alkalinate into a glue made from casein and calcium hydroxide.

Kragh and *Wootton* (1965) gave a very clear explanation as follows: "A glue may be made from casein and sodium hydroxide or ammonium hydroxide which has a relatively long working life but poor water resistance. A glue made from casein and calcium hydroxide has good water-resisting properties but gels irreversibly within an hour or two. The working life may be extended to several hours by the addition of sodium hydroxide. Casein-calcium hydroxide-sodium mixtures have their limitations. Small variations in sodium hydroxide change the solution properties appreciably."

There exists a long list of formulations for more or less water resistant casein glues. It is not in the scope of this book dealing with principles of wood science and technology to give a survey over the various formulas, but one example may be mentioned according to U. S. Patent No 1,456,842 by *Butterman* and *Coopeider* (19 23):

Ingredients	Parts of weight	Sequence of mixture	
Casein	100	⎫ 1	⎫
Water	220···230	⎬	⎪
			⎬ 3
Calcium hydroxide (hydrated lime)	20··· 30	⎫ 2	⎪
Water	100	⎬	⎭
Alkali silicate	70	⎫	
Copper chloride	2··· 3	⎬ 4	
Water	30	⎭	

For further details about the formulation of casein glues the chapter by *Kragh* and *Wootton* (1965) and the article by *Browne* and *Brouse* (1939, p. 247 to 253) should be consulted.

The rough casein powder should be stored dry and — if possible — air-tight. For hardwood per 1 kg (2.2 lb.) glue powder 1.6 to 1.9 kg (3.7 to 4.2 lb.) of water are necessary. For softwood the proportion is 1.8 to 2.2 (4.0 to 4.8 lb.) water. The specific weight of the loosely poured glue powder amounts to 1.4 to 1.7 g/cm³; therefore the proportion in volume parts ranges between about 1:1 to 1:1.6. The influence of the volume proportion of glue to water for butt joints of pine woods is illustrated in Fig. 1.39 (*Koch* and *Sachsenberg*, 1932, p. 89—91). The curve is instructive insofar as the bending strength of the butt joint increases up to a critical point of solids content (1:1 glue to water). Further it increases at a reduced rate.

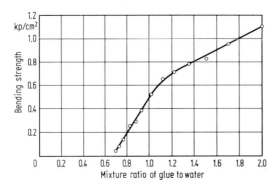

Fig. 1.39. Dependence of bending strength of casein glue joints across the grain (pine) on the ratio of glue to water. From *Koch* and *Sachsenberg* (1932)

Apparently the mixture is "oversaturated". Another very important point is that pot-life is a function of the relative alkali content (Fig. 1.40). Prolonged pot-life is due to the presence of colloidal silica acid. The relationships are complicated and not yet quite clear. *Narayanamurti* (1954, p. 40) published a table showing the approximate changes in viscosity with time for a casein-sodium hydroxide solution. The jelly strength reaches a maximum within two or three hours. Then hydrolysis reduces the viscosity but after a few days jellies are formed which remain strong for a long time.

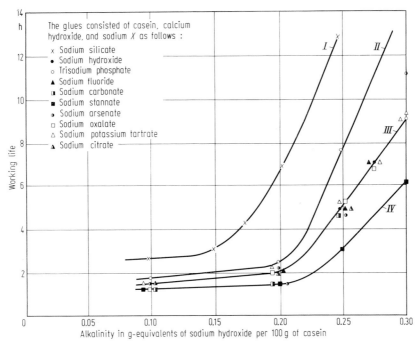

Fig. 1.40. Dependence of working life (pot-life) of casein glue on the alkalinity. Content of Ca(OH)₂ 16 g per 100 g casein plus the chemical equivalent of the sodium. Water content 400 g in glues of low alkalinity and 300 g in glues of 0.2 g equivalent NaOH or more. From *Sutermeister* and *Browne* (1939)

Casein glues are stirred with cold water (15 to 20 °C). Heating of the glues is not allowed. The pots for the glue should not be corroded under the influence of alkali. Zinc, aluminum, copper and iron are to be excluded, but vats made of wood or earthenware and enamel containers are suitable.

Cleanliness is imperative, since otherwise the glue will be rapidly decomposed. Decomposition reduces the viscosity and lowers the joint strength.

The casein glues should be spread evenly, but neither the open waitin/g time (between 1 and 25 min) nor the amount of spread (between 200 and 550 gm² or 0.04 and 0.11 lb./sq. ft.) has a major influence on joint strength. The average shear strength amounts to about 240 kp/cm² (3,400 lb./sq. in.) ± 18%, indicating a very acceptable variance for glue joints. *Truax* (1929) published for 40 wood species the following average values: Shear strength 84 to 238 kp/cm² (1,190 to 3,390 lb./sq. in.), wood failure 16 to 100%. Fig. 1.41 shows a casein glue joint with rather thin glue line especially between latewood layers. The wood was Douglas fir and the dark zones near the joint in the springwood are glue stains.

Pressing of casein glue joints should not be done suddenly but gradually in about three stages with a difference of 3 min. Pressure may range from 4 to 14 kp/cm² or 56 to 198 lb./sq. in.; in the average of 6 to 8 kp/cm² or 84 to 112 lb./sq. in. is recommended, although the influence of gluing pressure on joint strength is relatively small (Fig. 1.42). Quite another thing is the influence of wood moisture content on joint strength. Fig. 1.43, taken from a publication by *Kraemer* (1932

Fig. 1.41. Photomicrograph of a glue joint between two Douglas fir blocks glued with casein glue, very little penetration, thin glue line, weak joints between summerwood portion, dark areas along glue lines in springwood produced by staining from the glue. From *Truax* (1929), Plate 9

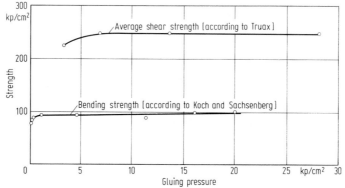

Fig. 1.42. Effect of gluing pressure on bending strength and shear strength of woods glued with casein glue. From *Truax* (1929), p. 39 and *Koch* and *Sachsenberg* (1932)

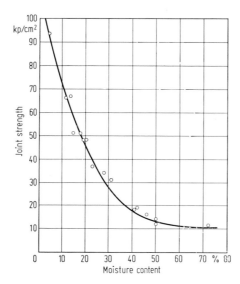

Fig. 1.43. Dependence of joint strength of casein glue joints on the moisture content. From *Kraemer* (1932)

p. 443) shows the hyperbolic effect of moisture content up to 75%. Joint strength at 10% moisture content is about 70 kp/cm² (1,000 lb./sq. in.), at 20% approximately 45 kp/cm² (640 lb./sq. in.) and at 30% about 30 kp/cm² (430 lb./sq. in.). For still higher moisture contents the decrease of strength is less, but the minimum for about 75% moisture content is only 10 kp/cm² (142 lb./sq. in.).

1.10.2.3 Blood Albumin Glues. In most countries blood albumin glues are only rarely used. The soluble albumin is a byproduct in slaughterhouses. The blood, first freed from fibrin, is evaporated as liquid residue at a temperature below the coagulation point (about 60 °C) (*Knight*, 1946). Coagulated albumin is insoluble and has practically no adhesive properties.

Fresh blood in liquid form has a storage life of only a few days. Addition of alkali to the albumin-water mixture improves the adhesive properties (*Kragh* and *Wootton*, 1965). The following are the maximum allowable impurities: fat up to 1%, ash up to 10% and water up to 10%.

Blood albumin glue is comparatively scarce. As an example of preparing blood glue one formula given by the U. S. Forest Products Laboratory (after *Lindner*, U. S. Patent No 1,459,541, 1923) may be quoted:

Ingredients	Part by weight	Procedure
Blood albumin (90% soluble)	100	1. 1···2 h slowly soaking and gently stirring
Water	140···200	
Ammonium hydroxide		2. Add the ammonia with more stirring
(0.88 sp. g.)	5.5	
Paraformaldehyde	15	3. Sift in the paraformaldehyde stirring the mixture rapidly

The mix is at first very thick, but will regain a normal consistency within about an hour. Hot pressing is recommended. For all practical purposes blood albumin glue is confined to plywood production by hot-pressing. Roller-type machines are used for spreading, usually at a rate which varies in a wide range. The following figures are taken from the literature and rounded off (1,000 g/m² = 0.2048 lb./sq. ft.).

Table 1.5. Glue Spread for Albumin Glues

Source	Glue spread	
	g/m²	lb./sq. ft.
Truax (1929, p. 50)	330···400	0.067···0.082
Mora (1932, p. 99)	158···488	0.032···0.10
Knight (1946)	320···365	0.066···0.074
Marian and *Fickler* in: *Kollmann* (1955)		
Table VI	200···350	0.041···0.072

There are significant differences in the glue spread and properties, especially compared with synthetic glues. The reasons for these differences can be explained as follows:

1. The viscosity of blood albumin glue increases rapidly with time. *Knight* (1952) has published very instructive figures:

Changes in viscosity with time for a blood albumin glue

Time (h)	025	1	2	4	6
Viscosity (poises)	185	320	485	545	550

2. The properties of blood albumin glues vary according to manufacture, especially the removal of fibrin and the processing temperature.

3. Not all glue spreading machines are equally effective for blood albumin glues. Generally double-roller machines are most suitable. It is essential to spread an adequate amount of glue. It is better to spread too much glue than too less.

4. Veneer needs less glue than thick wood pieces.

5. Higher speed in spreading is necessary where the consistency of the glue is rapidly increased. The working life of blood glues can be considerably increased by the addition of ammonia, lime or caustic soda (up to several days).

6. Closed assembly time is between 1 and 25 min.

7. Blood albumin glues can be mixed with casein, soybean-meal as well as synthetic resins. The German "AWF" (Ausschuß für wirtschaftliche Fertigung, Committee for Economic Production) recommends the following formula:

a) 40 parts by weight of albumin, dissolved in 60 parts of water.

b) To this solution may be added 100 parts of urea-formaldehyde resin (e.g. Kaurit) and 15 parts by weight of hot hardener; it is possible to increase the proportion of urea-formaldehyde up to threefold. In this case the bad smell of the albumin glue disappears.

8. Blood albumin glues may be pressed cold or hot. In the plywood industry hot pressing is the rule. Temperatures between 80 and 120 °C (176 and 248 °F, are common, but generally temperatures above 100 °C (212 °F) should be avoided since steam-blisters in the glue line can arise.

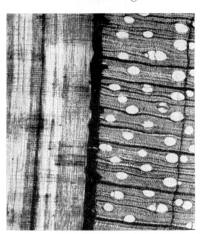 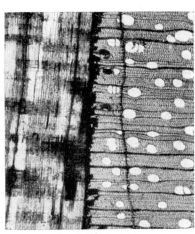

a   b

Fig. 1.44a, b. Photomicrograph of glue joint, birch plywood, glued with blood glue.
a) 3.5 kp/cm² (50 lb/sq. in.) gluing pressure; b) 14 kp/cm² (200 lb/sq. in.) gluing pressure

9. Pressures range from 3 to 30 kp/cm² (42.5 lb./sq. in. and 425 lb./sq. in.) Fig. 1.44a shows a photomicrograph of birch plywood glued with blood albumin glue under a pressure of 3.5 kp/cm² (50 lb./sq. in.) Fig. 1.44b illustrates the same material manufactured at 14 kp/cm² (200 lb./sq. in.) (*Truax*, 1929). There is an apparent difference of glue joints in thickness and penetration of glue into wood tissues as a function of pressure. For the low pressure one can see a rather thick

coherent glue line and scarcely penetration of glue into the adjacent veneer. For the high pressure the glue line is inhomogeneous, but much more glue has penetrated into the wood perpendicular to the grain. Nevertheless, *Truax* (1929), who published these photomicrographs, writes (explanations to Plate 9, after page 32): "In spite of differences in penetration and glue-line thickness, the joints gave practically the same strength when tested".

10. Blood glue gives a moderately strong bond with a high resistance even to boiling water. Microorganisms, however, can destroy the glue very quickly under wet conditions. The addition of paraformaldehyde only retards the decomposition.

11. In Western Europe, blood glues are not longer of importance. In the U. S. A. they are only used to some extent in the plywood industry on the East Coast. The blood albumin is combined with phenolic resin and a catalyst. The setting temperature is between 110 °C (230 °F) and 118 °C (244 °F) which is less than that required for pure phenolic resins. This is favorable when working with Douglas fir. The bond is both water and mould resistant.

1.10.2.4 Soyabean (or Soybean) Glues and Peanut-meal Glues.

Soyabean is a small erect herb of the bean family originally cultivated in China (Mandjuria), then in India, in Japan and to a large extent in the U. S. A. At first soyabeans were used and improved for many purposes, probably mainly as food stuff. The most valuable constituent of soyabeans were the oils. The oils have been extracted from the beans, but soon it was discovered that the crushed residues could be refined including addition of various chemicals.

The soyabean was originally exported from the Orient, and its oil-free residues were developed as a source of glue in the United States. Wood glues extracted from the soyabean found the attention of plywood manufacturers in the Pacific Northwest around 1923. They made good bonds and have a good resistance against the influence of water. The Douglas fir plywood industry is the largest consumer of soybean glue. Only after 1926 soybean glues became commercially important, but now this type of glue is one of the standard adhesives. The application turned from the Douglas fir and East Coast pine industries to other parts of the U. S. A. According to one of the most knowledgeable experts (*Perry*, 1948), *Laucks* (1923, 1939) was the pioneer in the field of developing suitable wood glues from soybean extracts. His developments have continued and the result is that his company is the leading producer of soybean glues in the U. S. A.

The most important details about manufacture, application, and properties of soybean glues are as follows:

1. In the U. S. A. there are now many important producers of a wide variety of soybean products.

2. Soyabean can be purchased as whitish powder in a dry state mixed before use only by the addition of water. More often the plywood factories have their own formulas, and they add lime, caustic soda, silicate of soda, and other special chemicals.

3. Prior to the introduction of urea-formaldehyde and phenolic resins, soybean glues had approximately the same water resistance as albumin glues.

4. Soybean glues are similar in their properties to casein adhesives.

5. Soybean glues are probably the lowest-cost, water resistant glues available. Of course, as stated by *Perry* (1948, p. 93): "There are a number of different grades, ranging from the minimum cost glue-base in the box-shook-industry to a highly refined chemically extracted soybean protein material that is not unlike casein in its physical and adhesive qualities".

6. Soybean glues can be used on hardwood veneer with a moisture content of up to 16%. Veneer with a relatively high moisture content gives reduced breakage and greater yield (cf. No 10).

7. The high caustic content of soybean glue mixtures has a tendency to stain various woods. If the pH value of the adhesives is too high, staining on the face veneer may occur in the form of bluish or dirty-brown stripes.

8. For cold pressing soybean glues should be spread approximately in the amount of 290 to 330 g/m² or 0.059 to 0.067 lb./sq. ft. (*Perry*, 1948, p. 94). *Kollmann* (1955, p. 1008), based on data given by *Marian* and *Fickler* (1953, p. 37—40), recommends quite similar values, namely 250 to 300 g/m² or 0.051 to 0.062 lb./sq. ft. For hot pressing 190 g/m² or 0.039 lb./sq. ft. is appropriate for a single glue joint. In this case the use of spirally grooved double-rubber rolls is useful.

9. The allowable assembly time for soybean glues varies with the water content of the mixture and for cold pressing may be from 15 to 25 min (*Perry*, 1948, p. 95) and for hot pressing from 5 to 15 min.

10. In the case of Douglas fir and other softwoods in the manufacture of plywood on the Pacific Coast, it is important to dry the veneer to the minimum value of 0 to 5% moisture content before spreading with soybean glue and then assembling into plywood.

11. Glue pressure for soybean glues is between 8 and 10 kp/cm² (114 and 142 lb./sq. in.) nearly in accordance with the recommendations of *Wood* and *Linn* (1950, p. 88) (125 to 150 lb. /sq. in.).

12. Pressing procedure:

a) Hydraulic cold pressing as soon as possible after glue spreading, the assembly time may be from 15 to 25 min between the application of the glue and the completion of pressure.

b) In hot pressing the permissible assembly time is shorter, as a rule between 5 and 15 min. In modern plants automatic loaders for hydraulic presses with 15 to 20 openings make this possible. The temperature is about 120 °C (248 °F).

c) The plywood can be stacked in a pile immediately after removal from the hot press and covered with an insulating cover for a postcuring treatment.

Properties and application of peanut meal glues are similar to those of soybean glues.

1.10.3 Shellac

Shellac is a resinous substance on the young twigs of various bushes and trees in East-India formed due to the sting of a small female insect (*Laccifer lacca* Kerr). The insects become fixed to the bark. They transform, by digestive processes, the tree sap into a resinous matter which they excrete over their whole bodies. The resin covers the brood of the insect. The resin with the carmine colored insects will be cooked with a weak solution of soda to produce the yellow-brown seed-lac. Shellac is molten and cast in form of plates and cakes.

Today shellac is used nearly exclusively for making varnishes and special cements. Formerly it was the main constituent of marine-glues for wooden chips. At present shellac plays no more a roll as a glue for wood.

1.10.4 Asphalt (Bitumen) and Mastic

Asphalt is any of various black or black-brown solid bituminous substances, occurring native and as a by-product of petroleum refinement. Asphalt is soluble in turpentine, petroleum and benzene and melting at 100 °C (212 °F). It is used nearly exclusively for building purposes (flooring, roofing as a barrier against earthmoisture) and for street pavements. In the woodworking industry asphalt as a glue has no importance, but wood paving blocks (spruce, fir, pine, for heavy loads beech and oak, impregnated with coal tar) are glued to concrete floors with special bituminous mixtures.

1.10.5 Natural and Synthetic Rubber (Neoprene)

Rubber is found widely in the vegetable kingdom. It is an elastic material, free of protein, always amorphous, but swells in water to a mucous mass. The swelling capacity varies from apparent solutions to more or less firm gellies. The main constituent of all gummy types is arabonic acid ($CH_2OH(CHOH)_3COOH$).

Caoutchouc, though called elastic gum or India rubber, does not belong to the family of rubbers. Nevertheless Caoutchouc (mixed with sulfur and heated to about 140 °C (284 °F)) has special properties: It is possible to produce tubes, plates e.g. pump flaps, rubber band, tennis balls, and tires. Adhesives from natural rubber can produce good joints between wood parts by hot pressing.

The joints made with commercial products are strong, both dry and wet, not when they are exposed to high relative humidity, but the adhesives are too expensive (*Truax*, 1932). Joints of wood to metal at room temperature are possible, but they do not have the resistance of hot-setting or two-step adhesive joints. Casein-rubber-latex adhesives have been used successfully at room temperature for bonding wood to metal at low stress levels. Some at room temperature setting adhesives have been developed with epichlorohydrin resins ($H_2C-CH-CH_2-Cl$).

$$\overset{}{\underset{\diagdown O \diagup}{}}$$

Practically nothing is known about the behavior of these adhesives under the influence of stress, moisture, and elevated temperatures.

In addition to the room temperature setting adhesives for wood to metal bonds there are some mastic and solvent type adhesives. They need only a low initial pressure, but the bond strength is rather low. The mastic and solvent-base adhesives are usually liquids or pastes consisting of synthetic or reclaimed rubber, asphalt, or certain low-cost resins that are dispersed either in inexpensive organic solvents or in water (*Eickner* and *Blomquist*, 1951). Neoprene is a linear polymer of chloroprene ($CH_2=CHOCl=CH_2$). It is stiff and slightly rubbery. Vulcanized with heat it becomes thermosetting.

Neoprene is used mainly in handicraft and small wood industries (for special purposes) and in combination with phenolic resin for bonding wood to metal. Neoprene has a remarkable compatibility with most resins.

As an example the following formulation may be given (*C. A. A. Rayner*, 1965, p. 331):

	Parts by weight
Neoprene	100
Phenolic resin	20···50
Magnesium oxide or zinc oxide	4··· 8
Small amount of sulfur	
Accelerator (for instance thiocarbanilide)	
Antioxidant	
Fillers (sometimes)	up to 50

4*

It may be that Neoprene glues in combination with phenolic resins as two-polymer adhesives, at present predominantly used for metal joints, become more important for wood gluing in the future, especially for bonds of wood to metal. Neoprene cements are extremely tacky, and therefore the open assembly time should be kept as short as possible. In this connection a few words about "tack" are necessary. Tack is a very complex phenomenon of decisive importance for adhesion. *Wake* (1961, p. 193) some time ago gave an analysis of rheological phenomena related to tacky behavior of adhesives. Tack is a property only of viscoelastic materials. Newtonian liquids, e.g. water, are not tacky. Tack contributes to adhesive bonding. A cross-linked rubber is not tacky because it does not flow. It is the energy required and not the force applied that is the source of tack.

Finally, it must be said that two-phase polymer systems consisting of rubber and thermoplastic are extremely interesting. *Salomon* (1965, p. 75/76) writes the following: "Organic adhesives consist frequently of two partly immiscible components differing widely in rheological properties. Nitrile rubbers and nylons are added to high modulus epoxy or phenolic resins as toughening agents. As the term implies, the brittleness of the polymer matrix is considerably reduced by this second phase. The dispersed polymer produces a seperate damping peak in the

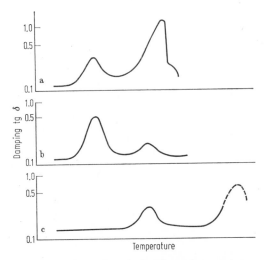

Fig. 1.45 a – c. Damping in two-phase polymer systems.

a) Rubber added as toughener to hard thermoplastic matrix; b) Rubber matrix stiffened by addition of hard thermoplastic material; c) Thermoplastic toughener added to thermohardening, brittle resin; the latter becomes thermally instable before damping maximum is reached. From *Salomon* (1965, p. 76)

mechanical spectrum; in other words, the damping properties are roughly additive. Stiffening of a rubber matrix by dispersion of a high-modulus thermoplastic filler can also be produced, and a corresponding damping peak for the stiffening agent is found at higher temperatures" (Fig. 1.45).

Rubber and latex have a wide variety of applications in industry and certainly to a limited extent in the wood industry. Table 1.6 gives a survey according to *Simonds, Weith,* and *Bigelow* (1949).

Table 1.6. Adhesives on Rubber and Cellulose Base and their Application
(From *Simonds*, *Weith*, and *Bigelow*)

Type	Nature of cement	Adheres to:	Typical industrial applications
Rubber Type A	Air-drying compounded rubber base solvent cement	Metals Rubber *Wood* Glass Plastics Stone Fabrics Cork Asbestos sheets	Plastic, metal, and linoleum covered tables Metal letters to glass, *wood*, etc. Clock construction Typewriter sound deadening Bonding anti-drumming materials Bonding mirrors in vanity cases Attaching handles to cardboard cartons *Furniture construction* Aircraft and auto sound insulation Foil-lacquered labels to cellophane
Rubber Type B	Air-drying rubber solvent cement	Rubber Leather Rubberized fabrics *Wood* Cloth Paper	*Sports goods* Felt insulation of metal, wood, etc. Cardboard box construction Automobile and aircraft Upholstering adhesive Blueprint adhesive *General household adhesive* Photo-mounting Attaching protective coatings to highly polished metal surfaces
Rubber Type C	Air-drying aqueous suspension of rubber	Leather Rubber Fabrics Paper *Wood*	Label adhesive Airplane and automobile upholstery *Suit case construction* Coardboard containers *Rubber sheeting to wood and plaster board*
Chloroprene selfvulcanizing	Air-drying rubberlike synthetic cement	Leather Rubber Rubberlike synthetic materials Cloth	Seaming coated fabrics Cable sheaths insulation Gasoline-proofing riveted metal seams Butt-jointing rubber and rubberlike synthetic materials Bonding oil-and-gasoline-resistant linings to outer sheaths, or to frictioned fabric Seaming garments and aprons for oil field use Seaming fabrics for gas bag
Thermosetting chlorinated rubber-chloroprene combination	Air-drying rubberlike synthetic cements	Rubber Metal Synthetic rubberlike materials Fabrics Braid	Bonding synthetic rubberlike materials to metal cable-armor Rubber-covered steel roll Bonding metal inserts to rubber and rubberlike synthetic materials Bonding resilient covering to steel keg and other containers Bonding steel and aluminium nipples and couplings to hose without mechanical attachment
Thermosetting cyclo-rubber chloroprene combination	Air-drying rubberlike synthetic cements	Rubber Metal Synthetic rubberlike materials Fabrics Braid	Bonding rubber to metal Bonding synthetic rubberlike material to fabel or braid Bonding rubber to synthetic rubberlike material Bonding certain synthetic rubberlike oil-resisting materials to the same material

Table 1.6 (Continued)

Type	Nature of cement	Adheres to:	Typical industrial applications
Cellulose Ester	Air-drying solvent nitro- cellulose cement	Leather Metals Cloth Wood Rubber	General office adhesives *Toy and novelty construction* *Furniture construction*

1.10.6 Glues on Cellulose Basis

1.10.6.1 Cellulose Acetates. Cellulose acetates are produced by acetylation of cotton linters with the aid of glacial acetic acid or acid anhydride in a suitable mixing process at a temperature of 5 °C (41 °F) before the catalyst, sulfuric acid, is added. There are three stages of acetylation (*Simonds, Weith*, and *Bigelow*, 1949). Cellulose acetates have been suggested for various industrial applications included veneering. Fig. 1.46 shows the flow of cellulose materials through various manufacturing processes to fibers, sheet plastics, films, and molding powders (*Simonds, Weith*, and *Bigelow*, 1949). In the modern plywood industry cellulose acetates are of no importance.

1.10.6.2 Cellulose Esters. The mechanism of esterification of cellulose is a complex one. Cellulose cements have been suggested often as adhesives for wood, and joints have been tested. In most cases the glues are too thin and need more than one application. The glue lines are slow drying and relatively expensive. *Gerngross* (1930, p. 428—433, 1949, p. 968—972) reported that with cellulose esters, joints of high strength and good water resistance are possible. Cellulose cements can be cold pressed, but their high cost and difficulties in application make them impracticable for woodworking purposes (*Truax*, 1932).

1.10.7 Silicate of Soda. Silicates of soda make rather good joints in wood of a fairly high dry strength. The silicates of soda are cheap, are easily applied, and are used to a large extent in the manufacture of fiber container. They are also suitable for the gluing of low grade plywood, such as the type used for tea and rubber chests, where only a short life is required. Sodium silicate glue lines lose their bonding power in a year or even less, especially if exposed to damp and wet conditions. Therefore, silicates of soda are not to be classified as permanent binders.

Nevertheless, there are sodium silicate cements available commercially. For use in the plywood industry the following formula is recommended by *Knight* (1946):

Soda to silica ratio   1 to 3.2
Baumé specific gravity   41° to 42°

The sodium silicate cements are clear syrupy liquids. In use they are spread at a rate of about 365 g/m² (0.075 lb./sq. ft.) and pressed cold. The assembly time should be as short as possible.

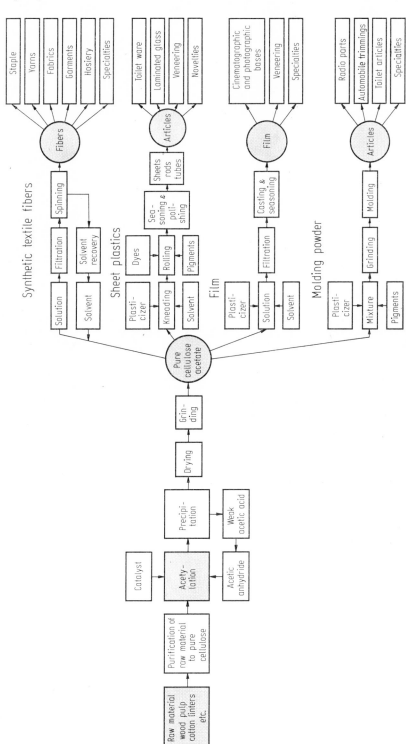

Fig. 1.46. Flow of cellulose materials through various manufacturing processes to the finished articles

1.11 Synthetic Glues

1.11.0   General Considerations

In the foregoing sections glues on the basis of natural products have been discussed. Their advantages and disadvantages were dealt with. Doubtlessly some of these glues have been considerably improved but even the best types are not fully waterproof and can be attacked by microorganisms. Therefore, these glues mainly are used for interior woodwork, in the open air danger of failures exists after some time.

Modern woodworking, structural glued wood members, as are laminated beams and arches, wooden aircraft, radar towers etc., would not be successfully possible without use of synthetic glues. Also the rapid development of particle-board industry is unthinkable without synthetic glues. These glues became available in the thirties. The startpoint was the manufacture of a phenolic film glue, the TEGO-film by the Theo Goldschmidt AG in Essen, Germany. Shortly later urea-formaldehyde resins (e.g. the Kaurit-glue) have been developed. All these glues, including the resorcinol and melamine glues, are thermosetting. Their bonding is irreversible.

Another family of synthetic glues are the thermoplastic types, mainly on the basis of polyvinyl chloride or polyvinyl-acetate. It may be referred to Table 1.7 and to the remark that often glues are offered in rather sophisticated mixtures. There is even no more distinct limit between natural and synthetic glues because both may be combined. The technique of the manufacture of various glues and of their use in practice is so complex that it is impossible to give a survey in details.

Reliable new statistical data on the production or consumption of synthetic adhesives in the whole world are not available. But Table 1.7 shows the production of synthetic adhesives in the U. S. A. in the year 1962. Doubtlessly in the meantime the development has proceeded, but the ratio of the production figures, roughly calculated, is instructive.

Phenolic and other tar acids, adhesive resins	60 %
Urea and melamine adhesive resins	27.3%
Polyvinyl acetate adhesives	10.2%
Epoxy resin adhesives	2.5%
	100 %

Table 1.7. Production of Synthetic Adhesives in the U. S. A. (1962[1])

Type	Production (Dry basis)	
	thousands pounds	metric tons
Phenolic and other tar acids, adhesive resins for:		
Laminating	90,997	41,276
Coated and bonded abrasives	17,874	8,108
Friction materials	24,745	11,224
Thermal insulation	91,566	41,534
Foundry and shell moulding	42,474	19,266
Plywood	72,607	32,934

Table 1.7. (Continued)

Type	Production (Dry basis)	
	thousands pounds	metric tons
Fibrous and granulated wood	15,754	7,146
All other adhesive uses	28,812	13,069
Urea and melamine adhesive resins for:		
Laminating	51,320	23,278
Plywood	97,186	44,083
Fibrous and granulated wood	63,816	28,946
All other adhesive uses	11,892	5,394
Polyvinyl acetate adhesives	79,215	35,931
Epoxy resin adhesives	8,941[2]	4,056

[1] U. S. Tariff Commission, S. O. C. Series P-63-1, March 25, 1963.
[2] Sales figure (production not available).

1.11.1 Phenol-Formaldehyde Resin Glue

1.11.1.1 History of Phenolic Resin Glues. The formation of resins based on the linkage between phenol and formaldehyde has been observed by *Baeyer* as early as 1872. An industrial application of these products was announced in the year 1909 by *Baekeland* (1910, 1912) who mentioned in his early patents the possibility of using condensation products of phenol and formaldehyde for the gluing of wood. Approximately 10 years later *McClain* (1919) applied for a patent of gluing wood with paper or similar sheets impregnated with phenolic resins. Some experts tried to use either phenolic resin dispersions or phenolic resin powders for the gluing of wood. Alcoholic solutions of phenolic resins were also investigated. Practically none of these procedures were successful, perhaps because it was impossible to achieve uniform glue spread with the material available (*Carswell*, 1947). After 1929 this problem was solved by the development of a technically useful, appropriately elastic phenolic resin glue film (TEGO-film) with an adequate shelf life. The brittleness of the phenolic resin film was reduced by Goldschmidt AG (1929) by the addition of glycol ($CH_2OHCHOHCH_2OH$). Approximately 1935 water soluble phenolic resins came into use for impregnation and as binding agents for veneer, plywood, and (compressed) laminated wood. During World War II liquid phenolic resin glues were developed.

1.11.1.2 Manufacture and Chemistry of Phenol-Formaldehyde Resins. For the manufacture of phenolic resins, phenol and formaldehyde are necessary. Besides phenol some other members of the homologous series such as cresols, xylenols, and resorcinols are used.

Phenol and the other compounds in this homologous series are present in coal tar, and they are manufactured from it. There are many industrial processes to produce phenol on an industrial scale synthetically, e.g. by alkali-melting of benzene sulfonamide. Pure phenol (C_6H_5OH) consists of colorless rhombic crystals which are strongly corrosive and poisonous. The same is true for cresol ($CH_3C_6H_4OH$) which can be produced synthetically. Water and occasionally acetone are used as solvents for phenol and cresol in the production of artificial resin glues.

The condensation of phenols with formaldehyde is at first slow and therefore must be sped up by alkali and by elevated temperatures. The formation of the resin progresses then rapidly with loss of water. Probably the first description of the reaction has been published by *Baeyer* (1872, p. 280), but the first industrial utilization of phenol-formaldehyde resins began not earlier than 1909. At this time *Baekeland* (1909, p. 149, and 1913, p. 506) developed a new technique for this type of reaction. The reaction between phenol and formaldehyde is much better known than that between urea and formaldehyde (*Ellis*, 1935; *Martin*, 1956). There are internationally applied terms, such as Novolak, Resol, Resitol and Resite, for the products obtained at various stages of condensation. Novolaks are thermoplastic phenolic resins, meltable and soluble in various organic solvents; they do not lose these properties even during prolonged heating. They are manufacutred mostly from phenol (used in molar excess) and formaldehyde through condensation under acid conditions. Strong acids, e.g. para-toluene sulphonic acid, oxalic acid, and sulfuric acid are used as catalysts.

Resols are phenolic resins produced when formaldehyde is used in molar excess under alkaline conditions. Resols in the "A" stage are soluble in organic solvents mainly lower alcohols and ketons, but in contrast to the novolaks they are irreversibly hardened at higher temperatures or at lower temperatures by the application of strong acids.

If "A" stage resol is heated it is converted to resitol or "B" stage. In this stage the resin is swollen by solvents and becomes rubbery on being heated, due to molecular branching and cross-linking. On further heating the resitol passes to the nearly cross-linked resite or "C" stage. In this stage the resin is practically infusible and insoluble.

Resitols are an intermediate stage of reaction product of phenolic resins. Resites are resins of a high molecular weight, they are completely insoluble in organic solvents, they do not swell any more, and it is impossible to melt them.

In this connection it is necessary to describe the reaction of phenol and formaldehyde. It would, of course, go far beyond the scope and task of this book to go into details. The following schemes elucidate the reactions. In the symmetrically built molecule of benzene (benzol C_6H_6) no position is favored in chemical reactions. Phenol is characterized by an asymmetry due to the presence of an OH group which increases the reactivity with respect to formaldehyde (methanol HCHO). The positions in phenol, resorcinol, and cresol able to react with formaldehyde are shown and marked by arrows below (from *Holzer*, 1962, p. 340):

Phenol   Resorcinol   m-cresol   p-cresol   o-cresol

II

Resorcinol has a second OH group in the metaposition and has therefore a much higher ability to react with formaldehyde than phenol and hence is cold-hardening without the addition of acid catalysts. Also m-cresol reacts better than phenol whilst p-cresol and o-cresol have a remarkably reduced ability to react. They form only linear chains but not crossed-linked systems of the condensation type. In the preparation of thermosetting (resol) adhesives a trifunctional phenol is necessary in order to produce such a three-dimensional cross-linked molecule network.

Phenol-formaldehyde resins first came into use in gluing plywood about 1930. The structure of hardened phenol-formaldehyde resins is not entirely clarified. The alkaline condensation is characterized by the formation of phenolic ethyl alcohol (CH_3CH_2OH) or derivates of methanol (CH_3OH).

Condensation products of phenol and formaldehyde can be either thermosetting (known as "resols") or thermoplastic (known as "novolaks"). The following formulae (IV to VII) show the possible reactions of phenol and formaldehyde according to *Rayner* (1965, p. 210).

The product of the first reaction between phenol and formaldehyde is either ortho- or para-monomethylol phenol. According to *Rayner* (1965) the above survey shows the different types of reaction. Formula III shows the result of the reaction between phenol and formaldehyde, formulae IV and V make clear that a methylol group of one molecule can react either with the nucleus of a second phenolic molecule splitting off water, or (VI) it may react with a methylol group (CH_2OH), attached to another phenolic molecule, losing again water. "The extent to which these reactions take place depends on the ratio of phenol for formaldehyde, the temperature, the pH and the catalyst" (*Rayner*, 1965).

For the novolak formation the formulae IV and VII are illustrative, for the resol formation V and VIII. Based on the scheme II, *Holzer* (1962) discussed the ability of various phenols to react with formaldeyde. In the manufacture of water soluble resols, commonly used for gluing purposes, the ratio of formaldehyde to phenol ranges from 1:1 to 3:1. Both chemicals react in alkaline solution under application of heat. Normal catalysts are strong alkalis such as sodium hydroxide or ammonia. In the first stage of the reaction mainly phenol alcohols are formed (IX). These further react forming ether-or methylol groups (X).

Under alkaline conditions the formation of methylol groups is rapid but can be slowed down by cooling if the resin has a useful concentration for practicable application. In such a way glue solutions are obtained which can be either concentrated or dried by spraying. As for nearly all solutions of thermosetting resins polymerization continues slowly, and therefore storage life is limited. Heating or the addition of acid catalysts starts again the reaction of condensation until a completely cross-linked resite ("C" stage) resin is obtained (XI).

IX

X

XI

Hultzsch (1940) has given an instructive and simple review on the formation of phenol resins (Fig. 1.47). The chemistry of phenolic resins is comprehensively described in the English literature by *Carswell* (1947), *Martin* (1956), and *Megson* (1958). Knowledge of the chemistry of phenolic resins is important for understanding the types of resins used as adhesives.

Methylphenol-formaldehyde resin as an adhesive for wood was investigated by *Guiher* (1970). The adhesive bonds in Douglas fir plywood with this type of resin are equal to those formed by phenol-formaldehyde. Higher specific gravities of veneer produced stronger plywood. Walnut shell flour improved glue joint strength when added in an amount of 10 to 20% based on the solids content of the resin adhesive. Since methylphenol is more expensive than phenol little thought was given towards practical use.

Gumprecht (1969) discussed the "tailoring" phenolic adhesives. He starts with the remark that the chemistry of phenolic adhesives is extremely complex and that there are many reasons for phenol and formaldehyde to combine. In order to be able to "tailor" phenolic adhesives it is first necessary to have some knowledge of the reactions occurring and the products formed. Thin layer chromatography is useful to determine phenolic intermediates.

Chow and *Mukai* (1969) described a method for directly determining the percentage of phenol-formaldehyde resin in cellulose and wood by the combination of standard transmission and differential techniques of infrared spectrometry.

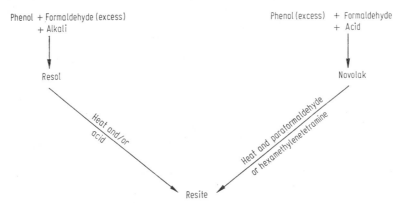

Fig. 1.47. Formation of phenolic resins

Chow and *Hancock* (1969) developed a simple and rapid spectrophotometric method for measuring the degree of cure of phenol-formaldehyde resin glues. The effect of wood extractives and glue additives is negligible. The influence of pressing temperature and time on degree of cure of a phenolic plywood glue can be determined by this method. The quality of a glue bond measured by percentage of wood failure and shear strength of bonds is very sensitive to the degree of cure.

1.11.1.3 Application of Phenolic Resin Glues. Phenolic resin glues are not only used for the manufacture of composite wood panels like plywood and laminated products where the veneers or wood layers are combined with a parallel orientation of fibers, but also for the impregnation of such laminated products even in moulded forms. Phenolic glues are used in the woodworking industry, e.g. for the manufacture of furniture. If odors are not allowed (e.g. for kitchen- and bedroom furniture, panels, and flooring) free phenol or cresol is not admitted in the resin glues. This is also desirable to avoid skin troubles of workers. The best phenolic glues are required for the following applications: sandwich constructions in the production of rolling stock, cars, planes, boats, and ships.

It is possible to control the formulations and the properties of phenolic glues in a very wide range. The following factors may be mentioned:

1. Ratio of formaldehyde to phenol or cresol;
2. Type and dosage of hardener;
3. Amount of polymerization of the resin prior to hardening (*Dietz*, 1949).

Generally, the phenolic glues available on the market are adapted to special uses, and this needs to be considered by the user. It is also possible to impregnate paper, textiles and similar materials and bond these to boards or panels in a two stage process to improve the surface quality of wood and wood based materials. Finally, it has to be mentioned that phenolic resin glues are important for the gluing of metals to wood.

For the application mentioned above resols are used nearly exclusively. The following types of phenolic resin glues are supplied by the chemical industry:

1. Liquid resins, mostly dissolved in water or alcohol;
2. Powdered resins;
3. Resin films.

The hot-setting phenolic glues are of the greatest importance for the plywood industry. They are usually produced with a molar ratio of about 1.5 to 2.25 formaldehyde to 1 of phenol. Cresol resins need a higher concentration of alkali. For plywood manufacture water soluble resols with a low condensation are used. The content of dry resin in the solution amounts to 40 to 50%. These glues are reddish-brown. For assembly work viscous amber-colored liquid phenolic resin glues are used together with liquid hardeners. Their storage life amounts to 2 to 3 months. They have a rather high content of dry resin (65 to 75%), they are not miscible with water, but they may be thinned with organic solvents. As hardeners strong acids, mostly para-toluene sulfonic acid, are used. The normal solution of para-toluene sulfonic acid has a pH of less than 0.5. *Rayner* (1965) gives a summary on the effect of pH on cold-setting phenolic resins. He points out that a "barrier pH" exists, above which curing does not take place at room temperature. This barrier pH depends on the type of phenol and may vary between 1 and 3. According to *Little* and *Pepper* (1947), the barrier pH with ordinary phenol is about pH = 1. The alkaline phenolic glues, used for plywood, may have a pH of 12 or even more. Comparing the extremes of the pH values for liquid phenolic glues, it is remarkable that most species of wood tolerate such a wide pH range. Nevertheless, it is well known that too high and especially too low a pH can have a long-term deleterious effect on wood fibers in close vicinity to the glue line. Water soluble resols were developed for thermosetting. After the glue has been spread, the veneer has to be redried to 6 to 10% moisture content. This rather low moisture content is essential because the pressing temperature was in the range between 135 and 160 °C (275 and 256 °F), nowadays in the plywood industries between 115 and 150 °C (239 and 320 °F). High temperatures over 130 °C frequently cause steam blisters if the moisture content of the veneer is not sufficiently low. An excessive penetration of resin into the wood is also possible. The following points must be observed:

1. The density, i.e. porosity, of the wood;
2. The moisture content of the wood;
3. The initial viscosity of the glue (low viscosity is dangerous);
4. Rate of gelation of the glue;
5. The average molecular weight which for hot setting phenolic glues is low compared with other polymers.

After glue application and redrying veneer can be stored prior to pressing for days or even months. Nevertheless, the whole process is too complicated and the high temperature may have an unfavorable effect on the wood. This was the reason why phenolic resin glues have been developed, setting at much lower temperatures and without the expensive and time consuming predrying. Resins setting by the combined effect of heat and acid catalysts were not successful. Another group of resins used hardeners containing paraformaldehyde. Such resins allow application without any predrying at pressing temperatures between 105 and 115 °C (280 and 239 °F). They have been well received in practice. Some of these products have a long shelf life and in utilization are to some extent similar to urea-formaldehyde resins.

These new types of phenolic glues are classified as medium-temperature ply-wood glues. The explanation is easy: The new resins have a higher molecular weight than formerly, a considerable part of the "curing" is already done in the plant producing the resins. The molecular weights of the medium temperature plywood glues are, as already mentioned, relatively high. The resins are approaching an initial "B" stage. According to *Rayner* (1965, p. 214): "For solutions to have good a storage-life, a dilute (with respect to resin), but rather strongly alkaline solution is necessary; a resin content of 40 to 50% and pH of 12 to 13 are therefore common." To raise the pH of a phenolic glue to 12 to 13 a arge addition of a strong alkali is necessary, much larger than that required to give the same pH value in water alone. Probably, a buffering effect is characteristic of many resinous polymers. When a more rapid hardening of the resin is wanted (moisture content of the veneers to be glued above approximately 10%) it may be favorable to add a reactive accelerator, e.g. resorcinol, resorcinol novolak resin, and paraformaldehyde. *Rayner* (1965, p. 216) gives instructive values for the effect of the addition of paraformaldehyde on the gelation time at 100 °C. If the resin solution is free of paraformaldehyde the gelation time amounts to 36 min, if 2 parts by weight paraformaldehyde are added to 100 parts resin solution the gelation time is shortened to 24 min, if 4 parts by weight paraformaldehyde are added the gelation time decreases to 15 min.

After World War II cold setting phenolic glues were developed. They were applied predominantly in assembly gluing of wood, especially in building construction. The cold-setting phenolic glues are two-part adhesives, consisting of the resin and of a hardener. The addition of the acid hardener causes generally an exothermic reaction. The lower the pH value and the higher the resin concentrations are, the higher the temperature rises. This means a remarkable shortening of the pot-life.

Cold-setting phenolic adhesives deliver good bond strength on curing at low temperatures, even down to the freezing point. The hardeners usually are sulfonic acids. With unsulfonated resins para-toluene sulfonic acid is commonly used as hardener.

The effect of cold-setting acid-catalysed phenolic glues on the wood tissues has been studied in Germany by *Plath* (1953), *Müller* (1953), and *Sodhi* (1957), in the United Kingdom by *Knight* (1959), *Chugg*, and *Gray* (1960), and also in the U. S. A. (U. S. Forest Products Laboratory, 1956). The general opinion at present is the following: Most acid phenolic glues give durable joints over long periods. Under the influence of low as well as high temperatures the wood fibers are attacked by the acid hardeners in the long run. Experiments to eliminate the danger of acid attack on wood by the addition of "protective solutions". were not conclusive. The acid-setting phenolic glues gained only a limited application. Their price lies between urea-formaldehyde and resorcinol-formaldehyde glues. In water-resistance they are near to the latter. In assembly gluing their gap-filling properties are valuable. They are not applicable to dielectric heating since arching often occurs.

Interesting are sulfonated phenolic resin adhesives. The sulfonation converts the water-insoluble resin into one which is soluble. Sulfonated resins can be produced by many different methods. Sulfonated resols may be sprayed or delivered as powder glues which can be easily redissolved in water. Curing can be effected by addition of a suitable acid (hydrochlorid acid).

In the plywood industry fillers and extenders are commonly used with phenolic glues. The glue line becomes relatively thick, starved glue lines are avoided if uneven veneers are flatted by suitable means, e.g. pairs of rollers or pressure, especially after having spread the veneer with water. Fillers reduce absorption and penetration of the glues into the wood tissues. Liquid phenolic glues cured at high temperatures tend to penetrate into the wood. Fillers change the rheological behavior, facilitate spreading, and, at last, make the glue mixture cheaper.

The distinction between fillers and extenders is not clear. The terms often are used synonymously. *Knight* (1959) distinguishes as follows: "'Fillers' are non-reactive components of the mixed adhesives, usually added on the manufacturer's instruction, and extenders are added by the glue user (often without the manufacturer's knowledge or consent)".

Commonly used fillers or extenders are flour of coconut, walnut, or pecan shell, soya bean, and wood. Also low-cost inorganic fillers such as gypsum, powdered chalk, clays, and some oxides and silicates (including magnesia) may be used. There is no limit to the list of insoluble substances to be used as fillers.

The amount of filler to be added varies; it depends on the chemical constitution of the filler, its density, and its particle geometry. One has to distinguish between extenders which participate in the binding process and those which do not. Dried blood albumin and starches, for example, are extenders which are themselves adhesives. Wood flour in the untreated condition contains approximately 30% lignin, which belongs to the phenolic-type of polymers, but it is uncertain whether the lignin contributes as extender or filler to the binding process. A study of the usefulness of various extenders has been published by *Williamson* and *Lathrop* (1949).

The main purpose of extenders is to lower costs, but there are some components, such as grain flours or soluble dried blood which contribute to adhesion. Fillers by contrast are rather inert components added to the resin adhesives to influence flow, consistency, or other qualities. According to *Perry* (1948), extenders are often used in ratios up to 100% or more of the resin content, while fillers normally are limited to 25%.

In the U. S. A. (American Specifications MIL-A-397 B, 1953 and MIL-A-5534 A, 1951) and in Canada (Can. Standard Assoc. 0112-7-1960), some specifications prohibit farinaceous and protein fillers because of their capacity to absorb moisture and to reduce resistance to bacteria and fungi. In addition to fillers and extenders, thickening agents also have to be mentioned. Their main purpose is to increase the viscosity of the glue mixture. They are mainly used for hot-setting plywood glues. The increase in viscosity enables the mixture to absorb more water, lowering the cost of the glue. Thickening agents include starches and dextrins, boric acid, polyvinyl alcohol, methyl cellulose, and polyethylene glycol. Usually, very small amounts of thickening agents (less than 2%) are used.

Rayner (1965) discusses possible modifications to phenolic adhesives. He mentions that "no revolutionary advance has been made". The addition of urea resins usually led to disappointing results. A number of patents claimed mixtures of phenolic adhesives with melamine, sodium silicate, and blood. In Finland, mixtures of blood albumin with 10 to 30% of phenolic resins have long been used for the manufacture of certain types of plywood. *Fischer* and *Bensend* (1969) investigated the gluing of southern pine veneer with blood modified phenolic resin glues.

Phenolic glue films — as already mentioned — were the first phenolic resins used for the manufacture of plywood. Special paper was impregnated with resol and dried. The resin could contain some cresols in addition to phenol. To keep the film elastic prior to hot pressing, softeners were added. The film rolls, stored dry and cool, have a storage life of approximately 1 year. The pressing temperature is at least 140 °C (284 °F), the pressure to be applied varies according to the wood species between 6 and 25 kp/cm² (85 and 355 lb./sq. in.). Under the effect of heat, the resol of the film becomes liquid if a sufficient amount of moisture is available. Therefore, veneer with a moisture content lower than 6%, based on the dry weight of the wood, cannot be glued with such films satisfactorily. This is a prerequisite, and its control is occasionally difficult in practice. Therefore, many manufacturers prefer liquid phenolic resin glues.

1.11.1.4 Properties of Phenolic Resin Glue Joints. Under the assumption that the gluing process has been effected in a proper way, and that especially the hardening conditions were correct, the resistance of the phenolic glue joints is very good, even under extraordinarily bad conditions. Proper glue joints will not delaminate under the following conditions (*Phinney*, 1950):

1. Long exposure to either cold or hot water;
2. Cyclic moistening and redrying;
3. Extreme values of temperature or relative humidity, acting permanently or intermittently;
4. Temperatures near, or even to somewhat above the charring temperature of wood;
5. Attacks by bacteria, fungi, and other microorganisms as well as termites;
6. Exposure to many chemicals such as oils, alkalis, and wood preservatives including fire retardants.

In summary, one can state that properly made glue joints with phenol and/or resorcinol glues are stronger and have a higher durability than the wood itself. The joints cannot be destroyed without destroying the wood adjacent to the glue line. Therefore, the phenolic glues in the broader sense of this glue family are widely used in the woodworking industries. They are used for the manufacture of plywood, laminated wood and in recent years of particleboards, and for furniture and prefabricated constructions. If under some conditions the glue joints must be odorless, residuals of free phenol or cresol must be avoided. This is also necessary to avoid skin diseases of the workers.

The formulations and properties of phenolic glues can be varied in a wide range. The main factors are the ratio of formaldehyde to phenol or cresol, the kind and the amount of the hardener, and the polymerization prior to hardening (*Dietz*, 1949). Generally, the commercial phenolic resins are adapted to special uses; therefore, it is necessary to use them with discrimination. *Plath* (1951) distinguishes the following principal types:

1. Assembly glues: rather viscous resins, highly condensed, generally containing water, to be thinned with organic solvents and usable even at low temperatures, down to 20 °C (68 °F) with acid catalysts;
2. Glues for plywood and laminated products: The resins are less condensed, can be diluted with water only to a small extent and harden at high temperatures between 135 to 160 °C (275 to 320 °F);
3. Glues for particleboards and fiberboards: resins of low condensation, containing alkalis, miscible with water, which harden with acid catalysts at high temperatures;
4. Hot-setting phenol or cresol resin glues are supplied either as film glues, occasionally as powders, and predominantly as solutions in water or alcohol. They are generally used for plywood and laminated products. These glues were mainly used for thin veneer for aircraft.

Liquid phenolic glues are predominant at present. As a rule the glues react with alkalis. The moisture content of the veneer should be between 1 and 3%, since above 5 to 6% blisters may occur during hot pressing. The assembly times vary, but most liquid phenolic glues need pressing temperatures between 115 and 140°C (239 and 284°F). Higher temperatures, as mentioned above, should be avoided if possible.

There are some types of phenol and cresol resin glues which harden at temperatures between 95 and 100°C (203 and 212°F). It is possible to shorten the setting time by increasing the temperature up to 150°C (302°F). The glues have a pH value approximately neutral or slightly alkaline, and thus deterioration of wood fibers is avoided. The pot-life of these glues at the temperature of 24°C (75°F) varies between 2 and 8 h. It is common to use a spread on both sides of the veneers of 200 and 300 g/m² (40 to 62 lb. p. 100 sq. ft.). Setting times depend on the glue, the species of wood, and the desired glue joint strength.

The durability of phenolic wood glue joints has been thoroughly investigated by many technologists. Various types of tests have been used involving exposure of samples in the laboratory, in the open air, embedded in soil, or dipped into water. Considering the remarkable amount of work done only a few investigations can be quoted: *Wangaard* (1946), *Narayanamurti* and *Pande* (1948), *Blomquist* (1949), *Knight* and *Newall* (1957) and U. S. For. Prod. Laboratory (1956).

The durability of phenolic wood glue joints also depends on the wood species and on glue line thickness (*Müller*, 1953). The rapid decrease of glue joint strength with increasing glue line thickness is probably caused by the following phenomena:

1. Stresses due to shrinkage within the glue joint;

2. Higher concentration of acid catalysts when these are used which enhances the danger of attack to wood fibers;

3. Relatively high brittleness of hardened, especially overhardened, phenolic glue joints.

1.11.2 Resorcinol-Formaldehyde Adhesives

Resorcinol-formaldehyde resins have been used as glues since approximately 1943. Resorcinol is a phenolic substance, but its reactivity is higher than that of phenol. The manufacture of resorcinol is represented, by formula (XII).

XII

It is important to see that there are two hydroxyl groups (OH) in the meta-position; they are much more active than the CH_3 groups found in m-cresol (XIII). The manufacture of resorcinol is rather complicated, and therefore the product is high in cost.

XIII

It is a special property of resorcinol-formaldehyde glues that they set quickly at low temperatures. Resorcinol-novolaks can be hardened to the resite state by the addition of paraformaldehyde at room temperature even without acid catalysts. This is important since "cold hardening" without strong acids is not possible for normal phenolic glues. To reemphasize: The novolak resin is converted to a thermosetting resin through the addition of formaldehyde, and herein lies the most important feature of resorcinol adhesive as compared to ordinary phenolic resin (*Rayner*, 1965).

Later it was found that the most important properties of resorcinol-formaldehyde adhesives can be retained, and the costs can be reduced, if some of the resorcinol is replaced by ordinary phenol. The usual replacement is up to one half the molecular proportion of total phenolic constituents. It may be possible to replace more than one half but the process of room-temperature setting after gelation becomes progressively slower. Cost reduction is the motive for admixing phenol to resorcinol resin glues, but the principal point is that the resins remain adequately active for hardening at low temperatures. Commonly, it is supposed that resorcinol-formaldehyde adhesives are neutral. This should be understood within certain limits for commercial products. These limits are between about pH 6 and 9, a range which is not at all harmful to wood.

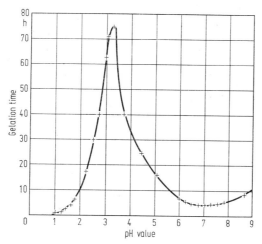

Fig. 1.48. A typical relationship between pH and gelation time for a resorcinol-formaldehyde resin. From *Little* and *Pepper* (1947)

The pot-life of resorcinol adhesives is greatly influenced by pH value (Fig. 1.48) (*Little* and *Pepper*, 1947). In the literature there is no distinction between gelation-time and pot-life. By altering the pH value, it is possible to alter also the pot-life. Normally, glues consisting of resorcinol and phenol mixtures in the ratio 1:1 cure more rapidly than the usual phenol-formaldehyde resins. The pH value of the liquid glue should be nearly neutral, i.e. between pH 7 and 8. Thus glues with rather long storage life (approximately one year) can be obtained (*Baumann*, 1967). The amount of formaldehyde necessary for setting usually is added immediately before application and mostly as a powder of paraformaldehyde.

Baumann (1967, p. 255) published the following data for mixtures of resorcinolphenol glues. They are given in Table 1.8.

Table 1.8. Pot-life, Assembly Times and Duration of Pressure for Resorcinol-Formaldehyde Resins
(aqueous solution 70% pH = 8···9)

Temperature °C	Pot-life h	Open assembly time (max) min	Duration of pressure[1]
10	10	—	—
15	6	—	—
20	3 1/2	15	10 h
30	1 1/2	10	6 h
40	—	—	2 h
60	—	—	20 min
80	—	—	5 min

[1] Valid for the temperature only of the glue joint, not of the surrounding air.

1.11.3 Urea-Formaldehyde Resin Adhesives

1.11.3.1 History of Urea Resin Glues. In the late twenties, the Badische Anilin- und Soda-Fabrik (BASF) at Ludwigshafen/Rhein (Germany) first synthetically produced urea on a commercial scale as fertilizer. The history of urea resins begins with various laboratory efforts and early patents. *Goldschmidt* (1896) investigated the reaction of urea and formaldehyde in neutral and acid media and obtained white granular condensation products. Later *Goldschmidt* (1897, 1898) produced mono- and dimethylurea with alkaline reactions. In the United States, *John* (1920), in his patent on urea-formaldehyde resins, pointed out their advantage as adhesives. In Britain, *Pollak* (1925, 1927) recognized the utility of urea- and thiourea-formaldehyde resin adhesives, and the salts of strong acids as accelerators. Among these were ammonium chloride, ammonium nitrate, and ammonium sulfate. In Germany the former I. G. Farbenindustrie AG in 1929 obtained a patent (I. G. Farbenindustrie AG, 1929) for a process of gluing wood, especially plywood, with aqueous solutions of condensation products from urea, thiourea or their derivatives and aldehydes or their polymers with an addition of acids, acid salts, or chemical splitting of acids. The first Kauritglues arrived on the market in the summer of 1931, but their introduction into the wood working industries, bound to tradition, was not easy. The Kaurit-glues had good properties but at that time the customary casein glues were very cheap.

The mechanism of urea-formaldehyde resin formation was then investigated thoroughly by *Walter* and *Gewing* (1931, 1935). *Delmonte* (1947) pointed out that "the wide scale application of urea-resins was preceded by the development of ample supplies of low-cost raw materials" such as "by-product gases, ammonia and carbon dioxide".

The chemical reaction of urea with formaldehyde will be dealt with in Section 1.11.3.3. Returning to the history from *Holzer*'s (1962) report the following may be summarized: The export of the new liquid urea-formaldehyde resins was difficult since their storage life generally amounted to only 3 to 5 months. After exhaustive experiments, BASF succeeded in drying the resin solutions by means of sprayers. The powder produced in this manner could be stored at least one year provided moisture was completely excluded. At present urea-formaldehyde resin glues are used throughout the wood based materials and woodworking industries. Production and application of these types of glue have been improved, but the complicated chemical proceedings in the production of urea-formaldehyde resins as well as the phenomena in binding are not yet entirely explained. Therefore, the manufacture of products with uniform predetermined properties needs considerable empirical data and personal experience.

1.11.3.2 Manufacture and Chemistry of Urea-Formaldehyde Glues. The initial raw materials for urea-formaldehyde resins are coal, water, and air. Water generator gas is produced from coal and may be converted into hydrogen, carbon monoxide, and carbon dioxide. Nitrogen from the atmosphere is obtained by liquefaction of purified air and subsequent fractional distillation. Ammonia is formed from nitrogen and hydrogen according to the Haber-Bosch process under the influence of high pressure (200 atmospheres), proper temperatures (500 to 600 °C), and catalysts from urea

$$2NH_3 + CO_2 = H_2N-\underset{\underset{O}{\|}}{C}-NH_2 + H_2O$$

XIV

The whole manufacturing process is technically not a simple one. Molten ammonium carbonate, at the high reaction temperatures employed, is very corrosive, and therefore the quality of the autoclave material must be extraordinarily high. There is also an economic problem. The process is carried out with an excess of ammonia and ammonium carbonate. Therefore, the excess of ammonia and the unused ammonium carbonate must be either returned to the reaction in a complicated rotation process or converted into ammonium salts for fertilizers (Fig. 1.49).

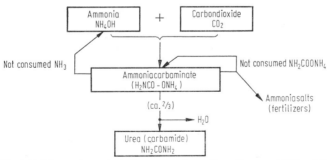

Fig. 1.49. Scheme of manufacturing process of urea. From *Baumann* (1967, p. 214)

Methyl alcohol (methanol) CH_3OH is prepared commercially from carbon monoxide and hydrogen under pressure with catalysts (BASF process) or in the largest quantities from petroleum. Steaming methanol is oxydized under the effect of a platinum net to formaldehyde gas

$$CH_3OH + 1/2\,O_2 = CH_2O + H_2O$$

XV

Formaldehyde is a gas easily soluble in water. Paraformaldehyde (Polyoxymethylene, CH_2O) is a polymer of formaldehyde.

Urea and formaldehyde condense provided that molar ratios, concentration, temperature, time, and pH values are adequate. By the variation of these main factors, urea glues can be made that differ widely in their properties. Solid urea is dissolved in formaldehyde solution of about 40% concentration at pH = 7 and at room temperature. In the solution 1 mole urea and 1.5 to 2 moles formaldehyde combine at first to a mixture of monomethylol, dimethylol, trimethylol and tetramethylol urea. As an example the reaction to dimethylol urea is taken (from *Holzer*, 1962, p. 318):

$$H_2N—C—NH_2 + 2\,CH_2O \qquad HO—CH_2—N—C—N—CH_2—OH$$

with the structural formula showing:

$$\underset{O}{\overset{\|}{}} \qquad\qquad \underset{H\ \ O\ \ H}{\overset{|\ \ \|\ \ |}{}}$$

XVI

The methylol ureas are not adhesives. Condensation does not take place yet. Methanol present in the diluted aqueous solution must be removed by means of distillation at about 90 °C since it is toxic and could retard the hardening of the glue. Subsequently an acid condensation at pH = 4.5 and in the range of temperatures between 85 and 90 °C (185 and 194 °F) is effected. This reaction is arrested by neutralization (pH = 7 to 8). Liquid glue is obtained by means of vacuum distillation to a concentration of 67% or dry glue powder by spray drying. The storage life of the liquid UF-glues amounts to approximately 3 months (at 18 °C ≈64 °F temperature), that of the glue powders to about one year under normal storage conditions. Liquid UF-glues have the tendency of "aging", which is recognizable by the increase in viscosity (Fig. 1.50). The increase in viscosity is based on a chemical reaction which is irreversible.

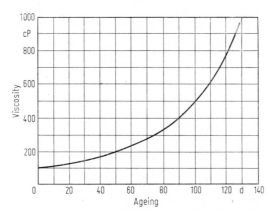

Fig. 1.50. Increase in viscosity of a urea-formaldehyde glue solution as a function of natural aging at 20 °C. From *Holzer* (1962)

"Unlike phenolic resins, urea-formaldehyde resins remain substantially soluble in water during resinification. Their hydrophilic qualities may be explained in part by the presence of the free methylol groups. At the time of setting or gelling, however, transformation takes place rapidly as the cross-linking of molecules takes place, with the formation of the infusible and insoluble structure. In studies of adhesion it is important to identify the pressure of these free unreacted groups, which may show particular affinity for certain surfaces due to their polar characteristics. The free methylol groups appear to possess an unusually good affinity toward cellulosic structures" (*Delmonte*, 1947, p. 55/56).

1.11.3.3 Hardening of Urea-Formaldehyde Resins. Precondensed neutral UF-glues are hardened by a sufficient reduction of the pH value. The free methylol and amino groups of the glue molecules, still in a low polymer state, react together and condense to a cross-linked network of macromolecules. Their exact structure is not known, but one suggested formula for hardened urea may be shown (Formula XVII, *Baumann*, 1967):

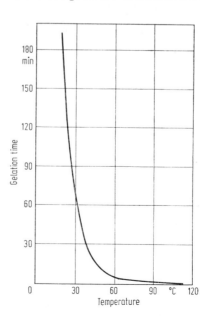

XVII

There are many hardeners, also called curing agents, catalysts, or accelerators. It is their task to reinitiate the interrupted action of condensation and to finish it as mentioned above. The rate of reaction during the hardening process is increased considerably by acids or by substances capable of liberating acid when mixed with the resin. In the same sense an increase in temperature is effective 1.51) while the reaction will be retarded if the temperature decreases. For the

Fig. 1.51. Dependence of the time of gelation on the temperature of a urea resin solution. From *Holzer* (1962)

usual cold hardeners the lowest admissible limit of temperature is about 15 °C (64 °F) and only for very effective "rapidly acting hardeners" complete setting may be achieved at temperatures as low as 5 °C (41 °F).

a) Hot hardeners. The direct addition of acids to UF-glues is in most cases unsuitable. The pH value falls too rapidly, pot-life becomes too short and the strenght of glued joints deteriorates with time. These troubles can be avoided by the use of ammonium salts of strong acids. "They are cheap, convenient to handle, and give a high ratio of pot-life to setting time" (*Rayner*, 1965, p. 191). A frequently used ammonium salt is ammonium chloride which reacts with the free formaldehyde or — more slowly — with methylol groups available in the resin solution and forming hexamethylene tetramine ("hexamine" or "hexa"), hydrochloric acid and water (Formula XVIII):

XVIII

"The rate of liberation of formaldehyde and consequently the rate of fall in pH is sharply increased by an increase in temperature (Fig. 1.52) which is one of the reasons why ammonium salts are excellent hardeners" (*Rayner*, 1965, p. 191).

Urea resins mostly used at elevated temperatures should have a pot-life as long as possible at room temperature and they should set very rapidly at high temperatures. From this aspect the free formaldehyde in the urea resin solution, which governs the amount of liberated acid and with it the reaction rate, should be bound in such a way that it is quickly freed again when the temperature rises. This is done by adding buffers such as ammonia, urea, hexamethylene tetramine. Fig. 1.53 shows that the pH value of a urea resin solution mixed with a buffered hardener decreases more slowly than that of an equal resin solution with an un-buffered hardener.

b) Cold hardeners. Efficient gluing at temperatures below 100 °C (212 °F) requires aggressive hardeners. An increase in the addition of ammonium chloride above 1% is not useful. Therefore, usually the amount of buffering substances is reduced but this means a decreased pot-life.

For cold hardeners, designed for room temperature cures within one half or one hour, free acids are preferred to ammonium salts. Strong acids effect a rapid condensation of the UF-glues, and therefore the cold hardeners are frequently applied to one wood surface and the resin to the other (I. G. Farbenindustrie AG, 1933). The most suitable cold hardeners are organic acids which are highly soluble in water with dissociation constants between 1.20^{-2} and 1.10^{-3}, e.g. tartaric acid or citric acid. Their pH lies in the range of 1 to 2, even in high concentrations, and thus there is no danger for the wood. Strong acids like hydro-

chloric acid may be applied only in very dilute solutions but do not reliably harden
the glue in this case, while in high concentrations they attack the wood (*Baumann*,
1967, p. 219).

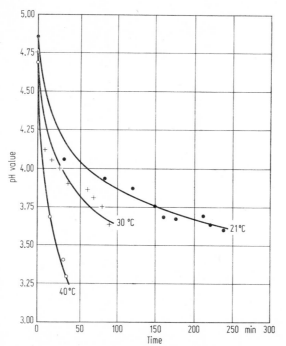

Fig. 1.52. Decrease in pH with time and temperature after addition of ammonium chloride to a UF-resin adhesive.
(After Ciba Aero Research Ltd, unpublished work.) From *Rayner* (1965, p. 191)

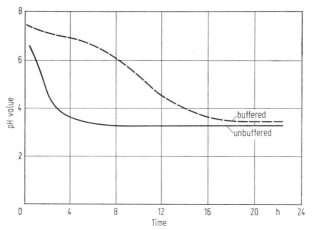

Fig. 1.53. Change of pH value in a urea resin solution mixed with an unbuffered and a buffered
hardener in dependence on time. From *Holzer* (1965, p. 322)

c) Organic HO-groups retard but not impede the condensation of UF-glues.

d) The setting of UF-glues consists of chemical curing and migration of the
water which is partly contained in the liquid glue and partly chemically liberated
from the resin during the condensation process. Both phenomena must occur

simultaneously. If the water is removed too quickly or if the wood is too dry, the reaction of hardening cannot be completed due to lack of solvent. If the decrease of water is too slow, induced by too high a moisture content of the wood or by too much water in the glue mixture, the hardening will be terminated too early, and instead of an adherent glue film a crumbling mass without cohesion will be formed.

Another very complex problem is the concentration of hardener in the drying glue joint. At the beginning of the hardening process the wood layers adjacent to the joint are not completely dried out. The formation of water during condensation has been mentioned already. Condensation and liberation of water continue during the pressing process. A moisture content gradient is present in the glue joint on the one hand and in the adjacent wood layers on the other hand. This gradient disappears slowly. The outer layers dry more quickly than the core. Due to this uneven drying there is also a gradient of acid concentration within the glue joint. The setting is therefore heterogeneous, and as a result high stresses develop within thick hardened glue joints of UF-glue. The stresses and strains in relatively thick UF-glue lines can become critical and delamination may occur. If volatile or unstable acids — e.g. formic acid HCOOH — which are not so strong as citric or tartaric acid are used, shrinkage is low but corrosion due to the acid vapor can be a serious problem.

e) Gap-filling hardeners. The original urea-formaldehyde glues cured by acids and applied in thicker layers had the tendency to shrink during the hardening process and to craze. Joint gaps in the order of a millimeter were dangerous. The first patent on improving gap-filling properties was granted to *Klemm* (1938).

1.11.4 Melamine-Formaldehyde Resin Adhesives

1.11.4.1 History, Chemistry, and Manufacture of Melamine-Formaldehyde Adhesives. The possibility to produce thermo-setting synthetic resins by condensation of melamine with formaldehyde was discovered independently and approximately at the same time at the end of the thirties by the firms Henkel (1935), Ciba (1938, 1939, 1965), and the former I. G. Farbenindustrie AG (1936). Melamine-formaldehyde resin is used as a sizing and bonding resin in the textile, paper, and wood products industries. Ultraviolet spectrophotometric method for the determination of melamine-formaldehyde adhesive in such products is described by *Chow* and *Troughton* (1969).

The melamine resins resemble the urea resins but offer the following advantages:

a) Greater resistance to water (withstanding the three-hour boiling water test);

b) Higher heat stability;

c) Ability to set at lower temperatures;

d) Suitability for impregnating as well as for gluing. Superiority to the urea resins in baked surface coatings. Development of exterior water-repellent (and decorative) faces for plywood.

Upgrading (fortifying) of urea adhesives by small percentages of melamine (Fig. 1.54) after *McHale* (1944) is done with the aim to reach hot-water-resistance and to some extent boiling-water-resistance (Fig. 1.55, after Ciba, *Houwink* and *Salomon*, 1965).

Disadvantageous is the fact that melamine-formaldehyde resin adhesives are more expensive than urea and phenolic resins. This is why melamine-formaldehyde resins have not attained great importance as glues. Melamine cannot be produced

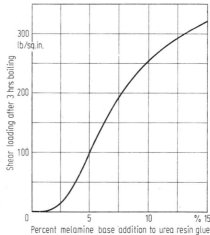

Fig. 1.54. Influence of percent melamine base addition to urea resin glue on shear loading after 3 h boiling. From *Delmonte* (1947)

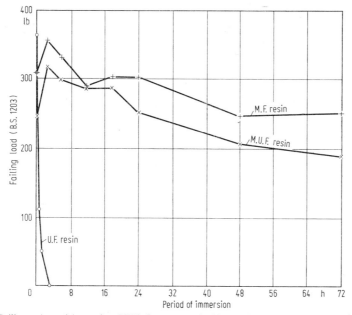

Fig. 1.55. Boiling-water-resistance of an MUF-glue compared with straight UF- and MF-glues. From *Houwink* and *Salomon*, after Ciba (1965)

on a technically simple way. There exist numerous patents regarding the manufacture of melamine (*Kaess* and *Vogel*, 1954). In the industry the following method is used (*Holzer*, 1962):

First calcium carbide is produced from coal and lime which requires a great deal of energy

$$4C + CaCO_3 = CaC_2 + 3CO.$$

Calcium carbide is transformed by the addition of nitrogen into calcium cyanamide

$$CaC_2 + N_2 = CaCN_2 + C.$$

Calcium cyanamide is reacted with sulfuric acid to form dicyandiamide.

The dicyandiamide, together with ammonia and under the influence of heat and pressure, turns to melamine in a reaction not yet completely clarified:

Melamine
XIX

The condensation reactions of melamine with formaldehyde are principally similar to those of urea with formaldehyde. One mole of melamine is condensed with 3 to 4 moles of formaldehyde at a pH value of 5 to 6 to form formaldehyde resins. Aqueous solutions of melamine adhesives are not as stable as solutions of urea adhesives. In consequence, melamine adhesives are usually shipped as spray-dried powders in tightly sealed drums, ready for suspension in cold water before their application as adhesives.

1.11.4.2 Application of Melamine-Formaldehyde Adhesives. The dry melamine resin powders are easily dissolved in cold water. The glue mixture, rapidly stirred, at first thickens greatly, but then becomes again less viscous. The duration of stirring determines the viscosity. After stirring the glue mixture is sufficiently thin to use. The best mechanical spreaders have ribbed rubber rollers.

The melamine-formaldehyde adhesives need spreads of only 100 to 150 g/m² (about 20 to 30 lb./1,000 sq. ft.); this is substantially lower than for animal (140 to 250 g/m², corresponding to 28 to 50 lb./1,000 sq. ft.), casein (240 to 300 g/m², corresponding to 48 to 60 lb./1,000 sq. ft.), vegetable (250 to 300 g/m², corresponding to 50 to 60 lb./1,000 sq. ft.), or urea glues (100 to 200 g/m², corresponding to 20 to 40 lb./1,000 sq. ft.). With respect to resistance against light (good), required moisture content of the wood (4 to 12%), consistency of the glue mixture (like syrup), skin irritation during handling (only possible with repeated contact), formaldehyde odor, staining of the wood (no), and color of the glue joint (colourless), melamine resin adhesives behave like urea resin adhesives. The open assembly time ranges from zero to 24 h. Press temperatures are between 70 °C (158 °F) and 120 °C (248 °F).

Pressure requirements vary with different formulations, with wood species and quality of wood surfaces in the range between 5 and 15 kp/cm² (about 70 and 210 lbs./sq. in.). The basic pressing time may be (at 100 °C or 212 °F) 3 to 5 min, plus 1 min for every millimetre of thickness, measured from the innermost glue line.

Melamine-formaldehyde resin glues have been used to some extent for production of waterproof or even boil-proof plywood. In woodworking industries they are still scarce. A promising use of pure melamine-formaldehyde adhesives is edge-jointing of veneers. Various types of tapeless veneer splicing machines are mentioned in Section 3. Melamine resin glues are especially suitable for splicing machines in which two veneers are brought together tightly without overlapping and pass parallel to the direction of the automatic travel (feed) with light pressure adjusted to veneer thickness. Heating is done by steam, hot water, electrical resistance, or high frequency fields. In the latter case the temperatures of the heating rods may reach 218 °C (424 °F) according to veneer thickness (Table 1.9).

Table 1.9. Gluing Conditions in a High Frequency-Heated Veneer Splicing Machine

Veneer thickness		Feed speed		Temperature of heating rods	
mm	in.	mm/min	ft./min	°C	°F
0.4	1/64	26	85	121	250
0.9	1/28	26	85	149	300
1.3	1/20	21	68	163	325
1.6	1/16	18	59	177	350
2.1	1/12	16	52	191	375
2 5	1/10	15	49	191	375
3.5	1/8	12	39	204	398
4.8	3/16	6	19	218	426

Recently, veneer splicers have been developed in which the veneer edges (glue joints) and the fibre direction are orientated perpendicular to the transport direction. Exact dosing of the glue is necessary to avoid squeeze-out.

In veneer splicing melamine resin glues offer the following advantages

1. Economical pot-life of the glues ready for use;
2. Setting within seconds;
3. Complete hardening;
4. Invisible or at least light glue joints.

The most disadvantageous phenomenon is that ocassionally the glued joints become brittle and the veneers, after having passed the machine, will be separated again. Tests made in close cooperation with the industry led to insights into the relationships between press temperatures and setting times (that is feed speeds) for various wood species (*Enzensberger*, 1952) (Fig. 1.56). Only press temperatures and feed speeds within the hatched regions lead to satisfactory glue joints: Lower press temperatures or higher feed speeds cause insufficient gluing. If the temperature is too high or the feed speed is too slow the joints will become brittle.

For the manufacture of boil proof plywood, *Plath* (1954) recommends three formulations:

a) Extended melamine-formaldehyde resin solutions:
Melamine-formaldehyde resin glue powder is dissolved in cold water and extended with rye flour under addition of water;

b) Mixture of melamine and urea-formaldehyde resin:
Into liquid urea-formaldehyde adhesive (Kaurit W) a solution of melamine-formaldehyde adhesive in water is poured; the melamine-formaldehyde resin acts as hardener;

c) Mixture of melamine-formaldehyde and urea-formaldehyde resin extended with a blood albumin solution.

The proportion of water in the glue mixture influences the shear strength of the glued joints (Fig. 1.57 from *Plath*, 1954). As mentioned earlier in this section, melamine adhesives are not very sensitive to moisture content prior to gluing (4 to 12%), but the best results for pure melamine-formaldehyde resin glues will be obtained for 10%, for MF and UF mixtures for 5 to 8% (Fig. 1.58 from *Plath*, 1954).

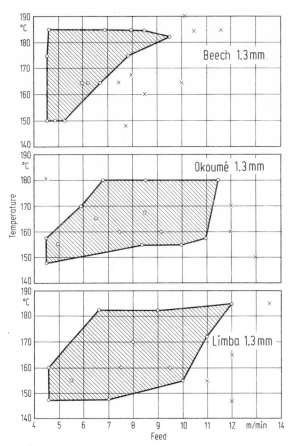

Fig. 1.56. Relationship between temperature and feed speed in veneer jointing with
melamine-formaldehyde adhesives. Points within the hatched areas indicate
good working conditions. From *Enzensberger* (1952)

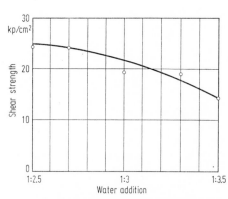

Fig. 1.57. Dependence of shear strength after 96 h
immersion in water on the proportion of water in a
mixture of melamine-urea resin (1:1) in gluing beech
veneer plywood. From *Plath* (1954)

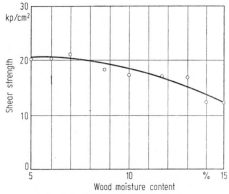

Fig. 1.58. Dependence of shear strength after 96 h
immersion in water from moisture content in gluing
beech veneers with a melamine-urea resin (1:1).
From *Plath* (1954)

1.11.4.3 Hardening of Melamine Adhesives. Unlike UF-resins, MF-resins can be cured without acid catalysts by heat treatment. A MF-resin can be hardened at room temperatures "by lowering the pH value sufficiently, but the cured product has poor cohesive and adhesive strength" (*Rayner*, 1965) and is very brittle. The brittleness can be reduced by an excess of formaldehyde (DRP 721 240; Henkel, 1942).

The hardening reaction of MF-resins can take place — as that of UF-resins — by the formation of ether-bridges (Formula XX) at temperatures up to 150°C (302°F) according to *Köhler* (1941, 1943) or at higher temperatures by the formation of methylene-bridges (Formula XXI).

$$H_2N-C\underset{N=C}{\overset{N=C}{}}C-N-CH_2OH + HOH_2C-N-C\overset{N=C}{}C-NH_2$$

XX

XXI

Under the influence of heat or catalysts (acids or ammonium salts of strong acids) finally an insoluble and infusible three-dimensional network comes into being which has a higher thermal stability than cured UF-resins. For this reason MF-resin adhesives yield boil proof and weather stable glue joints though they do not equal phenolic resins in this regard.

1.11.4.4 Properties of Melamine-Formaldehyde Resin Glue Joints. The principal advantages of MF-resins are summarized in Section 1.11.4.1. Pure MF-resins are used for the production of plywood and laminated board if boil proof joints must be obtained and if cheaper phenolic resins cannot be used. An example is the production of plywood for boat building with light wood species. In this case the dark glue layers of phenolic resins could shine through the outside veneers. But in most cases MF-resins, because of their high price, are either extended or blended with UF-resins (cf. Fig. 1.54 and Section 1.11.4.5).

MF-resins "can well fulfil the simultaneous role of an impregnating as well as an adhesive agent" (*Delmonte*, 1947, p. 73) (cf. Section 1.11.4.6).

1.11.4.5 Fortifying or Upgrading of Urea Adhesives. Normal unfortified UF-glues have very high resistance to cold water and appropriate resistance to hot water (up to a critical temperature between 70°C and 80°C — 158°F and 176°F). Besides the temperature the period of immersion is important. The resistance to boiling water is negligible. Glue joints in plywood, laminated products or con-structions are not used in contact with boiling water. The resistance to boiling water gives, however, according to experiments at the Forest Products Research Laboratory at Princes Risborough, England, a reliable measure of durability of the joint when exposed to weather.

Hot-water-resistance and to a lesser extent boiling-water-resistance of UF-resins can be improved by adding melamine (or resorcinol). The melamine acts as an accelerator of hardening, lowers the acidity of the UF-resins, and increases their water resistance. An example of the improved boiling-water-resistance of glued wood joints obtained with a urea- (melamine-) formaldehyde co-condensate MUF-resin) having a molecular ratio of $1:1:4$ is shown in Fig. 1.55 after Ciba (1938, 1939 a, b). The optimum amount of additive depends on the formulation of the urea glue, but on the average 10 to 20 parts per 100 parts by weight of urea resin solid will suffice. Ranges and limits of upgrading by the addition of melamine or melamine resins are discussed in more detail by *Rayner* (1965, p. 198—200). The influence of moisture content of veneer on shear strength of glued wood joints made with MUF-resins is shown in Fig. 1.58.

1.11.4.6 Surface Coating of Wood Based Materials. Surface coating (improving the surfaces) of wood based materials has been done formerly by lacquering, varnishing, or covering with plastic foils. During the last two decades new methods have been developed.

In the simplest case the solution of a thermosetting resin, e.g. a melamine resin in water, is spread on the surfaces of the board, dried, and then using smooth aluminium sheets under the simultaneous effect of pressure and heat converted into a film. The coating is transparent, scratch proof, of high lustre, boil proof, and will not be affected by alcoholic beverages, soap, ink, oil, etc.

In practice the water-soluble melamine resins will be cured to relatively brittle films, similarly to urea and phenol-formaldehyde resins. It has been tried to produce etherified resin films which are elastic, but these resins — important raw materials for lacquers — are soluble only in organic solvents. The results of these experiments are not yet encouraging (*Holzer*, 1962, p. 336). Overlay films without reinforcement tend to the formation of fine cracks either immediately after pressing or even after weeks of use. Cracks may not appear if the thickness of the resin films is less than 0.1 mm (4/1,000 in.), but this can scarcely be ob-tained. Nevertheless, melamine resin films have the following characteristics (*Simonds, Weith,* and *Bigelow,* 1949):

1. Excellent retention of color and lustre at abnormally high temperature;
2. Fast curing rates, even at the lower commercial temperature;
3. Higher hardness and brittleness than urea resin films (therefore higher liability to crack);
4. Comparing films of equal hardness and durability, melamine coatings actually have less chemical resistance than the urea type;
5. Less adhesion than urea films to some surfaces.

It is much better to use resin-treated films instead of resin solutions. Paper or some other fabrics serve as reinforcements for the resin. The sheets are impregnated with the resin solution and then carefully dried, avoiding curing, in a progressive kiln (*König*, 1958). This film will be pressed on the surface of a wood based material under high and uniform pressure (up to 14.5 kp/cm² corresponding to 250 psi) and a temperature of about 140 °C (284 °F). Using polished, chromium hardened cauls of brass or stainless-steel, flat or glossy surfaces can be obtained. The reinforcement of the film takes up the stresses developed below the surface and compensates them. If the amount of impregnating resin is not too high, cracking will not occur. One has to distinguish between:

a) Overlay-films or overlay-papers consisting of heavy paper on the basis of alpha cellulose with a weight of 20 to 50 g/m² (about 0.5 to 1.2 lb. p. sq. ft.) and a high resin coat of about 200%; after pressing the result is a clear transparent surface;

b) Decorative overlays consisting of filled, absorptive generally printed paper with a weight of 150 to 200 g/m² (about 30 to 40 lb. p. 1,000 sq. ft.) and a resin coat of 70 to 100%. Decorative overlays are generally 0.75 to 1.5 mm (about 0.03 to 0.06 in.) thick.

Remarkable advances have been made in the past twenty-five years in the development of decorative overlays for veneer, plywood, particleboard, and fiberboard (*Seidl*, 1947). The goal was to produce highly serviceable and attractive surfaces of varying color or design (e.g. for tables, bars, and counters). The most popular designs have the appearance of decorative veneer, fabric surfaces or consist of geometrical or artistic patterns. The decorative overlay usually masks the base material. *Seidl* (1947) describes a typical assembly of paper for the decorative overlay with sheets arranged in the following order: "A melamine-impregnated 'balancing' sheet, several sheets of phenolic-impregnated 'core-stock', a melamine-impregnated 'barrier' sheet, a printed paper carrying the design, and a thin transparent high-melamine content surface sheet. The latter serves to relieve the printed surface of abrasion that would result in wearing away of the pattern or color. A thin metal foil is sometimes molded under the decorative sheet to impart 'cigarette-proofness'. In order to minimize warping of the final assembly, another thin panel of phenolic plastic is usually pressed independently for use as a balancing sheet on the opposite side of the wood core."

Decorative laminated board (*König*, 1958; Holz-Zentralblatt-Verlag, 1958) with at hickness of 0.6 to 1.6 mm (about 1/42 to 1/16 in.) is produced by hot-pressing resin-impregnated paper. A typical sandwich assembly of laminated board is shown in Fig. 1.59.

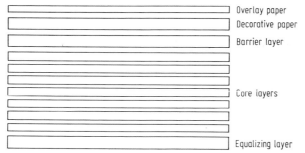

Overlay paper
Decorative paper
Barrier layer

Core layers

Equalizing layer

Fig. 1.59. A typical sandwich assembly of a laminated board. From *Holzer* (1962 b)

1.11.5 Thermoplastic Resin Adhesives

1.11.5.1 General Properties and Manufacture. Thermoplastic resin adhesives are synthetic high polymers which melt or soften when heated and reharden when cooled. They do not undergo any chemical change. The setting is effected physically. Thermoplastic resin adhesives are used as solutions or dispersions in water. Adhesion follows evaporation or absorption of the liquid constituent.

For thermoplastic adhesives there are different chemical classes, but typically all are linear macromolecules. This common feature influences the behavior in dependence on the molecular weight and distribution, the concentration of plasticizer and other additives. Thermoplastic adhesives may be used for various materials: wood, paper, and leather, a few bond metals, plastics and rubber.

The joint strenght of thermoplastic adhesives is not as important as that of thermosetting resins for structural uses. In the glue lines of thermoplastic adhesives the stresses are generally low. Typical is the tendency to large creep deformations.

1.11.5.2 Cellulose Adhesives. Cellulose is the most important chemical constituent of wood and other plant materials and belongs to the natural products. Therefore cellulose adhesives have been dealt with in Section 1.10.6. It may be repeated that one has to distinguish between cellulose esters and cellulose ethers.

1.11.5.3 Polyvinyl Acetates (PVA). "The vinyl group comprises the reactive part of a large number of compounds. They produce thermoplastic polymers, most of which are extremely useful as adhesives. They include polyvinyl esters, ethers and acetyles, polyvinyl alcohol and polystyrene" (*Rayner*, 1965). *Clark* Jr. (1968) measured the speed of set of room temperature setting wood adhesives. Prior to the introduction of polyvinyl acetate emulsion adhesives, the rapid setting glues for wood were various kinds of hot animal glue. Relative speed of set was predictable, based upon the "gelly strength", dilution, temperature, and moisture content of the wood to be bonded. The modern polyvinyl glues are fast setting polymers which can be modified to give various speeds of set as well as different assembly periods. Reduction in solids, addition of fillers, and other changes in formulation can affect the general performance of the glue.

Polyvinyl acetate is the most important polyvinyl ester adhesive. It is used to a large extent for gluing of wood as an aqueous dispersion without or almost without plasticizer. It is used for the gluing of solid wood as cold binder and has nearly replaced animal and casein glues for the following reasons:

1. Easy handling,
2. Cleanliness;
3. Practically unlimited storage time;
4. Resistance to microorganisms;
5. No staining of wood;
6. Approximately equal contact and gap filling properties as animal glues;
7. Low pressure for setting.

On the other hand there are disadvantages:

1. High sensitivity to water and therefore no suitability for exterior wood work;
2. Rapid reduction of joint strength under the influence of heat and moisture;
3. Unfavorable visco-elastic properties, therefore large creep and low fatigue resistance.

Polyvinyl acetate glues are not recommended for the gluing of smooth, non-porous surfaces.

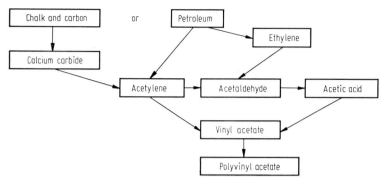

Fig. 1.60. Scheme for the manufacture of polyvinyl acetate. From *Baumann* (1967)

Fig. 1.60 illustrates the scheme for the manufacture of polyvinyl acetate. The following formula shows in a simplified manner how polyvinyl acetate is formel by radical polymerization:

$$nH_2C=CH$$

Polyvenylacetate
XXII

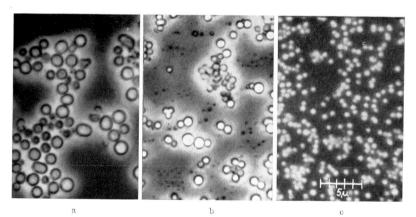

Fig. 1.61 a—c. Electron-microscopic pictures of three polyvinyl acetate emulsions. From *Baumann* (1967)

In reality there is always a branching. The number of branches increases with higher degree of polymerization.

The properties of polyvinyl acetate dispersions depend to a large extent on the size of the macromolecules. For wood glues a degree of polymerization of about 1,000 to 2,000 is usual. Of importance is also the size of the dispersed macroscopic particles which consist of a very large number of macromolecules. Fig. 1.61 shows electron-micrographs of three polyvinyl acetate emulsions. A coarse disperion is reproduced in part a, a medium dispersion in part b (approximate size 1:1.5 μm), and a fine dispersion in part c. Experience has demonstrated that the dispersions a and b are well suited for the gluing of wood, whilst finer dispersions penetrate too deeply into the wood or other porous surfaces (*Baumann*, 1967, p. 164).

The particles in the polyvinyl acetate dispersions are separated by water. After spreading the glue line dries, the water evaporates and the dispersed particles come into touch with one another. Flowing effects a coherent film. The viscosity of polyvinyl acetate wood glues can be influenced by the concentration, i.e. the proportional amount of water or by the temperature. A filling of this glue type is possible and diminishes costs. Occasionally filling increases the setting rate. Organic fillers such as starch should not be used for polyvinyl acetate dispersions since they reduce setting and joint strength. Good results can be obtained with powders of chalk and light spar.

Carroll and *Bergin* (1967) pointed out that since 1961 new types of PVA emulsion adhesives have been introduced with the aim to retain the quick setting and long pot-life properties of the emulsion form and to achieve or at leas tapproach the heat and moisture resistance and durability of the resorcinol formaldehyde wood adhesives.

As mentioned before, polyvinyl acetate glues are very durable if they are protected against moisture. Since they are resistant against bacteria, residues of glues may be mixed with new blends. According to explanations by manufacturers of modern polyvinyl acetate glues, they can be stored several years without deterioration. The resistance to freezing has been remarkably improved, the glues do not coagulate at temperatures as low as $-18\,°C$ ($-0.4\,°F$).

Nevertheless, polyvinyl acetate glues should be protected against freezing, since thawing of dispersions is not easy. The melting process should not be enforced by use of steam or hot water.

Polyvinyl acetate glues may be used like animal glues. Due to their relatively high water content (about 55%) polyvinyl acetate glues are not gap filling. The best results can be obtained with a spread of about $200\,g/m^2$ ($0.04\,lb./sq.\,ft.$) and a pressure of some kp/cm^2 ($1\,kp/cm^2 = 14.7\,psi$). The open assembly time for polyvinyl acetate glues depends on their ability to form a film. A simple reliable method is to touch the glue line in the work shop with the finger tips; a characteristic tackiness should be observed. Open assembly times between 5 and 10 min are usual.

The curves for development of glue joint strength with pressing time are comparable with those for casein glues. Within the first few hours joint strenght increases rapidly, for polyvinyl acetate glues quicker than for casein glues. For gluing solid, dry, stressfree wood parts at room temperature pressing times between 5 and 15 min are usual, and veneer requires 30 to 60 min. Any increase in pressing temperature reduces pressing time. If the temperature in the joint is $50\,°C$ ($122\,°F$), the pressing time is shortened by about 40%.

Polyvinyl acetate glues swell in contact with water. In atmospheres of very high relative humidity their joint strength may be reduced due to the absorption of water vapour, but after re-drying normal strength values are reached again. Between various glue formulations marked differences exist in behaviou under long-term loads (Fig. 1.62). The U.S. Forest Products Laboratory published a Progress Report in 1956. The consumer urgently needs more knowlegde on fatigue and long-term strength of very flexible glues such as polyvinyl acetate glues.

Some combinations of polyvinyl are not produced by polymerization of monomers, but by a chemical conversion of polyvinyl acetate. Polyvinyl is not soluble in water; it plays an important role in the gluing of metal to metal. During World War II, in England the "Redux"-process was developed by Aero Research Limited for metal surface bonding. The metal surfaces were degreased and pickled, then coated with a low viscosity phenolic resin and finally dipped into, or sprinkled with, a thermoplastic polyvinyl acetate powder. Assemblies are pressed for about 15 min at 14 kp/cm² (∼200 psi) and 149 °C (300 °F) or for longer times at lower temperatures. The "Redux"-process is also adapted for bonding wood to metal, provided the curing conditions are not detrimental to the wood (*Chapman*, 1951)

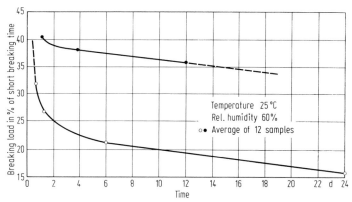

Fig. 1.62. Behavior of polyvinyl acetate glues under long duration loads. From U. S. Forest Products Laboratory

Polyacrylates and polymethacrylates. This group of chemicals is the basis for some cements and for synthetic rubber. In the future glues for wood might possibly be produced with these chemicals.

Polyethylene. The following formulae give a survey on various derivates of polyethylene (structural units $-CH_2-CH_2-CH_2-CH_2-$):

a by polymerization of monomers

Polyethylene

Polyisobutene

Polytetrafluor ethylene

Polyvinyl chloride

Polyvinyl acetate

Polyvinyl ether

Polyacrylate

Polymethyl acrylate

Polystyrene

b by chemical transformation of polymeres

Polyvinyl alcohol

Polyvinyl acetals

XXIII

Polyethylene is a relatively inexpensive raw material, its physical properties can be varied within wide ranges, it is chemically very resistant and will not be attacked by water. A problem is its non-polar nature. *Bikerman* and *Huang* (1959) proved that pollution reduces adhesion. The experiments of *Bikerman* contradict the hypothesis that conditions for adhesion between polar and non-polar substances are unfavorable.

Polyvinyl chloride (*PVC*). This is one of the most frequently applied and least expensive polymer plastics. Extruded and pressed articles, sheets and foils are produced, but as a glue polyvinyl chloride is restricted to metals.

1.11.6 Other Synthetic Organic Adhesives

1.11.6.1 Epoxy Glues. During Wood War II a new class of glue came into use: epoxy (or epoxide) resin glues. Due to their highly reactive basic components, epoxy resin glues are suitable for cold and hot setting. They are mainly used for gluing of metals and occasionally of metals to wood. Originally the manufacture was restricted to the firm Ciba, Basel.

Epoxy resin glues usually consist of the following components:

Linear low polymeric combination which has on both ends an epoxy group as follows:

$$-HC-CH_2$$
$$\diagdown O \diagup$$

The term epoxy was used first by *Berthelot* (1827—1907) who described the following formula:

$$H_2C-CH-CH_2Cl$$
$$\diagdown O \diagup$$

Baumann (1967) in detail describes manufacture and setting of epoxy components. He also discusses use and properties. Epoxy resin glues have many advantages, but the cold setting epoxy resin glues have a low resistance against delamination.

1.11.6.2 Polyurethane Adhesives. *Bayer* discovered at Leverkusen/Germany the principle of diisocyanate polyaddition. Its fundamentals were disclosed in 1937 in the German patent 728981.

This principle allows to produce plastics starting from aliphatic diisocyanates such as hexamethylene-1,6-diisocyanate and glycols such as butylene-1, 4-glycol.

$$HO-R-OH + OCN-R-NCO + HO-R-OH + OCN-R-NCO + HO-R-OH$$
$$\downarrow$$
$$-O-R-O-CO-NH-R-NH-CO-O-R-O-CO-NH-R-NH-CO-O-R-O-$$

Shaped articles, films, fibers, and bristles have been made from these plastics.

Foams, elastomers, lacquers and adhesives require higher molecular weight polyols as starting materials. Four adhesives in particular are used at ambient temperature in liquid form: hydroxyl-terminated polyesters, prepared from aliphatic and aromatic dicarboxylic acids (adipic acid, phthalic acid) and polyols like propylene-1,2-glycol, diethylene glycol or trimethylol-1,1,1-propane.

"Reactive polyurethane adhesives" are formed, if polyisocyanates either liquid or dissolved in an organic solvent are added to these hydroxyl polyesters (Desmocoll).

Originally the readily available and low cost toluene diisocyanate was used as component. Its pungent vapors, however, concentrated on the production line frequently caused respiratory irritations. In order to avoid health-hazards a special toluene diisocyanate-polyol-prepolymer was developed, commercially available as a 75% product dissolved in ethyl acetate (Desmodur L 75). Furthermore a solvent-free mixture of aromatic polyisocyanates (Desmodur VL), mainly consisting of diphenylmethane diisocyanate, was offered for the manufacture of adhesives. Unfortunately this low cost polyisocyanate shows a dark brown color which is objectionable in many applications.

Wood bonds of high strength result when "reactive polyurethane adhesives" are applied and the surfaces have been treated carefully before bonding. Softwoods and porous hardwoods should be roughened by sandpaper along the fibers. Nonporous wood from deciduous trees has to be treated by a toothed plane.

"Reactive polyurethane adhesives" are two-part adhesives. As a rule an excess of polyisocanate is added to the polyester, frequently up to 50% based on the stoichiometric amount required. The pot-life lasts several hours. For instance, by means of a doctor roll, an amount of approx. 200 g/sq. m adhesive is spread on both surfaces. If required, the adhesives' viscosity can be reduced by adding solvents such as ethyl acetate. If these adhesives, containing solvents, are applied, the solvents must be allowed to evaporate to a great extent from the coats before joining.

The bonds have to be pressed during setting which is produced by chemical reaction for instance crosslinking. In general few hours pressing is required. This time can be reduced by adding catalysts like urea. However, this leads to a reduction in pot-life.

As compared with other adhesives for wood "reactive polyurethane adhesives" offer advantages. Toughness and elasticity of the adhesive films can be varied over a wide range. The bonds withstand the influence of cold and boiling water

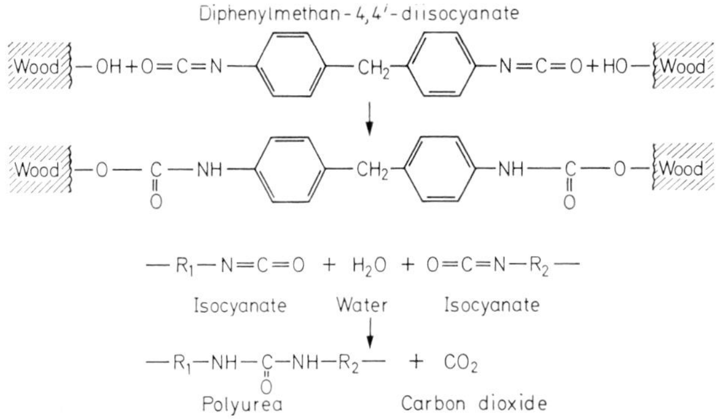

Fig. 1.63. Reaction between diisocyanates, wood, and water. From *Deppe* and *Ernst* (1971)

and present a remarkable outdoor resistance due to their stability in a damp climate. They are not attacked by oils, greases, aromatic, and aliphatic hydrocarbons.

Processing bonded parts involves no difficulty as the soft adhesive film does not wear the tools more than the wood.

Great attention is actually paid in this industry to polyisocyanates, which can act instead of alkaline setting phenolic resins as a bonding agent for wood in the production of particle board. Such polyisocyanate-bonded particle board is becoming important in building constructions. The hydrophobic effect of the polyisocyanate is of particular advantage in this application. The major technical problem which must be overcome is a too strong adhesion of the particle board to the iron moulds. As Fig. 1.63 shows, not only the urethane bridges but also a part of the polyurea is contributing to the bonding of the wood particles. The formation of the polyurea is due to a reaction between diisocyanates and water.

Literature Cited

American Specification MIL-A-397 B(1953) Adhesive: Room temperature and intermediate temperature setting resin (phenol, resorcinol and melamine base).
Anonymous, (1943) Penetration and Starvation. Aero Res. Techn. Notes, Bull. No. 7, Duxford, Cambridge.
Armonas, J. E., (1970) Gel permeation chromatography and its use in the development of resins. For. Prod. J. **20**: 22—28.
Baekeland, L. H., (1909, 1913) USP 939966, 942699, 942700 and 942809, Ind. Eng. Chem. **1**: 149, **5**: 506.
—, (1910) Condensation products of phenol and formaldehyde. USP 942809.
—, (1910) Condensation products of phenol and formaldehyde. USP 942700.
—, (1910) Insoluble product of phenol and formaldehyde. USP 942699.
—, (1910) Molded condensation products of formaldehyde and phenol. USP 939966.
—, *Thurlow, N.*, (1912) Wood finished with veneer attached by and coated with phenol-formaldehyde condensation product. USP 1019408.
Baeyer, A., (1872) Über die Verbindung der Aldehyde mit den Phenolen. Ber. d. Dt. Chem. Gesellsch. **5**: 289—282.
Barnes, D., (1970) Glue spreading efficiency of roll spreaders. For. Prod. J. **20**: 24—29.
Baumann, H., (1967) Leime and Kontaktkleber. Springer-Verlag, Berlin, Heidelberg, New York.
Bayer, O., (1937) Verfahren zur Herstellung von Polyurethanen, bzw. Polyharnstoffen. DRP 728981.
Bikerman, J. J., (1941) Strength and thinness of adhesive joints. J. Soc. Chem. Ind. **60**: 23—24.
—, (1947) The fundamentals of tackiness and adhesion. J. Colloid Sci. **2**: 163—175.
—, (1961) The science of adhesive joints. Academic Press, New York.
—, *Huang, C.-R.*, (1959) Polyethylene adhesive joints. Trans. Soc. Rheol. **3**: 5—12.
Black, J. M., (1943) The effect of fire retardant chemicals on glues used in plywood manufacture. U. S. For. Prod. Lab. Rep. No R 1427, Madison/Wisc.
Blankenstein, C., (1937) Leimen von Holz. AWF-Schrift No 164, Leipzig, Berlin.
Blomquist, R. F., (1949) Effect of alkalinity of phenol- and resorcinol-resin-glues on durability of joints in plywood. U. S. For. Prod. Lab. Rep. No R 1748, Madison/Wisc.
—, (1963) Adhesive-Past, Present and Future. U. S. For. Prod. Lab., Madison/Wisc.
BP 347242 (1929) Th. Goldschmit AG.
BP 435041 (1933) I. G. Farbenindustrie AG.
Brouse, D., (1938) Serviceability of glue joints. Mech. Eng. **60**: 306—308.
Browne, F. L., (1931) Adhesion in the painting and in the gluing of wood. Ind. Engng. Chem. **23**: 290—294.
—, *Brouse, D.*, (1929) Nature of adhesion between glue and wood. Ind. Engng. Chem. **21**: 80—84.
—, —, (1939) Casein glue. In: *Sutermeister, E., Browne, F. L.*, Casein and its industrial application. A. C. S. Monograph No 30, p. 247—253. Reinhold, New York.
—, (1947) quoted after: *Delmonte*, The technology of adhesives New York, p. 260.

Browne, F. L., Truax, T. R., (1926) The place of adhesion in the gluing wood. Colloid Symposium Monograph **4**: 258—269.

—, —, (1929) Significance of mechanical wood-joint tests for the selection of woodworking Glues. Ind. Engng. Chem. **21**: 74—79.

de Bruyne, N. A., (1944) The strength of glued joints. Aircraft Engng. **16**: 115—140.

—, (1947 a) The physics of adhesion. J. Scientif. Instruments, **24**: 29—35.

—, (1947 b) The physics of adhesion. Aero Res. Techn. Notes No 54, Duxford, Cambridge.

—, (1951) Some basic ideas in structural adhesives, London.

—, *Houwink, R.*, (1951) Adhesion and adhesives. Elsevier Publ. Comp., New York, Amsterdam, London, Brussels, p. 18.

Buttermann, S., Cooperider, C. K., (1923) Process of manufacturing waterproof adhesives. USP 1456842. U. S. Pat. Office Gaz. **310**: 1129.

Canadian Standard Association, Can. Standard 0112-7-1960 (1960) Phenol and resorcinol resin adhesives for wood (room and intermediate temperature curing).

Carroll, M. N., Bergin, E. G., (1967) Catalyzed PVA emulsions as wood adhesives. For. Prod. J. **17**: 45—50.

Carswell, T. S., (1947) Phenoplasts: Their structure, properties and chemical technology. Interscience, New York.

Chapman, F., (1951) Synthetic resin adhesives. In: *de Bruyne, Houwink*. Adhesion and adhesives, p. 224—226. Elsevier Publ. Comp., New York, Amsterdam, London, Brussels.

Chen, Ch.-M., (1970) Effect of extractive removal on adhesion and wettability. For. Prod. J. **20**: 36—40.

Chow, S.-Z, Hancock, W. V. (1969) Method for determining degree of cure of phenolic resin. For. Prod. J. **19**: 21—29.

—, *Mukai, H. N.*, (1969) An infrared method for determining phenol-formaldehyde resin content in fiber and wood products. For. Prod. J. **19**: 57—60.

—, *Troughton, G. E.*, (1969) Determination of melamine-formaldehyde content in wood and cellulose products by infrared spectrometry. For. Prod. J. **19**: 24—26.

Chugg, W. A., Gray, V. R., (1960) The durability of acid-setting phenolic resin adhesives. Timber Developm. Assoc. Ltd., London, Inform. Bull. 5.

Ciba AG (without date) Unpublished work.

—, (1938) DRP 718342. Melaminharz.

—, (1939 a) DRP 702890 (Swed. Pat. 202245). Melaminharz.

—, (1939 b) DRP 726855 (FP 863284). Reinigen von Melaminharz.

Clark, L. E. Jr., (1968) Measurement of "speed of set" of room temperature setting wood adhesives. For. Prod. J. **18**: 52—53.

Delmonte, J., (1947) The technology of adhesives. Reinhold Publ. Comp., New York, p. 325, 326, 349.

Deppe, H.-J., Ernst, K., (1971) Isocyanate als Spanplattenbindemittel. Holz als Roh- und Werkstoff **29** (2): 45—50.

Dibuz, J. J., Shelton, F. J., (1967) Glueline identification. For. Prod. J. **17**: 20—22.

Dietz, A. G. H., (1949) Engineering laminates. New York, London, p. 104.

DRP 550647 (1929) I. G. Farbenindustrie AG.

DRP 647303 (BP 455008) (1935) Henkel, Harzartige Kondensationsprodukte.

DRP 680707 (FP 817539, USP 2211709, USP 2211710) (1936) I. G. Farbenindustrie, Harzartige Kondensationsprodukte.

DRP 702890 (Swiss Pat. 202245) (1939) Ciba AG, Melamin.

DRP 718342 (1938) Ciba AG, Melaminharz.

DRP 721240 (1942) Henkel, Kaltverleimung.

DRP 726855 (FP 963284) (1939) Ciba AG., Reinigen von Melaminharz.

DRP 728981 (1937) *Bayer, O.*, Verfahren zur Herstellung von Polyurethanen, bzw. Polyharnstoffen.

Eickner, H. A., Blomquist, R. F., (1951) Adhesives for bonding wood to metal. U. S. For. Prod. Lab. Rep. No R 1768, Madison/Wisc.

Ellis, C., (1935) Chemistry of synthetic resins. Vol. I. Reinhold Publ. Corp., New York.

Enzensberger, W., (1952) Probleme der Furnierfugenverleimung. Holz-Zbl., **78**: 1137.

Farrow, C. A., Hamley, D. H., Smith, E. A., (1946) Phenolic resin glue line as found in yellow birch plywood. Ind. Eng. Chem. Anal. Ed. **18**: 307—310.

Fischer, C., Bensend, D. W., (1969) Gluing of southern pine veneer with blood modified phenolic resin glues. For. Prod. J. **19**: 32—37.

Freeman, H. G., Kreibich, R. E., (1968) Measurement of glue line viscosity. For. Prod. J. **18**: 15—17.

Frey, K., (1947) Über die Eigenschaften und die Anwendung von Melocol. Schweiz. Arch. f. Angew. Wiss. u. Techn. **13**: 83—93.

Gerngross, O., (1930) Über Sperrholzleime. Deutsche Versuchsanst. f. Luftfahrt e. V., Berlin-Adlershof, Rep. No. 192. DVL-Jahrbuch, p. 428—433.

—, (1949) Glue and gelatine. A paper for the Congress of the New International Association for Testing Materials. Ztschr. f. Angew. Chem. **42**: 968—972.

Gibbs, J. W., (1876/1877) On the equilibrium of heterogeneous substances. Trans. Connecticut Acad. **3** (1876): 103—243; (1877): 343—524.

Goland, N., Reissner, E., (1944) The stresses in cemented joints. J. Appl. Mech. **11**: A17—A27.

Goldschmidt, C., (1896) Über die Einwirkung von Formaldehyd auf Harnstoff. Ber. d. Dt. Gesellsch. **29**: 2438/2439.

—, (1897) Über die Einwirkung von Formaldehyd auf Harnstoff. Chem. Ztg. **46**: 460.

—, (1898) Action of formaldehyde on carbamide. J. of the Chem. Society **74**: 178—180 (Abstract).

Goldschmidt AG, (1929) BP 347242.

Guiher, J. K., (1970) Methylphenol-formaldehyde resin as an adhesive for wood. For. Prod. J. **20**: 21—23.

Gumprecht, D. L., (1969) "Tailoring" phenolic adhesives. For. Prod. J. **19**: 38—40.

Hahn, K. F., (1959, 1961) Symposium on adhesion and adhesives 1959. ASTM Spec. Tech. Publ. No 271.

Halligan, A. F., (1969) Recent glues and gluing research applied to particleboard. For. Prod. J. **19**: 44—51.

Henkel (1935) DRP 647303 (BP 455008). Harzartige Kondensationsprodukte.

—, (1942) DRP 721240. Kaltverleimung.

Hoekstra, J., Fritzius, C. P., (1951) Rheology of adhesives. In: *de Bruyne, N. A., Houwink, R.*, Adhesion and adhesives, p. 79. New York, Amsterdam, London, Brussels: Elsevier Publ. Comp.

Holzer, K., (1962a) Leime für die Herstellung von Lagenholz. In: *Kollmann, F.*, Furniere, Lagenhölzer und Tischlerplatten. Springer-Verlag, Berlin, Göttingen, Heidelberg, p. 285 to 355.

—, (1962b) Harnstoff-Formaldehyd-Harzleime. In: *Kollmann, F.*, Furnier, Lagenhölzer und Tischlerplatten. Springer-Verlag, Berlin, Göttingen, Heidelberg, p. 316—332.

Holz-Zentralblatt-Verlag (1958) Holz und Kunststoffe. Holzwirtschaftl. Jahrbuch No. 8, Stuttgart.

Houwink, R., Salomon, G., (1965) Adhesion and adhesives, Vol. I: Adhesives. Elsevier Publ. Comp., Amsterdam, London, New York.

Hse, Chung Y., (1968) Gluability of southern pine earlywood and latewood. For. Prod. J., **18**: 32—36.

Hultzsch, K. (1940) Chemie der Phenolharze. Springer-Verlag, Berlin, Göttingen, Heidelberg.

I. G. Farbenindustrie A. G. (1929) DRP 550647.

I. G. Farbenindustrie A. G. (1933) BP 435041.

Jarvi, R. A. (1967) Exterior glues for plywood. For. Prod. J., **17**: 37—42.

John, J., (1920) USP 1355834.

Johnson, J. W., (1970) Flexural tests of large glue-laminated beams made of nondestructively tested lumber. For. Prod. J. **20**: 20.

Jones, W. L., (1927) Some data on glues from raw starch. Wood Working Industries, Milwaukee, **3**: 21—27.

Kaess, F., Vogel, E., (1954) Herstellung und Verwendung von Melamin. Chem. Ind. Techn., **26**: 380—382.

Kirby, K. W., (1965) Vegetable adhesives. In: *Houwink, R., Salomon, G.*, Adhesion and adhesives. Vol. I: Adhesives. Elsevier Publ. Comp., Amsterdam, London, New York, p. 167—185.

Kitazawa, G., (1946) A study of adhesion in the glue lines of twenty-two woods of the U. S. N. Y. State College of Forestry, Syracuse, Techn. Bull. No. 66.

Klemm, H., (1938) Neue Leimuntersuchungen mit besonderer Berücksichtigung der Kaltkunstharzleime, München.

Knight, R. A. G., (1946) Requirements and properties of adhesives for wood. For. Prod. Res. Bull. No. 20. His Majesty's Stationery Office, London.

—, (1952) Adhesives for wood. Chapman & Hall, London.

—, (1959) The efficiency of adhesives for wood. For. Prod. Bull. No. 38, 2nd Ed., p. 21. Her Majesty's Stationery Office, London.

—, *Newall, R. J.*, (1957) Investigations into glues and gluing. For. Prod. Res. Lab. Rept. No. 107. Her Majesty's Stationery Office, London.

Koch, P., (1967) Minimizing and predicting delamination of southern plywood in exterior exposure. For. Prod. J., **17**: 41—47.

Koch, H., Sachsenberg, E., (1932) Untersuchungen über die Biegefestigkeit kaltgeleimter Hölzer. Sperrholz, **4**: 89—91.

Köhler, R., (1941) Kondensationskunststoffe aus Melamin und Formaldehyd. Kunststoff-technik, **11**: 1—4.

—, (1943) Über einige Versuche zur Härtung von Melamin-Formaldehyd-Kondensations-produkten. Koll.-Z., **103**: 138—144.

König, H. J., (1958) Herstellung, Prüfung und Eigenschaften melaminharz-vergüteter Schicht-stoffe für dekorative Verwendung. Kunststoffe, **48**: 513—522.

Kollmann, F., (1955) Technologie des Holzes und der Holzwerkstoffe. 2nd Ed., Vol. 2. Springer-Verlag, Berlin, Göttingen, Heidelberg.

—, *Teichgräber, R.*, (1961) Beitrag zur Prüfung der Brandeigenschaften, insbesondere der Schwerentflammbarkeit, von Platten aus Holz und Holzwerkstoffen. Holz als Roh- und Werkstoff **19**: 173—186.

Kraemer, O., (1932) Fragen der Holzleimung. Z. VDI., **76**: 443.

Kragh, A. M., *Wootton, J.*, (1965) Animal glue and related protein adhesives. In: *Houwink, R.*, *Salomon, G.*, Adhesion and adhesives, Vol. I: Adhesives., p. 141—165. Elsevier Publ. Comp., Amsterdam, London, New York.

Kreibich, R. E., *Freeman, H. G.*, (1968) Development and design of an accelerated boil machine. For. Prod. J., **18**: 24—26.

—, —, (1970) Effect of specimen stressing durability of eight wood adhesives. For. Prod. J., **20**: 44—49.

Laucks, I. F., *Davidson, G.*, (1923) Vegetable glue and method of making same. USP 1 689 732.

—, —, (1939) Glue and method of making. USP 150175.

Lindauer, A. C., (1923) Blood albumin glue. USP 1 459 541.

Little, G. E., *Pepper, K. W.*, (1947) The influence of pH on the setting time of phenolic resins. XIth Intern. Congr. Pure and Appl. Chem., London.

Marian, J. E., (1953) Trä och plast. Aktuella problem för limnings- och yttbehandlings-forskning. Svenska Träforskn. Inst., Träteknik, Medd. 48 B, Stockholm.

—, *Fickler, H. H.*, (1953) Eigenschaften von Leimtypen. Holz als Roh- und Werkstoff, **11**: 18—27.

Mark, H., (1943) Elasticity and strength. In: *Ott, E.*, Cellulose and cellulose derivates. Inter-science Publ. Inc. p. 990—1043.

Martin, R. W., (1956) The chemistry of phenolic resins, p. 298. Wiley, New York.

Matting, A., *Ulmer, K.*, (1963) Metallklebeverbindungen. III. Spannungs- und Dehnungs-verhalten. IV. Grenzflächenreaktionen u. Spannungsverteilung. Kautschuk, Gummi, **16**: 334, 387.

Maxwell, J. W. (1945) Shear strength of glue joints as affected by wood surfaces and pressures. Transactions Am. Soc. Mech. Eng. Easton, **67**: 104—110.

McBain, J. W. and coworkers, (1922) First Report of the Adhesives Research Committee, London. Dept. Sci. Ind. Res., Adhesive Res. Committee, London.

—, —, (1926) Second Report of the Adhesive Research Committee, London. Dept. Sci. Ind. Res., Adhesive Res. Committee, London.

—, —, (1932) Third Report of the Adhesive Research Committee, London. Dept. Sci. Ind. Res., Adhesive Res. Comm., London.

—, *Lee, W. B.*, (1927) Adhesives and adhesion: gums, resins and waxes between polished metal surfaces. J. Phys. Chem. **31**: 1674—1680.

McClain, I. R., (1919) Wood Veneer. USP 1 299 747.

McHale, W. H., (1944) Melurac — a boil-resistant, durable adhesive. Plastics, **1**: 40, 11, 112.

McNamara, W. S., *Waters, D.*, (1970) Comparison of the rate of glue-line strength development for oak and maple. For. Prod. J. **20**: 34—35.

Mees, H., (1930) Sperrholz. Vol. 2, p. 20 and 365.

Megson, N. J. L., (1958) Phenolic resin chemistry. Butterworths Sci. Publ., London.

Müller, A., (1953) Kaltleime auf Phenolharzbasis und die Gefahr der Holzschädigung. Holz als Roh- und Werkstoff, **11**: 429—435.

Mylonas, C., (1948) On the stress distribution in glued joints. Proc. VIIth Intern. Congr. Appl. Mech., London, **4**: 137—149.

—, *de Bruyne, N. A.*, (1951) Static problems. In: *de Bruyne, N. A.*, *Houwink, R.*, Adhesion and adhesives. Elsevier Publ. Comp., New York, Amsterdam, London, Brussels, p. 91—105.

Narayanamurti, D., *Handa, B. K.*, (1954) Rheology of Adhesives, Part I. Kolloid-Z., **138**: 40.

—, *Pande, J. N.*, (1948) Durability trials on glues and plywood. Ind. For. Bull. No 139, For. Res. Inst., Dehra Dun.

Northcott, P. L., (1964) Specific gravity influences wood bond durability. Adhesives Age **7** (10): 34—36.

Pauling, L., *Corey, R. B.*, *Branson, H. R.*, (1951) The structure of proteins: Two hydrogen bonded helical configurations of the polypeptide chains. Proc. Nat. Acad. U. S. **37**: 205 to 211.

Perry, Th. D., (1948) Modern plywood. Pitman Publ. Corp., New York, London.

Phinney, H. K., (1950) Phenol-formaldehyde and resorcinol-formaldehyde resin glues in gluing of wood. North Carolina State College, School of Forestry, Techn. Bull. No 4, p. 37.

Plath, E., (1951) Die Holzverleimung. Wiss. Verlagsgesellschaft, Stuttgart.

—, (1953) Studien über Phenolharzleime. I. Hitze- und Säurehärtung. Holz als Roh- und Werkstoff **11**: 392—400.

—, (1954) Melaminharze in der Holzindustrie. Holztechnik. **34**: 9—12.

Poletika, N. V., (1943) Rep. W. E. 170 M 2, Res. Lab. Curtis Wright Corp. (cited according to *de Bruyne, N. A., Houwink, R.*, 1951).

Pollak, F., (1922) BP 181014. Chem. Astr. **16**: 3531.

—, (1925, 1927) BP 261409. Chem. Abstr. **21**: 3432.

Rayner, C. A. A., (1965) Synthetic organic adhesives. In: *Houwink, R., Salomon, G.*, Adhesion and adhesives. Vol. I: Adhesives, p. 209—210, 221—222, 225, 296. Elsevier Publ. Comp., Amsterdam, London, New York.

Redfern, D. V., (1950) Gluing operations and industry trends. For. Prod. Soc., Reprint No 119.

Reinhart, F. W., (1954) Nature of Adhesion. J. Chem. Education, **31**: 128. .

Rinker, R. C., Kline, G. M., (1946) A general theory of adhesion. Symposium on Adhesives, A. S. T. M., p. 12.

Roš, M., (1945) Die Melocol-Leime der Ciba Aktiengesellschaft, Basel. Eidgen. Material-prüfungs- und Versuchsanstalt Zürich, Bericht No 152.

Salomon, G., (1965) Adhesives. In: *Houwink, R., Salomon, G.*, Adhesion and adhesives. Elsevier Publ. Comp., Amsterdam, London, New York.

Sauer, E., (1958) Chemie und Fabrikation der tierischen Leime und der Gelatine. Springer-Verlag, Berlin, Göttingen, Heidelberg.

Scales, G. M., (1951) Extension in "Structural Adhesives". London, p. 89.

Seidl, R. F., (1947) Paper and plastic overlays for veneer and plywood. For. Prod. Res. Soc., Proc. 1, p. 23—32.

Simonds, H. R., Weith, A. J., Bigelow, M. N., (1949) Handbook of Plastics. 2nd. Ed., p. 462, 663. D. van Nostrand Comp. Inc., Toronto, New York, London.

Sneddon, J. N., (1961) The distribution of stress in adhesive joints. In: *Eley, D. D.*, Adhesion, p. 46. Oxf. Univ. Press.

Sodhi, J. S., (1957) On the acid corrosion of wood by gluing with phenolic resin adhesives. Holz als Roh- und Werkstoff, **15**: 261—263.

Spellacy, J. R., (1953) Casein dried and condensed whey.Lithotype Process Co., San Francisco.

Strickler, M. D., (1968) Specimen designs for accelerated tests. For. Prod. J. **18**: 84—90.

Stumbo, D. A., (1965) Historical Table. In: *Houwink, R., Salomon, G.*, Adhesion and adhesives. Vol. I: Adhesives, p. 534—536. Elsevier Publ. Comp., Amsterdam, London, New York.

Sutermeister, S., Brühl, E. (1932) Das Kasein. Springer-Verlag, Berlin.

Truax, T. R. (1929) The Gluing of Wood. U. S. Dept. Agric. Bull. No 1500, p. 8, 39, 41. U. S. Gov. Printing Office, Washington.

—, (1932) Development of wood adhesives and gluing technics. Transact. A. S. M. E.

U. S. For. Products Laboratory, (1956) Summary of information on the durability of water-resistant woodworking glues. Rep. No 1530, Madison, Wisc.

USP 150175 (1939) *Laucks, I. F., Davidson, G.*, Glue and method of making.

USP 939966 (1910) *Baekeland, L. H.*, Molded condensation products of formaldehyde and phenol.

USP 942699 (1910) *Baekeland, L. H.*, Insoluble product of phenol and formaldehyde.

USP 942700 (1910) Condensation products of phenol and formaldehyde.

USP 942809 (1910) Condensation products of phenol and formaldehyde.

USP 1019408 (1912) *Baekeland, L. H., Thurlow, N.*, Wood finished with veneer attached by and coated with phenol-formaldehyde condensation product.

USP 1299747 (1919) *McClain, I. R.*, Wood veneer.

USP 1355834 (1920) *John, J.*

USP 1456842 (1923) *Butterman, S., Cooperider, C. K.*, Process of manufacturing waterproof adhesives.

USP 1459541 (1923) *Lindauer, A. C.*, Blood Albumin Glue.

USP 1689732 (1923) *Laucks, I. F., Davidson, G.*, Vegetable glue and method of making same.

Volkerson, O., (1938) Die Nietkraftverteilung in zugbeanspruchten Nietverbindungen und konstanten Laschenquerschnitten. Luftfahrtforschg. **15**: 41—47.

Voss, W. C., (1947) Engineering laminates. Fundamentals underlying the problems of their inhomogeneity. Edgar Marburg Lecture, A. S. T. M.

Wake, W. C., (1961) The rheology of adhesives. In: *Eley, D. D.*, Adhesion. Oxford University Press.

Walter, G., Gewing, M., (1932) Zur Konstitution der künstlichen Harze. II. Die theoretischen Grundlagen und quantitativen Untersuchungen der Harnstoff-Formaldehyd-Kondensation. Kolloid-Beih. **34**: 163—217.

Wangaard, F. F., (1946) Summary of information on the durability of woodworking glues. U. S. For. Prod. Laboratory, Rep. No 1530, Madison, Wisc.

Webb, D. A., (1970) Wood laminating adhesive system for ribbon spreading. For. Prod. J. **20**: 19—23.

Westphal, W. H., (1952) Physikalisches Wörterbuch. Vol. II, p. 700—701, Springer-Verlag, Berlin, Göttingen, Heidelberg.

Whistler, R. L., (1959) Industrial gums. Academic Press Inc., New York.

Williamson, R. V., Lathrop, E. C., (1949) Agricultural residue flours as extenders in phenolic resin glues for plywood. Mod. Plastics **27**: 111—112, 169—174.

Wood, A. D., (1963) Plywoods of the world. Johnston and Bacon Ltd., Edinburgh, London.

—, *Linn, T. G.*, (1950) Plywoods, their development, manufacture and application. W. and A. K. Johnston Ltd., Edinburgh, London.

Zoller, H. F., (1920) Casein Viscosity Studies. Journ. General Physiology, **3**: 635—651.

Zschokke, H., Montandon, R., (1944) Einfluß der Neigung von Schweißnähten zur Beanspruchungsrichtung auf die Zugfestigkeit. Schweiz. Archiv Angew. Wiss. Techn. **10**: 137.

2. SOLID MODIFIED WOODS

2.0 Introduction

Wood can be modified by either physical or chemical means, or their combination. Modifications to be considered in this chapter involve either impregnation with chemicals, densification under heat and pressure, or their combination. Impregnations involve depositing chemicals in the microscopically visible void structure or within the cell walls, or reacting the chemicals with the cell wall components without breaking down the wood structure. These impregnation treatments are made to impart decay resistance, fire retardance and/or dimensional stability to the wood, or to improve specific strength properties. The first two of these objectives can be accomplished by merely depositing either toxic or fire retardant chemicals in the void structure. When water borne chemicals are used they are partially deposited within the cell walls of the wood. Permanent dimensional stability, on the other hand, is imparted to wood only when the chemicals are deposited within the cell walls or chemically react with the cell wall components.

Only a few strength properties can be appreciably increased by impregnation treatments alone, and these only at high degrees of loading. Practically all of the strength properties can, however, be materially increased by densifying either normal wood or wood pre-impregnated with polymer-forming chemicals under temperature conditions so that mechanical damage to the wood is avoided.

These modifications of wood, involving only chemical treatments, thermal treatments, and densification treatments, that retain the original integrity of the wood throughout the processing, differ from reconstituted woods, that are broken down either mechanically or chemically and then reformed into solid products, with or without the use of adhesives. These reconstituted wood products comprising wood laminates, plywood, particleboard, fiberboards, and paper are covered in other chapters.

2.1 Fundamentals of Impregnation

The most important, and frequently the only, step involved in making the various modified wood products covered here involve the impregnation and distribution of the various chemicals throughout the structure. This has proven to be a difficult task in the heartwood of sizable specimens of many species. Table 2.1 lists important North American species divided into four groups according to increasing difficulty of impregnating the heartwood (*MacLean*, 1935, 1962). Even the more permeable sapwood of different species vary considerably in permeability. A short description of the mechanism of penetration and the reasons for variations under different conditions is hence of considerable practical importance.

Table 2.1. Classification of American Woods with Respect to Heartwood Penetrability
(*MacLean*, 1935)

Softwoods	Hardwoods
Group 1. Heartwood easily penetrated	
Bristlecone pine, *Pinus aristata*	Basswood, *Tilia glabra*
Pinon, *Pinus eduils*	Beech (white heartwood), *Fagus grandifolia*
Ponderosa pine, *Pinus ponderosa*	Black gum, *Nyssa sylvatica*
	Green ash, *Fraxinus pennsylvanica lanceolata*
	Pine cherry, *Prunus pennslyvanica*
	River birch, *Betula nigra*
	Red oaks (except blackjack), *Quercus spp.*
	Slippery elm, *Ulmus fulva*
	Sweet birch, *Betula lenta*
	Tupelo gum, *Nyssa aquatica*
	White ash, *Fraxinus americana*
Group 2. Heartwood moderately difficult to penetrate	
Douglas fir (coast type), *Pseudotsuga*	Black willow, *Salix nigra*
taxifolia	Chestnut oak, *Quercus montana*
Jack pine, *Pinus banksiana*	Cottonwood, *Populus spp.*
Loblolly pine, *Pinus taeda*	Largetooth aspen, *Populus gradidentata*
Longleaf pine, *Pinus palustris*	
Norway pine, *Pinus resinosa*	Mockernut hickory, *Hicoria alba*
Shortleaf pine, *Pinus echinata*	Silver maple, *Acer saccharinum*
Western hemlock, *Tsuga heterophylla*	Sugar maple, *Acer saccharaum*
	Yellow birch, *Betula lutea*
Group 3. Heartwood difficult to penetrate	
Eastern hemlock, *Tsuga canadensis*	Hackberry, *Celtis occidentalis*
Engelmann spruce, *Picea engelmannii*	Rock elm, *Ulmus racemosa*
Lowland white fir, *Abies grandis*	Sycamore, *Platanus occidentalis*
Lodgepole pine, *Pinus contorta*	
Noble fir, *Abies nobilis*	
Sitka spruce, *Picea sitchensis*	
Western larch, *Larix occidentalis*	
White fir, *Abies concolor*	
White spruce, *Picea glauca*	
Group 4. Heartwood very difficult to penetrate	
Alpine fir, *Abies lasiocarpa*	Beech (red heartwood), *Fagus grandifolia*
Corkbark fir, *Abies arizonica*	Blackjack oak, *Quercus marilandica*
Douglas fir (mountain type), *Pseudotsuga*	Black locust, *Robinia pseudoacacia*
taxifolia	Chestnut, *Castanea dentata*
Northern white cedar, *Thuja occidentalis*	Red gum, *Liquidambar styraciflua*
Tamarack, *Larix laricina*	White oaks (except chestnut oak), *Quercus spp.*
Western red cedar, *Thuja plicata*	

Movement of fluids in wood is controlled by the complex solid and void structure of its components. These structures are described in some detail in Chapter 1, Vol. I, of this book. The movement of fluids through wood can be by either of two mechanisms or their combination. Flow can occur through all parts of the natural void structure under an externally applied pressure or internally imposed capillary forces (*Stamm*, 1967 (1)). Movement of dissolved materials can occur by diffusion under a concentration gradient through a combination of the natural voids and the bound liquid held within the cell walls (*Stamm*, 1967 (2)). Diffusion is also involved in the drying of wood (*Stamm*, 1964, Chapters 23 and 24).

2.1.1 Flow through True Capillaries

The velocity of flow V, of a liquid through a true capillary tube, in cubic centimeters per second, is dependent upon its radius r, and length L in centimeters, the pressure drop across its length ΔP, in dynes per square centimeter, and the viscosity of the liquid η in centipoises according to the equation of Poiseuille

$$V = \frac{\pi r^4 \Delta P}{8\eta L}. \tag{2.1}$$

When n similar capillaries are connected in parallel with the same pressure drop across each, then

$$V = \frac{n\pi r^4 \Delta P}{8\eta L} = \frac{q r^2 \Delta P}{8\eta L}, \tag{2.2}$$

where q is equal to the combined cross section of the capillaries, $n\pi r^2$. Flow is thus dependent on the total capillary cross section and also the square of the radius. If the capillaries have different radii that do not vary greatly from the mean, the mean radius can be used as a first approximation in Eq. (2.2).

If a number of capillary tubes with different radii are connected in one series path, the rate of flow through each is of necessity the same. The portion of the total pressure drop through each capillary size will, however, vary as the reciprocal of the rate of flow through each when tested separately under constant pressure. Capillary flow is thus analogous to electrical conductivity in which the resistances or pressure drops are additive in series and the current or liquid flows are additive in parallel. The same analogy will later be shown to hold for diffusion (2.1.3).

2.1.2 Flow through Softwoods

2.1.2.1 Structure Involved. Assume for the moment that the various void water filled structures of softwoods act, as a first approximation, as if they were capillaries. Neglecting the minor effect of ray cells, passage of liquids, through softwoods in either the longitudinal or transverse direction will be through the lumen of the fibers in series with passage through the pit membrane openings in parallel with passage through the cell walls. The velocity of flow through the lumen will be equal to the combined velocity of flow through the pits and the cell walls. Thus

$$V = (q_f r_f^2) P_f = [(q_p r_p^2 / L_p) + (q_w r_w^2 / L_w)] P_m, \tag{2.3}$$

where q is the fractional void cross section, r the effective capillary radius and L the effective length. The subletter f is for the fibers or tracheids, p for the cell walls. P_f is the pressure drop across the lumen of the fibers, and P_m the pressure drop across the combined communicating structure. First, consider flow of a liquid in the fiber direction through air free liquid saturated wood. When the best approximations for the q, r, and L values for the heartwood of a softwood with a swollen volume specific gravity of 0.365 are substituted in Eq. (2.3) (*Stamm*, 1946),

$$[0.65(1.52 \times 10^{-3})^2 / 0.286] P_f = [0.0038(2.8 \times 10^{-6})^2 / 10^{-4} +$$
$$[0.10(2 \times 10^{-8})^2 / 6.4 \times 10^{-4}] P_m,$$
$$52{,}500 P_f = (3.0 + 0.00062) P_m,$$
$$P_f / P_m = 5.7 \times 10^{-5}.$$

The calculations show that flow through the cell walls is negligible compared to that through the pits, and that the pressure drop through the lumen of the fibers is negligible compared to that through the pits. The q, r, and L values for the different structures of the heartwood of various softwoods may vary appreciably from the aforegoing values without affecting the conclusions. However, the q_p and r_p values for highly permeable sapwood may be larger by a sufficient amount to make the pressure drop through the lumen a significant portion of the total pressure drop.

Similar calculations for transverse flow of a liquid through the heartwood of a softwood show that the pressure drop through the lumen of the fibers constitutes a still smaller fraction of the total pressure drop and that flow through the fiber walls is still negligible

$$2.5 \times 10^{10} P_f = (4.1 + 0.00035) P_m,$$

$$P_f / P_m = 1.64 \times 10^{-10}.$$

In this case the pressure drop through the lumen of the most permeable sapwood is still negligible compared to that through the pits.

It is thus apparent that the permeability of softwoods to liquids is almost entirely dependent upon the permeability of the pits. Unfortunately no equivalent structural data were available for hardwoods, but in all probability the permeability is dependent upon a combination of the permeability of the pits with that of the sieve plates and tyloses in the vessels.

2.1.2.2 Permeability to Liquids and Gases. The permeability of porous materials to liquids is usually expressed in Darcys

$$K = \frac{QL\eta}{A \Delta P}, \tag{2.4}$$

where Q is the rate of flow in cubic centimeters per second, A is the flow area in square centimeters, L is the effective flow length, ΔP is the pressure drop in atmospheres, and η is the viscosity in centipoise. This equation is equivalent to the Poiseuille equation (2.2) except that the radius variable is included in K, thus avoiding the need for its determination, which is both difficult and uncertain in most porous materials. K also involves a constant for conversion of the pressure drop from dynes/cm² to atmospheres.

The permeability of a porous material to a gas involves a more complicated relationship than for permeability to liquids. This is due to the fact that the volume of a gas varies with the pressure, and the fact that molecular or slip flow occurs through pit pores, the mean radius of which is of the same order of magnitude as the mean free path of the gas. The superficial permeability in Darcys (*Scheidegger*, 1960)

$$K_g = \frac{QL\eta}{A \Delta P} p/p', \tag{2.5}$$

where p' is the mean absolute pressure within the specimen and p is the pressure at which the rate of flow is measured. The other symbols are the same as those in Eq. (2.4). When χ in the equation

$$\chi = \frac{\lambda}{2r} \sqrt{\frac{8}{\pi}} \tag{2.6}$$

(*Perry* et al., 1963), where λ is the mean free path of the gas in centimeters and r is the effective pore radius in centimeters, is greater than unity, flow is essentially molecular and when it is less than 0.014 flow is essentially viscous. Between $X = 0.014$ and $X = 1.0$ both types of flow occur. According to *Adzumi* (1937),

$$\frac{Qp}{\varDelta P} = \frac{\pi N r^4}{8 \eta L} \, p' + 12 \, \sqrt{\frac{2\pi RT}{M}} \, \frac{N}{L} \, r^3, \tag{2.7}$$

where N is the number of parallel capillaries per unit area, R is the gas constant, T is the absolute temperature in degrees Kelvin, and M is the molecular weight of the gas. The other symbols are the same as given in Eqs. (2.5) and (2.6). Independently *Klinkenberg* (1941) derived the equation

$$K_g = K(1 + b/p'), \tag{2.8}$$

where $b =$ is $4c\lambda p'/r$ and c is a constant. The other symbols have the significance given in Eqs. (2.4), (2.5), and (2.6). This equation is the same as *Adzumi*'s when $K = N\pi r^4/8$. K for flow of a gas through wood can be obtained by plotting K_g against $1/p'$, which is virtually linear, and extrapolating to a $1/p$ value of zero. *Comstock* (1967) has shown that virtually the same K value is obtained for the flow of gas free iso-octane and amyl alcohol, that do not swell oven dry wood, and the flow of helium or nitrogen through the same eastern hemlock specimen in the fiber direction. Comparative measurements were made for a number of different specimens of both heartwood and sapwood with K values ranging from 0.05 to 8.0. *Comstock* thus showed that the K values are functions of the void structure of the wood and not of the nature of the fluid used. The K value for the permeability of water through wood will of course not be the same as that for a gas through the same dry wood specimen as pit membrane pores are known to decrease in size by a significant amount when the wood swells (*Stamm*, 1935).

2.1.2.3 Longitudinal versus Transverse Flow. *Comstock* (1970) and others (*Smith*, 1963; *Resch* and *Ecklund*, 1964; *Oshnach*, 1961) have shown that the longitudinal flow of gases through dry softwoods, expressed in Darcys, may range from 12,500 to 80,000 times greater in the fiber direction than in the tangential direction for various North American woods. A number of the highest ratios are probably too high due to the fact that too short specimens were used. On the average, half a fiber length at each end involves open lumina with little if any resistance to flow. An average fiber length should thus be subtracted from the thickness of the specimen used to obtain the effective specimen length in the fiber direction.

Theoretical permeability ratios have been calculated (by *Comstock*, 1970) on several different structural bases. The most realistic, involving an average fiber overlap of one quarter fiber length at each end and a concentration of the pits in the overlap, gives a permeability ratio of longitudinal to tangential flow

$$K_l/K_t = \frac{4(0.75l)^2}{w^2}, \tag{2.9}$$

where l is the average fiber length and w is the average fiber width. The author, assuming the same overlap but without assuming a concentration of the pits in the overlap, obtained

$$K_l/K_t = t_l/t_t \cdot \frac{l}{w} \tag{2.10}$$

where t_l/t_t is the ratio of the thicknesses traversed in passing through the same number of pits in the two directions corrected for end effects. Using an experimental t_l/t_t value of 111.7 an l value of 0.56 cm, and a w value of 0.003 76 cm for redwood, a K_l/K_t value of 16,000 was obtained (2.1.2.4) using Eq. (2.10).

In general longitudinal flow through wood in the absence of air menisci will be 10,000 or more times the tangential flow. Treating methods involving flow are thus almost entirely limited to longitudinal flow.

2.1.2.4 Mean Pit Membrane Pore Numbers and Radii.

The large natural variation in the permeability of softwoods and the reduction in permeability due to air drying are due to variations in both the number of effective pit membrane pores and their mean hydrodynamically effective radii. It would be expected that the aspiration of pits, caused by drying, will greatly reduce the number of effective flow paths through the pits in parallel without seriously affecting the mean effective radii. This has been shown to be the case by *Petty* and *Puritch* (1970) using equations developed by *Petty* and *Preston* (1969). They found that the longitudinal permeability to dry air, expressed as the volume of flow per unit time per unit pressure drop, increased with an increase in the mean pressure. The increase per unit increase in the mean pressure, however, decreased as the mean pressure increased. They reasoned that this decrease is due to the dual nature of the structure. If the permeability of the lumen and of the pits could be measured separately, the increase in permeability should be directly proportional to the increase in mean pressure. Then the theoretical lumen permeability x and the theoretical pit permeability y can be expressed as

$$x = l\overline{p} + a \tag{2.11}$$

and

$$y = m\overline{p} + b, \tag{2.12}$$

where $\overline{p}$ is the mean pressure and a, b, l, and m are constants. When z is the combined series permeability

$$\frac{1}{z} = \frac{1}{x} + \frac{1}{y} \tag{2.13}$$

and therefore

$$z = \frac{lm\overline{p}^2 + (lb + ma)\overline{p} + ab}{(l + m)\overline{p} + (a + b)}. \tag{2.14}$$

When four different experimental values for z are set into Eq. (2.14), values for the four constants can be obtained by the method of successive approximations. This fixes the straight lines of Eqs. (2.11) and (2.12) and the theoretical rates of flow through the lumen and the pit membranes taken separately. The Adzumi equation (2.7) is then combined with the x values for different values of p, and the number of effective lumen in parallel and the effective mean lumen radius are calculated. In the case of the pit permeability, where the mean effective radius is of the same order of magnitude as the membrane thickness, a modification of the Adzumi equation is used (*Petty* and *Puritch*, 1970) to obtain the number of effective pit membrane pores in parallel and their mean effective radius.

Measurements and calculations were made on *Abies grandis* sapwood that had been either air dried or solvent dried to minimize pit aspiration. The average number of the effective lumen, relative to the total number, was 32% for the air dried wood and 83% for the solvent dried wood. Aspiration of pits eliminated only a few lumina from being effective for solvent dried wood but eliminated about two thirds of the lumina in the air dry wood from being effective.

7*

The average number of pit membrane pores per effective tracheid was 600 for the air dried wood and 27,000 for the solvent dried wood. If, on the average, each tracheid has 100 bordered pits, half of which involve flow into the fiber and half out, there would be on the average only twelve effective pores per pit for the air dried wood and 540 effective pores per pit for the solvent dried wood. Aspiration thus reduces the number of effective pit membrane pores by 45 fold.

Mean effective radii are affected much less by the aspiration that occurs on drying. The average effective lumen radius for the air dry wood was 9.9 μm and for the solvent dried wood was 14.0 μm. The latter value agrees exactly with the average lumen radius calculated from a dry volume specific gravity of 0.42 and the fractional void volume (*Stamm*, 1964, Chapter 3) on the basis of the average lumen being square. The lower value for the air dried material is probably due to aspiration predominating on the pits of the larger earlywood tracheids.

The mean effective pit membrane pore radius was 0.12 μm for the air dried wood and 0.09 for the solvent dried wood. Somewhat smaller mean radii have been obtained by other flow methods (*Stamm*, 1964, Chapter 22). In the next section (2.1.2.5) it will be shown that maximum effective pit membrane pore radii for the passage of air through 50 pits in series of the sapwood of several different water saturated softwoods are of this order of magnitude (*Stamm*, 1970). The calculations of *Petty* and *Puritch* (1970) thus appear quite reasonable. Further measurements of this kind should be made on other species including longer specimens.

2.1.2.5 Effect of Gas-Liquid Interphases. When water or other liquids are made to flow through wood under constant pressure, the rate of flow tends to decrease with time unless extreme precautions are taken to insure that the liquid is entirely free from suspended particles that may tend to be filtered out on the pit membranes and thus tend to clog the openings. Even when the liquid is scrupulously freed of particulate matter, the rate of flow of the liquid with time may still decrease. This has been shown to be due to the release of dissolved air in the specimen as the pressure drops, forming air bubbles that tend to clog the pit membrane openings in the same way as particles (*Kelso*, *Gertjejansen*, and *Hossfeld*, 1963). Preboiling of water does not entirely free it of dissolved air. Passing preboiled water through a fine millipore prefilter or a bed of highly compressed cotton under high pressure causes the dissolved air to be removed in the prefilter. These precautions have been taken only in recent investigations. Because of this most of the older data for the permeability of wood to liquids might be questioned (*Erickson*, 1970).

It requires considerably more pressure to force a gas-liquid interface through a capillary tube than is needed to cause the flow of gas-free liquid. The surface tension forces effective at a gas-liquid interface have to be overcome for the gas to displace the liquid from the smallest opening with radius r in any single series path. The pressure to accomplish this is

$$P = \frac{2\sigma}{r} \cos \theta, \tag{2.15}$$

where σ is the surface tension of the liquid in dynes per cm, θ is the angle of wetting, P is the pressure in dynes and r is the opening radius in cm. This equation has been shown to hold for thin membranes as well as capillaries even when r is much greater than the thickness of the membrane. It also holds for triangular or slit like openings in the membrane (*Stamm*, 1966). In the first case the effective radius is the radius of the largest inscribed circle, and in the latter case it is the mean radius of the largest inscribed ellipse. Eq. (2.15) reduces to

$$r = \frac{1.5\sigma}{P_1} = \frac{20.4\sigma}{P_2} = \frac{0.29\sigma}{P_3} \tag{2.16}$$

for liquids that wet solids as water wets wood when above the fiber saturation point, when r is in microns, P_1 is in cm of mercury, P_2 is in grams per cm^2 and P_3 is in psi.

Eq. (2.16) has been recently applied to softwood cross sections of different species varying in thickness from 0.05 to 7.5 cm (*Stamm* et al., 1968; *Stamm*, 1970, 1 and 2) and to thin quarter and flat sawn specimens of redwood 0.10 cm thick (*Stamm*, 1970 (2)). Only small increasing pressures are required to displace water or water containing a wetting agent from cross sections with increasing thicknesses up to the maximum fiber length. The displacement occurs through the fiber lumen with the largest minimum radius of those that were cut across twice. The r values decrease with increasing thickness of the cross sections due to the decreasing number of available paths and the fact that the cuts have to be nearer to the tapered ends of the fibers.

When the maximum fiber length is exceeded, displacement of liquid has to occur through pit membranes. The gas pressure required to displace the liquid from a specimen of wood increases with an increase in the number of pits that have to be traversed in series. This is due to the fact that pit pore openings vary considerably in size not only in a single pit but between pits. For cross sections just exceeding the maximum fiber length only one pit has to be traversed. There are 60,000 to 120,000 softwood fibers per cm^2 of cross section of softwoods (*Stamm*, 1946; *Stamm*, et al., 1968), each with 50 to 250 bordered pits, containing at least 100 pore openings per pit membrane. Water displacement measurements were made over an area of 0.314 cm^2. In the case of cross sections just exceeding the maximum fiber length, displacement occurred through the largest pit membrane opening out of at least 94 million openings ($0.314 \times 60,000 \times 50 \times 100$). The probability of this largest opening resulting from a natural or mechanically caused flaw in the structure is quite great. As the thickness of the specimen is increased and several pits have to be traversed in series, the probability of such large openings occurring in any one path diminishes rapidly. The maximum effective pit pore radius would thus be expected to decrease with an increase in the number of pits traversed in series and tend to approach the average pit pore radius. In fact, the logarithm of the maximum effective pit pore radius was found to decrease linearly with an increase in the logarithmn of the number of pits traversed in series up to the highest pressures of about 500 psi that could be used without splitting the specimens (*Stamm* et al., 1968; *Stamm*, 1970 (1) and (2)).

The average number of pits n_l traversed in series was calculated from the thickness of the specimen t_l and the average fiber length l in cm as follows

$$n_l = \frac{t_l - l}{3/4l}. \tag{2.17}$$

The equation is based on the fact that on the average half a fiber length at each end of the specimen can be traversed without passage occurring through a pit. It is also based on the fiber overlap at both ends of a fiber being one quarter fiber length. Passage through a lumen may thus range from one half of a fiber length to a full fiber length and average three quarters of a fiber length.

Pressures required to just displace water from specimens of the sapwood and the heartwood of several never dried softwoods are given in Table 2.2. Heartwood requires pressures ranging from five to twenty times that required to displace water and allow the gas to pass slowly through the sapwood.

Table 2.2. Pressure Required to Just Displace Water by a Gas from Green Softwood Specimens 3 in. (7.6 cm) Long in the Fiber Direction (*Stamm*, 1970)

Species		Heartwood psi	Sapwood psi
Ponderosa pine	*Pinus ponderosa*	170—195[1]	17
Loblolly pine	*Pinus taeda*	160	15···34[2]
Douglas fir Coast. 2nd growth	*Pseudo tsuga menziesii*	275	57
Bald cypress	*Taxodium distichum*	128	36
Eastern larch	*Larix laricina*	350	18
Northern white cedar	*Thuja occidentalis*	330	44
Incense cedar	*Libocedrus decurrens*	580	85
Eastern red cedar	*Juniperus virginiana*	510	86
Redwood, buttlog	*Sequoia sempervirens*	145···163[1]	45
Redwood, 80 ft. up	*Sequoia sempervirens*	97···103[1]	30

[1] Variation from outer sapwood to pith.
[2] Variation from outer to inner sapwood.

Only limited measurements of this type have been made in the transverse directions (*Stamm* et al., 1968; *Stamm* 1970 (2)). Measurements usually have to be confined to specimens 1 mm thick or less due to the extremely high pressures required to cause displacement of water, which approach the tensile strength of the specimens perpendicular to the grain. It is also necessary to adequately seal the end grain to prevent lateral displacement.

The number of pits traversed in series in the tangential direction is

$$n_t = t_t/w - 1, \qquad (2.18)$$

where t_t is the thickness of the specimen in the tangential direction in cm and w is the average fiber width in cm in the tangential direction. Values for $1/w$ from microscope count on softwoods range from 250 to 350 (*Stamm*, 1946; *Stamm* et al., 1968). It has been shown that $\log r$ vrs $\log n$ plots for tangential displacement fall on the straight line plots of $\log r$ vrs $\log n$ for longitudinal displacmement (*Stamm* et al., 1968; *Stamm*, 1970 (2)). When n_l and n_t are sizeable and equal, such as 100 and $1/w$ is 300 and l is 0.35 cm, which is the case for a number of softwoods, $t_l/t_t = 78.7$. When $1/w$ is 266 and l is 0.56 cm, as in the case of redwood, $t_l/t_t = 111.7$. These ratios are practically equal to the ratio of three quarters of the fiber length to the fiber width. They represent depth of penetration not relative rates of flow. No unit cross section is involved as in ratios of flow. Taking the cross sections into account gives flow rate ratios $K_l/K_t = 8,350$ for the normal case where $1/w$ is 300 and l is 0.35 cm and $K_l/K_t = 16,600$ for the case where $1/w$ is 266 and l is 0.56 cm. These values are similar to the lower values given under 2.1.2.3. The high values are at least partially due to use of too short longitudinal specimens in which the open end effects were neglected.

Table 2.2 shows that gas pressures of several hundred psi have to be applied to the heartwood of a number of different softwoods to displace condensed water held in the pit membranes of specimens only 7.6 cm long in the fiber direction and only 0.1 cm thick in the tangential direction. It is because of this that it is frequently necessary to apply pressures of more than 500 psi to impregnate the heartwood of a number of species with preservatives. Wood is rarely predried below the fiber saturation before treatment. Even if it is throughly predried, vapor can condense in pit membranes in advance of the already impregnated zone and trap air in between. The smaller the pit membrane openings, the more readily the impregnating liquid will condense. Water will condense in openings with radii of 0.05 microns or less when the relative vapor pressure p/p_0 exceeds 0.98, according to the Kelvin equation

$$r = - \frac{2\sigma M}{RT \ln p/p_0}, \tag{2.19}$$

(where σ is the surface tension of water, M its molecular weight R is the gas constant and T is the absolute temperature). The pressure required to displace the condensed film is 290 psi $\left(\text{see Eq. (2.16)}\right)$. Even if the wood is completely predried and highly evacuated, the situation is not improved as the surface tension of liquids against their own vapor is essentially the same as against air. It is thus evident that impregnation of wood with a liquid is quite a different phenomenon from that, causing a liquid that completely saturates a sapwood specimen to flow through the specimen in the fiber direction.

The only presently known approaches for avoiding the high impregnating pressures now required for treating low permeability woods are to either increase the permeability by some pretreatment, or resort to a gas phase treatment under such conditions that condensation cannot occur in the pit membrane pores and the vapor will react with the cell wall substance to give the property change desired.

The pit pore permeability has been increased by passing chlorine or hydrochloric acid gas through partly dried wood in the fiber direction (*Stamm*, 1932). Unfortunately such reactions are difficult to control. Extensive embrittlement of the wood always accompanies significant increases in permeability. Predrying resinous woods by solvent seasoning (Vol. I, 8.4.3) (*Anderson*, 1946; *Anderson* and *Fearing*, 1960) with acetone or other solvents increases the permeability by dissolving resins and by preventing aspiration of the pits. Extraction and steaming improve the permeability of resinous woods (*Benvenuti*, 1963; *Nicholas* and *Thomas*, 1968), the latter presumably by a hydrolytic action on the pit membranes. Attack of wood by bacteria or molds such as *Trichoderma* (*Lindgren*, 1952; *Ellwood* and *Ecklund*, 1959) increase the permeability of sapwood greatly by breaking down the pit membranes. Unfortunately they do not attack heartwood which needs an increase in permeability, whereas sapwood does not.

Recent researches (*Resch*, 1967; *Prak*, 1970) have shown that gases and uncondensible vapors penetrate wood far more rapidly than liquids where liquid-gas interfaces are usually involved. Advantage is being taken of this in reacting gases or vapors with the cell walls of wood (*Baird*, 1969; *Stamm* and *Tarkow*, 1947; *Tarkow* and *Stamm*, 1953; *Schuerch*, 1968). Some of those reactions are described in Sections 2.4.5, 2.4.6.7, and 2.4.6.8.

2.1.3 Diffusion in Wood

Diffusion through a porous structure differs from flow through the structure in two important respects. Diffusion is dependent only on the effective cross section. Flow on the other hand is dependent upon the square of the average pore radius as well as the effective cross section (2.1.1). Diffusion is motivated by either a concentration gradient or a vapor pressure gradient, whereas flow takes place in response to an applied hydrostatic pressure or internal capillary forces. It is shown in 2.1.2.1 that flow through the cell walls of wood is negligible in combination with the other structures. Diffusion on the other hand takes place through the cell walls in series with the lumen and in parallel with the pit system. Diffusion through these three structures has been shown to be analogous to the flow of an electric current through the system (*Stamm*, 1946, 1964 Chapter 23, 1962 (2), and 1967 (2)), that is diffusion coefficients are additive in parallel and reciprocals of diffusion coefficients are additive in series.

Diffusion that occurs in the treatment of wood is always dynamic, occurring under a continuously changing concentration or vapor pressure gradient. This type of diffusion, frequently referred to as unsteady state diffusion, can be expressed mathematically by Fick's second law of diffusion:

$$dc/dt = D(d^2c/dx^2),\qquad\qquad(2.20)$$

which states that the rate of change of concentration with time, dc/dt, at any point x distance into the diffusion medium is proportional to the rate at which the rate of variation of concentration with distance changes, that is, the second derivative of concentration with respect to distance, d^2c/dx^2. There is no single general solution of this equation in integrated form. Simplified solutions have been developed, however, for specific boundary and other imposed conditions (*Barrer*, 1941; *Crank*, 1956; *Jost*, 1952).

Although diffusion of solutes can occur through the cell walls of wood, it is limited to molecules small enough to penetrate the solvent filled spaces within the swollen cell walls. *Tarkow* et al. (1966) have found for water swollen wood that polyethylene glycol molecules with molecular weights exceeding about 3,500 are excluded from entering the cell walls. The limiting molecular weight will vary with the density, degree of coiling, and shape of the particular molecule. The size of molecule that will enter the cell walls will also be greatly affected by the degree of swelling attained by the solvent. Amyl alcohol will not swell dry wood significantly but it can readily enter the cell walls of wood swollen in ethyl alcohol. It will be shown under 2.4.6.4 that high molecular weight waxes can diffuse into the cell walls of wood swollen in Cellosolve (ethylene glycol monoethyl ether).

Diffusion is involved in two different ways in the impregnation of wood, namely the diffusion of solutes from solution into liquid saturated wood under a concentration gradient or of vapors into dry wood under a vapor pressure gradient. When aqueous solutions of inorganic salts or of organic compounds of limited molecular size are impregnated into the void structure of wood by anyone of the methods considered under Section 2.2, diffusion of solute will occur readily from the void structure into the cell walls. Although diffusion is a slow process over appreciable distances it will be quite rapid into preswollen cell walls as diffusion times vary with the square of the thickness. For example, if it took 100 days to attain a given level of diffusion into a homogeneous gel 1 cm thick, it would require only about 4 sec to get the same level of diffusion into a thickness of gel equal to the average double cell wall thickness of a softwood, namely 7 microns. If the cell walls are not preswollen, the time will be increased appreciably as the slower rate of swelling, involving stress relaxation, (*Christensen*, 1960) will control the rate of take-up of solute rather than the rate of diffusion. When the lumen of thin cross sections of wood are rapidly filled with water, they normally come to 90% or more of swelling equilibrium in a few minutes (*Hittmeier*, 1967). They may, however, take hours to come to complete swelling equilibrium.

Diffusion into the cell walls is not a mass movement, but involves a molecular jump phenomenon (*Stamm*, 1964, Chapter 23). Molecules of polar vapors are adsorbed on the hydroxyl groups of the cellulose and lignin at the solid void interfaces. Attractive forces are operative, in the case of water, up to about seven molecules thick (*Stamm* and *Smith*, 1970) and decrease rapidly from the first to the seventh molecule. Molecules held in multilayers can readily jump to primary sites a short distance into the wood and then to secondary sites. This jump process is thus motivated by a concentration gradient. It is initially rapid but slows down greatly as the gradient diminishes.

Diffusion into the cell walls of dry wood may be much slower from concentrated solution than into preswollen wood. This is illustrated by an experiment in which matched 3 mm thick cross sections of a softwood were used. One was oven dried and the other saturated with water. Both were placed in saturated solutions of magnesium chloride with an excess of the salt present. A vacuum was applied over the immersed, originally dry, specimen to accelerate the take-up of saturated solution within the voids. In several days, the preswollen specimen came to a swelling equilibrium somewhat greater than the swelling in water alone, indicating that more of the salt was taken up within the cell walls than water was displaced. The originally oven dry specimen required 58 days to reach the normal swelling in water, and in 100 days had not quite reached the final equilibrium swelling of the other specimen (*Tarkow* and *Stamm*, 1950). This can be explained on the basis of water having to precede the salt in entering the cell walls (*Stamm*, 1934). In the case of the originally dry wood considerable energy is used in taking water from the solution so that it can diffuse into the cell walls. Presumably, water is taken up by the cell walls in small increments followed by diffusion of small amounts of salt into the adsorbed water. Such a stepwise process requires additional work on the solution so that the process is very slow. For this reason it is desirable in impregnating wood with concentrated fire retardant salt solutions to start with wood near the fiber saturation point.

When diffusion has to occur through the entire structure of swollen wood, it is usually too slow a process for general commercial use. It is, however, suitable for the treatment of thin green veneer (2.4.6.5.1) and is used in the treatment of precarved objects from green wood with polyethylene glycol (2.4.6.3). Diffusion of water soluble solutes into green wood has been shown to be 10 to 15 times as fast in the fiber direction of wood as transversely (*Stamm*, 1946, 1964, Chapter 23). Precarving is hence desirable as it exposes more end grain to accelerate diffusion. Diffusion is also an effective factor in the double diffusion process for treating green fence posts (2.2.2).

Vapor phase treatments are normally made under sufficient pressure to cause ready flow of vapor into the void structure. Diffusion from the voids into the cell walls of dry wood involves adsorption on the surface followed by molecular jump under a concentration gradient into the cell walls in a manner similar to that involved from the liquid phase.

2.2 Treating Methods

Wood treating methods fall into three different groups, namely, sap displacement, capillary absorption and diffusion, and pressure methods.

2.2.1 Sap Displacement

Sap displacement is the simplest means for obtaining a satisfactory treatment of green poles that are chiefly sapwood. Heartwood is too impermeable and generally has too low a moisture content for sap displacement to occur through its voids. The process consists of the longitudinal displacement of the sap with an aqueous treating solution under a small hydrostatic head (the Boucherie process, Vol. I, 5.2.3) or under a high pressure (2.3.2.2) avoiding any entrapment of air in the entering liquid (2.1.24). However, small amounts of air in the wood structure can be tolerated if properly distributed. In the case of softwoods, each fiber can contain an amount of air that when forced to the exit end under a small pressure will not expose all of the pits in the fiber overlap zone to liquid-air interfaces (2.1.2.4). This treatment has been most successfully applied to green, freshly cut southern pine poles (2.2.2.2). These species are almost entirely sapwood up to ages of 20 to 30 years, and are highly permeable and have high moisture contents.

2.2.2 Capillary Absorption and/or Diffusion

Superficial to moderately effective treatments can be made by a combination of capillary absorption and diffusion. Brush-on, spray, dip and soak methods, when applied to air dry wood, cause the take-up of solution by capillary absorption into the void structure from which diffusion occurs into the cell walls. When applied to water saturated wood, the entire process depends upon diffusion.

Capillary take-up of a solution by dry wood is about 100 times as far, or fast, in the fiber direction as in the transverse directions. Diffusion of a solute into water saturated wood is 10 to 15 times as rapid longitudinally as transversely (*Stamm*, 1946). It thus pays, in treating dry wood by capillarity or green wood by diffusion, to have as much end grain of the wood as possible exposed to the solutions.

Brush-on spray treatments are effective only when the surface of the wood is rough or the viscosity of the liquid is high, so as to attain a thick surface layer. Low viscosity solutions can be incorporated in a starch paste which acts as a thick surface reservoir for diffusing chemicals. This technique has been applied to green wood in chemical seasoning, a process that requires the take-up of chemical only in the outer layers of wood (*Loughborough*, 1939). Multiple applications of molten polyethylene glycol to green wood, held under non-drying conditions between application, is practiced as a means of minimizing the surface checking of wood (2.5.6.3).

A 3 to 5 min dip of air dry wood in water-repellent preservatives is frequently used on window sash and frames. Adequate take-up of the hydrocarbon solution, for the temporary protection sought, is attained with the commonly used ponderosa pine sapwood in this short period of time (2.5.3) (Vol. I, 5.2.12).

A combination of capillary absorption and diffusion is involved in the double diffusion process for treating green freshly pealed round fence posts with preservatives. This simple process involving the soaking of the post alternately in two different toxic salt solutions, that react with each other to form a relatively insoluble precipitate, is covered under (2.3.2.1).

Capillary absorption can be accelerated in two different ways, both involving the creation of a partial vacuum within the wood. One of these consists in heating air dry wood or wood just below the fiber saturation point in an oven or kiln, or while immersed in a tank of the treating liquid or solution. While still hot, the wood is rapidly transferred to a tank of the liquid at room temperature. On heating the wood, the air in the void structure expands and is partially driven from the wood. After transfer of the wood to the liquid at room temperature, the remaining air in the wood contracts and the treating liquid penetrates as a result of the pressure difference. This method is normally referred to as the hot-and-cold bath treating method (Vol. I, 5.2.1.4). Excluding pressure treating methods, this is the most effective means of treating dry or partially dry wood with oil borne preservatives. It can also be applied to advantage in introducing the bulking type of dimensional stabilizing agents into veneer or even into solid wood to obtain face protection (2.5.6.5.1).

The second accelerated capillary absorption method applies only to veneer just below the fiber saturation point. The veneer is compressed to one-third to one-half of its original thickness between compression rolls, the nip of which is just below the surface of a treating solution. Air is expelled from the veneer as it is compressed. As it immerges from between the rolls it loses its compression, thus sucking in the solution (2.5.6.5.1). This method of treatment is very effective in treating low to medium density veneer with aqueous bulking agents.

2.2.3 Pressure Methods

Pressure methods are the most effective means of treating solid wood containing sizeable amounts of air with preservatives, fire retardants, and dimension stabilizing chemicals. As was shown under heading 2.1.2.5, pressures of 500 psi and more may be needed to force air-liquid minisci through the pits of resistant woods. Such treatment requires expensive treating cylinders. However, treating times are usually reduced sufficiently below those required by other methods so that investment in pressure equipment pays wherever treatments are made on a large scale. Treating procedures involving both the so-called full-cell and empty-cell method are covered in Vol. I, 5.2.2, and by *Hunt* and *Garratt* (1953).

2.3 Wood Preservation against Attack by Organisms

2.3.1 General

Wood is treated with various chemicals to retard attack by organisms. The nature of attack by different organisms is covered in some detail in Vol. I, 4 under the title "Biological Deterioration of Wood" (p. 96—135). This is followed by Chapter 5 on "Wood Preservation" (p. 136—159), which includes sections on the effect of the fine structure of wood upon the treatment, preservative treating methods, and preservative chemicals. Treating methods are adequately covered, ranging from the simplest capillary absorption, diffusion, and sap replacement methods through pressure impregnation methods.

Conventional wood preservation involves the introduction of toxic chemicals into wood, in the form of low volatile slightly water soluble, organic liquids such as creosote, solutions of highly toxic chemicals such as pentachlorophenol in a petroleum solvent, or single or mixed inorganic salts dissolved in water. Decay resistance can also be imparted to wood by any treatment that permanently reduces the water sorption below the normal 20 per cent level required to support decay. This is shown to be true for all of the more effective dimension stabilizing treatments under headings 2.5.4.1, 2.5.5, 2.5.6.3, and 2.5.6.5.2.

Decay resistance can also be imparted to wood by destroying its thiamine content required to support decay. This has been done by heating wood under alkaline conditions such as in the presence of both water and ammonia vapor (*Bachelor*, 1959). Wood sterilized in this way, when in contact with damp soil and roots of plants, can, however, readsorb sufficient thiamin from them to again become susceptible to decay.

Although toxicity is not an essential property of wood preservatives, most of them are toxic. The toxic chemicals in present use are covered in sufficient detail by *Hunt* and *Garratt* (1953), *Van Groenou* et al. (1951), and in the Wood Handbook (1955, p. 399—428) so that they will not be discussed further here.

A wood preservative is not accepted for commercial use on the basis of toxicity tests alone. It must pass a series of service tests, usually made on fence posts or stakes driven into the ground in decay and/or termite infested areas. The U. S. Forest Products Laboratory maintains such test plots in southern Mississippi and in the Panama Canal Zone that are inspected annually, and the findings are published in the Proceedings of the American Wood Preservers' Association. Untreated fence posts usually lose practically all of their serviceable strength in 2 to 3 years in those areas whereas well preserved posts may be serviceable for 15 or more years.

The amount of preservative chemical needed to impart adequate decay resistance to wood varies greatly with the nature and toxicity of the chemical, the size, type and species of wood, and the exposure conditions. Often of equal importance is the extent to which the chemical is distributed throughout the structure. Only the sapwood of permeable species can be impregnated with a reasonable degree of uniformity in lumber and round pole sizes. In the latter case uniformity of treatment is fortunately not necessary. It is important to penetrate the outer sapwood to a depth only somewhat greater than the deepest check that may subsequently occur in the poles. This eliminates species that have only a thin layer of sapwood and resistant heartwood (see Table 2.1). In the case of lumber that is subsequently cut, uniformity of treatment is highly important for drastic exposure conditions. This requirement is best met by the southern pines that are mostly sapwood and have reasonably permeable heartwood. Coast grown Douglas fir and western hemlock lumber can be impregnated under high pressure with reasonable uniformity whereas mountain grown Douglas fir lumber is unsuitable for treatment (see Table 2.1).

When the distribution of chemical is reasonably uniform in southern pine fence posts, service tests have shown that a retention of about 6 lb./cu. ft. (16 to 18% of the dry untreated weight of the wood) of creosote is required to give a service life of about 15 years (Wood Handbook, 1955). Only 0.3 lb./cu. ft. of pentachlorophenol dissolved in spent crankcase oil per cubic foot of wood was needed to give the same life. Comparable preservative salt requirements vary from 0.1 lb./cu. ft. for mercuric chloride to 0.9 lb./cu. ft. for zinc chloride (*Hunt* and *Garratt*, 1953).

2.3.2 New and Modified Treating Methods

The older methods for treating wood with preservatives are adequately covered in Vol. I, 5.2, and by *Hunt* and *Garratt* (1953). For this reason only the newer methods and important modifications of older methods are considered here.

2.3.2.1 Double Diffusion. The double diffusion method for treating round, green fence posts is of importance because it is sufficiently simple that farmers can treat their own posts. The process consists of immersing freshly cut debarked posts on end, in old oil drums, to about half of their length in an aqueous solution of a toxic salt for one to three days. This is followed by a quick water rinse and immersion of the posts for an equal or somewhat longer period of time in a second toxic salt solution that reacts with the first to form an only slightly soluble precipitate (*Baechler*, 1953, 1954, 1958). The object of the water rinse is to avoid premature precipitation of the salts on the surface of the wood or within the drum. After the treatment is complete it is important that the posts be close piled and covered with a sheet of water impermeable film, such as polyethylene, to avoid premature drying. The posts are held in this non-drying condition for two to four weeks to allow diffusion of salts to continue deeper into the wood prior to complete precipitation. Final air drying can take place either before or after the posts are set in place.

Various salt combinations have been successfully used. Copper sulfate followed by sodium arsenate and copper sulfate followed by disodium phosphate are good combinations. Repeated treatments can, however, not be made in uncoated steel oil drums because iron tends to replace copper in solution leading to loss of copper and leaking of the drums. Wooden drums should hence be used in these cases. Combinations that can be used in steel drums are nickel sulfate followed by sodium arsenate, and zinc chloride to which small amounts arsenic acid have been added followed by sodium chromate. The latter combination has the advantage over the former in that it is cheaper.

Treatment is not limited to the portion of the posts that is immersed in the salt solutions. Capillary action along the surface of the posts carries some salt solution to the top of the posts. As capillary rise is about one hundred times greater along the fiber than across it, inward capillary absorption is small. Diffusion, on the other hand, is only 10 to 15 times as rapid in the fiber direction compared to the radial direction. Inward diffusion from the surface is thus appreciable. Adequate treatment of sapwood above the normal ground line is usually attained even when only half of the length of the posts is immersed.

The extent of capillary absorption and diffusion into the immersed portion of the posts will vary with the species, thickness and nature of the sapwood, the diameter of the posts, and their moisture content. Green yellow poplar posts, ranging in diameter from 7 to 12 cm, showed an average visual depth of penetration of 3.5 cm when soaked for one day in a 20% weight solution of zinc sulfate containing one sixth arsenic acid followed by soaking for one day in a 20% solution of sodium chromate, and then holding under non-drying conditions for three weeks (*Baechler*, et al., 1959). The total take-up of salt was about 2 lb./cu. ft. (9% of the dry weight of the wood). Loblolly pine posts gave about half as deep penetration and half as much take-up as the yellow poplar under the same conditions. Sweet gum gave almost the same take-up as the yellow poplar but only one-third the depth of penetration. These woods were almost entirely sapwood. Red and white oak posts with only 1.2 to 2.5 cm of sapwood required one and three days' soaking respectively to penetrate the sapwood completely.

Rough theoretical calculations for the take-up of a 20% by weight solution of zinc sulfate in the sapwood of water saturated Loblolly pine, on the basis of diffusion only, gave take-up values 65% of the measured values (*Stamm*, 1964, p. 747 to 775). Considerable capillary take-up must thus have accompanied diffusion. For capillary take-up to occur, the moisture content would have to be well below saturation. This would in turn reduce the theoretical diffusion considerably.

Probably less than half of the take-up can be attributed to diffusion. It thus appears that the initial moisture content of the wood is not as critical as it would be if the process was entirely dependent on diffusion.

2.3.2.2 Modified Boucherie Process. The Boucherie process, involving the longitudinal displacement of the sap from the green sapwood of freshly cut round poles by aqueous preservatives (see 2.2.1), dates back to 1838 (*Hunt* and *Garratt*, 1953). End cap seals are attached to the butt end of freshly cut trees with the bark intact. Aqueous treating solutions entering the cap displace the sap under a head of about 10 m of liquid (Vol. I, 5.2.3). The process has been used, with various small modifications, for years in Europe. The process was never extensively used in the United States because of excessive labor costs in attaching and maintaining the caps so that leaks did not occur, together with the long time required for the displacement.

Recently end seals or caps have been developed that are easy to attach and which will withstand pressures up to 200 psi (*Hudson*, 1969). Measurements made on 20 ft. (6.1 m) long, freshly cut southern pine poles gave flow rates of 6.4 gallons per hour (24.2 l) per square foot of butt surface under pressures of 50 to 200 psi. This is about 20 to 30 times as much as can be displaced by the conventional Boucherie process. The fact that the rate of flow increased only slightly in raising the pressure from 50 to 200 psi led to stopping the process, removing the cap and cutting off a thin disc of wood from the butt followed by replacing the cap and again applying pressure. The flow rate rose to 97 gallons per hour per square foot. This was explained on the basis of a negative pressure in the freshly cut poles drawing in air and blocking subsequent flow. It is thus highly advantageous to cut off a disc exceeding the maximum fiber length in thickness from the butt end of the pole just before attaching the cap. Green poles 20 foot long that were stored for a month or less after felling with the bark intact were adequately treated in five hours at 200 psi when a disc was cut off of the butt end just before treatment.

It has been recently demonstrated that longitudinal sap displacement can be carried out in a special treating cylinder on a full load of freshly cut green poles (*Hudson*, 1968). The principle was first demonstrated with a small vertical treating cylinder with a perforated bottom plate. A pad of filter cloth was placed on the plate and the poles, with freshly cut ends, placed on end on the pad of filter cloth. A slurry of very finely ground sand in the aqueous treating solution was run into the cylinder. The sand deposited on the filter cloth around the ends of the poles as the liquid passed through the pad and the perforated plate, giving a depth of several inches of slurry around the ends of the poles. Treating solution containing no sand was then pumped into the cylinder under pressure. This packed the sand seal around the poles to the point that passage of treating solution through the seal diminishes rapidly to zero. Sap then started to flow through the poles as the sap was displaced by preservative solution entering at the top of the poles. Flow under pressure was continued until the composition of the effluent was the same as that of the original treating solution.

This treating procedure can also be carried out in conventional horizontal treating cylinders. In this case more sand must be used for the seal to be effective as the angle of repose of the seal is about 60 degrees. The procedure works both on pealed or unpealed poles and green sapwood lumber. Green southern pine poles 20 ft. long have been treated in this way in one hour. The distribution of preservatives between the two ends of the poles is better for the peeled than for the unpeeled poles indicating that some lateral redistribution must be taking place. Green southern pine lumber, 2×4 in. $\times 8$ ft. long, gave little difference in the retentions of preservative determined 1.5 ft. from each end after a one hour treatment. Sixteen feet long lumber was adequately treated in three hours. The treating time by this process is not increased with an increase in thickness of the lumber as is the case with normal pressure treatment.

2.3.2.3 Predrying Methods. In order to obtain adequate penetration of freshly cut wood with preservatives by the most effective pressure treating methods, it is necessary to reduce the moisture content considerably prior to impregnation with preservatives. This has been done in the past by a presteaming in the treating cylinder for several hours at about 125 °C followed by applying a vacuum, or the wood is heated to 82 to 105 °C in the treating cylinder while immersed in a low volatility oil and pulling a vacuum over the oil (Bolton process) (*Hunt* and *Garratt*, 1953) — These procedures are quite effective in drying the sapwood but not the heartwood. A new procedure known as "vapor drying" is now in commerical use for predrying railroad ties and poles (*Hudson*, 1942, 1947). It consists of passing steam over the wood in a treating cylinder to remove the air. Hot vapor of a water-insoluble organic compound such as xylene or a high-boiling petroleum fraction at about 300 °C is admitted to the cylinder. Vapor in contact with the colder wood condenses and gives up its latent heat to the wood, thus raising the temperature of the wood to the boiling point of water causing evaporation of water. The condensate, together with the uncondensed organic vapor and water vapor, is drawn off and completely condensed. The organic liquid and the water are separated mechanically and the former is again vaporized and returned to the treating cylinder where the cycle is repeated. When the moisture content of the wood is reduced to the desired level of about 35% (oak and gum cross ties in 12 to 16 h and southern pine poles in about 10 h), the cylinder is evacuated to recover the absorbed organic liquid. The hot wood is then treated under pressure by one of the conventional methods (Vol. I, 5.2.2.). Even though the moisture content of the wood is not reduced below the fiber saturation point, some checking accompanies the drying. The checks are much finer than those occurring on air drying and tend to close on subsequent cooling of the wood. Because of checking the process is not suitable as a substitute for kiln drying of wood to low moisture contents.

2.3.2.4 Cellon Process. Another new, entirely different approach to wood preservation developed and extensively used by the Koppers Co. of Pittsburgh, Pa., was developed to attain a deeper and more uniform treatment of dry wood with such highly toxic organic chemicals as pentachlorophenol. The method, known as the Cellon process, involves the impregnation of pre-evacuated air dry wood in a pressure cylinder with a liquified hydrocarbon gas selected from saturated hydrocarbons containing 3, 4, or 5 carbon atoms per molecule (either normal or iso) in which the toxic chemical such as pentachlorophenol is dissolved together with a solubilizing agent (*Bescher*, 1965). The presently used liquified gas is n-butane. The solubilizing agent is normally iso-butyl ether, that is recoverable at the end of the treating cycle together with the butane. High boiling co-solvents that remain in the wood are also used.

The air dry wood is evacuated in the treating cylinder followed by admitting the liquified gas containing about 5% of pentachlorophenol and about 2% of the solubilizing agent under pressures ranging from 60 to 175 psi. The pressure is entirely controlled by the temperature of the liquified gas. The pressure is maintained for a period of time depending on the permeability and dimensions of the wood (Table 2.2). The viscosity of the liquified gas is about 1/5 that of water, and the surface tension is about 1/4 that of water. Flow into the structure should be about five times as rapid as the flow of water. Further the much lower surface tension should greatly reduce the pressure required to overcome the effect of condensation of vapor in pit openings in advance of the entering liquid (2.1.2.4).

When the treatment is complete the liquified gas is allowed to expand and vaporize into a storage tank where it is again liquified under pressure. The penta-chlorophenol, being non-volatile under the operating temperature, is deposited in place in the wood. There is no tendency for "bloom" to occur. This is a migration of preservative to the surface that occurs with less volatile solvents that migrate to the surface before evaporation. As yet unpublished experiments of *Resch* (Univ. California Forest Products Lab.) indicate that some pentachlorophenol actually penetrates the cell walls of the wood in spite of the fact that liquid butane is not a swelling agent for wood. The solubilizing agent together with residual water must open up the cell wall sufficiently for the pentachlorophenol to enter.

More polar co-solvents than the iso-butyl ether should, however, further in-crease the tendency for pentachlorophenol to enter the cell walls of wood.

Tests indicate that the heartwood of coast grown Douglas fir is almost com-pletely impregnated, but that the heartwood of inland grown Douglas fir is only superficially treated. Service tests of the treated wood look favorable to date. Chief uses for wood treated in this way are for posts, poles, and telegraph pole cross arms. The process has also been successfully used for treating elm stadium seats and maple gymnasium floors (*Goodwin* and *Hug*, 1961; *Henry*, 1963).

2.4 Wood Preservation against Fire

2.4.1 Fire Hazards

The combustibility of wood and the mechanisms of burning and charring are briefly covered in Vol. I, 5.4. Wood, because of its high carbon and hydrogen contents, is combustible. There is no known means of making it incombustible. The best that can be done is to reduce the rate of burning and fire spread by minimizing flaming in favor of charring (*Brendon*, 1965). Wood char is fortunately an excellent insulator, when sufficiently thick to remain intact. Because of this, large wooden beams have continued to support their load during intense fires. The Forest Products Newsletter (1961) cites informative examples. At a distance into a beam of only 0.6 cm beyond the charred zone temperatures as low as 182 °C have been recorded during a severe fire. Charring to depths of 2 to 4 cm in intense fires have occurred in one hour. A laminated beam of Douglas fir with a cross section of 23 by 70 cm exposed to an intense fire for one hour, during which the temperature rose from 250 to 870 °C, retained 70% of its original modulus of elasticity. In this test 6,000 lb. of dry wood was burned to attain a temperature of 870 °C in one hour at the surface of a single beam. In the process only 150 pounds of the beam was charred over a length of 160 cm to an average depth of 2.2 cm. According to this, the beam itself furnished only 2.5% of the fuel needed to support the fire. In contrast, steel beams will have only about one tenth of their original load bearing strength when an optimum temperature of 760 °C is reached. Under such conditions they are subject to buckling and collapse. Roofs of buildings are thus far less subject to collapse in intense fires when large wooden beams are substituted for steel beams.

The fire hazard of wood in buildings is chiefly confined to thinner material where the surface to mass ratio is large. The fire hazard in ordinary wood-frame construction can be materially reduced by incorporating fire stops (Wood Handbook, 1955, p. 337—346). These obstruct the chimney-like spaces between wall studs, sheathing and plaster, and between floors and floor joists. They are merely wooden cross structures that block possible circulation of hot air and flames through the encased channels. Additional fire protection is seldom required in single family homes.

In recent years building codes have been enacted in large cities, and in most states of the United States and provinces of Canada, restricting the use of wood and wood products in public buildings to materials that have been treated to meet specific flame spread requirements. Unfortunately, these requirements vary from one locality to another (*Larsen* and *Yan*, 1969). Practically all United States and Canadian codes are based upon a large scale flame spread test in a 25 ft. long tunnel (ASTM Standard E 84) in which a 25 ft. long specimen, 20 in. wide, forms the top of the horizontal tunnel with a heavy closely fitting lid on top with an asbestos-cement face contacting the specimen. A flame is applied to the end of the specimen. A fixed draft causes the flame to advance along the lower surface of the specimen. A flame spread rating is given the specimen on the basis of the time taken for the flame to reach the other end of the tunnel, or the distance traveled in more than 10 min. Temperatures attained at the vent end of the tunnel are used to calculate a fuel contribution factor. Smoke density ratings are also determined from the smoke density at the exhaust. Flame spread ratings are based on an asbestos-cement board having a rating of zero and a panel of 3/4 inch red oak flooring having a rating of 100. Most untreated wood or wood base panels have ratings in the range of 90 to 200.

Most of the codes are based on the hazard to the life of the occupants of a building in case of fire and not the monetary losses that might accompany a fire. For this reason fire spread ratings are usually more strict for hospitals, schools, auditoriums, and prisons, than for apartments and industrial buildings.

A different surface flame spread test is used in Great Britian involving a smaller 9 by 36 in. vertical specimen (British Standard 476, Part 1). Unfortunately an exact correlation with the American test cannot be made. The uniform British code, however, is stricter than the American codes requiring a larger proportion of plywood, particle board, and fiber board used in paneling, to be treated than in the United States. The "Building materials list" of the Underwriter's Laboratories (207 Ohio Street, Chicago, Illinois 60611) are available which specify wood and wood base materials that meet various United States and Canadian building codes. Fire Note No. 9 "Surface spread of flame tests on building products" January, 1965, is available from the Fire Research Station, Borcham Wood, England.

2.4.2 Fire Retardant Treatment

The rate of burning of wood and the spread of fire can be reduced by either surface coatings or impregnation treatments. The former depend primarily on forming of an envelope around the wood that excludes oxygen. Thick coatings of plaster or cement also give the wood some additional insulating value. Thin paint-like coatings of sodium silicate (water glass) give the cheapest form of protection. The glass-like film is quite effective in excluding oxygen while fresh. When subjected to intense heat it forms a film of froth-like bubbles that adds further insulation to the system. Unfortunately sodium silicate tends to hydrolyze as it ages and dries to form silica which tends to lose its adhesion to wood. It is thus necessary to recoat wood structures frequently.

Several fire retardant paints are on the market that contain ammonium phosphates, or phosphoric acid, and or chlorinated rubber-like compounds (*Van Kleeck*, 1956). None of the surface coatings are as effective as impregnation treatments. Once the film or coating is broken it tends to peal away from the wood, thus losing its protective power.

Fire retardance is imparted to wood by salt impregnants chiefly through their ability to increase the formation of char and decrease the formation of tars and volatile gases. This is shown from pyrolysis measurements on ponderosa pine shavings in which the char, tar, and water fractions were measured and the amount of non-condensible gas was calculated from the difference between the weights of the original salt free samples and the sum of the three measured products (*Brenden*, 1965). Part of the results are given in Table 2.3. The presence of each of the salts in all three concentrations increases the char considerably and the water produced was somewhat less while the tar production decreased appreciably. The first salt in the table, sodium chloride, has little if any fire retardant effect whereas borax and diammonium phosphate are good fire retardants. These produce almost twice the char and half of the tar produced by sodium chloride. Char will not propagate burning although it will maintain burning. On the other hand, tar, containing volatile combustible material, will propagate the spread of fire.

Table 2.3. Major Products of the Pyrolysis at 350 °C of Dry Ponderosa Pine Shavings Presoaked in Three Different Concentrations of Salt Solutions (*Brenden*, 1965)

	Concentration of soaking solution %	Loading of shaving[1] %	Products of pyrolysis[1]			
			Char %	Tar %	Water %	Gas %
None	0	0	19.83	54.90	20.87	4.41
Sodium Chloride	15	27.79	26.20	35.21	24.53	14.06
	5	10.34	25.91	35.69	25.18	13.21
	1	4.55	26.77	34.94	29.99	8.31
Borax	15	24.12	43.06	14.90	31.92	10.11
	5	4.28	48.39	11.75	30.42	9.43
	1	—	44.13	20.39	29.33	6.15
Diammonium phosphate	15	24.13	45.14	14.30	37.53	3.02
	5	6.69	45.49	16.76	32.01	5.73
	1	0.51	33.85	35.21	25.07	5.87

[1] Weight % for average of three measurements on salt free basis.

Other salts, besides borax and diammonium phosphate, that have been shown by earlier experiments to impart fire resistance to wood are boric acid, monoammonium phosphate, sodium phosphate, phosphoric acid, ammonium chloride and sulfate, and zinc and magnesium chlorides (*Truax* and *Baechler*, 1935). Of these, boric acid lacks the solubility in water to be used alone. It is, however, used in combination with other salts. The chlorides and sulfates, in general, are too acidic to be used alone in the high concentrations of 5 to 8 lb./cu. ft. (14 to 21% of the dry untreated weight of wood with a dry volume specific gravity of 0.6) needed to obtain a high degree of protection. These acid salts have a slow hydrolytic effect on wood which will significantly reduce the strength properties in time. They are also excessively corrosive towards various metal fittings.

Salts are frequently used in various mixtures. For example, a mixture of two parts of borax to one of monoammonium phosphate has been shown to be more effective than the phosphate alone (*Truax* and *Baechler*, 1935). As borax is cheaper than the phosphate salt, a series of tests were more recently made using borax alone, in which the product was evaluated by more modern methods (*Middleton* et al., 1965). Southern pine lumber, 1 and 2 in. thick, impregnated by the full cell method to a borax loading of 5 to 7.5 lb./cu. ft. (approximately 14 to 20% of the dry weight of the untreated wood), gave ASTM E-84 10 min flame spread values of 20 or less, fuel contribution values of 15 and less, and zero smoke development. Two inch thick western hemlock gave comparable values for the same range of loading. Two inch thick Douglas-fir when incised also gave similar results and only slightly poorer when not incised. These satisfactory loadings were obtained by evacuating for 30 min, running in of aqueous borax solutions ranging from 15 to 21% in concentration and applying a pressure of 145 psi at 150 °F (65 °C) for 10 to 30 h depending upon the species and thickness, followed by kiln drying. Completely filling of lumen with solution should be avoided to insure that collapse does not occur on drying. Exterior grades of Douglas fir plywood in 3/8 and 3/4 in. thicknesses were satisfactorily impregnated to loadings of 5 to 7 lb./cu. ft. Only in one instance did the ASTM E-84 flame spread exceed the satisfactory value of 20.

Fire retardant treated wood is more hygroscopic than untreated wood. Because of this it is important that the treated wood be quite dry before gluing or painting (*Gardner*, 1965). Better glue bonds were obtained with hot press glues than with cold or intermediate setting glues. As fire retardant treatments bulk the cell walls of wood (see 2.5.6) some strength is invariably lost on the bulked dimension basis. It is hence reasonable to calculate strengths on the basis of the untreated dimensions. Further strength losses may result with time due to the hydrolytic action of acidic salts such as zinc chloride. The corrosiveness of fire retardant treated wood varies not only with the fire retardant used but also with the particular metal with which it is in contact (*Gardner*, 1965). Fire retardants, with the exception of zinc chloride, are insufficiently toxic to impart protection against attack by fungi and termites. Hence fire retardant treated wood, even with a good paint coating to minimize leaching, should not be used out-of-doors in contact with the ground.

The chief shortcoming of fire retardant salt-treated wood is that it is not suitable for out-of-door uses where leaching can occur. Because of this, considerable effort has been made in recent years to develop non leaching treatments. An interesting treatment of this type (*Lewin*, 1968) involves treatment of wood with bromine, in which lignin is believed to be brominated in the same sense as wood pulp is bleached by chlorination. Confirmatory tests (*Larsen* and *Yan*, 1969) on wet-formed, dry pressed hardboards gave no better than a 75 ASTM-E 84 flame spread rating with a smoke density in the range 75 to 200.

Another relatively new approach to developing a non-leachable fire retardant treatment is to catalytically polymerize phosphorous containing monomers in the wood (see 2.5.6.6) or to add monomer soluble organic phosphorous compounds to a vinyl monomer with the object of occluding them in the polymer as it is formed (*Kenaga*, 1963). The Koppers Company, Pittsburgh, Pennsylvania, has recently gone into production in treating western red cedar shingles and shakes in this way. The company contemplates also treating structural lumber by this method. It is felt that this sound approach to fire retardant treatments will expand greatly during the next decade in spite of the fact that it is quite costly. Cost is the big barrier in all fire retardant treatments. In spite of this, the production of fire retardant treated wood and wood products in the United States increased from 0.67 to 4.7 million cu. ft. from 1954 to 1966 an increase of seven fold (Am. Wood Preserver's Assoc. Proceedings, 1955, 1967).

2.5 Dimensional Stabilization

Experience has shown that shrinking and swelling of wood can be materially reduced in rate, or in final magnitude, in any one of five different ways, namely by:

1. Cross-laminating veneer;
2. Applying external or internal water resistant coatings;
3. Reducing the hygroscopicity of wood components;
4. Chemically cross-linking structural components of the wood;
5. Bulking the cell walls of wood with chemicals.

2.5.1 Cross-Laminating

Wood swells, on the average, about fifty times as much in the transverse directions as in the fiber direction (*Stamm*, 1964, Chapter 13). When veneer is assembled into plywood, the lateral swelling of each ply is mechanically restrained from being its normal amount due to the fact that longitudinal swelling of adjacent plies is so much less. Swelling of plywood in the sheet directions is, in general, only slightly greater than the longitudinal swelling of separate plies. Each ply is thus under a high compressive stress in the transverse sheet direction. This results in a significant reduction in the hygroscopicity of the wood (*Barkas*, 1947), but far from enough to account for the reduction in external swelling. There is some increase in the swelling in the thickness direction of the plywood due to the relief of stresses in that direction, but the chief effect is to force the lateral swelling of each ply to occur internally into void structure.

Although this means of obtaining dimensional stability in the sheet directions is very effective, it has the shortcoming of promoting face checking (*Lloyd* and *Stamm*, 1958). When swollen plywood is again dried, each ply tries to shrink in the normal external manner, but it is held under tension from doing so by the adjacent plies. If and when this tensile stress exceeds the tensile strength of wood perpendicular to the grain, checking of the plies will occur.

The bulk strength properties of the plywood are not materially reduced by swelling and shrinking. The chief draw back is that unsightly face checks appear. It will be shown later (2.5.6.3 and 2.5.6.5.2) that this face checking can be practically eliminated by dimensional stabilization treatments of the face plies prior to assembly of the plywood, or by applying paper or plastic overlays to the plywood surfaces.

2.5.2 External Coatings

Coating the surface of wood with a water resistant finish materially reduces the rate of swelling, but it has little if any effect upon the final equilibrium swelling under long exposure conditions. The effectiveness of coatings varies greatly with the nature of the coating and the exposure conditions. Unfortunately all known finishes that will adhere to wood are somewhat permeable to water.

Table 2.4 gives the range of moisture excluding efficiencies of different types of finishes applied to all surfaces of 8 by 4 by 5/8 in. specimens with rounded edges and corners (*Browne*, 1933; *Hunt*, 1930). Specimens coated and dried at 60% relative humidity were weighed and then exposed together with weighed end matched controls for one week at 97% relative humidity, and again weighed. The moisture excluding efficiency is the gain in weight of the control minus that of the coated specimen divided by the gain in weight of the control expressed in percent. As swelling is directly related to the moisture sorption, the moisture excluding efficiency is also a rough measure of the temporary antiswell efficiency. The measurements summarized in Table 2.4 were made considerably before the advent of the modern emulsion finishes. They would be expected to have considerable lower moisture excluding efficiencies than oil base paints.

Table 2.4. Moisture-excluding Efficiency (MEE) of Various Classes of Finishes on Wood Specimens after Exposure for One Week at 97% Relative Humidity (*Hunt*, 1930)

Coating	MEE %
Aluminium foil between coats of varnish or oil base paint	99
2 coats of aluminium powder dispersed in varnish, or oil base paint	90···95
2 coats of pigmented oil base paint over a suitable primer	60···90
2 coats of varnish, enamels or cellulose nitrate lacquers	50···85
5 coats of linseed oil and 2 coats of wax	about 8

The moisture excluding efficiencies of finishes decrease appreciably with the time of exposure to high relative humidity, and still more when exposed to a cycling relative humidity or weather exposure. When the specimens are exposed to 97% relative humidity for two weeks, followed by exposures to 60% relative humidity for two weeks, and outside weather exposure for two weeks and this cycle is repeated five times, the moisture excluding efficiency falls below 70% even for aluminium paints. When the cyclic test is extended for a period of a year or more, most finishes lose practically all of their moisture excluding efficiency. This is due to the fact that moisture content cycling, accompanied by swelling and shrinking and ultraviolet light exposure tend to cause checking of the wood and chemical degradation of the finish, both of which lower the adhesion of the finish to the wood and the integrity of the coating.

2.5.3 Internal Coatings

Although external coatings are subject to being weathered or abraded away, which is not the case for internal coatings, they can be made more continuous and thus more effective than internal coatings. This can be readily demonstrated by applying two coats of varnish to a wood specimen and impregnating a matched specimen with several times as much varnish diluted with a varnish solvent. Subsequent moisture excluding efficiency tests will give higher values for the surface coated specimen than for the impregnated specimen. Nevertheless, internal coating by so called water repellents are useful for temporary protection of millwork, especially against absorption of liquid water (*Browne*, 1949; *Browne* and *Downs*, 1949). Water-repellents are natural resins, waxes or drying oils dissolved in clear volatile hydrocarbon solvents containing a toxic agent, usually pentachlorophenol. They are usually applied to predried mill-work by a short three minute dip technique. Take up of the solution by capillary action may be only a few millimeters deep in heartwood faces with somewhat deeper penetration into sapwood. Penetration in the fiber direction may, however, be several centimeters (*Browne*, 1950). The chief attribute of water repellents is their preservative action against decay. They will protect window sash and frames against absorption of liquid water over short periods of time prior to painting. They will not prevent the swelling , shrinking, and sticking of windows that occur as a result of relative humidity changes that occur from one season of the year to another. They will, however, reduce face checking and grain raising as a result of the reduced rate of swelling and shrinking.

2.5.4 Reduction in Hygroscopicity

Moisture adsorption occurs in wood and in other cellulosic materials as a result of the attractive force of the polar hydroxyl groups of the cellulose and the lignin for water. When these groups that are not active in holding the structure together are replaced by less polar groups, the affinity for water can be materially reduced. The theoretically ideal situation would be to replace hydroxyl groups with hydrogen. No method for hydrogenating only the groups that are not involved in bonding the various structures together is known. Hydrogenation of wood unfortunately breaks down both the cellulose and lignin structures into various smaller molecules (*Stamm* and *Harris*, 1953, Chapter 17). Wood and paper can, however, be acetylated to a high degree without breaking down the structure. This should theoretically reduce swelling and shrinking to about half of normal. It is, however, reduced to about 20% of normal. The fact that the reduction is greater than would be expected on the basis of changed hygroscopicity will be shown under 2.5.6.7 to be due to the larger acetyl than hydroxyl groups acting as bulking agents within the cell walls. Stabilization of wood by heat alone is the only presently known method that can be entirely attributed to a change in hygroscopicity.

2.5.4.1 Heat Stabilization. When wood is heated, preferably in the absence of oxygen, under temperature-time conditions that cause a small loss of water of constitution together with other breakdown products (*Stamm*, 1964, Chapter 19) its equilibrium swelling and shrinking is materially reduced. When the logarithm of the heating time is plotted against the heating temperature for constant weight loss, a series of almost parallel straight lines is obtained. The same is true for constant reductions in strength and in swelling and shrinking (*Stamm*, *Burr* and *Kline*, 1946). When air dry specimens of different softwoods, ranging in thickness from 1 to 24 mm, were heated beneath the surface of a low fusion metal, to minimize oxidation, a permanent reduction in swelling and shrinking of 40% was attained by heating for one minute at 315 °C, one hour at 255 °C, 1 day at 210 °C, one week at 180 °C, one month at 160 °C and one year at 120 °C, as is shown in Fig. 2.1.

It has been shown that the weight loss on heating is roughly proportional to the square of the accompanying reduction in swelling (*Stamm*, 1959, (3)). The activation energy of the thermal reaction is about 28,000 calories per mole of water lost (*Stamm*, 1956 (2) and 1959 (3)) irrespective of the absence or the presence of an acid catalyst. Acids, however, greatly increased the rate of the reaction.

Table 2.5 shows that weight loss and losses in strength properties become excessively high in attaining a reduction of swelling and shrinking of 40 percent and more. Abrasion resistance is the most adversely affected strength property. Abrasives actually gouge out full fibers rather than abrading away fibers at the high levels of dimensional stabilization, indicating that the bond between fibers is weakened. This is probably due to a break down of chains of hemicellulose molecules that pass through the middle lamella of the fibers and that help to hold the fibers together as the hemicelluloses are the most heat sensitive components of the wood (*Stamm*, 1965). Any applied use of this heating technique to attain dimensional stability of wood will be limited by the acceptable strength losses.

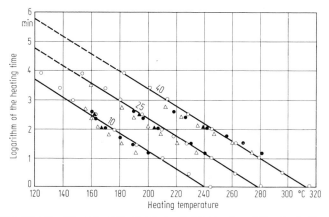

Fig. 2.1. Logarithm of heating time versus temperature required to give three different reductions in swelling and shrinking when the heating was done beneath the surface of a molten metal to exclude oxygen. From *Stamm, Burr* and *Kline* (1946)

Open circles: 1/16 in. thick Sitka spruce veneer, Shaded circles: 1/2 in. thick cross sections of western white pine, Open triangles: 3/8 in. thick flat sawn western white pine, Shaded triangles: 15/16 in. thick western pine boards. Numbers on plot indicate the reduction in swelling and shrinking (antishrink efficiency) in percent

Table 2.5. Weight and Strength Loss Accompanying Permanent Reduction in Swelling and Shrinking Resulting from Heating Dry Softwood Specimens beneath the Surface of a Molten Metal for 10 min at Different Temperatures (*Stamm, Burr* and *Klime*, 1946; *Stamm*, 1959 (3))

Temperature °C	Weight loss %	Modulus of rupture loss %	Hardness loss %	Toughness loss[1] %	Abrasion resistance loss[2] %	Reduction in swelling and shrinking %
210	0.5	2.0	5.0	4.0	40	10
245	3.0	5.0	12.5	20.0	80	25
280	8.0	17.0	21.0	40.0	92	40

[1] Forest Products Lab. toughness test, 1956.
[2] Heated in air.

The wood darkens in color as it is heated. Light colored softwoods attain about the color of walnut when heated to give a reduction in swelling and shrinking of about 25 percent.

It was first believed that heat stabilization of wood was the result of water molecules being split out between hydroxyl groups on adjacent cellulose chains with the formation of ether linkages between the chains, thus tying them together with cross links that limit separation of the chains by water (*Stamm* and *Hansen*, 1937). Although ether linkages may be formed within a cellulose chain as water of constitution is lost, it is improbable that they form between cellulose chains. If they did form, swelling by all media that are incapable of breaking other linkages should be reduced. Concentrated sodium hydroxide solutions and pyridine are such media. Both of them cause a greater swelling of heat stabilized wood than of matched unheated specimens. The fact that heat stabilization reduces the swelling in water, but not in these other media can be explained as follows. Hemicelluloses which are the most hygroscopic component of the wood are the most subject to thermal degradation to furfural and various sugars. These tend to react with each other to form water insoluble polymers, thus reducing the hygroscopicity. These polymers are, however, swollen and partially dissolved by strong alkalies and by pyridine (*Seborg*, *Tarkow* and *Stamm*, 1953). This is in keeping with the fact that hardboards, that are low in hemicelluloses, when subjected to heat stabilization, retain much more of their original abrasion resistance than wood.

Wood that has been given considerable dimensional stability by heat also has considerable decay resistance (*Stamm* and *Baechler*, 1960). Block culture tests, with *Trametes serialis* as the wood-destroying organism, over a period of two months gave weight loss of unheated white pine blocks of 28.4%. Matched blocks heated to give reductions in swelling and shrinking of 30 to 33% lost 12% of their weight, those heated to give a reduction in swelling and shrinking of 33 to 38% lost only 4.5% of their weight while those heated to give at least 40% reduction in swelling and shrinking lost no weight due to decay (*Stamm* and *Baechler*, 1960). Tests using *Lenzites trabea* as the wood destroying organism gave similar results. Cross-linking and bulking treatments that impart considerable dimensional stability to wood also impart considerable decay resistance to wood. It thus appears that all types of treatments that permanently reduce swelling and shrinking will also improve the decay resistance. This would be expected if the moisture sorption by the cell wall is reduced below that required to support decay, namely 20 to 25%.

2.5.5 Cross-Linking

Forming chemical cross-links between the molecular structural units of wood should, from a theoretical standpoint, be one of the most effective means of imparting dimensional stability to the wood. Nature has already provided some degree of stability to cellulose as a result of crosstying the amorphous water sorbing zones together through the crystalline zones. If these crystalline zones. were not present, cellulose should disperse in water as do a number of starches.

A good example of the effect of introducing cross-links into polymers is given by *Staudinger* (1937). He showed that incorporating small amounts of divinyl benzene in the vinyl benzene, in making a styrene polymer, converted the product from a benzene soluble material to a benzene insoluble material with the introduction of only one cross-link per several thousand carbon atoms in each polymer chain.

Formaldehyde has long been known as a cross-linking agent for cellulose (*Eschalier*, 1906). It has been used to impart crease resistance to cotton textiles by soaking them in a formalin solution containing a low concentration of mildly acidic salt as catalyst followed by drying (*Gruntfest* and *Geyliardi*, 1928). Similar experiments by *Tarkow* and *Stamm* (1953) with wood showed that no dimensional stabilization occurred until the wood was almost dry and then only when the acidity was quite high. It thus appeared desirable to carry out the reaction with formaldehyde in the vapor phase, generating the vapor by heating paraformaldehyde (*Walker*, 1944).

Dry Sitka spruce cross sections 1/8 in. thick were heated in sealed glass jars over paraformaldehyde at 50 to 60 °C for various lengths of time. The specimens increased not only in weight, but also in cross sectional dimensions indicating that condensation of paraformaldehyde must have occurred within the cell walls. This would give a first cycle reduction in swelling due to the bulking (see 2.5.6). As the paraformaldehyde in the cell walls of the wood can be lost both by heating in circulating air or by leaching with water, the dimensional stabilization obtained is, by necessity temporary unless some permanent weight increase occurs. Such permanent weight increases occurred only when volatile acid catalysts were present. Just detectable weight increases occurred with organic acid catalysts such as formic and acetic. When strong volatile mineral acids such as nitric and hydrochloric were used, up to 4 to 5% permanent increases in the weight of the wood occurred. No permanent reduction in swelling and shrinking in terms of anti-shrink efficiency,

$$\text{ASE} = \frac{S_c - S_t}{S_c} \times 100 \tag{2.21}$$

was obtained, where S_c is the shrinkage of the control and S_t that of the treated specimen expressed in percent, in the absence of a catalyst. When formic or acetic acid was used as the catalyst, antishrink efficiencies of less than 10% were obtained. When nitric or hydrochloric acids were used as catalysts, permanent antishrink efficiencies up to 70% were obtained with as little as 4% weight increase (*Tarkow* and *Stamm*, 1953). It was found that high permanent antishrink efficiencies could be obtained only when the pH of the system was maintained at 1.0 or less.

Unlike heat stabilized wood, wood that is reacted with formaldehyde swells less than the untreated controls in strong alkali solutions and in pyridine as well as in water (2.5.4.1). This, together with the following findings, indicate that the reaction is one involving cross-linking rather than bulking. Stabilization is accomplished by reducing the amount that the originally dry specimens can swell. In the next section (2.5.6) it will be shown that bulking agents reduce the amount that the originally swollen specimens can shrink. That is, formaldehyde reacted specimens have virtually the same dry dimensions as the controls whereas bulked specimens have the same swollen dimensions as the controls. Further a high dimensional stability can be obtained with only one sixth as much chemical by the formaldehyde reaction as by bulking.

Mildly acidic salts were tried as catalysts with the hope of carrying on the reaction under less acidic conditions (*Stamm*, 1959 (3)). These were introduced into the specimens by a presoak followed by drying. The square of the antishrink efficiencies obtained were roughly proportional to the permanent weight increases resulting from the treatments. In no case did mild catalysts give a significant reaction or antishrink efficiency even when the reaction was carried out at temperatures as high as 160 °C.

Optimum ASE values were obtained when the moisture content of the wood was between 5 and 10% under constant temperature-time and catalyst content conditions (*Tarkow* and *Stamm*, 1953). Evidently some moisture is needed to open up the cell wall structure. Larger amounts of water present tend to hydrate the formaldehyde and reduce its needed carbonyl content (*Walker*, 1944).

Table 2.6 gives the antishrink efficiencies corresponding to various permanent weight increases together with the relative abrasion resistance and relative toughness to that of untreated controls. It is obvious that these two chemically sensitive strength properties are so adversely affected by the treatment to almost nullify any potential usefulness of the treatment on wood. Specimens subjected to the same temperature-time combinations with catalyst, but no formaldehyde present, gave virtually the same strength losses. The major part of the strength loss can thus be attributed to hydrolysis of hemicelluloses together with lesser amounts of cellulose. The same explanation given for the loss in abrasion resistance of heat stabilized wood will apply here (2.4.5.1).

Table 2.6. Antishrink Efficiencies (ASE) Obtained with Different Permanent Weight Increases due to Reaction with Formaldehyde Vapor together with the Resulting Abrasion Resistance and Toughness Relative to the Values for Controls (*Stamm*, 1959 (3))

Weight increase of dry wood	ASE	Relative abrasion resistance	Relative[1] toughness
%	%	%	%
0.10	10.0	0.4	0.73
0.55	25.0	0.2	0.55
2.20	50.0	0.09	0.30
4.20	70.0	0.05	0.16

[1] Forest Products Lab., 1956.

The formaldehyde cross-linking treatment of wood to the 50% ASE level or above virtually eliminates loss in weight due to decay in the three month block culture decay tests (*Stamm* and *Baechler*, 1960).

Formaldehyde cross linked paper should have better strength properties than the cross linked wood because of its lower and more favorably distributed hemicellulose content. This was found to be the case. The formaldehyde cross-linking of Kraft paper carried to a 40% increase in the ASE value had wet tensile strength ranging from an initially negligible value to about one third of the dry strength value. The treatment increased the wet ring crush values about eight fold to about two thirds of the dry test values. The dry tensile strength was increased by 10% and the dry ring crush by about 5%. Burst was reduced to 85% of normal, tear to 67% of normal and fold to 50% of normal (*Cohen, Stamm* and *Fahey*, 1959 (1)). Dry tension was not affected by a cross fold in the paper at the 32% ASE level. At the 55% ASE level, a single cross fold reduced the dry tensile strength to 83% of normal. The paper gained only 0.7% in weight of chemical for an ASE value of 70% in contrast with the 4.2% weight increase for softwoods (*Stamm*, 1959 (3 and 4)). It thus appears that the formaldehyde cross-linking treatment could be used to advantage in imparting both some dimensional stability and appreciable wet strength to papers that are not subjected to fold or tear, such as punch cards or corrugated paper cores.

Aldehydes other than formaldehyde have been tested as to their cross-linking ability towards wood (*Tarkow* and *Stamm*, 1953). None were as effective or permanent as formaldehyde and all required a high degree of acidity. No acid catalyst had to be added when chloral was used as the cross-linking agent as it developed its own acidity which was sufficiently high to cause similar strength losses to those obtained with the formaldehyde reaction. Other types of cross-linking chemicals have been sought that will react with wood under practically neutral conditions. All compounds of this type that have been tested up to the present time have failed to impart a degree of dimensional stability to wood comparable to that obtained by bulking. The principle of cross-linking is, however, sufficiently sound to make it desirable to watch for new means for cross-linking wood.

2.5.6 Bulking Treatments

2.5.6.1 Salt Treatments. The principle of bulking the cell walls of wood to attain dimensional stability is best demonstrated by experiments in which it was first observed (*Stamm*, 1934). End matched cross sections of Sitka spruce, 1/8 in. thick by 2 by 2 in., with the annual rings parallel to two opposite faces, were oven dried, weighed and their tangential and radial dimensions measured to 0.001 in. They were then filled with water and with a series of different 25% saturated salt solutions, followed by oven drying. Weights and dimensions were determined at the conclusions of each step. Table 2.7 gives the relative vapor pressure over each saturated salt solution. Fig. 2.2 is a plot of the external volu-

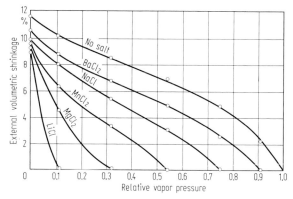

Fig. 2.2. External volumetric shrinkage versus the relative vapor pressure for thin Sitka spruce cross sections containing originally different 25% saturated salt solutions. From *Stamm* (1934)

metric shrinkage versus the equilibrium relative vapor pressure. In all cases shrinkage begins at the relative vapor pressure over the respective saturated salt solution that corresponds to the one within the cell walls of the specimens. On drying, water is first lost from the solution in the coarse capillary structure thus building up the concentration. Salt then diffuses into the cell walls until they become saturated. Saturation of the solution within the cell walls will occur whenever the original fraction of saturation exceeds the ratio of the volume of liquid required to saturate the cell walls to the total volume of liquid in the wood.

Table 2.7. Relative Vapor Pressures over Saturated Salt Solutions
with an Excess of Salt Present at 25 °C

Salt	Relative vapor pressure
Barium chloride	0.916
Sodium chloride	0.758
Manganese chloride	0.543
Magnesium chloride	0.331
Lithium chloride	0.117

Shrinkage to the oven dry condition is reduced by an amount equal to the volume of salt retained within the cell walls. The salt thus bulks the cell walls so that they can no longer shrink their normal amount.

Fig. 2.3 is a plot of the progressive shrinkage of the specimens of Fig. 2.2 versus their moisture content. Virtually parallel straight lines are obtained, indicating that the same shrinkage occurs per increment of water lost for all of the specimens. There is, however, less water to be lost, the higher the extent of bulking.

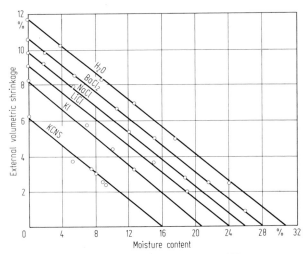

Fig. 2.3. External volumetric shrinkage versus moisture content for thin Sitka spruce cross sections containing originally different 25% saturated salt solutions. From *Stamm* (1934)

Shrinkage is thus reduced by two different mechanisms, namely by reducing the relative vapor pressure at which shrinkage begins, and by bulking the cell walls. Shrinkage of wood containing a saturated solution of lithium chloride will not start until the prevailing relative humidity falls below 12%. This occurs so rarely that such specimens will be practically immune from shrinking and swelling. This, however, is not a practical treatment as the specimen will be continuously wet and can not be properly glued or finished.

The principle of reducing the effective relative humidity with salts has been commercially applied to drying wood under less than normal stress conditions. This procedure, commonly known as "salt seasoning", (*Loughborough*, 1939; *Stamm*, 1964, Chapter 24), though successful in accomplishing its objectives, has been virtually abandoned because of the corrosiveness of the sodium chloride used on tools and fastenings.

2.5.6.2 Sugar Treatment. It is apparent from Figs. 2.2 and 2.3 that the ideal bulking agent would be a nonvolatile solid almost infinitely soluble in water that does not reduce the relative vapor pressure materially, thus avoiding the specimens being continuously wet at high relative humidities. These conditions are reasonably well met by sugar solutions, perferably invert sugars (*Stamm*, 1937) as is shown in Fig. 2.4.

Treating wood with sugars was commercially practiced in England for a short period (*Bateson*, 1939). As sugars are food for organisms, toxic agents had to be added. The chief shortcoming of the treatment was cost, and the facts that the wood was damp at relative humidities above 80% and that difficulty was encountered in obtaining good adhesion of wood finishes.

2.5.6.3 Polyethylene Glycol Treatment. Polyethylene glycol-1000 was found to be still better bulking agent as shown in Fig. 2.5 (*Stamm*, 1956). The figure shows that glycerine and the low molecular weight polyethylene glycols can virtually replace all of the water in the water swollen cell walls of small specimens of wood. The effectiveness of the polyethylene glycols drop off appreciably when their average molecular weights exceeds 1,500. *Tarkow*, *Feist* and *Southerland* (1966), have shown that polyethylene glycols having molecular weights exceeding about 3,500 are too big to penetrate the cell walls of wood. Thus, only the lower molecular weight portion of the high molecular weight polymers are effective. The high molecular weight fractions also have limited solubility in water.

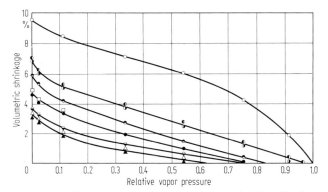

Fig. 2.4. External volumetric shrinkage versus the relative vapor pressure for thin white pine cross sections presoaked in different concentrations of sucrose or invert sugar. From *Stamm* (1937)
Open circles: water only, Left shaded circles: 6.25% sucrose, Top shaded circles: 12.5% sucrose, Open squares: 25.0% sucrose, Shaded circles: 80.0% sucrose, Right shaded circles: 6.25% invert sugar, Bottom shaded circles: 12.5% invert sugar, Open triangles: 25.0% invert sugar, Shaded triangles: 50.0% invert sugar

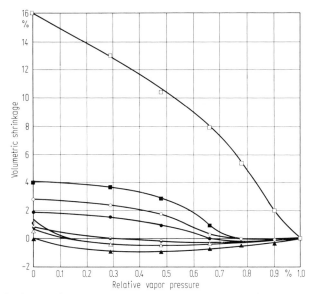

Fig. 2.5. External volumetric shrinkage versus the relative vapor pressure for thin Sitka spruce cross sections presoaked in 25% by weight aqueous solutions of glycerine and polyethylene glycol having the average molecular weights given in parenthesis in the legend. From *Stamm* (1956)
Open squares: water only, Right shaded circles: glycerine, Shaded triangles: polyethylene glycol (200 and 400), Open triangles: polyethylene glycol (600). Bottom shaded circles: polyethylene glycol (1,000), Shaded circles: polyethylene glycol (1,540), Open circles: polyethylene glycol (4,000), Shaded squares: polyethylene glycol (6,000)

Polyethylene glycol treatment can be best applied to green wood. The simplest technique merely involves soaking the green wood in about a 30% by weight aqueous solution. Caution must be taken in using more concentrated solutions or heating the solution to avoid faster outward diffusion of water than inward diffusion of chemical (*Kenaga*, 1963). The time of soaking varies somewhat with species. It increases appreciably with an increase in the specific gravity of the wood and the size of the wood specimens, and decreases greatly with an increase in the amount of end grain exposed (*Stamm*, 1964, Chapter 23). The rate of diffusion into water saturated wood varies inversely as the square of the thickness, and is ten to fifteen times as high in the fiber direction as in the transverse directions (*Stamm*, 1946). Fig. 2.6 shows the effect of soaking green Loblolly pine tree cross sections 3 cm thick in a 30% solution of polyethylene glycol-1000 for

Fig. 2.6. Adjacent originally green loblolly pine tree cross sections (27 cm diameter and 3 cm thick). Left: soaked in a 30% by weight solution of polyethylene glycol (1,000) for 24 h. Right: soaked in water for 24 h, both followed by air drying. The treated specimen, on the left, developed no peripheral radial checks. The control, on the right, developed a large peripheral radial check extending almost to the pith. The treated specimen took up on the average 16% of polyethylene on a dry weight basis. From *Stamm* (1959)

24 h followed by air drying compared with an end matched control specimen that had been soaked in water. The control developed a large wedge shaped check extending from the pith to the surface due to the stresses developed as a result of the tangential shrinkage being considerably more than the radial shrinkage. The treated specimen, on the other hand shrank sufficiently less than the control to avoid any similar checking. Eventually, a small internal check occurred at the pith of the treated specimen, presumably due to heartwood, which was just beginning to form, being inadequately treated in the 24 h soak. Subsequent oven drying of a matched pair of specimens showed that wedge shaped checks were avoided when the treated specimen took up only 16% of the chemical on a dry weight basis.

An alternate method for treating green wood is to apply a liberal coating of molten polyethylene glycol-1000 to the surfaces, store the specimen for a day in a sealed container or a polyethylene bag to avoid drying of the specimen, and repeating the application followed by a nondrying storage at one day intervals for several days before air drying.

Non refractory softwoods can be simultaneously treated and dried by immersing the specimens in a boiling solution of polyethylene glycol-1000. The method, however, does not appear practical as polyethylene glycol gradually decomposes when heated for extended periods of time in the presence of water (*Stamm*, 1967) and would be overly expensive.

Checking of the face plies of plywood resulting from relative humidity cycling can be virtually eliminated by treating the face plies prior to assembly with 20 to 30% of polyethylene glycol-1000. (The treatment can be made either by the soaking method starting with a green veneer or by the "hot-cold bath" or pressure treating methods (see 2.2.2, 2.2.3) starting with dry veneer, followed after storage for a day under nondrying conditions, by drying in a veneer drier.

Fig. 2.7 shows two pieces of Douglas fir plywood one with treated faces containing 25% of polyethylene glycol-1000 and the other without treatment both of which had been exposed to four cycles of two weeks at 90% relative humidity and two weeks at 30 % relative humidity. The figure shows that the controls checked badly while the treated specimens are entirely free from checks (*Stamm*, 1959 (2)).

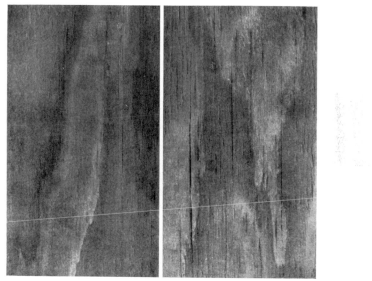

Fig. 2.7. Douglas fir three ply plywood. Left: with faces treated with 25% by weight of polyethylene glycol (1,000). Right: faces untreated. Both specimens were subjected to two weeks at 90% relative humidity followed by two weeks at 30% relative humidity four times before photographing. The face treated specimen at the left is free from face checks whereas the untreated control to the right is badly checked. From *Stamm* (1959)

Polyethylene glycol treatment virtually maintains wood in its green condition. Because of this the strength properties are close to those of green wood. Unlike formaldehyde cross-linking and part of the bulking treatments to follow (2.5.5, 2.5.6.5.2) the toughness of the wood is not adversely affected (*Stamm*, 1967 (1)).

Wood treated with moderate amounts of polyethylene glycol and carefully predried can be glued satisfactorily with all common types of glue (*Stamm*, 1959). The same is true for applied finishes over polyethylene glycol treated wood. A presurface treatment of Loblolly pine house siding with polyethylene glycol improved the weathering properties of applied alkyd emulsion house paints and two-can clear polyurethane finishes (*Campbell*, 1966, 1970). Finishes tend to seal the leachable polyethylene glycol in the wood (*Stamm*, 1959 (2)).

Treatment of wood with polyethylene glycol imparts considerable decay resistance to wood when it is not subjected to leaching (*Stamm* and *Baechler*, 1960) in spite of the fact that it is nontoxic. This can be explained on the basis of insufficient water being present in the cell walls to support decay.

The diffusion treatment of green wood with polyethylene glycol-1000 is being successfully utilized in the dimensional stabilization, and the avoidance of surface checking of gunstocks, various decorative carvings and tree cross section plaques (*Mitchell* and *Wahlgren*, 1959, 1961; *Stamm*, 1962). The objects are roughed out while the wood is still green, thus exposing considerable end grain, and then soaked in an aqueous solution of the polyethylene glycol or the alternate surface coating method is applied using the best empirically determined treating and storage times under nondrying conditions. The carving is then completed and the objects air dried, followed by a final sanding. One of the important features of this procedure is that the most vulnerable part of the carving to face checking such as the legs of a carved animals, take up the maximum amount of polyethylene glycol while the thicker less vulnerable parts, such as the body, take up appreciably less.

This method of treatment is ideal for the restoration of artifacts, that have been buried in a water logged condition for long periods of time. The most noted case to date is the restoration of the Swedish wooden battleship, Vasa, which was sunk in the harbor of Stockholm in 1628 and raised in 1961 (*Franzen*, 1962). As much as possible of the vulnerable water logged surface of this oak ship with pine masts is still in the process of being preserved by a polyethylene glycol treatment.

The most remarkable case of preservation of wood with polyethylene glycol is that of a pine log hermetically sealed in a bog in a glacial moraine in northern Wisconsin, which according to carbon 14 dating technique was buried for a period of 31,000 years. Small samples of the broken log, that were allowed to air dry in the laboratory, broke up into a pile of splinters. Evidently, hydrolyses of hemicellulose chains passing through the middle lamella weakened the fiber to fiber bonds to the extent that drying stresses caused random breakdown of the structure (see 2.5.4.1). Special precautions were taken in soaking specimens from the log in 10% polyethylene glycol-1000 solution and gradually building up the concentration to 30%. On air drying, the specimens remained perfectly sound with no new flaws. In this case the small drying stresses were insufficient to break up the structure. The slight shrinkage that did occur was probably sufficiently great to draw the structure together to the point that hydrogen bonds between the structural units replaced the originally broken covalent bonds. Not enough material was available to make strength tests. However, the treated and dried specimen was sufficiently hard that an end grain surface could not be scratched with a finger nail.

There would be many more applications for polyethylene glycol treatment if its price could be substantially reduced.

2.5.6.4 Wax Treatment. The chief physical shortcoming of the polyethylene glycol treatment of wood is that the chemical can be readily leached from the wood. It thus appeared desirable to deposit water insoluble chemical within the cell walls of wood. This can be accomplished by a replacement process (*Stamm* and *Hanson*, 1935). Water in the swollen wood is replaced by cellosolve (ethylene glycol monoethyl ether) by immersing the wood in this chemical and slowly boiling off the water, which has a lower boiling point than that of the cellosolve. Replacement occurs without shrinkage taking place. The specimen is then trans-ferred to a molten wax or natural resin and the cellosolve slowly distilled off. Wax replaces up to 80% of the cellosolve in the cell wall, thus giving an antishrink efficiency of about 80%. The replacement has been successfully carried out with paraffin wax, beeswax, stearin, and rosin. A mixture of beeswax and rosin seemed to work best. The replacement procedure worked well only on small wooden objects. Considerably more wax was introduced into the wood than was needed to bulk the cell walls. The wax further interferes with subsequent gluing and finishing. This treatment could be used on swollen wooden artifacts. No commercial application of the treatment is forseen.

2.5.6.5 Phenolic Resin Treatment (Impreg). A simpler approach to depositing water insoluble materials within the cell walls of wood is to treat the wood with water soluble resin forming chemicals that will diffuse into the cell walls followed by drying to remove the water, and then heating in the presence of a suitable catalyst to polymerize the resin. A number of different resin forming systems have been successfully polymerized within the cell walls of wood; namely, phenol, resorcinol, melamine- and urea-formaldehydes, phenol furfural, furfuryl aniline, and furfuryl alcohol (*Stamm* and *Seborg*, 1936). The most successful of these has been phenol-formaldehyde (*Stamm* and *Seborg*, 1939). It is cheaper than the resorcinol and melamine-formaldehydes, gives higher antishrink efficiencies than urea-formaldehyde (*Millett* and *Stamm*, 1946) and is a more weather resistant product. Further, less chemical is lost on drying than in the case of furfuryl aniline and furfuryl alcohol systems. In early experiments, unreacted mixes of phenol and formalin together with a mild acid or base catalyst were used. Too much chemical was lost in drying the wood after treatment. It was soon learned that virtually a nonvolatile "A" stage resin that is still soluble in water in all proportions is equally suitable (*Burr* and *Stamm*, 1943). A number of suitable "A" stage resins are commercially available. They range in resin-forming solid content from 33 to 70%, and in pH from 6.9 to 8.7. Their relative viscosities in 33% solids content, range from 3.5 to 4.7. Wood treated with these "A" stage phenol-formaldehyde resins has been given the name Impreg.

2.5.6.5.1 Treating Processes. Difficulty was encountered in adequately distributing the "A" stage resins in sizable specimens of solid wood by pressure impregnation. A small amount of solid wood Impreg is being made in this way from such readily impregnated woods as ponderosa pine and basswood. The bulk of the Impreg made at the present time is made from veneer and subsequently laminated to give panels of any desired thickness.

Veneer can be treated with "A" stage phenol-formaldehyde resins in a number of different ways. Green furniture veneer 1/32 in. or less in thickness will take up 25 to 30% of its dry weight of chemicals by diffusion when soaked in a solution containing 30 to 60% solids for an hour or two (*Stamm* and *Seborg*, 1939). Soaking times become excessive for thick veneer as diffusion times vary as the square of the thickness. Diffusion into green wood has the disadvantage that water diffuses out as the chemical diffuses in, making it necessary to frequently fortify the soaking solution with fresh chemical. Thin dry furniture veneers often require no longer soaking time than green veneer due to the presence of considerable cross grain which readily takes up the solution by capillarity. Frequently sufficient resin is taken up by capillarity when passing the thin veneer through a glue spreader one or two times in which a 60 to 70% resin forming solution is applied.

Thicker straight grain veneer usually requires a more drastic treatment. Veneer of medium to low specific gravity in thicknesses up to 1/8 in. and preferably with moisture contents of 20 to 30% can be treated with compression roll equipment (*Stamm*, 1944). The veneer is passed between compression rolls beneath the surface of the solution where it is compressed to about half of its original thickness. The veneer, on emerging from between the rolls, tends to recover its original thickness and in doing so sucks in the treating solution.

The chief method for treating 1/16 in. thick and thicker veneer is by pressure impregnation in a treating cylinder. The normal procedure is to immerse the veneer one sheet at a time in a tank that is filled with the resin forming solution. After the sheets of veneer are wetted on both faces they can be packed tightly without fear of forming dry pockets. The sheets of veneer are then completely submerged with a weight on top and sufficient treating solution added so that they will remain submerged after impregnation. The tray is then rolled into the treating cylinder, 20 to 200 psi pressure applied for 10 min to 6 h depending upon the species, whether sapwood or heartwood, and the thickness of the veneer. Basswood or cottonwood heartwood veneer 1/16 in. thick will take up its own weight of a 30% solution in 15 min at 30 to 40 psi. Birch heartwood veneer will require 2 to 6 hours at 75 psi (*Stamm* and *Seborg*, 1939).

The treated veneer should be covered in a close pile for one to two days to allow the resin forming components to equalize through the structure by diffusion. The veneer can then be dried in a continuous veneer drier or in a dry kiln. Very high drying rates should be avoided to prevent excessive migration of the resin forming components to the surface. Reasonable schedules are, drying for 8 h at 60 °C, or for 3 h at 72 °C. The resin is not appreciably polymerized under these conditions. To insure polymerization of the resin, kiln temperatures should subsequently be raised to 95 °C for a day, or the veneer drier temperature raised to 150 °C for half an hour to avoid compression of the veneer in the subsequent laminating step in a hot plate laminating press following normal laminating procedures (*Stamm* and *Seborg*, 1939).

2.5.6.5.2 Properties of Impreg. The reduction in swelling and shrinking, or antishrink efficiency, of Impreg made from Sitka spruce, and sugar maple increase with the resin content of the cell walls of the wood up to an optimum of about 70% at a resin content of about 35% (*Stamm* and *Seborg*, 1939) as shown in Fig. 2.8. A similar plot is also given for the urea-formaldehyde resin treated wood. The optimum antishrink efficiency in this case is only 50%. The difference is due to the limited solubility of the urea-formaldehyde resin precursor in water (*Millett*

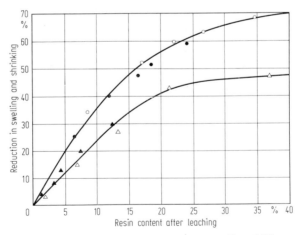

Fig. 2.8. Reduction in cross section swelling and shrinking of thin cross sections of Sitka spruce and sugar maple versus their content of phenol formaldehyde and urea-formaldehyde resins relative to the dry weight of the untreated specimens after leaching and drying. From *Millett* and *Stamm* (1946)
Open circles: spruce containing phenolic resin, Shaded circles: maple containing phenolic resin, Open triangles: spruce containing urea resin, Shaded triangles: maple containing urea resin

and *Stamm*, 1946). The fact that the antishrink efficiency of Impreg is limited to 70%, whereas that attained with polyethylene glycol approaches 100%, can be explained by the fact that the condensation reaction of phenol with formaldehyde involves the splitting-off of water, resulting in about a 25% contraction. Even if water in completely swollen cell walls was entirely replaced by the resin forming components, the optimum antishrink efficiency would be 75%. This is verified by the fact that repeated treatments followed by drying and curing of thin easily treated cross section of wood increased the antishrink efficiency to almost 100%. Voids left upon curing were filled to the extent of about 75% in each succeeding cyle of treatment (*Stamm* and *Seborg*, 1939). Such multiple treatments would not pay commercially.

Face checking of plywood and parallel laminates is practically eliminated on indoor exposure when the faces are treated with 25% to 30% of resin. Under out-of-doors weathering conditions face checking and erosion are materially reduced (*Lloyd* and *Stamm*, 1958).

Phenolic resin treatment, like other dimension stabilizing treatments, imparts considerable decay resistance to wood (*Stamm* and *Baechler*, 1960; *Stamm* and *Seborg*, 1939). The treatment also increases the electrical resistance materially (*Weatherwax* and *Stamm*, 1945). It also imparts considerable acid resistance to wood (*Stamm* and *Seborg*, 1939) but no alkali resistance. The latter can be obtained by forming a furfuryl alcohol resin within the wood structure (2.5.6.8) (*Goldstein*, 1955).

Phenolic resin treatment does not impart a true fire resistance to wood, but it does improve the integrity of the char thus cutting down on fire spread (*Stamm* and *Seborg*, 1939). Impreg is considerably more heat resistant then normal wood. Treated specimens have been subjected to the cycle of heating to 205 °C followed by cooling more than 50 times without visual harm (*Seborg* and *Vallier*, 1959),

Table 2.8. Strength Properties of Parallel-laminated Birch and Birch Impreg Made from 17 Plies of 1/16-in.-thick-rotary-cut Veneer (*Erickson*, 1947)[1]

Property	Untreated controls[2]	Impreg[3]
Specific gravity	0.67	0.9
Tension parallel to grain:		
Proportional limit (psi)	14,800	13,800
Ultimate (psi)	22,200	17,000
Modulus if elasticity (10^6 psi)	2.3	2.5
Compression parallel to grain:		
Proportional limit (psi)	6,400	7,600
Ultimate (psi)	9,800	15,600
Modulus of elasticity (10^6 psi)	2.3	2.6
Flexure, grain parallel to span	11,500	14,600
Proportional limit (psi)	11,500	14,600
Modulus of rupture (psi)	20,400	20,700
Modulus of elasticity (10^6 spi)	2.3	2.5
Shear parallel to grain and perpendicular to laminations (psi)	2,800	2,580
Toughness (Forest Products Lab. Test, 1956) (in.-lb.)	215	151
Izod impact (ft.-lb. per in. of notch)	10···12	1···3

[1] All values, average of 12 tests.
[2] Assembled with TEGO-film glue at 175 psi, moisture content at test 9 to 10%.
[3] Resin content, 56% on basis of dry weight of untreated wood.

whereas untreated controls charred and disintegrated badly under the same conditions after only a few cycles.

The strength properties of phenolic resin impregnated, parallel laminated, birch veneer together with untreated controls are given in Table 2.8. The treatment causes a slight loss in tensile properties, an increase in compression properties and virtually no change in flexural properties. The treatment causes a slight loss in shear and a large loss in both toughness and Izod impact. The latter is the only serious loss (*Erickson*, 1947). When the phenolic resin content of the wood is reduced to 25 to 35% the Izod impact will be raised to 2 to 5 feet pounds per inch of notch. Impreg hence should not be used in applications where the impact strength is critical.

2.5.6.5.3 Uses for Impreg. Impreg is chiefly used for automobile die models (*Seborg* and *Vallier*, 1954; *Stamm*, 1962). These models of all of the body surfaces are made in full size and must fit together regardless of the relative humidity. Parallel laminates of phenolic resin treated cativo veneer are hot pressed to one inch thick panels. These are assembled with cold setting glue to the desired thickness and then carved. Fig. 2.9 is a photograph of a car top model that was put through several relative humidity cycles without significant dimension changes, grain raising or roughening of the surface. Impreg is also used for various shell molding dies where its good resistance to heat is utilizied. The heated models are imbedded in sand containing a heat setting resin. The model is removed from the sand bed after curing and cooling, and metal cast in the sand cavity. The treated model can be reused up to 50 times.

Fig. 2.9. Automobile top die model made from mahogany impreg that was subjected to relative humidity cycling together with a model made from normal mahogany lumber. The Impreg model gave far less dimension changes than the conventional model and maintained a for better surface. Impreg die models are almost exclusively used in the United States at the present time. From *Seborg* and *Vallier* (1954)

2.5.6.6 Vinyl Resin Treatment. Considerable interest has developed in recent years in polymerizing various vinyl resins in cellulosic materials. The interest arose largely because these resins can be formed from their monomers by gamma ray irradiation which generates free radicals that act as exitation reaction sites in the system (*Chapiro* and *Stannett*, 1960; *Chapiro*, 1962). The term "graft polymerization" is frequently used to indicate a copolymerization reaction of the monomer or growing polymer with the hydrophilic material treated. The term is misleading when applied to cellulosic materials that have large microscopically visible void volumes. Polymers formed in this void structure are readily extracted with polymer solvents indicating that they are in the homopolymer form (*Huang* and *Rapson*, 1963; *Kent* et al., 1962, 1965). Not even in the case of polymers formed within the cell walls from swelling monomers, such as acrylonitrile where 82% of the polymer could not be extracted with a polymer solvent, is there assurance that grafting took place. If polymerization within the cell walls was carried to the stage in which the molecules were too big to diffuse out from the cell walls (*Tarkow*, *Feist* and *Southerland*, 1966) the same insolubility could have ocurred without chemical attachment to the cellulosic material.

Although most of the fundamental research on forming vinyl polymers in cellulosic materials has been confined to cotton fibers, textiles, cellophane and paper, considerable applied research has been carried out on wood in an attempt to produce an improved modified wood product (*Kenaga*; *Kenaga* et al.; *Kent* et al.; *Ramalingham* et al.; *Siau* et al.). The chief monomer used was styrene which swells wood only slightly on prolonged soaking (*Siau*, 1969). Because of this, when styrene was the sole impregnant, it imparted a negligible dimensional stability to wood after polymerization in the structure. Monomeric methyl methacrylate swelled the wood somewhat more than styrene (*Siau*, 1969) but still imparted only a small dimensional stability to the wood after polymerization. Acrylonitrile proved to be a much better swelling agent for wood and imparted considerable dimensional stability to the polymerized product. When swelling agents for wood that are miscible with the monomer are added to the monomer, such as methyl and ethyl alcohol, dioxane, formamide, dimethyl formamide and diethylene glycol, a wood-polymer product with antishrink efficiencies of as high as 74% was obtained (*Kenaga* et al., 1962; *Siau* et al., 1965). Mixtures of monomers have also been used. A mixture of styrene and acrylonitrile gave a high degree of swelling (*Loos* and *Robinson*, 1968) and a product at high loading with an antishrink efficiency of 65% (*Ellwood* et al., 1969). Phosgard, a fire retardant phosphorous compound, has been added to styrene, to acrylonitrile (*Kenaga*, 1963 (2)) and to methyl methacrylate (*Ellwood* et al., 1969) successfully.

Impregnation of solid wood has in general been accomplished by evacuating the wood specimens in a treating cylinder, then running in the pure monomer or monomer plus swelling agent, followed in the case of large specimens by the application of pressure. Preevacuation is desirable not only from the standpoint of aiding penetration but also from the standpoint of eliminating oxygen from the system as it acts as an inhibitor for subsequent polymerization by radiation. Modification of this procedure to attain virtually full loading, and reasonably more uniform distribution at partial loadings, are discussed by *Loos* et al. (1967), and *Meyer* and *Loos* (1969). They used partial vacuum, atmospheric pressure or an initial gas pressure in the wood followed by pressure impregnation, various holding times and in some cases post evacuation, or draining the cylinder followed by a post application of nitrogen pressure to attain different levels of loading and improved distribution. The conditions will vary with the size, shape, and extent of end grain, the species, whether sapwood or heartwood, the specific gravity of the wood and the viscosity of the treating liquid. Even when all of these factors are taken into account, only roughly constant degrees of loading can be obtained with solid wood due to the high degree of variability of wood.

Practically all of the monomers used are volatile. Because of this, some of the investigators wrapped their impregnated specimens in aluminum foil prior to irradiation. Avoiding resin depleted surfaces of the wood has been more effectively accomplished by using less volatile monomers such as tributyl styrene in place of styrene (*Kenaga*, 1970).

Irradiation by gamma rays has been carried out with cobalt-60 sources of different intensity. Dose rates and times of irradiation varied with the size and shape of the sample and its position relative to the source. Care should be exercised to avoid over irradiation, as the higher total doses of irradiation have a deleterious effect upon the mechanical properties of the product. *Loos* (1962) has shown that the toughness of dry unimpregnated irradiated wood increases slightly with an increase in the irradiation dose up to a total dose of 0.1 megarad, then decreases to the control toughness at a total dose of 1.0 megarads and decreases drastically in going to a total dose of 10 megarads. Even though the presence of monomer may decrease the degradative effect on the wood, it appears desirable to limit the total dose to only a few megarads when toughness is critical. Tensile tests on matched specimens of yellow poplar solvent exchanged from ethanol to styrene and cured at total dose of 4 to 9.8 megarads increased in both tensile strength and modulus of elasticity up to a total dose of 7 megarads followed by no increase indicating that wood degradation is probably showing its effect. Toughness being more sensitive to thermal degradation than tensile strength and modulus of elasticity in tension, it is not surprising for degradative effects to appear at lower levels of irradiation in the toughness test.

Free radicals, like those needed to initiate the polymerization of vinyl resins, are produced on heating benzoyl peroxide thus breaking the -0-0- bond giving two free phenyl radicals and carbon dioxide. These free phenyl radicals have been shown to promote vinyl resin polymerization in a manner similar to gamma ray initiated polymerization (*Meyer*, 1965). Matched specimens of soft maple were impregnated with methyl methacrylate monomer, part of which contained 0.2% of benzoyl peroxide. The specimens containing the catalyst were heated in an oven at 68 °C for 12 h. Part of these were initially brought to 68 °C by radio frequency in order to bring them to the reaction temperature more rapidly, followed by oven heating. The uncatalyzed specimens were irradiated with a total dose of 5 megarads. Compression parallel to the grain tests gave practically the same increase of about 18% for the proportional limit by both methods of heating the catalyzed specimens and for the irradiated specimens. Likewise the modulus of rupture was increased by about 19% and the end hardness by about five fold in all three cases. Use of 0.2 % of catalyst must thus be practically equivalent to irradiation with 5 megarads of radiation.

2.5.6.6.1 *Strength Properties.* Most of the available strength properties of wood-vinyl resin composites are limited in scope. Because of this the U. S. Atomic Energy Commission sponsored an extensive testing program in which the University of West Virginia treated and irradiated specimens matched and prepared by the Department of Wood and Paper Science of North Carolina State University. The latter agency made all of the physical and mechanical tests and the Research Triangle Institute statistically analysed the results. This three-institution project involved four species of wood namely, Loblolly pine (*Pinus taeda*), eastern white pine (*Pinus strobus*), yellow poplar (*Liriodendren tulipifera*) and northern red oak (*Quercus rubra*), four treating monomers namely methyl methacrylate, styrene 60% — acrylonitrile 40%, methyl methacrylate 88% — phosgard 12%, and ethyl acrylate 80% — acrylonitrile 20%. The first three set to hard resins, the fourth to a rubbery resin. Swelling agents were not added to any of the systems. A third variable was the extent of loading of the wood with polymer, expressed as the actual loading divided by the maximum loading. It was originally planned to make all tests at zero, 1/3, 2/3 and full loading. The loading turned out to be so variable that it complicated the analysis of the data. Eight different tests were made on the material, namely static bending, vibration (dynamic modulus of elasticity), compression perpendicular to grain, tension perpendicular to grain, toughness

(U. S. Forest Products Lab., 1956) hardness (both end and side), abrasion (U. S. Navy method) and rate and equilibrium dimension changes with changes in relative humidity. Either 512 or 1,024 mechanical tests were made by each method. All specimens, except controls, were irradiated at the rate of about 0.25 megarads per hour for 20 h.

Table 2.9 gives a summary of the results for Loblolly pine and yellow poplar treated with both styrene-acrylonitrile and methyl methacrylate at one quarter and full loading. Except for toughness, there is an increase in strength properties resulting from forming of polymer within the structure which in all cases increases with the extent of loading (*Ellwood* et al., 1969). The loss in toughness at an irradiation dose of 5 megarads would be expected on the basis of the effect of irradiation on the toughness of unimpregnated wood just given (*Loos*, 1962). The compression perpendicular to the grain is increased up to 5 fold, and the side hardness up to 7 fold. Good antiswell efficiencies were obtained at high levels of loading with the styrene-acrylonitrile mixture as would be expected from the degree of swelling in the monomer mixture (*Loos* and *Robinson*, 1968). The antiswell efficiencies obtained with methyl methacrylate were better than would be expected on the basis of swelling in the monomer. In both cases the appreciable improvement in antiswell efficiency in going from one quarter to full loading was not anticipated. The difference between the values at these two levels of loading cannot be due to bulking. True grafting at lumen surfaces may, however, greatly retard entrance of moisture into the fiber far better than more deposition of solids in the lumen.

The chief advantage of vinyl resin treated wood over phenol-formaldehyde treated wood is in the toughness and abrasion resistance of the product at high levels of loading. Because of the cost of high loading, the potential usefulness of the treatment appears to be in the treatment of only the surface plies for furniture and flooring. The step of impregnating with monomer would be greatly simplified in the case of veneer. The use of low volatility monomers in a volatile swelling solvent would avoid depletion of resin at the surfaces and insure dimensional stability. Curing of the resin with benzoyl peroxide and heat could be carried out in a press simultaneously with assembly of the completed panel.

2.5.6.7 Acetylated Wood. Dimensional stabilization of wood by the various resin treatments is not dependent upon the resin actually reacting with the wood. Merely depositing water insoluble resins within the cell walls is needed. Acetylation, on the other hand, depends upon a chemical reaction in which hydroxyl groups of the wood components, that are not active in holding the structure together, are chemically replaced by acetyl groups. The reaction was first carried out by soaking 1/8 in. thick cross sections of wood in a mixture of acetic anhydride and pyridine (*Stamm* and *Tarkow*, 1947) or acetic anhydride and dimethyl formamide (*Clermont* and *Bender*, 1957) followed by heating in sealed glass jars to 90 to 100 °C for several hours, leaching the specimens to remove unreacted chemical and then drying. Weight increases of 18 percent for hardwoods and 25% for softwoods due to the acetylation were readily attained giving antishrink efficiencies up to 70 percent.

In a similar manner wood has been butyrylated with butyric anhydride (*Stamm* and *Tarkow*, 1947). The data, when compared with the data for acetylation, show that the degree of dimensional stability attained is not a function of the number of hydroxyl groups replaced by ester groups, but rather a function of the partial specific volume of the ester groups added within the cell walls. This finding puts acetylated wood into the bulking type of treatment.

Table 2.9. Properties of Vinyl Resin Impregnated Wood

	Loblolly pine (sp. gr. 0.59 to 0.62)					Yellow poplar (sp. gr. 0.50 to 0.58)				
	Loading[1]	SA[2]		MMA[3]		Loading[1]	SA[4]		MMA[5]	
	Control	0.25	1.0	0.25	1.0	Control	0.25	1.0	0.25	1.0
Static bending										
Proportional limit (10^4 psi)	0.63	0.95	1.11	0.80	0.93	0.75	0.97	1.40	1.05	1.85
Modulus of rupture (10^4 psi)	1.37	1.62	2.55	1.84	2.16	1.41	1.59	2.22	1.59	2.34
Modulus of elasticity (10^2 psi)	1.82	1.97	2.41	2.08	2.26	1.94	1.96	2.23	1.94	2.18
Compression perpendicular to grain										
Stress at proportional limit (10^2 psi)	0.94	1.56	4.29	1.47	3.84	0.64	1.02	3.46	1.01	3.58
Tension perpendicular										
Av. radial and tangential (20^2 psi)	3.97	3.53	5.45	3.72	5.43	5.75	4.62	6.82	5.08	7.21
Hardness[6]										
End grain (lb.)	411	592.0	1,067	629	1,454	389	573	1,388	565	1,440
Side grain (lb.)	177	255.0	709	275	1,260	177	305	1,079	265	1,220
Toughness (in.-lb.)[7]	324	257.0	283	253	308	167	125	183	183	278
Relative abrasion resistance[8]	1	1.06	3.52	1.21	1.79	1	1.07	3.10	1.06	2.29
Antiswell efficiency (%)[9]	0	27.5	64.5	16.5	38.5	0	31.0	61.0	22.5	41

[1] Loading: fraction of void volume filled.
[2] SA: 60% Styrene 40% acrylonitrile: multiply by 0.83 to convert to relative weights.
[3] MMA: methyl methacrylate: multiply by 0.91 to convert to relative weights.
[4] Multiply by 0.96 to convert to relative weights.
[5] Multiply by 1.05 to convert to relative weights.
[6] Load to penetrate 0.444 in. diameter ball to depth of 0.050 in.
[7] Forest Products Lab. method, 1956.
[8] U. S. Navy abrador.
[9] Av. for end and center 1/8 in. thick cross sections, 30 to 85% relative humidtity.

Other ester forming reactions have been carried out in wood including allylation, crotylation, and phthaloylation (*Risi* and *Arseneau*, 1947, 1958). None of these reactions show any advantage over acetylation, and the groups tend to hydrolyze off more readily.

The liquid phase acetylation of wood with acetic anhydride proved to be wasteful of chemical, so vapor phase treatments were tried. Birch veneer 1/16 in. thick gave a 20% permanent weight increase on heating the veneer over a mixture of 20% pyridine and 80% acetic anhydride at 90 °C for six hours, resulting in a 70% antishrink efficiency. Sitka spruce veneer required heating for ten hours to attain the acetyl content of 26% required to give the same antishrink efficiency of 70% (*Stamm* and *Tarkow*, 1947).

Vapor phase acetylation was carried out on one foot square sheets of veneer in a sealed circulating tunnel made of thin gauge stainless stell. Heated air was passed over a shallow pan filled with the acetylating mix, then across the faces of the vertically hung sheets of veneer and back over the pan. Good acetylation occurred, but it was slower than desired as the equipment could not be operated under pressure. Unfortunately the equipment became so badly corroded in the few months that it was used that it had to be discarded. Corrosion of equipment is thus a critical factor in the process. It was later found that aluminium equipment can be used (*Stamm* and *Beasily*, 1961). It rapidly builds up an oxide coating that is quite resistant to further attack.

Small scale vapor phase acetylation has been carried out in an aluminium cell that could withstand several atmospheres of pressure (*Baird*, 1969). White pine cross sections when heated over acetic anhydride for one hour at 130 °C gave a permanent gain in weight of 20.5% and an antishrink efficiency of 56%. Original moisture contents of the wood ranging from 0 to 11% had little effect upon the antishrink efficiencies. When 15% of dimethyl formamide was added to the treating solution, weight increases up to 28% and antishrink efficiencies up to 72% were obtained in one hour.

Tests were made to determine if the acetyl groups could be hydrolyzed off from acetylated wood in service (*Stamm* and *Tarkow*, 1947). Ten cycles of relative humidity change between 30 and 90% at 27 °C over a period of four months showed no loss in antishrink efficiency. Soaking in a 9% solution of sulfuric acid for 18 hours at 25 °C did not reduce the dimensional stability. When soaked in the acid at 40 °C, the antishrink efficiency dropped from 75 to 65%. Exposure of acetylated birch panels in the warm salty water of the Gulf of Mexico for a year showed no attack by *Teredo* and no loss in antishrink efficiency whereas the untreated controls were badly attacked. Acetylated birch stakes inserted in termite infested soil for five years showed no sign of attack. Sitka spruce acetylated to give a 70% antishrink efficiency when exposed to *Lenzites trabea* in a soil-block culture test for 3 months showed a negligible loss in weight compared to a 47% loss in weight of the untreated controls (*Stamm* and *Baechlor*, 1960). Acetylated wood thus shows a high degree of permanence in its stability and a high resistance to attack by organisms.

Douglas fir plywood with acetylated faces, when exposed to the weather on a test fence for two years without a surface finish, showed only a very slight roughening and checking. The untreated controls weathered and checked badly (*Tarkow*, *Stamm* and *Erickson*, 1955). The weathering of out-of-doors paints over acetylated wood was considerably better than over controls and appear somewhat better than over phenolic resin treated wood. Presurface acetylation also appears to improve the weathering properties of painted wood (*Campbell*, 1966, 1969).

Table 2.10 gives some of the strength properties of acetylated veneer. Most of the strength properties are unaffected by acetylation or slightly increased (*Tarkow, Stamm* and *Erickson*, 1955). In another study (*Baird*, 1969), the toughness of white pine, having a 23.3% weight increase due to acetylation and an antishrink efficiency of 67%, gave a relative toughness of 0.93 and a relative abrasion resistance of 1.09.

Table 2.10. Effect of Acetylation on the Properties of 1/16 in. Thick Rotary Cut Veneer
(*Tarkow, Stamm* and *Erickson*, 1955)

Property	Sitka spruce		Yellow birch		Basswood	
	Control	Acetylated	Control	Acetylated	Control	Acetylated
Acetyl content	0	31	0	22	0	20
Antishrink efficiency	0	71	0	72	0	73
Specific gravity (dry volume)	0.36	0.42	0.62	0.63	0.32	0.34
Tension parallel to grain[1]						
Ultimate strength (10^4 psi)	1.39	1.53	1.44	1.50	—	—
Modulus of elasticity (10^6 psi)	1.65	1.84	1.84	2.27	—	—
Static bending parallel to span[2]						
Modulus of rupture (10^4 psi)	0.81	1.14	1.56	1.92	0.74	0.34
Modulus of elasticity (10^6 psi)	1.54	1.94	2.64	2.75	1.60	1.56
Relative toughness[3]	1.0	1.72	1.0	1.04	1.0	1.28

[1] Average of 3 specimens.
[2] Average of 6 to 10 specimens.
[3] Forest Products Lab., 1956: average of 12 specimens.

Acetylation of unbleached kraft paper using pyridine and various salt catalysts gave antishrink efficiencies up to 50%. Wet breaking lengths were greatly increased by acetylation. Dry breaking lengths, burst and tear were adversely affected only when the antishrink efficiency exceeded 50%. Fold was drastically reduced in some cases at considerably lower antishrink efficiencies. (*Stamm* and *Beasley*, 1961), again indicating that the fold factor is the most seriously affected strength property of paper subjected to a dimension stabilizing treatment.

Liquid phase acetylation of wood was carried out on a pilot plant scale by the Koppers Co., Pittsburgh, Pa., for several years but was finally abandoned for economic reasons. It is felt that this most favorable means of dimensionally stabilizing wood from the standpoint of the accompanying properties deserves further commercial consideration both for liquid and vapor phase methods. Substituting aluminum alloy equipment for stainless steel is also well worth considering.

2.5.6.8 Other Treatments. Furfuryl alcohol resin has been successfully formed in the cell walls of wood to give the wood dimensional stability and alkali resistance (*Goldstein*, 1955). Furfuryl alcohol swells dry wood very slow so that it is desirable to have some water or other wood swelling chemical present. The reaction requires an acid catalyst and should be carried out in a confining treating cylinder to avoid excessive loss of the volatile chemical. Great precautions have to be taken to avoid the use of an excess of acid as the polymerization may take place even at room temperature with explosive violence. The product is always dark brown to black. Wood was commercially treated this way for a short period of time, but was finally abandoned for economic reasons.

Wood has been dimensionally stabilized by a vapor phase bulking reaction with ethylene oxide with trimethylamine present to open up the structure and serve as catalyst (*McMillin*, 1963). Wood specimens 1.76 by 1.75 by 1.0 in. in the fiber direction were placed in an autoclave heated to 95 °C. Air was removed from the system by evacuation, trimethyl amine preheated to 65 °C was introduced into the autoclave up to a pressure of 1 psi absolute and held for 15 min. Ethylene oxide preheated to 65 °C was then introduced into the system under a pressure of 50 psi and held for various periods of time followed by venting and applying a vacuum to remove unreacted gases. Weight increase up to 20 to 30% were obtained at the higher levels of treatment giving antishrink efficiencies up to 65%.

Reactions of wood with various isocyanates in the vapor phase have been investigated as a means of imparting dimensional stability to the wood (*Baird*, 1969). Isocyanates swell wood but slightly so that it is necessary to use an accompanying swelling agent to open up the structure. Dimethyl formamide proved to be suitable for this purpose. Butyl isocyanate proved to be more suitable than ethyl, t-butyl and phenyl isocyanates giving antishrink efficiencies up to 78% for a weight increase of 49% in two hour exposure at 130 °C. Toughness and abrasion resistance, however, were reduced to 72 and 75% of the values for the controls.

2.6 Densified Wood

2.6.1 Densifying by Impregnation

Wood can be densified and otherwise modified by impregnating its void structure with any substance that can be solidified within the structure. This can be done not only by polymerizing phenol-formaldehyde resins (see 2.8.6.5) and liquid vinyl impregnants within the structure (see 2.5.6.6) but also by impregnating the voids of the wood with molten natural resins, waxes, sulfur and even low fusion metals, followed by cooling to solidify the impregnant. Such impregnations improve chiefly the compressive strength and the hardness of wood. In some cases the hardness of wood is increased to such a high degree, when the voids are filled with impregnants, that the normal hardness test is not applicable. Instead of determining the load to imbed a 1.11 cm diameter steel ball to one-half of its diameter, a modulus of hardness is determined by measuring the area of penetration under different loads, and calculating the load per unit of cross section penetrated (*Weatherwax*, *Erickson* and *Stamm*, 1948). Table 2.11 gives the relative face modulus of hardness of four different woods loaded with different impregnants compared with phenolic resins impregnated wood, Impreg (see 2.5.6.5), phenolic resin impregnated compressed wood, Compreg (see 2.6.2.2), and stable untreated compressed wood, Staypak (see 2.6.2.3). Although solidified impregnants alone increase the hardness of wood much more than the increase in specific gravity, they are not as effective in increasing the hardness as by compressions.

Table 2.11. Dry Volume Specific Gravity and Relative Face Modulus of Hardness of Four Species of Wood Highly Impregnated with Vinyl Resins, Molten Sulfur and Molten Metal Compared with Impreg and Compreg Containing 30% of Phenolic Resin and Stable Untreated Compressed Wood, Staypak (*Ellwood* et al., 1969; *Stamm*, 1964, Table 20.3)

Species	Treatment	Sp. gr.	Relative face modulus of hardness
Douglas fir	none	0.60	1.0
	Impreg	0.67	1.3
	Compreg	1.35	19.8
	Staypak	1.32	11.5
	Sulfur	1.30	3.5
	Low melting alloy[1]	6.14	7.4
Yellow birch	none	0.61	1.0
	Impreg	0.79	2.0
	Compreg	1.30	10.0
	Staypak	1.36	11.6
	Sulfur	0.98	2.6
	Low melting alloy[1]	4.31	2.9
Loblolly pine	none	0.60	1.0
	Methylmethacrylate		7.0
	Styrene acrylonitrile[2]		4.0
Yellow poplar	none	0.57	1.0
	Methylmethacrylate		6.9
	Styreneacrylonitrile[2]		6.0

[1] Low melting alloy of bismuth, 50%, lead, 31.2%, and tin, 18.8%.

[2] 60% styrene-40% acrylonitrile.

2.6.1.1 Metalized Wood. Metalized wood was first made in Germany based on the proposal of *H. Schmidt* of the former Kaiser Wilhelm Institute for Iron Research at Düsseldorf. German patents No. DRP 493,905 and 506,477 were granted and made public by *Naeser* (1930) and *Martell* (1930). Specimens of permeable wood as large as $40 \times 10 \times 5$ cm were impregnated to various degrees. The process worked best on the sapwood of such ring porous species as beech. The vessels treated most readily. Wood ray cells did not become impregnated. Walnut, with a normal dry volume specific gravity of 0.6 or less, was increased in specific gravity to between 0.95 and 3.83 depending on the treating conditions and specimen size. The specific gravity of pine was increased to as much as 4.83. Hardness increases of two to three fold were obtained, in fact hardnesses, in some cases, exceeded those of the pure metals. This appears to be due to the wood structure minimizing creep of the softer metals.

High degrees of impregnation of small specimens of wood have been attained with an alloy consisting of 50% bismuth, 31.2% lead and 18.8% of tin that melts at 97 °C. This particular alloy was used because it expands about 2% upon solidification from the molten state. It was used to seal off the voids of wood in bound water diffusion studies (*Stamm*, 1959). More than 90% of the voids in specimens $5 \times 5 \times 0.3$ cm were filled by the following technique. Oven dry specimens were placed on the solidified treating metal in a pressure bomb with a non-melting metal weight on top. The bomb was sealed, highly evacuated and then heated to 130 to 150 °C to melt the treating metal. The weight on top of the specimen forced the specimen below the surface of the molten metal. The vacuum was released and gas pressure ranging from 600 to 2,500 psi was applied for 20 min up to one hour. The pressure was released, the bomb opened and partly cooled. Just before the molten metal solidified, the specimen was removed and the surfaces scraped free of adhering metal. This technique was used in impregnating the specimens of Table 2.11 with both molten metal and with sulfur.

Metalized wood will char when exposed to fire but will not burn with a flame until a large part of the metal has been melted and expelled from the structure as a result of expansion of the contained air. During this initial period the temperature is kept down as a result of the heat of fusion of the metal and its high thermal conductivity.

The anisotropic properties of metalized wood are changed considerably when the wood is only partially impregnated. The thermal conductivity is increased considerably more in the fiber direction than in the transverse direction. Ratios of longitudinal conductivity to transverse conductivity have been increased from the normal wood values of 2:1 to as much as 10:1. The electrical conductivity is quite similarly affected.

During World War II metalized wood, containing about 3% of lubricating oil, was used in Germany for ship screw bearings in place of lignum vitae, which is the densest of woods high in lubricating resins. Even without a lubricant, frictional adhesion is minimized when the shaft and bearing are made of dissimilar metals. Lead would be an ideal bearing material for steel shafts if lead did not tend to creep. The wood shell prevents this creep. Present uses for metalized wood are quite limited, but it has interesting possibilities for future specialty uses.

2.6.2 Densifying by Compression

Wood can be densified and its properties modified not only by filling its void volume with polymers, molten sulfur, or molten metals (see 2.6.1) but also by compressing it under conditions such that the structure is not damaged. Compressed solid wood has been made in Germany since the early nineteen thirties (*Kollmann*, 1936) and marketed under the trade name of Lignostone. Laminated compressed wood is marketed under the trade name Lignofol. A resin treated laminated compressed wood known as Kunstharzschichtholz has also been in commercial production. These densified materials are used for textile shuttles, bobbins and picker sticks, for mallet heads, for forming jigs and for various tool handles.

Early United States patents for compressed wood are as follows: *Sears*, 1900; *Walch* and *Watts*, 1923; *Olesheimer*, 1929; *Brossman*, 1931; *Esselen*, 1934; and *Olson*, 1934. The latter differs from the others in that the wood is compressed circumferentially by driving cylindrical pieces of wood through tapering dies. These patents are concerned almost entirely with the mechanics of compression and did not adequately consider plasticization of the wood or stabilization of their product. Because of this none of the patents were put into continuous use.

2.6.2.1 Plasticizing Wood. Both moisture and heat act as plasticizers for wood. This is illustrated by measurements of the compression of hickory discs, that just fit into compression molds so as to avoid lateral movement. Fig. 2.10 gives the fractional thickness resulting from compression versus the applied pressure at three different moisture contents and three different pressures (*Seborg* and *Stamm*, 1941). The compression, per unit of applied pressure, increases to a maximum at the inflection points (maximum yield points) and then decreases to zero as the maximum compression is approached. Compressing at 6% moisture content and 160 °C, and at 26% moisture content and 26 °C give quite similar curves with the maximum yield point for each occurring at 1,700 psi. Increasing either the moisture content or the temperature shifts the maximum yield point to lower pressures. From a practical use standpoint it is desirable to make compressed wood at a

moisture content close to that of average use conditions (6 to 8% for in-door use
and 6 to 12% for out-of-door use) to avoid subsequent shrinkage and face checking.
It is thus preferable to increase the temperature rather than the moisture content
to obtain increased plasticity. It is, however, undesirable to press at temperatures
much above 160 °C as heating dry wood to 186° C for one hour in a press in the
presence of air will cause about 0.5 % weight loss from the wood substance, (see
Fig. 2.1) resulting in an appreciable loss in toughness (*Stamm*, 1964, Chapter 18,
Fig. 5). All of the curves for heated hickory specimens in Fig. 2.10 are approaching
the ultimate compression at 4,000 psi.

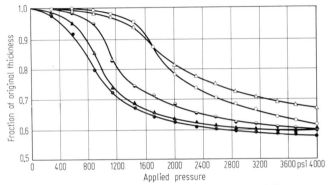

Fig. 2.10. Fraction of the original thickness to which 1 cm thick hickory sapwood discs were compressed at different moisture contents and temperatures versus the applied pressure. From *Seborg* and *Stamm* (1941)
Open triangles: 26% moisture content and 26 °C, Open circles: 6% moisture content and 160 °C, Bottom shaded circles: 12% moisture content and 160 °C, Shaded circles: 26% moisture content and 160 °C, Shaded triangles: 26% moisture content and 130 °C

Veneer of five different softwoods and four hardwoods at 6% moisture content
was hot-pressed at 149 °C in making parallel laminates and plywood at four different pressures ranging from 250 to 1,000 psi. Plots of the specific gravity of the
product against the applied pressure gave sigmoid curves (*Lloyd* and *Stamm*,
1958). The data show that both the pressure to obtain maximum plastic yield,
and to obtain maximum compression, increase with an increase in the original
specific gravity of the wood.

When the veneer was impregnated with a cell wall penetrating phenol-formaldehyde resin forming mix, followed by drying under non-curing conditions
(see 2.5.6.5.1), compression of the assemblies in a hot press was appreciably increased.
The maximum plastic yield occurred almost at the initial application of pressure
with all five of the softwood assemblies. At 500 psi the softwood laminates compressed to a dry volume specific gravity of 1.2, and at 1,000 psi they compressed
to a specific gravity of about 1.3, where 1.4 is approximately the ultimate compression. At 500 psi the compression of the four hardwoods was appreciably less
than for the softwoods, but at 1,000 psi it was practically the same. If, however,
the resin in the plies was precured by heating, the softwoods compressed to specific gravities of only 0.85 to 0.95 at 1,000 psi. It is thus seen that uncured
phenol-formaldehyde resin forming mixes plasticizes the wood at elevated temperatures far more than water, while precured resin within the plies has the
reverse effect. This has been taken advantage of in making resin treated laminates
with compressed faces on uncompressed cores in a single compression and assembly
operation (*Stamm* and *Seborg*, 1944).

The plasticizing effect of various organic liquids and aqueous solutions on
Norway pine has been investigated by measuring their effect upon the compressive
strength (*Erickson* and *Rees*, 1940). Benzene and other non-wood-swelling hydro-

carbons have a negligible effect upon the compressive strength of drywood. Liquids that swell wood less than it is swollen in water decrease the compressive strength of wood by virtually the same amount that it is decreased by a moisture content that produces the same amount of swelling. Liquids and aqueous solutions that swell wood more than it is swollen in water (*Stamm*, 1964, Chapter 15), cause a greater loss in compressive strength than by water alone. Apparent exceptions are chloral hydrate and tannic acid. This could be due to their reacting chemically with the wood. In general the compressive strength of wood decreases as the extent to which it is swollen increases. This would be expected on the basis of swelling causing the breakage of hydrogen bonds between the structural units of wood.

2.6.2.1.1 Bending Wood. Wood is normally bent, after presteaming, in various types of strapping devices that minimize the effect of tension on the convex surface and maximize compressive yield on the concave surface, thus minimizing breakage. This subject is covered in Vol. I, 9.8 and also in the Wood Handbook (1955, pp. 299—309). Hardwoods, in general bend more readily than softwoods. This is probably due to the fact that the lignin in hardwoods is more plastic and solvent soluble than in softwoods (*Guss*, 1945).

Chemical plasticization of wood has been studied as a means of facilitating the bending of wood (*Loughborough*, 1942, 1943). Wood impregnated with a concentrated aqueous solution of urea is swollen about 10 percent more than it is swollen in water, on an external volume basis and considerably more on a cell wall volume basis (*Stamm*, 1964, Chapter 15, Fig. 1). Impregnated specimens of hardwoods, one centimeter square, are very plastic at oven temperatures following drying and can be twisted or bent quite easily to severe curvatures. This technique was applied for a short period in bending 2-in. thick oak ribs for wooden mine sweeper boats. Although the bending could be done at lower pressures than steam bending at the same temperature, more failures occurred than with steam bending. This is believed to be due to an uneven distribution of the urea in the wood resulting in non-uniform plasticization.

More recently it has been found that impregnation of wood with liquid ammonia causes a high degree of swelling (*Stamm*, 1955) and a high degree of plasticity to wood (*Schuerch*, 1964; *Bariska, Skaar* and *Davidson*, 1969). Liquid ammonia, unlike water enters the crystallites of cellulose, modifying their crystal lattice, as well as being sorbed in the amorphous regions. Liquid ammoina plasticizes isolated lignin at about −30 °C in contrast to heat plasticization that starts at about 125 °C (*Goring*, 1962, 1963). Because of this plasticizing action of liquid ammonia, small specimens of wood have been given very severe bends while still saturated at temperatures as low as −30 °C. Upon evaporation of the liquid ammonia, the wood shrinks, loses its plasticity, and becomes set in the bent form.

It is still uncertain whether this method of plasticizing wood for bending can be economically applied on a commercial scale in bending of boat ribs or furniture parts. A few preliminary unpublished tests by *Stamm* at North Carolina State University in making a cold pressed veneer by this method did not show any advantages over simple heat plasticization at 150 °C (see Fig. 2.11) and would be considerably more expensive.

2.6.2.2 Resin-Treated Compressed Wood (Compreg)

2.6.2.2.1 Making Compreg. The plasticizing effect of phenol-formaldehyde resin forming mixes on wood has been taken advantage of in making a resin treated compressed wood with specific gravities ranging from 1.2 to 1.35 at pressures of 1,000 psi or less at 125 to 150 °C. Fortunately when heat and pressure are simultaneously applied, the wood responds to compression more rapidly than the resin is cured (*Stamm* and *Seborg*, 1941). Phenol-formaldehyde resin treated compressed wood, frequently called Compreg, is normally made from treated veneer as described in 2.5.6.5.1, rather than solid treated wood not only to avoid treating difficulties, but also to insure subsequent drying without curing of the resin prior to pressing. Parallel laminated Compreg can be made from veneer without the application of glue between the plies when the resin content of the veneer is 35 percent or more as sufficient resin exudes from the plies for assembly (*Stamm* and *Seborg*, 1941). When making cross banded Compreg or when the resin content of the plies is less than 30%, a somewhat reduced glue spread below the normal 15 pounds per 1,000 square test is needed. It is important to predry the impregnated plies to 2 to 4% moisture content after any glue spread and prior to assembly as the equilibrium moisture content of Compreg in use is in this range. The drying temperature-time combination must be such that the resin does not cure. Such conditions range from drying over night at 55 °C in an oven or for 45 minutes at 85 °C in a veneer drier. Relatively thin Compreg panels can be pressed and cured at temperatures up to 150 °C in 10 to 20 min. Thick panels may require temperatures as low as 125 °C to avoid exceeding the exothermic reaction temperature of the resin, above which temperatures have spontaneously risen, without dissipation of heat, to the charring point. Because of this difficulty Compreg panels should not be made in thicknesses exceeding 2 cm. When greater thicknesses are sought, they should be obtained by subsequent lamination of thinner panels.

Compreg can be molded by a so-called "expansion molding" process. Single sheets of dry but uncured phenolic resin treated veneer are rapidly heated to 120 to 150 °C in a hot press or dielectric field without significant further cure of the resin. The hot plies are transferred to a cold press where they are rapidly pressed to one-half to one-third of their original thickness under a pressure of 500 to 1,000 psi. As soon as the plys have cooled to room temperature, they can be removed from the press without springback occurring as the thermosetting resin has set in a thermoplastic sense. This compressed veneer, when kept dry, can be stored for months at room temperature without recovery from compression. The compressed veneer is cut to template sizes, and fitted into a metal mold so as to fill the mold. The mold is locked in a closed position, and heated. The heat softens the resin that has been holding the plies in the compressed condition. The plies thus tend to recover their original thickness but they are restrained from doing so by the mold, thus developing an internal recovery pressure equal to about half of the original compressing pressure. When the temperature time combination suitable for cure of the resin in a thermosetting sense is attained, the mold is cooled and the molded product removed (*Stamm* and *Turner*, 1945). No parting line lip has to be removed as in ordinary compression molding.

2.6.2.2.2 Properties and Uses of Compreg. Compreg is much more dimensionally stable than the best untreated compressed wood. It approaches the dimensional stability of Impreg in the sheet directions but it swells appreciably more in the thickness direction because of the reduced thickness on which dimension changes are based.

Compreg has a natural lustrous finish that can be restored by merely sanding and buffing when cut or scratched. It can be readily cut or turned using metal-working tools operated at slower than normal speeds. Compreg can be glued to Compreg or to normal wood with both hot press phenolic and room temperature-setting resorcinol glues. When thick Compreg panels are face glued to each other, the surfaces should be lightly machined before assembly to assure good contact and a longer open assembly period should be used because of the slower penetration of the glue solvent into the wood.

Compreg is quite resistant to decay and attack by termites and marine borers (*Stamm* and *Seborg*, 1941). Its electrical resistance, like that of Impreg, is far better than that of normal wood. Its acid resistance is better than that of Impreg because of the reduced permeability of the wood. It is more flame resistant than Impreg, largely because of its greater density.

Most of the strength properties of Compreg are increased over those of the wood from which it was made about in proportion to the increase in specific gravity, as shown in Table 2.12. The table gives data for Compreg made both with a fiber penetrating water soluble phenol-formaldehyde resin-forming system and an alcohol soluble more advanced resin that penetrates the cell walls of the wood to a lesser degree. It also gives data for Compreg compressed to two different degrees. The only strength property that is adversely affected by the combined resin treatment and compression is the Izod impact strength. The same is true for toughness determined by the U. S. Forest Products Lab. method (1956) where the reduction was from 215 to 161 in.-lb. The impact and toughness values for Compreg are, however, considerably better than those for Impreg (Table 2.8). Table 2.12 shows that less embrittlement occurs when the less fiber penetrating spirit soluble resin is used. This, however, results in a product with greater swelling. The type of resin to be used will thus depend upon whether dimensional stability or toughness is more important for the particular application. The compressive strength perpendicular to the grain and the hardness of Compreg are improved over the values for the original wood far more than the increase in specific gravity as shown by the modulus of hardness values in Table 2.11. The hardness increase is, in fact, increased considerably more than it is increased by full impregnation with methyl/methacrylate, and styrene-acrylonitrile resins, sulfur or molten metals.

Compreg was used during World War II largely for the roots of wooden aeroplane propellers and for ship screw bearings. Since then its use has been largely limited, because of the cost for forming dies and jigs, weaving shuttles, knife handles, glass door pulls, and railroad track connectors where electrical resistance is needed for an automatic signaling system.

A similar resin treated compressed wood to Compreg has been made in Germany since the early nineteen thirties using either a phenol formaldehyde or cresol formaldehyde resin. Uses have been quite similar to those in the United States. A comparable English product with the trade name "Permali" has been made since World War II largely for electrical insulating connectors for electrical signaling. Spirit soluble phenol-formaldehyde and cresol-formaldehyde resins rather than water soluble resinoids are used.

2.6.2.3 Stable Untreated Compressed Wood (Staypak). The lack of toughness of dimensionally stable Compreg can be attributed to the hardening and stiffening effect of the treating resin within the cell walls of the wood. This suggests eliminating the resin when high toughness is sought. Ordinary compressed wood, however, tends to lose its compression under swelling conditions. It thus appeared desirable to determine the pressing conditions under which springback or recovery from compression is a minimum for untreated compressed wood. Fig. 2.11 is a plot of the maximum permanent recovery from compression of untreated compressed birch laminates under different moisture content and pressing time con-

Table 2.12. Strength Properties of Parallel-laminated Birch Compreg and Semicompreg Made from both Spirit and Water-soluble Resin treated 1/16 in. thick Veneer (*Findley* et al., 1946) and the Corresponding Untreated Compressed Wood, Staypak, pressed at 6% Moisture content and 150 °C (*Erickson*, 1947)

		Untreated uncompressed control	Spirit-soluble resin		Water-soluble resin		Staypak
			Compres	Semi-Compreg	Compreg	Semi-Compreg	
Resin content	%	0	28.5	28.0	29.7	30.2	0
Specific gravity	%	0.67	1.36	1.22	1.36	1.24	1.40
Pressing pressure	psi	175.0	1,500.0	600.0	1,500.0	600.0	2,000.0
Tensile strength parallel to grain	10^4 psi	2.22	5.37	4.36	5.29	4.57	4,37
Modulus of elasticity in tension	10^6 psi	2.30	4.12	3.72	4.21	3.90	4.6
Ultimative compressive strength parallel to grain	10^4 psi	0.98	2.42	2.27	2.56	2.40	2.21
Modulus of elasticity in compression	10^6 psi	2.30	4.21	3.81	4.25	4.0	4.6
Shearing strength parallel to laminations	10^3 psi	2.80	4.84	3.64	4.40	3.92	5.67
Izod impact ft.-lb. per in. of notch		10...12	10.0	8.3	8.6	6.3	11...14
Swelling[1] %		—	14.2	13.9	10.6	11.0	—
Recovery[2] %		—	2.90	2.97	2.02	2.11	—

[1] Of cross section 1/8 inch thick, measured in pressing direction after soaking in water for 48 h.

[2] Increase in thickness of cross sections beyond original oven-dry thickness after drying subsequent to swelling.

ditions at 2,000 psi pressures versus the pressing temperature (*Seborg, Millett* and *Stamm*, 1945; *Stamm, Seborg* and *Millett*, 1948).

The maximum permanent recovery from compression in percent

$$r = \frac{t_r - t_c}{t - t_c} \cdot 100, \qquad (2.22)$$

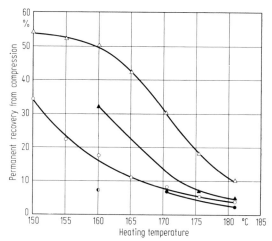

Fig. 2.11. Permanent recovery from compression of yellow poplar parallel laminates made from 21 plies of 1/16 in. thick veneer pressed at different moisture contents and times under pressure of 2,000 psi, versus the heating temperature.
Open triangles: 6% moisture content, 5 min, Shaded triangles: 6% moisture content, 30 min, Open circles: 9% moisture content 30 min, Shaded circles: 12% moisture content, 30 min

where t_r is the thickness of the compressed product after maximum recovery, t_c is the thickness of the compressed product and t is the thickness of the original assembly prior to compression. The thickness after recovery t_r was obtained from a 2 mm thick cross section cut from the specimen that was immersed in water until swelling in the compression direction was complete, followed by air drying and then oven drying and again measuring. The recovery in Fig. 2.11 is chiefly affected by the temperature. Raising the pressing temperature from 150 to 180 °C when the initial moisture content was 6 percent and the time in the press was 5 minutes reduced the maximum permanent recovery from 54 to 10%. Holding the specimens under heat and pressure for 30 rather than 5 minutes caused a further reduction in recovery. Also pressing at a higher initial moisture content had the same effect. All of the specimens approached being completely compressed at the high pressure of 2,000 psi. The recovery thus cannot be a function of the degree of compression. It appears to be a function of the residual stress in the product. When a thin specimen of the compressed wood is immersed in water, both swelling and moisture plasticization occur. The latter allows the internal stresses to be relieved resulting in an additional dimension increase to that due to swelling. On drying, the dimension increase due to reversible swelling, is lost in the form of shrinkage. The remaining irreversible increase in dimensions is due to recovery from compression or springback. Less recovery will, of course, occur if the compressed wood is subsequently exposed to less severe swelling conditions.

Fig. 2.11 shows that when dimensional stability is sought untreated compressed wood should be made at as near use moisture content conditions as possible and at as high a temperature as is compatible with the allowable loss in toughness, holding the product under heat and pressure for a minimum of 30 min. As the moisture needed to adequately plasticize the wood is rapidly lost from the end grain of the wood at these high temperatures, it is advisable to rapidly compress the wood at a temperature near the boiling point of water to seal in the moisture and then raise the temperature to 160 to 170 °C for a minimum of 30 min in making a product 1.5 to 2.0 cm thick (*Stamm, Seborg* and *Millett*, 1948). At temperatures above 125 °C lignin is sufficiently thermoplastic (*Goring*, 1962, 1963) that it is necessary to cool the product in the press to 100 °C or less before releasing the pressure and removing the product. Side restraint should be applied while pressing to prevent lateral spread on the thicker products.

Compressed wood made under conditions that minimize springback are invariably darkened in color except at the ends where some moisture is always lost. Thin cross sections cut from the light colored ends invariably spring back badly after swelling in water and oven drying. Thin cross sections cut from the darkened center of the specimen show little springback. The darkening in color is thus an index of stability. This darkening has been attributed to a slight possible plastic flow of lignin in situ, which tends to relieve the pent up stresses in the wood caused by the compression. Isolated lignin will begin to show appreciable plastic flow at temperatures above 125 °C (*Goring*, 1962, 1963). It is hence not unreasonable to expect lignin in situ to yield sufficiently under pressure at only slightly higher temperatures so as to relieve the pent up stresses. This stable form of compressed wood has been given the name, Staypak. It has the same natural lustrous appearance as Compreg. Its strength properties are in general comparable with those of Compreg, as shown in Table 2.12, except that the toughness (Izod impact) is considerably greater.

Almost all of the untreated compressed wood commercially manufactured at the present time is made at an appreciably lower temperature than Staypak. It is hence not suitable for use under conditions that lead to appreciable springback. It is proving satisfactory, however, for such indoor uses as picker sticks in weaving mills where the prevailing relative humidity is reasonably controlled.

The chief disadvantage of Staypak is the fact that the panels require considerable end and some side trim to remove the unstable light portion. This can be avoided in the case of a laminated product by pretreating the plies with as little as 5 percent of phenolic resin forming mix. This compromise material between Compreg and Staypak is virtually free from springback over the whole panel. It will, however, swell appreciably more than Compreg.

Surface densification of untreated wood has recently been accomplished by a continuous belt press developed by Lam-H-Nard Div., Hoover Ball Bearing Co. of Ann Arbor, Michigan (*Tarkow* and *Seborg*, 1968) in which the wood surface is rapidly heated to as high as 260 °C, immediately pressed at several thousand psi and cooled while still under pressure. The heating, pressing and cooling occur so rapidly that heat, enough for plasticization of the wood, penetrates the wood only a few millimeters. Specific gravities up to 1.0 near or at the surface have been obtained with redwood and maple with a negligible increase in the core. Abrasion resistances of the surface have been increased up to 20 fold. A somewhat similar surface densification method has been developed by the Elmendorf Laboratories of Palo Alto California. These processes impart a natural lustrous surface to wood and should find considerable commercial application.

Literature Cited

Adzumi, H., (1937) On the flow of gases through a porous wall. Bull. Chem. Soc., Japan **12** (6): 304–312.

Am. Soc. Testing Materials, (1961) Surface burning characteristics of building materials. Part 5, Standard E 84: 1178–1184.

Am. Wood Preserver's Assoc. Proc., (1955) Wood preservation statistics **51**: 324.

Am. Wood Preserver's Assoc. Proc., (1967) Wood preservation statistics **60**: 246.

Anderson, A. B., (1946) Chemistry of western pines. Ind. Eng. Chem. **38**: 450–454, 759–761.

—, *Fearing, W. B. Jr.,* (1960) Solvent seasoning of tanoak. For. Prod. J. **10** (5): 234–238.

Baechler, R. H., (1953) Effect of treating variables on absorption and distribution of chemicals in pine posts treated by double diffusion. J. For. Prod. Res. Soc. **3** (12): 170–176.

—, (1954) Double diffusion treatment of wood. Chem. Eng. News **32** (Oct. 25) 4288.

—, (1958) How to treat fence posts by double diffusion. U. S. Dept. of Agr. For. Prod. Lab. Report, 1955.

—, (1959) Improving wood's durability through chemical modification. For. Prod. J. **9** (5): 166–171.

—, *Conway, E., Roth, H. G.,* (1959) Treating hardwood posts by the double diffusion method. For. Prod. J. **9** (7): 216–220.

Baird, B. R., (1969) Dimensional stabilization of wood by vapor phase chemical treatments. Wood and Fiber **1** (1): 54–63.

Bariska, M., Skaar, C., Davidson, R. W., (1960) Studies of the wood-anhydrous ammonia system. Wood Sci. **2** (2): 65–72.

Barkas, W. W., (1947) A discussion of the swelling stresses and sorption hysteresis of plastic gels. Great Brit. Dept. Sci. Ind. Research, For. Prod. Special Report, No. 6.

Barrer, R. M., (1941) Diffusion in and through solids. Cambridge University Press.

Bateson, B. A., (1939) Chem. Trade J. **105** (8) (2724): 93, 98.

Benvenuti, R. R., (1963) An investigation of methods of increasing the permeability of Loblolly pine. M. S. Thesis Dept. Wood Science and Technology, North Carolina State University, Raleigh, N. C.

Bescher, R. H., (1968) Process for impregnating wood with pentachlorophenol and composition therefore. U. S. Patent No. 3200003.

Brenden, J. J., (1965) Effect of fire-retardant and other inorganic salts on pyrolysis products of ponderosa pine. For. Prod. J. **15** (2): 69–72.

Brossman, J. R., (1931) Laminated wood product. U. S. Patent No. 1834895.

Browne, F. L., (1933) Effectiveness of paints in retarding moisture absorption by wood. Ind. Eng. Chem. **25**: 835–842.

—, (1949) Water-repellent preservatives for wood. Architectural Record, Mar: 131–133.

—, *Downs, L. E.,* (1945) A survey of the properties of commercial water repellants and related products. U. S. For. Prod. Lab. Mimeo. R 1495.

Burr, H. K., Stamm, A. J., (1943) Comparison of commercial water-soluble phenol formaldehyde resinoids for wood impregnation, U. S. For. Prod. Lab. Mimeo 1384.

Campbell, G. G., (1966) An investigation of improving the durability of exterior finishes on wood. M. S. Thesis, Dept. Wood and Paper Science, North Carolina State University, Raleigh, North Carolina.

—, (1970) The effect of weathering on the adhesion of selected exterior coatings to wood. Ph. D. Thesis, Department of Wood and Paper Science, North Carolina State University, Raleigh, North Carolina.

Chapiro, A., (1962) Radiation chemistry of polymer systems. Interscience Publishing Co., New York

—, *Stannett, V. T.,* (1960) Radiation grafting to hydrophilic polymers, International J. Applied Radiation and Isotopes **8**: 164–167.

Choong, E. T., Barnes, H. M., (1969) Effect of several wood factors on the dimensional stabilization of southern pine. For. Prod. J. **19** (6): 55–60.

Christensen, G. N., (1960) Kinetics of sorption of water vapor by wood. Australian J. Applied Sci. **11**: 294–304.

Clermont, L. P., Bender, F., (1957) The effect of swelling agents and catalysts on acetylation of wood. For. Prod. J. **7** (5): 167–170.

Cohen, W. E., Stamm, A. J., Fahey, D. J., (1959) Dimensional stabilization of paper by catalized heat treatments. TAPPI **42**: 904–908.

—, —, —, (1959) Dimensional stabilization of paper by cross-linking with formaldehyde. TAPPI **42**: 934–940.

Comstock, G. L., (1963) Moisture diffusion coefficients in wood as calculated from adsorption desorption, and steady state data. For Prod. J. **13** (3): 97–103.

—, (1965) Longitudinal permeability of green eastern hemlock. For. Prod. J. **15** (10): 441–449

Comstock, (1967) Longitudinal permeability of wood to gases nonswelling liquids. For. Prod. J. **17** (10): 41—46.

—, (1968) The relationship between the permeability of green and dry eastern hemlock. For. Prod. J. **18** (8): 20—23.

—, (1970) Directional permeability of softwoods. Wood and Fiber. **1** (4): 283—289.

—, *Côte, W. A.*, (1968) Factors affecting permeability and pit aspiration in coniferous softwoods. Wood Sci. and Tech. **2** (4): 279—291.

Crank, J., (1956) The mathematics of diffusion. Clarendon Press, London.

Daniel, J. H., More, S. T., Segro, N. R., (1962) Graft polymerization of acrylonitrile on cellulose. Tappi **49** (1): 53—57.

Ellwood, E., Gilmore, R., Merrill, J. A., Poole, W. K., (1969) An investigation of certain physical and mechanical properties of wood-plastic combinations. U. S. Atomic Energy Commission Report ORO-638 (RTI-2513-T13).

Erickson, E. C. O., (1958) Mechanical properties of laminated modified wood. U. S. For. Prod. Lab. Memeo, No. 1639, revised.

Erickson, H. D., (1970) Permeability of southern pine wood: a review. Wood Science **2** (2): 149—158.

—, *Rees, L. W.*, (1940) Effect of several chemicals on the swelling and the crushing strength of wood. J. Agr. Research **60**: 593—603.

Eschalier, X., (1906) French Patent No. 374,724 additions 8422 (1906); 9904 (1908); 9905 (1908); 10760 (1909).

Esselen, G. J., (1934) Wood Treatment and product. U. S. Patent No. 1952664.

Findley, W. N., Worley, W. J., Kacalieff, C. D., (1946) Effect of molding pressure and resin on results of shorttime tests and fatigue tests of compreg. Trans. Am. Soc. Mech. Eng. **68**: 317—325.

Forest Products Laboratory's toughness testing machine, (1956) U. S. For. Prod. Lab., Report 1308.

Forest Products News Letter ,(1961) C. S. I. R. O. (Australia) No. 281.

Franzen, A., (1962) Ghost from the depths: the warship Vasa, National Geographic **121** (1): 42—57.

Gardner, R. E., (1965) The auxiliary properties of fire-retardant treated wood, For. Prod. J. **15** (9): 365—368.

Goldstein, I. S., (1955) The impregnation of wood to import resistance to alkali and acid. For. Prod. J. **5** (4): 263—267.

Goodwin, D. R., Hug, R. E., (1961) A new wood preserving process. For. Prod. J. **11** (11): 504—507.

Goring, D. A. I., (1962) The physical chemistry of lignin. Butterworths, London.

—, (1963) Thermal softening of lignin, hemicellulose and cellulose. Pulp and Paper Mag. Can. **64**: T517—527.

Gruntfest, I. J., Geyliardi, D. D., (1948) The modification of cellulose by reaction with formaldehyde. Textile Research J. **18**: 643—649.

Guss, C. O., (1945) Acid hydrolysis of waste wood for use in plastics. U. S. For. Prod. Lab. Mimeo R1481.

Henry, W. T., (1963) A new method of impregnating wood with preservatives. Proc. Am. Wood Preserver's Assoc. **59**: 68—76.

Hittmeier, M. E., (1967) Effect of structural direction and initial moisture content on the swelling rate of wood. Wood Sci. and Tech. **1** (2): 109—121.

Huang, R. Y. M., Rapsen, W. H., (1963) Grafting polymers onto cellulose by high energy radiation. Effect of swelling agents on the gamma-ray induced direct radiation grafting of styrene onto cellulose. J. Polymer Sci. Part C (2): 169—188.

Hudson, M. S., (1942) Treating wood and wood products, apparatus and method for drying wood. U. S. Patents No. 2273039; 2435218; 2435219; 2535925.

Hudson, M. S., (1947) Vapor drying: the artificial seasoning of wood in vapor of organic chemicals. For. Prod. Research Soc. Proc. **1**: 125—146.

Hudson, M. S., (1968) New process for longitudinal treatment of wood. For. Prod. J. **18**(3): 31—35.

—, *Shelton, S. V.*, (1969) Longitudinal flow of liquids in southern pine poles. For. Prod. J. **19** (5): 25—32.

Hunt, G. M., (1930) Effectiveness of moisture-excluding coatings on wood U. S. Dept. Agr. Circular No. 128.

—, *Garratt, G. A.*, (1953) Wood preservation. McGraw Hill Book Co. Inc., New York.

Jost, W., (1952) Diffusion in solids, liquids and gases. Academic Press Inc., New York.

Kelso, W. C., Jr., Gertjejansen, R. A., Hossfeld, R. L., (1963) The effect of air blockage upon the permeability of wood to liquids. University of Minnesota Agr. Exp. Station, Tech. Bull. 242.

Kenaga, D. L., (1963) Effect of treating conditions on the dimensional behavior of wood during polyethylene glycol soak treatments. For. Prod. J. **13** (8): 345—349.

—, (1963) Stabilization of wood products with acryliclike compounds, U. S. Patent No. 3077417.

—, (1963) Stabilization of wood products with styreneacrylonitrile bis (2 chloroethyl) vinyl phosphate. U. S. Patent 3077419.

—, (1970) The heat cure of high boiling styrene-type monomers in wood. Wood and Fiber **2** (1): 40—51.

—, *Fennessey, J. P., Stannett, V. T.*, (1962) Radiation grafting of vinyl monomers to wood. For. Prod. J. **12** (4): 161—168.

Kent, J. A., Winsten, A., Boyle, W. R., (1962) Preparation of wood-plastic combinations using gamma radiation to induce polymerization. U. S. Atomic Energy Commission Report O. R. O-612.

—, *Loos, W. E., Ayres, J. E.*, (1965) Preparation of wood-plastic combinations using gamma radiation to induce polymerization. U. S. Atomic Energy Commission Report O. R. O.-628.

Klinkenberg, L. J., (1944) The permeability of porous media to liquids and gases. Drilling Prod. Pract. 200—213.

Kollmann, F. F. P., (1936) Technologie des Holzes. Springer-Verlag, Berlin.

Larsen, M. L., Yan, M. M., (1969) Meeting fire hazard requirements with wood-base panelboards. For. Prod. J. **19** (2): 12—16.

Lewin, M., (1968) Israel Patent.

Lloyd, R. A., Stamm, A. J., (1958) Effect of resin treatments and compression upon the weathering properties of veneer laminates. For. Prod. J. **8** (8): 230—234.

Loos, W. E., (1962) Effect of gamma radiation on the toughness of wood. For. Prod. J. **12** (6): 261—264.

—, *Robinson, G. L.*, (1968) Rate of swelling of wood in vinyl monomers. For. Prod. J. **18** (9): 109—112.

—, *Walters, R. E., Kent, J. A.*, (1967) Impregnation of wood with vinyl monomers. For. Prod. J. **17** (5): 40—49.

Loughborough, W. K., (1939) Chemical seasoning of overcup oak. Southern Luberman, Dec. p. 137.

—, (1942) Process of plasticizing lignocellulose materials. U. S. Patent No. 2298017.

—, (1943) Process for resinifying lignocellulose materials. U. S. Patent No. 2313953.

MacLean, J. D., (1935) Manual of preservative treatment of wood by pressure. U. S. Dept. Agr. Misc. Publ. No. 224, Washington, D.C.

—, (1962) Preservative treatment of wood by pressure methods. U. S. Dept. Agr. Handbook No. 40, Washington, D. C.

Martell, P., (1930) Journal for Applied Chem. Vol. 1930, 257.

McMillin, C. W., (1963) Dimensional stabilization with polymerizable vapor of ethylene oxide. For. Prod. J. **13** (2): 56—61.

Meyer, J. A., (1965) Treatment of wood-polymer systems using catalyst-heat techniques. For. Prod. J. **15** (9): 362—364.

—, *Loos, W. E.*, (1969) Process of and products from treating southern pine wood for modification of properties. For. Prod. J. **19** (12): 32—38.

Middleton, J. C., Dragaov, S. M., Winters, F. T. Jr., (1965) An evaluation of borates and other inorganic salts as fire retardants for wood products. For. Prod. J. **15** (12): 463—467.

Millett, M. A., Stamm, A. J., (1946) Treatment of wood with urea resin-forming systems: dimensional stability. Modern Plastics **24**: 150—153.

Mitchell, H. E., Iversen, E. S., (1961) Seasoning green-wood carvings with polyethylene glycol-1000. For. Prod. J. **11** (1): 6—7.

—, *Wahlgren, H. E.*, (1959) New chemical treatment curbs shrink and swell of walnut gunstocks. For. Prod. J. **9** (12): 437—441.

Naeser, G., (1960) Umschau **34**, 250.

Nicholas, D. D., Thomas, R. J., (1968) Influence of steaming on ultrastructure of bordered pit membranes in loblolly pine. For. Prod. J. **18** (1): 57—59.

Olesheimer, L. J., (1929) Compressed laminated fibrous product and process of making the same. U. S. Patent No. 1707135.

Olson, A. G., (1934) Process of shrinking wood. U. S. Patent No. 1981567.

Osnach, N. A., (1961) The permeability of wood. Derev. Prom. **10** (3): 11—13.

Perry, R. H., Chilton, C. H., Kirkpatrick, S. D., (1963) Perry's Chemical engineering Handbook. Ed. 4. McGraw-Hill, New York.

Petty, J. A., Puritch, G. S., (1970) Effect of drying on the structure and permeability of the wood of *Abies Grandis* Wood Sci. and Tech. **4** (2) 140—155.

—, *Preston, R. D.*, (1969) The dimensions and number of pit membrane pores in conifer wood Proc. Roy. Soc. B**172**, 137—151.

Prak, A. L., (1970) Unsteady-state gas permeability of wood. Wood Sci. and Tech. **4** (1): 50—69.

Ramalinghan, K. V., Werezak, G. N., Hodgins, J. W., (1963) Radiation induced graft polymerization of styrene in wood. J. Polymer Sci. Part C Polymer Symposium No. 2: 153 to 167.

Resch, H., (1967) Unsteady-state flow of compressible fluids through wood. For. Prod. J. **17** (3): 48—54.

—, *Eckland, B. A.*, (1964) Permeability of wood exemplified by measurements on redwood. For. Prod. J., **14** (5): 199—206.

Risi, J., Arseneau, D. F., (1957) Dimensional stabilization of wood. For. Prod. J. **7** (6): 210—213.

—, (1957) Dimensional stabilization of woods II crotonylation and crolylation. For. Prod. J. **7** (7): 245—247.

—, (1957) Dimensional stabilization of wood: III butylation. For. Prod. J. **7** (8): 261—265.

—, (1957) Dimensional stabilization of wood: IV allylation. For. Prod. J. **7** (9): 293—295.

—, (1958) Dimensional stabilization of wood: V Phthaleylation. For. Prod. J. **8** (9): 252—253.

Scheidegger, A. E., (1960) The physics of flow through porous media. Ed. 2. University of Toronto Press, Canada.

Schuerch, C., (1964) Wood plasticization. For. Prod. J. **14** (9): 377—381.

—, (1968) Treatment of wood with gaseous reagents. For. Product J. **18** (3): 47—53.

Sears, C. U., (1900) Preparing wood matrices. U. S. Patent No. 646547.

Seborg, R. M., Millett, M. A., Stamm, A. J., (1945) Heat-stabilized compressed wood. Staypak. Mech. Eng. **67**: 25—31.

—, *Stamm, A. J.*, (1941) The compression of wood. Mech. Eng. **63**: 211—213.

—, *Tarkow, H., Stamm, A. J.*, (1953) Effect of heat upon the dimensional stabilization of wood. J. For. Prod. Research Soc. **3** (3): 59—67.

—, *Vallier, A. E.*, (1954) Applications of Impreg for patterns and die models. J. For. Prod. Research Soc. **4** (5): 305—312.

Siau, J. F., (1969) The swelling of basswood by vinyl mensmers. Wood Sci. **1** (4): 250—253.

—, *Meyer, J. A., Skaar, C.*, (1965) Wood-polymer combinations using radiation techniques. For. Prod. J. **15** (10): 426—434.

Smith, D. N., (1963) The permeability of wood to liquids and gases. Fifth F. A. O. Conference on Wood Technology, U. S. For. Prod. Lab., Madison, Wisconsin.

—, *Lee, E.*, (1958) The longitudinal permeability of some hardwoods and softwoods. Div. Sci. Ind. Research, For. Prod. Res. Spec. Report No. 13. Her Majesty's Stationary Office, London.

Stamm, A. J., (1932) Effect of chemical treatment upon the permeability of wood. Ind. Eng. Chem. **24**: 51—52.

—, (1934) Effect of inorganic salts upon the swelling and shrinking of wood. J. Am. Chem. Soc. **56**: 1195—1204.

—, (1937) Minimizing wood shrinkage and swelling: treatment with sucrose and invert sugar. Ind. Eng. Chem. **29**: 833—836.

—, (1944) Wood impregnation. U. S. Patent No. 2350135.

—, (1946) Passage of liquids vapors and dissolved materials through softwoods. U. S. Department of Agriculture Tech. Bull. No. 929.

—, (1955) Swelling of wood and fiberboards in liquid ammonia. For. Prod. J. **5** (6): 413—416.

—, (1956) Dimensional stabilization of wood with Carbowaxes. For. Prod. J. **6** (5): 201—204.

—, (1956). Thermal degradation of wood and cellulose. Ind. Eng. Chem. **48**: 413—417.

—, (1959) Bound water diffusion into wood in the fiber direction. For. Prod. J. **9** (1): 27—32.

—, (1959) Effect of polyethylene glycol treatment upon the dimensional stabilization and other properties of wood. For. Prod. J. **9** (10): 375—381.

—, (1959) Dimensional stabilization of wood by thermal reactions and formaldehyde cross-linking. Tappi **42**: 39—44.

—, (1959) Dimensional stabilization of paper by catalyzed heat treatment and cross-linking with formaldehyde. Tappi **42**: 44—50.

—, (1962) Stabilization of wood: a review of current methods. For. Prod. J. **12** (4): 158—160.

—, (1962) Wood and cellulose-liquid relationships. North Carolina State Agr. Exp. Station Tech. Bull. 150, Raleigh, North Carolina.

—, (1964) Wood and Cellulose Science. Ronald Press Co., New York.

—, (1966) Maximum pore diameter of film materials. For. Prod. J. **16** (12): 59—63.

—, (1967) Heating dry wood and drying greenwood in molten polyethylene glycol. For. Prod. J. **17** (9): 91—96.

—, (1967) Movement of fluids in wood: Part I, flow of fluids. Wood Science and Tech. **1** (2): 122—141.

—, (1967) Movement of fluids in wood: Part II, diffusion. Wood Sci. and Tech. **1** (3): 205—230.

Stamm, A. J., (1970) Maximum effective pit pore radii of the heartwood and sapwood of six different softwoods as affected by drying and resoaking. Wood and Fiber 1 (4): 263—269.

—, (1970) Variation of maximum tracheid and pit pore dimensions from pith to bark for ponderosa pine and redwood before and after drying determined by liquid displacement. Wood Sci. and Tech. 4, 81—96.

—, *Baechler*, R. H., (1960) Decay resistance and dimensional stability of five modified woods. For. Prod. J. 10 (1): 22—26.

—, *Beasley*, J. N., (1961) Dimensional stabilization of paper by acetylation. Tappi 44 (4): 271—275.

—, *Burr*, H. K., *Kline*, A. A., (1946) Heat stabilized wood, Staybwood. Ind. Eng. Chem. 38: 630—637.

—, *Clary*, S. W., *Elliott*, W. J., (1968) Effective radii of lumen and pit pores in softwoods. Wood Sci. 1 (2): 93—101.

—, *Hansen*, L. A., (1935) Minimizing wood shrinkage and swelling: replacing water in wood by non-volatile materials. Ind. Eng. Chem. 27: 148—152.

—, —, (1937) Minimizing wood shrinkage and swelling: effect of heating in various gases. Ind. Eng. Chem. 29: 831—833.

—, *Harris*, E. E., (1953) The Chemical Processing of Wood. Chemical Pub. Co., New York.

—, *Seborg*, R. M., (1936) Minimizing wood shrinkage and swelling: treatment with synthetic resinforming materials. Ind. Eng. Chem. 28: 1164—1170.

—, —, (1939) Resin-treated wood Ind. Eng. Chem. 31: 897—992.

—, —, (1941) Resin-treated, laminated, compressed wood. Trans. Am. Inst. Chem. Eng. 37: 385—397.

—, —, (1943) Process for making an improved wood. U. S. Patent No. 2321258.

—, —, *Millett*, M. A., (1948) Method for forming compressed wood structures. U. S. Patent No. 2453679.

—, *Smith*, W. E., (1969) Laminar sorption and swelling theory for wood and cellulose. Wood Sci. and Tech. 3 (4): 301—323.

—, *Tarkow*, H., (1947) Dimensional stabilization of wood. J. Phys. and Colloid Chem. 31: 493—505.

—, —, (1951) Method of stabilizing wood, U. S. Patent No. 2570070.

—, *Turner*, H. D., (1954) Method of molding. U. S. Patent No. 2391489.

Staudinger, H., (1936) The insoluble polystyrene. Trans. Faraday Soc. 32: 323—335.

Tarkow, H., *Feist*, W. C., *Southerland*, C. F., (1966) Interaction of wood and polymer materials: penetration versus molecular size. For. Prod. J. 16 (10): 61—65.

—, *Seborg*, R. M., (1968) Surface densification of wood For. Prod. J. 18 (9): 104—107.

—, *Stamm*, A. J., (1953) Effect of formaldehyde treatments upon the dimensional stability wood. J. For. Prod. Research Soc. 3 (3): 33—37.

Tarkow, H., *Stamm*, A. U., *Erickson*, E. C. O., (1955) Acetylated wood. U. S. Dept. Agr. For. Prod. Lab. Mimeo. No. 1593. revised.

Truax, T. R., *Braechler*, R. H., (1935) Experiments in fire proofing wood. Proc. Am. Wood Preserver's Assoc. 31: 231—248.

U. S. Forest Products Lab., (1941) Forest Products Laboratory's toughness testing machine. U. S. For. Prod. Lab. Report 1308.

Van Groenou, H. B., *Rischen*, H. W. L., *Van den Borge*, J., (1951) Wood preservation during the last 50 years. A. W. Sijthoff's Uitgoversmentschappij, N. V. Leiden, Holland.

Van Kleeck, A., (1956) Fire-retardant coatings. U. S. For. Prod. Lab. Mimeo 1280, revised.

Walker, J. F., (1944) Formaldehyde. Reinhold Pub. Corp., New York.

Walsh, F. L., *Watts*, R. L., (1923) Composit lumber. U. S. Patent No. 1465383.

Weathermax, R. C., *Erickson*, E. C. O., *Stamm*, A. J., (1948) Modulus of hardness test. Am. Soc. Testing Materials. Bull. No. 153.

—, *Stamm*, A. J., (1945) The electrical resistivity of resin-treated wood and laminated hydrolized and paperbase plastics. Elect. Eng. Trans. 64: 833—839.

Wood Handbook, (1955) U. S. Dept. Agr. No. 72, Washington, D. C.

3. VENEER, PLYWOOD AND LAMINATES

3.1 Introduction

3.1.1 History, Status and Trends in Germany and North America

Probably the first veneer was manufactured in ancient Egypt around 3000 B.C. These veneers were relatively small pieces of valuable woods selected for the manufacture of costly furniture for kings and princes. The small veneer pieces were produced by hand sawing, then smoothed with suitable grinding materials (for example pumice) and subsequently combined with thin pieces of other materials such as metals or ivory in an artistic manner to produce articles ranging from bedsteads to coffins. The types of glues used are unknown, but they were probably basically albumin. The throne found in the tomb of Tut-anch-Amun 1361 until 1352 B.C.) (Fig. 3.1) is made from cedarwood overlaid with thin sheets of ivory and ebony. Queen Cleopatra VII (69 until 30 B.C.) sent as a gift a precious table with inlaid work to Julius Caesar. It is known from publications by Pliny the Elder (24 until 79 B.C.) that during his time the masters in Rome were well acquainted with veneering techniques and some principles of plywood-effects.

After a long stagnation during the Middle Ages, in the Renaissance (in Europe during the 14th, 15th and 16th centuries) the artistic woodwork and veneering revived. Primarily, the principle of creating a mosaic of small pieces of tinted and natural woods was guiding. Subsequent to the Italian Renaissance the art of inlaid work attained remarkable heights in France, Spain, the Netherlands and England. During the reign of the French kings Louis XV (1710 to 1774) and Louis XVI (1774 to 1793) extraordinarily beautiful furniture was produced, as for example the famous "Bureau du Roi" now located in the Louvre in Paris. "It is a prototype of the modern roll-top desk and is said to have cost the king (Louis XV) more than a million francs" (*Perry*, 1947, p. 21). In the 18th century the German *David Röntgen* (from Neuwied) manufactured masterpieces highly appreciated by kings and sovereigns.

The first patent for a veneer saw in Europe was granted to a French mechanic in 1812. Such a saw was not used in the industry before 1825. Subsequently, saws of this type were improved and manufactured in Hamburg. The amount of sawdust produced was high. The development of the first veneer-slicing machine came from France, where a patent was granted to *Charles Picot* in the year 1834. Approximately thirty years were necessary to make this machine sufficiently reliable for industrial use.

Much progress in the manufacture of veneer as the basis of the coming plywood industry resulted from the construction of the rotary lathe; this led to economical mass production of veneer. The first machine for producing a continuous sheet of veneer with a knife from rotating round logs was invented in 1818. The arrangement for holding the log as it was cut were threaded (screw-like) spindle ends which fitted into holes bored into the ends of the log. In the year 1840 U.S. patent No.

1758 for a veneer lathe was granted to *Dresser*, another patent in 1844 in France to *Garand*. The logs were as much as 2 m (6.6 ft.) long, and the peeling velocity was between 4 and 5 m/min (13 to 16 ft./min). The knife of the peeler was vertically adjustable, and a pressure bar was already in use.

The first factories for the manufacture of veneer were built in Germany in the middle of the 19th century. The peelers were mostly of French origin. American-made peelers were imported by Germany also. After 1870 the firm of *A. Roller*, Berlin, delivered simple peelers. The quick development and improvement of the rotary peelers gave an impetus to the plywood industry prior to World War I.

Fig. 3.1. Throne of Tut-anch-Amun. From *Hamann* (1962)

The first commercial manufacture of plywood was in Germany, other European countries and in the U.S.A. primarily for special purposes such as laminated rims of grand pianos (1860), sewing-machine cabinets (1867), perforated chair seats and backs (1875), tops and fronts of furniture (1885 to 1900), "yet the all-around advantages of plywood were little appreciated until around 1890" (*Perry*, 1947, p. 31). The quality of plywood became better in the last decade of the 19th century, and the new product slowly found a good market. Many plywood factories were erected. Cross banding proved to be an effective means of reducing changes in dimensions, warping and checking. The term "plywood" did not really become a trade name (in the U. S. A.) until the war period of 1914 to 1918 (*Perry*, 1947, p. 28).

Thicker plywood panels were produced in an expensive manner. Cores were produced by edge gluing heartwood-free boards which had been sawn from logs. The core then was covered on both sides with thick veneer whose grain direction was perpendicular to the grain direction of the core. This type of primitive plywood was called "core stock". It was manufactured by hand and was therefore expensive.

Under these circumstances the economic situation of the plywood industry was not favorable. Nevertheless this type of plywood could be sold to the furniture industry, which could avoid in this manner the purchase of costly machinery. But, unfortunately, this crude core stock gave reasons for complaint by the user.

In the year 1910 *Kümmel* developed a new method for manufacturing core stock. He face-glued planed boards into a block which was then cut perpendicular to the glue lines into thin boards using a horizontal or a vertical gangsaw. The thickness of the thin boards corresponded to the desired thickness of the core. In this way the dimensional stability of the plywood was improved remarkably. The smaller the pieces in the core, the less the warping in the completed core stock. In 1923 a factory (J. Brüning & Son at Rehfelde, West Germany) made the cores not from boards but from peeled veneers. The higher cost due to the higher glue requirement was compensated by the higher yield for peeling instead of sawing.

Core stock was developed step by step as a handicraft while the manufacture of plywood was from the beginning a matter of industry. Beech wood, originally sold mainly as a fuel in Germany, came into use as a veneer and plywood species, especially after the improvement of belt-dryers in 1907. During World War I plywood was utilized in the belligerent countries for military purposes. Plywood of high quality was an important material for the construction of aircraft. The firm Schütte-Lanz at Rheinau, West Germany, built lighter-than-airships with the internal skeleton made entirely of plywood. The first airships of this type built in the years 1909 and 1910 had a maximum diameter of 18.5 m (60.7 ft.) and a length of 131 m (430 ft.), and could carry a load of approximately 21,000 kg (21.3 long tons). Animal (hide, leather or bone) glues could not be used because they are not resistant against moisture, but a glue mixture of casein and albumin was suitable.

The original difficulties of the German plywood industry after World War I were overcome quickly. The main consumer was again the furniture industry, but the ship-yards also purchased large volumes of plywood for the reconstruction of the merchant fleet. The export of the German plywood industry was important, since at this time the value of the German Mark was low, the quality of the plywood was good, and Russia was not able to deliver plywood. Monetary inflation in Germany had caused severe losses of capital, but after 1924 the situation became better, and within five years the capacity of the German plywood industry had increased approximately fourfold. To avoid detrimental competition, the three biggest plywood factories decided to cooperate.

The cooperative program was as follows:

1. Specilization in types and dimensions of plywood, establishment of standard sizes;

2. Exchange of products;

3. Exchange of information on methods of fabrication;

4. Uniform calculation of production costs;
5. Common marketing and terms of delivery;
6. Demarcation of market areas;
7. Common advertising, combined with similar activities;
8. Common measures against "dumping".

Even though all points of the program could not be realized, the benefits for the three factories were remarkable. This led in July, 1933 to the founding of a German Society (Interessengemeinschaft Deutscher Sperrholzfabriken, IDS) with at first 15 and later 18 plywood factories as members.

Veneer and plywood manufacturing in North America is described by *Lutz* and *Fleischer* (1963). The following information is a more or less literal extract from their report.

Veneer and plywood manufacturing in North America may be divided into two distinct major industries:

a) The large and highly automated plants located in the Western United States and Canada manufacture primarily construction plywood from Douglas fir and related softwood species, such as ponderosa pine, white pine, sugar pine, redwood and recently western hemlock, white fir, western larch, spruce, cedar, and noble fir.

Western larch is equivalent in strenght to coast-type Douglas fir and is used with it in construction plywood. The other species named are lower in density and strength than coast-type Douglas fir and, as a result, are converted into thicker plywood for equivalent structural uses.

In recent years, the demand for suitable raw material has resulted in the use of Douglas fir logs containing white pocket, a decay caused by *Fomes pini*. The commercial standard now permits limited white pocket in some grades of veneer. This change was made only after extensive processing and engineering research at the U. S. Forest Products Laboratory in cooperation with the Douglas Fir Plywood Association.

b) Many small plants scattered throughout the Eastern United States and Canada cut primarily hardwood and make a wide variety of veneer and plywood products. In 1961, the softwood veneery industry of North America consumed about four times as much timber as the hardwood veneer industry. Approximately 11% of the softwood veneer and 12% of the hardwood veneer were produced in Canada. The remainder was produced in the United States.

In 1970, there were 164 western softwood plywood plants of which 140 were in the United States and 24 in Canada. These western plants are large in size and use straight-line production. Many of the plants are capable of producing 100 to 200 million square feet (9 to 18 million m²) of 3/8 in. (9.5 mm) thick plywood per year. This production capacity has been accomplished by use of new, highly productive equipment and through modernization of older plants. Automatic lathe chargers, retractable chucks, air jet replacing conventional dryers, mechanical veneer dryer feeders, electronic moisture detectors, more efficient and economical patchers, automated jointing and edge gluers, mechanical press loaders and unloaders, automatic trim saws, and automatic feeding of panels to improved wide belt sanders are now commonly used in the softwood plywood industry.

There is currently (Forest Industries, 1971, Jan. issue) in the U.S.A. a large and rapidly growing softwood plywood industry in the Southeast and in the South, based on southern pine species. There are 44 plants in this group (southern pine plywood), producing from 1/5 to 1/4 of the total softwood plywood production in the U.S.A.

There are also three softwood plywood plants in Eastern and Central Canada (Ontario and Quebec).

c) Most softwood veneer is cut one-eighth to three-sixteenths of an inch (3.1 to 4.7 mm) in thickness and 8 ft. (2.5 m) long on lathes equipped with roller nosebars. Some softwood veneer is produced in 1/10 and 1/12 in. (2.5 and 2 mm) thicknesses and may be cut as thin as one-sixteenth of an inch (1.6 mm) for special uses. Commercial standards limit the maximum thickness to one-fourth of an inch (6.3 mm).

Softwood veneer is generally cut, dried, and glued into plywood in the same plant. However, in the past decade more than 50 small plants have been built to cut and ship green or dry veneer to plywood plants for further manufacture.

Another development in the softwood plywood industry has been the use of No. 2 sawlogs rather than high-grade peelers. This became possible because a growing volume of plywood is being used for sheathing purposes, made of C-grade and D-grade veneers. C-grade permits sound, tight knots up to 1 1/2 in. (38 mm) in diameter, knotholes not over 1 in. in diameter (25.4 mm), splits not wider than 3/16 in. (4.8 mm), and similar characteristics. D-grade veneer, used only in interior-type panels, permits knotholes not exceeding 2 1/2 in. (64.5 mm) in diameter and equivalent characteristics.

The increased use of knotty material has stimulated industry interest in the practice of heating the logs before cutting. Research has shown that heating makes it possible to produce smooth and tight veneer from low-grade logs. Several older plants and some new plants now steam Douglas fir and other softwood bolts before peeling.

Some hardwood veneer and plywood is produced in the Western United States and Canada, mainly from cottonwood and alder. Other species used to a lesser extent include tanoak, madrone, walnut, bigleaf maple, Oregon myrtle, and the various western oaks. These hardwoods are used primarily as face veneers on softwood plywood panels. Western manufacturers also import a large amount of hardwood veneer from the Eastern United States as well as lauan veneer from the Philippines.

The hardwood veneer and plywood industry is scattered throughout the eastern half of the United States and Canada. It is estimated that there are between 600 and 700 mills producing veneer, and about one-fourth of these also produce plywood. Veneer production may be broken into three broad groups — 1. face veneer, 2. commercial veneer for use in standard hardwood plywood, and 3. container and packaging veneer.

1. The manufacture of fancy face veneer is a specialty operation in about 50 plants scattered throughout the eastern half of the United States with a large concentration in Indiana and Kentucky. There are a few such plants operating in Eastern Canada. Most face veneer is cut by the slicing process, using vertically operating slicers. Some is cut on lathes equipped with a stay-log or is cut "full-round". Practically all of the veneer cut and dried at these plants is sold to secondary manufacturers for conversion into finished products. Face veneer plants specialize in such species as walnut, cherry, oak, mahogany and maple. Many of these plants also have sawmills. The logs are sawn through the pith and the sawyer then decides whether to prepare flitches for veneer or to saw the log into lumber. The bolts or flitches are almost always heated before slicing; heating in water vats is preferred. Face veneers are generally relatively thin, one twenty-eighth of an inch (0.9 mm) being the most common thickness.

2. Commercial veneers are cut almost entirely on the lathe and are used to make standard hardwood plywood. They are generally cut in thicknesses ranging from one-twentieth to three-sixteenths of an inch (1.3 to 4.8 mm). Grades and dimensions are often chosen to suit the buyer's specifications, and veneers are graded on the basis of such defects as knots, wormholes, splits, and discoloration. The biggest volume of commercial-grade veneer is cut from gum, including tupelo, sweetgum, and black gum. The second major species used is birch, primarily yellow birch, followed by yellow poplar, red oak and white oak. Other species cut in volume include basswood, maple and walnut. Used as the face the major species is birch, followed by gum, oak, and maple. Ash and elm have recently been used for commerical hardwood plywood. Other species used include cherry, pecan, aspen, cottonwood, beech and sycamore. Imported woods used for the manufacture of standard grades of hardwood plywood include cativo, the Philippine hardwoods (particularly the lauans), mahogany and "African mahogany".

3. Container and packaging veneer. Approximately one third of the total veneer cut from hardwoods is made into containers. Principal species used are gum, beech, sycamore, cottonwood, and the oaks. Some container veneer is also made from such softwoods as southern pine, ponderosa pine, and Douglas fir. Container veneers are usually produced by rotary cutting, or they may be produced on special shok slicers. Existing grading rules are not well defined and are based chiefly on such properties as appearance and printability of surface, and the absence of defects that may interfere with nailing or may damage the container contents.

The veneer and plywood industry of North America is in a state of transition. Timberland and mill holdings are being consolidated into larger ownerships, and in the larger companies, plywood production is being integrated with the production of lumber, pulp and paper, hardboard, and other forest products. This move will result in noticeable changes, particularly in the hardwood industry, toward greater automation and quality control in production, and greater application of technical knowledge and results of research.

In spite of the unfavorable import situation, the hardwood veneer and plywood industry continues to show a degree of confidence in the future. Hardwood plywood manufacturers have been working towards the establishment of an industry-wide quality control program, controlled out of a central industry testing laboratory, that is aimed at better customer satisfaction with its product.

The softwood plywood industry shows no signs of having reached a production plateau. Its aggressive promotion program is currently informing the public of the advantages of plywood structural components and of the desirability of having a second house for vacation purposes in every family. On the whole, therefore, it may be concluded that veneer and plywood production in North American can be expected to show continuing growth in the future.

3.1.2 World Production and Analysis of Consumption

Plywood is a material which is relatively light in weight, which can be machined easily and which is reasonably priced. All these factors have increased steadily the consumption of veneer and plywood. New uses are being found continually. This explains the rapid growth in production of veneer and plywood in industrialized countries. Fig. 3.2 shows the development of veneer and plywood production prior and subsequent to World War II. The production of plywood in the entire world and its most important areas and countries after 1955 are shown

Table 3.1. World Plywood Production 1955 to 1970 (in 1,000 m³) (After FAO World Forest Products Statistics, Rome)

Country	1955	1956	1957	1958	1959	1960	1961	1962	1963	1964	1965	1966	1967	1968	1969	1970²
Total¹	10,720	11,280	11,780	13,010	14,880	15,346	16,510	18,142	20,196	22,302	24,295	25,348	26,435	29,672	30,660	32,600
United States and Canada	6,250	6,710	6,750	7,720	8,820	8,910	9,684	10,573	11,888	13,141	14,511	14,840	14,927	16,469	15,639	15,969
Europe and USSR	3,200	3,030	3,260	3,360	3,650	4,020	4,158	4,392	4,760	4,944	5,075	5,118	5,251	5,427	5,876	6,076
Asia	860	1,080	1,260	1,390	1,840	1,860	2,076	2,510	2,866	3,509	3,938	4,650	5,502	6,992	8,231	9,559
Central and South America	200	220	240	270	250	283	326	368	366	375	389	398	404	415	530	571
Africa	110	140	150	160	190	115	125	162	187	194	241	204	224	229	249	270
Australia and Pacific Area	120	120	130	130	140	157	141	137	130	140	142	137	128	141	135	154
United States	5,560	5,910	5,980	6,830	7,950	7,910	8,580	9,329	10,528	11,628	12,809	13,037	13,059	14,510	13,635	14,119
Japan	680	850	980	1,670	1,290	1,286	1,499	1,833	2,073	2,453	2,627	3,101	3,778	4,744	5,784	7,008
USSR	1,050	1,120	1,160	1,230	1,300	1,354	1,428	1,486	1,544	1,659	1,711	1,772	1,819	1,832	2,050	2,158
Canada	680	800	760	886	870	1,000	1,104	1,244	1,360	1,513	1,702	1,803	1,868	1,959	2,004	1,850
Germany, Fed. Republic	650	470	510	470	500	665	637	653	666	654	670	632	584	598	615	569
Finland	360	370	310	290	350	414	406	426	468	511	545	550	573	616	686	706
France	230	260	300	340	350	380	382	414	439	482	477	494	536	547	614	639
Philippines	60	70	80	100	176	191	154	187	236	277	353	319	308	410	309	341
Italy	150	140	140	150	150	150	200	200	400	300	300	280	300	340	380	420
Rumania	50	50	60	60	80	95	123	171	201	229	236	251	264	271	285	—
Brazil	100	100	90	100	90	124	141	185	181	147	130	118	77	43	169	—
Taiwan	20	20	40	22	28	33	60	84	100	213	251	315	336	423	499	537
Czechoslovakia	130	100	110	110	170	184	185	187	183	169	163	164	162	167	149	161
Poland	160	160	170	190	180	158	160	165	168	168	166	156	168	174	184	205
Spain	50	60	80	80	80	75	92	130	115	108	131	172	210	230	221	228
Yugoslavia	30	40	50	60	80	102	115	120	118	145	155	149	132	135	156	161
Australia	90	90	90	100	110	123	109	100	97	106	100	96	98	103	94	106
China (Mainland)	—	10	10	10	10	110	105	85	90	90	100	110	120	130	140	150

¹ Addition of Regions may not exactly match with total world due to rounding.
² Provisional.

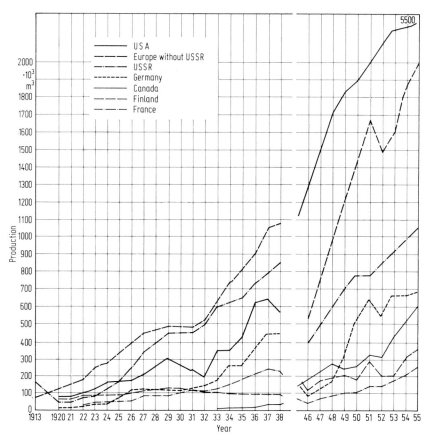

Fig. 3.2. Production of plywood in various countries between 1933 and 1955. From Bois et Scieries, Paris (January 1957)

in Table 3.1. A comparison of the growth in the production of plywood, particleboard and fiberboard (Fig. 3.3) is very instructive. Apparently within the last decade the rate of growth for particleboard is the greatest and demonstrates the very severe competition between this material and plywood and fiberboard. It can be assumed that this trend will not change in the near future but a small and gradual increase in the production of veneer and plywood may be expected up to 1985 (Fig. 3.4) (*Grosshennig*, 1971), using FAO-Data from 1962 until 1985.

Unfortunately, no exhaustive or reliable statistics on the consumption of plywood for different uses exist. Therefore it is possible only to list the most important uses:

1. Furniture (cabinets, tables, desks, bedsteads, bars, certain types of chairs, auditorium seating, etc.);

2. Residential construction, especially parts for pre-fabricated houses, roofs, ceilings, partitions, subflooring, wall panelling, shopfitting, stage and studio joinery;

3. Engineering constructions (e.g., structural members, curved shell roofs, hollow tubes for antennas, etc.);

4. Doors (flush doors, panel doors);

5. Concrete shuttering;

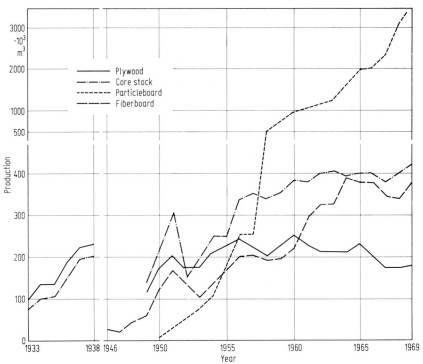

Fig. 3.3 Comparison of the growth in the plywood, particleboard and fiberboard production from 1933 until 1969.
FAO World Forest Products Statistics (Rome)

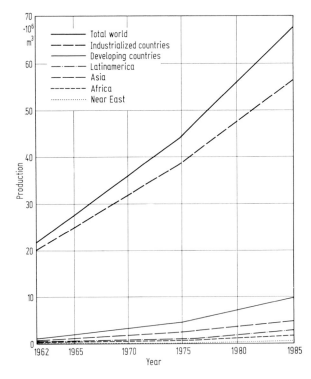

Fig. 3.4. Development and estimated
trend of growth of production of
plywood from 1962, extrapolated until
1985. From *Grosshennig* (1971)

6. Musical instruments (pianos, cabinet organs, stringed instruments, radio and television cabinets, instrument cases);

7. Means for land transportation (trucks, station wagons, passenger cars, bus interiors, trailers, freight cars, railway passenger cars, unit carriers);

8. Boats and ships, built-in equipment for large vessels (with consideration of fire safety requirements, especially partitions and bulkheads in passenger-carrying vessels), rescue boats, pontoons;

9. Aircraft construction (e.g., small sport and training planes, gliders, propellers, helicopter blades);

10. Containers (packages for shipping, storage and distribution), pallets, crates, export cases, tool chests, drums, boxes and barrels, trunks, bags and suitcases, trays, reels for electrical cables and wires;

11. Coffins and caskets;

12. Sporting goods (tennis rackets, laminated skis, sleds, toboggans, paddles, etc.);

13. Miscellaneous uses, such as patterns for foundries, forming dies (made from high-density plywood and laminated wood), handles for tools, silent gears made from high-density plywood, parts for equipment in chemical factories (for example tubes, vats and vessels, filter frames etc.), toilet seats.

3.2 Veneer

3.2.1 Wood Species

Some wood species used in the U.S.A. for the manufacture of veneer are listed with their predominantly used commercial names below (trade names in the "U. K. Standards"). The statistics to which extent these species are sliced or peeled in various countries are incomplete and not comparable.

Wood species used for veneer and plywood manufacture in North America

A. *Softwood*

Douglas fir (Oregon pine)	White fir (Grand fir)
Ponderosa pine	Western larch
White pine (Yellow pine)	Spruce
Sugar pine	Western red cedar
Redwood (Sequoia)	Noble fir
Western hemlock	

B. *Hardwood*

1. *For fancy-face veneer*

Walnut, American	Mahogany, African
Cherry, American	Maple
Oak	

2. *For commercial veneer for standard hardwood plywood*

Gum (Sweet gum, American red gum)	Cherry, American
Tupelo, Black gum	Pecan (Hickory)
Birch (primarily Yellow birch)	Aspen

11*

Yellow poplar (American whitewood)	Cottonwood (Canadian poplar)
Red oak and white oak	Beech, American
Basswood	Sycamore (American plane)
Maple	Cativo
Ash, American	Lauan
Elm (various species)	Mahogany, African

3. Container and packaging veneer

Gum (sweet gum, American red gum)	Oak
Beech, American	Southern pine (American pitch pine)
Sycamore (American plane)	Ponderosa pine
Cottonwood (Canadian poplar)	Douglas fir (Oregon pine)

C. Softwoods used for decorative plywood

Western white pine	Redwood (Sequoia)
Ponderosa pine	Western red cedar

D. Softwoods used for construction of plywood

Douglas fir	Western hemlock
Southern pine (American pitch pine)	White fir (Grand fir)
	Engelmann spruce

E. Hardwoods used for face veneer

Walnut, American	Maple
Cherry, American	Rek oak
Pecan (Hickory)	White oak
Birch	
Gum (Sweet gum, American red rum)	

Wood species used for veneer and plywood manufacture in Western Europe

In Western Europe the wood species listed below are widely used or have a good chance of future in veneer manufacture.

The most significant references for species nomenclature are the Nomenclature of the Association Technique Internationale des Bois Tropicaux and the British Standards 881 and 589. It would exceed the scope of this book to survey all commercial and native designations of the various origins of individual wood species. Therefore, the international indications or standard names as the trade designation recommended for easy reference are used.

A. Wood species used for core veneer in Western Europe

1. Softwood

Parana pine (Brazilian pine)	European spruce
Douglas fir	Podo

2. Hardwood

From Europe

Birch	Alder	Poplar
Beech	Lime	

From North America

Whitewood (in U. S. A. known as Yellow poplar)

The import of North American wood species into Europe for core veneers is generally too expensive; whitewood-core, however, is particularly used in the piano industry.

From Africa

Abura	Fromager (Ceiba)
African mahogany (Acajou	Ilomba
d'Afrique)	Kapokier (Bombax)
Aiélé	Kumbi
Ako (Antiaris, Kirundu)	Limba (Afara)
Bossé	Obeche (Wawa, Samba, Ayous)
Dabéma	Okoumé (Gaboon)
Dibétou	Olon
Emien	Onzabili
Essessang (Erimado)	Tchitola
Faro (Ogea)	Tola (Tola-Branca, Agba)
Framiré	Tiama

From American Tropics

Andiroba	Guácimo
Assacú (Possentrie)	Jequitiba (Abarco)
Baboen (Virola)	Marupa
Cativo	Prima vera
Cedro (Cedar)	Quaruba
Espavel	Saman
Fromager (Ceiba)	Tepa
Gobaja (Coupaya)	

From Asia

Dipterocarpaceae family

This wood family is of great importance as a supplier of timber in the Asiatic tropics. Although the trees occur in mixed forests, in most cases they amount to more than half of the usable stand and distinguish themselves by strong, straight and knot-free trunks. Approximately 380 species are known, belonging to 19 genera. The exported lots are found to be very heterogeneous and the timber for commercial use is sorted rather arbitrarily according to wood density and color. The lower density timber of this family is well suited for peelers. The appearance of this wood is, however, very plain, dull and often brownish-gray. The best-known peeler species are:

Red Lauan	Kaunghun
Meranti	Krabak
Red Seraya	Mersawa
White Lauan	White Seraya

Further species:

Jelutong	Ramin

B. *Wood species for face veneers*

1. *Softwood*

Alerce	Larch, European
Douglas fir	Thuya burls
Yew	Wavona burls (Burls of Redwood,
European spruce (Whitewood)	Sequoia)
Scots pine (Redwood)	Alpine stone pine

2. Hardwood

From Europe

Maple	Cherry, European
Birch	Walnut, European
Pear	Plane, European
Boxwood, European	Ash, European
Sweet chestnut	Elm (various species)
Oak	

From North America

Madrone burls	Gum (Sweet gum, American red rum)
California-laurel (Laurel burls)	Bird's-eye maple
Walnut, American	

From Africa

African mahogany (Acajou d'Afrique, Khaya)	Limba (Afara)
	Makoré
Ako (Kirundu, Antiaris)	Moabi
Avodiré	Movingui
Bété (Mansonia)	Mukulungu
Bilinga	Mutenye
Bossé	Okoumé (Gaboon)
Bubinga (Kevazingo)	Ovoga
Dibétou	Sapelli
Douka	Sipo (Utile)
Ebiara	Tiama
Evino	Tola (Tola-Branca, Agba)
Framiré	Wengé
Iroko (Kambala)	Zingana (Zebrano)
Kokrodua (Afrormosia)	
Kosipo	

From American Tropics

Angelin (Partridge wood)	Imbuia
Andiroba	Mahogany
Araribà (Amarillo)	Palissandre Brésil (Brazilian rosewood)
Cedro (Cedar)	Rauli
Cocuswood	Roble
Courbaril (Copalier)	Urunday (Tigerwood)
Espenille (West Indian Satinwood)	Wacapou
Freijo (Laurel)	Zapatero

From Asia

Cickrassy	Padauk (Amboyna burls)
Chuglam	Palissandre Asie (Indian rosewood)
Ebène Asie (Ebony, black, Ceylon ebony)	Sen
	Teak
Japanese ash (Tamo)	

From Australia

Oak, Silky	Walnut, Australian	Eucalyptus

The majority of the eucalyptus species are heavy construction woods; however, some species of brighter color and lower density (probably *Eucalyptus gigantea* Hook. and *E. regnans* F. v. M.) were brought to Germany and successfully used as veneer. As veneer Eucalyptus is very light-colored, lemon-white, with a natural luster and regular stripes due to alternating spiral grain.

It is interesting to compare the figures for the export of main species from Africa as an important source of veneer logs. The outstanding position of Okoumé, Obeche, Sipo and Limba is evident.

Table 3.2. Main Species Exported from Africa in 1963, 1964 and 1965 (in 1,000 cu. meters). After O. E. C. D. (Organization for Economic Cooperation and Development) cf. Bois Tropicaux, p. 28—31. Paris 1968

	Logs			Sawn		Timber	
	1963	1964	1965	1963	1964	1965
Okoumé (*Aucoumea klaineana* Pierre)	1,247	1,451	1,372	6	7	5
Obeche (*Triplochiton scleroxylon* K. Schum.)	1,226	1,326	1,024	64	75	71
Sipo (*Entandrophragma utile* Sprague)	470	698	739	62	78	106
Limba (*Terminalia superba* Engl. et Diels)	331	339	348	18	21	20
Acajou, Mahogani (*Khaya ivorensis* A. Chev. and *Khaya anthoteca* C. DC.)	247	262	244	60	66	65
Sapelli (*Entandrophragma cylindricum* Sprague)	188	206	200	42	56	62
Makore (*Dumoria heckelii* A. Chev.)	123	157	174	11	12	13
Tiama (*Entandrophragma angolense* C. DC.)	108	135	127	7	9	10
Kokrodua (*Afrotmosia elata* Harms)	91	91	64	39	49	37
Iroko (*Chlorophora excelsa* Benth. et Hook. f.)	62	97	81	12	21	17
Ilomba (*Pycnanthus angolensis* Exell)	123	116	139	—	—	2
Azobé (*Lophira procera* A. Chev.)	51	72	56	11	12	12
Niangon (*Tarrietia utilis* Sprague)	55	72	71	9	10	10
Doussié (*Afzelia* spp.)	38	60	72	8	14	13
Tola (*Gossweilerodendron balsamiferum* Harms)	38	53	50	14	14	13
Mansonia (*Mansonia altissima* A. Chev.)	68	64	111	2	3	3
Ozigo (*Dacryodes Buettneri* H. J. Lam.)	66	68	62	—	—	—
Dibetou (*Lovoa trichilioides* Harms)	49	62	83	4	4	4
Abura (*Mitragyne cilta* Aubrev. et Pelleriagr. and *Mitragyne stipulosa* O. Kuntze)	66	62	56	2	2	1

Table 3.2. (Continued)

	Logs		Sawn	Timber		
	1963	1964	1965	1963	1964	1965
Andoung, Ekop (*Monopetalanthus Heitzii* Pellegr. and *Tetraberlinia bifoliolata* Hauman)	43	50	34	—	—	—
Tali (*Erythrophleum guineense* G. Don and *Erythrophleum micranthum* Harms)	8	12	1	27	26	15
Famiré (*Terminalia ivorensis* A. Chev.)	24	30	25	9	9	10
Douka (*Dumoria africana* A. Chev.)	29	39	31	—	1	—
Bossé (*Guarea cedrata* Pellegr.)	21	30	20	7	4	6
Fromager (*Ceiba pentandra* Gaertn.)	19	34	17	—	—	3
Kirundu, Ako (*Antiaris Africana* Engl.)	41	33	20	1	1	1
Wengé (*Millettia Laurentii* De Wild.)	—	—	—	2	27	17
Copalier (*Buibourtia coleosperma* J. Leonard)	18	17	19	2	10	10
Tchitola (*Oxystigma oxyphyllum* J. Leonard)	25	28	26	—	—	--
Mecrussé (*Androstachys Johnsonii* Prain)	—	—	—	19	24	14
Kosipo (*Entandrophragma Candollei* Harms)	20	25	8	—	1	—
Banga-Wanga (*Amblygonocarpus obtusangulus* Harms)	—	—	—	6	12	6
Bilinga (*Sarcocephalus Diderrichii* De Wild. and *Sarcocephalus Pobeguini* Hua)	3	1	1	6	9	8
Total of all species	5,130	6,040	5,597	674	734	690

3.2.2 Pretreatment of Logs Prior to Slicing and Peeling

3.2.2.1 Protection of Logs against Decay and Checking. Veneer is produced from a great variety of hardwood and softwood species as has been shown in the foregoing section. Normally the face veneer consists of valuable woods. The form of tree stems and their dimensions must be suitable for veneer manufacturing. The grain should be straight; deviations are permitted only within limits fixed by economic considerations (yield) and quality requirements. Logs for the production of veneer are expensive. For this reason it is necessary that logs on route to the veneer and plywood plants should not spoil. Logs are subject to deteriorating influences after felling in the forest, as cargos in transit and in the wood yards, in the temperate zones, especially during the spring and summer season. Felling, peeling and subsequent operations produce small but nevertheless dangerous spots for the entry of wood staining or destroying fungi. The moisture in a green stem is unequally distributed throughout its length and diameter (Fig. 3.5). The stems dry — at least in their outer zones — rather rapidly. If logs have not been debarked the water diffusion occurs mainly parallel to the grain, and drying is therefore restricted to the cross sections at the ends of the logs. Hence it follows that shorter logs are more endangered. *Mayer-Wegelin* (1932) found that beech logs which were not debarked had not changed in moisture content at a distance of 1 m (3.28 ft.) from the ends during three months after felling.

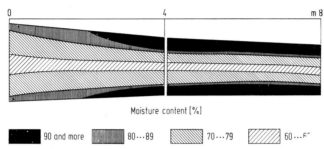

Moisture content [%]

■ 90 and more   ▥ 80···89   ▨ 70···79   ▩ 60···6⁻

Fig. 3.5. Distribution of moisture content in a beech stem cut into two sections after felling in March. From *Mayer-Wegelin* (1932)

Deterioration from various causes is most pronounced if the wood species have little natural decay resistance and if the climatic conditions, including during transport, are unfavorable.

Plywood factories should preserve the round logs by the best possible methods.

Early attempts have been made to retard the loss of moisture from the end grain by suitable chemicals; examples are hot coatings, such as paraffin, resin and lampblack, coal-tar pitches, asphalt, and cold coatings such as pigment and oil pastes, emulsions of asphalt, wax, paraffin, polyvinyl and latex with a pigment of aluminium and drying oils. *McMillen* (1950) at the U. S. Forest Products Laboratory published a report about methods and results on end coatings for preventing end checking of logs during storage, handling and drying. He summarizes: "End coatings must be inherently of high water resistance and must be applied in a comparatively thick film to eliminate checking" ... "For practical use, a coating must adhere to the wood, remain intact during handling, and be flexible enough to adjust to changing wood dimensions, yet not too messy to handle."

Lambert and *Pratt* (1955) listed end check preventives in the order of their merits as follows:

1. Emulsions of wax;
2. Latex with aluminium pigment;
3. Transparent hardening oils, filled and probably based on copal resin.

It is essential to treat all open spots on the logs. The coating should be applied generally not later than nine weeks after felling, in March not later than six weeks after felling, and in May not later than three weeks after felling. During the winter season no coating should be applied during frost or on frozen surfaces. To prevent stain and decay, it is necessary to apply a toxic substance to the logs within 24 h of cutting (*McMillen*, 1950).

Checks at the end surfaces are always dangerous and should be avoided as far as possible. A rather primitive, but to some extent effective means is the use of iron clamps (Fig. 3.6). It is essential that these clamps are inserted prior to drying and that as many medullary rays as possible be crossed at a right angle (Fig. 3.7). The efficiency of these clamps is limited. The clamps must be removed before the wood is steamed. The work is troublesome and time-consuming. Plastic antichecking clamps are available and in use. These do not have to be removed before cutting the log. If log diameters are not too great, occasional end checks may be prevented as shown in Fig. 3.8.

Round logs for veneer manufacture are transported to the factories either by vessels or by train or trucks. In Scandinavia the logs are mainly floated by combining about fifty round logs into a raft. Round logs for veneer must be carefully treated on unloading and piling; otherwise damage occurs.

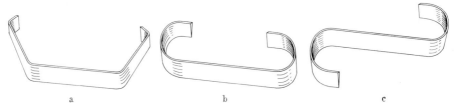

a   b   c

Fig. 3.6a—c. Antichecking irons.
a) Beegle iron; b) C-iron; c) S-iron. From *Hunt* and *Garratt* (1953)

Fig. 3.7. Arrangement of S-irons on the cross section of a round log. From *Kollmann* (1962)

Fig. 3.8. Arrangement of bandages and tension screws on a round log.
a Round log, *b* Iron sheet bandage, *c* Tension screw. From *Klotz* (1940)

3.2.2.2 Steaming and Boiling (Cooking) of Round Logs. Veneer can be manufactured by sawing, slicing or peeling. Only a few very expensive woods are still converted to veneer by sawing. For sawing a pretreatment by steaming or boiling is not necessary, which has the advantage that the natural color will not be changed. The waste, of course, is large. Most veneer (approximately 98%) is produced by slicing or rotary peeling. In Section 3.2.1 an incomplete survey on wood species for veneer manufacturing in the world is given. The decision as to whether a veneer should be produced by slicing or by rotary peeling depends on the structure of wood and the decorative effects required.

With the exception of a few very wet and soft woods, such as basswood (*Tilia glabra* Vent.), eastern cottonwood (*Populus deltoides* Bartr. ex Marsh.), Port Orford cedar (*Chamaecyparis lawsoniana* Parl.), Sitka spruce (*Picea sitchensis* Carr.), and yellow poplar (*Liriodendron tulipifera* L.), steaming or boiling of logs is very desirable.

Actually the boiling temperature of water is seldom used. Almost always the temperature is less than 100 °C (212 °F).

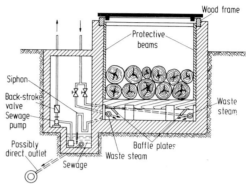

Fig. 3.9. Schematic cross section through a steaming pit for direct heating. From *Doffiné* (1956)

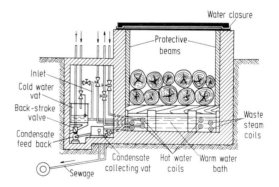

Fig. 3.10. Schematic cross section through a steaming pit for indirect heating. From *Doffiné* (1956)

Steaming gives plasticity to the logs for slicing and peeling and facilitates the subsequent process of veneer drying. The production rate of veneer dryers is higher when veneer from steamed logs is dried. Steaming is done by heating in steaming pits or vats either directly or indirectly. Direct steaming is always perilous for the logs, because direct contact of the steam with the logs greatly increases the formation of cracks and checks. Furthermore, direct steaming very often pollutes the logs with oil mixed with steam. Indirect steaming avoids this risk completely, but the installation is more expensive. In the long run indirect steaming is cheaper than direct steaming. Figs. 3.9 and 3.10 show schematically both installations. *Doffiné* (1956) compares direct and indirect steaming as follows:

	Direct steaming	Indirect steaming
Advantages	1. Simple arrangement of heating coils 2. Utilization of exhaust steam with lowest possible pressure (1.08 to 1.1 atm)	1. Very mild treatment of wood with lowest possible losses 2. Possibility to regain condensation water, therefore efficient management 3. No difficulties in removing condensation water 4. No necessity to use oil-free steam 5. Very high economy in combination with high pressure hot water systems
Disadvantages	1. Necessity of thorough control in order to avoid wood losses 2. No possibility to regain condensation water, therefore unfavorable economy 3. Need to use completely oil-free steam	1. Necessity of somewhat higher steam pressure (1.3 to 1.4 atm) 2. Extensive and expensive heating coil system

Kollmann and *Hausmann* (1955) found that the higher investment costs of reinforced concrete steaming pits in the case of indirect steaming are more than compensated for by the lower steam consumption. In the direct steaming of beechwood the steam consumption varied between 181 and 222 kg/m³ (11.3 and 13.9 lb./cu. ft.) whereas in indirect steaming only 131 to 142 kg/m³ (8.2 to 8.9 lb./ cu. ft.) were necessary. The temperature increase in indirect steaming is slower and more unform (Fig. 3.11) than in direct steaming (Fig. 3.12) for the same steaming periods. *Perry* (1947, p. 99) is of the opinion that boiling "is usually preferred to steaming as this gradual and better controlled process is less likely to damage the wood structure than the more abrupt steam treatment", but this remark is valid only for direct steaming. The range of steam temperatures is normally between 50 and 100 °C (122 and 212 °F). Stumps and crotches (for slicing) require very low temperature (38 °C ≈ 100 °F) for several days, "as so much of the cutting is across the grain" (*Perry*, 1947, p. 99). *Kuhlmann* (1960) ascertained that for Okoumé a temperature of 60 °C (140 °F) is most favorable

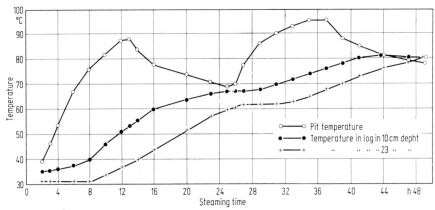

Fig. 3.11. Curves of temperature for direct steaming of wood with 12 h interruption of steam supply during night. From *Kollmann* and *Hausmann* (1955)

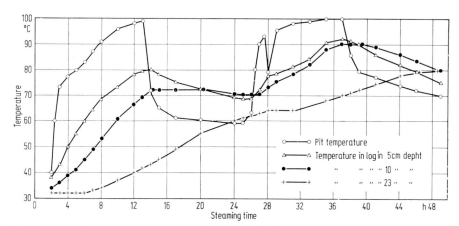

Fig. 3.12. Curves of temperature for indirect steaming of wood with 12 h interruption of steam supply during night.
From *Kollmann* and *Hausmann* (1955)

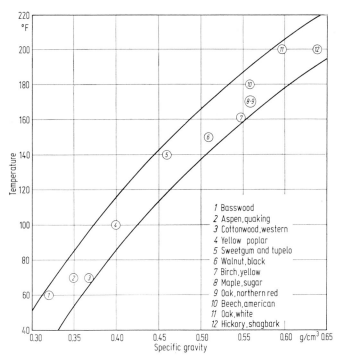

Fig. 3.13. Favorable temperature range of rotary cutting veneer of hardwood species of various specific gravities.
From *Fleischer* (1959, 1965)

and that an average log temperature of 40 °C (104 °F) is sufficient to peel smooth veneer.

Optimum temperatures can be determined and adjusted by taking into consideration the quality of cutting desired and species factors, such as specific gravity, presence of hard knots, tendencies for end-splitting and color changes in the wood (*Fleischer*, 1959—1965). In most species good cutting can be obtained over a moderately wide range of temperatures. Hardwoods of low specific gravity[1] ($\leq$0.40) such as aspen or cottonwood can be peeled at room temperature. Hardwoods, such as sweetgum with an average specific gravity of 0.46, cut well when at a temperature of 60 °C (140 °F), while yellow birch, with a specific gravity of 0.55, requires a temperature of 71 °C ($\approx$160 °F). Very dense woods such as white oak (0.60) or shagbark hickory (0.64) cut best at a temperature of about 93 °C ($\approx$200 °F). Fig. 3.13 illustrates the favorable temperature range for rotary cutting 1/8-in. (3.2 mm) thick veneer of various hardwood species.

The temperatures required for proper cutting of softwoods are generally higher than those required for hardwoods of comparable density. This may be due to the structure of many softwood species which are characterized by alternate bands of soft springwood and dense summerwood. Recommended cutting temperatures for some coniferous species (1/8-in. thick veneer) are shown in Table 3.3.

The proper temperature for other softwood species must be determined by trial.

Table 3.3. Proper Veneer Cutting Temperatures for Some Softwood Species in the U. S. A. (*Fleischer*, 1969—1965)

Species	Favorable temperature range	
	°C	°F
Douglas fir (coast type)	60···77	140···170
Fir, true (*Abies* sp.)	54···71	130···160
Hemlock, western	54···71	130···160
Larch, western	60···77	140···170
Pine, ponderosa	49···66	120···150
Pine, southern yellow	71···88	160···190
Redwood	66···82	150···180

Heat transfer in green logs has been studied at the U. S. Forest Products Laboratory by *MacLean* (1930, 1946, cf. Vol. I, p. 250—255, reference: his publications and *Fleischer*'s report and study 1959—1965). Charts have been developed that are useful for quickly calculating the time required for heating veneer bolts or logs to any desired cutting temperature at several depths, at mid-length. An example of such a chart is given in Fig. 3.14. The curves were calculated for wood having an approximate specific gravity (ovendry weight/green vol.) of 0.5 and a diffusivity of 0.00027 sq. in./sec = 0.00063 m²/h (cf. Vol. I, p. 251) when heated in water. Fig. 3.15 contains two curves, the upper of which indicates the diffusivity ratio when green wood is heated in water, the lower when green wood is heated in steam. The relative heat balance in steaming is unfavorable due to heat losses mainly through losse heating pit covers (Fig. 3.16). Tightening covers should reduce the heat losses in practice. *Kuhlmann* (1960) calculated the figures in Table 3.4.

The required duration of steaming periods depends on several factors, such as wood species, log diamter, log length and condition of the wood before the steaming, especially moisture content and moisture content gradients. Steaming not only gives the wood the necessary plasticity for peeling, but also facilitates the subsequent drying of veneer. The output of veneer dryers is remarkably higher when handling steamed rather than unsteamed veneer.

---

[1] Based on volume when green and weight when oven-dry.

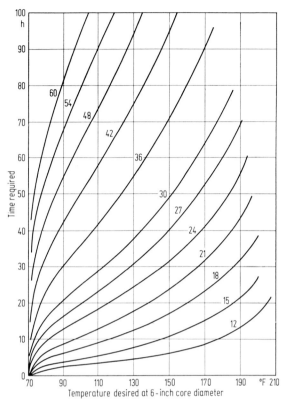

Fig. 3.14. Time required to attain desired temperatures in veneer bolts at mid-length and with 6″ core diameters for wood with an approximate specific gravity of 0.5 and a diffusivity of 0.000 27, heated in water. From *Fleischer* (1959, 1965)

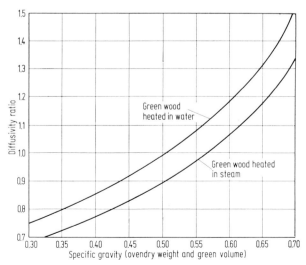

Fig. 3.15. Relation between specific gravity and the diffusivity ratio used to adjust heating times for woods of different density, heated in steam of hot water. From *Fleischer* (1959, 1965)

Table 3.4. Steam Consumption in kg/h (lb./h) for Heating Okoumé Veneer Logs by Indirect
Steaming
(heat content 643 kcal/kg or 1,157 BTU/lb.)

Log diameter	cm (in.)	60 (23 5/8)	70 (27 9/16)	80 (31 1/2)	90 (35 7/16)	110 (43 5/16)
		Consumption of steam per hour				
	60 °C (140 °F)	102 kg (224 lb.)	96 (211 lb.)	94 (207 lb.)	—	—
Average steam temperature	80 °C (176 °F)	168 kg (370 lb.)	155 kg (341 lb.)	147 kg (324 lb.)	144 kg (317 lb.)	140 kg (309 lb.)
	99 °C (210 °F)	—	268 kg (591 lb.)	262 kg (578 lb.)	240 kg (529 lb.)	232 kg (511 lb.)

1 kcal/h = 3.968 BTU/h

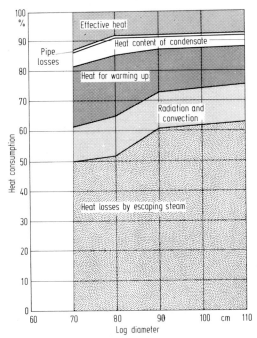

Fig. 3.16. Relative heat balance for steaming of Okoumé veneer bolts of various diameters.
From *Kuhlmann* (1960)

3.2.2.3 Debarking or Cleaning. In the need to improve the economics of
veneer and plywood production simple and reliable debarking machines in
addition to labor-saving equipment for transporting logs before and after cross-
cutting are necessary. The steamed and cut to length logs are passed to the
cleaning machine. The main function of this machine is to remove bark, bast
and dirt. Knife-edge life of slicers and rotary peelers is remarkably extended with
properly cleaned logs. The output of the most commonly used debarking machines
is approximately equal to the capacity of two normal peeling lathes.

Fig. 3.17 shows schematically a debarking machine which can handle practically all species of wood, including non-cylindrical and irregularly grown logs. Loading and removing is effected rapidly and continuously. The logs to be debarked are conveyed to the feeding chain of the debarker. The chain transports each log to the center of the machine. Two hydraulic lift cylinders raise the log to spindle height, and the spindles clamp the log and set it in an adjustable rotation.

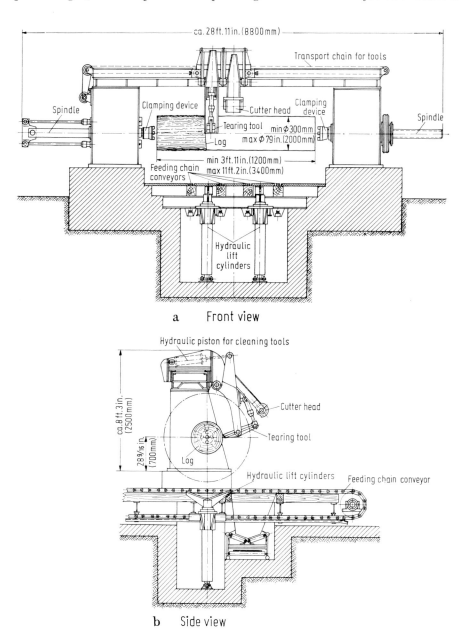

Fig. 3.17a, b. Debarking and cleaning machine.
a) Front view; b) Side view. RFR, Hamburg, now: Keller & Co., Laggenbeck

The bark is removed from the rotating log by a tearing tool, while a following cutter-head cleans the log surface. Either immediately after debarking or while the tools return to their initial position, the cutting depth of the following cycle can be adjusted. A complete debarking and cleaning cycle with the illustrated machine, including loading, clamping and discharging, for a log of 1 m (3.3 ft.) diameter and 2.7 m (8.0 ft.) length requires only about 3 min. Maximum log diameter is 2 m (6.6 ft.). The power requirement is approximately 30 kW. In the case of a highly productive veneer peeling plant the cleaned logs are transported on chain conveyors to a log magazine in front of the feeding and centering devices of the peeling lathes.

3.2.3 Veneer Sawing

3.2.3.1 Introduction. Veneer in ancient times and after the Middle Ages was produced in an inefficient and wasteful manner. The ancient Egyptians probably used edge-tools for shaping or smoothing wood. Another tool was a two-handled saw, approximately 90 cm (3 ft.) long. Later the pit saws, still used in many regions of South East Asia, were usual. This primitive pit sawing technique is as follows: A pit is dug and a log is supported over this pit in which a worker stands. A second man stands over the log and the pit. Both workers saw simultaneously. The waste is very high and the procedure does not produce veneer of uniform thickness, but rather wavy boards.

3.2.3.2 Circular Sawing of Veneer. The first circular saw for the production of veneer (thickness 1.6 mm ~ 1/16″) was used in England in 1805. In 1860 a circular veneer saw with a diameter of 350 mm (1 ft. 2 in.) was installed in a veneer factory and worked well. Later, in England, the construction of circular veneer saws was improved by the invention of segment saws. The cutting edges of these saws consisted of 16 special steel-segments. These steel-segments were securely bolted to a circular blade which at the maximum had a diameter of 2,050 mm (6 ft. 9 in.). The whole blade was stiffened by a heavy cast-steel support (Fig. 3.18). The (hickness of the teeth at the cutting diameter was reduced by grinding to 0.9 mm ~1/32 in.). The set of the teeth averaged only 0.25 mm (1/100 in.). It was possible to cut veneer about 1 mm (~0,04 in.) thick, but the yield was extremely low. Sawing of thin veneer is a very wasteful process. With the figures given above the yield is only about 42%. The economic conversion of valuable woods to veneer by this technique is impossible. The cutting velocity was rather high; for instance, the largest circular blades operating at 480 rpm had a cutting velocity of about 51 m/sec. The feed speed was as much as 2 m/min (6.6 ft./min). In comparison

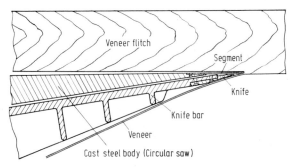

Fig. 3.18. Section through a veneer-circular sawblade with cutting segments, turn-away knife and knife support. From *André*, in *Kollmann* (1962)

with the veneer gang saw the efficiency of the veneer circular saw was approximately threefold. This figure is misleading, however, because the resharpening of the teeth of the circular sawblade required a rather long time, during which the circular saw could not be operated. The exchange of dull gang saw blades required only a few minutes, while the sharpening of a veneer circular saw with inserted teeth required an average of 120 min. Directly behind the circular saw teeth the veneer was turned away from the saw by means of a so-called splitter so as to produce minimum stresses in the weak steel segments.

Fig. 3.19. View of a modern veneer gang saw with horizontal movement of the sawblade.
From *André*, in *Kollmann* (1962)

Maximum flitch dimensions had a length of 6.5 m (21.3 ft.), a width of 0.7 m (2.3 ft.) and a height of 0.6 m (2 ft.). The thickness of the veneer was adjusted mechanically to an accuracy of approximately 0.1 mm (~0.004 in.). At the beginning of the 20th century the technology of veneer circular saws was kept private. In England about 1935 the use of veneer circular saws was discontinued. In Germany only few machines of this type were in use, in the U. S. A. some veneer mills occasionally sawed very hard species whose color would be spoiled by cooking. The following wood species may be mentioned: satinwood, black ebony, rosewood, kingwood, mahogany curls and quartered oak. The flitches to be cut were approximately in the green state. Circular veneer saws produced more ruptured grain and similar defects than other types of veneer saws. In general the structure and mechanical properties of sawn veneers are excellent, but woods containing tannic acid are extremely sensitive to its effect. They must be treated with chemicals to avoid staining as a result of extended contact between the wood and the steel of the tools. It is important to remove carefully sawdust from the surfaces of the sawn veneer, otherwise discolorations are unavoidable.

12*

3.2.3.3 Band Sawing of Veneer. Band saws could be manufactured of much thinner steel plates than ordinary circular saws. *Perry* (1947, p. 24) reports that the first patent was granted in England to *William Newberry* in 1808. The principle of construction was quite similar to that of modern band saws. The use of band saws for the manufacture of veneer was very limited and practically disappeared around 1870.

3.2.3.4 Horizontal Gang Saws for Veneer Manufacturing. The horizontal frame saw or gang saw was created about 1880. Fig. 3.19 shows a modern machine. The horizontal reciprocating movement of the frame with the saw blade follows a slight curve to facilitate "free-cutting" of the saw blade and smooth feeding of the wood block. The crankshaft motion is transmitted by a connecting rod to the frame. The usual length of the connecting rod is about 3.5 times the stroke. The horizontal veneer gangsaw has an average cutting velocity of 7 m/sec (1,380 ft/ min), feed speed is as much as about 0.7 m/min (2.3 ft./min), and power consumption amounts to about 4 kW. The volume of sawn veneer is very low since the waste due to the saw kerf is too high in comparison with the kerfless slicing and peeling. Even so horizontal veneer gang saws have some advantages. The movement of the saw blades is very regular, with no vibrations. To reduce the waste the saw blades should be as thin as possible (0.9 mm, 1/28 in.) with a set of 0.25 mm (1/100 in.). The yield for veneer of 1 mm (1/25 in.) thickness is only about 41.6%. An important feature is the so-called "knife" (Fig. 3.20) which keeps the saw blade in correct position and relieves the pressure on the saw teeth

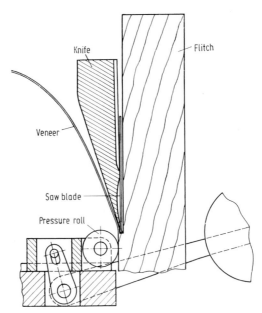

Fig. 3.20. Position of knife and pressure roll on a veneer gang saw. From *André*, in *Kollmann* (1962)

by turning aside the veneer immediately after the cut. A pressure roll gives exact guidance to the block. Checks in the veneer are completely avoided. Check-free veneer produced with horizontal gang saws is used in the manufacture of musical instruments, thicker oak veneer and veneer of Alpine stone pine for panelling.

3.2.4 Veneer Slicing

3.2.4.1 Introduction. As mentioned before sawn veneer is now limited to special purposes such as musical instruments. Exact statistics on veneer production do not exist, but probably at least 95% of veneer is produced by rotary peeling, perhaps 3 to 4% by slicing, and the balance by sawing.

The methods of manufacturing veneer depend on the use of veneer. *Grosshennig* (1962) reported on the methods of manufacture and use of veneer:

a) Decorative veneer for furniture and panelling is produced either by slicing or rotary peeling;

b) Eccentric clamping with special devices in rotary lathes produces interesting figures in veneer;

c) Veneer for cores and for the core stock and outer layers of softwood plywood is manufactured on rotary lathes only;

d) Face veneer for core stock can be produced either by rotary peeling or by slicing;

e) Veneer for construction uses may be manufactured by rotary peeling or by slicing;

f) Very thin veneer, e.g. for paper-hangings, 0.05 to 0.2 mm thick, can be produced only by rotary peeling.

The requirements in veneer quality depend upon the methods of manufacture and have an influence on the construction of the machines. The prerequisites are:

a) Uniform thickness of the veneer over the whole surface;

b) No cracks or breaks on either surfaces;

c) Smooth surfaces on both sides.

It is difficult to meet these requirements for all wood species. Grain orientation has an important influence, as has the pretreatment by boiling or steaming. Only a few wood species can be manufactured into plywood with the veneer in green condition. In these cases difficulties during the subsequent drying may occur. Some species such as maple are very sensitive to higher temperatures, therefore they are sliced or peeled at room temperatures to produce unstained veneer.

The production of high quality veneer requires that each of the following conditions be considered:

a) Suitable wood species;

b) Adequate quality of the log, correct grain structure, and growth rate;

c) Proper pretreatment of the wood by boiling or steaming (temperature and time);

d) Adequate cutting velocity;

e) No vibrations within the machine;

f) Proper geometry of the cutting knife, especially between knife and nosebar;

g) Replacement of cutters at the right time;

h) Protection of iron parts of the machine by lacquer, chromium plating or heating etc. against chemical effects of condensation water containing acids (e.g. tannin) to avoid staining of the veneer.

3.2.4.2 Horizontal Slicing of Veneer. Decorative veneer as well as veneer for face layers are manufactured by slicing and by rotary peeling. For this reason most veneer plants have both types of machines. Therefore it is desirable to compare the various types of machines in the industry.

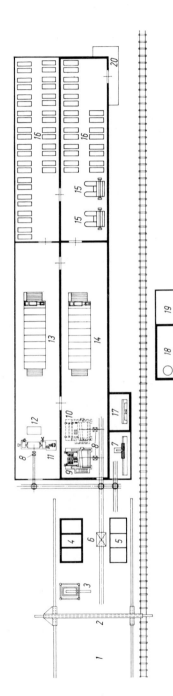

Fig. 3.21. Schematic layout of a veneer plant:

1 Log yard, *2* Bridge crane, *3* Cross-cut saw, *4* (*5* enlargement) Steaming pits, *6* Truck, *7* Log band saw, *8* Electro-crabs, *9* (*10* enlargement) Veneer slicer, *11* Lathe for eccentric peeling, *12* Offload table for veneer, *13* and *14* Veneer dryers, *15* Veneer pack jointer, *16* Veneer piles, *17* Knife grinder, *18* Boiler house, *19* Machine hall (Electric generators), *20* Loading ramp. From *Grosshennig*, in *Kollmann* (1962)

Veneer slicers should be constructed in such a manner that they can be operated easily and that the movement of the veneer produced is easy. Fig. 3.21 shows the layout of a modern veneer plant where a rotary lathe and slicers are combined. The logs from the woodyard 1 are transported by a travelling crane to the cross cut saw 3; from there they usually are moved to the steaming pits 4 and 5, but occasionally on a plant truck 6 via a debarking station to a flitch band saw 7. A chain saw or a horizontal gang saw may be used. The round logs are converted to flitches, using cutting methods which vary with wood species, growth properties, wood figure desired, and experience in the factory. The common methods of log preparation are illustrated in Fig. 3.22.

After this pretreatment the logs are steamed and carried on trucks or the more modern roller conveyors to the slicers 9 and 10. Electric hoists precisely position the logs in the slicer.

In a similar way logs are transported to the rotary peelers 11. The veneer passes through the dryers 13 and 14, after which its volume is determined, and it is cut into packs on automatic clippers (joinders or guillotines) and bundled for storage.

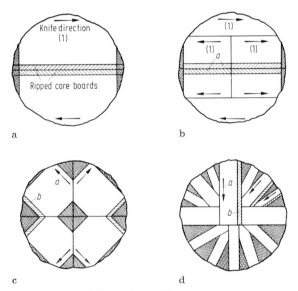

Fig. 3.22a — d. Type of flitch preparation.
a) Simple bisectioning for production of crossband veneer; b) Oak logs for heart ripping; c) Preparation for quarter sawing; d) Sectioning in flitches. The arrows indicate schematically the movement of the sawblades.
From *Grosshennig*, in *Kollmann* (1962)

In Europe high-grade veneer is manufactured primarily on crank-driven horizontal slicers. Compared with rotary veneer peelers the number of slicers is small.

The two major types of slicers are either horizontally or vertically oriented, with the latter originally constructed in the U. S. A. Both types of machines have advantages and disadvantages. Each slicer consists of a sturdy bed to which the flitch is fixed by dogs; the powerful knife is carried on a rigid frame. In the European or horizontal machine the bed is fixed and the knife is driven across the flitch, whereas in the American or vertical slicer the knife is fixed and the bed is movable (*Wood*, 1963). Fig. 3.23 shows the operating principle and the general construction of a horizontal slicer. The knife beam and the pressure bar move jointly to and fro in the machine, reciprocated by a crank drive mechanism. The table carrying the flitch to be sliced is elevated by feed screws after each stroke, the amount of lift applied being equal to the required veneer thickness.

Because of the very great inertial forces involved special attention has been given to the design of the crank drive. The crank wheels are separate from the drive gearing and between them an additional mainshaft bearing is located.

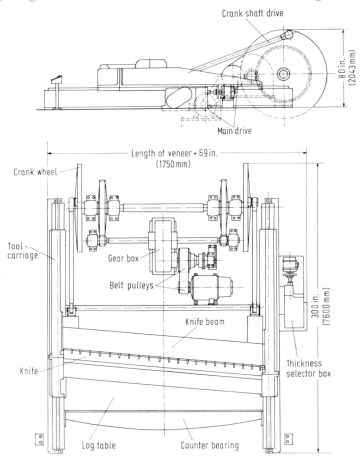

Fig. 3.23. Heavy duty veneer slicer. Above: side view, below: top view. RFR, Hamburg, now: Keller & Co., Laggenbeck

For best results the machine should operate at the speed best suited to the type of wood being handled, and this calls for a variable-speed main drive.

Continuously variable speed control is, of course, preferable and is obtainable by using a D. C. motor or by installing a Ward Leonard set. Cuts per minute vary from 30 to 45.

The veneer thickness is set by means of a selector box. This box operates via a cam and gives perfectly smooth and troublefree operation even when slicing very thick veener. For veneer thicknesses up to 0.10 in. (2.5 mm) the incremental progression is 0.001 in. (0.025 mm); up to 0.20 in. (5.0 mm) increments of 0.002 in. (0.05 mm) are obtainable, while up to 0.4 in. (10 mm) the thickness can be changed in steps of 0.004 in. (0.1 mm).

The heavy duty veneer slicer with a crank drive was originally developed for bolts with lengths not exceeding 4,000 mm (13 ft.). With a better utilization of forests bolts of greater length and diameter became available, and larger machines were necessary. The factories constructed veneer slicers able to handle flitches up to a maximum length of about 5,100 mm (16 ft. 9 in.). Machines of this size meet practically all demands.

3.2.4.3 Vertical Slicing of Veneer. In most models of veneer slicers in use in the U. S. A. the flitch moves downward over the knife to cut the veneer (Fig. 3.24). Occasionally, however, the knife is moved by power, and the flitch moves only to regulate the veneer thickness. Machines of the second type are usually relatively small machines.

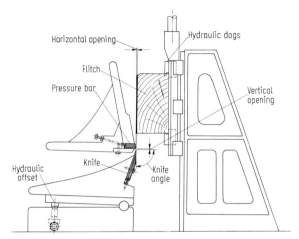

Fig. 3.24. Diagram of the cross section of a veneer slicer.
From *Lutz* and *Fleischer* (1963)

In operating a slicer with a stationary knife the flitch is held on a heavy steel frame by hydraulic dogs. The dogs grip, as Fig. 3.24 shows, the flitch at the top and bottom and hold it firm. The flitch table is given an alternating up-and-down movement on rigid supports. On each downward stroke the flitch is pressed against the edge of the knife. A nosebar insures proper alignment. On the up-motion the knife automatically recedes, so that there is no interference with the flitch; the next step is an automatic advance of the knife for regulating the thickness of the next veneer. The veneer passes through a slot between the knife and the nosebar. It is turned over by hand, and each successive slice is stacked in order. Large slicers operate at speeds of 35 to 50 strokes/min, but some machines operate at 100 strokes/min, and very small slicers for the manufacture of box shook and battery separators operate at up to 240 strokes/min (*Lutz* and *Fleischer*, 1963).

The most important and critical task in cutting good quality veneer either on the slicer or the lathe is proper positioning and alignment of the knife and the nosebar. These settings can be facilitated by the use of suitable instruments (*Fleischer*, 1956). For satisfactory cutting, slicer knives are generally ground to a bevel of 20° to 22°. The slicer knife is set with ahead of about one third of a degree into the flitch. After the knife is set, the nosebar is aligned to it. The edge of the rigid nosebar is generally set about 0.030 in. (0.75 mm) above the knife edge and the horizontal opening is about 0.007 in. (0.17 mm) less than the veneer thickness. Lower bar settings may cause excessive splintering at the end of the flitch, smaller horizontal openings may cause bumping and vibration of the knife carriage. Some features of a typical veneer slicer (Fig. 3.25) are as follows:

a) Sizes up to 4,000 m (∼13 ft.) flitch length;

b) The base is made up of four main members, two of cast iron and two of fabricated steel, cross ribbed and stiffened inside, machined, and bolted together;

c) There are three A-frames, fabricated from steel;

d) The cast iron flitch bed is made as light as possible. The steel dog screws have hardened points;

e) New machines have electro-magnetic brakes instead of hand brakes;

f) The connecting rods are fabricated from steel; counter balances are provided in the drive gears and on each main gear shaft;

g) The knife bar is of heavy massive construction. The knife is held in place with cap screws placed on 5 in. centers along the entire length of the knife;

h) The nosebar cap is of a design similar to the knife bar;

i) The feed is effected by means of a ratchet operated from an adjustable crank and a rod on one of the main drive gear shafts. The slicer cuts approximately 30 to 40 slices per minute;

k) Veneer passes off the knife over a brass plate on to a conveyor which is a part of the machine. The conveyor consists of a number of belts which are individually adjustable. The smaller type has a total power consumption of 34 kW (46 HP), the larger of 41 kW (56 HP).

Fig. 3.25. Side view of a modern American vertical veneer slicer. Johnson City Foundry & Machine Works Inc., Tennessee

3.2.5 Veneer Rotary-Peeling

At present 95% or more of the veneer is produced by rotary-peeling. Rotary-peeling or rotary-cutting is the basis for the modern plywood industry.

There is a similarity between turning and rotary-peeling, but the residues in turning curled chips are more or less worthless and in most cases they are burned. Rotary-peeling has the aim to produce an endless "chip" which is the veneer. The veneer should be smooth on both faces. This is, of course, not completely possible because during peeling the veneer is bent to cause different stresses on the two sides. Modern rotary peelers must be very sturdy and well bedded so that shocks or vibrations are avoided. This is necessary since different wood species with various densities, structures and contents of knots are to be peeled.

In any peeling machine the log is mounted between two centers and is revolved against a knife fixed over the whole length of the log. Considering the
technology of rotary-peeling it is necessary to distinguish two actions which are
difficult to separate from each other. The two actions are cutting and splitting.
With material of an anisotropic and heterogeneous character such as wood, any
cutting action is accompanied by splitting. Splitting causes undesirable conse-
quences.

At the instant when the knife engages the wood and the cutting action begins,
the wood tends to split along the grain and in advance of the knife edge. With
respect to the cross section of the veneer this advance splitting causes the formation of vertical cracks (known as lathe checks) which reduce the smoothness of
the surface of the veneer (Fig. 3.26 and 3.27). The risk of this checking is avoided
or reduced by using a nosebar.

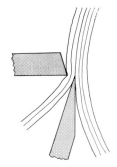

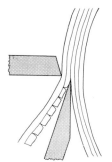

Fig. 3.26. Scheme of veneer
peeling. Nosebar correctly
adjusted. RFR, Hamburg, now:
Keller & Co., Laggenbeck,
W.-Germany

Fig. 3.27. Scheme of veneer
peeling. Nosebar wrongly
adjusted, no pressure applied.
RFR, Hamburg, now:
Keller & Co., Laggenbeck,
W.-Germany

Terms and principal activities in veneer rotary-peeling are as follows:

1. On *rotary lathe carriage:*

a) The *nosebar* is the device which bears on and applies pressure on the log at
the knife edge;

b) The *pressure bar* is the metal casting which holds the nosebar. Unfortunately
due to the lack of standardized terminology in some publications the term "pressure
bar" is used as being identical with the term "nosebar".

2. On the *rotary lathe spindles:*

a) The *chucks* are the "holders" which are on the ends of the spindles, and
which support the veneer bold and deliver rotational torque to it;

b) *Dogging* is the process of driving the chucks into the veneer bolt (by spindle
movement). This is also sometimes referred to as *chucking.*

The purpose of the nosebar is to compress the wood ahead of the knife so
that the knife can make a clean cut without the advance splitting. The pressure
applied by the bar should be adjusted to the species of wood being cut. Figs. 3.28
and 3.29 show clearly the importance of correct adjustment of the nosebar and the
knife. The first figure demonstrates that an incorrect adjustment is accompanied
by the development of severe lathe checks, while the second figure shows that
a proper adjustment produces smoother veneer with much less severe checks.

Fig. 3.28. Cutting 5/16 in. (7.9 mm) yellow birch veneer without nosebar pressure. The lathe checks and the roughness of the veneer surfaces are pronounced. From *Fleischer* (1949)

Fig. 3.29. Cutting 5/16 in. (7.9 mm) yellow birch veneer with an opening 0.27 in. (6.8 mm), 68% of the veneer thickness. Thickness and smoothness of the veneer were excellent, but the wood was overcompressed. From *Fleischer* (1949)

The position chosen for the knife and the nosebar has been studied thoroughly over many years. Some of the developments are illustrated in Fig. 3.30 in their chronological order (*Schreve*, 1937). In the original construction (position *a*) a peeling machine had a fixed vertical knife with the flat side oriented to the log; both tools, knife and nosebar, were set to the horizontal centerline of the log. This construction was unsatisfactory since the flat side of the knife was exposed to wear. In this case regrinding becomes highly uneconomical. Position *b* shows the knife on top above the horizontal centerline of the log. The direction of log-rotation is opposite to that in position *a*. The solution is poor because the rotation of the log causes all the resulting thrust loads to act upwards. A further disadvantage is the difficulty in handling the veneer when discharged in the vertical direction. The position *c* in Fig. 3.30 is unsatisfactory since the essential parts of knife and nosebar are covered by the log and are camouflaged to the operator's view.

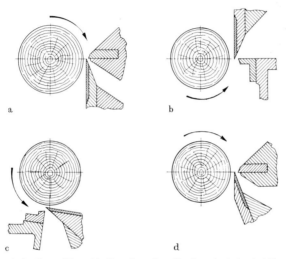

Fig. 3.30. Developments for the position of knife and nosebar. (In chronological order) From *Schreve* (1937)

Position *d* in Fig. 3.30 is correct from the design point of view and gives excellent results in practice. The veneer is discharged according to the law of gravity, all loads are distributed evenly over the main structural parts of the machine, and the knife presents its beveled surface to the log. This is the principle on which most modern veneer cutting machines are now based.

The relative position of knife and nosebar is shown in more detail in Fig. 3.31.

The following considerations are based on a report by *Fleischer* (1949). The quality of rotary cut veneer depends mainly on the knife and on the nosebar. "These are mounted in a carriage that moves ahead a distance equal to the thickness of the veneer being cut each time that the bolt revolves once in the lathe." Fig. 3.31 shows a diagrammatic drawing of veneer bolt, veneer knife and nosebar in their relative position during cutting operation and clarifies also the terminology used.

As the veneer is cut it must bend rather sharply because the knife acts as a wedge. The wedge angle must therefore be kept as acute as possible but sufficient support for the knife edge must be provided and undue deflection and breakage prevented.

The knife is set in the lathe with its edge horizontally in line with the centers of the spindles that support the bold (cf. Fig. 3.31), and its bevel side approximately vertical and tangential to the bolt at the point of cutting. The wood generally is somewhat compressed at this point by the nosebar (Fig. 3.32a, right side).

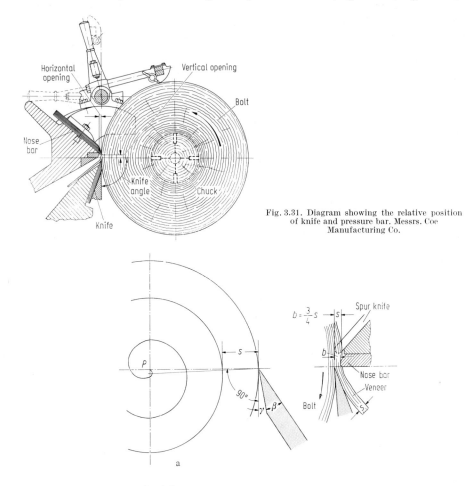

Fig. 3.31. Diagram showing the relative position of knife and pressure bar. Messrs. Coe Manufacturing Co.

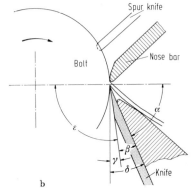

U S A Symbol	Term		
	Amer.	Brit.	Germ.
α	Rake angle	Cutting angle	Spanwinkel (γ)
β	Bevel or wedge or lip angle	Grinding (sharpness) angle	Keilwinkel (knife chamfer angle (RFR)(β)
γ	Clearance angle	Clearance angle	Freiwinkel (α)
δ (Δ)	Chip deflection (also cutting) angle		Schnittwinkel
$\varepsilon = 90° \pm \gamma$	Knife setting angle		

Geometrical relationships 1. $\beta + \gamma = \delta$ 2. $\alpha + \beta + \gamma = 90°$

Fig. 3.32. Setting of cutting angle

Due to the bending at the knife edge the veneer has a tendency to break or split to produce lathe checks. These characteristic breaks always extend inward from the under side, or "loose side", of the veneer (Figs. 3.27 and 3.28). The opposite side of the veneer is the "tight side".

The amount of pressure applied on the wood by the nosebar can be determined and measured in terms of both the horizontal and the vertical openings between the knife edge and the nosebar edge (Fig. 3.31). *Fleischer* (1949) found that suitable cutting may be obtained in most cases by varying the vertical opening o_v of the bar approximately proportional to the thickness s of the veneer, for instance $o_v = 0.005$ in. (≈ 0.13 mm) for $s = 1/100$ in. (≈ 0.25 mm) veneer to a maximum of $o_v = 0.30$ in. (almost $1/32$ in. (≈ 0.76 mm) for $s = 1/8$ in. (≈ 3.2 mm)). With respect to the horizontal nosebar opening experiments at the U. S. Forest Products Laboratory demonstrated that when the nosebar is withdrawn to the point where no pressure is applied, the veneer is often rough and generally contains severe lathe checks. When the bar is moved ahead the horizontal opening is reduced and a pressure is applied. In this case the lathe checks are almost completely eliminated, and the veneer is relatively smooth (Fig. 3.26 and 3.29). The thickness of the veneer may be reduced slightly because of overcompression. Excessive compression in the case of some softwoods may produce "shelling" or "slivering" due to a separation between the thin-walled springwood and the thick-walled summerwood. "The degree of tightness and smoothness, however, may also be markedly affected by the temperature of the wood at the time of cutting and by other factors such as the sharpness of the knife" (*Fleischer*, 1949).

The following experimental results are presented by *Fleischer* (1949) in his report:

1. Knife angle adjustments or knife setting angles varying from 89 degrees, 30 min, to 90 degrees, 30 min, according to the species (Fig. 3.33), were found to be most satisfactory;

2. There is a definite trend toward higher knife setting angles as the veneer thicknesses decrease;

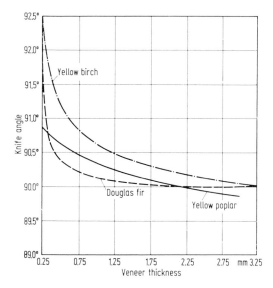

Fig. 3.33. Knife angle for rotary cutting veneer of various thicknesses for 3 wood species.
From *Fleischer* (1949)

3. If the knife setting angle is too small the veneer produced varies in thickness in waves that may be several feet long.

4. If the knife setting angle is too high for a given thickness of veneer of a particular species, a limit will soon be reached beyond which a further increase results in corrugated veneer. In this veneer very short waves occur, generally only 1/4 in. (6.4 mm) to 3/8 in. (9.5 mm) from crest to crest, and apparently result from a rhythmic vibration of the knife edge.

The diameter of the log decreases continuously during the peeling, and the effect of this is a reduction of the bearing area of the knife against the log (Fig. 3.34). Modern veneer lathes, therefore, are provided with devices to automatically adjust the knife setting angle, so that a constant bearing area for the knife results. If the knife setting angle becomes too high the veneer thickness becomes nonuniform and the log surface resembles a washboard. If the bearing area of the knife on the log is too large, the veneer quality is poor. A prerequisite for high veneer quality without thickness variations is also ruggedness of the lathe.

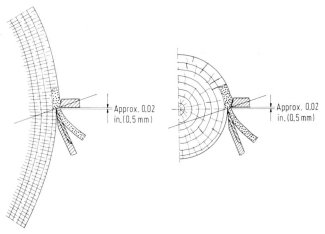

Fig. 3.34. Scheme of the reduction of log diameter in rotary peeling. RFR, Hamburg, now: Keller & Co., Laggenbeck

The kinematics of veneer peeling is complex. The veneer is peeled from the round log in a tangential direction in an archimedrian spiral with equidistant windings. The distance between the singular windings is equal to the veneer thickness (Fig. 3.32 a left side). The meaning of knife setting angle, wedge or sharpness angle and clearance angle is illustrated in Fig. 3.32[1]. The clearance angle is according to *Andresen* (1934) especially important. Each species of wood has an optimum cutting velocity and an optimum cutting angle. It may be repeated:

---

[1] There is no internationally acknowledged terminology for angles in veneering. The author designed the sketch Fig. 3.32b after studies of the literature (*Andresen*, 1934, in *Kollmann*, 1936, p. 608/609; *Fleischer*, 1949; *Harris*, 1946; *Reinmuth*, 1950; *Sacht*, 1951; U. S. For. Prod. Lab., Rep. No D 1766-6, 1951, Vol. I of this book, 1968, p. 478) and after a correspondence with Mr. *J. C. Beech*, For. Prod. Res. Lab., Princes Risborough; Prof. *Fred E. Dickinson*, For. Prod. Lab., Richmond, Cal.; Mr. *John F. Lutz*, U. S. For. Prod. Lab., Madison, Wisc.; Mr. *W. M. McKenzie*, For. Prod. Lab., South Melbourne; Prof. Dr. *G. Pahlitzsch*, Inst. f. Werkzeugmaschinen und Fertigungstechnik, TU Braunschweig; Mr. *Torsten Englesson*, Swed. For. Prod. Res. Lab., Stockholm). Reference is made also to Mr. *McKenzie*'s proposals for a uniform terminology (I. U. F. R. O. Sawing and Machining Group, 1965); CIRP-Wörter-buch der Fertigungstechnik, Vol. 4, Grundbegriffe des Spanens, Verlag W. Girardet, Essen 1969; DIN 6581 (Mai 1966) Geometrie am Schneidkeil des Werkzeugs.

If the cutting angle is too high the knife edge develops vibrations and due to the interaction of summerwood and springwood within the annual rings the veneer becomes corrugated. If the cutting angle is too small the bearing surface between the cutter and the log becomes too large, the power consumption of the machine increases and the veneer thickness varies in long waves.

Andresen (1934) investigated the theoretical interrelationships between clearance angle, log diameter and vertical distance h between knife edge and spindle axis. He found that in peeling logs with small diameters the knife angle may be kept constant. For logs with diameters from 100 to 300 mm (about 4 to 12 in.) the knife setting angle should be 91 to 92 degrees. The distance h (taken as positive above the spindle axis and as negative below the spindle axis) has an influence on the optimum clearance angle (Fig. 3.35 a for lathes with common support, b for lathes with automatic control of clearance angle).

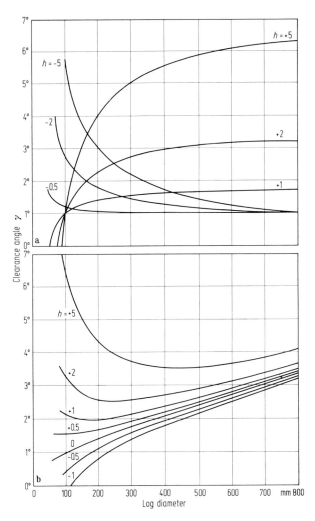

Fig. 3.35a, b. Dependence of the clearance angle on the log diameter and on the vertical distance of the cutting edge from the log axis.
a) Common support; b) Support with automatic control of the clearance angle. From *Andresen* (1934)

The theoretical consideration by *Andresen* (1934) and the results of *Fleischer*'s (1949) experiments are well confirmed by practical experience in the U.S.A. and in Europe. One important point is that apparently the optimum cutting angle does not depend upon the temperature of the wood.

The usual practice today is to adjust the cutting angle to the properties of the wood species concerned. Table 3.5 gives a few values accepted in practice.

Table 3.5. Cutting (Chip Deflection) Angles for Rotary Peeling Lathes for Various Wood Species

Wood species	Cutting angle $(\delta = \beta + \gamma)$ in degrees
Walnut	15···17
Oak	17
Birch	18···20
Beech	20···21
Okoumé	22
Poplar	22···23
Spruce	20···21 $\left.\begin{array}{l}\\ \gamma = -1°\end{array}\right\}$

Another point is that the veneer knives must have adequate hardness; the Brinell-hardness-values should range between 550 and 625 kp/mm². Hardness is defined as the resistance of the surface of a material against penetration of a harder test body. There exists a great number of test methods that use independent hardness scales. The values of hardness obtained, for instance, by static impression of a metal as hard as steel or diamond ball, after *Brinell* or *Martens*, or of a pyramid (opening angle of the surfaces 136°) after *Vickers* are calculated as the ratio of the force applied to the area of impression. The Rockwell-C-hardness is determined as the depth of penetration of a diamond-cone (opening angle 120°, rounded point with 0.2 mm radius) after application of a pre-load of 10 kp and then an additional load of 60, 90 or 150 kp. The arithmetical difference between the depth of penetration caused by the major load and that caused by the minor load as substracted from an arbitrary constant, and the resultant figure expresses the hardness of the specimen (*Hengemühl*, in *Siebel*, 1939, p. 360). A dynamic hardness test consists of the measurement of the height of resilience of a dropping test hammer (after *Shore*).

Fig. 3.36. View of a veneer lathe. RFR, Hamburg, now Keller & Co., Laggenbeck.

In Europe and in the U. S. A. peeling lathes[1] are available covering log diameters ranging from 400 to 2,000 mm (16 to 79 in.) and veneer lengths ranging
from 800 to 3,800 mm (32 to 150 in.) (Fig. 3.36). Economic reasons limit the size
of the machines. If the peeling lathe is not adequately massive in cutting a long
log, the pressure of the knife on the log midway between the chucks may cause
bending and "chattering" of the log. At one time such a log had to be removed
from the peeling lathe and reduced in length. At present, hold-down or anti-
chattering devices (back-up rolls) are available which guarantee a fixed position
of the log even for reduced diameters. The principles of back-up rolls are illustrated
and explained in the legend in Fig. 3.37 (*Grosshennig*, 1971). A leading firm in
the U.S.A., the Coe Manufacturing Company, Painesville/Ohio, and Portland/
Oregeon, explains:

"In order to control core deflection at smaller diameters, particularly when
cutting soft species, and to minimize variations in veneer thickness, a back-up
roll mechanism is required. This can either apply a pre-set pressure or be controlled by a follower roll.

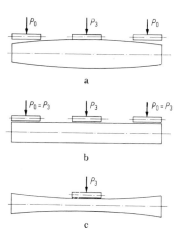

Fig. 3.37 a — c. Principles of working methods o
back rolls with the aim to avoid flexure of the
cores.
a) The back roll in the middle exerts a load P_3 until
the other both back rolls are in contact with the
core; in this case $P_0' = P_3'$; b) If the 3 back
rolls from the beginning are in contact with a
cylindrical peeling block, then, supposed the load
P_3 is adequate, no flexure occurs; c) The single
back roll in the middle workssalone and with a
load P_3 too high. From *Grosshennig* (1971)

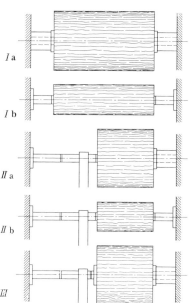

Fig. 3.38. Scheme of arrangement and working
of telescopic spindles for rotary veneer
cutting lathes. RFR, Hamburg, now: Keller &
Co., Laggenbeck

With a 'Pilot Operated Back-Up Roll', automatic follower contact with the
log occurs after it has been reduced to approximately 10 in. (250 mm) in diameter.
A tracer arm and wheel assembly located near the chuck end of the log is used
to 'follow' the reducing core diameter. As the core tends to bend away from the
knife additional pressure is automatically applied to maintain the back-up roll
assembly in the same relative position as the tracer arm. The tracer arm feature
eliminates over- and under-pressure on the log, resulting in flat veneer from the
minimum core diameter.

---

[1] The author is indebted to two firms for information: in Germany, RFR, W. Ritter,
C. L. P. Fleck and Sons, and A. Roller, now C. Keller & Co, Laggenbeck/Westphalia; in the
U. S. A., Coe Mfg. Company.

Several different types of back-up rolls are available to suit individual requirements. Among the most popular is the single roll with automatic pilot operation. In some applications double back-up rolls are preferred.''

Nevertheless, for every log cut a core remains to be discarded. The diameter of this core depends on the original diameter before peeling and on the defects in the log. Formerly there were special rotary cutters, designed for peeling small-diameter logs. They were fitted with relatively small chucks, and the cores coming from these lathes were small in comparison with those from lathes cutting large logs. In order to take advantage of the modern veneer peeling line and to increase the yield of veneer produced from logs with bigger diameters the invention of telescopic spindles was important. Telescopic spindles consist of two spindles, one sliding into the other. The outer spindles are hollow-bored and carry a pair of large chucks, whereas on the inner spindles the smaller chucks which are attached can be retracted into the large spindles. Both spindles are movable along the axis of the lathe and are mounted on the ends of the machine. At the beginning the log is peeled with the larger chucks and spindles; then having attained a predetermined log diameter the outer spindles are retracted automatically and the log is peeled to the smallest possible core diameter without stopping the machine. These telescopic spindles (Fig. 3.38) allow the peeling of large diameter logs down to smallest possible core diameters without changing the chucks or stopping the machine. Dogging and spindle movement are done hydraulically. The log is mounted initially between the two larger chucks (Fig. 3.38, I a, Fig. 3.38, I b). When the log is shorter than the minimum peeling length determined by the range of normal spindle movement a spindle extension must be inserted. When the core of the log is defect free, the spindle extension using the smaller chuck and the stay bracket (spindle support) can be mounted on one end and on the other end the outer and inner chucks can be fixed (3.38 II a). When the log is peeled to a predetermined diameter, the outer chuck is retracted, and the log can be peleed to the diameter of the smallest chuck (Fig. 3.38 II b).

When the core, however, is fairly large the bigger chuck must be fixed onto the extension. With this arrangement the retraction of the outer chucks on the other side would serve no useful purpose (Fig. 3.38, III). In this instance the minimum core diameter is fairly large.

For proper peeling correct pressure of the nosebar is important. The higher the pressure the greater becomes the power consumption and the wear of the lathe spindles bearings. Round logs with a uniform structure can be peeled with higher pressures than logs with irregularities. According to *Fleischer* (1949) sharp edges of the nosebar are more favorable than round edges.

The cutting speed v (peripheral speed of the log, which in turn is equal to the discharge speed of the cut veneer) of modern peeling lathes is as follows

$$v = \frac{r \cdot n \cdot \pi}{30} \text{ [m/sec]} \tag{3.1a}$$

(where r = radius of the log in m, n = number of revolutions per minute and $\pi = 3.1416$). The cutting velocity, if a constant number of revolutions is assumed, increases in direct proportion to the log diameter ($D = 2r$). If the radius r of the log is measured in ft., then the equation for the cutting velocity v' in usual British or American writing holds:

$$v' = 2 \cdot r \cdot n \cdot \pi \text{ [ft./min]}. \tag{3.1b}$$

As mentioned before, each wood species has an optimum cutting velocity. If the cutting velocity is too low, the band of veneer will tear.

For large diamter logs, the cutting velocity must be thoroughly controlled. Otherwise the output of the lathe decreases and the quality of the veneer becomes lower. The best economy in peeling depends greatly on correct setting of the knife and nosebar and certain other relevant factors.

To obtain veneer of high quality and a high rate of output the cutting speed should be maintained constant as much as possible. This means that the rotational speed of the log must be increased with decreasing log diameter. At present the maximum cutting speed under normal working conditions amounts to 200 m/min (660 ft./min); the actual ("average") cutting speed, however, will depend on the general conditions prevailing, and ranges from 30 m/min to 50 m/min (98 ft./min and 164 ft./min). Again the quality of the veneer will suffer if cutting speeds are too low or too high.

Early rotary lathes operated at the same speed throughout the cutting of the entire log. The efficiency obtained in this way was low. Increases in cutting speeds and greater interest in veneer quality led to the use of variable speed drives for the log spindles, at least on larger veneer lathes.

Table 3.6 shows types of drives available for standard peeling lathes produced in Germany. The least expensive method of obtaining a stepwise control of speed is by using a pole-changing motor or a multispeed gearbox located between the main drive and the machine. For the latter type of drive a change from one speed to another is possible only when the machine is stopped. A multi-speed pole-changing motor provides less difficulty, but such a motor is complicated in design and can cause trouble. The best solution to the drive problem is doubtless a continuously variable speed control. If available, the best type of drive for this purpose is a DC shunt motor.

On smaller veneer peeling machines it is possible to employ a continuously variable PIV-transmission (*Positive Infinitely Variable* gear). For larger types the most satisfactory drive is a continuously variable Ward Leonard set. A speed control in the range of 1:6 is normally adequate, but for special applications a ratio of 1:8 can be reached by this method. In each case the main drive must be adapted to the requirements of the factory and especially to the available supply of electricity.

Standard veneer lathes are able to cut by the half-round method. The stresses involved are rather heavy and special attachments are necessary, e.g. a hydraulic braking system. The purpose of this brake is to cushion the shock-loads which are inevitable in half-round cutting. Many types of fancy veneer are cut by the half-round method. Such veneers are needed in the form of separate sheets for the matching of texture, e.g. birch and maple.

The log can be dogged either by electric or hydraulic means. The design and diameter of the chucks are governed by the type of wood peeled and the log diameter. The smallest diameter is that of the spindle or of the spindle extension.

For the total power consumption N_t of veneer lathes only empirical equations exist. An equation developed by *Birett* (1933, p. 29) is as follows:

$$N_t = N_i + N_c = N_i + l \cdot v \cdot s \, \frac{R_c}{10.2} \, [\text{kW}], \tag{3.2}$$

in which

N_i = the idling consumption [kW], v = cutting velocity [m/sec],
N_c = the cutting consumption [kW], s = thickness of veneer [mm],
l = the length of the log [m], R_c = specific cutting resistance $\approx 20 \, \text{kp/cm}^2$.

3.2.6 Veneer Handling from the Slicer or Lathe

In the U. S. A. veneer is transported from the slicer or lathe by hand-pulling into a table, by reeling, or more commonly by a mechanical tray system. Pulling

Table 3.6. Types of Drive Available for Standard Peeling Lathe Types P (RFR)

	Speed control facilities	Remarks
Constant speed	None	Adequate if only small output required, for greater output not advisable
Pole-changing motor 4-speed gearbox	2 to 4 speeds 4 speeds	Automatic speed control possible / Unnecessary for small output For straightforward operating conditions not involving high production rates For highest output not suitable
Direct motor current shunt	1:3 stepless	Offers advantages if D. C. supply available
PIV (*positive infinitely variable*) gear	1:3 up to 1:6 stepless	Satisfactory drive giving high output / Automatic speed control possible / Medium types: PIV transmission for largest peeler types not suitable owing to high input power required
D. C. motor with duplex clutch	1:9 with one step	Speed control range of D. C. motor sufficient without duplex clutch Very large speed control range for larger types but change-over necessary half-way
Ward Leonard set	1:6 stepless	Efficient drive enabling full advantage to be taken of the capacity of the machine

veneer by hand is limited by economic reasons to small hardwoods plants. In the larger hardwood plants mechanical reeling is used, but in practically all American softwood mills the tray system is used. Mechanical reeling has the advantage of less space requirement and waste. Veneer from 1/100 in. (0.25 mm) to 1/8 in. (3.2 mm) in thickness can be reeled.

Veneer reels (bobbins) are mounted in a storage rack, full reels in an upper track and empty reels in a lower track. For continuous cylindrical peeling an empty reel is brought down by the reeling unit in descending from its upper position; it is then ready for immediate operation in receiving the veneer from the lathe. A reel can be wound to a diameter of 69 cm (27 in.). A full reel is hoisted to the upper track of the storage rack; while the reeling unit descends it automatically carries down an empty reel, to complete the cycle.

Two reel-storage racks are necessary if a modern veneer lathe running at a very high speed is used. The storage rack takes a total of four reels. The rails of the upper track are inclined towards the unreeling system so that the loaded reels are driven by gravity and no motor is needed. The unreeling systems is driven by a special gear. The veneer is automatically fed from the unreeling system into the feeding device of the clipper.

Modern tray conveyors or deck systems used to handle green veneer require less personnel between lathe and clipper. The veneer leaves the lathe on a tipple which is automatically adjusted to feed the veneer waste conveyor or to a tray leading to the clippers. Belts on the tipple carry the veneer to the tray. When a tray is filled, a light flashes, the veneer ribbon is broken by an air jet or by hand and the tipple feeds another tray. The tray systems normally are between 100 and 200 ft. (30 to 60 m) long, permitting adequate storage of the green veneer. In Europe the tray system is not used due to lack of space. In the United States probably the general adoption of the tray system depends on the desire for high-speed production with minimum labor. Storage of a single sheet of veneer in flat position, as is common on the tray system, assists rapid and accurate clipping.

Sliced face veneer is, as mentioned before, turned over by hand and piled in order. Occasionally a belt conveyor is used to move the veneer away from the slicer. Hand-stacking at the end of the conveyor is necessary (*Lutz* and *Fleischer*, 1963).

3.2.7 Trimming and Clipping

Veneers produced on a slicer should be trimmed lengthwise in order to give them an attractive appearance and to cut out defects. This can be done on veneer guillotines which are able to take a pack of veneers. Height of the pack is up to 80 mm (3 1/4 in.) provided the cut is performed parallel to the fiber; if the cut is across the fiber the height of the pack is limited to 50 mm ($\sim$2 in.). These figures are valid if the thickness of a single veneer does not exceed 1.5 mm (1/16 in.). In case of thicker veneer the total height of the veneer pack should be reduced to avoid damage to the machine. The veneer pack is held down during the cutting process. The cutting knife performs a swinging movement whereby a very clean cut is achieved. The counter-knife is placed on the machine table and adjusted by a large number of screws. Cutting lengths vary from 2,600 mm (102 in.) to 5,100 mm (201 in.).

If the veneer packs, which often are of considerable length, are cut by a veneer guillotine both in longitudinal and transveral direction to the grain, loss of time and frequently damage occur. For this reason cross-cut guillotines were developed. These machines are suitable for trimming veneer packs up to a width of 1,200 mm (47 in.) and a height of 50 mm ($\sim$2 in.) and veneer thicknesses up to 1.5 mm (1/16 in.). The knife is inclined to execute a drawing cut. The speed is about 30 cuts/min; if a pole-changing motor is provided 30 and 60 cuts/min may be obtained. There are various possibilities to fit both types of guillotines in a cutting line. An example is shown in Fig. 3.39.

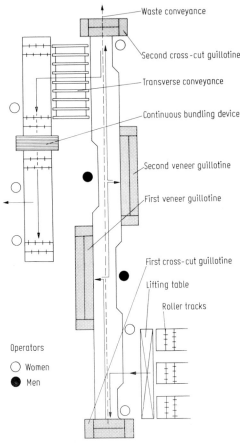

Fig. 3.39. Scheme of a cutting line consisting of two guillotines for longitudinal cut and two guillotines for cross cut in connection with a continuous bundling machine for the finished veneer packets. RFR, Hamburg, now: Keller & Co., Laggenbeck

The veneer leaves the lathe normally trimmed on both sides during the peeling process by means of a pair of trimming or spur knives. The two knives, which can be adjusted for position, are attached to the nosebar holder. The depth of cut suits the veneer thickness (Fig. 3.4). For certain purposes the veneer is slit lengthwise into two or three narrower strips. Slitting knives with eccentric withdrawal mechanism are furnished for this case.

For the clipping of the veneer different types of clippers are available:

a) The most efficient clipper is operated by compressed air. The pressure is about 6 to 8 kp/cm² (85 to 114 lb./sq. in.) depending upon the width and thickness of the veneer. The time for a cut is extremely short (about 1/10 to 1/20 sec). The clipper may be used for wet or for dry clipping. The veneer draw-in speed is adjustable between 0 and 60 m/min (0 and 197 ft./min);

b) Electric clipper suitable for heavy cuts of veneers of poor quality which can not be cut at high speed. Time for a cut approximately 1/3 sec;

c) Hydraulic clippers for speed of the veneer up to 20 m/min (65 ft./min); normal speeds 10 to 15 m/min (∼33 to 49 ft./min).

Many types of veneer clipper are in use in various countries. The older ones were controlled by hand or foot, but the modern types are partially or wholly automatic. A typical clipper or guillotine consists of a sturdy iron frame, carrying at each end vertical knife guides, a cutting table, a shear plate and a movable knife. The knife is counterbalanced by a weight which in the case of hand control is raised by depressing a foot pedal. The automatic clippers in operation in Europe as well as in the U. S. A. are speed-controlled by switch buttons.

Most clippers are equipped with an automatic lengthening device. Several different lengths can be set, provided several contact rolls are used. The contact rolls are operated by switch buttons. Individual cuts can be performed by hand in case veneer defects must be cut out.

Clipping is done with an addition of 5 to 7 cm (2 in. to 2 3/4 in.) to allow for losses in the subsequent trimming of the pressed plywood and with a further width allowance of 5 to 9% for shrinkage during veneer drying (*Irschick*, 1949, p. 2).

Summarizing it may be said that one of the major problems in the veneer and plywood industry is the streamlining, the work flow subsequent to the peeling machine. The average production rate of the veneer band by the rotary peelers is 50 m/min (164 ft./min). This is a very large quantity of veneer, and it follows that the capacity of the veneer clippers together with the reeling and unreeling devices and the conveyors necessarily is of decisive importance in the whole operation.

3.2.8 Veneer Peeling Lines

The productivity of veneer peeling plants can be substantially increased by mechanizing the processing equipment surrounding the peeling lathe. The amount of veneer peeled depends on the time that the lathe is actually running and on the speed of the machine in relation to the speed of the removal of the peeled veneer. A German firm has developed the several items of equipment that together form the peeling line illustrated in Fig. 3.40. The idea was to make the whole production process flow smoothly and continuously. The steamed and cut-to-length logs are passed to the debarking and cleaning machine. The capacity of the cleaning machine is approximately equal to the capacity of two peeling lathes. Subsequently the logs are moved on chain conveyors to the two log storage units of the lathe feeding and centering unit. These are located in front of the peeling lathes. For greater convenience the log magazines can take up two log lengths. Centering of the ingoing log is accomplished during the peeling of the preceding log. The centered log is fed into the peeling lathe and promptly dogged.

The hydraulic dogging devices on the lathe, acting on both ends, speed up the dogging operation; hydraulically operated telescopic spindles allow peeling to small core diameters. In this way peeling speeds up to 200 m/min (660 ft./min) can be attained.

Special attention has been paid to ensure efficient removal of waste, cores and continuous veneer strips from behind the peeling lathe. Waste is transported by a conveyor belt to a waste-hog unit. Cores are collected on pallets and pass on roller type runways to the chipper. Pallet return is facilitated by elevated tables.

Veneer reeling and unreeling are dealt with in Section 3.2.6, veneer trimming and clipping in Section 3.2.7.

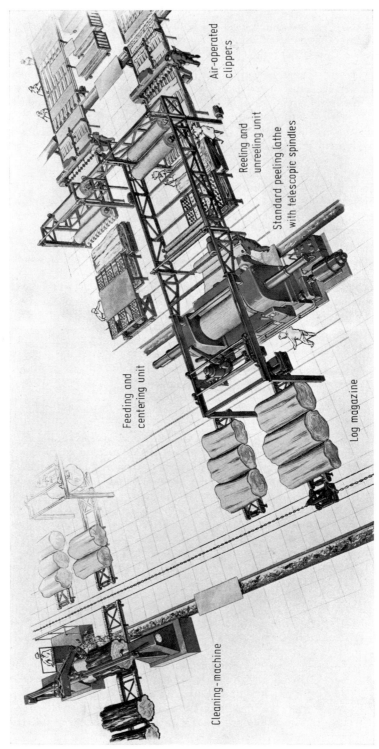

Fig. 3.40. Perspective view of a complete veneer peeling line. RFR, Hamburg, now: Keller & Co., Laggenbeck

Following the clippers are sorting tables which facilitate the separation and stacking of different widths and qualities of veneers.

The plants can also be operated efficiently when a continuous dryer is used. In this case the wet veneers are fed into the dryer by a special unreeling unit, and after drying they are transported on special conveyors to an air-operated clipper.

The output of a veneer peeling plant of conventional design averages approximately 4,000 to 6,000 m (13,000 to 20,000 ft.) of veneer per shift while the described peeling line increases the output to approximately 12,000 m ($\sim$40,000 ft.) per shift. Peeling lathe utilization is thus increased from about 20% to about 50% of total operating time, at the same time requiring less labor.

3.2.9 Veneer Drying

3.2.9.1 Introduction. Immediately after slicing or rotary peeling the moisture content of veneer is very high. It ranges influenced by wood species, ratio of heart to sapwood, pre-treatment by boiling or steaming from about 30 to 110%, based on the oven-dry weight. The high moisture content causes high density of fresh veneer. Birch veneer coming from the peeling lathe has a density of about 910 kg/m³ (57 lb./cu. ft.), okoumé veneer about 600 kg/m³ (38 lb./cu. ft.), pine veneer about 830 kg/m³ (52 lb./cu. ft.). Quick drying is necessary with respect to the following facts:

a) Only properly dried veneer can subsequently be processed, for instance, glued and machined;

b) After drying, staining or infection of veneer by mold or wood destroying fungi and chemical reactions of wood constituents with the surrounding atmosphere are eliminated.

Sliced decorative veneer especially such having rich, curly figure cut from sound roots of a tree as butt veneer or stump veneer are still very carefully dried in drying sheds with natural air circulation. Veneers for plywood in Central Europe and North America are exclusively artificially dried. The originally in Eastern Europe and some Balkan countries practised gluing of wet veneer has been abandoned in these countries due to severe technical imperfections.

The final moisture content of sliced veneer shall range from 8 to 12%, of rotary peeled veneer for plywood as a rule from 6 to 8%. Equal distribution of moisture content in the veneer is necessary for dimensional stability of plywood made from the veneer. With respect to the economy of operation the veneer should be dried as quickly as possible, of course, without suffering damage. It is necessary that veneer during drying do not become too wavy. Some species lose their characteristic color when drying conditions (temperature, relative humidity, time) are not properly controlled.

3.2.9.2 Types of Veneer Dryers, Conventional Drying Times

A. Drying-Sheds, racks and lofts are still used for highly sensitive sliced veneer that do not tolerate elevated temperatures. In large veneer plants these sheds require much space. Their outer surfaces are painted in a dark color so that as much as possible sun radiation is absorbed for heating the shed interior. The walls are perforated by slits on all sides or furnished with blinds opened during warm seasons to accelerate the drying. In winter artificial heating is necessary, but the temperature should not exceed 30 °C (86 °F), and the relative humidity should not decrease below 50%. It is impossible to estimate the drying times because they depend on too many factors such as wood species, veneer thickness, climatic conditions. Air drying of veneer is always "a time-consuming operation that largely has passed out of use" (*Perry*, 1947, p. 115).

B. *Drying Kilns and Channel Dryers.* Veneer drying kilns are in construction similar to usual lumber drying kilns (Fig. 3.41), but with the following differences:

1. The capacity of the fans must be about twice that for drying lumber because veneer with low thicknesses and loosened structure due to peeling exhibit a high diffusion rate. Also the circulation speed of the drying agents should be higher than that for lumber drying;

2. The heating system must be amply dimensioned with respect to higher heat consumption due to relatively greater water evaporation.

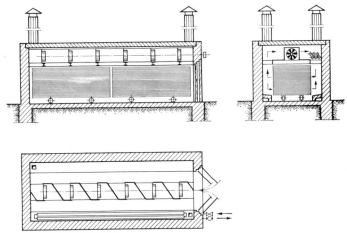

Fig. 3.41. Vertical projection, horizontal projection and cross section of a veneer drying channel.
Büttner-Schilde-Haas AG, Krefeld

The veneer are piled on kiln-trucks. Occasionally channel dryers or progressive kilns are in use. Progressive kilns are built in the form of a tunnel at least 20 m (about 65 ft.) long in which both temperature and relative humidity can be controlled throughout the entire length. Practice has demonstrated that 30 m (∼100 ft.) is the optimum length and that 52 m (∼170 ft.) (*Henderson*, 1951) is the maximum: "At the loading end of the kiln the temperature is low and the humidity high, but as the truck proceeds through the channel it meets with a steadily rising temperature and a gradual reduction in the humidity until at the point of discharge the temperature is at its highest and humidity at its lowest. The channel should be kept full of loaded trucks: as fresh trucks are pushed into the kiln a corresponding number are discharged" (*Wood* and *Linn*, 1950, p. 59).

A cross section through a typical channel dryer is illustrated in Fig. 3.42. Drying times for veneer of various species and different thicknesses is shown in Fig. 3.43.

At present veneer drying kilns and channel dryers are widely discarded in consequence of the following reasons:

a) Low veneer quality, warping due to inadequate flat-holding;
b) Defects, especially staining at the spots of support;
c) Unequal drying;
d) Much labor for piling and unpiling.

Nevertheless progressive kilns are used in factories producing blockboards.

C. Hot-plate and Breather Dryers. Hot-plate Dryers are in their construction similar to that of hydraulic hot presses. The veneer — in up to five layers a pack — is put into the daylights of the press. The hot-plate dryer is most frequently used to flatten face veneer and to bring it from an air-dry condition to a low moisture

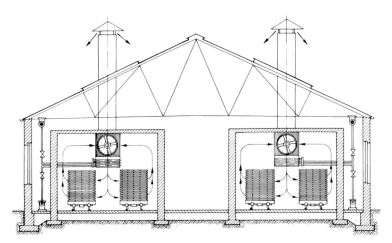

Fig. 3.42. Section through a shed containing two typical channel dryers. From *Wood* and *Linn* (1950)

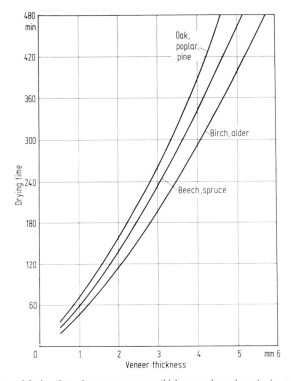

Fig. 3.43. Dependence of drying time of veneers on veneer thickness and wood species in veneer drying channels.
Büttner-Schilde-Haas AG, Krefeld

content just before it is glued (*Lutz* and *Fleischer*, 1963). The closed time is equal to the drying time. *Keylwerth* (1953) found that for a drying temperature of only 110 °C (230 °F) there is practically no difference in the drying time using a contact-dryer or a convection dryer, but at 145 °C (293 °F) the drying time in the contact-dryer was reduced to approximately 36% of that in a convection dryer.

The disadvantages of the hot-plate dryers are as follows:

1. For greater veneer sizes the transport of water mainly perpendicular to the direction of pressing, consequently from the center of veneer to its edges, is difficult;

2. Intermittent working;

3. Manual loading is expensive; automatic feeders and unloaders increase the capacity but investment and production costs become too high.

Therefore hot-plate dryers generally are no more in use.

Breather Dryers are used for core stock over 2 mm (5/64 in.) in thickness, or for re-drying veneer. They consist of a number of hollow steam-heated steel plates which move together and apart at regular intervals. The plates are so coupled that when one series of pairs is closed the alternative series of two particular plates is open. The veneer is smoothed under a slight pressure between the closed platens, and at the same time moisture in the veneer is converted into steam which escapes as the platens open. Veneer drying is done quickly down to 6 to 9% moisture content without buckling. But breather dryers are economic only if rather high temperatures are used. This may cause case-hardening even of thin veneer.

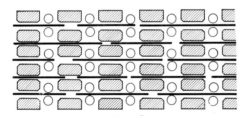

Fig. 3.44. Scheme of construction and mode of operation of a progressive plate veneer dryer. From *Merritt-Monsanto*

D. *Progressive Plate Dryers* were developed in the U. S. A. (by the firm Merritt-Monsanto) on the basis of breather dryers for drying hardwood veneer from Northern states (birch, walnut, maple, but also mahogany and cedar). Fig. 3.44 shows that this type of veneer dryer has a series of alternate pairs of rollers and steam-heated platens, operating intermittently. "When the platens are open the veneer is forwarded by the rollers, but only for the distance of one platen. The rollers stop, the platens close for a short interval: the platens open, the rollers advance the veneer again, and so on." (*Perry*, 1947, p. 114/115; *Kollmann*, 1962, p. 227/ 228). This process flattens the veneer (working breadth 3 m (10 ft.)). The progressive plate dryers do not reach by far the capacity of roller or conveyor dryers but they are still used in the U. S. A. with respect to their low space requirement, relatively low steam consumption and their flattening effect.

E. In *Roller or Roll Dryers* the veneer stock to be dried is propelled through the dryer by a series of rolls. The pairs of rollers are driven with variable speed transmission, the distance between two pairs of rollers is mostly between 40 mm (1.58 in.) and 85 mm (3.35 in.), the lower the thinner the veneer. Veneer travels between top and bottom rollers with weight of top rollers (pinching effect) on veneer. Both top and bottom rollers are positively driven. Veneer is free to

shrink as it travels from one pair of double rollers to the next pair. The dryers
are built in sections, the number of which (for instance 5 to 18) is determined by
the desired capacity. The effective dryer length is adapted to the production
conditions, and is in Central Europe usually between 8 and 30 m ($\sim$26 and 98 ft.),
in the U.S.A. up to 47 m ($\sim$154 ft.). The longer the dryer the better — except
very thin veneer — the drying quality.

The width of rollers is for standard roller dryers in Germany between 4 and
4.5 m, in the U.S.A. between 12 ft. and 14.5 ft. ($\sim$3.7 and 4.4 m). The dryers are
built with 2, 3, 4, 5, 6 and even more decks. The vertical space between the decks
is ample to accomodate the steam coils and permit uniform circulation of air over
both sides of the veneer. The whole heating chamber is enclosed in a well insulated
framework to prevent heat losses by conduction or radiation.

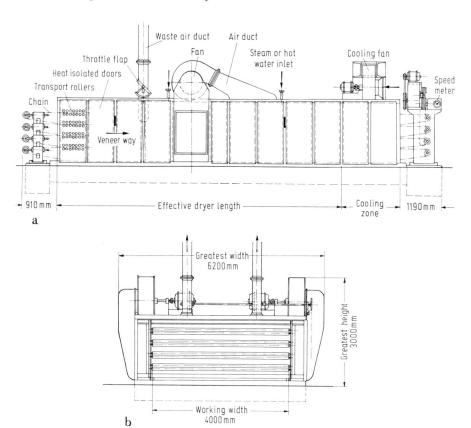

Fig. 3.45 a, b. Schematic vertical projection and cross section of a veneer roller dryer. Siempelkamp & Co., Krefeld

There are two circulation systems:

1. Air movement along the axis of the dryer (for instance Siempelkamp & Co.,
Fig. 3.45). This system operates on the counterflow principle. The wet, green,
boiled or steamed veneer stock is fed into the zone of lowest temperature and
highest humidity. As the veneer dries, it moves toward the discharge and into
progressively hotter and drier air. This eliminates danger of case-hardening and
produces flat, pliable veneer;

2. Air circulation across the axis of the dryer (for instance Büttner-Schilde-Haas). The hot air is blown into the dryer on one side, passes over the veneer and is sucked on the opposite side. Commonly guiding sheets or chests make the circulation uniform. Fig. 3.46 illustrates the cross section through a dryer of this type.

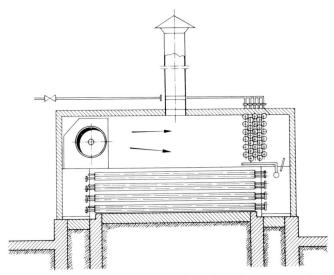

Fig. 3.46. Cross section through a roller veneer dryer with air circulation across the axis of the dryer. Büttner-Schilde-Haas AG, Krefeld

The dependence of drying time on veneer thickness for various wood species is shown in Fig. 3.47. As has been mentioned already the initial moisture content of the veneer may vary between 30 and 110%, the final moisture content after drying between 6 and 9%. The temperature in roller dryers may be between 100 and 165 °C (212 and 329 °F). Only for oak there is a limit of 85 to 90 °C (185 to 194 °F). In the U.S.A., Oregon pine is dried with very high temperatures between 165 and 205 °C (329 and 401 °F). The drying times are very short, for instance for heartwood veneer (*Gebhardt* et al., 1951):

2.5 mm thick at 165 to 180 °C (329 to 356 °F) and 5 to 8% final moisture content 5 min

2.5 mm thick at 165 to 205 °C (329 to 401 °F) and 1.5 to 4% final moisture content 7 min.

The capacity C of a roller dryer can be calculated as follows:

$$C = l \cdot b \cdot s \cdot c \cdot n \cdot t \, [\text{m}^3/\text{h}], \tag{3.3}$$

where

l = effective dryer length in m,
b = width of rollers in m,
s = thickness of veneer in m,
c = covering factor for the lines in the dryer = 0.65 to 0.80,
n = number of decks,
t = drying time in h.

The specific steam consumption per 1 kg evaporated water amounts from 1.75 to 2.0 kg.

F. Endless Belt Dryers consist of usual insulated heating chambers with fans producing air circulation and efficient heaters. Pairs of endless belts of wiremesh type serve as conveyors, driven at each end of the chamber by sprocket wheels. The bottom surface of the upper of two belts moves in the same direction through the dryer as the top surface of the lower, and the sheets of veneer are carried

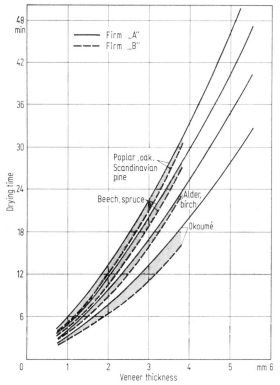

Fig. 3.47. Dependence of drying time of veneers on veneer thickness and wood species in a roller veneer dryer, manufactured by two firms

between the two surfaces and held down firmly by special devices such as spring-levers (Fig. 3.48). As the figure (marked by arrows) shows the veneer is counter-running through the dryer. Feeding and unloading are done on both ends of the dryer. If unidirectional movement of veneer is desired then a more complicated construction (Fig. 3.49) is necessary; the total number of belts is twice the number of veneer ribbons. This type of dryers has a greater height and is more expensive.

Belt dryers can be used for nearly all wood species and veneer thicknesses down to 1 mm (~0.04 in.).

The air circulation is mostly effected by fans located on one side of each dryer section beside the decks (Fig. 3.50). The temperature in belt dryers is between 80 and 110 °C (176 and 230 °F), in special cases (drying of birch veneer) up to 140 °C (284 °F). Recirculation not only assures proper drying but also reduces heat supply. Drying time versus veneer thickness for various species is shown in Fig. 3.51. The drying time is about 25 to 100% longer than in roller dryers; this is partly due to the relative humidity used as high as practicable. Belt dryers

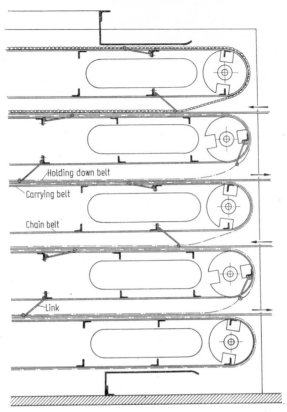

Fig. 3.48. Scheme of the belt arrangements in an endless belt dryer with five decks. Büttner-Schilde-Haas AG, Krefeld

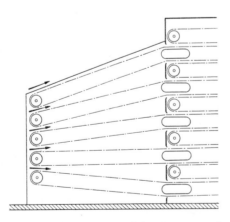

Fig. 3.49. Scheme of the belt arrangement in an endless belt veneer dryer with unidirectional movement of the veneer. Büttner-Schilde-Haas AG, Krefeld

are suitable for very sensitive veneer and usable for nearly all drying tasks in the veneer and plywood industry. Capacity and length of dryers may be guessed using formula 3.3.

G. A nearly revolutionary development in veneer drying was the introduction of *Jet Dryers* in the fifties. The method was known from drying textiles and many other sheetlike materials such as paperboard. Heated air is blown onto both sides

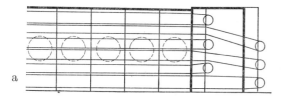

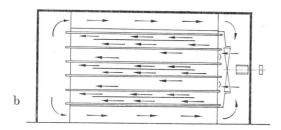

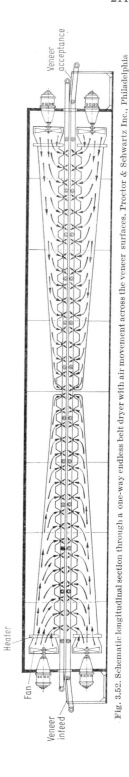

Fig. 3.50. a) Longitudinal, and b) Cross section of a veneer belt dryer, with endless belts in three decks. Fans are located beside the decks. Proctor & Schwartz Inc., Philadelphia

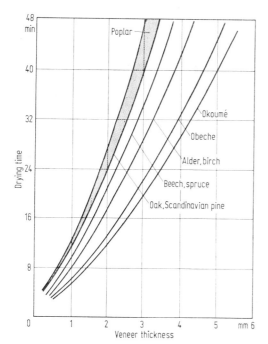

Fig. 3.51. Dependence of drying time of veneers on veneer thickness and wood species in endless belt veneer dryers. Büttner-Schilde-Haas AG, Krefeld

Fig. 3.52. Schematic longitudinal section through a one-way endless belt dryer with air movement across the veneer surfaces. Proctor & Schwartz Inc., Philadelphia

of the veneer at high velocity through nozzles in air ducts located immediately above and below the conveyor (Fig. 3.52). Air strikes each side of the veneer with an impinging action at right angles to both surfaces. By this method every square centimeter (or square inch) of each sheet receives an intensive, equal volume and velocity of air simultaneously. In conventional systems the hot air is blown in a horizontal parallel to the surfaces of the veneer.

The great advantages of jet dryers listed below are proved by the fact that most modern high-capacity veneer dryers use this system consisting predominantly of convection drying. The advatages are as follows:

1. Much higher drying rate, drying time only 25 to 50% of that in conventional roller or belt dryers. This allows not only less lines or decks but also shorter dryer lengths;

2. Simpler feeding due to the reduction of the number of decks (for belt dryers only two instead of formerly in most cases 5 decks), simple operation, labor saving, minimum maintenance;

3. Handling veneer with any thickness from 0.3 to 1.3 mm (about 1/100 to 1/20 in.). Handling small pieces and pieces up to 4.85 m (~16 ft.) long;

4. Handling crotches, stumps, burls, butts and clusters. Drying certain kinds of face veneer that could not be dried fully successfully by other mechanical means. The firm Proctor mentions: African mahogany; Oriental wood; walnut; sapeli; walnut butts, rotary cut; myrtle cluster, rotary cut; African mahogany crotches, sliced; prima vera; lace wood;

5. Less heat consumption due to compact construction;

6. Extremely uniform drying;

7. Maintenance of the desirable bright, flat, resilient condition of the veneer, no discoloration;

8. Saving of 3.5 to 6.5% material. The veneer is cut to size after drying, thus eliminating the high shrinkage allowance resulting from uncontrollable shrinkage. The veneer reaches the clipper only after drying.

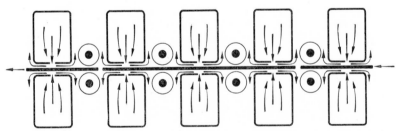

Fig. 3.53. Scheme of a jet veneer dryer with sequences of air ducts with slit nozzles and roller pairs for veneer transport. Büttner-Schilde-Haas AG, Krefeld

For the production of stripped veneer of approximately uniform quality and thickness a one-deck dryer with the clipper at the receiving end may be recommended. For higher capacity requirements two (or more) decks provide the best space-saving solution. It is possible to run the various decks at different speeds in order to dry different qualities or thicknesses at the same time. High output capacity on very limited space can be achieved by using two or more level dryers. The deck is automatically reversed and runs through the next level.

Because of the large radius of the deflection pulleys this system is applicable also for thick veneer.

There are many approved construction details: for instance nozzle boxes with slits as nozzles (Fig. 3.53), or jet tubes which can be individually removed, 76 mm (3 in.) thick exterior doors and panels, easy opening roller cam latches on dryer and duct doors, radial type or axial type fans, streamlined inlet and outlet, each fan operated by its own motor, air cooled fan bearings, cleaning section for removal of all loose debris (knots, loose fibers, etc.) before drying, combination of a roller and a conveyor dryer in one housing, sieve drums with sucking effect as veneer supports automatic temperature control in a wide (e.g. 6:1) range, variable speed transmission for rollers.

The air velocities in jet veneer dryers range from 15 to 60 m/s (3,000 to 12,000 ft./ min), the temperatures between 210 and 290 °C (410 °F and 554 °F) (*Hildebrand*, without date, *Milligan* and *Davies*, 1963). The drying time is low as Fig. 3.54 shows. In many U. S. mills steam is the lowest cost source of heat. Table 3.7

Table 3.7. Pressure and Temperature of Saturated Steam, Possible Drying Temperatures

Saturated steam Pressure	kp/cm²	8	10	12	16	20	25	30
	lb./sq. in.	113.8	142.2	170.7	227.6	284.5	355.6	426.7
Temperature	°C	169.6	179.0	187.1	200.4	211.4	222	233
	°F	338	354	369	395	412	432	451
Drying temperature	°C	155	160	170	180	190	200	210
	°F	311	320	338	356	374	392	410

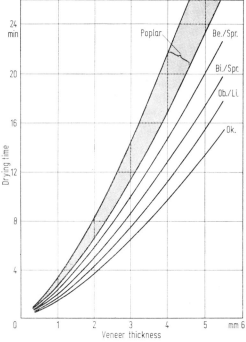

Fig. 3.54. Dependence of drying time of veneers on veneer thickness and wood species for a roller and endless belt veneer dryer with jet ventilation. Büttner-Schilde-Haas AG, Krefeld

shows the interrelationships between pressure and temperature of saturated steam. Highest possible temperatures in veneer drying are desirable but usually the energy reserve of the boiler house is limited and cannot be increased. The risk of selfignition has to be considered and possible trouble in subsequent gluing if the moisture content of the veneer is lower than 4%. Wet veneer ($u > 30\%$) can be dried with temperatures above 200 °C (392 °F) because in this case the veneer temperature remains constant at nearly 100 °C (212 °F) as long as free water is present in the wood tissues up to the "critical point" where fiber saturation is reached.

With the aim to use high temperatures in veneer drying *Directly Heated Dryers* were developed. Electricity as source is scarce with respect to its high costs. Burners for natural gas or light oil are highly efficient. The combustion products from the burner and re-circulated gases are forced into the dryer by the blower(s). The hot gases then pass out through the nozzles. Liberal high capacity burners permit rapid temperature recovery when needed. A wide range of flame turn-down permits economical utilization of fuel when limited load conditions are encountered. Certified safety control equipment is used. Safety organs are built into the heaters to provide maximum protection against un-expected interruption of gas (or oil), air, or electrical power for fans. The gas or oil burners are operated with compressed air of 4 to 8 at. Directly fired burners are independent from the boiler house, guarantee shortest possible heating times, high temperatures, low operation costs, no special service and maintenance (*Tiefenbach*, 1965).

H. Veneer Dryer Feeders and Unloaders, Grading. Veneer feeders must work in harmony with the veneer dryer. Wasteful gaps between sheets of veneer as they are fed to the dryer should be reduced to a minimum. Feeders must be capable of handling all sizes and thicknesses of veneer. They should be synchronized to the drying speed. Many dryers are specifically intended to handle long thin sheets of hardwood veneer. In order to obtain maximum capacity from a veneer dryer, it is highly desirable to insure that the machine is kept full by having a mechanism to accomplish the feeding of veneer in exact cycles. Preferred by many operators is compactness of the feeder when equipped with a load positioner. In this case manual effort by the operator is held to a minimum. Smooth functioning hydraulic system guarantees precise positioning of the feed decks for each cycle. When there are certain space limitations a "short tipple" can be applied, these feeders are able to handle the majority of requirements encountered in mills today. Other feeder types are capable of handling all sizes of veneer. At the entering end the pinch rolls are vertically adjustable to suit load height.

An electrical interlock with the dryer drive provides automatic cycling. Dryers of 3, 4 and 5 decks can be served.

Where space is limited a lift is used in conjunction with a load positioner. The load level is regulated manually by the operator. Fork trucks put the green veneer on a chain. When the lift is in its lowered position the piles of veneer can be moved into it with minimum waste time.

For feeding thin hardwood veneer a special arrangement of two main components is designed: a conveyor to transfer the veneer in a direction at right angles to the dryer where the second component, a pinch hole assembly, grips the veneer and delivers it to the dryer feed section.

All modern veneer dryer unloaders should work with automatic timing and electronic controls. Accessibility of all parts of a complete unloader is highly desirable. Sheets move through the pinch rolls driven at dryer speed until each particular deck has timed out and is ready for discharge. At the proper instant the high speed drive and magnetic clutches accelerate the rolls of a given deck and pass the sheets to the discharge conveyor. The following factors assure good operation:

1. No extra "spotter" is required;
2. Overlapping of sheets does not occur;
3. Veneer is uniformly delivered;
4. Handling damage is a minimum;
5. Correct moisture content readings;
6. Better granding of veneer.

I. Special Veneer Drying Methods. Occasionally experiments were carried out to dry veneer by new methods with the aim to improve the drying quality or to increase the drying rate. The results of these experiments were discouraging and the two main principles are only briefly mentioned. More details are reviewed by *Kollmann* (1955, 1962), *Krischer* and *Kröll* (1959), *Kneule* (1959), U. S. Forest Products Laboratory (1947).

Vacuum Drying needs a steel container; its ground area must be adapted to the maximum veneer size. The vacuum in the container is produced by special pumps. The drying temperatures (40 to 80 °C or 104 to 176 °F) are low. Therefore the drying is very mild and the veneer remains flat, smooth and flexible. But the installation and the drying process is much too expensive and the capacity is too low.

Infrared Radiation is successful for hardening paints and enamels on metal. Seasoning of wood, however, presents a basically different problem because:

1. The heat must penetrate the wood without raising the surface temperature to such a degree that drying defects occur;

2. Infrared radiation penetrates wood only to a slight depth (mostly up to about 1 mm $\sim$ 1/24 in.). Therefore it is not advantageous to use infrared radiation as a source of heat for seasoning lumber;

3. It seemed possible to dry thin veneer in this way but restraint must be provided against wrinkling and buckling by drying between mechanisms (such as plates, closely spaced, iron rolls, springs, heavy wire mesh) which interfere with the infrared radiation equipment. Drying times for beech veneer when infrared radiation exerted undisturbed influence upon both surfaces were found by *Keylwerth* (1951), as indicated in Table 3.8.

Table 3.8. Drying Times for Beech Veneer
(250 W-radiator, 20 cm (8 in.) distance, initial moisture content 70%)

Moisture content	Veneer thickness	
%	1.5 mm ($\sim$1/16 in.) Drying times in min	2.5 mm ($\sim$1/10 in.)
50	3	5
30	7	10.5
20	9.5	14
10	13	19
5	15.5	23

These drying times are not favorable as compared with those in jet dryers;

4. The consumption of energy is too high (between 1.5 and 5 kWh per 1 kg water evaporated). Modern veneer jet dryers suppressed further interest in infrared radiation drying of veneer.

High Frequency Drying of veneer has been investigated by *Gefahrt* (1966, 1967). He remarks that veneer usually leaves the peeler with no uniform moisture content. There are zones in the veneer of different diffusion resistance. This fact may cause some difficulties in the case of convection drying. The electric conductivity in veneer along the grain is about 8 to 10 times greater than across the grain (Fig. 3.55). This difference in electrical conductivity causes a relevant raise in temperature (Fig. 3.56). With respect to a high drying rate the electrical field

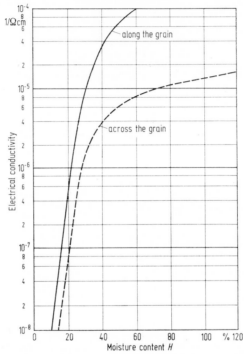

Fig. 3.55. Dependence of electrical conductivity of alder veneer along the grain and across the grain on the moisture content. From *Gefahrt* (1966)

should be applied along the fiber with smallest possible airspace between veneer and electrodes. Veneer with higher initial moisture content are desiccated more rapidly than drier veneer. Nevertheless, high frequency is only applicable if uniform final moisture content alone by convection drying cannot be reached. Perhaps a combination of convection drying and high frequency drying may be used, but one is still far away from this goal.

More promising is the application of HF-heating for constructive wood elements (*Gefahrt*, 1970), especially in the production of glued-laminated structural members. Due to the relatively high Ohmic conductivity of the liquid glue under the influence of the HF-field Joulean heat is mainly generated in the glue joint. This energy concentration in the glue joint occurs if the glue lines and wood layers are parallel to the HF-field lines. The glue joints with the highest electrical conductivity are heated most. Too much heating of wood is thus avoided and the least amount of energy is consumed (*Kollmann*, 1951, 1955). Optimum values may be reached by adding common salts to the liquid glue. Prerequisite for good results is correct dimensioning of the electrodes.

Recently *Radiation Techniques* using γ-rays produced by a Co-60-source in combination with HF-heating serve for the treatment of wood-polymer combinations (*Meyer*, 1965; *Siau* and coworkers, 1965). Similar experiments were published by the Forest Products Research Laboratory at Princes Risborough, U. K.

3.2.9.3 Physics of Veneer Drying. The drying of veneer is less difficult than the drying of lumber with its greater thickness. The loosening of wood structure during slicing or rotary peeling reduces the internal resistance to moisture diffu-

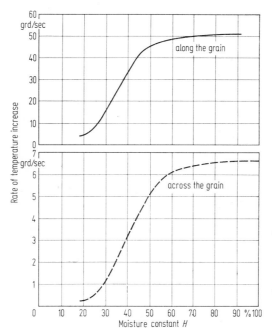

Fig. 3.56. Comparison of heating rate of alder veneer along the grain and across the grain. From *Gefahrt* (1966)

sion between air and wood substance, as *Martley* (1926) pointed out when he introduced an empirical external diffusion resistance to calculate moisture movement.

The physics of veneer drying was neglected thereafter until about 1952. Scientific investigations of veneer drying were carried out by *Keylwerth* (1952, 1953), *Fleischer* (1953), *Milligan* and *Davies* (1963), *Fessel* (1964).

Under the assumption that drying is done principally by convection, the following three phases (Fig. 3.57) occur:

1. Phase: A short heating time; below the dew-point, water vapor condenses on the surfaces of the veneer;

2. Phase: Constant rate period, during which capillary water is evaporated at constant drying rate, at constant veneer temperature;

3. Phase 3: Final drying below the fiber saturation point; the temperature of the veneer increases rapidly and approaches the temperature of the drying medium. In this phase drying rate decreases greatly.

Studying Fig. 3.56 and considering that drying time is plotted on a logarithmic scale the following may be concluded: For drying red beech veneer, 2 mm thick, 14 minutes drying time reduces the moisture content from 97% to 5% on the

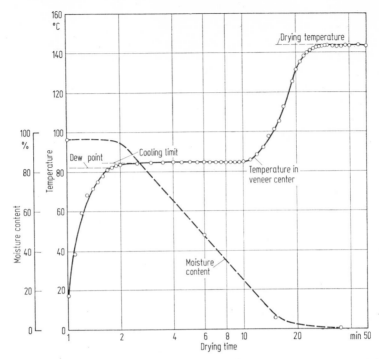

Fig. 3.57. Dependence of the temperature in the veneer center and of the moisture content on the drying time in constant climate. From *Keylwerth* (1953)

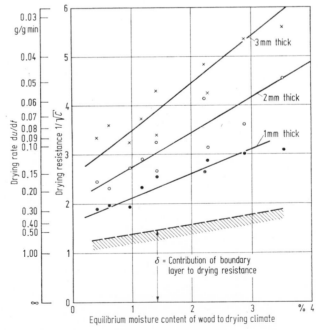

Fig. 3.58. Relationship between drying rate, drying resistance and equilibrium moisture content of wood in veneer drying. From *Keylwerth* (1953)

average. The time is distributed as follows: 1 min for heating, 8 min for evaporation of capillary water and 5 min for the final phase. Apparently the diffusion laws are not valid for thin veneer.

The heat transfer to the veneer is of prominent importance, especially in the boundary layer between drying medium and wood (Fig. 3.58). *Keylwerth* (1953) pointed out that intense turbulence in the boundary layer improves the heat

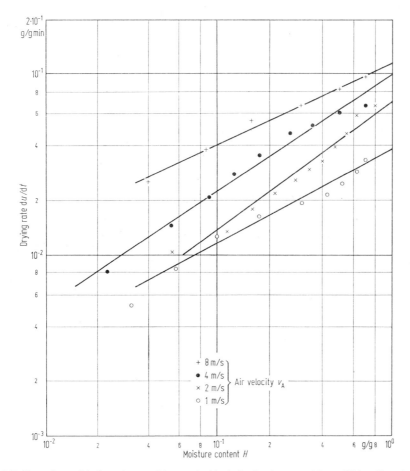

Fig. 3.59. Dependence of drying rate on moisture content in drying beech veneer, 3.2 mm thick, with various air velocities. From *Keylwerth* (1953)

transfer. *Fleischer* (1953) found that the rate of heat transfer to the veneer decreases parabolically with decreasing moisture content. Therefore the rate of evaporation of water during drying also decreases parabolically, when plotted against decreasing moisture content. In a double logarithmic scale the interrelationship between drying rate and moisture content is approximately linear (Fig. 3.59). *Fleischer* (1953) showed that at temperatures above 100 °C (212 °F) the drying rate (expressed as evaporation constant S of poplar veneer, 1/8 in. (3.2 mm) thick or thinner, can be estimated empirically as follows:

$$S = 2.98 \,(\log \vartheta) + 0.78 \,(\log v_A) - 1.34 \,(\log s) - 11.34, \qquad (3.4)$$

where

S = evaporation per sq. ft. of surface area [g/min],
ϑ = drying temperature [°F],
v_A = air velocity [ft./min],
s = thickness of drying specimen [in.].

Data calculated using the foregoing equation are only partially in agreement with experimental results obtained by *Keylwerth* (1953). A short discussion is necessary.

Peck, Griffith and *Nagaraja Rao* (1952) found, drying balsa wood boards, the formula

$$\frac{dH}{dt} = -CH^n, \tag{3.5}$$

where

$\dfrac{dH}{dt}$ = drying rate $\left[\dfrac{g}{g\ \min}\right]$,

C = coefficient of drying,

H = moisture content $\left[\dfrac{g}{g}\right]$,

n = exponent.

Keylwerth (1953) confirmed this formula for veneer drying. He calculated, based on experimental data, the drying resistance as the quotient $\dfrac{1}{\sqrt{C}}$ which is

$$\frac{1}{\sqrt{C}} = R_{iD} + R_{eD}, \tag{3.6}$$

where

R_{iD} = internal diffusion resistance of the wood,
R_{eD} = external diffusion resistance in the boundary layer between drying agent and wood.

It is evident that with increased veneer thicknesses (1, 2, 3) the drying resistance does not increase in proportion to the thickness of veneer, but follows approximately the relationship $(1 + R_{eD}):(2 + R_{eD}):(3 + R_{eD})$.

This is shown in Fig. 3.58. One can derive from this figure:

1. The external diffusion resistance R_{eD} decreases with increasing velocity v of the drying medium;
2. For an infinitesimal drying rate the boundary layer must disappear which means that the external diffusion resistance R_{eD} becomes zero;
3. The higher the velocity of the drying medium the better the heat transmission. The experiments by *Keylwerth* (1953) show that the rate of heat transmission is proportional to $v_A{}^{0.8}$ where v_A is the velocity of the drying medium (normally a mixture of air and steam).

The rate of heat transmission can be improved greatly by the use of nozzles in jet dryers. *Kröll* (1959) has investigated the conditions in jet dryers. Fig. 3.60 shows the dependence of the drying rate on the air velocity and thereby on heat transmission change when a stream of the drying medium flows perpendicular to the veneer surface. Apparently the distance of nozzle opening from the veneer to be dried should not be too large. The lateral distance between nozzles should be less

than 10 times the diameter of the nozzles. Fig. 3.60 shows for felt board the relationship between drying velocity and air velocity after passing round or slit
nozzles. Similar conditions can be assumed for the drying of veneer.

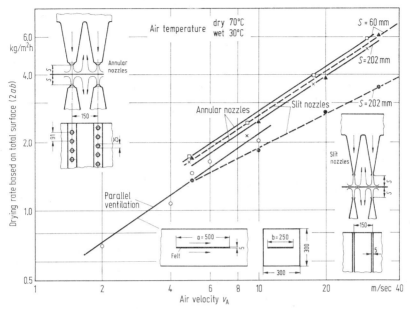

Fig. 3.60. Dependence of drying rate of felt board on the air velocity in the first phase of drying using various
nozzle types. From *Kröll* (1959)

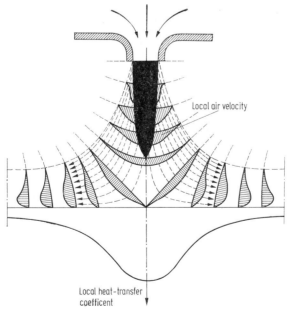

Fig. 3.61. Local air velocity in a plane jet hitting a horizontal plate and distribution of local heat transfer coefficient, semi-schematical. From *Kröll* (1959)

In any veneer dryer the drying rate is inversely proportional to the humidity of the drying air. The drying resistance decreases if the difference $\Delta\vartheta$ between readings of dry-bulb (ϑ_d) and wet-bulb (ϑ_w) thermometers (psychrome tric difference) becomes greater. The time for heating up the veneer (Fig. 3.57) is an exponential function of the psychrometric difference. Industrial experience has shown that the tendency of the veneer to develop waves increases with decreasing relative humidity in the dryer; the veneer also becomes more brittle if the relative humidity is too low.

Veneer is heterogeneous with respect to density, grain orientation, texture and presence of knots. These variations in structure cause variations in drying resistances. This is especially valid for thick veneer or if the nosebar in veneer slicers or rotary peelers is not correctly adjusted. Another point to be considered is the faster drying rate which occurs at the ends of the veneer. This faster drying causes the waves to be very severe at the ends of the veneer. Superimposed is a warping of the veneer since often the surfaces are influenced by different drying conditions.

It is important to keep the veneer flat and smooth during the drying. Belt dryers are better than roller dryers if the veneer is sensitive or has high shrinkage. The jet dryers brought about remarkable improvement in this respect. The flatiron-effect of contact in roller dryers has been mentioned.

The kinetics of veneer jet drying has been investigated recently by *Comstock* (1971). His studies indicate that internal diffusion of moisture certainly does not control drying rate. A mathematical relationship between drying rate and moisture content consisting of two linear portions is proposed. The experimental data demonstrate that individual specimens follow the same relationship regardless of initial moisture content. Utilizing the proposed rate equations, the relationships between the rate coefficients and variables occurring in practice, drying times between any two moisture contents can be calculated for a given dryer for a broad range of veneer thickness, density or temperature and velocity. Several softwood species (Douglas fir, Loblolly pine and Yellow birch) have been dried: The drying rates have been essentially identical when corrected to a common density and thickness. Drying times for different species appear to be predictable purely on the basis of the density and green moisture content of the species, other things being equal. A constant drying rate was never observed in the studies by *Comstock* on veneer drying. The rate of drying began to decrease immediately after the warm-up period ended.

If the veneer to be dried are to be kept flat, then the drying temperatures must be high enough that plastic deformations in the wood can occur. In this way differences in shrinkage can be equalized. *Kübler* (1956) investigated the minimum temperature for flattening wavy veneers between two hot-press plates through plastic deformation. If the veneer are not to become wavy then the temperatures during the drying process should not go below the boundary curve in Fig. 3.62. This figure also indicates that the optimum temperatures vary inversely with the relative humidity (and the moisture content of the wood), therefore with progressive drying the minimum temperatures should be increased. Jet drying produces flatter surfaces but it should be considered that near the end of the drying process certain valuable veneer is very sensitive to high temperatures. Oak and beech are examples.

According to *Lutz* (1955) end waviness in dried veneer can be avoided by using the following methods:

1. In continuous dryers the veneer (for example birch 1/16 in. in thickness) should be overlapped at the ends at a distance of about 6 to 12 mm (1/4 to 1/2 in.).

The double thickness of the veneer produces, in effect, a reduced drying rate near the veneer ends.

2. In drying veneer perpendicular to the grain the ends should be moistured for a width of 6 mm (1/4 in.) by appropriate means.

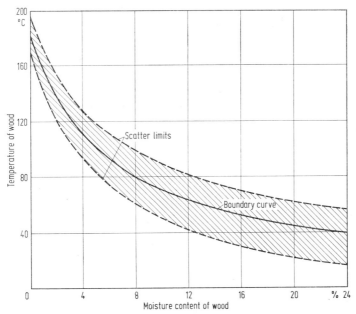

Fig. 3.62. Minimum temperature for the plastic smoothening of wavy veneers between hot iron plates under a pressure of 1 kp/cm² (14.2 lb./sq. in.). From *Kübler* (1956)

3.3 Plywood

3.3.0 General Considerations

The principal disadvantage of solid wood as structural material is its heterogeneity and anisotropy and its absorptive capacity for water and vapors (and therefore its low dimensional stability). It is possible to overcome this disadvantage by crossbanding to a large extent. These improvements, of course, are not adequate for special purposes. If high tensile strength is necessary, plywood cannot be used because the strength along the grain in the face veneers is remarkably reduced by the adjacent veneers orientated perpendicular to the face grain. Most constructions made of plywood have high resistance against buckling. The valuable properties of wood, including plywood, are expressed by the favorable ratio of strenght properties to density. Favorable too are high elasticity, low thermal conductivity, and in plywood practically no influence of notches. Plywood can easily be glued. In comparison to metals it is workable with less power consumption. Its price is relatively low and the raw material for veneer and plywood is continuously produced by nature.

Another advantage is that by the production of veneer and gluing it into plywood, veneer plywood, core plywood or composite plywood defects inherent in solid wood are better distributed. This knowledge led to the creation of laminated wood in about 1930 with the aim to produce board with high tensile strength along the grain and small variations in mechanical properties across the grain.

The principal types of plywood are shown in Fig. 3.63. In the U. S. A. and in the U. K. the term "plywood" covers not only boards consisting of cross banded glued veneers but also corestock.

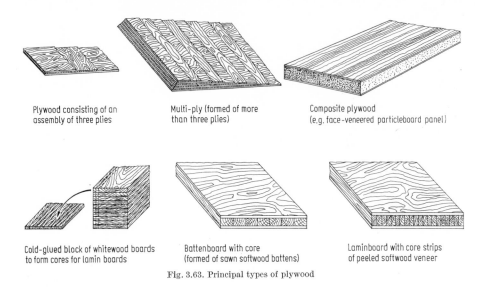

Plywood consisting of an assembly of three plies

Multi-ply (formed of more than three plies)

Composite plywood (e.g. face-veneered particleboard panel)

Cold-glued block of whitewood boards to form cores for lamin boards

Battenboard with core (formed of sawn softwood battens)

Laminboard with core strips of peeled softwood veneer

Fig. 3.63. Principal types of plywood

The new terminology for plywood is as follows:

1. *Plywood*, a panel consisting of an assembly of plies bonded together with the direction of the grain, in alternate plies usually at right angles;

2. *Veneer plywood*, a panel in which all the plies and possibly the core are made of veneers orientated parallel to the plane of the panel;

3. *Core Plywood:*

a) Wood core plywood (core of solid wood or veneers);
b) Battenboard (core made of strips of solid wood more than 30 mm wide, glued or not be glued together);
c) Blockboard (as above strips 7 mm to 30 mm wide);
d) Laminboard (core made of strips of solid wood or veneer not wider than 7 mm);
e) Cellular board (core consists of a cellular construction);
f) Composite plywood (core made of materials other than wood).

ISO-Recommendation R 1098 (June 1969) provides the general requirements for plywood panels for general use. This ISO-Recommendation does not concern panels covered with decorative veneers, faced panels and panels for special use. The following requirements are recommended:

1. Construction of panels:

a) The grain of two successive plies should form a right angle;
b) Plies symmetrical to one another relative to the center ply or the core should have the same thickness and should consist of the same species of wood or of species having similar physical characteristics. They should be produced either by rotary cutting or by slicing;
c) The tight side of veneers constituting the face band back should be on the outside;

2. When the panels leave the factory the moisture content should be between 12 and 14%;

3. Workmanship:

a) The panels should be rectangular with straight edges and sharp corners;

b) Tolerance limits fixed by ISO-Recommendation No 1954.

Table 3.9. Dimensions for Veneer Plywood for General Purposes
(Draft ISO-Recommendation No 1954, Part 1, 1971)

Veneer plywood of Broadleaved species		Coniferous species	
Minimum number of plies	Nominal thicknesses mm	Minimum number of plies	Nominal thicknesses mm
3	3	3	6.5
	4		8
	5		9.5
	6		13
5	(7)[1]	5	16
	8		19
	(9)[1]		
	10		
	12		
			22
7	15	7	26
	19		29
9	22	9	32
	25		

[1] The thicknesses in parentheses are infrequently used.

The lengths and widths are not yet internationally standardized. In Germany according to DIN 4073 veneer plywood has lengths in the range between 1,700 mm and 3,050 mm, widths between 1,000 mm and 15,250 mm. For core stock lengths are between 1,525 mm and 1,830 mm, widths between 3,500 mm and 5,100 mm.

In Germany still the following terminology is used (DIN 68705, Dec. 1958):

Veneer boards have a thickness of 0.3 mm up to 15 mm (0.012 in. to 0.59 in.). The thicker ones are called *Furniture boards*. The number of the veneer layers varies. The lowest figure is 3, then follow 5, 7, 9 ... layers. According to the German Standards veneer boards consisting of five or more layers (maximum thickness 0.5 mm) are classified as *"Multiplex boards"*.

The principal terms concerning plywood are defined in ISO 2074-1972. The vocabulary is English and French.

Irschick (1949) recommended for economical production of plywood the following rules:

1. The maximum length and width of panels has to be adapted to the available logs. European logs do not compare favorably with logs from overseas which always are larger in diameter and contain practically no knots;

2. The volume of logs on the yard should be limited so that the store of dried veneers can be kept small and well supervised. Veneer logs exceeding 2 m (6 ft.) in length should be crosscut, if during the peeling defects are observed;

3. The manufacturer should consider how the expectations of the consumer to avoid large waste can be met.

3.3.1 Principles of Manufacture

3.3.1.1 Veneer Jointing and Repairing. In the manufacture of large multiply panels, battenboard, blockboard and laminboard the outer plies are formed by jointing together several narrow widths of sliced or rotary cut veneer. With respect to the finishing of the boards either by painting, lacquering or coating it is imperative that all joints be perfectly accurate.

Good joints cannot be achieved unless the veneer to be joined together has been cut with perfectly straight edges free from all defects. The cutting is done on the veneer jointer, a very important machine in veneer or plywood factories (*Wood*, 1963, p. 56).

Dried veneer normally has buckled edges and therefore a joint can only be made if the veneers are held throughout their length by a pressure adequate to produce a plane surface.

Modern veneer trimmers possess a rigid bed with accurately machined top and a heavy carriage which travels on a track or on guides which are machined into the solid bed, and which carries regularly two cutters, one for rough and the other for fine cutting. Formerly the cut was made by saws, but now cutter-heads are usual. For instance a modern machine has eight-knife cutter heads working in tandem.

In plywood mills all veneers after jointing to be spliced on a tapeless splicer are coated at the freshly cut edges with a film of hide glue or a thermosetting resin adhesive before the clamping pressure is released. With this technique it became possible to joint the veneers without tape.

Older tapers were typical of the veneer and plywood industries. Two leaves of the veneer to be joined together were fed into leading rollers. These rollers are set so that they press the edges of the veneer closely together. A roll of pre-gummed tape is positioned above the bed with veneer and is led towards the veneer. The principles are as follows:

a) Leading the veneers to be joined;
b) Pressing the edges of the veneer together;
c) Moistening the glue at the edges;
d) Automatical laying of the tape over the joint;
e) Pressing the tape against the veneer by an electrically heated roller, causing setting of the glue.

The tape is generally made from very tough, flexible and thin (0.05 mm thick) Kraft paper, occasionally of a perforated type. It is desirable that tapes can be easily removed. Tapes are also used for repairing splits in face veneers (*Kollmann*, 1962, p. 257—261, *Wood*, 1963, p. 58).

Trouble may occur if tapes lie between face veneer and crossbanding. Therefore the tapes should always be laid to the outer side of the face veneer and removed after pressing. There are tapes which can be used internally; they are combinations of thermosetting resins and paper. These tapes should be used only in conjunction with resin adhesives which polymerize under the same conditions (moisture content, temperature, pressure) as those indicated by the manufacturer of the tape.

Tapeless jointers for veneers were developed for along-the-grain and later for across-the-grain jointing. Machinery and the jointing process are explained in the technical literature and in advertisements of manufacturers. Some machines are very efficient and have found their way into the plywood mills throughout the world. A survey was given by *Kollmann* (1962, p. 262—276).

In preparing veneer for the cores of large plywood panels it may be necessary to joint together two or more pieces of veneer of adequate width to obtain the lengths required. There are two principal possibilities, illustrated by Fig. 3.64:

a) Dovetail joints, produced by special cutterheads;
b) Scarf joints.

Occasionally veneers during the drying process are split or damaged. In this case they must be repaired either by re-cutting into two or more pieces of narrower veneer on a guillotine and thereafter by jointing in the normal way. Larger knots and defects in color and texture should be repaired by plugging or patching. In the U.S.A. the manufactures of Douglas fir veneer cut the plugs in the shape of a gun-boat. By this method the patch gets the opportunity of blending well into the surrounding veneer. All plugged boards should be perfectly sanded prior to painting or veneering.

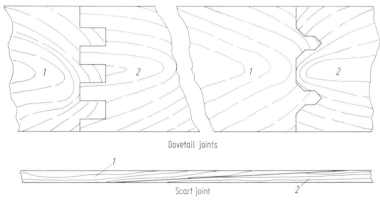

Dovetail joints

Scarf joint

Fig. 3.64. Types of veneer joints. From *Wood* (1963)

3.3.1.2 Applying the Adhesive. Originally in the trade and workshops the adhesive was applied by brushing. This method is not applicable to industrial operation. The spray-gun method is also rarely used in large-scale operation because:

a) The solvent necessary to reduce the viscosity of the glue for spraying must be evaporated subsequently;
b) Waste of adhesive from overspray is inevitable;
c) Ventilating equipment is necessary to provide disposal of noxious vapors.

For veneer and plywood the use of roller spreaders is now most common. The adhesive is distributed by means of two or four rollers. The thickness of the spread is controlled by doctor rolls. A modern spreader is shown in Fig. 3.65.

3.3.1.3 Pressing. Some pressure is necessary to make a good glue joint. Hand pressure is still used, but only in small veneering shops, whilst clamps are useful for edge veneering, repairing and other minor cabinet-making jobs and for curved work. For the majority of plywood the bond is produced to a small extent in cold presses, but primarily in hot presses. The cold press has been used in the decade before World War II. It does not need highly skilled labor, and the risk of errors is relatively small (*Wood* and *Linn*, 1950, p. 103).

Adhesives for cold pressing are applied in a fluid state and need more time to set than those which utilize heat. Cold-pressed plywood must be carefully re-dried after the setting of the adhesive.

Until about 1939 cold pressing was the method used by the majority of manufacturers of softwood plywood in the U.S.A. and of plywood manufacturers in Australia, Brazil, Europe, and Japan for the production of cores for block-boards and laminboards. After about 1936 hot-plate presses became more and more popular in plywood manufacture in many parts of the U.S.A. In Europe hot-pressing is generally adopted in the production of plywood of the species alder, beech, birch and Okoumé. The scope of this book does not encompass a description or discussion of cold presses. According to *Wood* (1963, p. 90) the original method of manufacturing plywood on a hot-plate press was described in a patent granted in 1896 to *Christian Luther* of Reval. The development of the hot-plate presses proceeded rapidly after World War I. New glues called for greater precision.

Fig. 3.65. Modern roller spreader with two spreading and two doctor rolls. Bürkle & Co., Freudenstadt

There are several types of hot-plate presses, mainly massive machines. They consist of a heavy structural head supported by four or more columns and a press table. When the press is open, there are spaces between the plates, known as "openings" or "daylights". Pressure is applied by raising (hydraulically) the press table which in turn lifts the platens.

Heat is supplied to the plates by steam or hot water. Platen temperatures of 160 to 163 °C (320 to 325 °F) should be reached. Pressure is produced by rotary or piston pumps, the former being used for closing the press rapidly and the latter for maintaining the high pressures. As a general rule, pumps operate on a mixture of water and water-soluble emulsion to avoid possible packing troubles when working on oil.

The output of the factory is determined by the number of "daylights" in the hot presses and by the type of adhesive used. Production systems with automatic lay-up stations with mechanical loaders and unloaders have been developed which by the shortest possible press-charging times and by the adaption to specific technological requirements guarantee high productivity and consistent product quality. There are, for example, press lines with tray-system loaders for the production of plywood. In some factories two press lines operate in parallel. The hot platen size may vary; examples are 3,850 mm × 1,750 mm (152 in. × 68 in.), 3,500 mm × 1,750 mm (138 in. × 69 in.) and 5,100 mm × 183 mm (200 in. × 72 in).

In W.-Germany hot presses with 16 to 40 daylights are installed. Mechanical loading can be effected by the following principal types of equipment:

1. Conventional method with metal cauls which circulate each plywood panel which is placed between an upper and a lower caul. Advantages: High safety of operation, possibility to charge the openings with several smaller sizes. Disadvantages: Higher investment costs, more floor space required, higher heat consumption;

2. Caulless systems:
a) With pallets or traybelts;
b) With push-loading.

Caulless systems need cold prepressing of the veneers glued and assembled in the usual manner. Prepressing guarantees high safety of hot press operation, better utilization of wood (less rejects through open joints/gaps/and overlaps), smaller oversize between net size and gross size of panels, adjustment of moisture content in assembled plies, prevention of veneer waviness. This last fact is especially important if phenolic resin glues are used because in this case the open assembly or waiting time up to hot pressing may reach 20 and more minutes.

For a high-efficiency 40-daylight press line, built in W.-Germany, the following data can be given:

Platen size 3,200 mm × 1,900 mm (126 in. × 75 in.),
Specific pressure 15 kp/cm² (213 lb./sq. in.),
Opening rate 600 to 700 mm (23.6 to 27.6 in.) per second,
Output 20 m³ (706 cu. ft.) per hour, 5 mm thick boards,
800 boards per hour, pressing time 2 min., idle time 1 min.

In Finland 20 daylights are usual, the specific pressure on the birch-plywood is rather high: 20 kp/cm² ($\sim$285 lb./sq. in.). In the U. S. A. a 40-daylight press with a tray system loader for the production of plywood is adequate. Hot platen size is 2,700 mm × 1,375 mm ($\sim$106 in. × 55 in.). The specific pressure is 14 kp/cm² (199 lb./sq. in.). Fig. 3.66 shows a modern 30-daylight press for the production of plywood[1]. The assemblies are conveyed to the cold pre-press with its oilhydraulic drive. Then the possibility exists that the assemblies can rest until they come to the lifting platform with rolls. For the efficiency of the press line the infeed device is essential. The Figure shows that there is a tray system loader and that the hot press is oil-hydraulically driven. Unloader, outfeed and stacking device, and lifting platform with rolls are shown.

Specific pressures may vary between 12 and 20 kp/cm² (171 and 285 lb./sq. in.). *Wood* (1963, p. 95) discusses the application of heat and pressure and writes that "in all hot pressing three rules should be kept in mind:

1. The quicker the setting-time the greater the accuracy required;
2. Heat applied and time of heating should be kept to a minimum;
3. Never attempt to speed up production by using excessive heat or pressure."

---

[1] By courtesy of Messrs. Siempelkamp & Co., Maschinenfabrik, Krefeld, West-Germany.

Excessive or prolonged heat will reduce the quality of the plywood. The plasticity of wood increases rapidly above the boiling point of water, so that too high temperatures applied cause undesirable compression.

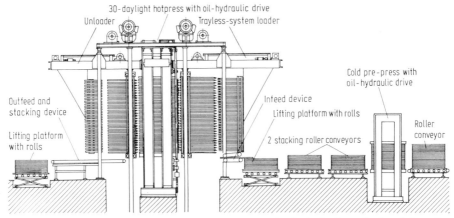

Fig. 3.66. Scheme of a thirty-daylight pressline with trayless-system loader for the production of plywood

3.3.1.4 Conditioning. Plywood panels on removal from the hot press are in an instable condition. The outer plies are overdried but not the core. This produces temperature and moisture content gradients which must be reduced or removed by one of the following methods:

a) Close-stacking the hot plywood panels (without stickers) as they come from the press. During a period of about a fortnight the pile is brought to approximately a uniform temperature and moisture content. This procedure is not applicable from the economic point of view in large factories;

b) The best method is to pass the panels as they come from the hot press through an air-conditioned channel; this method was practised during World War II in Germany for plywood and especially veneer plywood (Schichtholz) for aircraft manufacture, but it is too expensive for the industry in normal times. About conditioning of plywood *Schneider* reports in *Kollmann*, 1962, p. 525 to 555;

c) The panels may be sprayed with, or plunged into, water. Afterwards the panels are stacked in a tight pile for some hours. This enables the moisture to approach an equilibrium in the plywood and to relieve the strains set up during pressing (*Wood*, 1963, p. 100 to 101). In this case experience is most important because correct procedures are not available and accurate measurements are not possible.

3.3.1.5 Panel Finishing (Sizing and Thicknessing)

3.3.1.5.1 Sizing or Trimming[1]. If the plywood panels are cooled off in piles and equalized with respect to temperature and moisture content, they must then be trimmed to the exact size. For this purpose a double circular saw is used which trims two parallel edges in one operation. The rolling carriage of the saw can carry several panels on top of each other at one time.

---

[1] The author thanks Messrs. Anthon GmbH, Flensburg, W.-Germany; Becker & van Hüllen, Krefeld, W.-Germany; Greenlee Bros. & Co., Rockford, Ill., U.S.A.; Meyer & Schwabedissen, Herford, W.-Germany, for information.

The trimming of the third and fourth edge can be done either on the same machine after having adjusted the distance of the saws accordingly. If a bigger capacity is desirable two different saws have to be installed which are placed advantageously one behind the other and at a right angle to each other. In this case one machine cuts the width and the other one the length of the panel. If the machine is equipped with an additional cross cut support all four edges can be trimmed on one machine which lowers however the capacity of the machine.

A very high capacity on trimming two edges simultaneously can be attained with a trimming saw equipped with endless feed chains, as on this machine no carriage has to be returned. Stops and shadow line guides take care of low waste of material and of proper rectangular trimming of the panels.

The manufacturers in various countries build a variety of panel machines to fully meet the requirements of plywood and other panel plants. Only a few more details can be given.

Sizing machines installed in the mill of a large U.S. Western plywood factory produce size and trim 4 ft. by 8 ft. plywood panels at the rate of 30 per minute, maintain high standards of accuracy and cut smoothly due to high saw speed (6,000 rpm). Spring-loaded hold-downs in a scattered pattern apply uniform pressure over the entire panel. The stub saws are driven by a common arbor or either a direct-coupled or a belted-motor drive. Variable speed drive provides any feed required from 90 to 270 ft./min.

Flattening arms depress warped panels as they move over the outfeed belt units of the transfer device. Refuse strips drop to a floor conveyor as the panels move from one machine to the other.

A W.-German automatic panel sizing saw can be used either:

a) In a production line, feeding being from a lifting platform or the like and conveying after the cut into downstream sanding machines (cutting of the panels in the cold condition) or

b) As individual machine immediately behind the press discharge/bin, the panels being stacked on a lifting platform after cutting (cutting of the panels in the warm condition).

Another W.-German automatic panel cutting saw (Schwabedissen) offers the following facilities:

1. Trimming on four sides;
2. Cross and length-cutting parting cuts;
3. Breaking down the material into irregular sections (Fig. 3.67).

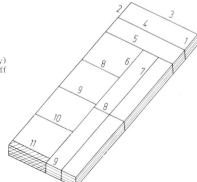

Fig. 3.67. Sequence of operations for an automatic panel cutting saw with program control.

Untrimmed panels:
1. — Forward run: length-trimming cuts 1 and 2 (simultaneouly)
1. — Return run: trimming cut 3, head-cuts 4 and 5 (take off strips)
2. — Forward run: length-cutting parting cuts 6 and 7 (simultan.)
2. — Return run: cross-cutting parting cuts 8 to 11 incl. cross-trimming cut

Trimmed panels:
1. — Forward run: no cuts
1. — Return run: head cuts 4 and 5 (take off strips)
2. — Forward run: length-cutting parting cuts 6 and 7 (simultan.)
2. — Return run: cross-cutting parting cuts 8 to 10

Meyer & Schwabedissen, Herford

All this is achieved with a single loading of the panel stack and by an automatic working cycle.

The electro-mechanical program-control system with micro-switches provides a very high output with minimum operating effort, optimally with electronic size pre-selection or punched card control for frequently changing cross-cutting program.

Some characteristic technical data are listed in Table 3.10.

Table 3.10. Technical Data for Automatic Panel Cutting Saws

A

Panel Sizing and Parting Saw (Schwabedissen, Fed. Germ. Pat. No 1 080 287)

Cutting width: 1,550···1,850···2,150 mm 　　　　　51 ···　　73···　　85 in.	Sawblade diameter:　　330 mm (13 in.)
	Tool speed:　　　　　3,000 rpm
Cutting length　2,500 mm 　　　　　98 in.	Table carriage drive:　　　　　2.2 kW (3 HP)
Further lengths in steps of 500 mm ($\sim$ 20 in.) depending on press dimensions)	Feeds:
Cutting height: 60 mm (2.4 in.)	Forward:　　18 m/min (59 ft./min)
Lenght-cutting motors: 7.5 kW (10 HP)	Return —
(normally 3 — maximum 6)	fast speed:　　10 m/min (33 ft./min)
Cross-cutting motor: 7.5 kW (10 HP)	

The machine can be supplied with a charging and discharging mechanism.

B

Automatic Panel Saw (Anton)

Max. panel length	2,700···3,700···4,100···5,200 mm 　　106···　　146···　　161···　　205 in.
Max. panel width	1,900···1,900···1,900···1,900 mm 　　75···　　75···　　75···　　75 in.
	Extended (length) 6,000 mm (236 in.)

Max. cutting width of length saw	950 mm (37 1/2 in.)	Tool speed
Min. cutting width of length saw	75 mm (3 in.)	
Min. cutting width of cross-cut saw	220 mm (9 in.)	Length saw　　4,500 rpm
Max. cutting height	60 mm (2 1/2 in.)	Cross-cut saw 5,500 rpm

Motor ratings:　　　　　　　　　　　　　　　　　Saw blade diameters:

Length saw	5.6 kW (7.5 HP)	Length saw	350 mm (14 in.)
Cross-cut saws (standards 4 units)	4.1 kW (5.5 HP)	Cross-cut saw	300 mm (12 in.)
Cross-cut saws (reinforced, optional)	5.6 kW (7.5 HP)		
Length saw feed	0.19 kW (0.25 HP)		
Table feed	0.16···0.25 kW (0.22···0.34 HP)		
Fence adjustment	0.24 kW (0.33 HP)		

Compressed air demand 5 at.
Compressed air volume
with 4 cross-cut saws 50 l/min

Rates of feed:　　　　　　　　　　　　　　Exhaust pipe connections with
　　　　　　　　　　　　　　　　　　　　　4 cross-cut saws:

Length saw feed	21 m/min ($\sim$ 70 ft./min)	Manifold	250 mm dia. (10 in.)
Length saw return	21 m/min ($\sim$ 70 ft./min)	Length saw	150 mm dia. (6 in.)
Table feed	12 m/min ($\sim$ 40 ft./min)	Cross-cut saws	100 mm dia. (4 in.)
Table return	18 m/min ($\sim$ 60 ft./min)	Air velocity	25 m/sec (82 ft./sec)
		Air volume	75 m³/min (350 cu. ft./min)

Heavier machines are necessary for the production of wood core plywood (battenboard, blockboard, laminboard as defined by ISO 2074).

Sizing of particleboard is dealt with in paragraph 5.2.10.1.

3.3.1.5.2 Thicknessing (Sanding or Scraping). Trimming to dimensions of plywood panels and squaring up on double cut-off saws is also called "epualizing". The equalized plywood sheets are passed for final processing to sanders or scrapers on which surface imperfections are removed and a smooth even finish is given to the panels.

There are the following principal machines for final processing in use in plywood factories:

A. *Drum sanders;*	C. *Wide belt sanders;*
B. *Belt sanders;*	D. *Scrapers.*

The principal properties of abrasives for wood and the technology of sanding processes (volume of abrasion, influences of wood species, grain size, sanding time, belt pressure and speed, temperature etc.) are dealt with in Volume I of this book (p. 528—533). Reference is also made to the contributions by *Kollmann* (1955, p. 740—757, where the U.S.A. grain numbers are compared with the corresponding German values according to DIN 1971); *Doffiné*, in *Kollmann*, 1962, p. 494 to 516; *Wood*, 1963, p. 103—109; *Koch*, 1964, Chapter 11; *Wehner*, in *Kollmann*, 1966, p. 380—388; *Koch*, 1972, p. 894—899.

Ad A. Drum sanders in plywood plants are similar to those used by other wood-working factories and by the particleboard industry. They may consist of two, three or four drums which, as a rule, are positioned above the table and are given a lateral and oscillating movement (e.g. 9.5 mm $\sim$3/8 in. amplitude at 20 to 25 cycles per minute) in addition to the rotary action. The plywood panels are carried by an automatic roll-feed for endless bed-feed through the machine beneath the rotating and with sandpaper covered drums. Each drum (roller, cylinder) is covered with felt which acts as a cushion between drum and abrasive paper. The latter is spirally wound over the periphery of the drum. The first drum carries the roughest paper, the following a finer one and the last a finishing grade. Silicon carbide, the hardest artificial abrasive, is most commonly used for primary heads. *Ferguson* (1968), however, states that for rapid wood removal aluminium oxide is best. Secondary heads commonly use silicon carbide in grits from 80 through 120 (*Stevens*, 1966).

The larger modern sanders are massive machines which run at all speed without vibration. Occasionally the drums are placed below the table.

In the U.S.A. panel sanding machines are usually double deck; they simultaneously machine both top and bottom of the panel in one pass. Six heads are not uncommon. Some usual technical data are as follows:

Belt speeds: 20 to 30 m/sec ($\sim$3,950 to 5,900 ft./min), for Southern pine plywood (*Koch*, 1972, p. 894) 5,000 to 6,750 ft./min ($\sim$26 to 35 m/sec);

Feed speed: for hardwood plywood in W.-Germany 12 to 18 m/min ($\sim$40 to 60 ft./min), for softwood plywood in the U.S.A. 80 to 200 ft./min ($\sim$24 to 61 m/min);

Depth cut per primary heads: up to 0.03 in. (0.76 mm) for softwood;

Running time of belts: 50 to 100 h;

Moisture content of the plywood not exceeding 8%;

Width of abrasive belts: in the U.S.A. commonly 50 to 53, 63, or 67 in. (1,270 to 1,350, 1,600 or 1,700 mm);

Power demand: Cutting force is positively correlated with specific gravity of wood, depth of cut, width of belt and feed rate (*Kollmann*, 1955, p. 749, *Ward*, 1963);

> Example: 53-in. wide four-head panel sander, capable of feed speeds to 200 ft./min, each sander head carries 125 HP (94 kW), total feed rolls drive 25 HP (18.8 kW) (*Koch*, 1972, p. 895);

Direction of feed against direction of belt travel, possibly across the grain (*Stewart*, 1970).

Ad B. Belt sanders are useful to remove odd pieces of tape or scratches after drum sanding, for finishing fine veneer work or curved surfaces. One or two abrasive belts run over two or more pulleys. Belt speeds 12 to 30 m/sec (2,400 to 5,900 ft./min). With two belts the second should be of finer grit than the first. Recommendations for type of abrasive and grit size for various work are given by *Wood*, 1963, p. 106. The table which carries the plywood panel can be moved forward and backward across the direction of travel of the belt by the operator who also maintains contact between the abrasive belt and the surface of the panel via a flexible pad and a lever.

Ad C. Wide belt sanders are heavy, sturdy built machines free from vibration for processing up to 2,500 plywood panels 8 ft. × 4 ft. (2,440 mm × 1,220 mm) per shift. Maximum sanding widths of W.-German wide belt contact sanders (Anthon & Söhne, Flensburg): 1,100 mm — 1,300 mm — 1,650 mm — 1,950 mm (44 in. — 51 in. — 65 in. — 77 in.), belt speed 25 m/sec. (∼4,900 ft./min), feed speed 8.5 to 27 m/min. (28 to 89 ft./min). Power demand as shown in Fig. 3.68 and Fig. 3.69. In the U.S.A. for softwood plywood up to 120 ft./min (∼37 m/min).

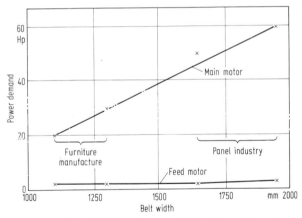

Fig. 3.68. Power demand of wide belt sanders versus belt widths used in various industries

There is a variety of wide belt sanders available either with top belts or with bottom belts, also in tandem-arrangement (Fig. 3.70). In order to install the longest possible belt (for instance, 2,620 mm (103 in.) or up to 3,650 mm (144 in.) the belt runs, in a triangular way, over an idler roll and a contact roll (tensioning roll) which retains it against the panel to be sanded. Great belt lengths increase operational life, reduce the unproductive time required for changing belts, improve cooling of belts, assist chip extraction and prevent clogging of dust and chips.

Top wide belt sanders and bottom belt sanders complete a production line. Without being turned the workpieces pass through both machines and are succes-

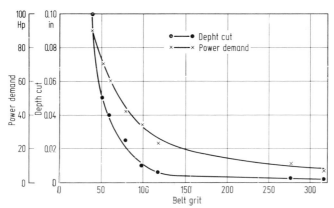

Fig. 3.69. Interrelationships between power demand, depth cut and belt grit sanding 4 ft.-wide Southern pine plywood at a feed speed of 80 ft./min with a 50 in. belt. Curves designed by *Kollmann*, using data published by *Koch* (1972, p. 899)

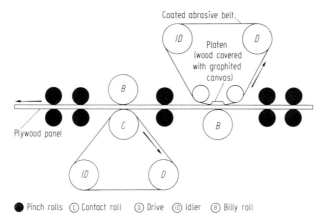

Fig. 3.70. Scheme of a wide belt sander in tandem arrangement for sanding plywood panels. Smithway Machine Co., Seattle, Wash., and Behr Manning Co., Troy, New York

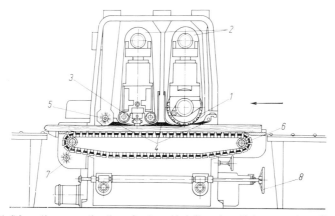

Fig. 3.71. Schematic cross section through a top wide belt sander with two separate sanding belts.
1 Contact roll for presanding, *2* Tension and control roll, *3* Sanding pad for finish sanding, *4* Flexible pressure lip, *5* Brush roll, *6* Feed chain belt, *7* Brush roll for cleaning the conveyor belt, *8* Thickness adjustment of workpiece.
Anthon & Söhne, Flensburg

sively sanded on both surfaces. With conventional single-belt sanders rough and fine sanding cannot be carried out in a single pass, as both processes require their own specific abrasive grits. Fine sanding can only be effected with a machine fitted with two separate sanding belts (Fig. 3.71). The feed system consists either of flexibly mounted rubber coated rolls or of transport chains.

Some manufacturers have combined in one machine both drum and belt sanders, the former for preliminary cut, the latter for final finish.

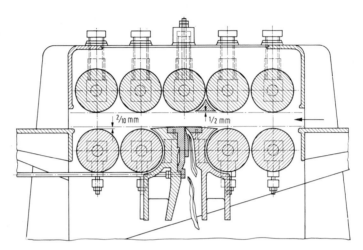

Fig. 3.72. Schematic cross section through a scraping machine. Böttcher & Gessner, Hamburg

Ad D. Scraping machines work in such a way that the plywood panels are transported by strong feed rollers across a stationary scraping knife (Fig. 3.72), which scrapes a thin (0.1 to 0.15 mm thick) chip — according to the adjustment — from the whole width of the panel. Feed direction is parallel to grain direction, feed speed is 20 to 25 m/min (66 to 82 ft./min). Machine scrapers are not suited for plywood manufactured from coniferous species (with considerable differences in densities of springwood and summerwood, cf. Vol. I, Table 6.4, p. 174) but well suited for veneer produced from diffuse-porous woods of broadleaved trees. Best suited are Okoumé (if not swirl-grained) and other African veneer woods, to some extent also oak. Scraping machines are commonly used in U.S.A. and French plywood mills whereas the application in W.-Germany decreased. Under certain circumstances scraping machines offer the following advantages:

1. Absolutely smooth and — contrary to sanded panels — bright and shiny surfaces;
2. Less wood losses (in sanding 4 mm thick plywood 10 to 15% of the total panel volume are lost, in scraping only about half of his amount);
3. No sander-dust;
4. Higher output as compared with sanding;
5. Low power demand;
6. Cheap maintenance.

Nevertheless, the poorer quality of okoumé-logs recently imported suspended the scraping machines in Western Europe.

3.4 Properties of Plywood

3.4.0 Preamble

Wood-based panel products (plywood, particleboard and fiber building board) are used mainly in furniture manufacture and in construction. The increased production of prefabricated houses has opened up new and extended fields of application for these materials. They are also preferred for reconstruction, modernization and partitioning of old houses and administrative buildings. All these sheet materials, produced in an ever-growing volume (cf. Fig. 3.3), offer specific advantages such as the following:

1. Almost unlimited market availability and the possibility of their in-plant conversion into secondary products by the panel manufacturers themselves. (As an example the production of extruded Okal-particleboard and the construction of ready-for-use prefabricated Okal-houses may be quoted);

2. Large dimensions standardized and adapted to consumer requirements;

3. Dry construction methods;

4. Moisture resistance better than that of solid wood;

5. Reduction in swelling and shrinkage as compared with solid wood and consequently better dimensional stability;

6. Greater isotropy and homogeneity than solid wood;

7. Easy workability with hand tools and machines;

8. Wide range of densities, thicknesses, structural strength and weights per unit area, meeting most practical requirements;

9. Greater stiffness, rigidity, resistance to buckling, strength, hardness, abrasion resistance may be imparted if desired. The mechanical properties of wood-based panel products are especially favorable in relation to specific weight ("order of merit" in comparison to other building materials);

10. Low thermal conductivity at right angles to the panel surfaces;

11. Adequate acoustical properties, sound absorption improved in special acoustic boards by holes or slits;

12. Possibility of gluing and fastening by nails, screws and other convectors;

13. Paintability;

14. If desired fire resistance for specific building requirements;

15. Generally aesthetic appearance;

16. Reasonably priced.

The foregoing list indicates that about the same physical, mechanical and technological properties are essential for all wood-based panel products. They are all tested, and the test results evaluated, by the same or by very similar methods. Therefore it is only necessary to consider the details of one product group and to be informed about the properties of the other two material groups in a more condensed manner. The most comprehensive coverage has been chosen for particleboard (Chapter 5) because of its relative and growing importance and because of its special interest in relation to technical development and versatility in application.

3.4.1 Density, Weight per Unit Area

Density is the ratio of the mass of a test piece to its volume, calculated to the nearest 0.01 g/cm³ in accordance with the following formula:

$$\varrho = \frac{M}{V},\qquad\qquad(3.7)$$

where

ϱ = the density in [g/cm³],
M = the mass in [g],
V = the volume in [cm³].

The test pieces should be square in shape, with sides measuring 100 mm. The test pieces should be conditioned to a constant mass in an atmosphere of a relative humidity of 65% ± 5% and a temperature of 20 °C ± 2 °C. The thickness is measured at four different points, shown as circles in Fig. 3.73. The mean arithmetical value of the four measurements is considered to be the thickness of the test piece.

The density of plywood depends on the species used for the manufacture. There are some types of plywood with a relatively low density, for instance Okoumé and poplar-plywood. The density is higher for plywood manufactured from alder, birch and beech. The pressure applied in the hot presses ranges between 6 and 30 kp/cm² (86 to 428 lb./sq. in.). In Europe only occasionally coniferous species such as spruce and pine were converted to plywood. Softwood plywood is produced in great amounts in the U.S.A. Early studies of the U.S. Forest Products Laboratory (*Fleischer* and *Lutz*, 1963) played an important part in developing the necessary technology. Remarkable was the establishment of the Southern pine plywood industry. *Koch* (1972, p. 1175) quotes that in 1968 weight per unit area W_A can be calculated as follows:

$$W_A = \varrho \cdot s = \frac{M \cdot s}{V}, \tag{3.8}$$

where symbols and dimensions may be used as above and s = thickness in cm; for getting W_A in kg/m² the result must be multiplied by the conversion factor 10.

About 15% of the softwood plywood used in the U.S.A. is southern pine and *Holley* (1969) predicted 30% for 1975. For a number of reasons most Southern pine plywood is hot pressed with phenol-formaldehyde resin glues.

Glue bond quality is affected by: 1. Variations in growth ring per inch, 2. Specific gravity, 3. Roughness of the veneer rotary cut, 4. Wood moisture content (near 4%), 5. Pitch content, 6. Surface contamination, 7. Mold, 8. Resin formulation, 9. Uniformity of glue spreading, 10. Glue line thickness, 11. Assembly time.

Prepressing is practised in some softwood plywood plants. In hot-pressing southern pine plywood (using phenol-formaldehyde glues) specific pressures may range from 100 to 300 lb./sq. in. (~7 to 21 kp/cm²), but are usually 175 to 225 lb./sq. in. (~12.3 to 15.8 kp/cm²). Temperatures are usually 275 °F to 315 °F (135 °C to 157 °C). Press times depend on panel thickness and glue formulation (for instance 6.5 min for 3/8-in. (9.5 mm) three-ply panel).

With respect to the low pressure applied the shrinkage or compression does not exceed usually 5%. Therefore the density of normal plywood is only a little higher than that of the original wood. *Kollmann* (1951) collected data as reproduced in Table 3.11.

Increases in pressure cause a rapid increase in density, but only up to a pressure of about 100 kp/cm² (1,424 lb./sq. in.) (Fig. 3.74) (*Küch*, 1951). It becomes clear that pressures higher than about 100 kp/cm² (1,424 lb./sq. in.) are not effective in increasing board density.

Table 3.11. Structure, Density and Moisture Content of Plywood

Plywood		Density	Moisture content
Species	Thickness mm	g/cm³ (lb./cu. ft.)	%
Beech	1	0.77 to 0.88 (48 to 55)	6⋯8
Beech	2.5	0.73 to 0.81 to 0.95 (∼46 to 51 to 56)	6⋯8
Beech	6	71 to 0.77 (∼44 to 48)	6⋯8
Birch	1 2.5 6	0.82 to 0.92 (51 to 57)	6⋯8
Douglas fir	1/4 in. ∼ 6.4 mm 3 plies 3/8 in. ∼ 9.5 mm 5 plies 1/2 in. ∼12.7 mm	0.50 (31) 0.48 (30) 0.48 (30)	

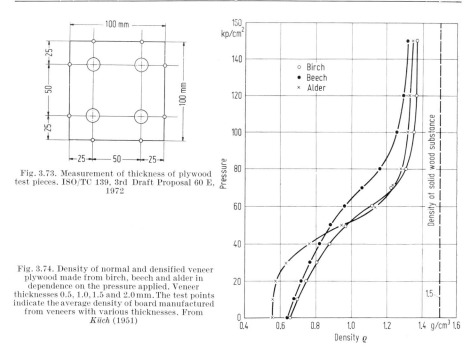

Fig. 3.73. Measurement of thickness of plywood test pieces. ISO/TC 139, 3rd Draft Proposal 60 E, 1972

Fig. 3.74. Density of normal and densified veneer plywood made from birch, beech and alder in dependence on the pressure applied. Veneer thicknesses 0.5, 1.0, 1.5 and 2.0 mm. The test points indicate the average density of board manufactured from veneers with various thicknesses. From *Küch* (1951)

3.4.2 Moisture Content, Absorption and Swelling

The moisture content H of a plywood test piece is determined by the same method which is applied to particleboard and fiber building board. Sampling and cutting of test pieces should be carried out in accordance with International Standards (ISO) or with national standards.

ISO recommends test pieces of any shape (possibly rectangular) and dimensions with a minimum area of 25 cm². Each test piece is weighed at the time of sampling and dried at a temperature of $103 \pm 2\,°C$ to constant mass after cooling in dry atmosphere. The balance used must allow a reading to an accuracy of 0.01 g. The moisture content should be calculated to the nearest 0.1% in accordance with the following formula:

$$H = \frac{M_H - M_0}{M_0} \cdot 100, \qquad (3.9)$$

where

H = the moisture content of the test piece in [%],
M_H = the mass of the test piece at the time of sampling in [g],
M_0 = the mass of the test piece after drying in [g].

Kollmann (1962, p. 782—789) reports, based on extended mechanical tests of plywood produced by ten manufacturers (with 3 to 9 plies, nominal thicknesses between 0.8 and 15 mm, densities between 430 and 794 kg/m³, various wood species), that the moisture content of the plywood after storage in normal climate (relative humidity 65% ± 2%, temperature 20 °C ± 1 °C) varied between 7.3% and 12.7%. The average value was about 10.0 %. *Wood* (1963, p. 444) published for 3/4 in. Douglas fir plywood (7-ply) $H = 10.4\%$, for 18 mm okoumé plywood (7-ply) $H = 11.8\%$ and for 18 mm blockboard (birch faces, Baltic redwood core) $H = 10.7\%$. The values are quite similar. The influence of the glue line on density and on moisture content is evident.

Wood is a hygroscopic ligno-cellulosic material. For adsorption and for desorption of condensable vapors from the surrounding atmosphere sorption isotherms can be determined by tests (cf. Vol. I, p. 189—195). *Kollmann* (1963) developed an equation which mathematically describes hygroscopic isotherms in the total range of relative humidity. The application of high temperatures used in veneer drying and hot-pressing of plywood reduces the sorption capacity (*Kollmann* and *Schneider*, 1963). Cured thermosetting artificial resin glues are not hygroscopic. The interrelationships are complicated because the very fine (even submicroscopic) capillaries in the system wood/glue line play a role.

Fig. 3.75 (*Küch*, 1939) showed that the hygroscopicity of beech veneer plywood (15 and 45 plies per 1 cm thickness) slightly increased as compared with

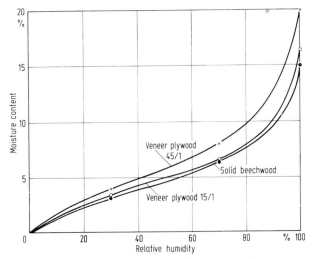

Fig. 3.75. Hygroscopic equilibrium of veneer plywood made of beech veneer, 45 plies per 1 cm thickness of board or 15 plies. From *Küch* (1939)

solid wood. Surprising are the results of investigations made by *Seifert* (1972). Fig. 3.76 reproduces the 25 °C-isothermes for plywood with the following structure and history of manufacture:

Structure: **1.3 mm okoumé — 2.0 mm spruce — 2.8 mm spruce**
    2.0 mm spruce — 1.3 mm okoumé

Peeler logs steamed at 70 °C to 90 °C (158 °F to 194 °F)

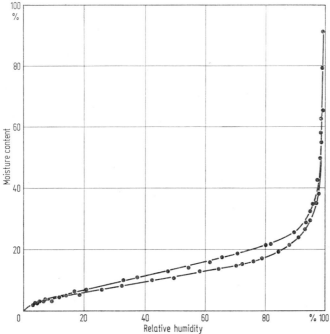

Fig. 3.76. Adsorption and desorption curves for plywood. Structure: 1.3 mm okoumé — 2 mm spruce — 2.8 mm spruce — 2 mm spruce — 1.3 mm okoumé, at 25 °C (77 °F). From *Seifert* (1972)

Veneer dried to about 10% moisture content, based on oven-dry weight.
Glued with phenol-formaldehyde resin glue (boiling proof according to German specification "AW 100"), about 4% resin based on oven-dry weight of wood.
Pressing conditions: temperature 120 °C (248 °F), specific pressure 12 kp/cm² (171 lb./sq. in.), press time 12 min.

Above 90% the hygroscopic curves change their character and above about 95% relative humidity a steep increase in moisture content occurs. This certainly does not come from the wood, but from the glue line. The phenol-formaldehyde resin glues need hardeners, normally on the basis of acid salts. Therefore it may be assumed that these hardeners absorb rapidly moisture near to the saturation point. Probably there is a difference between plywood glued with such phenol-formaldehyde glues and plywood glued with urea-formaldehyde resins. Further investigations are necessary.

Compressed veneer plywood (Compreg) absorbs less moisture than solid wood (*Vorreiter*, 1942). *Winter* (1944) found the same results. Both authors attributed their results to the increased density and to the presence of water-repellent artifical resins. Doubtlessly the pressing temperature has a great influence. *Seborg, Millett* and *Stamm* (1944) found reductions of water absorption versus immersion time (*Staypak* consisting of birch veneer, specimen 2.5 cm × 1.9 cm × 7.6 cm) for various temperatures and hot-pressing times as shown in Fig. 3.77. The influence of temperature is great up to about 157 °C (315 °F). Further raise of temperature looses more and more its stabilizing effect. This is testified also by the observation that above this temperature the influence of the heating duration (between 5 and 45 min) may be neglected. *Seborg, Millett* and *Stamm* (1944) explain the heat stabilization of densified wood with a plastic flow of the cementing substances in the wood tissues. This flow diminishes the internal stresses set up during the

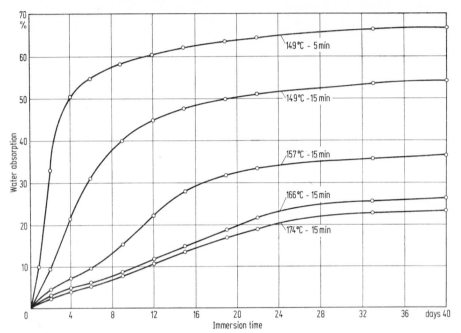

Fig. 3.77. Water absorption versus immersion time for specimen 2.5 cm × 1.9 cm × 7.6 cm from compressed birch veneer plywood. Veneers preconditioned at 50% rel. humidity, temperature and time used for stabilization as indicated. From *Seborg, Millett* and *Stamm* (1944)

compression and reduces the tendency of recovery or "spring-back" from compression usual for plywood after release from the hot press. *Stamm* and *Harris* (1954) write: "A decrease in both equilibrium swelling and recovery from compression, and in the permanent recovery from compression after drying, occurs with increasing moisture content, temperature, and time of heating. The latter is shown in Fig. 3.78 for cross sections of 1/8 in. in the fiber direction that were immersed in water for 40 h. The data show that it is possible to reduce the spring back or recovery from compression from values over 50% at 300 °F to values as low as 10% for wood at 6 percent moisture content (equilibrium with 30% relative humidity) merely by raising the temperature of pressing to 360 °F. It can be further reduced to 2 to 5% by using wood with a slightly higher moisture content (9 to 12%). At intermediate temperatures, the permanent recovery can be appreciably reduced by increasing the time of pressing."

Cross-banding or cross-laminating does not eliminate swelling and shrinking, but it does change their directions. Rotary-cut veneer with a density (ovendry weight based on dry-volume) of 0.45 will swell about 3.5% in the tangential direction between 30 and 90% relative humidity. In general it will swell only

0.1% in the fiber direction under the same conditions (*Stamm*, 1964, p. 313). When an odd number of plies is cross-banded to form normal plywood, the corresponding swelling will be only a few tenths of a percent in both face sheet directions. Due to the restraint of one ply upon another, there will, in general, be a slight tendency for the thickness swelling to increase. This tendency may, however, be neglected. The restraint reduces the hygroscopicity of the wood to a measurable amount (*Barkas*, 1947), which in turn reduces the total swelling by a corresponding extent. *Gaber* and *Christians* (1929) compared the swelling or shrinking of Okoumé-plywood glued with casein perpendicular to the grain in the face ply (Fig. 3.79)

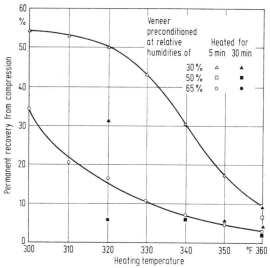

Fig. 3.78. Effect of relative humidity at which veneer was conditioned, final heating temperature and time on the permanent recovery from compression of Yellow poplar-Staypak, pressed at 1,000 lb./sq. in. (≈ 70 kp cm²). From *Stamm* and *Harris* (1954)

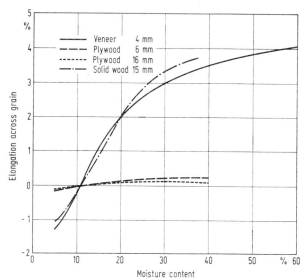

Fig. 3.79. Elongation perpendicular to grain in face veneer (in % of specimen width in air dry condition) for Okoumé solid wood, veneer and plywood. From *Gaber* and *Christians* (1929)

with the corresponding values for 4 mm thick veneer and 15 mm thick solid board. The initial moisture content was 11%. The variations in thickness under the same test conditions are shown in Fig. 3.80.

When swollen plywood is re-dried, shrinkage occurs and tends to restore the original unswollen dimensions. There is always the possibility, however, that the primary swelling causes some compression failures in the fiber direction of the plies. As the compression is relieved face-checking may be produced which is worse for softwoods than for hardwoods and considerably worse than for solid wood or veneer plywood (parallel laminates) (*Lloyd* and *Stamm*, 1958).

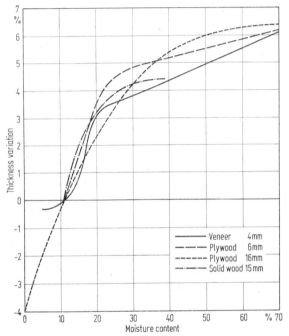

Fig. 3.80. Thickness variation, radial to the annual rings in percentage of the thickness in air dry condition of okoumé solid wood, veneer and plywood. From *Gaber* and *Christians* (1927)

Therefore *Stamm* (1964, p. 313—342) states "that cross-laminating alone will give dimensional stabilization in the sheet directions, but at the possible expense of face-ckecking." There are the following principal methods to improve dimensional stability of plywood:

1. External coatings (e.g. two or three coats of aluminum powder dispersed in oil-paints, varnishes or enamels giving water-excluding efficiencies of 90 to 95%;

2. Facing with resin-impregnated paper, non- or slow swelling plastic films, aluminium foils, etc.;

3. Resin treatments of veneer, mainly with phenol-formaldehyde resins (Impreg), but also with resorcinol-formaldehyde, melamine-formaldehyde, and furfural alcohols;

4. Polyethylene glycol (PEG) treatment of (Douglas fir) plywood faces which virtually eliminates checking.

3.4.3 Thermal Conductivity, Acoustical Properties

Thermal conductivity is an important property of plywood in building constructions.

In this regard plywood behaves quite similar to solid wood. It may be referred to Vol. I, p. 246—257, and to Section 5.4.5 and to Figs. 5.147 and 5.148 in this book. Normal plywood has an average density of 0.5 if it is made from veneer from coniferous species and of 0.8 if it is made from hardwoods. Considering some scattering one can assume on the average for softwood plywood a thermal conductivity perpendicular to the grain of 0.10 kcal/m h °C and for hardwood plywood 0.145 kcal/m h °C. Conversion factors into English units are given on pp. 673—684.

The acoustical properties of solid wood and wood based materials are dealt with in Vol. I, p. 274—285, and in Section 5.4.6 of this book.

The *sound transmission loss* is directly proportional to the logarithm of the weight of a wall per unit area. Recent investigations proved that weights per unit area below 800 kg/m² (∼160 lb./sq. ft.) are remarkably more unfavorable than formerly was assumed. *Meyer* (1931) measured for plywood with a thickness of 5 mm (about 3/16 in.) a sound transmission loss of only 17 decibels (dB), the corresponding value for a 1/1 brick wall, plastered, 27 cm (∼10 5/8 in.) thick is about 48 decibels (cf. Fig. 6.143 in Vol. I). According to German Standard Specification (DIN 4110) airtight single walls must have a minimum transition loss of 53 dB in the frequency range between 100 and 3,000 Hz. The sound transmission loss increases (with some irregularities) with increasing sound frequency; there is a limit frequency which is characteristic of an extraordinary reduction of sound insulation. This limit frequency is determined by the quotient of density to dynamic modulus of elasticity of the material in the form of a shell (*Cremer* and *Zemke*, 1964, p. 93). With respect to high sound insulation it must be demanded that the limit frequency is either very low (<100 Hz, flexural stiff shells) or possibly high (>3,000 Hz, flexible shells). Plywood panels, 1 cm thick, with a weight of 8 kg/m², have a limit frequency of 1,800 Hz. They are not suited for single shell walls with regard to appropriate sound insulation. Double or multi-shell (or skin) walls can provide efficient sound-insulation. The architect has to pay careful attention to the manifold problems of sound insulation in his designs. Calculations are not always possible or reliable; therefore experience and experiments are necessary. The problem of sound absorption is quite different from that of sound insulation. Sound absorption requires soft, porous materials, such as carpets, heavy fabrics, wood wool-board and acoustic tiles. Plywood has under some circumstances a maximum of sound absorption of about 26% at a frequency of 512 Hz, above about 1,000 Hz the sound absorption is nearly constant with 10% (Fig. 3.81). *Sabine* (1927) found for fir board, 20 mm thick, in the frequency range from 100 to 5,000 Hz sound absorption values between 8 and 11%.

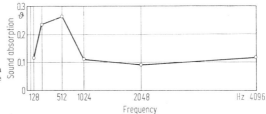

Fig. 3.81. Relationship between sound absorption and frequency for plywood, 3 mm thick, board size 1.21 m × 1.52 m, distance from wall 2.5 cm. From *Meyer* (1931)

3.4.4 Mechanical Properties

3.4.4.1 Elasticity and Rigidity. In Vol. I (p. 292—321) of this book elasticity, plasticity and creep of solid woods have been dealt with. *Norris* (1937/1939, 1942, 1945) has introduced to the elastic theory of wood and plywood.

In the more general case of stresses applied in any direction there are 36 elastic constants (cf. Vol. I, p. 293). The method of dyadics (or tensors) can be applied according to *Norris* to plywood assuming that many of the problems are simple because they are "two-dimensional". Reference may be made, e.g. to *Love* (1927), *Sokolnikoff* (1956), *Timoshenko* and *Goodier* (1951), *Greene* and *Zerna* (1968).

The method of tensors leads in the case of plain stress to 9 elastic constants which again are not all independent and can be reduced to 5 as mentioned below:

E_L　　= Young's modulus of elasticity in the direction of the grain,
E_{TR}　= Young's modulus of elasticity perpendicular to the direction of the grain,
l/m_L　= μ_1 = Poisson's Ratio in the direction of the grain,
l/m_{RT} = μ_2 = Poisson's Ratio perpendicular to the direction of the grain,
G　　= Modulus of rigidity.

Since about 1920 many investigations on the elastic theory of the anisotropic wood have been published. It may be referred to Vol. 1 of this book, p. 294—296, Table 7.1, systems of elastic constants for different species of wood, condensed from *Keylwerth* (1951), and p. 298, Table 7.2, Poisson's ratios for wood, tabulated by *Hörig*.

Norris (1945) quotes values of elastic properties of spruce, mahogany, ash and walnut, given by *Jenkin* (1920, p. 105), and remarks (p. 94): "It has been found that for the softer woods such as poplar and pine average values of the elastic constants can be used with success ... The average values which have been found to be useful are the following":

$$E_L = \left(\frac{1}{s_{22}}\right) = 1,800,000 \text{ lb./sq. in.} \approx 127,000 \text{ kp/cm}^2$$

$$E_{TR} = 90,000 \text{ lb./sq. in.} = 6,400 \text{ kp/cm}^2,$$

$$G = \left(\frac{1}{s_{44}}\right) = 200,000 \text{ lb./sq. in.} = 14,100 \text{ kp/cm}^2,$$

$$m_L = 2.5 \text{ (abstract)},$$

$$m_{RT} = 50 \text{ (abstract)}.$$

For further considerations the following assumptions or approximations are made:

　　1. All stresses applied act in the plane of the plywood;

　　2. The strains are uniform through the plywood;

　　3. The strain at any point on one of the faces of the plywood is equal to the strain within the plywood directly under the point;

　　4. The arrangement of the plies and their number is not important, but the proportion of wood having its grain running in one direction and the proportion having its grain running in a direction at right angles to the first are very important;

　　5. The kinds of wood of which the various plies are made is very important.

Formulae can be derived which give the average elastic constants of the plywood if the structure of the plywood (kinds of woods, their elastic constants, thicknesses of the vairous plies) is known. The necessary equations and references are given by *Norris* (1945, p. 79, 84, 91, 92).

The relation between the principal strains and the strains in a direction at an angle α to the principal axis of extension is given by the formula

$$\varepsilon = \varepsilon_t \cos^2 \alpha + \varepsilon_c \sin^2 \alpha, \tag{3.10}$$

where

ε = strain in a direction at an angle α,
ε_t = principal strain of extension,
ε_c = principal strain of contraction,
α = angle between direction of principal axis of extension and direction in which the strains are required.

Formula (3.11) gives the rotational strain Φ in a corresponding manner:

$$\Phi = -(\varepsilon_t - \varepsilon_c) \sin \alpha \cos \alpha. \tag{3.11}$$

The last two formulae can be combined and reproduced by a single diagram known as Mohr's stress and strain circle (*Mohr*, 1835).

Fig. 3.82 is a diagram of this kind.

The relations between the principal strains and the distance c to the center and the radius r of the strain circle are given by the following formulae:

$$c = \frac{\varepsilon_t + \varepsilon_c}{2} \tag{3.12}$$

$$r = \frac{\varepsilon_t - \varepsilon_c}{2} \tag{3.13}$$

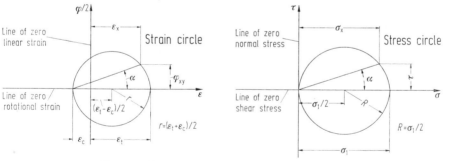

Fig. 3.82. Mohr's strain circle (left) and stress circle (right)

The stress circle is entirely analogous to the strain circle. The rotational strain is defined by *Norris* (p. 75) numerically equal to 1/2 of the shear strain defined in various books upon the mechanics of materials. The maximum rotational strain is directly proportional to the shear stress which causes it. The proportionality factor is called the modulus of rigidity.

It should be kept in mind that the average Young's modulus of elasticity of a plywood panel in the direction of the face grain is the average stress σ_m (extension) divided by the principal strain (elongation, extension) ε_t. The average value of Poisson's ratio for the panel is the contraction ε_c, divided by the elongation ε_t. In isotropic materials such as steel, there is a definite relation between the modulus of elasticity, Poisson's ratio, and the modulus of rigidity. No such relation exists for anisotropic materials such as wood. Computations of the elastic constants can be carried out with the assumption that an average stress is applied in a direction at any angle γ to the direction of the face grain of the plywood. The plywood panel (or specimen) will not retain its rectangular shape due to the rotational strain. It will move either to the right or to the left, depending upon in which sense the grain angle was taken (Fig. 3.83). The computations need proper numerical values for the elastic constants. Based on the values given above for softer woods, *Norris* (1945, p. 99) listed the average elastic constants for plywood with varying percentage of the wood having the grain parallel and across the face grain, respectively (Table 3.12).

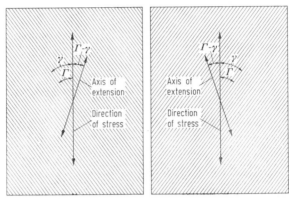

Fig. 3.83. Scheme of rotational strain for two possible angles between direction of stress and axis of extension. From *Norris* (1945)

Table 3.12. Average Elastic Constants of Plywood (*Norris*, 1945, p. 99)

Grain	Grain	E_L	E_{RT}	Poisson's ratio		Modulus of rigidity
				$1/m_L$	$1/m_{RT}$	G
%	%	face grain lb./sq. in. (kp/cm²)				lb./sq. in. (kp/cm²)
100	0	1,800,000 (126,500)	90,000 6,420	0.4000	0.0200	
90	10	1,637,000 (115,000)	262,000 (18,400)	0.1380	0.0221	
80	20	1,466,000 (103,000)	435,000 (30,500)	0.0834	0.0247	
70	30	1,298,000 (91,000)	606,000 (42,600)	0.0600	0.0280	
60	40	1,123,000 (79,000)	779,000 (54,700)	0.0462	0.0320	
50	50	945,000 (66,400)	945,000 (66,400)	0.0353	0.0363	200,000 (14,200)
40	60	779,000 (54,700)	1,123,000 (79,000)	0.0520	0.0462	
30	70	606,000 (42,500)	1,298,000 (91,000)	0.0280	0.0600	
20	80	435,000 (30,500)	1,466,000 (103,000)	0.0247	0.0834	
10	90	262,000 (18,400)	1,637,000 (115,000)	0.0221	0.1380	
0	100	90,000 (6,410)	1,800,000 (126,500)	0.0200	0.4000	

It is possible to construct curve sheets by plotting the values of c, r, and the angle $(\Gamma - \gamma)$ against the angle γ to the grain at which the stress is applied. Examples for pure tension are given in Figs. 3.84 and 3.85, and for the case of a simple tensile stress with an equal compressive stress applied at right angles to the tensile stess in Figs. 3.86 and 3.87. The curves designed by *Norris* do not correspond with the symbols used in modern theory of the elasticity of orthotropic

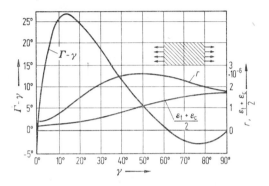

Fig. 3.84. Values of the angle $\Gamma - \gamma$ of r and $\dfrac{\varepsilon_t + \varepsilon_c}{2}$ for plywood in tension. $L = 90\%$, $S = 10\%$. From *Norris* (1945)

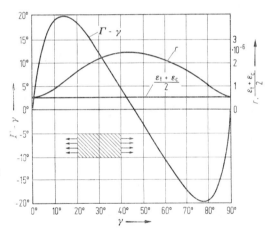

Fig. 3.85. Values of the angle $\Gamma - \gamma$ of r and $\dfrac{\varepsilon_t + \iota_c}{2}$ for plywood in tension. $L = 50\%$, $S = 50\%$. From *Norris* (1945)

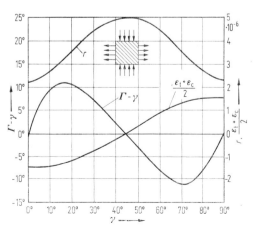

Fig. 3.86. Values jor the angle $\Gamma - \gamma$ of r and $\dfrac{\varepsilon_t + \varepsilon_c}{2}$ for plywood in tension + compression. $L = 90\%$, $S = 10\%$. From *Norris* (1945)

materials. Nevertheless they are very instructive because the angle $(\Gamma - \gamma)$ denotes clearly the deformation of a plywood panel or strip under the influence of unilateral forces.

More details about transformation of stress by Mohr's circle are published by *Hoff* (in *Dietz*, 1949, p. 6—32). He refers to further literature in this field (*Timoshenko*, 1940/1941; *Norris*, 1943; *Hoff*, 1945; *March*, 1942). *Hoff* computed

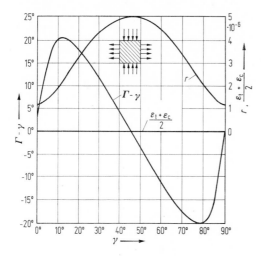

Fig. 3.87. Values for the angle $\Gamma - \gamma$ of r and $\dfrac{\varepsilon_t + \varepsilon_c}{2}$ for plywood in tension + compression. $L = 50\%, S = 50\%$. From *Norris* (1945)

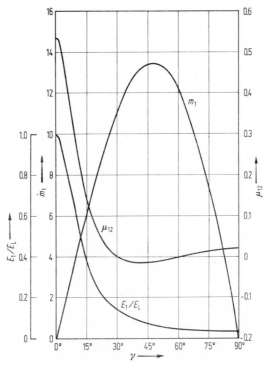

Fig. 3.88. Curves for the dependence of elastic constants for pure tension in a spruce veneer on the angle γ between principal stresses and grain. From *Hoff* (1945)

the elastic constants for pure tension in a spruce veneer for various angles γ subtended by the grain with the principal directions and plotted the results in the graph Fig. 3.88. The material is spruce with the following properties.

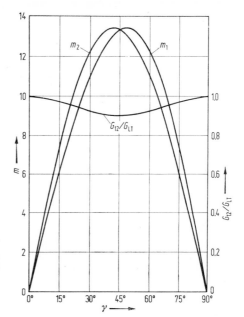

Fig. 3.89. Curves for the dependence of elastic constants for pure hear in a spruce veneer on the angle γ between principal stresses and grain. From *Hoff* (1945)

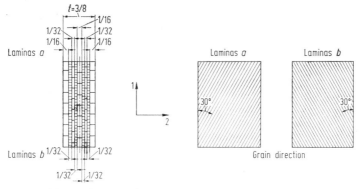

Fig. 3.90. Structure of a laminated spruce plate. From *Dietz* (1949)

$$E_L = 1{,}430{,}200 \text{ lb./sq. in.} \approx 100{,}000 \text{ kp/cm}^2,$$
$$E_T = 51{,}400 \text{ lb./sq. in.} \approx 3{,}610 \text{ kp/cm}^2,$$
$$G_{LT} = 52{,}800 \text{ lb./sq. in.} \approx 3{,}710 \text{ kp/cm}^2,$$
$$\mu_{LT} = 0.539,$$
$$\mu_{TL} = 0.0194.$$

Displacements caused by pure shear in a spruce veneer with the same properties are shown in Fig. 3.89. Other wood species exhibit a similar behavior, but the numerical values may differ considerably (cf. Vol. I, p. 294—296 and Table 3.12 in this book). For plywood with the structure as shown in Fig. 3.90 the apparent elastic constants were determined and plotted in Figs. 3.91 and 3.92.

A comparison with Figs. 3.88 and Fig. 3.89 shows that cross lamination reduces materially the variation in the value of the apparent Young's modulus. For the single spruce veneer the maximum is 1,430,000 lb./sq. in. and the minimum 51,400 lb./sq. in. while for the plywood the corresponding values are 747,000 lb./sq. in. and 186,000 lb./sq. in. The shearing rigidity is increased by cross lamination for every angle γ, and the maximum value amounting to 6.76 times that found for the single veneer is reached when $\gamma = 45°$.

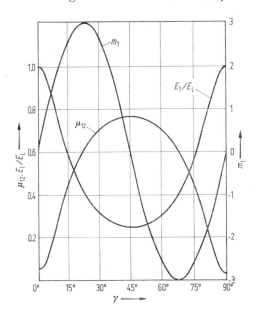

Fig. 3.91. Dependence of elastic constants in pure tension in a plywood plate according to Fig. 3.90 on the angle γ between stress and grain in the surface ply. From *Dietz* (1949)

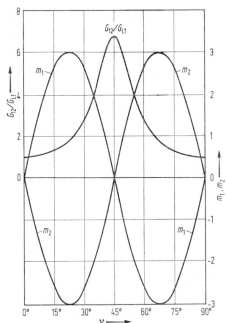

Fig. 3.92. Dependence of elastic constants in pure shear in a plywood plate according to Fig. 3.90 on the angle γ between stress and grain in the surface ply. From *Dietz* (1949)

Considering the great variety of various types of plywood with respect to kinds of wood (even different in various layers of laminae, cf. Fig. 3.90), structure, glue lines, dimensions especially thickness, etc., values for elastic constants as well as for strength properties published by many authors over the past fifty years can be taken only as a guide but not as representative. True values for engineering plywood must be determined in any case by experiments.

The modulus of elasticity E_t of plywood in tension should be determined following ISO/TC 139 proposals (July 1972):

1. Test pieces, having a reduced cross-section at the center of the length to avoid failure in the grip area (Fig. 3.93) are used; the region of reduced cross-section (L_1 in Fig. 3.93) should be of sufficient length to accomodate extensometers;

2. Load-elongation-curve within the range of elastic deformation has to be drawn (Fig. 3.94);

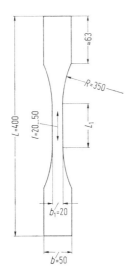

Fig. 3.93. Shape and dimensions
of plywood tensile test pieces.
ISO/TC 139 (1972)

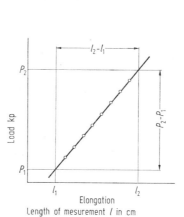

Fig. 3.94 Load-elongation curve within
the range of elastic deformation.
ISO/TC 139 (1972)

3. The modulus of elasticity should be calculated in accordance with the following formula

$$E_t = \frac{l \cdot (P_2 - P_1)}{s \cdot b \cdot (l_2 - l_1)},$$ (3.14)

where

E_t = modulus of elasticity of the test piece in kp/cm²,

s = thickness of the test piece in cm to the nearest 0.01 cm,

b = width of the test piece at the minimum cross-sectional area in cm, to the nearest 0.01 cm,

l = basis measuring length of the tensometers in cm, to the nearest 0.01 cm,

$P_2 - P_1$ = increment of load in kp, on the straigth line of the load-elongation-curve, determined with an accuracy of at least $\pm 1\%$,

$l_2 - l_1$ = increment of elongation of the basis length of measurement l of the test piece, corresponding to the difference of load $P_2 - P_1$, determined to the nearest 0.0001 cm.

The modulus of elasticity should be calculated to the nearest $1,000 \text{ kp/cm}^2$. Usually the modulus of elasticity E_b in bending should be determined following ISO/TC 139 (Aug. 1972):

1. Test pieces should be rectangular as shown in Fig. 3.95. Width (b) 7.5 cm, length (L) = 25 times the nominal thickness, but not less than 100 mm, plus approximately 50 mm.

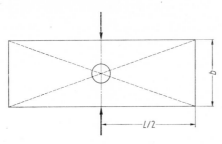

Fig. 3.95. Test piece for bending tests on plywood. ISO/TC 139 (1972)

Note: If the deflection of the test piece becomes too large so that rupture (failure) does not occur, the distance between the supports must be reduced for testing the bending strength.

The width and thickness of each test piece should be measured as follows:

a) The thickness at one point in the middle of the test piece shown as a circle in Fig. 3.95;

b) The width of the transverse axis, as shown by the arrows in Fig. 3.95;

2. Conditioning: The test pieces should be conditioned to constant mass in an atmosphere of a relative humidity of $65\% \pm 5\%$ and a temperature of $20\,^\circ\text{C}$ $\pm\, 2\,^\circ\text{C}$. Constant mass is considered to be reached when two successive weighing operations, carried out at an interval of 24 h, do not differ by more than 0.1% of this mass of the test piece.

3. Principle of the bending apparatus. A cylindrical loading head, placed parallel to the supports and equidistant to them, movable in the vertical direction and having the same length and diameter as those of the supports (Fig. 3.96). The load-deflection curve (similar to Fig. 3.94, but with $(f_2 - f_1)$ instead of $(l_2 - l_1)$) should be determined up to one third of the maximum load.

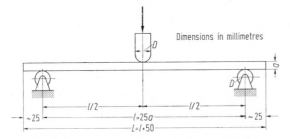

Dimensions in millimetres

Fig. 3.96. Principle of the bending apparatus for testing plywood. ISO/TC 139 (1972)

Increments of load of equal magnitude shall be chosen so that not less than 8 — and preferably 15 or more — readings of load and deflection are taken. Automatic load and deflection recordings apparatus may be used, provided the curves obtained are of sufficient magnitude to allow accurate measurement of load and deflection.

The deflection f in the middle of the test piece (below the loading head) should be measured with a suitable measuring instrument to the nearest 0.001 cm;

4. The modulus of elasticity should be calculated in accordance with the following formula

$$E_b = \frac{l^3 \cdot (P_2 - P_1)}{4 \cdot b \cdot a^3 \cdot (f_2 - f_1)} \tag{3.15}$$

where

E_b	=	modulus of elasticity of the test piece in kp/cm^2,
l	=	distance between the supports, in cm, to the nearest 0.05 cm,
b	=	width of the test piece, in cm, to the nearest 0.01 cm,
a	=	thickness of the test piece, in cm, to the nearest 0.001 cm,
$P_2 - P_1$	=	increment of load, in kp, on the straight line portion of the load-deflection-curve, determined with an accuracy of at least $\pm 1\%$,
$f_2 - f_1$	=	increment of deflection, in cm, at mid-length corresponding to $P_2 - P_1$, determined to the nearest 0.001 cm.

The modulus of elasticity of each test piece should be expressed to the nearest 1,000 kp/cm^2.

In cases, where the "true modulus of elasticity E^+" without shear effects is needed, it can be calculated in accordance with the following formula (*Timoshenko*, 1940):

$$E^+ = E_b \left[\frac{1}{1 - 1.2 \cdot \left(\frac{E_b}{G}\right) \cdot \left(\frac{a}{l}\right)^2} \right] \tag{3.16a}$$

G = Modulus of shear in kp/cm^2.

As an approximative solution, E_b/G of plywood can be considered as about 14, so in the case of $a/l = 1/25$ the true modulus of elasticity becomes:

$$E^+ = 1.03 \cdot E_b. \tag{3.16b}$$

Hearmon (1948) described the methods for measuring Young's and rigidity moduli as follows:

Brief description	Constant measured	Approximate frequency range (Cycles/sec. or Hz)
Slow flexural vibration, end loading, breadth/length plane of specimen horizontal	E_B	3···5
Slow flexural vibration, end loading, breadth/length plane of specimen vertical	E_B	0.75···5
Rapid flexural vibration induced electrically, specimen unloaded	E_B	10···200
Static bending, end loading	E_B	—
Static bending, 4 point loading (standard test method)	E_B	—
Longitudinal vibration (Kundt's tube)	E_C	2,500···5,000
Longitudinal vibration induced electrically	E_C	
Static compression (Standard test method)	E_C	—
Slow torsional vibration	G	0.5···5
Static torsion	G	

Hearmon stated that the dynamic method gives results consistently higher than the static methods, in general agreement with the findings of *Greenhill* (1942, 1944) and of *Griffith* and *Wigley* (1918). He investigated the effects of frequency, of temperature and of moisture content. To the latter point information published by *Jenkin* (1920), *Carrington* (1921, 1922) and by *Doyle, Drow* and *McBurney* (1945, 1946) is given. The influence of the grain angle on Young's modulus and rigidity modulus is discussed. In connection with the elastic theory of plywood plates *Hearmon* (1948) mentions the first attempts to apply mathematical methods to the solution of elastic problems of wood and plywood made by *Price* (1928), *Hörig* (1931) and *March* (1936) and he writes: "Owing to the greater number of independent elastic constants possessed by wood and plywood as compared with isotropic materials, the mathematics involved is often extremely complicated. Further, wood and plywood are less "perfect" from the point of view of elastic theory than are most isotropic materials, and the discrepancies between theory and experiment will tend to be increased by the imperfections."

The buckling of plywood plates and cylinders is especially important. This is, shortly speaking, the problem of elastic stability (*Hearmon*, 1961). *March* et al. (1942 to 1945) investigated the buckling of plywood. He suggested to introduce a "buckling stress coefficient" k_c and computed the critical buckling stress σ_{cr} as functions of elastic constants, geometrical conditions, grain angles and edge conditions (clamped and/or supported on all edges or variably on ends and sides). It would go far beyond the scope of this book to go into details of the applied anisotropic elasticity of plywood. Reference must be made to *Hearmon* (1948, 1961). Only to give the reader a very condensed review the equation for σ_{cr} may be given as follows:

$$\sigma_{cr} = k_c E_L \frac{a^2}{b^2}, \tag{3.17}$$

where

E_L = Young's modulus along the grain in dyn/cm² or kp/cm²,
a and b = lengths of the short and the long side of plate respectively, in cm.

Fig. 3.97 shows curves referring to $1:1:1:1:1$, 5 ply 0° or 90° plates made from sliced mahogany having the following elastic constants (*Goland* and *Aleck*, 1943):

$E_L = 9.3 \times 10^{10}$; $E_R = 0.73 \times 10^{10}$; $G_{LR} = 0.45 \times 10^{10}$ (all constants in dyn/cm²); $\mu_{RL} = 0.024$.

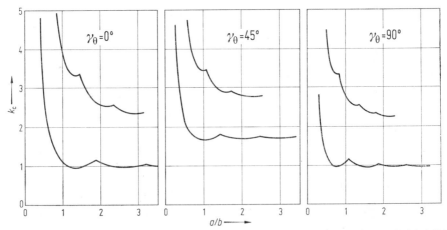

Fig. 3.97. Buckling stress coefficient k_c versus the ratio of short and long side of 5ply mahogany (1:1:1:1:1) in compression.

3.4.4.2 Strength Properties

3.4.4.2.0 General Considerations. The demand for data on the mechanical properties of plywood increased due to its extensive use as a structural material, but the wide variety of constructions and thicknesses available in different countries makes it impossible to obtain these data by testing procedures alone. Therefore single methods for calculating the tension, compression, bending and shear strength of any plywood construction have been developed (*Curry*, 1954).

3.4.4.2.1 Tensile Strength. The *tensile strength* is determined for a conditioned ($H \approx 12\%$) test piece according to Fig. 3.93. Wedge-type grips are used to hold the specimens and the rate of loading corresponds to crosshead movement of the testing machine of 0.6 to 2.5 mm per minute.

The tensile strength should be calculated in accordance with the following formula:

$$\sigma_{tB} = \frac{P_{max}}{a \cdot b_1},\qquad(3.18)$$

where

σ_{tB} = tensile strength of the test piece, in kp/cm²,

P_{max} = maximum load (failing load), in kp, to the nearest 1% of the load,

a = thickness of the test piece, in cm, to the nearest 0.001 cm,

b_1 = width of the test piece at the minimum cross-sectional area, in cm, to the nearest 0.01 cm.

The tensile strength of each test piece should be expressed to the nearest 5 kp/cm².

In the Forest Products Research Laboratory, Princes Risborough, England, 3/16 in. 3-ply plywoods made of 25 different species were tested. The block diagrams of Figs. 3.98, 3.99 and 3.100 (*Curry*, 1953) show that there is no regression between density and tensile strength.

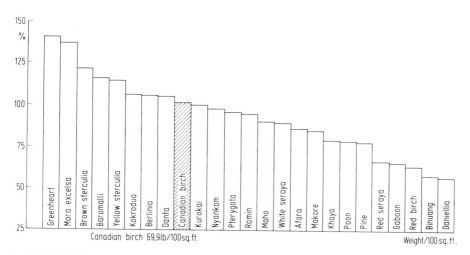

Fig. 3.98. Block diagram of the relative weight per unit area (lb. per 100 sq. ft.) for plywood made from 25 imported wood species. As indicator Canadian birch with the value 100%. From *Curry* (1953)

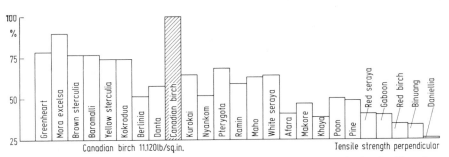

Fig. 3.99. Block diagram of the relative tensile strength perpendicular to the outer face for 25 species.
From *Curry* (1953)

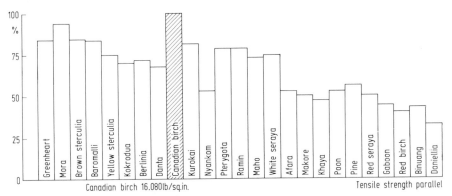

Fig. 3.100. Relative tensile strength parallel to the outer face for 25 species. From *Curry* (1953)

The influence of the grain angle between the principal stress direction in the face veneer and the grain direction can be clearly shown by polar diagrams (Fig. 3.101) (*Kraemer*, 1937). Multiply and star plywood have smaller differences in strength along, inclined to and across the face veneer.

3.4.4.2.2 Compressive Strength. The determination of *compressive strength* should be carried out in accordance with ISO/TC 139 proposals as follows:

1. The test pieces should be rectangular with the following dimensions: width (b): 50 mm,
thickness (a): it corresponds with the thickness of the plywood panel.
For the time being it is not mandatory to test plywood panels with a thickness < 3 mm in accordance with International Standards. For these boards particular methods may be agreed upon between purchaser and supplier;
length, height: 4 times the nominal thickness of the plywood panel;
Care should be taken in preparing the test specimens to make the end surfaces smooth and precisely parallel to each other at right angles to the length.

2. *Conditioning.* The test pieces should be conditioned to constant mass in an atmosphere of a relative humidity of 65% ± 5% and a temperature of 20 °C ± 2 °C.

3. The test piece should be placed centrally between the two steel-plates, so that the axis in the length-direction of the test piece corresponds with the axis of the loading system.

The load should be applied continuously at constant loading speed in such a way that the maximum load is reached within 90 sec $\pm$ 30 sec. The reading of the maximum load should be done to the nearest 1% of the load.

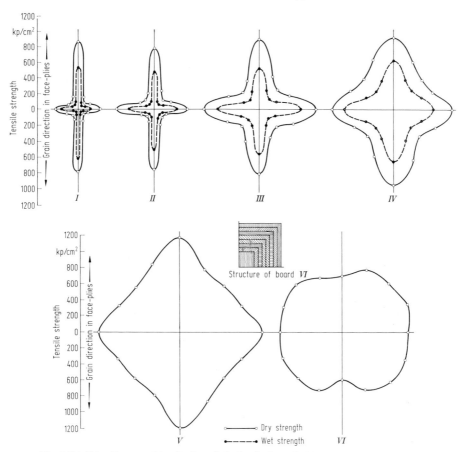

Fig. 3.101. Polar diagrams of tensile strength for beech plywood with the following properties:

| | Board No | | | | | |
	I	II	III	IV	V	VI
Board thickness mm	5	5	5	1.2	1.2	1.2
Structure	3 plies	5 plies	5 plies	5 plies	11 plies	8 plies 45°
Gluing	Casein-blood	albumin	TEGO-film	—	TEGO-film	—
Density g/cm³	0.75	0.76	0.74	0.85	0.95	0.85
Moisture content %	10.5	11.9	6.5	7.2	7.0	6.8

From *Kraemer* (1937)

4. The *compressive strength* should be calculated in accordance with the following formula:

$$\sigma_{dB} = \frac{P_{\max}}{A} = \frac{P_{\max}}{s \cdot b},$$
(3.19)

where

σ_{dB} = compression strength of the test piece in kp/cm^2,
P_{max} = maximum load (failing load) in kp, to the nearest 1% of the load,
A = cross sectional area of the test piece in cm^2,
s = thickness of the test piece, in cm, to the nearest 0.001 cm,
b = width of the test piece, in cm, to the nearest 0.01 cm.

"Long-term loading" raises the question, how long is meant by "long-term". For timber constructions "The Swedish Royal Board of Building and Planning" (BABS) in regulations governing the Building By-Laws distinguishes between long-term load, when the time is not limited (since 1967 "exceptional cases of load" comprise loading times limited to a maximum of one week), and two kinds of short-term load.

The least loading types are exemplified by impacts, breaking and acceleration forces (wind force, incidental load by human being for example on roofs, by vehicles, load on scaffolding, effect of atmospheric changes, etc.). Material stress may be increased in long-term load by the modification coefficient 1.4, for the above mentioned cases of short-term load by the factor 1.2 (*Lundgren*, 1969, Fig. 3.102). The strength of timber subjected to sustained loading has been studied by many research workers, some of them may be mentioned: *Kollmann* (1951, 1952), for plywood *Larsson* (1956), *Norén* (1964). For the time dependence of the tensile strength of particleboard and hard fiber building boards curves are reproduced in Chapters 5 and 6.

3.4.4.2.3 Bending Strength. The *bending strength* is determined on conditioned ($H \approx 12\%$) test pieces as used for tests of Young's modulus (Fig. 3.95) and with the same apparatus (Fig. 3.96). Where tests for the modulus of elasticity and bending strength are to be combined, then after approximately 1/3 of the ultimate failing load has been applied, the constant rate of loading should be such that the failure occurs in a further time of 60 seconds $\pm$ 20 sec.

The bending strength should be calculated in accordance with the following formula:

$$\sigma_{bB} = \frac{3 \cdot P_{max} \cdot l}{2 \cdot b \cdot a^2} \tag{3.20}$$

where

σ_{bB} = bending strength of the test piece in kp/cm^2,
P_{max} = maximum load, in kg, to the nearest 1% of the load,
l = distance between the supports, in cm, to the nearest 0.05 cm,
b = width of the test piece, in cm, to the nearest 0.01 cm,
a = thickness of the test piece, in cm, to the nearest 0.001 cm.

The bending strength of each test piece should be expressed to the nearest 5 kp/cm^2.

A comparison of bending strength (modulus of rupture) tests made on 25 different species of 3/16-in. 3-ply wood was published by *Curry* (1953) in a table and in block diagram form. With a rather wide scattering the modulus of rupture, for face grain parallel to span, showed a positive linear regression to density expressed as weight per unit area (lb./100 ft.^{2}).

A theoretical analysis of stresses and strains in plywood has been made by *Curry* (1954, p. 2—10). When a plywood strip is subjected to bending, some of the fibers will be strained in tension and some in compression as shown in Fig. 3.103a. Since the stress is directly related to the strain below the proportional

limit, the stress distribution across the section, assuming that the strip is composed
of veneers distributed in two directions at right angles and glue lines between,
may be as shown in Fig. 3.103 b. In a balanced plywood construction the neutral
Axis N-A will be identical with the geometrical axis going through the center of
the section. The stresses in the glue lines are higher than in the adjacent veneers,

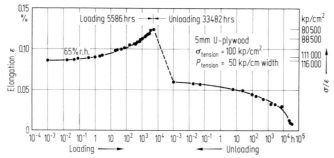

Fig. 3.102. Elongation as a function of loading and unloading time for 5 mm plywood glued with urea-formaldehyde
resin; tensile stress 100 kp/cm². From *Lundgren* (1969)

especially in those loaded perpendicular to the grain. An increase in the number of
glue lines in plywood comprising laminates less than about 1/20 in. (≈ 23 mm)
thick produces an increase in strength (*Kraemer*, 1934; *Kollmann*, 1936, p. 637 to
644, *Preston*, 1950). *Curry* (1954) mentions further facts or requirements:

　　1. The loose (or "slack") face with small cracks produced during veneer peeling
(cf. p. 188 and Figs. 3.27 and 3.28) should be placed towards glue line because the
adhesive can penetrate the cracks and on hardening will seal them. This procedure
also will improve finished appearance and stability of the laminated construction;

　　2. The perpendicular-to-span veneers contribute very little to the stiffness
and strength of the plywood.

　　Equations for the calculation of the modulus of elasicity for the full cross-
section, for the radius of curvature for simple bending unaccompanied by shear
and for the modulus or rupture of the extreme parallel-to-span veneer are developed, and their application is demonstrated by *Curry* (1954, p. 3—5, 22—25).

　　The reduction of permissible stresses under long-term load in Sweden based on
E-modulus values going down from 100,000 kp/cm² to 70,000 kp/cm² is reasonable
in practice because a maximum load never occurs continuously for decades of
time. *Lundgren* (1968) found in experiments:

　　"According to Fig. 3.102 a 50 mm wide strip of 5 mm phenolic resin glued
plywood has an E-modulus of 116,000 kp/cm² immediately on loading, that is, within

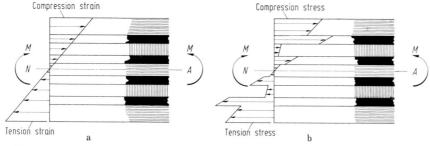

Fig. 3.103 a, b. Stress (b) and strain (a) distributions across a plywood strip subjected to bending.
From *Curry* (1954)

a few seconds. After an hour E has dropped to 111,000 kp/cm², after 1,000 hrs
to 88,500 and after 5,600 h to 80,500 kp/cm². The last value is 69% of the instantaneous value and 73% of the 1 h-value. (In all cases E has been calculated
as the mean value for the entire area of the plywood.)

However, the creep exhibits no tendency whatever to cease during the loading
time. If the curve is extrapolated to a tenth power, that is, to a loading time of
6.5 years, the E-moduli will be substantially lower, 69,000 or 62% of E_{1h}. If one
goes up a further tenth power, $E_{65\,years}$ is only about 51% of E_{1h}."

3.4.4.2.4 *Shear Strength.* The *shear strength* of a panel in different directions is
difficult to determine with certainty. Even if the tests are properly carried out
the obtained values are not generally applicable in practice. This is due to the
edge effects of various kinds and the size of the surface (*Lundgren*, 1969). There
are various types of shear tests:

A. *Panel shear test;*   E. *Tension shear test;*
B. *Plate shear test;*   F. *Torsion shear test;*
C. *Rolling shear test;*   G. *Glue shear test;*
D. *Two-rail shear test;*   H. *Punching.*

Ad A. Panel shear stresses are defined as stresses which tend to distort into the
shape of a rhombus any square element in a plane parallel to the surface of the
plywood while the surface of the plywood remains plane (Fig. 3.104). The shear
tests should be made on 9-in. (229 mm) square panels having 3 1/2-in. (89 mm)
tabs projecting from each side to which hardwood or plywood pads are glued
(Fig. 3.105). Suitable test apparatuses are shown in Fig. 3.106. In this figure the

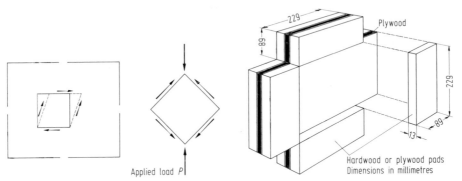

Fig. 3.104. Scheme of distortion of square element by   Fig. 3.105. Panel shear specimen. BSI (Dec. 1968)
shear stresses into a rhombus. BSI (Dec. 1968)

arrangement b will be incorporated in ISO regulations. Density of wood species
(low, medium, high) and plywood thickness have a remarkable influence on the
results. Therefore ASTM D 805-1963/66 (arrangement a in Fig. 3.106) gives for
three density groups and plywood thicknesses from 0.05 to 0.60 in (1.3 to 15 mm)
a table of dimensions for the panel shear specimens. Using method b for materials
greater than 13 mm in thickness, the dimensions of the reinforcing blocks and
the test frame would have to be increased, but at the moment it is not possible
to make specific recommendations.

The material shall be conditioned to standard humidity and temperature prior
to gluing on the pads (which should also have been conditioned) and then conditioned again before testing. The tests are usually made with the face grain of

the material parallel or perpendicular to the sides of the loading rig, but tests may also be made with the grain inclined to the sides.

The load shall be applied by compression along a diagonal. The movement of the crosshead of the testing machine shall be continuous at a rate of 0.0333 mm per sec $\pm$ 25%.

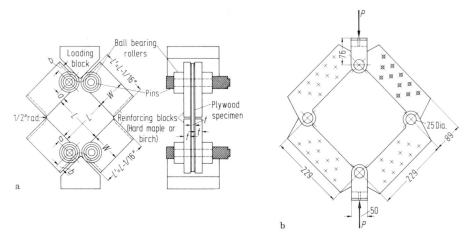

Fig. 3.106a, b. Arrangement for testing panel shear stresses.
a) ASTM 1966; b) BSI 1968

The maximum panel shear stress on the plywood shall be calculated as follows:

$$\tau_s = \frac{0.707P}{Ls}, \qquad (3.21)$$

where

τ_s = maximum panel shear stress in kp/cm²,
P = maximum applied load in kp,
L = side of square panel specimen in cm,
s = thickness of specimen in cm.

Ad B. The *plate shear* has been beveloped at the U.S. Forest Products Laboratory and is standardized in ASTM 805-63/66. The grain direction of the individual plies or laminations shall be parallel or perpendicular to the edges of the test pieces which shall be square and the length and width no less than 25 nor more than 40 times the thickness. The arrangement of the test is shown in Fig. 3.107. In order that the loads may be applied at the corners, metal plates shall first be attached as shown in Fig. 3.107. The load shall be applied with a continuous and uniform motion of the movable head at a rate of 0.003 times the length of the plate in inches (cm), expressed in inches (cm) per minute, within a permissible variation of $\pm$ 25%.

The deformation shall be measured to the nearest 0.001 in. (0.02 mm) at two points on each diagonal equidistant from the center of the plate. These measurements preferably shall be made at the quarter points of the diagonals. The plate shall not be stressed beyond its elastic range, and increments of load shall be chosen so that not less than 12 and preferably 15 load-deformation readings are taken.

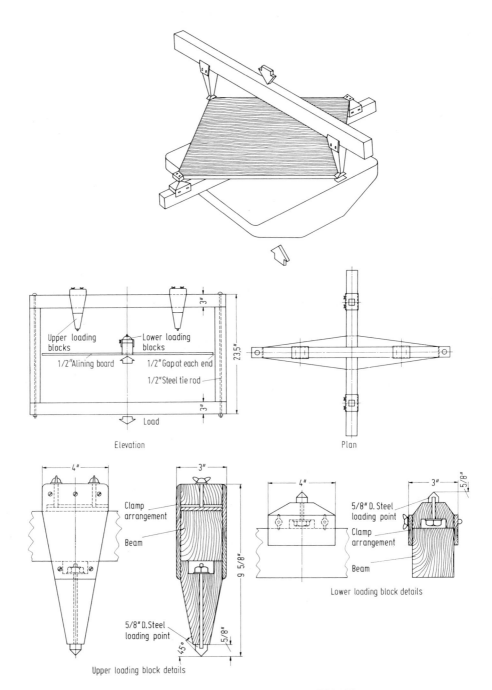

Fig. 3.107. Apparatus for plate shear test. ASTM (1966)

The shearing modulus of elasticity shall be calculated as follows:

$$G = \frac{3u^2P}{2s^3f} \tag{3.22}$$

where

G = shearing modulus, in lb./sq. in. (kp/cm²),
P = load applied to each corner, in lb. (kp),
s = thickness of the platen, in in. (cm),
f = deflection relative to the center, in in. (cm), and
u = distance from the center of the panel to the point where the deflection is measured, in in. (cm).

Ad C. Rolling shear is the term applied to a particular type of failure which occurs in certain veneers of plywood (BSI Dec. 68). The surface produced by the failure appears as though the fibers have rolled off each other (Fig. 3.108). Formally, rolling shear is defined in relation to planes perpendicular to the grain direction in the affected veneer; a square element in one of these planes distorts into a rhombus under the action of rolling shear stresses.

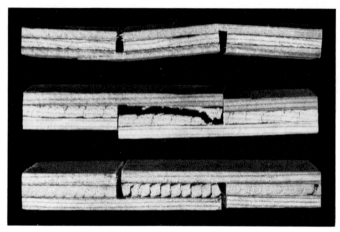

Fig. 3.108. Types of failure in rolling shear tests

Rolling shear stresses may be critical in some applications of plywood, for example in box beams, I-beams and stiffened panels where plywood is glued to solid wood or to other plywood. It is recognized that the apparent failing stress will depend on the geometry of the joint, and the following test has been selected for comparative purposes.

The test is made on the double lap tension specimen shown in Fig. 3.109. The specimens comprise two 25 mm wide central strips of plywood to which are glued two 25 mm wide cover plates having the face grain perpendicular to their length. The cover plates are positioned so that there is a 25 mm double lap joint at one end and a 50 mm double lap joint at the other; the gap between the ends of the central strips is 6 mm. The centre strips both have the face grain parallel to their lengths. The plywood strips are conditioned as required before assembly and the assembled specimens are again conditioned before testing.

The load is applied with the movable head of the testing machine moving at 0.01 mm per second, and the load at failure is recorded.

The result for any specimen which fails in a manner other than in rolling shear in the shorter lap shall be rejected.

The rolling shear stress τ_s at failure is given by

$$\tau_s = \frac{P}{2A} \text{ kp/cm}^2, \tag{3.23}$$

where

P = failing load in kp,
A = smaller area of overlap (625 cm²).

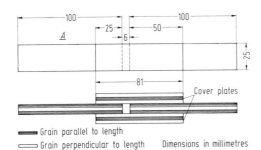

Fig. 3.109. Double lap tension specimen for rolling shear test. BSI (1968)

Ad D. The *two-rail shear test* is important in the U.S.A. for softwood plywood with respect to its structural applications where shear-through-the-thickness has greatly increased interest in determining possible effects of defects, such as knots, short grain and splits, and manufacturing characteristics, such as core gaps and repairs of various kinds, on shear strength. ASTM standards are not satisfactory for measuring grade effects. Therefore in the laboratory of the Douglas Fir Plywood Association (DFPA) a new test was developed to produce pure shear and to permit failure along the specimen's weakest plane (*Batey* jr., 1960, *Post*, 1961, *Batey* jr. and *Post*, 1962).

Test Specimens:

Length: 24 in. (610 mm), trimmed.
Width: 16 to 17 in. (406 to 432 mm) desirable, may be reduced to 14 inches (356 mm) for 3/8 in. plywood and about 15 in. (381 mm) for 1/2-in. if the full width is not available from specimen material. Edges need not be trimmed.
Thickness: 1/4 in. (6 mm) to 3/4 in. (19 mm) and probably thicker can be tested.
Grain Orientation: Face grain should be perpendicular to rails for all normal testing in which grain direction is not to be a factor under study or the purpose of the tests does not require the use of another orientation. Normal constructions of 3 plies must be tested with face grain perpendicular if lateral stability during test is to be assured.
Splits, Veneer Joints, and Gaps: Occurrence of these characteristics will have their minimum effect when located in the shear area if they are oriented perpendicular to the rails and their maximum effect when parallel to rails.
Rail Stock: Normally three rails are obtained from each piece with the following dimensions:

Length 31 in. (∼790 mm)
Width   5 in. (127 mm),

Thickness about 1/16 to 9/16 in. (1.6 to 14.3 mm). Both surfaces must be planed to uniform thickness.

Gluing Rails to Specimen Panel:

For gluing rails to plywood normally a casein adhesive is used.

Assembly:

The specimen panel and the rails with glue applied are assembled in a special jig which accurately clamps the rails and panel in position. Two 12 penny nails are driven in each rail to hold the parts in position (Fig. 3.110).

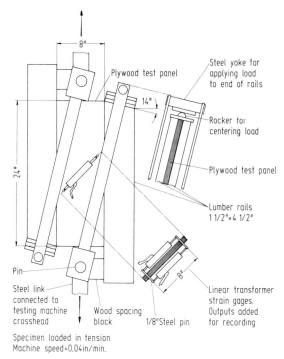

Fig. 3.110. Two-rail shear test for plywood. Douglas Fir Plywood Association, *Post* (1961)

Magnification:

The available magnifications obtainable with the Olsen 51 recorder together with the "K" constants to be used in computing modulus of rigidity are as follows:

Recorder magnification setting	Magnification (2 gages summing)	"K" constant gage length 8"
A	114.806	429.224
B	228.056	912.224
C	455.00	1,820.00

For most laboratory testing for modulus of rigidity, the "A" recorder setting has been found most useful.

Maximum shearing strength τ_s is calculated as follows:

$$\tau_s = \frac{P_{\max}}{L \cdot a},\qquad (3.24)$$

where

$P_{\max}$ = maximum load in lb. [kp],
L = specimen length in inches [cm],
a = specimen thickness in inches [cm].

Modulus of rigidity G_{LT}

$$G_{LT} = \frac{(P_u - P_l)K}{(R - R_l)Lt},\qquad (3.25)$$

where

P_u and P_1 = upper and lower load readings from line fitted to load deformation curve from recorder in lb. [kp],
R_u and R_l = deformation readings from upper and lower points of line fitted to recorder curve in in. [cm],
K = constant given in table of magnifications depending upon recorder magnification and gauge length.

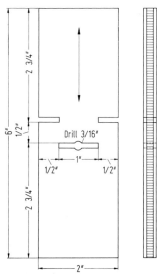

Fig. 3.111. Tension type shear specimen. ASTM D 805 (1966)

Arrow indicates direction of grain in face plies

Ad E. Tension shear strength may be determined according to ASTM D 805-1963/66.

The grain direction of all plies shall be at 0 or 90° to the length of the specimen. The grain direction of the face plies shall be parallel to the length of the specimen.

The test specimen shall be 6 in. ($\sim$15 cm) in length, 2 in. ($\sim$5 cm) in width, and of a thickness equal to the thickness of the material. The specimen shall be cut to provide a shearing length of 1/2 in. ($\sim$13 mm) between saw slots, and the distance between shearing lines shall be 1 in. (25 mm) as shown in Fig. 3.111.

The outer slots shall be cut, using a hollow ground grooving saw, and the inner slot shall be cut, after first drilling a pilot hole, using a jig saw or by any other method. The slots shall be carefully cut both as to length and position to provide the desired shear length.

The load shall be applied continuously throughout the test at a rate of motion of the movable head of the testing machine of 0.025 in. (~0.6 mm) per min until failure. Maximum load only will be obtained.

The shearing strength shall be calculated by dividing the maximum load by the product of the thickness of the specimen times twice the shear length.

Ad F. Torsion shear tests are carried out and evaluated as described in Section 5.4.7.7.3.

Ad G. Glue shear stress may be determined according to ASTM D 805-1963/66 but it may be referred to paragraph 1.3 and Fig. 1.21 in this book. For plywood glue shear tests the following shall be observed:

The direction of grain in each face ply shall be parallel to the length of the specimen and the grain direction of the core shall be perpendicular to the length of the specimen.

The test specimen shall be nominally 1 in. (2,5 cm) in width, 3 1/4 in. (8.25 cm) in length, and of a thickness equal to that of the three-ply plywood selected. When the plywood consists of more than three plies, it shall be stripped of all except any three selected plies prior to test. The length of shear area shall be 1 in. (2,5 cm) when the face ply thickness is greater than 1/20 in. (1.2 mm) and shall be 1/2 in. (13 mm) when the face plies are 1/20 in. (1.2 mm) or less in thickness as shown in Fig. 3.112, specimens A and B, respectively.

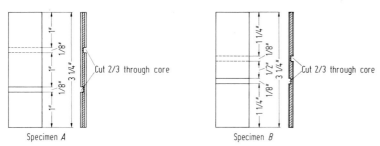

Fig. 3.112. Test specimens for plywood glue shear test. ASTM (1968)

The specimens may be tested dry, in which case they shall have a moisture content of 8 to 12% at the time of test, or they may be tested after a soaking or boil test, depending on the type of information desired.

The ends of the specimen shall be firmly gripped during test and the load shall be applied at a rate of 600 to 1,000 lb./min (~270 to 450 kp/min) until failure. The jaws for gripping the specimen are shown in Fig. 3.113. All strengths shall be expressed in lb./sq. in. (kp/cm²) of shear area.

Ad H. Punching is a dynamic test. Important theoretical and practical considerations are given in paragraph 5.4.7.6. Punching may be described also as a toughness test. Adequate or even high toughness of plywood is desired if it is used for constructional applications where shocks occur and must be sustained. The British panel impact test shall be carried out with specimens cut 305 mm square with the face grain parallel and perpendicular to the sides. The test apparatus shall be as shown in Fig. 3.114. The combined mass of the rod, cone and hemisphere shall be 4.5 kg.

The panel shall be freely supported along all four edges, and the hemisphere shall be made to strike the centre of the panel. The rod shall be dropped through heights from 13 mm increasing in increments of 13 mm until fracture of the panel occurs. Fracture is indicated when the spherical end has penetrated the panel and is arrested by the flange of the cone. The height of drop required to produce fracture shall be taken as the panel impact strength.

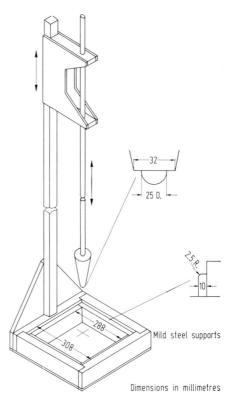

Dimensions in millimetres

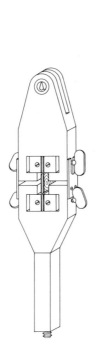

Fig. 3.113. Jaws for gripping specimen for plywood glue shear test. ASTM (1968)

Fig. 3.114. Apparatus for making panel impact (punching) tests BSI (Dec. 1968)

The ASTM D 805-1963/66 thoughness test uses a machine consisting of a frame supporting a pendulum, being so arranged that a measured amount of the energy from the fall of the pendulum may be applied to a test specimen. The pendulum shall consist of a bar to which is fastened a weight adjustable to different positions and shall carry at its upper end a drum or pulley whose center provides the axis of rotation. The force shall be applied to the specimen by means of a flexible steel cable passing over the drum. A stationary graduated scale or dial and a vernier operated by the moving drum shall be provided for reading the angles through which the pedulum swings. The machine shall be adjusted before test so that the pendulum hangs truly vertical and adjusted to correct for friction. The cable shall be adjusted so that the load is applied to the specimen when the pendulum swings to within approximately 15° of the vertical so as to produce complete failure by the time the downward swing is completed.

The test specimen shall be supported as a beam on two vertical pins which exert reactions that are perpendicular to the plane of the plies or laminations. These pins shall be adjusted to the span length taken from Table 3.13, which provides for a specimen overhang of 1 in. ($\sim$2.5 cm) at each end.

Table 3.13. Span Lengths for Various Specimen Lengths

Span Length in. ($\sim$cm)	Specimen Length in. (cm)
2 (5)	4 (10)
3 (7.6)	5 (12.7)
5 (12.7)	7 (17.8)
6 (15)	8 (20)
8 (20)	10 (25)
9 (23)	11 (28)
12 (30)	14 (35)

The load shall be applied at the center of the span perpendicular to the plane of the plies. The load shall be applied through a tup, attached to a flexible cable, having a radius of curvature approximately equal to one and one-half times the depth of the test specimen. The weight position and initial angle of the pendulum shall be chosen so that complete failure of the specimen is obtained on one drop. Most satisfactory results are obtained when the difference between the initial and final angel is at least 10°.

The initial and final angle shall be read to the nearest 0.2° by means of the attached vernier. The toughness shall then be calculated as follows:

$$J = WL\,(\cos \alpha_2 - \cos \alpha_1), \qquad (3.26)$$

where

J = toughness (work per specimen), in in.-lb. (cmkg),
W = weight of the pendulum, in lb. (kg),
L = distance from the center of the supporting axis to the center of gravity of the pendulum, in in. (cm),
α_1 = initial angle, in degrees,
α_2 = final angle the pendulum makes with the vertical after failure of the test specimen, in degrees.

3.4.4.2.5 Hardness. Hardness in its problems is discussed in Section 5.4.7.8 on modified woods (Impreg, Compreg, Staypak). According to ASTM D 1 324-1960 the Janka-hardness test is recommended, occasionally a Rockwell-hardness test is chosen.

3.4.5. Technological Properties

The technological properties of plywood are various, influenced by many factors (kinds of wood, structure, type of adhesive used, pressure applied in manufacture, sanding or scraping etc.). The results of tests which are mostly not standardized are scattered, usually not reproducible and hardly suited for comparisons.

General considerations are similar to those for particleboard (Section 5.4.8.0). This is also valid for *surface quality* (Section 5.4.8.1); accuracy of dimensions and *dimensional stability* (paragraph 5.4.8.2). In this connection the close correlation between strength and dimensions on the one hand, and moisture content and long-term loading on the other hand of wood based sheet materials must be pointed out. With respect to the variety of factors there is a sense of uncertainty among users though many studies have been carried out. The following subjects must be observed (*Lundgren*, 1969):

1. Atmospheric conditions outdoors and indoors, outdoor exposure to sunlight, rain, snow or in shade, ventilating indoors in dwelling houses (summer cottages, unheated garages, under-floor space, cold attics, bathrooms, bedrooms, heated living rooms etc.;

2. Moisture cycles causing residual dimensional changes due to hysteresis, release of internal stresses due to earlier drying and pressing;

3. Unequal exposure to moisture changes, for example one-sided wetting with water;

4. Painting or other surface treatment (oil-tempering of hard fiber building boards), edge-sealing; on small test pieces the edge effect proved negligible whereas painting for plywood is worthwile (*Lundgren*, 1969, p. 42);

5. Permeability. Diffusion across the grain through solid wood takes place two or three times faster than through plywood (*Lundgren*, 1969, p. 51);

6. Sorption rate is multiplied many times at high temperature;

7. Reversible and irreversible moisture movements in the plywood or other wood based sheet materials;

8. More or less direct relationship between changes in thickness and length;

9. Changes in stress and strain of panels subjected to short and long-term loads under various moisture conditions, creep and relaxation;

10. Fiber-raise;

11. Asymmetrical structure in panel cross-section;

12. Variations of quality in commercial plywoods;

13. Dimensions of panels or test pieces;

14. Edge conditions, for example clamping.

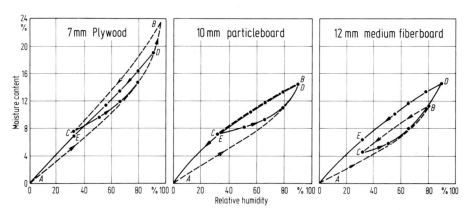

Fig. 3.115. Hysteresis loops for complete moisture cycles as indicated for plywood, particleboard and medium fiberboard. From *Lundgren* (1969)

The hygroscopic moisture equilibrium has been dealt with theoretically in Vol. I, p. 189—195, and with reference to plywood in Section 3.3.2.2 of this book. *Lundgren* (1969) published hysteresis loops for complete moisture cycles 32 — 90 — 32 — 0% atmospheric relative humidity (Fig. 3.115). Within the domain of hysteresis any point can be reached by suitable operations. The point A is the origin of adsorption and desorption isotherms at an imaginary wholly dry condition, the panels are humidified in some manner up to B. An experimental cycle (characterized by practical limits) starts at C and follows the loop $C - D - E$. The presence of adhesives, water repellents or preservatives increases the dry weight of panels and thereby reduces the moisture content and variations thereof if the relative humidity is below 80%. When the air humidity exceeds about 80% then the moisture content increases rapidly. An investigation by *Keylwerth* (1956) demonstrated the excellent dimensional stability of laminboard in comparison with other plywood panels. Further contributions to the problem of the structural behavior of plywood are published by *Peck* and *Selbo* (1950), *Gratzl* (1955—1963), *Mukudai* (1960), *Kufner* (1966), *Bacher* (1960), *Walter* and *Rinkefeil* (1961), *Neusser* (1962), *Mörath*, *Neusser* and *Krames* (1966).

The *machining properties* are similar to those of solid wood, and it may be referred to Vol. I, p. 475/541. The wear of cutting edges in machining plywood depends principally on the type of wood, structure of plywood, number, thickness and properties of glue lines, sharpness angle, material of cutter edges (for example high-speed steel, cobalt-chrome-tungsten-alloy, tungsten-carbide), output (cutting or milling length) in lin. m or ft. In general there is lower wear (measured as the average displacement (recession) dE of the edge throughout the length of the engaged portion of the cutter (Fig. 3.116). Fig. 3.117 shows the relation of sharpness angle to wear of cutter (Forest Products Research, 1955). The great effect of various adhesives in veneer plywood is illustrated in Fig. 3.118 (*Mang*, 1956). The blunting of cutter edges in machining plywood is many times increased as compared with that in machining solid wood without glue lines.

The capacity of plywood panels to provide a firm *hold for nails, screws and plugs* of different kinds, and conversely, the possibility of fastening panels with them, do not depend exclusively on the basic strength properties of the plywood

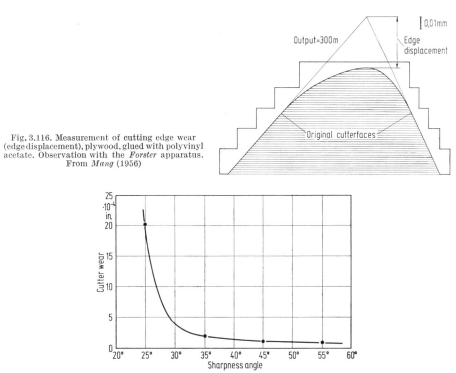

Fig. 3.116. Measurement of cutting edge wear (edge displacement), plywood, glued with polyvinyl acetate. Observation with the *Forster* apparatus. From *Mang* (1956)

Fig. 3.117. Relation between cutter wear and sharpness. For. Prod. Res. Report, Princes Risborough (1956)

(Lundgren, 1969, p. 133—135, Section 5.4.3.5 in this book). Nails as well as screws provide a good hold perpendicular to the surface, especially if suitable holes are pre-drilled, but plywood, mainly core plywood, is decidedly inferior if nails or screws are inserted in the edge surface. This is partly due to relatively

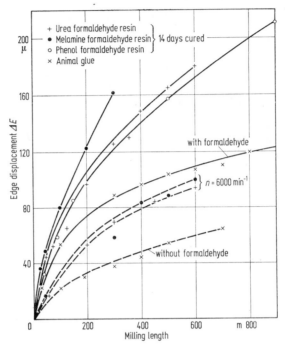

Fig. 3.118. Edge displacement in milling plywood, glued with various adhesives and using high-speed steel. From *Mang* (1956)

low transverse tensile strength in the panels, partly due to the wedge effect produced by the taper of nails and even more of ordinary wood screws. *Lundgren* (1969, p. 135) compared the screw holding power of timber and various wood based panels and boards (Table 3.14).

Table 3.14. Screw-Holding Power of Timber and Various Wood Based Panels and Boards

Type	Sheathing screw 19 mm × 10 withdrawal from surface kp	Wood screw 25 mm × 7 withdrawal from edge kp	pre-drilling
Plywood, 7 mm	85		
Particleboard, 10 mm	55	75	⌀ 2.6 mm
Core plywood, 12 mm	85	40	⌀ 2.8 mm
Hard fiber building board, 12 mm	50		
Oil-tempered hard fiber building board, 6.4 mm	80		
Pine timber, 10 mm	115	125	⌀ 2.0 mm

The ratio of withdrawal force from surface of pine timber and two types of plywood was 1:0.74, the ratio of withdrawal force from edge of pine timber board and core plywood was 1:0.32. These figures are not representative, but just examples. The conditions are similar to those for solid wood. The principles of withdrawal resistance of nails and screws has been dealt with by many experts. The results of their experiments and theories are published in a condensed manner by the U.S. Forest Products Laboratory (1935) and by *Kollmann* (1955, p. 849 to 915).

Between the specific nail or screw holding power perpendicular to grain P_a and the density ϱ_0 of the material (oven-dry weight based on volume when oven-dry) and the diameter d of the nail or screw exists the following function

$$P_a = C \cdot l \cdot \varrho_0{}^n \cdot d \qquad (3.27)$$

where

l = depth of penetration,
C = a constant (for example 485 for nails in air-dry wood, 720 for screws in air-dry wood),
ϱ_0 = density in g/cm³,
n = an exponent (about 5/2 for nails, about 2 for screws),
d = diameter of the nail or screw in cm.

Scholten (1950) has published experimental results for 18 wood species. Already the publications by *Trayer* (1928, 1933), *Cockrell* (1933), *Langlands* (1933), *Gaitzsch* (1934), *Gaber* (1935), *Stoy* (1935), *Fahlbusch* (1944), *Keylwerth* (1944), and *Kloot* (1944) created the fundamentals for properly designing nail, screw or bolt

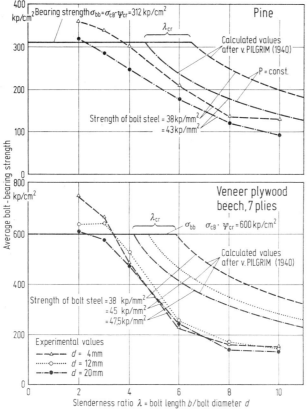

Fig. 3.119. Average bolt bearing strength of long bolts inserted in pine wood and veneer plywood (beech, 7 plies per 1 cm thickness of board) in dependence on a slenderness ratio. From *Fahlbusch* (1944)

joints. The bearing strength of plywood under bolts may be characterized by two typical diagrams (Fig. 3.119 and 3.120). *Fahlbusch* (1944) found that the bearing of bolts increases linearly with air-dry density. Moisture content and grain angle have less influence on bearing strength of bolts in plywood than in solid wood.

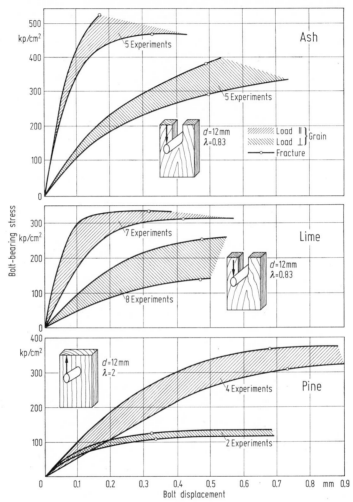

Fig. 3.120. Diagrams of bolt displacement for compressive and tensile load as indicated (short bolts). From *Fahlbusch* (1944)

3.4.6 Resistance against Destruction

Plywood panels if used as structural elements should exhibit adequate resistance to destruction. This resistance depends on the kinds of wood from which the plywood is manufactured, on its structure, on the adhesives used, on surface finish and/or protection, on load conditions, on exposure conditions (indoor or outdoor) etc.

Abrasion (wear) resistance is relevant only if plywood is used for floorings, for able tops and for bodies of cars, trucks and in railroad wagons. The various factors causing abrasion and the tests for abrasion are described in Section 5.4.9.1 in this book. Fig. 5.201 shows the average abrasion restistance indicated by experiments with the Taber-apparatus. Plywood is not incorporated in the diagram but it may behave similarly to normal (not oil-tempered) hard fiber building board.

The resistance to biological attack is closely related to the long-term durability of adhesives for plywood (*Knight* and *Soane*, 1962). The suitability over a 5 to 10 years period of certain adhesives for wood (and thus plywood) in relation to defined exposure environments was investigated in the *British Forest Products Research Laboratory* at Princes Risborough. *Knight* (1963) published the results summarized. Tests were made in the following environments:

1. Exposed to weather (W) in open fields;
2. Soaked (S) in fresh or salt wate runder outdoor conditions;
3. Protected (P) from direct rain and sun by a ventilated hut or roof, but exposed to outside air, unheated by artificial means;
4. Indoor (I) conditions with normal heating and ventilation;
5. Laboratory's tropical house in which the dry room (THD) maintains a daily cycle of temperature and humidity similar to those in Egypt and central India and the wet room (THW) following the climate of Lagos and Singapore.

Only the phenolic and resorcinol resins are completely reliable under severe exposure conditions. Confirmation is given over an average period of 17 1/2 years. Melamine-formaldehyde and urea-formaldehyde fortified with melamine-formaldehyde were reliable under the conditions S, I, THD, questionable under the conditions P and THW, and unsuitable if exposed to weather (W). Pure urea-formaldehyde was reliable under the conditions S, I, THD, questionable under the conditions P and THW, and unsuitable if exposed to weather (W). Casein and animal glues were reliable only under the conditions I and THD. *Knight* and *Doman* (1962) reported also about reducing the surface checking of plywood exposed to the weather. It must be said, that in the U.S.A. animal glues, among the first to be used, are still employed but like casein-based glues in limited quantities. Dried blood glues have a long tradition and are still utilized in substantial quantities in their modern form. Urea-formaldehyde resins introduced in the U.S.A. in 1937 (*Shelton*, 1969) are deficient in long-term and high-temperature-environment. Melamine-formaldehyde resins offer excellent properties for longterm durability but their present cost is high. The popularity of phenolic plywood adhesives is increasing; they provide the most durability for the money spent. There exists a myriad of combinations of phenols with aldehydes. Plywood properly manufactured with a phenolic glueline is resistant to temperature extremes, water (both hot and cold), many chemicals, and mold and fungi to the point that, on attack, the wood can disintegrate, leaving the glueline holding a mass of splinter.

The increasing use of phenol-formaldehyde resins has been made possible by three important trends:

1. Their steadily decreasing glueline cost;
2. The steadily decreasing basic press time requirement (Fig. 3.121), and
3. The increasing versatility.

Manufacture conditions for panels produced from 3 to 11 or more plies:

Pressures $100 \cdots 300$ lb./sq. in., usually $175 \cdots 225$ lb./sq. in.
                 $7 \cdots 21$ kp/cm^2,   usually $12.5 \cdots 16$ kp/cm^2

Temperatures $220 \cdots 325\,°$F,   usually $275 \cdots 315\,°$F
                 $\sim 105 \cdots 163\,°$C,   usually $135 \cdots 157\,°$C

Hse (1971) investigated properties of phenolic adhesives as related to bond quality in southern pine plywood. In the range tested, contact angle (57 to 105°), heat of curing reaction (95 to 235 cal/g) and glueline thickness (8 to 21 μm) were linearly and positively correlated with wet shear strength and percentage of wood failure. The most effective bonding occurred when the resin wetted the veneer surfaces. The optimum glue bonds were from resins with high chemical reactivities. Such resins appeared to produce a high degree of cross linking when cured and required short cure times.

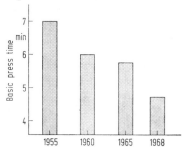

Fig. 3.121. Plot of the steadily decreasing press time requirement for western softwoods for phenol-formaldehyde systems. From *Shelton* (1969)

For plywood manufacture woods with a high starch content liable to *attack by mold or fungi or insects* are generally unsuitable (*Wood*, 1963, p. 10). Usual commercial plywood can be infested to a varying degree by the more prevalent wood-boring insects. *Hickin* (1963) gives an excellent survey on the insect factor in wood decay. Very valuable are the numerous references. Experts in the U.K. stated that damage by the common furniture beetle (*Anobium punctatum*) has increased. Previous studies have indicated that okoumé plywood is unsuitable for larval development. An experiment to ascertain the effect of making up plywood with 1/8 in. Okoumé veneers and blood casein glue, which is known to facilitate larvae development in birch or alder plywood, has been undertaken. Okoumé plywood made up with this glue, and with urea-formaldehyde glue for comparison, and exposed for egg-laying in 1957 had been cut up. Although a satisfactory number of eggs had been laid it was found that no larvae had survived in the plywood made up with either glue nor in the solid wood controls. It thus appears that the use of blood casein glue (which would increase the nitrogen content), in the case of unsuitable wood species, may not render them susceptible.

With *Anobium* beetle infested birch plywood showed exit holes, and X-ray examination ascertained active larvae (Forest Products Research Laboratory, 1962). *Schmidt* (1953, 1954) investigated the attack of termites (*Reticulitermes*) on plywood. He found that multi-ply panels manufactured of beech veneers (11 plies, veneer thickness 1.4 mm) with urea-formaldehyde glue were rapidly destroyed up to one half of the volume within 3 months, completely beneath the earth within 6 months. Then plywood (veneer thickness 0.6 mm) glued with phenolic resins was more resistant. The edges were not attacked, light injuries only on surfaces. Within the frame of building control regulations the chemical preservation of plywood building panels became mandatory. Preservatives that are mixed with the glue and then applied showed best homogeneous distribution (*Deppe* and *Kottlors* 1973) and herewith best efficiency.

Reaction of plywood *to fire* is similar to that of other sheetlike wood based materials. The problems are dealt with, and Figs. 5.204 and 5.205 show that thin normal plywood does not behave well as compared with solid spruce board, particleboard and especially excelsior board. After surface treatment with a foam-coating preservative (spread amount in the average 300 g/m²) the plywood panels showed remarkable improvement with respect to fire resistance.

Literature Cited

André, M.-A., (1962) Furnierherstellung durch Sägen. In: *Kollmann, F.*, Furniere, Lagenhölzer und Tischlerplatten. Springer-Verlag, Berlin, Göttingen, Heidelberg.

Andresen, A., (1934) Kinematics of peeling knives. Mechanical Woodworking (Russian), Moscow, No **12**: 29—32.

Bacher, F., (1960) Das Stehvermögen von glatten Türen und Paneelplatten bei laboratoriumsmäßiger Prüfung und im praktischen Einsatz. Holzforschung und Holzverwertung **12**: 22—30.

Barkas, W. W., (1947) A discussion of the swelling stresses and sorption hysteresis of plastic gels. Great Brit. Dept. Sci. Ind. Research, For. Prod. Spec. Report No 6.

Batey, T. E. Jr., (1960) The two-rail shear test. Douglas Fir Plywood Association, Tacoma, Wash.

—, *Post, P. W.*, (1962) Procedures for preparing and testing two-rail shear specimens. Douglas Fir Plywood Association, Tacoma/Wash.

Birett, H., (1933) Der Elektroantrieb von Rundschälmaschinen. Sperrholz **5**: 29—32.

Carrington, H., (1921) The moduli of rigidity for spruce. Phil. Mag. **41**: 848.

—, (1922) Young's modulus and Poisson's ratio for spruce. Phil. Mag. **43**: 871.

CIRP Wörterbuch (1969) Wörterbuch der Fertigungstechnik. Vol. IV. Grundbegriffe des Spanens. Verl. W. Girardet, Essen.

Cockrell, R. A., (1933) A study of the screw-holding properties of wood. Bull. New York State College Forestry **6** (3).

Comstock, G. L., (1971) The kinetics of veneer jet drying. For. Prod. J. **21**: (9): 104—110.

Cremer, L., Zemke, H. J., (1964) Bau- und Raumakustik. In: Bauen mit Spanplatten, p. 93. Triangel Spanplattenwerke, Triangel.

Curry, W. T., (1953) The strength properties of plywood. Part I: Comparison of 3-ply woods of a standard thickness. Dept. Sci. Ind. Res., For. Prod. Res. Laboratory, Bull. No 29, Her Majesty's Stationary Office, London.

—, (1954) The strength properties of plywood. Part II: Effect of the geometry of construction. Dept. Sci. Ind. Res., For. Prod. Res. Lab., Bull. No 33. Her Majesty's Stationary Office, London.

Deppe, H.-J., Kottlors, Ch., (1973) Verteilung von Holzschutzmitteln in Holzspan- und Baufurnierplatten. Holz als Roh- und Werkstoff **31**: 115—120.

Dietz, A. G. H., (1949) Engineering laminates. John Wiley & Sons Inc., New York, Chapman & Hall Ltd., London.

DIN 6581, (1966) Geometrie am Schneidkeil der Werkzeuge.

Doffiné, E., (1956) Dämpfgruben für die Furniererzeugung. Norddt. Holzwirtschaft **4**: 6—7.

Doffiné, H., (1962) Oberflächenglättung von Lagenhölzern. In: *Kollmann, F.*, Furnier, Lagenhölzer und Tischlerplatten. Springer-Verlag, Berlin, Göttingen, Heidelberg, p. 494—516.

Doyle, D. V., Drow, J. T., McBurney, R. S., (1945, 1946) The elastic properties of wood. U.S. Dept. Agric. For. Prod. Lab., Mimeo 1528, Suppl. A to H.

Fahlbusch, H., (1944) Ein Beitrag zur Frage der Tragfähigkeit von Bolzen in Holz bei statischer Belastung. Inst. f. Maschinenkonstruktion und Leichtbau d. Techn. Hochschule Braunschweig, Ber. No 49-09.

Ferguson, J. S., (1968) What you should know about today's coated abrasives. Ind. Woodworking **20**: 20—23, 28.

Fessel, F., (1964) Furnier-Durchlauftrocknung mit Düsenbelüftung. Holz als Roh- und Werkstoff **22**: 129—138.

Fleischer, H. O., (1949) Experiments in rotary veneer cutting. For. Prod. Res. Society. Proc. Ann. Meeting, **3**: 137—155.

—, (1953) Drying rates of thin section of wood at high temperatures. Yale University, School of Forestry, New Haven, Bull. No 59.

—, (1956) Instruments for alining the knife and nosebar on the veneer lathe and slicer. For. Prod. J. **6**: 1—5.

—, (1959/1965) Heating rates for logs, bolts, and flitches to be cut into veneer. U.S. For. Prod. Laboratory, Rep. No 2149, Madison, Wisc.

—, *Lutz, J. F.*, (1963) Technical considerations for manufacturing southern pine plywood. For. Prod. J. **13**: 39—42.

Forest Products Research Board, (1956) Report for the year 1955. Her Majesty's Stationary Office, London.

—, (1962) Report for the year 1961. Her Majesty's Stationary office, London.

—, (1968) Report for the year 1967. Her Majesty's Stationary Office, London.

Gaber, E., (1935) Statische und dynamische Versuche mit Nagelverbindungen. Versuchsanst. f. Holz, Steine u. Eisen, Karlsruhe, H. 3.

Gaber, E., Christians, G., (1929) Vergleichende Untersuchungen an Vollholz und Sperrholz aus Okumé. Maschinenbau **8**: 385—395.

Gaitzsch, F., (1934) Die Festigkeit maschineller Nagelverbindungen. Forsch. Ber. Holz (3): 43.

Gebhardt, H., v. Grudzinski, H., Matz, H., Herstellungsmethoden in amerikanischen Sperrholzfabriken. Holz als Roh- und Werkstoff **9**: 273—278.

Gefahrt, J., (1966) Hochfrequenzwärme in der Furniertrocknung. Moderne Holzbearbeitung **1**: 182—184

—, (1967) Die Verwendung der Hochfrequenzenergie in der Holzindustrie. Holz als Roh- und Werkstoff **25**: 125—129.

—, (1970) Die Anwendung der Hochfrequenzerwärmung im Holzbau. Holz als Roh- und Werkstoff **28**: 146—154.

Goland, M., Aleck, B., (1943) The veneer orientation for plywood plates with optimum resistance to buckling. Rep. Airpl. Div., Curtiss Wright Corp. No 5: 57.

Gratzl, A., (1955) Zur Meßmethodik bei der Prüfung des Stehvermögens plattenförmiger Körper. Mitt. Österr. Ges. f. Holzforschg. **8**: 69—72.

—, (1956) Zur Meßmethodik bei der Prüfung des Stehvermögens plattenförmiger Körper. Intern. Holzmarkt **54**: 13—17.

—, (1963) Einflüsse auf das Stehvermögen von Möbelteilen. Holz als Roh- und Werkstoff **21**: 149—153.

Green, A. E., Zerna, W., (1968) Theoretical elasticity. 2nd Ed. Clarenson Press, Oxford.

Greenhill, W. L., (1942) The damping capacity of timber. J. Counc. Sci. Industr. Res. Austr. **15**: 146.

—, (1944) A preliminary investigation of the comparative damping capacity of various timbers. Div. For. Prod. Austr. Proj. T. P. 11 (Unpublished Progr. Report).

Griffith, A. A., Wigley, C., (1918) A preliminary investigation of certain elastic properties of wood. Rep. Memor. Adv. Common. Aero., No 28, London.

Grosshennig, E., (1962) Furnierherstellung durch Messern. In: *Kollmann, F.*, Furniere, Lagenhölzer und Tischlerplatten. Springer-Verlag, Berlin, Göttingen, Heidelberg.

—, (1971) Die moderne Furniererzeugung II. Holz als Roh- und Werkstoff **29**: 169—177.

Hamann, R., (1962) Geschichte der Kunst. Droemersche Verlagsanstalt, Th. Knaur Nachf., München, Zürich.

Harris, P., (1946) A handbook of woodcutting. Dept. Sci. Ind. Res., For. Prod. Res. His Majesty's Stationary Office, London, p. 5.

Hearmon, R. F. S., (1948) The elasticity of wood and plywood. For. Prod. Res. Laboratory, Spec. Rep. No 7. His Majesty's Stationary Office, London.

—, (1961) An introduction to applied anisotropic elasticity. Oxford University Press, London.

Henderson, H. L., (1951) The air seasoning and kiln drying of wood. 5th Ed., p. 287. Albany, N. Y.

Hengemühle, W., (1958) Härteprüfmaschinen und -geräte. In: *Siebel*, Handbuch der Werkstoffprüfung I, p. 247. Springer-Verlag, Berlin, Göttingen, Heidelberg.

Hickin, N. E., (1963) The insect factor in wood decay. Hutchinson, London.

Hörig, H., (1931) The elasticity of spruce. Z. techn. Phys. **12**: 369.

—, (1935) Anwendung der Elastizitätstheorie anisotroper Körper auf Messungen an Holz. Ing.-Arch. **6**: 8—14.

Hoff, N. J., (1945) A graphic resolution of strain. J. Appl. Mech. **12**: A211—A216.

—, (1949) The strength of laminates and sandwich structural elements. In: *Dietz*, Engineering laminates, p. 6—89. John Wiley & Sons, Chapman & Hall, New York, London.

Hse, Ch.-Y., (1971) Properties of phenolic adhesives as related to bond quality in southern pine plywood. For. Prod. J. **21**: 45—52.

Hunt, G. M., Garratt, G. A., (1953) Wood preservation. New York.

Irschick, E., (1949) Furnier und Sperrholz und ihre fabrikmäßige Herstellung, p. 20. Stuttgart.

Jenkin, C. F., (1920) Report on materials used in the construction of aircraft and aircraft engines. Brit. Aeronautical Res. Comm., H. M. Stationary Office, London.

Keylwerth, R., (1944) Lochleibungsunterlagen an Schichtholz, Sperrholz und Vielschicht-Sperrholz des Flugzeugbaus. Techn. Ber. (ZWB) **11** (2).

—, (1951a) Die anisotrope Elastizität des Holzes und der Lagenhölzer. VDI-Forschungsheft No 430, VDI-Verlag, Düsseldorf.

—, (1951b) Infrarotstrahler in der Holzindustrie. Holz als Roh- und Werkstoff **9**: 224—231.

—, (1951) Der Verlauf der Holztemperatur während der Furnier- und Schnittholztrocknung. Holz als Roh- und Werkstoff **10**: 87—91.

—, (1953) Furnier-Trocknungsversuche. Holz als Roh- und Werkstoff **11**: 11—17.

—, (1956) Dimensionsstabilität und Gleichmäßigkeit von Möbel- und Türplatten. Holz als Roh- und Werkstoff **14**: 353—360.

Kloot, N. H., (1944) Tests on the holding power of titan plain and processed cement-coated nails. J. Counc. Sci. Ind. Res. (Australia) **17**: 156.

Knight, R. A. G., (1963) The efficiency of adhesives for wood. Bull. For. Prod. Res., No 38, 3rd Ed. Her Majesty's Stationary Office, London.

—, *Doman, L. S.*, (1962) Reducing the surface checking of plywood exposed to the weather. L. Inst. Wood. Sci. **10**: 66—73.

—, *Soane, G. E.*, (1962) The durability of glues for plywood manufacture. Series V: Investigations into glues and gluing. Progr. Rep. 131, For. Prod. Res. Laboratory, Princes Risborough.

—, (1964) Wood machining processes. Ronald Press Co., New York.

Koch, P., (1972) Utilization of the southern pines. U. S. Dept. Agric., For. Serv., Southern Experiment Station, Agriculture Handbook No 420, p. 894—899. Pineville, La.

Kollmann, F., (1936) Technologie des Holzes. 1st Ed. Springer-Verlag, Berlin.

—, (1951a) Herstellung von geformten Sperrholz- und Schichtholzteilen. Holz als Roh- und Werkstoff **9**: 416—422.

—, (1951b) Technologie des Holzes und der Holzwerkstoffe. Vol. I, 2nd Ed. Springer-Verlag, Berlin, Göttingen, Heidelberg.

—, (1952) Über die Abhängigkeit einiger mechanischer Eigenschaften der Hölzer von der Zeit, von Kerben und von der Temperatur. 1. Mitt.: Der Einfluß der Zeit auf die mechanischen Eigenschaften. Holz als Roh- und Werkstoff **10**: 187—197.

—, (1955a) Technologie des Holzes und der Holzwerkstoffe. Vol. II, 2nd Ed. Springer-Verlag, Berlin, Göttingen, Heidelberg.

—, (1955b) Some aspects of wood technology — Lecture Notes (being a series of lectures delivered at the Forest Research Institute, Dehra Dun), The Manager of Publications, Delhi.

—, (1962) Furniere, Lagenhölzer und Tischlerplatten. Springer-Verlag, Berlin, Göttingen, Heidelberg.

—, (1963) Zur Theorie der Sorption. Forschg. Ing. Wes. **29**: 33—44.

—, *Hausmann, B.*, (1955) Vergleichende Untersuchungen beim indirekten und direkten Dämpfen von Holz. Holz als Roh- und Werkstoff **13**: 365—371.

—, *Schneider, A.*, (1963) Über das Sorptionsverhalten wärmebehandelter Hölzer. Holz als Roh- und Werkstoff **21**: 77—85.

Kraemer, O., (1934) Aufbau und Verleimung von Flugzeugsperrholz. Luftf.-Forschg. **11**: 33—52.

—, (1937) Neue Wege bei der Beurteilung von Sperrholz. Holz als Roh- und Werkstoff **1**: 26—29.

Krischer, O., Kröll, K., (1956/1959) Trocknungstechnik. Bd. 1: Die wissenschaftlichen Grundlagen der Trocknungstechnik. Bd. 2: Trockner und Trocknungsverfahren. Springer-Verlag, Berlin, Göttingen, Heidelberg.

Kröll, K., (1959) Trockner und Trocknungsverfahren. Springer-Verlag, Berlin, Göttingen, Heidelberg.

Kübler, H., (1956) Plastische Formung und Spannungsbeseitigung bei Hölzern unter besonderer Berücksichtigung der Holztrocknung. Holz als Roh- und Werkstoff **14**: 442—447.

Küch, W., (1939) Untersuchungen an Holz, Sperrholz und Schichthölzern im Hinblick auf ihre Verwendung im Flugzeugbau. Holz als Roh- und Werkstoff **2**: 257—272.

—, (1951) Über die Vergütung des Holzes durch Verdichtung seines Gefüges. Holz als Roh- und Werkstoff **9**: 305—317.

Kufner, M., (1966) Entwicklung eines Verfahrens zur Prüfung des Formänderungsverhaltens von plattenförmigen Holzwerkstoffen. Holz als Roh- und Werkstoff **24**: 4—9.

Kuhlmann, A., (1960) Wärmeverbrauch und Wärmebilanz beim Dämpfen von Gaboon für die Furnierherstellung. Diss. Techn. Hochsch., München.

Lambert, G. M., Pratt, W. E., (1955) End check preservatives. For. Prod. Res. Soc., News-Digest, Madison, Wisc.

Langlands, L., (1933) The Holding Power of Special Nails, Counc. Sci. Ind. Res., Div. For. Prod., Techn. Paper No 11, Melbourne.

Larsson, G., (1956) Några hallfasthetsprov på plywood samt harda och härdade träfiberskivor. Byggmästaren No B3, Stockholm.

Lloyd, R. A., Stamm, A. J., (1958) Effect of resin treatment and compression upon the weathering properties of veneer laminates. For. Prod. J. **8**: 230—240.

Love, A. E. H., (1927) A treatise on the mathematical theory of elasticity. 4th Ed. Cambridge University Press, New York.

Lundgren, S. Å., (1969) Wood-based sheet as a structural material. Part I, p. 143. The Swedish Wallboard Manufacturers' Association, Stockholm.

Lutz, J. F., (1955) Causes and control of end waviness during drying of veneer. For. Prod. J. **5**: 114—117.

—, *Fleischer, H. O.*, (1963) Veneer manufacture in the United States. In: Plywood and other wood-based panels. Intern. Consultation on Plywood and Other Wood Based Panel Products, Vol. II, FAO, Rome.

Klotz, L., (1930) Aus der Praxis der Furnier- und Sperrholzherstellung. In: *Bittner-Klotz*, Furniere, Sperrholz, Schichtholz. II, p. 8.

Kneule, F., (1959) Das Trocknen. H. H. Sauerländer & Co. Aarau, Frankfurt/Main.

Maclean, J. D., (1940) Studies of heat conduction in wood. Results of steaming green round southern pine timber. Proc. Amer. Wood Preserv. Ass. **26**: 197—219.

Mang, W., (1956) Zerspanungsuntersuchungen über die Abnutzung von Fräserschneiden beim Gleich- und Gegenlauffräsen von Buchenschichtholz mit verschiedener Leimfugenbeschaffenheit. Holz als Roh- und Werkstoff, **14**: 339—352.

March, H. W., (1936) The bending of a centrally loaded rectangular strip of plywood. Physics **7**: 32.

—, (1942) Buckling of flat plywood plates in compression, shear or combined compression and shear. U. S. Dept. Agric., For. Prod. Laboratory, Mimeo 1316, Madison, Wisc.

—, coworkers, (1942/1945) The buckling of flat plywood plates in compression, shear and combined compression and shear. U. S. Dept. Agric., For. Prod. Laboratory, Mimeo 1316, Supplements A-I. Madison, Wisc.

Martley, J. G., (1926) Moisture movement through wood, the steady state. For Prod. Res., Techn. Paper 2, London.

Mayer-Wegelin, H., (1932) cited according to *Trendelenburg—Mayer-Wegelin* (1955) Das Holz als Rohstoff. 2nd Ed., Carl Hanser-Verlag, München.

Meyer, E., (1931) Grundlegende Messungen zur Schallisolation von Einfach-Wänden. Sitzg. Ber. Preuss. Akad. Wiss., Berlin.

Meyer, J. A., (1965) Treatment of wood-polymer systems using catalyst-heat techniques. For. Prod. J. **15**: 362—364.

McMillen, J. M., (1950) Coating for the prevention of end checks in logs and lumber. U.S. For. Prod. Lab. Rep. R 1435 (revised) Madison, Wisc.

Milligan, F. H., Davies, R. D., (1963) High speed drying of western stoftwoods for exterior plywood. For. Prod. J. **13**: 23—29.

Mörath, E., Neusser, H., Krames, U., (1966) Beitrag zur Kenntnis des Formverhaltens von Sperrholz. Holz als Roh- und Werkstoff **24**: 467—469.

Mohr, Ch. O., (1835, 1882) Über die Darstellung des Spannungszustandes und des Deformationszustandes eines Körperelements und über die Anwendung derselben in der Festigkeitslehre. Civilingenieur **28**: 113—156.

Mukudai, J., (1960) Study on lumber-core-plywood. The factors affecting its surface quality and warping. Bull. Governm. For. Exper. Stat. No 126: 1—18. Tokyo.

Neusser, H., (1962) Über die Dimensionsstabilität von Sperrholzprodukten. Holzforschung u. Holzverwertung **14**: 61—69.

Norén, B., (1964) Svensk furuplywood. Statens råd for byggnadsforskning, Handlingar No 45. Stockholm.

Norris, Ch. B., (1943) The application of Mohr's stress and strain circles to wood and plywood. U.S. Dept./Agric., For. Prod. Lab. Mimeo No 1317, Madison, Wisc.

—, (1945) Technique of plywood. 5th Ed. I. F. Laucks Inc., Seattle, Wash.

Peck, E. C., Selbo, M. L., (1950) Flatgrained yellow poplar for cores in furniture panels. U.S. For. Prod. Lab. Rep. No 1785, Madison, Wisc.

Peck, E. C., Griffith, R. T., Nagaraja Rao, K., (1952) Relative magnitudes of surface and internal resistance in drying. Ind. Eng. Chem. **44**: 665—669.

Perry, Th. D., (1944) Moulded plywood. Wood **9**: 107—113.

—, (1947) Modern plywood. Sir Isaac Pitman & Sons Ltd., London.

Post, P. W., (1961) The two-rail shear test. Final Report. Douglas Fir Plywood Association, Tacoma/Wash.

Preston, S. B., (1950) The Effect of fundamental glue line properties on the strength of thin veneer laminates. Proc. Nat. Mtg., For. Prod. Res. Soc. **4**: 228—240.

Price, A. T., (1928) A mathematical discussion of the structure of wood in relation to its elastic properties. Philos. Trans. A. **228**: 1.

Reinmuth, P., (1950) The manufacturing of pine veneer. Proc. For. Prod. Res. Soc. **4**: 332.

Sabine, W. C., (1927) Collected papers on acoustics, p. 93. Cambridge.

Sacht, J.-J., (1951) Moderne Maschinen für die Furnierherstellung. Holz als Roh- und Werkstoff **9**: 20—26.

Schmidt, H., (1953) Eigenschaften und Bewertung der Versuchstermiten. Holz als Roh- und Werkstoff **11**: 385—388.

—, (1954) Formen des Termitenangriffs an Furnier-, Span- und Faserplatten. Holz als Roh- und Werkstoff **12**: 44—46.

Schneider, A., (1962) Klimatisierung von Lagenhölzern bei Fertigung und Lagerung. In: *Kollmann, F.*, Furniere, Lagenhölzer und Tischlerplatten, Springer-Verlag, Berlin, Göttingen, Heidelberg, p. 255—558.

Scholten, J. A., (1950) Nail-holding properties of southern hardwoods. Southern Lumberman **181** (2273): 208.

Schreve, C., (1937) Schälen und Messern von Furnieren. In: Holzprobleme der Gegenwart. Mitt. Fachaussch. Holzfragen, No 17: 71—87, Berlin.

Seborg, R. M., Millett, M. A., Stamm, A. J., (1944) Heat-stabilized compressed wood (Staypak). U.S. For. Prod. Lab., Rep. No 1580, Madison, Wisc.

Seifert, J., (1972) Zur Sorption und Quellung von Holz und Holzwerkstoffen. 1. Mitt.: Einflüsse auf das Sorptionsverhalten der Holzwerkstoffe. Holz als Roh- und Werkstoff **30**: 99—111. 2. Mitt.: Das Quellungsverhalten von Holz und Holzwerkstoffen. **30**: 294—303. 3. Mitt.: Die Volumenkontraktion zwischen Holz und Wasser. **30**: 332—342.

Shelton, F. J., (1969) Plywood adhesives — developments and trends. For. Prod. J. **19**: 9—12.

Siau, J. G. and coworkers, (1965) Wood polymer combinations using radiation techniques. For. Prod. J. **15**: 426—434.

Skark, L., (1962) Lagerung, Aufbereitung und Auftrag der Leime. In: *Kollmann, F.*, Furniere, Lagenhölzer und Tischlerplatten. Springer-Verlag, Berlin, Göttingen, Heidelberg, p. 355—395.

Sokolnikoff, I. S., (1956) Mathematical theory of elasticity. 2nd Ed. New York.

Stamm, A. J., (1964) Wood and cellulose science. Ronald Press Company, New York.

—, *Harris, E. E.*, (1954) Chemical processing of wood. Thames & Hudson, London

Stevens, S. F., (1966) Care of belts and rolls — keys to good panel sanding. Ind. Woodworking **18**: 24—25, 47.

Stewart, H. A., (1970) Abrasive vs. knife planing. For. Prod. J. **20**: 43—47.

Stoy, W., (1935) Tragfähigkeit von Nagelverbindungen im Holzbau. Mitt. Fachaussch. Holzfragen, H. 11, Berlin.

Tiefenbach, J., (1965) Die Luftführung in einem neuzeitlichen Schnittholztrockner. Holz als Roh- und Werkstoff **23** (4): 152—154.

Timoshenko, S., (1940/1941) Strength of Materials. 2nd Ed., Part I (1940), Part II (1941). Van Nostrand, New York.

—, *Goodier, J. N.*, (1951) Theory of Elasticity. 2nd Ed. New York.

Trayer, G. W., (1928) Bearing strength of wood and other factors influencing fitting design. NACA Technical Notes No 296.

U. S. For. Prod. Laboratory (1951) Veneer cutting and drying properties. Rep. No D 1766-6, Madison, Wisc.

Vorreiter, L., (1942) Gehärtete und mit Metall oder Öl getränkte Hölzer. Holz als Roh- und Werkstoff **5**: 59—69.

Walter, F., Rinkefeil, R., (1961) Ein Beitrag zur Bestimmung der Formbeständigkeit von Holzwerkstoffen. Holztechnologie **1**: 67—72.

Ward, D., (1963) Abrasive planing challenges your knife cutting techniques. Hitchcock's Wood Working Digest **65**: 11, 29—32.

Wehner, E., (1966) Fertigbearbeitung. In: *Kollmann, F.*, Holzspanwerkstoffe. Springer-Verlag, Berlin, Heidelberg, New York, p. 372—387.

Winter, H., (1944) Richtlinien für den Holzflugzeugbau, Teil B III c, Berlin.

Wood, A. D., (1963) Plywoods of the world, their development, manufacture and application. W. & A. K. Johnston & G. W. Bacon Ltd., Edinburgh, London.

—, *Linn, Th. G.*, (1950) Plywoods, their development, manufacture and application. W. A. K. Johnston Ltd., Edinburgh, London.

4. Sandwich Composites

4.0 Introduction

A sandwich composite is a layered construction formed by bonding two thin facings to a relatively thick core. It is a "stressed skin" construction in which the facings resist nearly all of the applied in-plane, edgewise loads and flatwise bending moments. The thin, spaced facings provide nearly all of the bending rigidity to the construction. A thick core spaces the facings and transmits shear between them so they are effective about a common neutral axis. The core also provides most of the shear rigidity to the sandwich. The advantage of spaced facings to provide greater stiffness without much increase in amount of material needed was reported by *Timoshenko* (1953) to have been investigated by *Duleau*, circa 1820. This investigation led to the design of tubular bridges, I-beams and other stiff structural shapes.

Specific nonstructural advantages can be incorporated in a sandwich composite by proper choice of facing and core materials. An impermeable facing can be employed to act as a moisture barrier for various house panels; an abrasion-resistant facing can be used for floors; and decorative effects can be obtained by using panels with plywood, hardboard, particleboard, or plastic facings for walls, doors, tables, and furnishings. Core material can be chosen to provide thermal insulation, fire resistance, and decay resistance.

The component parts of sandwich composites should be compatible with service requirements. Moisture-resistant facings, cores, and adhesives must be used if the composites are to be exposed to adverse moisture conditions. Similarly, heat-resistant or decay-resistant facings, cores, and adhesives should be used in environments involving heat and decay. Properties for use in design of sandwich must also be for the composites and their components in the same invironment.

4.1 Basic Design Principles

The sandwich composite is analogous to an I-beam, with the facings carrying direct compression and tension loads as do I-beam flanges, and the core carrying shear loads, as does the I-beam web. A high ratio of stiffness to weight is achieved by using a thick, lightweight core that is strong enough to stabilize thin facings through a facing-to-core bond. For sandwich cores to be lightweight they are usually of low-density material, some type of cellular honeycomb material or of corrugated sheet material. Because of the lightweight cores their low effective shear modulus will require that design methods account for core shear deformations. The main difference in design procedures for homogeneous material is the inclusion of effects of low core shear properties on deflection, buckling, and stress for the sandwich.

Because thin facings are used to carry loads in a sandwich, prevention of local failure under edgewise direct or flatwise bending loads is necessary. Modes of failure that may occur in sandwich under edge load are shown in Fig. 4.1. Shear crimping failure shown in Fig. 4.1b appears to be a local mode of failure but is actually a form of general overall buckling as shown in Fig. 4.1a except that the wavelength of the buckles is very small because of low shear modulus of the core. If the core is of honeycomb or corrugated material the thin facings may buckle or dimple into spaces between core walls or corrugations as shown in Fig. 4.1c. Wrinkling of facings, as shown in Fig. 4.1d, may occur if the facing buckles as a plate on an elastic foundation provided by the core. The facing may buckle into or away from the core depending on the flatwise tensile strength of the core or bond between the facing and core. Since the facing is never perfectly flat, the wrinkling load will also depend upon the initial eccentricity of the facing or original waviness.

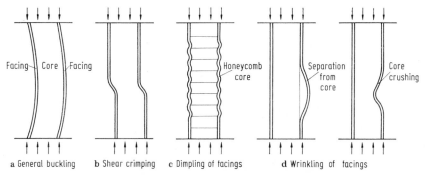

Fig. 4.1 a−d. Possible modes of failure of sandwich composites

In addition to general buckling and local modes of failure, sandwich is designed so that facings do not fail in tension, compression, shear, or combined stresses due to edgewise loads or normal loads, and cores and bonds do not fail in shear, flatwise tension, or flatwise compression due to normal loads.

The basic design principles can be summarized into four conditions as follows:

1. Sandwich facings shall be at least thick enough to withstand chosen design stresses at design loads.

2. The core shall be thick enough and have sufficient shear rigidity and strength so that overall sandwich buckling, excessive deflection, or shear failure will not occur under design loads.

3. The core shall have a high enough flatwise modulus of elasticity, and the sandwich shall have great enough flatwise tensile strength and compressive strength, so that wrinkling of either facing will not occur under design loads.

4. If dimpling of the facings is not permissible and the core is honeycomb or corrugated material, the cell size or corrugation spacing shall be small enough so that dimpling of either facing into the core spaces will not occur under design load.

In designing sandwich to meet these necessary conditions it is assumed that the construction is assembled so that the facings and core will respond, under load, as required. Thus it is assumed that the joint between facings and core is strong enough to hold the facings from wrinkling under design load and also that excessive deflection or shear failure at this joint will not occur. Moreover, joint integrity is assumed for the environment in which the sandwich structural component is to serve and for the life of the structure. The proper joint material and use of the material must be carefully considered if a successful sandwich structure is desired.

Formulas in the following were derived by fundamental mechanics.

4.1.1 Sandwich Bending Stiffness

A sandwich composite under forces normal to its facings has a bending stiffness, per unit width, given by the formula:

$$D = \frac{1}{\dfrac{E_1 t_1}{\lambda_1} + \dfrac{E_c t_c}{\lambda_c} + \dfrac{E_2 t_2}{\lambda_2}} \left[\frac{E_1 t_1}{\lambda_1} \frac{E_2 t_2}{\lambda_2} h^2 + \frac{E_1 t_1}{\lambda_1} \frac{E_c t_c}{\lambda_c} \left(\frac{t_1 + t_c}{2} \right)^2 \right.$$

$$\left. + \frac{E_2 t_2}{\lambda_2} \frac{E_c t_c}{\lambda_c} \left(\frac{t_2 + t_c}{2} \right)^2 \right] + \frac{1}{12} \left[\frac{E_1 t_1^3}{\lambda_1} + \frac{E_c t_c^3}{\lambda_c} + \frac{E_2 t_2^3}{\lambda_2} \right], \qquad (4.1)$$

where E is elastic modulus of facing; E_c is core elastic modulus; λ is one minus the product of two Poisson's ratios ($\lambda = 1 - \mu_{ab}\mu_{ba}$); t is facing thickness; 1 and 2 are subscripts denoting facing 1 and 2; t_c is core thickness; and h is distance between facing centroids. The sketch in Fig. 4.2 shows the notation. For many combinations of facing materials it will be found advantageous to choose thickness such that $E_1 t_1 = E_2 t_2$. For sandwich with facings of the same material and thickness formula (4.1) reduces to:

$$D = \frac{E t h^2}{2\lambda} + \frac{1}{12} \left(\frac{2 E t^3}{\lambda} + \frac{E_c t_c^3}{\lambda_c} \right). \qquad (4.2)$$

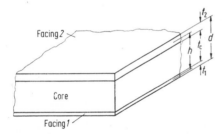

Fig. 4.2. Sketch showing notation for sandwich composites

It will usually be found that the second term on the right side of (4.2) is very small compared with the first term, and it will usually be sufficiently accurate to define sandwich bending stiffnes by:

$$D = \frac{E t h^2}{2\lambda}. \qquad (4.3)$$

4.1.2 Sandwich Extensional Stiffness

The extensional stiffnes of a sandwich, stretched or compressed

$$H = E_1 t_1 + E_2 t_2 + E_c t_c \qquad (4.4)$$

and for equal facings by:

$$H = 2 E t + E_c t_c. \qquad (4.5)$$

4.1.3 Sandwich Shear Stiffness

A sandwich with fairly thin facings on a thick core has a transverse shear stiffness per unit width given approximately by the following formula:

$$U = \frac{h^2}{t_c} G_c \approx h G_c, \tag{4.6}$$

where t_c is core thickness and G_c is the core shear modulus.

4.1.4 Facing Stresses

The facing stresses caused by edgewise (in-plane) forces are given by the formula:

$$\sigma_{1,2} = \frac{N E_{1,2}}{E_1 t_1 + E_2 t_2}, \tag{4.7}$$

where N is edgewise force per unit width of sandwich.

The average facing stress caused by bending moment is given by the formula:

$$\sigma_{1,2} = \frac{M}{t_{1,2} h} \tag{4.8}$$

and the maximum facing stresses (at outer fibers) caused by bending moment are given by

$$\sigma_{1\text{max}} = \frac{M}{t_1 h} \left[1 + \left(1 + \frac{E_1 t_1}{E_2 t_2} \right) \frac{t_1}{2h} \right], \tag{4.9}$$

$$\sigma_{2\text{max}} = \frac{M}{t_2 h} \left[1 + \left(1 + \frac{E_2 t_2}{E_1 t_1} \right) \frac{t_2}{2h} \right], \tag{4.10}$$

where M is bending moment, per unit width, causing curvature of the plane of the sandwich.

4.1.5 Core Stresses

The most important stress in sandwich cores is the shear stress caused by transverse shear loads. Because the cores have low elastic moduli compared with facing elastic moduli the resultant core shear stress distribution is essentially constant through the depth of the core and the shear stress is given approximately by the formula:

$$\tau_c = \frac{P}{h}, \tag{4.11}$$

where P is shear load per unit width of sandwich. This formula is satisfactory for sandwich with facings so thin that they carry little transverse shear. Thick, stiff facings can carry considerable transverse shear and thus lower the core shear stress. Details regarding this can be found in the literature and particularly in the derivation by *Norris, Ericksen* and *Kommers* (1952).

Slight core compression stress in a direction normal to the plane of the sandwich can be caused by bending moment causing curvature of the plane of the sandwich. An approximate formula for core compression normal to the sandwich facing is given by the formula:

$$\sigma_c = \frac{M^2}{Dh},$$ (4.12)

where M is bending moment per unit width of sandwich and D is bending stiffness per unit width of sandwich.

4.1.6 Minimum Weight Sandwich Composites

Often sandwich composites are used because they are efficient, lightweight constructions. Since the sandwich is comprised of facings (that have some weight) and a lightweight (not weightless) core it is of interest to determine a possible "optimum" or minimum weight sandwich for a given stiffness, D. If the facing density is denoted by ϱ and the core density by ϱ_c the weight of a sandwich with thin, equal facings is given by the formula:

$$W = 2\varrho t + \varrho_c h,$$ (4.13)

where W is weight per unit area of sandwich. Solving formula (4.3) for t and substitution of this into the weight formula W and minimizing by calculus with respect to h eventually results in:

$$h^3 = \frac{8\lambda D}{E} \cdot \frac{\varrho}{\varrho_c}.$$ (4.14)

After further manipulation it can be shown that:

$$\frac{t}{h} = \frac{\varrho_c}{4\varrho}.$$ (4.15)

Examination of sandwich proportioned for minimum weight shows that the total core weight must be two thirds the weight of the sandwich and the total facing weight one third the weight of the sandwich. It has been assumed that the bond between the core and facings is the same for all sandwich of the type considered. Thus the minimum weight sandwich for a given bending stiffness is a possibility but further work with the results will show that it is usually of academic interest because very thin facings may not be able to be used as required for minimum weight.

4.1.7 Local Failure Modes

The local modes of failure of sandwich are shown in Fig. 4.1 c and d. Failure b, in Fig. 4.1, is often called a local failure but it is really a form of general buckling in which shear instability causes the appearance of many small buckles that finally produce a shear crimp. Failures c) and d) are primartily facing failures in which the final buckled pattern is influenced by the core. Dimpling of sandwich facings can occur if thin facings are supported on a honeycomb with large cells. Dimpling that does not cause total structural failure may, of course, be severe enough so that permanent dimples remain after removal of load. The facing stress at which dimpling of the sandwich facing will occur was found by *Norris* (1964) to be given by the empirical formula:

$$\sigma_D = 2\,\frac{E}{\lambda}\left(\frac{t}{s}\right)^2, \tag{4.16}$$

where s is honeycomb cell size (diameter of inscribed circle) and the other symbols were defined previously.

Wrinkling of sandwich facings (Fig. 4.1 d) may occur if a sandwich facing buckles as a plate on an elastic foundation. Analysis of this localized buckling is complicated by unknown amplitude of waviness of the sandwich facing. It was found by *Norris, Ericksen,* and *March* (1949) that the wrinkling stress is given approximately by the formula:

$$\sigma_w = Q\left(\frac{EE_cG_c}{\lambda}\right)^{1/3}, \tag{4.17}$$

where Q is dependent upon the parameters q and K as presented in the graph of Fig. 4.3. Formulas for q and K are:

$$q = \frac{t_c}{t}\,G_c\left(\frac{\lambda}{EE_cG_c}\right)^{1/3}, \tag{4.18}$$

$$K = \frac{\delta E_c}{t_c F_c}, \tag{4.19}$$

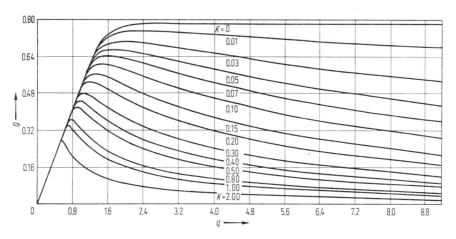

Fig. 4.3. Parameters for determining wrinkling of sandwich facings

where δ is initial amplitude of facing waviness and F_c is flatwise sandwich strength (the lesser of flatwise compression or tension). Unfortunately present state of the art does not permit a suitable choice for values of δ. If test values of wrinkling stresses are known, the graph of Fig. 4.3 can be used to determine a K that fits the data and then compute δ from formula (4.19). Changes in design for similar sandwich can then be made by assuming this δ and using the graph of Fig. 4.3 to redesign the sandwich.

4.2 Materials and Fabrication

Sandwich facings, cores, and bonds can be chosen to have certain structural and nonstructural characteristics that can be combined to advantage in producing effective constructions.

4.2.1 Facings

Almost any sheet material can be employed as a sandwich facing. Mechanical properties needed for structural design are modulus of elasticity, edgewise compression and tension design allowable stresses. The properties should be for the expected environment of the sandwich and should consider temperature (including fire resistance, if necessary), moisture, and any deteriorating atmosphere of conditions. Considerations should be given to impact resistance, abrasion resistance, creep behavior, and fatigue loading, nonstructural characteristics such as weight, permeability, dimensional stability, thermal conductivity, sound absorption and reflection, electrical characteristics, decorative qualities, or special features must be recognized in the final design. Availability and cost may, of course, dictate the eventual choice of facing material for many applications.

It is not the purpose of this book to give detailed design values for facing properties; however, values to indicate general data ranges are given for many different materials in Table 4.1. Additional data on plywood are given in Chapter 3, on particle boards in Chapter 5, and on fiberboards in Chapter 6.

Table 4.1. Properties of Facings

Material	Weight pci	Elastic modulus 10^6 psi	Facing stress comp. ksi	tens. ksi	shear ksi	Thermal expansion 10^{-6} in./ in./°F	Thermal conductivity BTU/h/ft.²/ °F/in.
Aluminum							
2024-T62	0.100	10.0	47	47	36	12.6	960
7075-T6	0.101	9.6	62	60	42	12.9	960
Magnesium							
AZ31B-H24	0.064	6.5	24	29	18	14.0	670
Titanium							
6Al-4V	0.160	16.0	126	120	76	4.6	45
Stainless steel							
301-1/2 H	0.286	26.0	58	110	80	9.2	120
Glass fabric plastic laminate	0.065	2.5	18	22	1.6	8.0	—
Plywood[1]	0.016	1.0	1.0	1.4	0.1	4.0	1.0
Birch Impreg[1]	0.029	2.5	10	10	—	—	—
Birch Compreg[1]	0.047	3.9	16	16	—	—	—
Papreg	0.051	3.1	7	7	—	—	—
Fibrous board	0.014	0.09	0.2	0.2	—	—	—
Hardboard	0.039	0.8	1.3	1.0	—	—	1.1

[1] Properties of plywood, Impreg, and Compreg are values for the grain parallel to load.

4.2.2 Cores

Cores for sandwich composites are usually of lightweight material; or constructions in themselves that are designed of a heavy material to form a lightweight core such as a honeycomb core or a corrugated core. Mechanical properties needed for structural design are primarily shear modulus and shear strength. Values of flatwise elastic modulus and flatwise compression and tension strength are also needed for design but are somewhat secondary to shear properties. If cores are fairly rigid their bending stiffness values are also needed to compute sandwich bending stiffness accurately. The properties should be for the expected environment of the sandwich and should consider temperature (fire resistance, if necessary), moisture, and any other possible deteriorating conditions. Impact resistance, fatigue loading, and creep behavior under load must also be included in designs where important.

Nonstructural characteristics such as weight, formability, permeability, thermal conductivity, electrical properties, moisture sorption, or special features must be accounted for in the final design. As for facings, availability and cost may, of course, dictate the final choice of core for many applications. The detailed properties of cores are beyond the scope of this book; however, the properties of

Table 4.2. Properties of Cores

Core material	Density	Flatwise strength			Shear modulus		Shear strength	
		E	Comp.	Tens.	TL	TW	TL	TW
	pcf.	ksi	psi	psi	ksi	ksi	psi	psi
Balsa wood	5.0	222	670	—	9.8	12.2	160	150
,, ,,	7.0	404	1,160	—	14.9	21.0	230	200
,, ,,	9.0	586	1,650	—	20.1	29.7	300	260
,, ,,	11.0	769	2,150	—	25.2	38.3	380	320
Aluminum honeycomb	3.0	122	230	—	29.1	12.8	150	80
Aluminum honeycomb	4.0	190	360	—	42.3	18.6	240	120
Aluminum honeycomb	5.2	252	570	—	57.0	25.6	330	170
Aluminum honeycomb	6.3	306	770		75.2	32.9	390	230
Aluminum honeycomb	7.7	330	870	—	98.4	33.8	540	290
Aluminum honeycomb	9.0	492	1,160	—	144.6	42.1	590	320
Glass fabric plastic honeycomb	3.5	64	290	—	11.9	6.6	160	80
	4.4	113	440	—	17.9	8.4	280	140
	6.3	118	790	—	23.0	10.6	460	240
	8.2	116	880	—	28.2	10.5	480	220
Paper honeycomb	1.7	—	80	—	—	2.5	—	30
	2.0	—	110	—	—	—	—	—
	3.1	—	450	—	—	7.7	—	110
Foamed glass	9.0	180	100	50	69	69	40	40
Cellulose acetate foam	6.2	30	165	310	44	—	120	—
Polyurethane foam	6.0	4	80	200	1.8	1.7	120	110
	9.0	12	165	150	3.6	4.0	95	100
	10.0	9	140	300	3.5	2.8	230	200
Polystyrene foam	1.8	1.2	20	—	1.2	—	32	—
	3.0	2.0	65	—	1.8	—	58	—
	4.5	3.0	130	—	2.6	—	88	—

many cores are given in Table 4.2. Notation used in conjunction with honeycomb cores is shown in Fig. 4.4.

Core properties such as rigidity and strength generally increase as core density increases, for any particular material. The increase is not great at low densities but gradually climbs and approaches linearity with density for many cores.

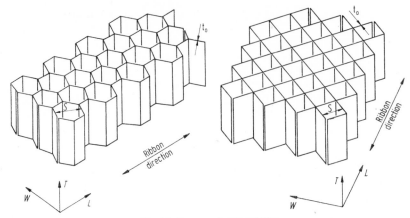

Fig. 4.4. Honeycomb core notation

Elastic properties of honeycomb cores can be estimated by considering the core density, the elastic properties of the ribbon material, and the core cell shape. Flatwise compression modulus of elasticty varies in proportion to the ratio of core density to ribbon material density times the modulus of elasticity of the ribbon material. Ribbon in this sense defines the strip of material that is bent to form the cell shapes of the final honeycomb with either hexagon or square cells. Effective shear modulus in a direction parallel to core cell walls can be estimated by considering only the portion of the walls parallel to the direction of stress to be resistant to shear. Shear properties of cores with hexagonal cells are greater in the core ribbon direction than perpendicular to the core ribbon direction. If the honeycomb cells are square the shear properties are more nearly alike in the two principal directions; in fact tests show that properties of square cell core are about the same at 45° to the ribbon direction as parallel to the ribbon direction.

Nonstructural properties of cores include the one of most interest in building construction design — thermal conductivity. Thermal insulation of honeycomb cores are not notably good because of direct radiation loss through the cell (in a direction normal to the facings) and convection circulation loss if cell sizes are larger than about 3/8 in. Thermal insulation of foam cores vary inversely as their density. This is roughly true of wood cores except that conductivity of wood cores is also dependent upon grain direction. Thermal values for a few cores are given in the following table:

Core material	Thermal conductivity BTU/h/ft.²/°F/in.
Balsa wood, 8 pcf. paral. to grain	1.00
Balsa wood, 8 pcf. perp. to grain	0.40
Foamed glass, 9 pcf.	0.39
Polyurethane foam, 9 pcf.	0.20
Polystyrene foam, 2 pcf.	0.25
Paper honeycomb	0.42

4.2.3 Adhesives

Special types of sandwich comprised of metals can be assembled by joining with welds, brazes, and diffusion bonding but by far the most sandwich are assembled by bonding with adhesives. This section is concerned with adhesive bonding which is essential to properly join many sandwich facings and cores.

An adhesive has been defined as "A substance capable of holding materials together by surface attachment". From this definition it is apparent that the adhesive materials and materials being joined must be compatible and that a proper surface must be prepared for bonding. In the fabrication of sandwich components adhesives are used not only for bonding facings to core but for bonding fittings, reinforcing plates, edge strips and inserts. The adhesive used is often a resin formulation especially developed to give high-strength bonds over a wide range of exposure and stressing conditions. The ever-expanding advances in chemical discovery and application in the field of adhesives prevents a complete presentation. The recommendations of manufacturers of adhesives and their evaluation of properties should be consulted for various applications.

The use of adhesives to create primary structure of sandwich composite demands the use of durable adhesives. For application in wood and wood base materials the types most useful are resorcinol or phenol-resorcinol adhesives. A few isolated applications at controlled moderate temperature and humidity may indicate acceptability of a urea-resin adhesive.

For bonding metals one of the principal types of adhesive is based on phenol resins modified with neoprene, the butadiene-acrylonitrile synthetic rubbers; or the polyvinyls. Adhesives of these types are available in a two-step process as tape with two liquid components, or as a combination neoprene, phenol, and polyamide resin supported as a single tape with liquid components to be added. These adhesives require elevated temperature and application of pressure for curing; however, special formulations curing at pressures as low as 25 psi have been successful. Shear strengths in standard 1/2 in. aluminium lap-joint specimens can range from 2,500 to 5,000 psi. Epoxy resin adhesives have been developed that have good adhesion to metals, glass, plastics, rubber, and other materials. These formulations are made to be used at curing temperatures of 70 to 200 °F and at relatively low bonding pressures. Strengths in standard 1/2 in. lap-joint specimens range from 3,000 to 3,800 psi. Epoxy-phenolic adhesives can be formulated to have good strength at temperatures as high as 500 °F. Strengths of 1,000 psi at room temperature in standard aluminium lap-joint specimens after 200 h at 500 °F have been attained.

Adhesives can be supplied in many forms: powders to be added to solvents, solvents to be added to tapes and films, or completely reactive thick viscous, liquids or films. Storage and mixing or other preparation of adhesives should be correctly done. Often temperatures and humidity should be controlled in a low range for optimum adhesive performance.

Lap-joint strength data are not considered of prime use for determining adequacy of adhesives for bonding sandwich facings to cores, particularly honeycomb cores. For honeycomb cores the adhesive needs to form strong fillets at the ends of the core cells. Sandwich flatwise tensile strength (if it is too low, face-wrinkling can occur) data have been converted into fillet strength in pounds per in. of fillet length so that flatwise tensile strength of sandwich with cores of any cell size or shape can be estimated. The fillet length for cores of hexagon and square cells is determined by dividing 4 by the cell size. Thus the sandwich tensile strength at core-to-facing bond failure is given by the formula:

$$\text{Tensile strength} = \frac{4}{s} \text{ (fillet strength)},$$

where s is core cell size.

The intrinsic elasticity and strength of adhesives have not been evaluated to any great extent; probably because design information for adhesive bonded joints have not been derived. However, several types of adhesives have been evaluated in bond lines between ends of aluminium tubes in torsion and tension by *Kuenzi* and *Stevens* (1963). Properties evaluated are given in Table 4.3.

Table 4.3. Mechanical Properties of Adhesives

Adhesive	Elastic modulus ksi	Shear modulus ksi	Poisson's ratio —	Shear strength psi	Tensile strength psi
Neoprene elastomer-phenolic	4.8	1.6	0.50	2,500	1,200
Nitrile elastomer-phenolic	6.3···18.0	2.1···6.0	0.50	4,200	2,500
Polyvinylphenolic, liquid + powder	500	184	0.36	8,200	8,300
Polyvinylphenolic on glass-fabric	325	118	0.38	3,700	4,400
Epoxy, liquid	508	180	0.41	6,000	—
Epoxy, phenolic-gl. fab.	400	160	0.25	5,500	2,600
Epoxy, polyamide film	180	64	0.41	7,900	7,000
Modified epoxy film	140	50	0.41	5,600	8,000

4.2.4 Sandwich Assembly

The principal operation in the manufacture of panels of sandwich composites is bonding the facings to the core. This can be done in a press, hot or cold, by bag-molding, or in continuous roller presses. The type of equipment must be chosen to effectively cure the bond while joining the sandwich. In contrast with other assembly procedures using presses which produce high pressures for laminating plastics or plywood, equipments for sandwich need exert only low pressures otherwise cores will be crushed. It is essential that core thicknesses be maintained within close limits and that press platens be accurately parallel otherwise even light pressure can cause progressive crushing of the entire sandwich core. It may be advantageous to use stops in the press if the sandwich has no edgings. Use of foamed-in-place plastic core may require no press but will need restraints so that the foam will press against the facings and bond to them without use of additional adhesive.

Facing materials may need to be cleaned and primed before bonding. Cleaning to remove waxes and grease or oil is essential for preparing many facings; particularly metal. The preparation of a metal facing will also involve chromic acid or similar acid etching solutions and thorough rinsing if durable bonds are to be obtained. Specific treatments recommended by the adhesive manufacturers should be followed. Inspection of the treated metal surface is accomplished by many fabricators with the use of a water-film test. This test consists of running cold water over the surface, allowing the excess water to run off, and then inspecting the surface for areas where the water film breaks due to the presence of greases, oils, or waxes. Instruments have also been developed for use in determining contact angles between a water drop and metal surface.

Surfaces that show areas with breaks in a water film or high water drop contact angles should be recleaned. Prepared surfaces must be protected from contamination before bonding by wrapping the parts in clean paper. If bonding is to be done at a much later date the cleaned metal surfaces should be primed with an appropriate adhesive primer and this should be cured. Subsequent bonding can be tailored toward the use of an adhesive compatible with the core and primer. Thus if a primed metal facing is to be bonded to a woodbase core it would not be necessary to choose an expensive metalbonding system but use a phenol or phenol-resorcinol instead.

Preparation of cores prior to assembly in sandwich composites usually comprises cutting to proper thickness, and expanding honeycomb cores to proper cell size if the core was purchased in the unexpanded condition. Cutting is most often done with a band saw and with care the tolerance on thickness can be held to 0.008 in.

Application of adhesives will depend on the form of adhesive. Thin solutions are formulated for use as sprayable primers or for application in multiple coats by spraying. The more viscous solutions are designed for application by brush, hand roller, scraper, roll glue spreader, and by direct extrusion of the adhesive. The film or tape adhesives are simply laid in place, and tacked in position if necessary by touching with a hot iron or moistening with a solvent followed by forced air drying.

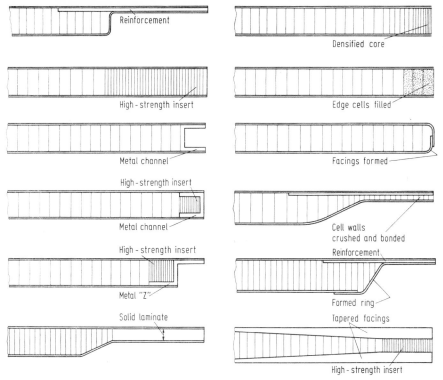

Fig. 4.5. Sandwich panel edge treatments

In many sandwich panels loading rails or edgings are placed between the facings before assembly. Special fittings or equipment, such as heating coils, plumbing, or electrical wiring can be placed more easily in the panel during manufacture than after it is completed. The most persistent difficulties in use of sandwich panels are caused by the edges, inserts, and connectors for the panels. Usually the simpler designs incorporating few if any inserts and doublers have been the most satisfactory. The need to have these pieces fit well so that good bonding can be accomplished is essential. A variety of edge treatments for attaching panels is shown in Fig. 4.5.

4.2.5 Inspection

The completed sandwich must have the facings well bonded to the core in order to function properly. Because all sandwich parts are internal and thus hidden from view it is exceedingly difficult to inspect and determine quality of the sandwich. Carefully controlled systematic inspection of raw materials and fabrication processes are the first steps necessary for obtaining a satisfactory panel.

Observations during manufacture may show whether a panel has loose facings if "blisters" appear under thin facings while the panel is removed from a hot press. The appearance of the blisters indicates areas where the facing is not bonded to the core but the absence of blisters does not indicate a defect-free panel.

One of the simplest and most effective methods of test for voids in the adhesive bond between facings and core is to tap the sandwich with a small metal piece such as a coin or a small, lightweight hammer. Proceeding around the panel while tapping and listening will result in good panel areas emitting a clear tone while unbonded areas produce a lower tone or a dull thud. If there is close contact between facings and core, no difference in tone quality can be detected between poorly bonded areas and well bonded areas. Therefore poorly bonded areas cannot always be detected by tapping.

The use of ultrasonic inspection has shown that hidden flaws, voids, and other defects can be located by their attenuating effect upon high-frequency vibrations.

Disturbance of expected uniformly distributed thermal conductivity of sandwich panels can be indicative of unbonded areas or inclusions. Detection of these areas is possible by use of thermocouple readings, infrared sensing photographic or television cameras, or color changes in liquid crystal coatings as uniformly distributed heating is applied to the opposite sandwich facing. Inversely, the frost pattern immediately formed when a cooled sandwich panel is brought into a warm, moist atmosphere may also show unbonded areas.

Various kinds of proof loading devices have been employed to evaluate sandwich used in critical areas. Proof loads are chosen high enough to stress the sandwich and cause failure if defects are significant and yet are low enough to leave defect-free panel unharmed.

4.2.6 Test Methods

Test methods and apparatus have been devised to evaluate sandwich and core materials — especially for evaluation of properties needed in structural design. Details of the methods and design of apparatus are available from the American Society for Testing and Materials, 1916 Race Street, Philadelphia, Pennsylvania (ASTM). Only pertinent material will be presented here. All the test methods are listed in the following but detailed remarks are given as needed.

ASTM C271 Density of Core Materials for Structural Sandwich Constructions

ASTM C272 Water Absorption of Core Materials for Structural Sandwich Constructions

ASTM C273 Shear Test in Flatwise Plane of Flat Sandwich Constructions or Sandwich Cores

The shear test determines shear properties of cores or sandwich by application of shear load through thick steel loading plates bonded to each side of the core or sandwich. Shear load is directed parallel to the plane of the facings. Complete load-deformation data can be obtained to furnish shear stress-strain curves if desired. The specimen length must be chosen to be at least 12 times the core thickness so that nearly pure shear will be applied. Loading can be through tension fittings for weaker cores and compression fittings for heavier, stronger cores.

ASTM C274 Terms Relating to Structural Sandwich Constructions.

The primary definition is that for structural sandwich construction which is defined as "A laminar construction comprising a combination of alternating dissimilar simple or composite materials assembled and intimately fixed in relation to each other so as to use the properties of each to attain specific structural advantages for the whole assembly".

ASTM C297 Tension Test of Flat Sandwich Constructions in Flatwise Plane.

This test stresses the sandwich in tension directed normally to the facings by bonding metal loading blocks to the facings. Tensile forces only are assured by utilizing swivel type loading fixtures such that bending cannot be introduced.

ASTM C363 Delamination Strength of Honeycomb Type Core Material.

The bond between core ribbons is evaluated in this test by exerting tensile forces to tear the core apart.

ASTM C364 Edgewise Compressive Strength of Flat Sandwich Constructions.

The correct procedure for loading small sandwich specimens in edgewise (in-plane) compression is described in detail. This test is most useful for checking the stresses at which local facing failures—dimpling and wrinkling occur. Adjustable apparatus is illustrated for applying concentric load to the sandwich so that unknown eccentricities are not included in the test. Essential to success is the requirement that deformations in each facing be adjusted to be equal in the early stages of testing to eliminate unwanted bending, hence unknown stresses in the facings. Emphasis is placed on the measurement of deformation in *each* facing—not the recording of testing machine head movement as a way of determining deformation data.

ASTM C365 Flatwise Compressive Strength of Sandwich Cores.

The usual precautions in loading compression specimens through a spherical bearing block are required. Preparation of specimens of honeycomb core include dipping of loaded ends in reinforcing resin or cement to avoid local crushing and failure at low loads. Crushing of bare core is allowed as an alternate test method for a rapid quality control or acceptance test.

ASTM C366 Measurement of Thickness of Sandwich Cores.

Two procedures are outlined for measuring the thickness of production samples of flat sandwich cores. A method utilizing a loaded roller with attached dial gage and a method employing a reciprocating disc and dial gage that give similar results are described.

ASTM C392 Flexure Test of Flat Sandwich Constructions.

Although a favorite type of test, the flexure test of sandwich, even though simple, is most difficult to analyze properly. Elastic behavior can be determined from load-deflection data as measured on the specimen, not from testing machine head movement. Shear deflection as well as bending deflection can be determined. Failure loads causing facing failure can be resolved into facing stresses if facings are thin and sandwich deflection at failure was not too large. Shear failures in the core or core-to-facing bond can be resolved into core shear stress if facings are thin. Thick facings required to cause failure of strong cores in shear carry considerable shear load themselves and division of load is impossible to compute beyond elastic limits. Therefore the flexure test is a poor one for determining core shear strenght.

ASTM C394 Shear Fatigue of Sandwich Core Materials.

Effects of repeated shear loads on shear strength of sandwich cores are determined by the "two-plate" shear test in a fatigue machine. Flexure tests of sandwich usually results in facing failure unless very thick facings or reinforced facings are used, and then core shear stress is not easily computed.

ASTM C480 Flexure-Creep of Sandwich Constructions.

Attempts to evaluate durability of sandwich by exposing specimens to high moisture and heat repetitively have resulted in this standard method. Many materials satisfactorily passing the various exposure cycles have also been found to have good durability in buildings. Of two cycles described the more drastic exposures are in Cycle A in which the sandwich is subjected to six cycles of the following set of six steps:

1. Immerse specimen in water at $49 \pm 2\,°C$ for 1 h;
2. Spray with steam at $93 \pm 3\,°C$ for 3 h;
3. Store at $-12 \pm 3\,°C$ for 20 h;
4. Heat at $99 \pm 2\,°C$ in dry air for 3 h;
5. Spray with steam at $93 \pm 3\,°C$ for 3 h;
6. Heat at $99 \pm 2\,°C$ in dry air for 18 h;

Repeat these steps for 6 cycles and inspect samples for failures. Also strength data obtained before and after aging can be compared to assess durability.

4.2.7 Repair

With use of sandwich construction as with any other type of construction it is inevitable that a certain amount of damage will occur. Proper precautions will minimize damage but if damage does occur, acceptable repair methods must be available.

During manufacturing, where hazards of dropped tools and equipment are encountered, serious damage can be eliminated by protecting exposed corners and by using temporary corner covers. Temporary splines for protection of edges during shipping and erection of panels have been found to be worthwhile.

Repair procedures are developed with the objective of equaling, as nearly as possible, the strength of the original part with a minimum of weight increase or appearance change. This can only be done by replacing damaged material with identical material or an equivalent substitute. Abrupt changes in cross-sectional area should be avoided by tapering joints, by marking small patches round or oval-shaped instead of rectangular, and by rounding corners of all large repairs.

Damaged facings can be repaired by carefully removing the portion of the facing in the vicinity of the damage and replacing with a patch-shaped and scarfed into the facing—and bonded in place. Scarf slopes shall be flat enough to develop good strength in the facing joint. Slopes of 12 or 20 to 1 are often used with plywood facings but slopes as flat as 100 to 1 must be used with plastic and metal facings if reasonably high loads are to be carried.

Core damage is repaired by replacing with similar core and then repairing with a proper facing. A stronger, heavier replacement of core material is used if weight and thermal insulation properties are not critical.

4.3 Structural Components

Basic structural properties of sandwich composites, presented in Fig. 4.1, are used in this section to describe behavior of various structural components.

4.3.1 Beams

Sandwich panels are often used as beams, particularly in floors, roofs, and walls of buildings.

By combining sandwich bending and shear stiffness in appropriate formulas the deflections of sandwich beams can be determined. For most sandwich with relatively thin facings and a moderately rigid core, the following formula will be quite accurate.

The deflection of a sandwich beam can be found by solving the differential equation:

$$\frac{d^2\delta}{dx^2} = -\frac{M_x}{D} + \frac{1}{U}\left(\frac{dS_x}{dx}\right), \tag{4.20}$$

where δ is deflection, x is distance along the beam, M_x is bending moment per unit beam width at point x, S_x is shear load per unit beam width at point x, D is bending stiffness and U is shear stiffness. This equation integrates to:

$$\delta = \frac{k_B P a^3}{D} + \frac{k_s P a}{U}, \tag{4.21}$$

where P is load on beam per unit beam width, a is span length, and k_B and k_S are constants that can be evaluated for any particular loading. Constants for a few typical loadings are given in the following table:

Loading	Beam ends	Deflection at	k_B	k_S
Uniformly distributed	Both simply supported	Midspan	5/384	1/8
Uniformly distributed	Both clamped	Midspan	1/384	1/8
Conc. at midspan	Both simply supported	Midspan	1/48	1/4
Conc. at midspan	Both clamped	Midspan	1/192	1/4
Conc. at 1/4 points	Both simply supported	Midspan	11/768	1/8
Conc. at 1/4 points	Both simply supported	Load pt.	1/96	1/8
Uniformly distributed	Cantilever	Free end	1/8	1/2
Conc. at free end	Cantilever	Free end	1/3	1

4.3.2 Columns

Behavior of sandwich under edge load (in-plane forces) depends upon bending and shear stiffness and sandwich size. From the derivation of the buckling loads of sandwich plates by *Ericksen* and *March* (1958) the following formula for columns is obtained after letting the plate width approach infinity:

$$N = \frac{n^2\pi^2 D}{a^2 \left(1 + \dfrac{n^2\pi^2 D}{a^2 U}\right)}, \tag{4.22}$$

where n is the number of halfwaves into which the column buckles. Formula (4.22) can also be written as:

$$N = \frac{\pi^2 D}{a^2 \left(\dfrac{1}{n^2} + \dfrac{\pi^2 D}{a^2 U}\right)} \tag{4.23}$$

and this formula has a minimum value when $n = 1$ resulting in:

$$N_a = \frac{\pi^2 D}{a^2 \left(1 + \dfrac{\pi^2 D}{a^2 U}\right)} \tag{4.24}$$

An upper limit for formula (4.23) is given for $n = \infty$. This limit is the shear instability limit and formula (4.23) reduces to:

$$N_i = U. \tag{4.25}$$

4.3.3 Plates

Many useful structural sandwich components can be constructed as plates. The relatively great stiffness of sandwich plates and their inherent resistance to deflection under normal load and buckling under edgewise (in-plane) forces are of prime importance in designing structural components. The buckling behavior of sandwich is such that buckling usually precipitates failure as was found by *Kuenzi* (1951) and therefore the buckling load of panels is not only indicative of loss of panel shape but of possible failure as well.

4.3.3.1 Plates under Normal Load. Since sandwich panels are utilized for their stiffness the design of plates of sandwich under normal load can begin by choosing thicknesses to limit deflections to allowable values. The maximum deflection of isotropic sandwich panels under uniformly distributed normal load was found by *Raville* (1962) to be given by the formula:

$$\delta = 2K_1 \frac{pb^4}{Eth^2}, \tag{4.26}$$

where δ is panel center deflection, K_1 is a constant given by the graph in Fig. 4.6, p is intensity of uniformly distributed load, b is panel width (dimension a in Fig. 4.6 is panel length), E is facing elastic modulus, t is facing thickness, and h is distance between facing centroids. K_1 is a function of the panel aspect ratio b/a and a parameter V relating shear and bending stiffness. The parameter V is defined by the formula:

$$V = \frac{\pi^2 D}{b^2 U}.$$

(4.27)

The facing stresses are maximum at the center of the panel and the stress at the facing centroid is given by:

$$\sigma = K_2 \frac{pb^2}{ht},$$

(4.28)

where K_2 is given by the graph of Fig. 4.6.

The core shear stresses are maximum at the middle of the panel edges and are given by the formula:

$$\sigma_{cs} = K_3 \frac{pb}{h}$$

(4.29)

where K_3 is given by the graph of Fig. 4.6.

4.3.3.2 Plates under Edge Loads. The solution of buckling problems is aided by the strain energy expressions for sandwich plates derived by *Libove* and *Batdorf* (1948). The total strain energy due to bending and shear is given by:

$$Q_1 = \frac{1}{2} \iint \left\{ D_x \left[\frac{\partial}{\partial x} \left(\frac{\partial w}{\partial x} - \frac{P_x}{U_x} \right) \right]^2 + 2 D_x \mu \left[\frac{\partial}{\partial x} \left(\frac{\partial w}{\partial x} - \frac{P_x}{U_x} \right) \right] \left[\frac{\partial}{\partial y} \left(\frac{\partial w}{\partial y} - \frac{P_y}{U_y} \right) \right] \right.$$

$$+ D_y \left[\frac{\partial}{\partial y} \left(\frac{\partial w}{\partial y} - \frac{P_y}{U_y} \right) \right]^2 + \frac{D_{xy}}{2} \left[\frac{\partial}{\partial x} \left(\frac{\partial w}{\partial y} - \frac{P_y}{U_y} \right) + \frac{\partial}{\partial y} \left(\frac{\partial w}{\partial x} - \frac{P_x}{U_x} \right) \right]^2$$

$$\left. + \frac{P_x^{\,2}}{U_x} - \frac{P_y^{\,2}}{U_y} \right\} dx\,dy.$$

(4.30)

The potential energy of the external forces is given by:

$$Q_2 = \frac{1}{2} \iint \left[-2pw + N_x \left(\frac{\partial w}{\partial x} \right)^2 + N_y \left(\frac{\partial w}{\partial y} \right)^2 + 2 N_{xy} \frac{\partial w}{\partial x} \frac{\partial w}{\partial y} \right] dx\,dy.$$

(4.31)

In these formulas x and y are the panel axes, w the deflection, P the shear load, D the bending stiffness, D_{xy} the twisting stiffness ($D_{xy} = G_{xy} th^2/2$ where G_{xy} is the facing shear modulus), U the shear stiffness, and N the applied edge loads per unit edge length, and p the intensity of uniformly distributed normal load.

These formulas are solved after assuming a deflection form for w. Values of P_x and P_y are not known but it can be assumed that:

$$g_x \frac{\partial w}{\partial x} = \frac{\partial w}{\partial x} - \frac{P_x}{U_x},$$

(4.32)

$$g_y \frac{\partial w}{\partial y} = \frac{\partial w}{\partial y} - \frac{P_y}{U_y}.$$

(4.32)

And then the final expressions are minimized with respect to g_x and g_y.

4.3.3.2.1 Flat Panels under Edgewise Compression or Bending. The facing
stress of sandwich plates under edgewise (in-plane) compression is given by:

$$\sigma = \frac{N}{2t},\tag{4.33}$$

where N is load per unit length of plate loaded edge and t is facing thickness.

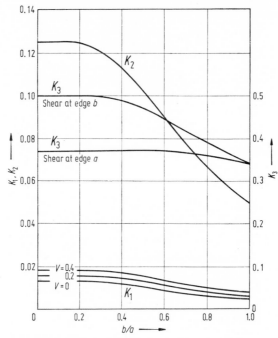

Fig. 4.6. Constants for determining deflection and stresses of uniformly loaded rectangular sandwich panels

The buckling load of a simply supported, flat, rectangular isotropic sandwich
plate under uniaxial compression is given by the formula derived by *Ericksen* and
March (1958):

$$N_c = K_c \frac{\pi^2}{b^2} D,\tag{4.34}$$

where D is bending stiffness of the sandwich, b is width (loaded edge) of the plate,
and K_c is a buckling coefficient dependent upon panel aspect ratio (a/b where a is
panel length) and the shear stiffness parameter given by the formula:

$$V = \frac{\pi^2 D}{b^2 U}.$$

Values of K_c for several choices of V are given in the graph of Fig. 4.7. In this
figure n denotes the number of half waves in the buckled form of the panel. The
horizontal asymptotes of K_c are given by the formula:

$$K_c = \frac{4}{(1+V)^2}.\tag{4.35}$$

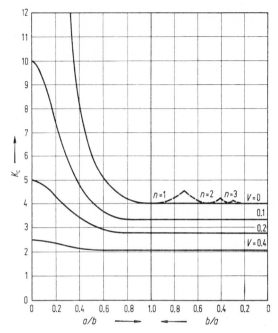

Fig. 4.7. Buckling coefficients for simply supported, flat, rectangular sandwich panels under uniaxial compression

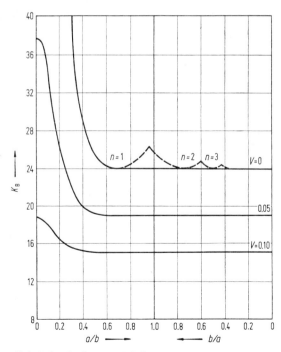

Fig. 4.8. Buckling coefficients for simply supported, flat, rectangular, sandwich panels under edgewise bending moment

The graph of Fig. 4.7 also indicates the number n into which the plates buckle with the dashed curve for $V = 0$.

Panels under edgewise bending can also buckle. The load at which these simply supported isotropic sandwich panels buckle is given by the formula derived by *Kimel* (1956):

$$N_B = K_B \frac{\pi^2}{b^2} D, \tag{4.36}$$

where K_B is given in Fig. 4.8, N_B is load per unit edge width. This load is related to the edgewise bending moment (M) by the formula:

$$N_B = \frac{6M}{b^2}. \tag{4.37}$$

4.3.3.2.2 Flat Panels under Edgewise Shear. The facing stress of sandwich plates under edgewise (in-plane) shear is given by the formula:

$$\tau = \frac{N}{2t}, \tag{4.38}$$

where N is load per unit length of plate edge and t is facing thickness.

The buckling load of a simply supported, flat, rectangular, isotropic sandwich plate under edgewise shear is given by the formula derived by *Kuenzi, Ericksen* and *Zahn* (1962) as:

$$N_s = K_s \frac{\pi^2}{b^2} D. \tag{4.39}$$

Values of K_s are given in the graph of Fig. 4.9.

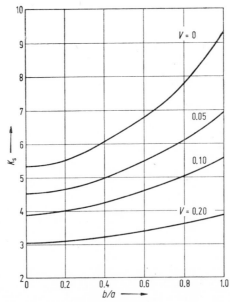

Fig. 4.9. Buckling coefficients for simply supported, flat, rectangular, sandwich panels under edgewise shear

4.3.3.3 Combined Loads. General buckling of sandwich panels under combined loads is given approximately by interaction formulas in terms of the ratios, R, wherein R denotes the ratio of applied load under combined loading to buckling load under separate loading ($R = N/N_{cr}$). Appropriate subscripts are given to R to denote load and direction. *Plantema* (1966) discusses these formulas and their limitations in detail.

Buckling of panels can be estimated from the formulas:

Biaxial compression	$R_{cx} + R_{cy} = 1$,	(4.40)
Bending and compression	$R_{cx} + (R_{Bx})^{3/2} = 1$,	(4.41)
Compression and shear	$R_c + (R_s)^2 = 1$,	(4.42)
Bending and shear	$(R_B)^2 + (R_s)^2 = 1$.	(4.43)

The combination of edge loads with loads directed normal to the plane of the sandwich can greatly magnify deflections and stresses due to normal load only. The deflections and stresses under combined edge and normal load can be closely approximated by the formula:

$$\Psi = \frac{\Psi_0}{1 - N/N_{cr}}, \tag{4.44}$$

where Ψ is deflection or stress due to combined loading, Ψ_0 is deflection or stress due to normal load only, N is edgewise loading (single or combined), and N_{cr} is general buckling load (single or combined).

4.3.4 Cylindrical Shells

The stiffness of sandwich makes it an ideal composite for the walls of cylindrical shells that may be required to withstand various loadings. Cylindrical shells under axial compression are difficult to design because of poor agreement between theoretical values and experiment, the difference being in the order of several times rather than a few percent. This is also the case for shells under axial bending. Cylindrical shells under external pressure (no axial load) or torsion do not buckle at loads much different than theoretical, however.

4.3.4.1 Cylindrical Shells under External Radial Pressure. A load of external radial pressure of intensity, p, on a cylindrical sandwich shell produces hoop compression stresses in the facings equal to:

$$\sigma_c = \frac{pr}{2t}, \tag{4.45}$$

where r is radius of middle surface of the shell and t is thickness of each facing. The stress at which buckling of the cylinder walls will occur is given by the formula:

$$\sigma_c = K \frac{E}{\lambda}, \tag{4.46}$$

where K is a coefficient dependent upon the shell length L (in a parameter L/r), a parameter

$$\Psi = \frac{h}{2r},$$

where h is the distance between facing centroids, and values of the shear stiffness parameter V where

$$V = \frac{D}{r^2 U}. \tag{4.47}$$

Values of K for isotropic sandwich shells were derived by *Kuenzi*, *Bohannan* and *Stevens* (1965) and are given in the graphs of Figs. 4.10, 4.11, and 4.12 for values of V of 0, 0.05, and 0.10, respectively. The curves dashed in the top curve of the figures shows the effect of numbers of buckles, n, in a circumference.

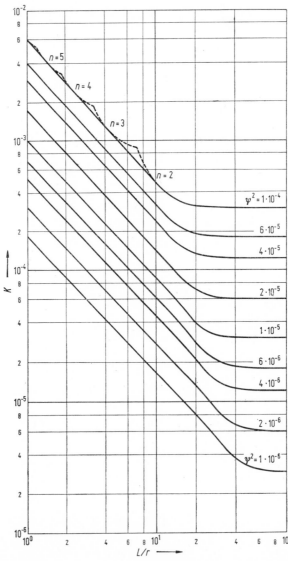

Fig. 4.10. Buckling coefficients for sandwich cylindrical shells under external radial pressure, $V = 0$

4.3.4.2 Cylindrical Shells under Torsion. The torque, T, applied to a circular cylinder in torsion produces a circumferential force, N, per unit circumference of:

$$N = \frac{T}{2\pi r^2}, \tag{4.48}$$

where r is the mean radius of curvature of the sandwich cylinder walls. The force N produces a facing stress of:

$$\sigma_s = \frac{N}{2t}, \tag{4.49}$$

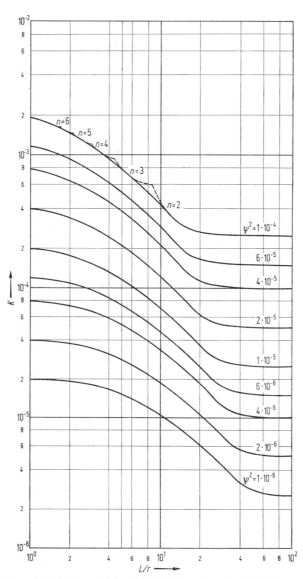

Fig. 4.11. Buckling coefficients for sandwich cylindrical shells under external radial pressure, $V = 0.05$

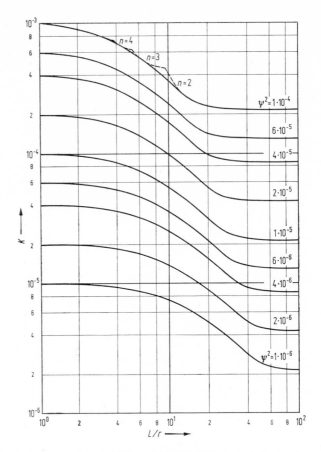

Fig. 4.12. Buckling coefficients for sandwich cylindrical shells under external radial pressure, $V = 0.10$

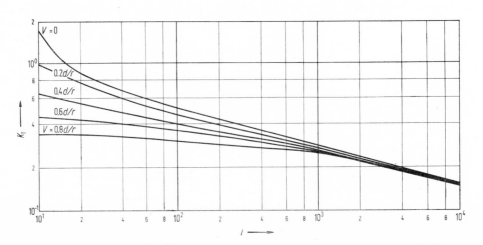

Fig. 4.13. Buckling coefficients for sandwich cylindrical shells under axial torsion.

where t is thickness of each facing of the sandwich. The facing stress at buckling of the sandwich wall is given by the formula derived by *March* and *Kuenzi* (1958) as:

$$\sigma_s = K_T E \frac{d}{r}, \tag{4.50}$$

where K_T is given in Fig. 4.13, and d is total sandwich thickness. Values of K_T are dependent on parameters:

$$J = \frac{L^2}{dr}, \tag{4.51}$$

where L is cylinder length, and

$$V = \frac{Eth}{2\lambda r^2 G_c}, \tag{4.52}$$

where G_c is core shear modulus associated with shear displacement in the radial and axial directions.

4.3.4.3 Cylindrical Shells under Axial Compression or Bending. Facing stresses are related to axial load, N, per unit circumference, by the formula:

$$\sigma_c = \frac{N}{2t}, \tag{4.53}$$

For axial compression $N = \dfrac{P}{2\pi r}$, where P is total axial load and r is mean radius of curvature of the cylinder.

For total bending moment, M, applied at the cylinder ends:

$$N = \frac{M}{\pi r^2}. \tag{4.54}$$

The cylinder wall will buckle at a facing stress derived by *Zahn* and *Kuenzi* (1963) given by the formula:

$$\sigma_c = \frac{kKE}{\sqrt{\lambda}} \cdot \frac{h}{r}, \tag{4.55}$$

where, for isotropic sandwich:

$$K = \frac{h}{4rV}, \tag{4.56}$$

$$V = \frac{Eth}{2\lambda r^2 G_c}$$

and k is a reduction factor determined by the appropriate curve of Fig. 4.14 as given by NASA (1965). The reduction factor attempts to account for the effects of initial shell irregularities.

4.3.4.4 Cylinders of Sandwich under Combined Loads. The buckling of the sandwich walls of cylinders under combined loads is given by interaction formulas in terms of the ratios, R, where R denotes the ratio of applied load or stress under

combined loading to buckling load or stress under separate loading ($R = N/N_c/$).
Axial compression and external lateral pressure

$$R_c + R_p = 1 \,, \tag{4.58}$$

Axial compression and torsion

$$R_c + R_T = 1 \,, \tag{4.59}$$

Torsion and external or internal pressure

$$R_p + (R_T)^2 = 1 \,, \tag{4.60}$$

where R_p is positive for external pressure and negative for internal pressure.

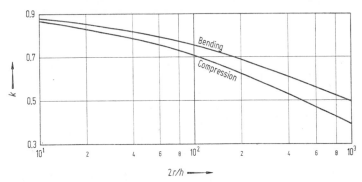

Fig. 4.14. Reduction factor, k, for the buckling of sandwich cylinders in axial compression or bending

4.4 Dimensional Stability

The dimensional movement of one sandwich facing with respect to the other causes bowing of an unrestrained panel. The dimensional movement may be due to stresses causing strains or due to changes in moisture content or temperature from one side of a panel to the other. If dimensional change is the same in both facings the length and width dimensions of the panel will change but bowing will not occur. The dimensional stability is primarily related to the facings because the core does not have enough stiffness to cause panel bowing or to cause the panel to remain flat. The amount of bowing depends, however, upon the core thickness.

It is possible to calculate the bowing of a sandwich panel if the expansion of each facing is known. The maximum deflection due to bowing caused by expansion of one facing resulting from temperature or moisture differential is given approximately by:

$$\Delta = \frac{ka^2}{8d} \,, \tag{4.61}$$

where k is the expansion of one facing compared to the opposite facing, a is the length of the panel, and d is the sandwich thickness.

Data accumulated on a test sandwich structure erected in 1947 are reported by *Anderson* and *Wood* (1964) on the bowing of sandwich panels as affected by seasonal variations in temperature and humidity. Bowing was found to follow a cyclic pattern, panels usually showing the same deflection in the same season year after year. Maximum bowing of 3″ thick panels with 1/4″ plywood facings was about 1/4″ in their 8 ft. length. Aluminium-faced panels 2″ thick had a bow of nearly 1/10″. Maximum bow in 3″ thick panels with 1/8″ or 1/4″ hardboard facings was about 1/2″. These various sandwiches had paper honeycomb cores.

4.5 Durability

Although abundant data and analytical information have been accumulated in the literature for the structural design of sandwich, relatively little information is available concerning durability of sandwich. In order for proper function of the sandwich to be maintained, the various parts — facings, core, bond — and inter-action of these in the environment must be evaluated or established through use. Choice of a durable facing material bonded to a durable core with a non-durable adhesive will not produce a durable sandwich composite. The key to success would be to utilize a good, durable adhesive as is available today for bonding of many different materials.

The durability of sandwich composites in an experimental structure was re-ported after 15 years of exposure by *Anderson* and *Wood* (1964). Losses in stiffness and strength were insignificant for certain combinations of materials. Wall panels of resin-treated paper honeycomb cores and plywood facings have demonstrated excellent performance. However, other combinations of facings and paper honey-comb cores have resulted in only fair to moderate performance. Composites exhibiting good behavior after being in the structure for several years were also good behavers in accelerated aging tests.

Literature Cited

Anderson, L. O., Wood, L. W., (1964) Performance of sandwich panels in FPL Experimental Unit. Forest Prod. Lab. Paper FPL 12.

Ericksen, W. S., March, H. W., (1958) Effects of shear deformation in the core of a flat rect-angular sandwich panel. Forest Prod. Lab. Report 1583 B.

Kimel, W. R., (1956) Elastic buckling of a simply supported rectangular sandwich panel subjected to combined edgewise bending and compression. Forest Prod. Lab. Report 1857 A.

Kuenzi, E. W., (1951) Edgewise compressive strength of panels and flatwise flexural strength of strips of sandwich constructions. Forest Prod. Lab. Report 1827.

—, *Bohannan, B., Stevens, G. H.*, (1965) Buckling coefficients for sandwich cylinders of finite length under uniform external lateral pressure. U. S. Forest Serv. Res. Note FPL 0104.

—, *Ericksen, W. S., Zahn, J. J.*, (1962) Shear stability of flat panels of sandwich construction. Forest Prod. Lab. Report 1560.

—, *Stevens, G. H.*, (1963) Determination of mechanical properties of adhesives for use in the design of bonded joints. U.S. Forest Serv. Res. Note FPL 011.

Libove, C., Batdorf, S. B., (1948) A general small-deflection theory for flat sandwich plates. NACA TN 1526.

March, H. W., Kuenzi, E. W., (1958) Buckling of sandwich cylinders in torsion. Forest Prod. Lab. Report 1840.

NASA, (1965) National Aeronautics and Space Administration. Buckling of thin-walled circular cylinders. NASA SP-8007.

Norris, C. B., (1964) Short-column compressive strength of sandwich constructions as affected by size of cells of honeycomb core. Materials. U.S. Forest Ser. Res. Note FPL 026.

Norris, C. B., Ericksen, W. S., Kommers, W. J., (1952) Flexural rigidity of a rectangular strip of sandwich construction. Forest Prod. Lab. Report 1505 A.

Norris, C. B., Ericksen, W. S., March, H. W. et al., (1949) Wrinkling of the facings of sandwich constructions subjected to edgewise compression. Forest Prod. Lab. Report 1810.

Plantema, F. J., (1966) Sandwich construction. John Wiley & Son, Inc.

Raville, M. E., (1962) Deflection and stresses in a uniformly loaded, simply supported, rect-angular sandwich plate. Forest Prod. Lab. Report 1847.

Timoshenko, S., (1953) History of the strength of materials. McGraw-Hill Book Co.

5. PARTICLEBOARD

5.0 History of Production, Consumption and Use of Particleboard

The idea to create particleboard has a long history, but the term "particleboard" was not used for a long time. Within about 90 years many patents were granted, predominantly for "artificial board". With respect to the large number of patents (DRP 967328 *Fahrni* (1942), ČSP 56350 *Pfohl* (1936), Schweiz. P. 193139 *Pfohl* (1937), DRP 692159 *Pfohl* (1940), ČSP 67763 *Dyas* (1940), USP 796545 *Watson* (1901), USP 2007585 *Satow* (1930), USP 2033411 *Carson* (1936), French.P. 679708 *Samsonow* (1929) the question arises why such sheetlike boards in large sizes suitable for "dry building" were not produced earlier. The reasons are as follows:

1. There was no proper idea about production of particles, flakes, splinters. Sawdust at the beginning was generally not suitable.

2. Glues for particleboard were available, but proper kind and necessary amount were not known.

3. Pressure cycles were not known.

4. Consumers were reluctant because they were accustomed to solid wood or plywood.

5. Many machines for the production of the new material (chippers, dryers, mixers, chip spreaders, mat prepresses, single and multi-daylight presses extrusion presses etc.) were lacking, they had to be developed.

6. Procedures for testing particleboard did not exist, strength values were underestimated, hygroscopicity overestimated, surfacing and gluing was not experienced.

7. Screw-holding and nail-holding power was not known or understeimated.

The patents granted to *Pfohl* 1936/37 were ahead of time. He proposed to produce uni- and three-layer particleboard made of flat thin prismatic wood chips or small sticks made from solid wood. The idea was to manufacture with a low content of binding agents light and stiff boards for the use in furniture manufacture, but the time was not ripe yet due to the reasons mentioned above.

With respect to some shortage of wood in Germany in the year 1941 the first plant for the production of particleboard in a technical scale was erected (Torfit-Werke at Bremen/Hemelingen). The raw material was dried spruce-sawdust width addition of 8 to 10% phenolic resin. The boards were produced in rather large sizes (3 m by 2 m ~9.5 ft. by 6.4 ft.) with thicknesses varying from 4 to 25 mm (5/32 to about 1 in.). Pressures (80 to 100 kp/cm², corresponding to 1,140 to 1,420 lb./sq. in.) and temperatures (160 °C = 320 °F) were high; therefore and due to the fine particles the density of the board was too high, ranging from 0.9 to 1.1 g/cm³ (56 to 69 lb./cu. ft.).

The plant was destroyed in World War II during an air raid. It was not reconstructed. But a progress in the technique was reached because for the first time special machines such as dryers, blenders, mat forming devices, hydraulic presses etc. were developed and used. In the following years urea-formaldehyde resins in colloidal solutions were used as binding agents. They were cheaper and could be cured at lower temperatures than phenolic resin glues. In the year 1942 *Roos* and coworkers in the Westdeutsche Sperrholzwerke AG, Wiedenbrück (W.-Germany), established a particle board plant in which beech veneer residues were chopped by wing beater mills to coarse splinters. These splinters were blended with 8 to 10% urea-formaldehyde resins — based on dry weight —, were spread in form boxes and then pressed to boards in a multi-platen hydraulic hot-plate-press. The boards produced had a thickness of 12 mm (about 1/2 in.) and a density 0.7 to 0.8 g/cm³ (44 to 50 lb./cu. ft.). The pressure applied was between 60 to 100 kp/cm² (~850 to 1,420 lb./sq. in.). This type of board was used for panelling and interior walls. A few very small works started in Germany in the years 1942 to 1943 the production of particleboard, but according to statistical data the yearly production in 1943 was only about 10,000 metric tons. Two firms tried to produce thin rigid particleboard on the basis of sawdust.

The results were not encouraging, but engineering, nevertheless, made some progress. In the year 1943 *Fahrni* described for the first time in a German technical scientific journal the status of the manufacture of particleboard and discussed the problems to be solved in the future. This publication gave an enormous impetus. *Fahrni* explained that the structure of the board in connection with the amount of binding agents, the density of the product and the special field of application is of greatest importance. He developed rather strong, light 3-layer-boards with a density around 0.6 g/cm³ (~37.5 lb./cu. ft.). The boards had a core consisting of splinters produced in mills and face or deck layers with higher density and strength in a thickness of only 1 to 1.5 mm (about 3/64 to 1/16 in.). For the face layers very thin flakes cut from round logs were used. The principle was that the "skin effect" of the face layers improved remarkably the strength of the boards. The new method was called Novopan and in Switzerland (at Klingnau) the first small plant on this basis was erected in the year 1944, but it was soon enlarged. Of importance was the development of many special machines by *Fahrni* for the particleboard industry.

Up to the year 1946 the basic problems for the manufacture of particleboard were not dealt with in a technical scientific manner. *Klauditz*[1] started such investigations in his institute at Braunschweig in the year 1946 and was in close cooperation with *Winter* (1949), who was interested on light constructions and aircraft. The main problem was the dependence of strength and quality of particleboard on the following factors: Shape and size of splinters or flakes, wood species, board density. The most important result of early investigations was the knowledge that especially the relationship between length, thickness and width of particles does not only determine the internal surfaces available for gluing, but also the strength properties of the board and the economics of manufacture (Fig. 5.1). The great influence of chip thickness on board quality (density and bending strenght) is shown in Fig. 5.2.

In the industry *Herdey* sen., *Himmelheber* (1948), *Steiner*, Interwood Ltd. (1947/51), Triangel (1947/62) and Behr (in this firm especially *Fischer* and coworkers) investigated also the effect of chip shape and size on manufacture and board properties. The influence of various wood species on the quality of particleboard is shown in Fig. 5.3.

---

[1] The many important publications of *Klauditz* cannot be quoted all in this book, but reference may be made to a list published by the author in Holz als Roh- und Werkstoff. Vol. 21 (1963), p. 121—123.

In the U. S. A. approximately in 1955 programs for investigations related to particleboard manufacture were established at the Forest Products Laboratory, Madison/Wisc., and the School of Engineering, North Carolina State College, Raleigh, N. C. (*Johnson*, 1956).

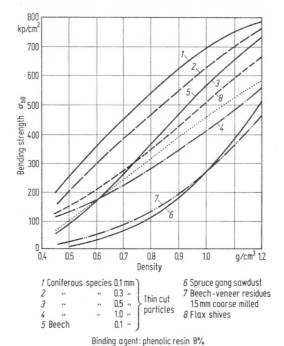

1 Coniferous species 0.1 mm		6 Spruce gang sawdust
2 " " 0.3 "		7 Beech-veneer residues
3 " " 0.5 "	Thin cut	1.5 mm coarse milled
4 " " 1.0 "	particles	8 Flax shives
5 Beech 0.1 "		

Binding agent: phenolic resin 8%

Fig. 5.1. Effect of chip size, wood species, and density (specific gravity) on bending strength (modulus of rupture) of particleboards. From *Klauditz* (1949, 1955)

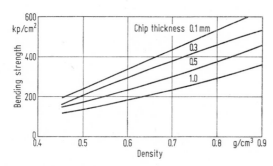

Fig. 5.2. Effect of chip thickness (flakes from coniferous species) and board density on bending strength, resin content 8% (based on oven dry wood). From *Klauditz* (1949, 1955)

The first larger modern particleboard plant with a daily production capacity of about 20 metric tons was erected in the year 1949 by *Himmelheber* and co-workers, the later Triangel-Holzwerkstoff GmbH. The daily capacity of newer plants, mainly based on the experience won by *Fahrni* increased in the years 1950/51 to 100 m³ (3,531 cu. ft.). Full mechanization of the whole manufacture process was reached.

Characteristics of the development of particleboard industries are the following particular points:

1. Utilization not only of round logs but in increasing amounts of industrial residues;

2. Increasing quality requirements;

3. Rapid development of new, highly efficient machines;

4. Quick development in the particleboard industries. The curves indicating capacity versus time are approximately parabolic. In West-Germany particleboard industry started in 1948. In 1952 the daily output of a plant was about 10 metric tons, in 1958 about 40 metric tons, in 1968 about 110 metric tons. In 1969 the capacity varied between 150 and 170 m³/day, the most modern works range in their capacity between 300 and 400 m³/day and recently between 600 and 800 m³/day with the possibility of enlargement up to 1,000 m³/day. Smaller plants with capacities below 200 m³/day become rare. They may be established only at locations where the conditions of raw material supply and marketing of ready board are particularly favorable.

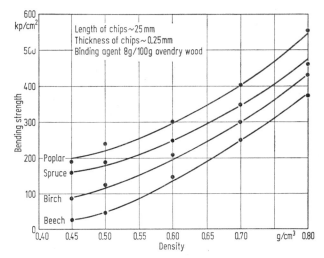

Fig. 5.3. Effects of wood species on bending strength in relation to density of particleboard (length of chips ~25 mm, chip thickness ~0.25 mm, 8 g binder per 100 g oven-dry wood). From *Klauditz* (1952)

It may be mentioned that already in 1948/49 *Kreibaum* developed his continously working extruding method. In the U.S.A. a similar vertical extrusion process a few years later was developed as Chipcraft system and a horizontal extrusion process as *Lanewood* process (*Johnson*, 1956). Particleboard became attractive in many European and other countries, e.g. in Belgium, France, Great Britain, Italy, Japan, the Netherlands, Austria and Czechoslovakia in the years from 1949 to 1953. Unilayer board on the basis of shavings were developed in Great Britain by the firm Airscrew Company & Jigwood Ltd. Remarkable was the development of the continously pressing Bartrev-process in England. Unfortunately the heavy and expensive Bartrev-press had some initial faults and therefore only a few plants were established but a push was given, and later on continuous presses of simpler construction were invented. It would go too far to mention more details of the development of the particleboard industry up to 1956. The following 15 years are characterized by:

a) Deepened and enlarged knowledge about the scientific principles of particleboard manufacture;

b) Improvement of production methods, especially by mechanization and automation;

c) Improvement of board quality;

d) Extension of use of particleboard in furniture manufacture, for interior work in houses, and beginning of exterior application.

Klauditz (1966) mentioned that the publications about particleboard increased from 1956 to 1962 by about 1,500. The author of this chapter estimates that in the whole world at least further thousand papers dealing with particleboard were published in the meantime.

FAO arranged in 1957 an International Consultation about Particle Board and Fiberboard in Geneva. A European organization for common exchange of experience and for cooperation was established in the year 1958: FESYP, Fédération Européenne des Syndicats de Fabricants de Panneaux de Particules.

In 1969 at the annual meeting of the Swedish Association of Pulp and Paper Engineers *Bengtson* (1970) presented a paper on the trends in consumption of fiber building board and other sheetlike materials for the building industry. This paper has been published in Defibrator News 1: 1970, Stockholm. *Bengtson* started his paper with the remark that at the beginning of the 1970s the world production of wood based sheet materials amounted to no less than 50 million m³ and he noted that the British trade press refers to "the panel products explosion". The development of the consumption of various types of wood based sheet materials must be viewed both worldwidely and locally. *Bengtson* was correct in saying that a survey is very difficult because sheets are used different thicknesses and with various in-service properties. The Economic Commission of Europe (ECE) therefore introduced a new unit, the "panel unit" (p. u.) on a trial basis. This unit may be rather arbitrary and certainly needs more studies for refinement. After all, the p. u. gives some opportunity for comparison of statistical data; one has to know that 1 m³ plywood = 1.1 p. u., 1 metric ton particleboard = 1.4 p. u., 1 metric ton hardboard = 1.6 p. u., 1 metric ton insulating board = 2 p. u. During the period from 1950 to 1965 world consumption of wood based sheet

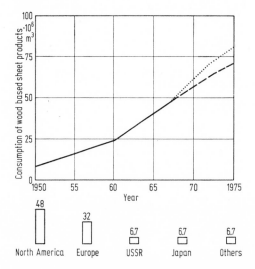

Fig. 5.4. World consumption of wood based sheet products during the period from 1950 to 1967 with subsequent extrapolated curves to the year 1975. Bar diagrams at the bottom show the high share of industrialized countries on total consumption. From *Bengtson* (1970)

products increased more than 5 fold (Fig. 5.4), and FAO expects that this trend will continue in the next future. A preliminary study (dotted curve) indicates an accelerated rate of increase, but later investigations let expect a steady rise up to 1975. The bar diagrams at the bottom of Fig. 5.4 show that (in 1967) the industrialized countries (North America, Europe, USSR, Japan) count for approximately 93% of total consumption. The greatest increase in the future probably will occur in the USSR, Latin America and Africa. Further studies showed that plywood had the greatest increase in the world and that particleboard increased faster than fiber building board. It is interesting to note that the regional pattern of consumption gives a different picture than that for the world. Thus Fig. 5.5, for example, shows that particleboard is increasing fastest in

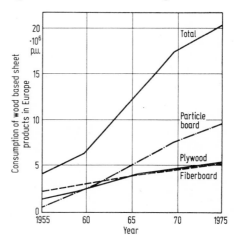

Fig. 5.5. Consumption of wood based sheet products in Europe from 1955 to 1975. Explanation of p. u. (= panel unit) is given in the text. From *Bengtson* (1970)

Europe. In the sketch the p. u. unit is used. According to data from the ECE Timber Committee particleboard dominate in Germany, Southern and Eastern Europe, whereas in Northern Europe fiber building board dominate and in the United Kingdom plywood. At a symposium, organized by the Timber Committee of the ECE in Geneva in June 1968, interesting material was presented which made clear that the consumption of wood based sheet materials in a country is correlated to its gross national product (GNP) (Fig. 5.6), which is a measure of the standard of living in a country. Table 5.1 shows a comparison of per capita consumption in 1965 of various types of sheet materials in major using countries.

The consumption of plywood per capita in the U.S.A. is greater than Sweden's per capita consumption of fiber building board, expressed in weight units, but of about the same order in square meters. Compared with North America and Sweden, Germany uses a relatively small amount of wood based sheet material,

Table 5.1. Comparison of per capita Consumption in 1965 of Various Types of Sheet Materials in Major Using Countries (*Bengtson*, 1970)

		Units/capita	m²/capita estimated
Plywood	North America	45 kg	9
Particleboard	Norway and Germany	20 kg	2
Fiber building board	Sweden	38 kg	10

especially expressed in m² per capita. *Bengtson* (1970) concludes that "taken as
a whole" the market for sheet materials is definitely confusing. A very thorough
study is needed to investigate the full reasons behind this heterogeneous pattern.

FAO has collected a great deal of interesting statistics concerning the uses
of various wood based sheet materials in different countries (FAO, 1968). As far as
particleboard is concerned it may be summarized as follows:

a) *Building*

Topping the list of consumers of particleboard for building purpose was Chile
(64%), followed by India (44%). Countries reporting use in building trade equal
to about a third of total consumption of particleboard were Belgium (mostly
flax shives boards), Canada, France, Sweden and the U.S.A. For building purposes
a considerable share of consumption was found in Poland (25%), Britain (27.5%)
and in Japan (22.5%); Austria and West Germany reported limited use: 12%
and 17% respectively. In Yugoslavia only 4% of the total consumption of
particleboards were going into the building trade.

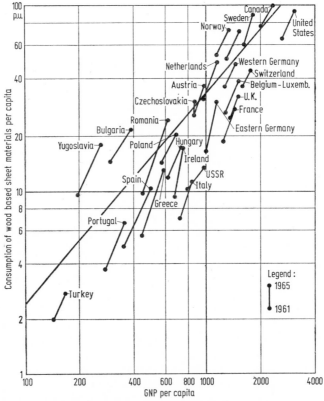

Fig. 5.6. Consumption of wood based sheet materials per capita in various countries (1961 and 1965) as correlated
to the gross national product (GNP). ECE Timber Committee (1968), *Bengtson* (1970)

The use of particleboard for flooring, mainly in domestic dwellings, offers the
following advantages and needs the following provisions (*Gomme*, in *Mitlin*, 1968,
p. 160—162):

1. Large and different sizes are available;

2. Very few joints compared with the multiplicity inherent in conventional floorings with tongued and grooved wood boards;

3. Acceptable quality, tested to conform with the flooring requirement of B. S. 2604: 1963, Amendment No 3. All edges laid on joists must be either supported by inserted noggins or must be tongued and grooved. This insures not only ample strength at the joints but complies with Fire Regulations which prohibit "straight-trough-joints";

4. Structures may be nailed or glued. In short term loading tests they indicate little difference, in the long term, however, glued joints will better withstand the conditions in use;

5. In multi-storey buildings the tongued and grooved particleboards may be laid as "floating platform", held to some extent by the surrounding skirting on the concrete sub-floor which is insulated for instance by expaned polystyrene or resin bonded glass-fiber. Suspended floors are strong enough when 19 mm (3/4") thick particleboard are nailed direct to joists at 16 in. to 18 in. (406 mm to 457 mm) centers about 16 in. (406 mm) apart;

6. Particleboard should be kept dry at all times to ensure that moisture pick-up is a minimum; in no circumstances particleboard should be stored under damp conditions or out-of-doors;

7. Resistance to warping and movement is remarkable;

8. Color and texture can be maintained with wax polish, oil stain, hard varnish;

9. Wearing quality is entirely satisfactory for use in place of floor boards of similar thickness (Section 5.4.9.1).

b) *Furniture*

FAO statistics show that in 11 countries the furniture industry had the relatively greatest consumption of particleboard. In Poland it is estimated at over 98%, in Yougoslavia at about 93%, while Belgium reported about 70%, Austria 71%, West Germany 68%, Sweden 61%, Canada 60%, France 56% and Britain 52%. It was only in Japan (46%), U. S. A. (44%), India (about 40%) and Chile (26%) that the quantities used for furniture manufacture were less than half of the total consumption of particleboard.

5.1 Raw Materials

5.1.1 Wood and Other Ligno-Cellulose Materials

5.1.1.0 General Considerations. In the total production costs for particleboard manufacture the costs of wood are very important. The data available are still contradictory. *Wyss* (1957) published figures between 8 and 23% for four plants in different countries, *Höchli* (1956) calculated 19% — in comparison with 72% for blockboard —, and FAO (1957/58) published for Europe and the U. S. A. rounded off costs of the wood in particleboard plants between 24 and 33%. In the same FAO study figures are given which demonstrate that wood costs are the higher the greater is the capacity of a plant. The reason is that with increasing capacity wages and general expenses contribute relatively less to the total production costs.

Over 90% of the total dry weight of particleboard is usually wood or equivalent fibrous ligno-cellulosic material. The high yield between about 75 and 90% in particleboard manufacture (calculated for ready cut to sizes and planed or sanded board) is favorable as compared with plywood manufacture where only values between 45 and 58% may be obtained.

The wood species used for particleboard manufacture vary. In Western Europe initially coniferous species (mostly spruce and pine, to a smaller extent fir, Sitka spruce and Douglas fir) were the preferred material, but beech, poplar, and birch were employed subsequently in an increasing proportion for reasons of both economy and availability. Other hardwoods, such as alder, horse-chestnut and willow are also used. FAO (1957/58) published surveys on the most important woods for particleboard manufacture in all parts of the world.

Kumar (1968) dealt with the manufacture of particleboard from tropical hardwoods. He points out that mixed species are economically used. With the rise of the standard of living and utilizing potentials of the new products in developing countries a further progress may be expected for production and application of particleboard.

Timber suitable for particleboard manufacture can be divided into five basic groups:

a) Unprocessed forest products such as thinning and thick branches;

b) Coarse industrial residues such as slabs, edgings, off-cuts from sawmills, peeler cores and rejects from veneer manufacture;

c) Fine industrial residues, especially planer mill shavings and sawdust;

d) Wood chips from machining of dry wood;

e) Residues such as slabs, edgings and off-cuts from furniture manufacture and other industrial operations on dry wood with a moisture content of 10 to 15%.

Unfortunately world — wide statistics are not yet available, but *Stegmann* and *Storck* (1963) made an investigation. They found that in West Germany for the year 1954 from the total amount of wood consumed by the particleboard industry, the fiberboard industry and the pulp and ground wood industry only 3% were used for the manufacture of particleboard. In the year 1961 the consumption for particleboard manufacture had increased to 21.5%. Typical was that especially the consumption of industrial wood residues had increased from 1952 to 1962 about forty times (Fig. 5.7).

Apart from wood the most important fibers and ligno-cellulose raw materials utilized for particleboard manufacture are bagasse and flax shives. They differ essentially from wood chips, basically from other ligno-celluloses, such as cotton, hemp and jute stalks. Straw is not suitable because its large capillaries collect inside too much resin and the adhesion is poor due to the high content of mineral substances. Experimental work has been done on ground nut shells and bark, but at the moment none of these materials is used in the industry.

5.1.1.1 Unprocessed Forest Products.

It is not known how much from the wood coming out of the forests and going into particleboard plants are thinnings, thick branches (limbs), round logs or billets. The highest commercial value has pulp wood, but it is mainly desired — as its name says — by the pulp and paper industry. It is evident that from round logs with greater diameters more chips can be produced than from logs with smaller diameters. Since only the latter are going into the particleboard industry the differences in the chip yield are not important. From an economic point of view the higher costs for transportation and handling of round wood with small diameters and thinnings has to be considered. Knotty or bent logs or billets are suitable for chipboard manufacture, but the content of solid wood expressed as a percentage of the piled raw material is only 0.6 to 0.7, whereas for straight round logs with diameters of 30 cm and more (about 12 in. and more) the proportion can increase up to 0.8 (in the German literature for forestry and forest products industries the term fm/rm is used, which means fm = cubic meter of solid wood and rm = cubic meter of piled wood containing air spaces). If the raw material has a greater length, a bigger diameter and a relatively high weight, then conveyors to the reduction units are necessary. Transportation methods (trucks, roller bands or a combination, e.g. with a bridge crane) are available and depend on the type of material and the capacity of the work. Manual conveying should be avoided. The conveying of chips ready for board manufacture is dealt with in Section 5.2.4.

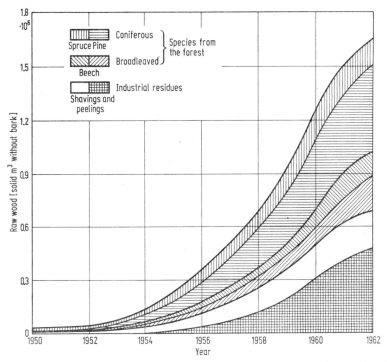

Fig. 5.7. Consumption of raw wood in the West-German particleboard industry from 1950 to 1962. From *Stegmann* and *Storck* (1963)

5.1.1.2 Industrial Wood Residues. The most important industrial wood residues for the particleboard industry are coarse residues from saw mills, such as slabs, edgings and off-cuts. These types of mill residues are a valuable raw material, not only for the particleboard industry, but also for the fiberboard and for the pulp and paper industry. Quality and grading are essential. Preferred are slabs, cut to uniform lengths and if possible bundled. The content of solid wood in one piled cubic meter of industrial wood residues varies in wide ranges due to the many types of wood species, sizes and dimensions. An average survey is given in Table 5.2.

As far as West Germany is concerned — and this may be valid for other industrialized countries, even for the U.S.A. — up to a few years ago hardwood and industrial residues (formerly called "wood waste", a term which is no more justified) were easily available and rather cheap. Industrial residues in the form of

Table 5.2. Content of Solid Wood in Percentage of Piled Wood for Various Types of Industrial
Residues
(*Vorreiter*, 1943; *Kollmann*, 1951)

	Solid content % or fm/rm (see above)
1. Residues from sawmills	
a) Off-cuts (in heaps)	~**0.37**
b) Slabs	0.51···**0.58**···0.64
c) Scantlings and edgings (length 1 m)	
middle thick (*Aaro*, 1961)	0.57···**0.56**···0.63
thin (*Aaro*, 1961)	0.47···**0.52**···0.57
spruce and fir (*Flatscher*, 1929)	0.44···**0.47**···0.49
d) Sawdust (*Levon*, 1931)	**0.33**
2. Residues from planer mills, shavings and chips from woodworking industries	0.18···**0.20**···0.25
3. Residues from veneer and plywood plants	
a) Edgings from peeled veneer	**0.45**
Edgings from sliced veneer	**0.55**
b) Peeler cores (*Jalava*, 1929), and (*Aaro*, 1961)	0.78···0.91
half split (*Aaro*, 1961)	0.69···**0.70**···0.71
4. Residues from woodworking	
a) Mixed pieces (e.g. from vehicle manufacture)	0.60···**0.65**···0.70
b) Sawdust	0.32···0.38

shavings, guarantee a high yield and offer, already mechanically desintegrated, advantages for further manufacture process. The amount of industrial residues will not remarkably increase and therefore in the future, at least in West Germany, the demand for hardwood and thinnings will increase.

The differing content of solid wood mass per piled or poured cubic meter (again the German term "rm" = literally translated "space meter") of wood residues or round wood determines the yield of chips. An important German firm, involved in the development of particleboard plants, found that 1 rm slabs or edgings delivers about 150 to 160 kg (~330 to 350 lb.) ovendry chips, 1 rm round logs 250 to 260 kg (~550 to 570 lb.) oven-dry chips.

Sawdust was used for the first time, as mentioned in Section 5.0, in the Torfit-Werke at Bremen-Hemelingen, Germany, for the production of particleboard. Another German chemical industry tried to produce hard chipboard for floorings on the basis of sawdust blended with phenolic resins. The experiments failed. *Klauditz* (1947) proved that it is not possible to manufacture particleboard of medium density with sufficient strength properties on the basis of sawdust with 8 to 10% binding agents (solid resin based on oven dry wood), but he pointed out that the problem is still important. Recent investigations and experiments in the industry showed that it is possible to add in a limited amount rather coarse sawdust to the raw material for particleboard manufacture. The new trend is to use tools which produce either coarse sawdust or better to employ tools which produce no sawdust, but various types of chips as by-products which are usable in the particleboard industry. Coarse grained sawdust commonly takes up too few glue, whereas too fine particles absorb too much glue. Therefore the technique of blending must be adapted to the special case if very small particles prevail.

5.1.1.3 Flax shives. Flax is an annual plant. There are about 90 species, all native in temperate zones. The common flax (*Linum usitatissimum* L.) is an important economic plant used

a) For the fibers obtained from its stem which are converted into linen, thread, cordage;

b) For the lin seed oil, extracted from its seed.

Flax, cultivated since prehistoric times, is now grown throughout the world and is second to cotton in importance as fiber plant. Flax fiber is produced extensively in the Soviet Union, Poland, France, Belgium and Holland. As a residue or by-product in the production of flax fibers the woody stems or flax shives are obtained. They are fine rectancular-shaped particles of ligno-cellulose material obtained by longitudinal division of the stalk of the flax plant during scrutching of the retted flax (FAO, 1958, 1959). Flax shives became an essential raw material for the production of particleboard. The technology of manufacture is different from that of producing particleboard on the basis of woods. A well developed process is the Belgian Linex-Verkor-process (*Swiderski*, 1960).

The flax shives are already particles and therefore chipping machines or mills are not necessary, but the shives are thoroughly cleaned for removing dust, fibers and too coarse particles. The shives come to the plant with a low moisture content of 11 to 14% — as compared with 30 to 50% in wood — a fact which favorably influences the heat economy. The dust free shives are fed into a machine which removes fibers and fiber bundles. Subsequently they pass a cyclone, a puffer silo and a pneumatic sorting device. A scheme of the whole process is given in Fig. 5.8.

5.1.1.4 Bagasse[1]. Bagasse is the fibrous ligno-cellulose residue, left after extraction of the sugar from sugar cane. Table 5.3 shows that the chemical composition of bagasse is similar to wood (*Hesch*, 1968, 1969, 1970, 1972).

Table 5.3. Chemical Composition of Bagasse and Wood

	Bagasse %	Beech %	Pine %
Cellulose	46	45	42
Lignin	23	23	29
Pentosans and hexosans	26	22	22
Other components	5	10	7

Bagasse is available in tropical countries. The yearly world output is about 65 million metric tons, dry matter. Most of the bagasse is still used as fuel for the sugar mills.

For a long period bagasse has been considered as low quality substitute for wood. During the recent years this attitude has changed. New developments proved that bagasse is competitive for pulp, paper and board manufacture. On long terms, experts expect that bagasse, which is today considered as by-product, may become the main product and sugar the by-product.

At present about 60 industrial plants use bagasse as raw material. Most of them are pulp and paper mills and particleboard plants, followed by a few fiberboard plants and one furfurol plant.

Compared with wood, bagasse is a cheap raw material. In many places a great surplus exists. Much bagasse is still burnt or is dropped into rivers or into the sea.

---

[1] The author is indebted to Dr. *Hesch* for this contribution.

Depending on climatic conditions of the various countries, fresh bagasse is available from 2 to 12 months per year. As world average a cane grinding season of 6 months can be assumed. For the "off-season" bagasse has to be stored to enable year-round production.

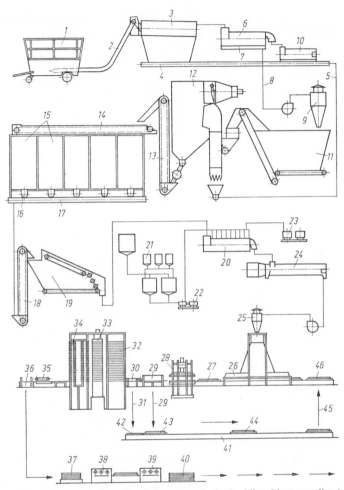

Fig. 5.8. Total scheme of the manufacture of particleboard on the basis of flax shives according to the Belgian Linex-Verkor process. From *Swiderski* (1960)

1 Carriage, *2* Rake conveyor, *3* Dust removal, *4, 7, 16* Screw Conveyor, *5* Pneumatic conveyor, *6* Fiber separator, *8* Suction pipe line, *9, 25* Cyclone, *10* Fiber cleaning, *11, 19* Dosing bin, *12* Pneumatic separator, *13, 18* Bucket elevator, *14* Belt conveyor, *15* Conditioning chamber, *17* Longitudinal conveyor, *20* Glue mixer, *21* Preparation of glues, *22* Pumps, *23* Compressor, *24* Drum (rotary) dryer, *26* Charging station, *27* Charging box, *28* Pre-press, *29* Conveyor for pre-pressed casting boxes, *30* Chain conveyor, *31* Transverse conveyor, *32* Loader, *33* Press, *34* Unloader, *35, 36,* Edging, *37* Conditioning, *38, 39* Cylinder grinders, *40* Storing, *41* Return, *42* Cauls, *43* Frame, *44, 46* Cauls and frames, *45* Transverse conveyor

Bagasse storage is connected with some problems. Fresh bagasse contains 2 to 4% residual sugars. Cane is harvested in immature condition due to the higher sugar content at this stage. The lignification is not completed, and therefore more low molecular components are contained in the bagasse than in wood. Together with the residual sugar, they are nutrients for the growth of bacteria and fungi.

The convention to store the raw bagasse is to bale it and to build the bales up to pyramids. In countries with high rain fall the pyramids have to be covered by sheets. However, baling and handling the bales are expensive operations. Therefore attempts were made to develop bulk storage systems which can be mechanized.

Similar to wood, attacks by bacteria and fungi can be kept within tolerable ranges either by storing the bagasse completely wet or by reducing the fiber moisture content below the fiber saturation point.

Artificial drying of the depithed and washed bagasse gives a good preservation but is too expensive. Artificial drying of fresh depithed bagasse prior to washing preserves not only the fibers, but also residual sugar. If bagasse with residual sugar is converted to boards they are attacked by fungi as soon as they come into humid environments. Correspondingly expensive preservatives have to be added.

Based on experiments in the tropics a new simpler storage has been developed (*Hesch*, 1970, 1972). The bagasse is depithed and screened quickly after leaving the sugar mill. The pith returns to the boilers. The prepared fiber only is conveyed to the storage area.

Immediate depithing offers saving in proportion from 25 to 30% of removed pith and fines. The second reason is: Fresh bagasse starts to ferment very soon. By the fermentation the nuisant residual sugar and other low molecular components are converted mainly to alcohol and acide acid. Fermentation is an exothermic process and generates heat. By the fermentation the strength properties of the bagasse fibers can be affected by hydrolysis. If, however, the bagasse is depithed before being stored, the absence of pith and fines results in a better aeration of the bales. The heat and vapor dissipate excessive temperature which could lead to hydrolysis. At the same time the humidity of the bagasse is reduced. Lower temperatures and reduction of humidity together worsen the growth conditions of the bacteria, and fermentation interrupts itself. After a storage time of 4 to 6 weeks a humidity of 30 to 25% is achieved, such a low humidity does not favor neither growth of bacteria nor of fungi.

The third reason for immediate depithing is connected with quality and physical properties of the particleboard. Strength, water absorption and surface quality depend on the pith content of the prepared bagasse. This applies to the pith of the inner parts of the cane stalk. Here the pith has voluminous cells with thin and weak walls. As long as the bagasse is wet, these spongy cells stand off and are easy to be gripped and shaved off by the tools of a mill. But once the bagasse is compressed without being depithed, the spongy pith cells are pressed against the strong fibers. With the loss of humidity they remain in this stage. Therefore during subsequent depithing the tools are not able to grip them, depithing will be less perfect. If much of the pith comes into the board, a "spring back" of individual particles and the board as a whole will be observed, as soon as such board comes into humid environments. This "spring back" is irreversible.

For depithing several varieties of hammer mills exist. Fig. 5.9 compares the principle of a conventional hammer mill with horizontal shaft (a), with a one-chamber type vertical shaft mill and (b) a two-chamber type vertical shaft mill (c).

Type b) does depithing only. Separating the loosened pith has to be done subsequently by a rotating screen. Type c) is designed to depith and fractionate in one operation. The fines are supposed to pass the perforations between the central chamber and the exterior chamber. Better separation of fines is achieved with the combination of type b) and a rotating screen.

The principle of the process for bagasse particleboard manufacture is shown in Fig. 5.10. The bagasse is dried first to 3 to 5% humidity. Then the raw material is distributed to the surface layer preparation and the core layer preparation line. For refining wing beater mills are employed. The unsuitable fines are eliminated by screens and can be used for the manufacture of cattle fodder. Gluing and spreading are done by similar machines as used for wood particleboard manufacture. This applies in principle to the press, too. However, a longer press cycle than with wood must be obeyed, due to the very homogeneous cross section of bagasse particleboard. This requires a longer time for the dissipation of the vapors from the board.

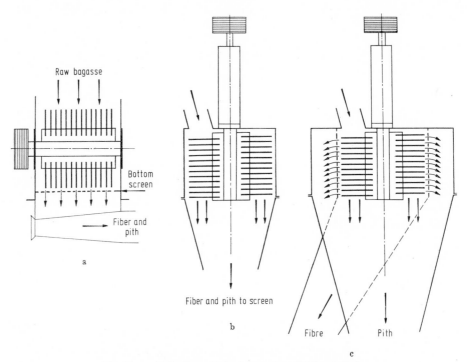

Fig. 5.9 a — c. Reduction units for bagasse in particleboard plants.
a) Hammer mill with horizontal shaft; b) Hammer mill with vertical shaft in one-chamber execution; c) Hammer mill with vertical shaft in two-chamber execution. From *Hesch* (1972)

Strength and surface properties of bagasse particleboard are similar to high quality wood particleboard, today. Compared with wood and other annual plants, bagasse offers a wider range of densities. *Hesch* (1968) mentions that bagasse boards can be produced with densities between 300 kg/m³ and 1,000 kg/m³ (∼19 and 62 lb./cu. ft.), generally with densities varying from 300 to 700 kg/m³ (∼19 to 44 lb./cu. ft.). The process of particleboard production is extraordinarily versatile with respect to thickness and density of board. The possible applications are numerous, such as for furniture, shelvings, built-in furniture, partitions, prefabricated doors, floors and subfloors, ceilings and roof deckings, prefabricated houses. For external application all particleboard has to be protected by waterproof coatings (aluminium, PVC-foils, enamelled asbestos cement, bituminous paper board etc.). If bagasse is compared with those species of wood that grow in cane areas, the comparison turns out expressively in favor of bagasse under technical as well as economical aspects. From all so-called "annual plants" bagasse is the best and most versatile one. On long term bagasse will become the most important source of ligno-cellulose fiber after wood.

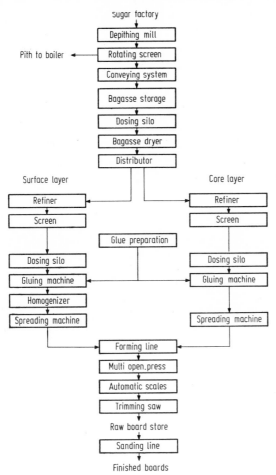

Fig. 5.10. Principles of the production for a three-layer bagasse particleboard plant established by Trinidad Bagasse Products Ltd. System *Hesch* (1972)

5.1.2 Adhesives

5.1.2.1 Urea- and Melamine-Formaldehyde Resins. The binding agents have besides the chips the main influence on nearly all properties of particleboard and therefore from the beginning of the particleboard industry many types of glues and cements were partly applied on an experimental scale and scientifically investigated. *Mörath* (1966) gave a survey. Only a few names of researchers can be mentioned: *Blomquist* (1960, 1961, 1962), *Herdey* (1958), *Klauditz* (1960), *Kollmann* (1958), *Marian* (1958), *Marra* (1960), *Meinecke* (1960), *Rackwitz* (1955), *Rayner* (1966, 1968). Of special importance is the question of costs because in the particleboard industry the costs of binding agents contribute most to the total production costs. At present practically the following types of artificial resin glues are used in particleboard manufacture: Urea-formaldehyde resins, melamine-formaldehyde resins and phenolic resins. All these glues are thermosetting. Thermoplastic resins are used only for the production of molded particle parts. In Chapter 1 at first the physical-chemical principles of gluing are dealt with, and then the various types of glues are discussed. The economy in gluing depends not only on the costs of the glue, but on many other properties such as permissible storage time, pot-life, necessary pressing times and pressures, resistance of the particleboard against effects of high relative humidity, dripping water and temperature especially under cyclic conditions.

Two reaction products of an amine and formaldehyde are used in the particleboard industry: Urea-formaldehyde resins and melamine-formaldehyde resins. Throughout the world urea resins are used more than any other resin type. In Central Europe at least 90% of particleboard is produced with these glues. Melamine resins would be principally attractive for chipboard, but they are applied only on a small scale with respect to their high costs. Manufacture and chemistry of urea-formaldehyde resins are treated in detail in Section 1.11.3.2 hardening in Section 1.11.3.3. Fig. 5.11 shows the relationship between viscosity

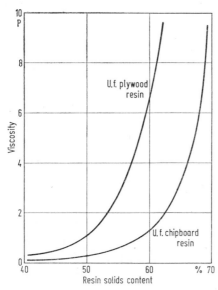

Fig. 5.11. Relationship between viscosity and resin solids content for a typical (UF) urea-formaldehyde chipboard resin compared with an average plywood resin. From *Rayner*, in *Mitlin* (1968)

and resin content for a typical urea-formaldehyde (UF) chipboard resin compared with an average plywood resin glue. It can be seen that UF plywood resins have (usually) a higher viscosity than UF chipboard resins.

For the development of the chipboard industry a decrease in curing times was essential in order to increase the output. Shorter curing times may be reached by three means:

a) Higher pressing temperatures;
b) Faster press closing;
c) Moistening of mat surfaces prior to hot pressing.

The percentage of resin content influences the curing time. Many investigations proved that for various types of typical chipboard resins there exists a nearly hyperbolic relationship between gelation time[1] and percentage of resin content. Table 5.4 gives a few figures based on a diagram published by *Rayner* (1968).

Table 5.4. Variation of Gelation Time at 100°C (212°F) in Minutes with Percentage of Resin Content

Resin content %	40 min	50 min	60 min	70 min
Glue				
A	6.2	4.3	3.25	2.5
B	5.8	3.85	2.4	1.95
C	4.45	2.8	2.25	1.9

The moisture content of the chip mat in contact with the hot press platens has a two-fold effect on the curing process:

a) Dilution of the resin by water retards gelation and thereby delays precuring;

b) Steam generated in the outer regions of the board improves heat transfer to the core.

Much information about this subject has been published. The most important references are *Fahrni* (1942, 1943), *Klauditz* (1952, 1955), *Rackwitz* (1954), *Kollmann* (1957). Fig. 5.12 shows the effect of surface water on heat transfer during

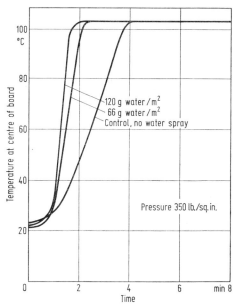

Fig. 5.12. Effect of surface water on heat transfer during board production in a hot press (board thickness 3/4", platen temperature 145°C, pressure 350 lb/sq. in. (25 kp/cm²). From *Rayner*, in *Mitlin* (1968)

[1] Gelation time is the time taken for a given mass of resin and a particular hardener to change from a liquid to an incipient gel under defined conditions (*Rayner*, 1968, p. 19).

board production in hot presses (*Rayner*, in *Mitlin*, 1968, p. 15). *Rayner* points out that in the case of urea-formaldehyde resins a high rate of heat transfer does not necessarily shorten the pressing time.

The commonly used hardeners for urea-formaldehyde resins are discussed in Section 1.11.3.3. For chipboard manufacture the best all-round hardeners are the ammonium salts of strong acids. The hardener may be applied to the wood chips separately, either before or after the blending with the resins. The latter method avoids too short pot-life. In chipboard manufature a reasonably long pot-life is required and therefore retarding agents such as ammonium hydroxide, which is unpleasant to handle, or hexamine may be added. Melamine is a very efficient retarding agent in hot hardeners. A quantity between 0.5% and 2% of the resin weight is sufficient.

The degree of cure is indicated by the thickness of the board after its removal from the press. The board becomes appreciably thicker than the spacers. A climax of this "growth" of the board is reached within a few — say three to four minutes — but the board cools down and the internal forces become equalized. Board thickness decreases again. Cooling down is accompanied by evaporation of moisture and expansion due to elastic recovery.

Melamine-formaldehyde resin adhesives are treated in Section 1.11.4. The chemical similarity of melamine and urea-formaldehyde resins is pointed out. — The main advantage of melamine-formaldehyde resins is a superior resistance to water, especially hot water. Hence follows for the manufacture, the particle-boards can be hot-pressed with a higher moisture content. An important use of melamine resins is upgrading or fortifying of urea cements (cf. Section 1.11.4.5). By this technique water resistance may be remarkably increased, swelling is reduced, but there is no substantial effect on dry strength.

The percentage of resin required in the manufacture of wood particleboard depends mainly on official strength specifications. In Europe mostly chips from softwoods are used; generally for single layer board 7 to 9% urea- or melamine resin solids are considered adequate, for three-layer board about 7% for the core and 10% for the outer layers. It is the general opinion that hardwood chips require more resin than softwood chips, also chips from flax shives and bagasse need a slightly increased amount of resin. In any case a uniform distribution of the resin over the chips is desirable.

The German chipboard manufactures once used fillers mainly to reduce the glue costs but during recent years fillers were not used any more. The presence of fillers can cause problems during spraying with conventional equipment.

5.1.2.2 Phenolic Resin Binders. Reference is made to Section 1.11.1 dealing with these types of resin binders. For the wide range of interior applications, usual for particleboard, the cheaper urea-formaldehyde resins are sufficient. They offer the advantage to cure very rapidly at elevated temperatures by the addition of a suitable hardener; a sudden transition from the fluid to the gel state occurs. Phenol-formaldehyde resins undergo a slower transition, and therefore they need higher pressing temperatures and longer pressing times. By this reason the moisture content of the chips to be glued is more critical with phenol-formaldehyde than with urea-formaldehyde resins. If the moisture content is too high too much steam is generated within the board during hot pressing.

This superfluous steam will disturb or even inhibit the curing process, within the board either blisters may occur or the board may delaminate on removal from the press.

5.1.2.3 General Rules for Chip Resination. Based on articles by *Mörath* (1966) and *Nicholls* (1968) and on own experience the following facts can be stated:

a) Artificial resin glues for particleboard manufacture are applied as solutions of between 40% and 65% (mostly 40% and 50%) solid content;

b) If the resin increases in viscosity during storage adjustment by addition of water (up to 20%) is possible in order to maintain the required spraying properties;

c) Minimum percentage of resin required as binder is approximately 7 to 10%, based on dry wood weight;

d) Maximum moisture content of chips prior to drying is 10% to 50%, based on the total weight of mat;

e) Re-drying of the chips after resination should be achieved rather by high air velocities in the dryer than by high air temperatures (not more than about 70 °C or 158 °F) within short drying time;

f) Good coverage of chips can be accomplished by "atomized spray" using a mixer equipped with a series of nozzles ("spray guns");

g) Press temperatures from 130 °C to 160 °C (266 °F to 320 °F) are suitable. Pressure for manufacture of medium density boards is usually in the range of 14 to 35 kp/cm² (~199 to 498 lb./sq. in.); the lower pressures are recommended for softwood chips, the higher pressures for hardwood chips;

h) Press cycles vary with various factors such as moisture content of chips, type of resin binders, thickness of board, press temperature etc. According to FAO (1958/59) for a board of 19 mm (3/4 in.) thickness, for instance the cycle may range from 8 to 30 min, the average being between 12 and 15 min;

i) As mentioned under d) maximum moisture content of chips prior to blending entering as mats the hot presses is about 10 to 50%, based on total weight. This moisture content is reduced by the pressing operation to about 5 to 12%;

k) Some pressing operations use a combination of hot pressing and high-frequency heating. This technique makes it possible to shorten the pressing time and so to increase the board production. The technique is particularly suitable for the manufacture of thicker boards. *Pungs* and *Lambertz* (1957) discussed the principles of different systems and the economic aspects. A critical examination of high-frequency heating in comparison with normal contact heating is necessary. Preheating by means of high-frequency up to 80 °C (176 °F) before entering the pressure zone of the press is usual in the continuous matpressing procedure.

5.1.3. Additives

5.1.3.1 Water-Repellents. Dimensional stability of particleboard is generally good and if water-repellents are properly applied; swelling, cupping or twisting are nearly excluded. These advantageous properties promoted the quick application of particleboard in the furniture manufacture. Only the thickness swelling under the influence of high relative humidity or still more of water is remarkable. Fig. 5.13 shows thickness swelling of particleboard in water and in moist air. Length swelling is relatively low due to uniform distribution of the chips in horizontal planes and due to the content of binding agents. The high swelling in thickness is not only caused by the high compression of the chip material in the hot press — from which also follows the "kick back" on the removal of the press — but also by the porosity of the particleboard. One has to bear in mind that the values are relative figures depending on the particular test procedures, especially on the size of the test specimens and on their position in water, vertical, inclined or horizontal.

Mineral waxes or paraffins are by-products in the petroleum industry where crude oil is first treated to separate and purify volatile fractions, such as gasolins, kerosene, naphta, and solaroils (high-boiling oil extracted in amount of about 1% in the distillation of lignite-tar oil). Wax distillate is the higher fraction from which paraffin wax and paraffin oil are separated.

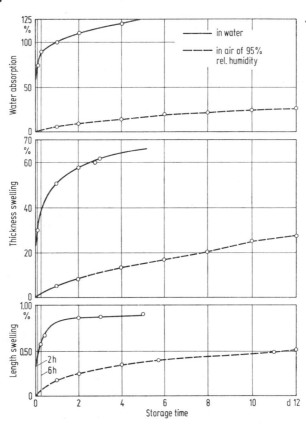

Fig. 5.13. Water absorption, thickness swelling, and length swelling of three-layer particleboard (UF-resin binder, board thickness 12 mm ~ 15/32 in., density 0.75 g cm³ ~ 47 lb./cu. ft.) stored in water (temperature 20°C ~ 68°F) or wet air (relative humidity 95%). From *Liiri* (1961)

Commercial paraffin graded according to melting points between 48°C and 56°C (~118°F and 133°F) is an important material for package sealing, for increase in moisture resistance of board, for polishes, cosmetics and many minor uses. Paraffin dispersions are used in the particleboard industry mainly for the following reasons:

a) High water-repellent effect;
b) Favorable range of melting points;
c) Favorable price in comparison with other possibilities to produce a water-repellent effect.

The maximum content of solids in emulsions reaches 65% based on total weight. The paraffin particles are very small (diameter approximately 1.5 μm) if homogenizers for the production of the emulsions are employed. If normal stirring machines are used the paraffin particles are coarser (about 14 μm diameter), and the storage time of the emulsion becomes relatively low.

With respect to particleboard manufacture the paraffin should be:

a) Homogeneous;
b) Compatible with binder, hardener, water, and preservatives;
c) Pumpable;
d) Suitable for dosing.

There are three different types of emulsifier systems and, therefore, anionic, non-ionic, and cationic paraffin or wax emulsions are used. Though most of these emulsions can be incorporated with the resin and hardener up to now the anionic emulsions have found the widest application.

It is possible either to blend the paraffin emulsion with the resin in the mixing vessel adjusted if necessary by addition of an ammonia solution to pH of about 8, and then to add water and hardener or to apply the anionic emulsion separately from the glue by providing a special metering and spraying system for the emulsion. Both methods are used in the industry. Most German particleboard plants purchase the waxes and produce the emulsions themselves. Wax emulsions ready for use are supplied by the mineral oil industry. They have an appreciably higher content of solids, and their particles are more homogeneous because sophisticated homogenizers are employed. Occasionally the question has been asked whether or not only by the incorporation of an artificial resin binder into the chips a water repellent effect may be obtained. This is not the case with urea-formaldehyde resins sprayed in economically justified amounts. If phenolic resin binders are used adding of water-repellents is necessary because the phenolic resins possess high alkalinity and are sprayed with a relatively high water content.

The effects of paraffin additions shows Fig. 5.14. The amount of paraffin or wax added to the mixture must be large enough to be efficient with respect to water repulsion and swelling resistance, but small enough not to reduce the bond between the chips and the glue applied. Usually 0.3 to 0.5% paraffin based on the ovendry weight of softwood chips, 0.5 to 1.0% of hardwood chips are employed. A mixture of the wax with the adhesive as an emulsion requires less wax to obtain about the same water resistance in the board as with the hot spraying method. It is essential — as it is with the adhesives — to obtain a uniform distribution of the wax. Because the amount of wax introduced into the mixture is small, the problem of distribution is difficult and "mixing wax with adhesive and introducing both materials jointly offers obvious advantages" (*Lyman*, in *Mitlin*, 1968, p. 35). Experiences in the industry proved that wax-treated board is suitable for veneering, is to some extent shower-proof; it is said that such boards due to the presence of wax are easier to be cut and machined.

More information about the influence of water repellents on the properties of particleboard is given in the literature (*Klauditz*, 1960, *Eisner* and *Kolejak*, 1960, *Walter*, 1960, *Buschbeck* and *Kehr*, 1960, *Müller*, 1962, *Wittmann*, 1971). The latter author summarized his results as follows:

If liquid paraffin meltings free of emulsifiers are sprayed, particleboard can be manufactured, satisfying the standards in thickness swelling. To be sure 20 to 50% more paraffin must be used to obtain the same water-repellent effect as with dispersions. It seems possible to reduce the amount of paraffin by improving the machine equipment, and it is economically interesting that no emulsifiers are necessary. If emulsions (dispersions) of paraffin are used, the kind of incorporation depends on the binder type. For phenol resins separate spraying is advisable because then the tensile strength across the board plane as well as the thickness

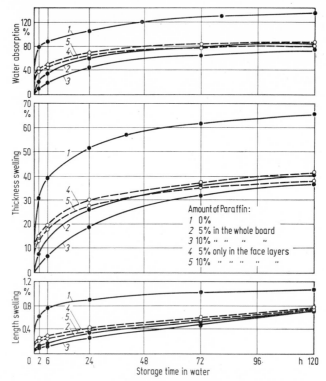

Fig. 5.14. Effects of additions of paraffin emulsions to the whole board or only to the deck layers (----) on water absorption, thickness and length swelling as a function of the duration of the storage in water (temperature 20°C ∼ 68°C). From *Liiri* (1961)

swelling will be improved. If the binders are UF-resins the paraffins do not interfere with the strength properties of the board. The reason may be relatively low concentration of paraffin in the UF bond board. The emulsion if separately sprayed should be incorporated into the chip material prior to the spraying of the binder. The hygroscopic behavior of board bond with phenol-formaldehyde (PF) resins is quite different from that of board bond with urea-formaldehyde (UF) resins; if for instance 1.0% solid paraffin, based on the weight of dry wood, are employed, PF-board in an absorption test takes up more than twice as much water as board bond with UF-resin and containing 0.5% solid paraffin, based on ovendry weight. The reason for this increased water absorption of board bond with phenol resins is partly the higher density of such board and mainly the mentioned high alkalinity of the phenol resins. Particleboard can be protected against penetration of water into their edges by application of a two-component lacquer, poor of solvents and chemically hardening. Such a lacquer has a content of solids of about 85% thus producing rather thick ("filled") layers which are elastic and resistant against many chemicals. This treatment has proved itself for parts of prefabricted houses, balustrades, schools, armories, halls for factories, store-houses and holidayhomes which may be exposed to the weather.

According to experience it can be stated:

a) The diffusion resistance of particleboard well treated with water-repellents is very high;

b) Thickness swelling of particleboard protected on all sides is negligible;

c) The absorption of moisture of particleboard, protected on all sides, is after 14 days storage in a climate of 40 °C (104 °F) and 100% relative humidity only about 1%;

d) Thickness swelling under the conditions mentioned above is less than 1%;

e) Attack of fungi on the protected particleboard is improbable.

5.1.3.2 Extenders. By a control of the flow of liquid binders and their pene-tration into the wood under certain conditions resin costs may be reduced. One of the tasks of extenders is "holding of the adhesives on the wood particle surfaces and to aid the distribution of the adhesive on the wood particle surface as a thin and uniform resin film" (*Johnson*, 1956). Usually flours are added as extenders to the resin mix. Wheat flour contributes some bonding power, but there is a limit of such extenders which can be introduced without seriously reducing the quality of the bond of the resin. Flours impart some plasticity to binders and reduce the inherent brittleness of urea resins (*Bornstein*, 1956). Proper extension, based on solid resin by weight, must be determined by experience or experiment. "It has been found that under certain conditions the extender greatly enhances the free flow of the chip mass because the moisture tends to remain on the surfaces and serves as a lubricant" (*Johnson*, 1956). Hardened, pulverized or chemically treated flours (bromated or phosphated) cause difficulties. High ash or protein contents and variations in the acidity of the flour are also undesirable. Blockage of spraying nozzles caused by the use of extenders is mainly responsible for the decreasing. use of these agents. Miller Hofft Inc. (1953) discuss characteristics of chipboard. They present graphs showing how most chipboard properties (not hardness) can be improved by resin extension (Fig. 5.15 and 5.16). *Johnson* (1956, p. 203) points out that the function of the extender can be achieved also by other means such as increasing resin mix viscosity or increasing chip moisture content or mixing or covering the chips with resin more thoroughly. Extension has about the same effect as raising the resin content because it improves the distribution and con-centration of existing resin.

5.1.3.3 Fungicides, Insecticides. During a long period the decay resistance of particleboard was not considered as a serious problem because the board delivered by the manufacturers had a low moisture content and were used exclusively for furniture manufacture and interior work. Longlasting influences of higher relative humidity or wetting with liquid water were excluded and therefore the conditions for fungi attacks were not actual. The situation changed with increased use of particleboard for exterior works, e.g. walls.

The decay resistance may be increased partly by a higher amount of binding agents. *Clark* (1960) remarks that an increase in the addition of urea resin from 4% to 8% nearly doubles the decay resistance. With phenol resins an increase

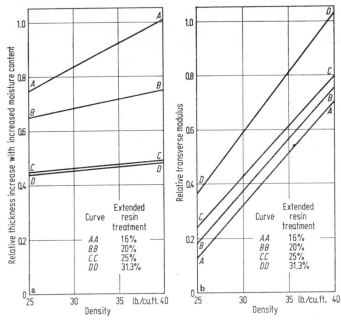

Fig. 5.15a, b. Relation of chipboard properties. a) Relative thickness increase with increased moisture content (dimensional stability); b) Relative transverse modulus to density at various resin extensions. Miller Hofft Inc. (1953), in *Johnson* (1956, p. 204)

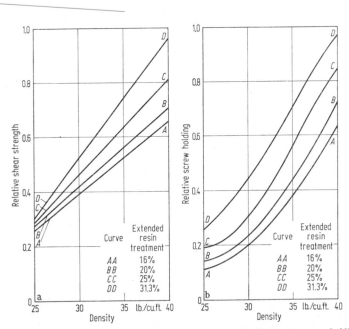

Fig. 5.16a, b. Relation of chipboard properties. a) Relative shear strength; b) Relative screw-holding power to density at various resin extensions. Miller Hofft Inc. (1953), in *Johnson* (1956, p. 205)

in the resin amount from 3 to 6% and an addition of 1% wax density from 0.55 to 0.66 g/cm³ (~34 to 41 lb./cu. ft.) has no influence.

Compatibility of the preservatives with the binding agents has to be considered. It may be mentioned that the manufacture of chips of such wood species which have a higher content of toxic extractives is especially effective. It has been observed that particleboard bond with urea resins and containing a blend of chips with 50% western red cedar chips has the highest decay resistance. In Europe mainly pentachlorphenol up to 1% (in the tropics up to 2%) and copper-pentachlorphenol (0.4 to 1%) or sodium-fluoride or sodium-silicofluoride (0.5 to 1%) are added (*Künzelmann*, 1960).

Klauditz and *Stolley* (1951) sprayed in their experiments aqueous solutions of sodium pentachlorphenol to the chips, and subsequently the binding agent was incorporated; another technique is to combine preservative and binder in a solution which then is sprayed. Separate incorporation of binder and preservative can be done in such a way that the preservative is sprayed to the wet chips after their preparation. The impregnation effect is satisfying but the following drying of chips causes losses of the preservative. The quoted researchers employed an aqueous solution with 20% of preservative, based on the weight of water, and the incorporated amount of the preservative was 1 to 2%, based on the weight of oven-dry chips. As binders either urea resins or phenol resins were sprayed in a solution of 50% concentration in an amount of 8% solid resin, based on the weight of oven-dry chips. Blending of the solution of binder and preservative caused no disadvantages. Table 5.5 elucidates this fact. Instructive are the figures of Table 5.6.

The results in both Tables 5.6 and 5.7 show that particleboard bond with urea as well as with phenol resin may be attacked by fungi. The binding agent alone exerts no protection which is worth mentioning, but already an addition of 2% pentachlorphenol results a sufficient protection against fungi. Attacks by termites are reduced but not to a satisfying extent (Table 5.8).

Table 5.5. Modulus of Rupture (Bending Strength) of Particleboard without and with Addition of Pentachlorphenol
(*Klauditz* and *Stolley*, 1951)

Binding agents, preservatives respectively	Density		Bending strength	
	g/cm³	lb./cu. ft.	kp/cm²	lb./sq. in.
8% phenol resin	0.54	33.7	310	4,410
8% phenol resin 2% pentachlorphenol (sprayed in a mixture)	0.53	33.1	290	4,130
8% urea resin	0.60	37.5	260	3,700
8% urea resin 2% pentachlorphenol (separately sprayed)	0.61	38.1	301	4,280
8% urea resin	0.57	35.6	261	3,710
8% urea resin 2% pentachlorphenol (sprayed in a mixture)	0.57	35.6	323	4,590

Spruce chips, thickness 0.1 to 0.2 mm.

5.1.3.4 Fire Retardants. Producing fire retardant particleboards, mostly phosphates are used for this purpose. *Stegmann* (1958) showed that these chemicals lessen the strength of urea-formaldehyde bonded boards. The conclusion is that

Table 5.6. Reduction of Decay of Particleboard by Use of Urea Resin without and with Pentachlorphenol as Preservatives Using Various Techniques of Application (*Klauditz* and *Stolley*, 1951)

Binding agents, preservatives respectively	Fungi	Loss of Weight %
8% urea A	*Coniophora cerebella*	52···54···56
	Poria vaporaria	0
	Merulius lacrymans	39
8% urea A 1% pentachlorphenol separately sprayed	*Coniophora cerebella*	3···22···41
8% urea A 2% pentachlorphenol separately sprayed	*Coniophora cerebella*	0···0···1
8% urea B	*Coniophora cerebella*	49···52···53
8% urea B 4% pentachlorphenol separately sprayed	*Coniophora cerebella*	0···1···1

Spruce chips, thickness 0.1···0.2 mm

Table 5.7. Reduction of Decay of Particleboard by Use of Phenol Resin without and with Pentachlorphenol as Preservatives Using Various Techniques of Application (*Klauditz* and *Stolley*, 1951)

Binding agents, preservatives respectively	Fungi	Loss of weight %
8% phenol resin	*Coniophora cerebella*	18···22···26
8% phenol resin 1% pentachlorphenol separately sprayed	*Coniophora cerebella*	3··· 9···15
8% phenol resin 2% pentachlorphenol separately sprayed	*Coniophora cerebella*	5··· 6··· 6
8% phenol resin 2% pentachlorphenol sprayed in a mixture	*Coniophora cerebella*	1··· 2··· 3
	Poria vaporaria	1
8% phenol resin 2% pentachlorphenol sprayed in a mixture	*Coniophora cerebella*	2··· 3··· 6

Spruce chips, thickness 0.1···0.2 mm

Table 5.8. Termite Resistance of Protected Particleboard (*Klauditz* and *Stolley*, 1959)

Protective agent mixture % Pentachlorphenol	Degree of destruction in 5 months % Sample		
	1	2	3
0	35	60	75
0.8	5	10	10
1.5	less than 5	0	0

apparently phenolic boards are more suitable for phosphate addition. Phenolic resins are more resistant to elevated temperatures than urea resins. *Schaeffer* (1970) investigated cure rates of resorcinol and phenol-resorcinol adhesives in joints of ammonium salt treated southern pine. Joints of the treated wood made with the mentioned adhesives passed the minimum requirements of commercial standards for shear strength and wood failure. Six normal phenol-resorcinol or resorcinol resin adhesives did not meet the commercial standard minimum requirements for glued, fire resistant treated wood. Shear strength and wood failure were higher with all adhesives studied for untreated wood than for treated wood. *Schaeffer's* results may be transferred to the conditions of chipboard.

5.2 General Technology

5.2.0 Introduction

The definition of particleboard, the rapid growth of the industry concerned and the manifold uses are discussed in Chapter 5.0. The raw materials and the additives and preservatives are dealt with in Chapter 5.1. Doubtlessly not only the development, but also the future growth of the particleboard industry depends upon the types of chips, the structure of board, the resin binders and on the manufacturing equipment. A wide variety of raw materials and particles of varying sizes and shapes may be used.

The capital investment required to establish a particleboard plant is lower than that for a fiber building board plant of comparable size. In the last decade there was a trend to increase the capacity of particleboard plants and thus to increase their efficiency.

Significant patent literature underlying the establishment of modern particleboard plants is mentioned in the introduction to Section 5.0.

A schematic outline of the principal operations in processing is shown in Fig. 5.17. FAO (1958/59, p. 66) remarks: "The technique of actual manufacture has become increasingly complex and calls for a great deal of specialized knowledge, scientific, technological, and economic. It follows that the establishment of a

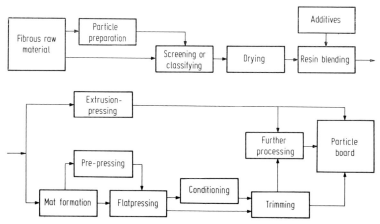

Fig. 5.17. Schematic outline for particleboard manufacture. From FAO (1958/59)

particleboard operation in countries where such an industry does not yet exist should be preceded by a most careful investigation and assessment of all relevant factors, e.g. raw materials and use, markets, economic conditions, wages, availability of capital, rates of interest, etc. Thorough investigations of these and other requirements are essential before a decision is taken to establish a new plant."

5.2.1 Particle Preparation

5.2.1.0 General Considerations. The properties of particleboard primarily depend upon the type of particles used. Their preparation, along with drying, screening, blending and mat formation are the most important factors in particleboard production. All steps of manufacture, mentioned above, are closely related to the type of particles. The resin requirement is also related to it.

Pfohl (1936/37) in his pioneer patent recommended to use small rectangular sticks. Suitable machines to produce such sticks were not available, blending was difficult, the raw material was too expensive, board properties were problematic. *Fahrni* (1942) distinguished between coarse splinters, produced by hammermills for the core of a three-layer board, and cut flakes produced by a special machine for the decklayers, consisting mainly of very thin particles in order to get smooth surfaces which can easily be painted, lacquered, covered with plastic foils or veneered.

Cutter-type particles ("flakes") are occasionally called "engineered" particles. With dry material more fines and dust are produced than with green material. Fines are not desirable for high quality board because they absorb more resin and they affect mechanical properties adversely (FAO, 1958/59), but the question should not be exaggerated because the amount of fines up to 6% does not interfere with the strength of the board (*Kollmann* and *Teichgräber*, 1962). In the industrial production of particleboard, however, the tendency of fines to sift to the bottom of the board during pressing may cause an unbalanced construction. In the use of the final board warping may occur, but one of the newest developments has changed this situation.

5.2.1.1 Wood Yard, Conveying of Raw Wood, Metal Detection. A plant, producing daily 200 m³ (about 7,000 cu. ft.) of finished particleboard with an average density of 620 kg/m³ ($\sim$39 lb./cu. ft.) and an average thickness of 19 mm ($\sim$3/4 in.), manufactures per year (with 280 working days) about 56,000 m³ ($\sim$two million cu. ft.) or approximately 35,000 metric tons of particleboard. Under the assumption that for the production of 1 m³ ($\sim$35 cu. ft.) of finished board 2.2 rm (= piled wood, cf. Section 5.1.1.1) of raw wood are needed, such a plant consumed per year 123,000 rm wood. As a rule the plants store a supply of wood for 3 to 6 months production. Larger storage is not advisable because:

1. The danger of wood losses by decay and stain increases;
2. The wood becomes too dry, checks occur reducing the yield and resulting in poor chips;
3. The handling costs become too high due to much storage labor;
4. In the case of fire extinguishing is impeded;
5. The elasticity in purchasing raw wood is restricted.

In particleboard plants frequently the raw material grades for the manufacture of face layer chips and for that of core layer chips are frequently separately stored, such as round logs, slabs and edgings or softwoods and hardwoods or indigenous and imported woods. The layout of the total wood yard and the dimensions of the particular piles — usual lengths, governed by the local conditions, in many cases between 30 and 60 m (~100 and 200 ft.), heights predominantly between 3 and 6 m (~10 and 20 ft.) — are determined according to fixed rules, experience or special requirements. One can guess that on each m² of wood yard (1 m² ≈ 10.8 sq. ft.) about 1.5 to 2 rm wood can be piled. These values are generally accepted for fixed piles (Fig. 5.18). For conveying the wood to these piles trucks are used

Fig. 5.18. Separate storage of round logs, slabs, and edgings

in combination with travelling crabs or travelling cranes equipped with "polyp-grabs" or field railroads, with rubber-tired trucks pulled by Diesel locomotives.

In order to avoid piling which requires high labor costs, so-called "wild-piles", discordered heaps, recently came into use (Fig. 5.19). The whole wood yard is controlled by a travelling bridge crane equipped with a polyp-grab which takes the wood from the pile and throws it into a log canal wherefrom the floated logs are taken over by a wire mesh belt forwarding them directly to the chip producing machines.

Many particleboard plants use chain or belt conveyors for log transportation. Interesting is the combination of such an inclined conveyor with a trough-carousel (rotating trough) for feeding five chipping machines (Fig. 5.20) surrounding the carousel (*Kollmann*, 1966, p. 127).

It is advisable to pass the raw wood through an electronic metal detector. The instruments must be very sensitive. The wood to be tested is transported on an endless rubber belt to an electric field. Metal, especially iron parts entering the aperture disturb the uniform electric field, a voltage is caused in the pick-up coils and amplified in order to give signals, to make markings, or to stop the conveyor. The schematic circuit diagram of a stationary metal detector is shown in Fig. 5.21 (*Schaffrath*, 1957; *Graham*, 1958; *Sorg*, 1959).

Fig. 5.19. "Wild pile". In the fore-ground on the left the wire mesh belt conveys logs, on the right the pump
house

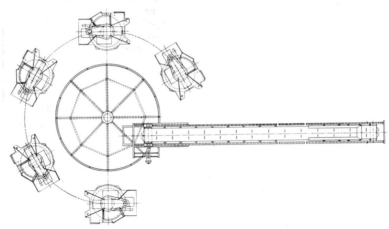

Fig. 5.20. Scheme of wood supply for 5 flake-shaving machines by means of an inclined conveyor with a trough-
carousel. Schenck, Darmstadt

5.2.1.2 Moisture Content. The moisture content of logs or slabs piled on the
wood yards of particleboard plants varies regionally, seasonally, according to
species, season of felling, storage in the forests, and conditions of transport in the
wide range between about 35 and 120%, based on oven-dry weight; in extreme
cases minimum values down to 16% and maximum values up to 200% may
occur.

Wet storage in natural or artificial log ponds — absolutely to be recommended
for the valuable veneer logs and blocks used by plywood factories, and usual in
Scandinavian sawmills — are not necessary for the raw wood in particleboard
industries, but under some circumstances wetting by spraying or sprinkling may
be useful. According to American experience small droplets (so-called "water-

dust") are effective and economic; nevertheless a precipitation of about 3.5 mm/h
(~0.14 in. per hour) is necessary. Only the storage area for the wood to be converted into face layer chips is sprayed.

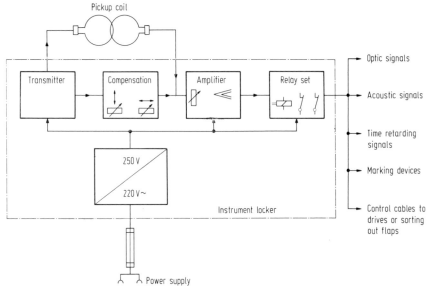

Fig. 5.21. Schematic circuit diagram of a stationary metal detector. From *Sorg* (1959)

If the logs are to be barked adequate moisture content is indispensable. Therefore as a rule the wood is pretreated with water or steam. Storage in cold water
or sprinkling does not considerably increase the moisture content but removes
adhering dirt. The effect of treatment with hot water of 95 °C (203 °F) temperature
is better than that by steaming at about the same temperature. *Bott* (1955) not
only describes the various methods of industrial wood debarking methods but also
the possibilities of plasticizing the wood by steaming processes.

From the aspect of wood cutting the moisture content (m. c. or recently H)
of the wood should not be lower than about 35%, based on the oven-dry weight.
Fig. 5.22 shows the rapid increase of specific cutting work E_c and specific cutting
power P_c/w below 35% m. c., and Fig. 5.23 shows the reduction of quality (expressed as inverse radius of chip bend) below 35% m. c. (*Pahlitzsch* and *Mehrdorf*, 1962). Furthermore, below the fiber saturation point screen analysis indicates
the percentage of flakes with large surface areas remarkably decreases whereas
the percentage of fines increases. With respect to chipping as well as with respect
to subsequent drying the raw wood should possibly have moisture contents
between 35% and 50%.

5.2.1.3 Barking (Debarking). The first step in particle preparation is the debarking of wood. The amount of bark, based on the total volume of the stem
with bark is about 10 to 17% for pine, for spruce 8 to 15%, for birch 7 to 15%,
for beech 6 to 10%. Manual debarking is rare; because it is expensive, slow, and
inefficient. Mechanical debarking is usual (cf. Section 6.5.1). Debarking machines
with cutting tools can be movable or stationary. The capacity for the usual types
is listed in Table 5.9. This table also gives a survey of the debarking machines

Table 5.9. Data for Debarking

Kind of debarking	Log diameter cm	Energy consumption HP	Men required
With cutting tools			
Conventional manual debarking	10···25	—	1
Manual debarking with mechanized handtools	10···50	2···4	1
Smallpeeler with milling or cutterheads and cutterdisks	10···20 10···30	1.5···2p. tool 4	1···2 1···2
Rapid-peeler (*Bezner*)	up to 25	4···7.5	1···2
Floating cutterhead peeler (*Bezner*)	7···40	13.5	1
By friction forces between wood			
Drum-peeler (*Paschke*)	10···30	Size O 8···10 Size I 20···25 ⎱ Size II 30···35 ⎰	1 1···2
„ „ (*Hanke*)	5···40 (0.5 m woodlength) (1.0 m woodlength)	13 20	1 1
By friction of tools against wood			
Rotor-peelers (*Cambio, Skoglund, Sund*)	5···30	⎰ 15···16 ⎱ 20 ⎰ 30	1 1 1
Hydraulic debarking			
In connection with friction forces (*Schongau-Entrindungsmaschine*)	pulpwood (partly barked)	310	8
Hydraulic ringbarker (*Hansel*)	5···50	15···50	1

---

[1] cf. explanation of „fm" and „rm" at the end of Section 5.1.1.2. Cord is a unit of volume used chiefly for pulp and fuel wood, now generally equal to 128 cu. ft.

Methods and Machines

Capacity[1] per hour (h) solid m³ = fm piled m³ = rm	Remarks	Loss of wood %	Sources
0.125···0.3···0.5 rm/h	—	5···8	
0.6···1.2···(5)		5···8	*Bott* (1955) *Zieger*, 1960, p.229
0.45···1.25 1···3		8···15 up to 26	*Bott*, 1955 *Zieger*, 1960, p.237
1···2 rm/h at 12 cm ∅		8···15 (for small diameters up to 25)	*Bott*, 1955 *Bezner*, letter 17.2.64
10 cm ∅ at 5···6.9 rm/h 16 cm ∅ at 8.7···11.2 rm/h 30cm ∅ at 16···17.5 rm/h		6···15	*Bott*, 1955 *Bezner*, letter 17.2.64
0.50···1.04 1.24···2.50 2.50···5.00	8···9 rm at charge: 100 l/min water consumption 12···15 rm at charge 150 l/min water consumption	0.5···0.8 (···2)	*Bott*, 1955
2···4 rm/h 4···8 rm/h	420 l/min water consumption (1 l = 0.264 U. S. gal.)		*Bezner* letter 17.2.64
12···30 17···36 24···48		} 0···2	
15···16 rm/h (theoret. maximum capacity 30 rm/h		Soaking in hot water <0.1	*Holzhey* letter 14.2.64
Feed speed 10 to 75 m/min (33 to 245 lineal ft./ min)	Water consumption 2.300···3,200 l/min Water pressure 75···115 kp/cm²		*Kollmann* (1955), p. 18—19

making use of friction forces between logs or slabs or of friction between tools and the wood. Only two hydraulic type machines are listed in Table 5.9, but there are numerous debarkers using high-velocity jets of water with velocities of 100 to 130 m/sec (300 to 400 ft./sec) against the surface of the log. An example is the Streambarker (Model D), manufactured by Allis Chalmers, Milwaukee, Wisc. It is reported that at one debarking operation, on southern pine logs, a rate of 7,850 lineal feet ($\sim$2,400 m) per hour was reached (*Johnson*, 1956, p. 112).

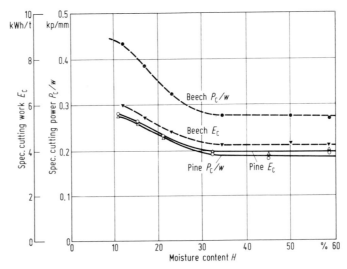

Fig. 5.22. Dependence of specific cutting work and specific cutting power on the moisture content of the wood. From *Pahlitzsch* and *Mehrdorf* (1952)

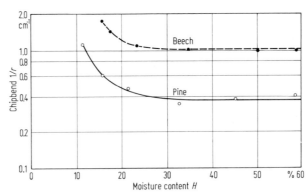

Fig. 5.23. Dependence of chip bend on the moisture content of the wood. From *Pahlitzsch* and *Mehrdorf* (1962)

Such huge installations are not yet used in the particleboard industry, but are used in the pulp industry. It may be that they will become of interest for the particleboard plants, especially if these are attached to pulp mills.

The friction log barker, type Waterous, as shown in Fig. 5.24, is a machine developed in Canada, originally for use at large sawmills. The method of operation is as follows: Continuous movement of the logs by means of three heavy feed chains to the top of the hopper — constant trembling action — water sprayed forcefully upon the logs washes away the removed bark — decision of the operator as to whether a log may be ejected or processed again by appropriate operation of the unloading deflector.

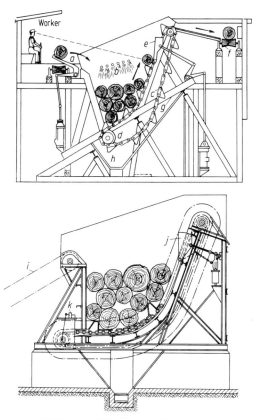

Fig. 5.24. Scheme of the Waterous friction log barker.
a Loading flipper, *b* Water spray (overhead), *c* Chains, *d* Idler wheels, *e* Driving sprocket, *f* Live discharge rolls, *g* Unloading deflector, *h* Water spray trough and bark sluice, *i* feed conveyor, *j* kickback arm, *k* lubrication.
Waterous Ltd. Brantfield, Canada

Debarking in trough (hopper) and drum machines is effected by conversion of external mechanical forces into friction between the logs (*Mitlin*, 1968, p. 54).

Fig. 5.25 illustrates a drum friction log debarker, type Paschke. As is known from experience, even dry pine pulpwood can be completely debarked within 2.5 to 3 h, partly peeled spruce pulpwood within 3 to 5 h. About 10 to 13 l water are consumed per 1 rm.

The Cambio-barker (Fig. 5.26) applies friction forces directly to the log surfaces by means of blunt edge hard metal teeth which are radially set and elastically pressed against the log. Fig. 5.27 shows that, for a Cambio-barker, a nearly parabolic function exists between average log diameter and practical output (capacity). The effect of feed velocity is insignificant. From the economic point of view large log diameters cannot be afforded because they are too expensive.

Fig. 5.25. View of a drum friction log barker, type Paschke

Fig. 5.26. View on the outlet-side of two Cambio-barkers in a German particleboard plant

More details about debarking machines supplementary to Table 5.9 have been published by *Smith* (1948), *Bott* (1955), *Johnson* (1956, p. 98—114) and *Kollmann* (1966, p. 101—112).

As a fuel, bark is no considerable gain for the boiler house because its calorific value varies only from 1,000 kcal/kg to 2,500 kcal/kg (1,800 BTU/lb. to 4,500 BTU/lb.). *Mitlin* (1968) quotes a wider range from 1,100 kcal/kg to 3,900 kcal/kg (~2,000 BTU/lb. to 7,000 BTU/lb.).

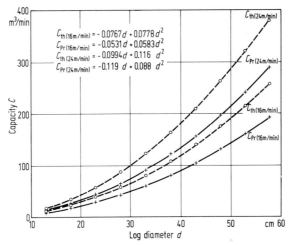

$$C_{th(16m/min)} = -0.0767d + 0.0778 d^2$$
$$C_{Pr(16m/min)} = -0.0531d + 0.0583d^2$$
$$C_{th(24m/min)} = -0.0994d + 0.116\ d^2$$
$$C_{Pr(24m/min)} = -0.119\ d + 0.088\ d^2$$

Fig. 5.27. Theoretical (th) and practical (pr) capacity C of a Cambio-barker as a function of log diameter d for feed speeds of 16 and 24 m/min respectively

5.2.1.4 Reduction of Wood[1]

5.2.1.4.1 Crosscutting, Splitting. In general flake-producing machines (for instance such equipped with magazines in which the raw material is placed) can be fed only with logs *crosscut to fixed lengths* (0.33 or 0.38, 0.50 or 0.58 and 1 to 1.16 m (~1.1 or 1.25, 1.64 or 1.90 and 3.3 to 3.8 ft.). For crosscutting, automatically working bandsaws in conjunction with chain conveyors are used. A unit may consist of one, two or three bandsaws (Fig. 5.28). If the diameters of the round logs are greater than about 24 cm (about 10 in.) the wood must be slit by means of sturdy *splitting* machines (Fig. 5.29).

5.2.1.4.2 Primary Reduction, Knife Hogs. In many cases a *primary wood reduction* is necessary. This is useful especially for industrial wood residues, limbs, veneer wastes, plywood edgings etc. The more common types of machines used for this primary reduction are *knife hogs.* The knives may be fastened either on disks (Fig. 5.30) or in a number of 2 to 36 knives on rotors (steel cylinders, Fig. 5.31). Table 5.10 gives data for two series of knife hogs as examples.

5.2.1.4.3 Secondary Reduction, Hammer-Mill Hogs, Toothed Disk-Mills, Impact Disk-Mills. Hammer-mill hogs are used as *secondary reduction* units. "The heart of

---

[1] The terminology is not yet internationally clear and standardized. In the following explanations, terms principally in accordance with *Johnson* (1956) are used: *primary reduction* (to about walnut-sized chips), *secondary reduction* (to, for example 6.5 mm ~1/4 in. × 19 mm ~3/4 in. sq. to 32 mm ~1 1/4 in. long and finer), *shaving or flake* (chip thickness 0.06 to 0.40 mm ~1/400···1/64 in.), *after-reduction* to extremely thin and fine particles for surface layers, using special mills or defibrators (defiberizers). Because the various reduction machines are made as versatile as possible there are no typical boundaries between them.

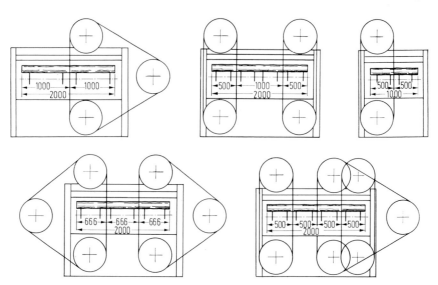

Fig. 5.28. Possible arrangements of crosscut-bandsaws for pulp and particleboard wood. (Figures are dimensioned in mm [100 mm = $3^{15}/_{16}$ in.])

Fig. 5.29 Splitting machine on the wood yard of a German particleboard plant

the shredder is the hammer, hammer drum, breaker plate, and screen bars or perforated plate" (*Johnson*, 1956, p. 118). Hourly capacity and power consumption depend on the desired degree of reduction, on wood species, on moisture content, on the size of material (for instance chips 180 to 300 mm ≈ 0.8 to 1.2 in. long) to be reduced, and on the desired plant capacity.

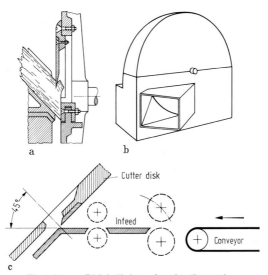

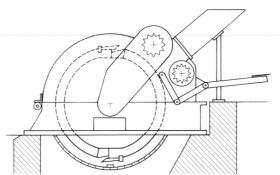

Fig. 5.30 a — c. Disk knife hogs chopping the wood.
a) Feed inclined from overhead; b) Horizontal feed, type Wigger-Hansel, Unna; c) With inclined disk, type
Gustin, Charleville-Mezières

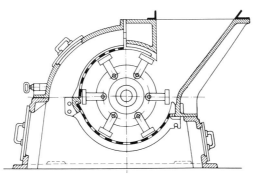

Fig. 5.31. Scheme of a rotor knife hog (type Maier)

Fig. 5.32. Schematic cross section through a hammer-mill hog ("Schlagrotor", type Maier, Brackwede)

There are many types of secondary reduction equipment manufactured in industrial countries. Fig. 5.32 shows a schematic cross section through a hammer mill. When the hammer drum (rotor) revolves at high speed (700 to 2,200 rpm), the hammers are extended from the drum by centrifugal force. A very small

Table 5.10. Knife Hog Data

Type	Diameter of rotor		Revolutions per minute	Power consumption		Approx. capacity per hour	
	mm	in.		PS (HP)	kW		
A. Sumner Iron Works, Everett, Wash., U. S. A.							
No 25	610	24	1,500	(25··· 50)	19··· 37	6 cord[1]	
No 35	860	34	1,200	(50···100)	37··· 95	10 cord	
No 45	1,310	51 1/2	900	(100···150)	75···112	16 cord	
No 65	1,530	60	600	(200···250)	150···187	20 cord	
B. Maier, Maschinenfabrik, Brackwede/Westf., W.-Germany							
00	400	15.7		5.4··· 8.2	4··· 6	300 kg[2]	600 kg
0	600	23.6		10.9···13.6	8···10	600 kg	1,000 kg
I	800	31.5		19.0···27.2	14···20	1,000 kg	2,000 kg
II	1,200	47.2		41··· 54	30···40	2,000 kg	4,000 kg
III	1,600	63		68···122	50···90	6,000 kg	12,000 kg
IV	2,000	78.6		95···164	70···120	8,000 kg	20,000 kg

[1] 1 cord = 128 cu. ft. piled wood = 5.63 m³ piled wood (rm) = 90 cu. ft. solid volume without bark = 2.55 m³ solid volume without bark (fm).   [2] 1 kg = 2.205 lb.

clearance exists between the — in many cases hardened — cutting corners of the hammers and the knives or screen bars, fixed on a heavy cast frame. By the interaction between hammers and screen bars the material is torn or shredded into small particles. The screen plates are mounted just outside and around three quarters or the lower half of the swing of the hammers.

Fig. 5.33 illustrates a commonly used two-stage wood particle preparation equipment set-up in the U.S.A. This set-up provides for the oversize chips, produced by the primary reduction, to be separated by a screen, and fed to a reshredder for final processing to the desired particle size. The output from the reshredder or secondary reduction unit is again screened. If oversized particles still exist, the screen sends them back to the reshredder. (*Johnson*, 1956, p. 123). Only a few limiting data for Williams-hogs and reshredders are listed in Table 5.11.

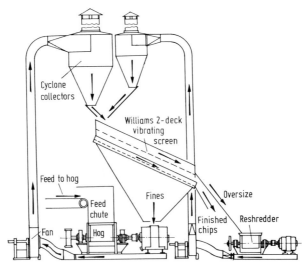

Fig. 5.33. Scheme of a two-stage wood particle preparation

Table 5.11. Limiting Data for Williams-Hogs and Reshredders

Type	Capacity	Revolutions	Power consumption	
	tons per hour	per minute	HP	kW
"No-Knife" hogs	1/3···7	1,800 → 750	20···250	15···187
Chip reshredders	1···6	3,600 → 1,200	40···150	30···112

The hammer-mills may be classified as one type of sieve-mills (5.34). The principal difference between sieve-mills and disk-mills is that the first-mentioned machines have an ample free grinding space which may not be completely charged because in this case the mill becomes blocked, whereas disk-mills have a restricted grinding space. Feeding may be practised either centrally (for instance in wing beater mills (Alpine, Condux)) or peripherally (for instance in hammer-mills).

A type of reduction unit that has largely been used for the manufacture of chipboard in W.-Germany is the Condux-toothed mill which belongs to the family of "disk-mills" (Fig. 5.35). Fig. 5.35a shows the schematic diagram of a Condux-mill. It consists of two large grinding disks with rather blunt teeth. The teeth in the unit decrease in size toward the outside of the machine, and the wood is

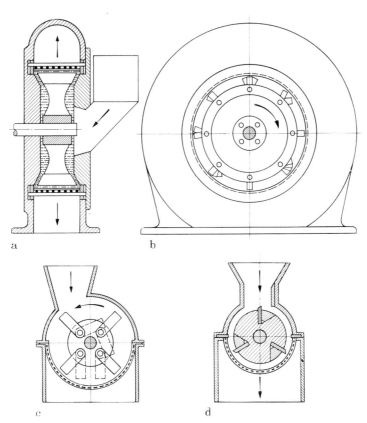

Fig. 5.34a—d. Scheme of sieve-mills. a) Wing beater mill; b) Hammer basket screen mill; c) Hammer-mill; d) Cutting mill

ground progressively finer as it approaches the periphery. The disks lie in horizontal planes. The wood (chips) is fed at the top of the upper disk. Since the lower disk rotates at high angular velocity, the particles are moved by centrifugal forces toward the outer edges of the disks. When the particles have been reduced to the proper size they are dispensed into a trough where they are finally discharged into a conveying system (*Kollmann*, 1955). The particles produced are relatively long and uniform.

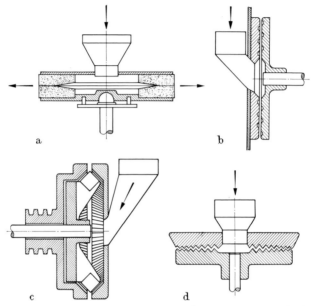

Fig. 5.35a—d. Scheme of disk-mills. a) Toothed disk-mill (Condux); b) Defiberizing mill, Condux, Wolfgang near Hanau; c) Impact disk-mill, Pallmann, Zweibrücken

The various types of mills are very versatile by proper choice of speed (rpm), of grinding tools and other adjustment possibilities. Species, size and conditions of woods to be reduced may vary considerably. Tests are informative. The effect of the moisture content of the wood should be realized.

High moisture content is beneficial for the quality of the chips to be produced (less fractures across the grain) but the throughput of the mill is considerably reduced (Fig. 5.36).

Kull (in *Kollmann*, 1966, p. 163) quotes for the choice of the particular mill types the following recommendations of the firm Condux:

a) Wing beater mills: For after-reduction of flakes to be used as face and core layers;

b) Hammer-mills: For veneer waste, long splinters, shavings, chips from planing operations;

c) Knife-mills: For after-reduction of flakes;

d) Hammer basket screen mills: For fine grinding or pulverizing wood or cork;

e) Toothed disk-mills: Especially for moist or wet material (shavings from peeling round logs);

f) Cutter ring mills (chippers): For the production of uniform, thin and fine chips from coarser chopped chips.

5.2.1.4.4 Shaving or Flake Production, Disk Type Chippers, Cutter Spindle and Cutter Cylinder (Cutter Head) Chippers. Originally only industrial residues in the form of chips (sawdust, shavings from peeling or planing etc.) were taken into consideration for the manufacture of "artificial board" (cf. Section 5.0). Even later after the first misfortunes to produce useful and firm particleboard the utilization of industrial residues predominantly was intended.

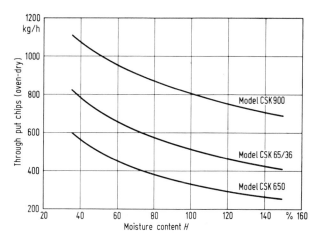

Fig. 5.36. Throughput of Condux wing beater mills in dependence on moisture content of the chips. The figures are valid for the after-reduction of face layer chips, 0.2 mm (0.0079 in.) thick. The throughput is approximately doubled for core layer chips, 0.4 mm (0.0157 in.)

Is was recognized soon that the board quality depends very much on kind and size of the chips. If they originate more or less by chance they are not ideal for particleboard manufacture. "Engineered" chips are a prerequisite, and *Fahrni* (1940/43), as has been mentioned, first came to the conclusion that three-layer boards are promising: two face layers consisting of thin areal flakes produced by a special machine and a core layer consisting of milled splinters and coarse particles.

The utilization of 0.25 to 0.50 mm (0.1 to 0.2 in.) thick flakes ("scales or laminae") for the manufacture of 4 to 12 mm (1/16 to 1/2 in.) thick board (Fig. 5.37) had been proposed by *Watson* (1901), but he did not specify how to produce such flakes economically. The successful development began in the year 1943 (*Fahrni*, Novopan-process) and in 1947 (Interwood AG patent, Behr-Himmelheber method.) Fig. 5.38 shows a schematic survey of some procedures of particleboard production. In consideration of the high importance of particleboard development (technical and economic) the survey in Fig. 5.38 may be completed by the block diagrams in Fig. 5.39. Details of chipping and flake producing, processes and machines are published by *Kollmann*, 1954, 1955, 1966, *Scheibert*, 1958, and *Steiner*, 1954.

The following principal groups of flake-producing machines may be distinguished:

1. Freely Cutting Machines
The knives are freely conducted across the clamped wood. Chipping without residues is scarcely possible.

2. Cutter Spindle Machines

In cutter spindle machines the wood is fed to the periphery of the rotating cutter shaft. Counter-bars control the feeding and prevent the wood from moving along the shaft. Feeding is done (*Kull*, in: *Kollmann*, 1966, p. 149) by:

a) Intermittently working levers (Fig. 5.40);
b) Pistons;
c) Endless chain conveyors (Fig. 5.40b, c, d).

Vertical or inclined feeding of the cutter spindle is favorable and accepted in the paper industry. The machines are suitable up to log lengths of 1 m (3.5 ft.). Fig. 5.41 shows the supply line for a cutter spindle flake producing machine with inclined endless belt conveyor and inclined feeding.

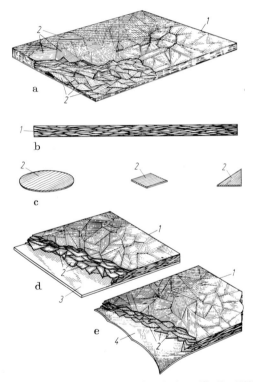

Fig. 5.37a—e. Sketches of a particleboard in *Watson's* patent specification (1901, cf. Section 5.0)

Most cutter spindle machines can be fed only with wood cut to fixed lengths. If thin and to some extent flexible wood assortments are used, special saws must be purchased, and the expenditure of labor is relatively high. "Long wood" may be chipped moving by gravity in tilted chutes or conveyed by endless belts in horizontal chutes.

The cylindrical tools have large diameters with respect to adequate internal supply. Fig. 5.42 shows some types of cutter cylinder machines. A few data are listed in Table 5.12 for one typical machine of this type. The cutting velocities are very high due to the opposite direction of rotation of the inner part and the cutter cylinder. The machine is driven by two electromotors.

Table 5.12. Data for an Impact Disk-Mill
(*Pallmann*, cf. *Kull*, in: *Kollmann*, 1966, p. 153)

Type	Pz 6	Pz 8	Pz 10	Pz 12
Capacity Flakes from chips kg/h	400/500	800	1,200	2,000···2,500
lb./h	880/1,100	1,760	2,650	4,350···5,560
Number of knives	20	26	42	52
Power consumption kW	12 + 24	15 + 30	20 + 40	25 + 55
PS	26 + 35	20 + 41	27 + 54	34 + 75
Dimensions of flakes	0.4···1.5 mm thickness, about 2 cm length (0.016···0.06 in., 0.84 in.)			

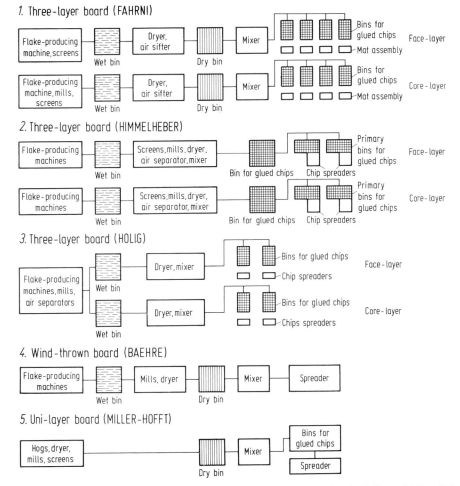

Fig. 5.38. Schematic survey on procedures of particleboard production. From *Steiner*, in *Kollmann* (1966, p. 221)

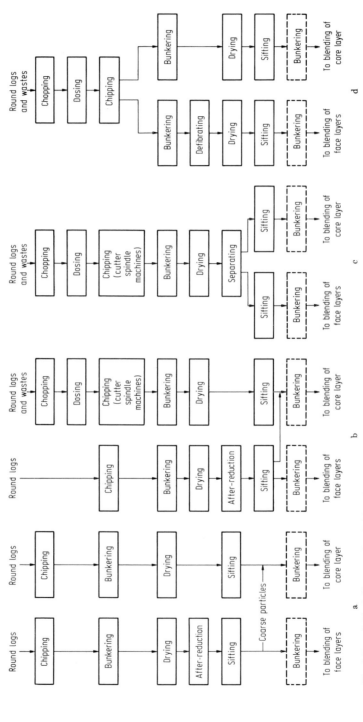

Fig. 5.39a—d. Block diagrams of particleboard manufacture. a) Chipping separated for face layers and core layer; b) Chipping of material for face layers, chopping for core layer; c) Chopping of total material; d) Chopping of total material, defibration for face layers. From *Trutter* and *Himmelheber* (1970)

3. Disk-Type Chippers

Disk-type flake-producing machines consist mainly of a large rotating disk mounted in a vertical or horizontal plane. The disks contain slots for the setting of knives, and these slots are placed radially in the disk. Adjustable scoring knives or teeth controlling the length of the flakes are also radially placed in the disks. In most instances the raw material must be cut to a maximum size in order to fit into the feeding mechanism of the machine. The increasing distance of the knives from the center of the driving shaft causes differences of rotational speed ($w = (n\pi r)/30$, where $\pi = 3.14$, $n =$ number of revolutions per min, $r =$ radius or distance of the knives from the center of the driving shaft and w has the dimension m/sec (1 m/sec $\approx$ 197 ft./min). These differences may influence the specific cutting capacity as well as (probably) the cutting quality. Practice, however, does not verify the latter assumption. The various feeding possibilities are illustrated in Fig. 5.43: pistons or endless conveyors, arranged horizontally, inclined or vertically.

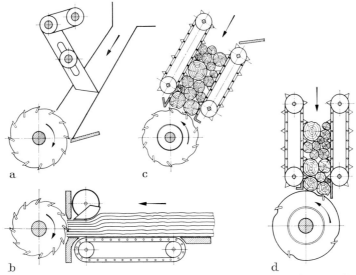

a   c

b   d

Fig. 5.40a—d. Cutter spindle chippers. a) Intermittently working, Rottmann, Wilhelmshaven, b) Continuous horizontal feeding of veneers (Rottmann), c) Inclined continuous feeding, Hombak, Bad Kreuznach, d) Vertical continuous feeding, Ortmann

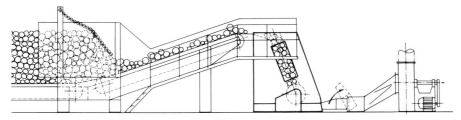

Fig. 5.41. Supply line of round logs in a particleboard plant, Bezner, Ravensburg

The capacity of the machines depends very much on the following conditions:

a) Thickness of flakes to be produced;
b) Wood species (density);
c) Shape of wood;

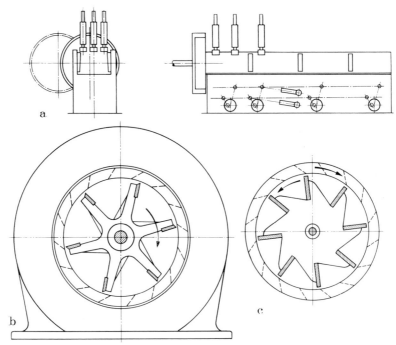

Fig. 5.42a – c. Cutter cylinder chippers. a) Universal and veneer flake producing machine equipped with cup wheel and conveyor for long wood; Krenzler, system Behr-Himmelheber, b) Cutter ring chipper with fixed knife holder, Condux, Wolfgang near Hanau, c) Impact disk mill with double drive, Pallmann, Zweibrücken.

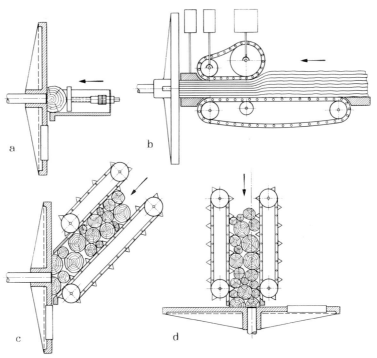

Fig. 5.43a – d. Disk-type chippers. a) Disk-planer with intermittent feeding, Fischer, system Novopan; b) Vertical disk with continuous feeding for veneer, Anton, Flensburg; c) Vertical disk with oblique continuous feeding, Wigger, Unna; d) Horizontal disk with vertical continuous feeding, Bezner, system Himmelheber

d) Condition of raw material (moisture content, cleanness);

e) Tool edges (sharpness);

f) Well organized wood supply;

g) Maintenance of the flake-producing machines;

h) Well-trained personnel;

i) Particular conditions of market.

It is not possible to give in this chapter a complete survey of chippers. In addition to the foregoing explanations and especially to the sketches in Fig. 5.34, 5.35, 5.40, 5.42 and 5.43 only a few machines are mentioned.

1. Cutter Spindle and Cutter Cylinder Chipper

a) Double acting, automatic, flake-producing machine, vertical ring-magazine (external diameter, 2,500 mm ∼ 8.2 ft.) with 9 receptacles for the wood, 0.13 to 0.8 rpm. Radial strutting pistons clamp the wood against the outside of the magazine. Below this magazine the logs are salient 120 mm (∼ 4 3/4 in.) and are chipped by means of two knife-turbines (diameter 700 mm (∼27 9/16 in.), each turbine with 12 knives, 900 rpm, deflectors lead the chips via a cone into a guide tube, total power consumption 61.5 kW or 82 HP (Behr, Torwegge).

b) Flake-producing machine consisting in the upper portion of the inclined hopper (occasionally divided into compartments) and feeder mechanisms in the lower portion of the revolving pan (2 to 8 rpm) housing two cutter heads driven by individual motors, output varies depending on length and thickness of chips desired and raw material, maximum about 6,000 kg/h (∼13,200 lb./h), average about 2,300 kg/h (∼5,070 lb./h), power requirement up to 50 kW (67 HP) (Guilliet Mfg. Co., France).

c) Circular saw chipper for the production of rather short curly chips ("S"-type chips) with the property of dry-felting. Raw materials: Thinnings, peeler cores, coarse industrial wood residues, crosscut 95 cm (∼37 in.) length, necessary moisture content about 70% (based on oven dry weight). Pressed down by chains moved back and forth in magazine. Tool: package of about 45 to 50 wobbling circular saw blades, power consumption at full load 60 kW (∼80 HP), output 600 kg (1,330 lb.) (Homogenholzwerke).

d) Cutter spindle chipper, 3 knives, 3 rows of scoring teeth, wood block feeding grain disposed parallel to cutterknife edge, energy requirement ca. 50 HP, utilization of end trims from kiln-dried lumber (Long-Bell Lumber Co.).

e) Cutter spindle machine for the production of flakes, cutter spindle with sets of scoring knives, revolving drum as feeder, wood: length up to 1 m (∼39 in.), diameters up to 200 mm (∼8 in.), exhauster fan for flakes, output about 1,000 kg (2,205 lb.) of green chips per hour, power consumption 28 kW (38 HP) (Meyer and Schwabedissen).

f) Principle of a veneer rotary peeler for the manufacture of chips (length ca. 25 mm ∼ 1 in., width 10 mm ∼ 3/8 in., thicknesses 0.06 to 0.30 mm (∼ 1/400 to 1/80 in.) from round logs and limbs, greatest wood diameter 800 mm (∼31 in.), minimum core diameter 50 mm (∼2 in.), maximum wood length 1,070 mm (∼42 in.) and 1,250 mm (∼49 in.) (larger machine-type) respectively, 4,500 rpm, average output 700 kg (1,540 lb.) per hour (Roller, type Ferd. Schenck).

g) Cutter spindle machine for chip production from moistened split billets, slabs, edgings, boards, crosscut to machine-width, 8 knives chipping along the grain, control of chip thickness by adjustment of planing knives, of chip length by scoring knives, starting power consumption 60 kW (~82 HP), output 1,110 kg (~2,430 lb.) wet chips per hour (Rottmann).

2. Disk-Type Chippers

a) Chipping machine with cup wheel, 7 knives, circulating gear rim, receptacle for the wood, at the top open, below partly open for the band conveyor, pressure beam motor. Chips produced are shown in Fig. 5.44 (Behr, Grupp).

Fig. 5.44. Chips produced with the machine, Behr, Grupp, Oberkochen

b) Vertical disk type flake-producing machine for utilization of veneer waste which is previously cut to a maximum length of 300 mm (~1 ft.) and bundled is placed upon a conveyor belt with grain direction across the movement direction of conveyor, clamping mechanism, circular saw, cutting disk chipping parallel to fiber grain, chip dimensions: length 25 to 200 mm (~1 to 8 in.), thicknesses 0.25 to 0.39 mm (~0.01 to 0.015 in.), width equal to veneer thickness. Output (Douglas fir veneer) about 1,000 kg (2,205 lb.) per hour (Elmendorf Research Inc., Palo Alto, California, U. S. A.).

c) Vertical disk-type flake-producing machine, great cutter disk, diameter 2,650 mm (~8 ft. 8 in.), 150 rpm, two rotating feeding and working chambers for wood lengths of 500 mm (19.7 in.), wood held by a hydraulically operated ram against the knife disk, knives and scoring knives easily accessible, output approximately 1,000 kg (2,205 lb.) of chips per hour, power requirement about 26 kW (~35 HP) (Ortmann).

5.2.1.4.5 After-Reduction. Though there exists a great variety of machines for the manufacture of chips, even in the case of perfect adjustment, it is not possible to produce a chip material suitable for exacting processes of particleboard production. Therefore after-reduction (called also after-chipping or chip improvement) is necessary in order to obtain chips with the desired properties. The usual shaving or flake-producing machines are not fit for purpose. Special sieve mills or toothed disk mills are necessary for further reduction; if the thickness also has to be diminished re-shredders or defibrators (defiberizers) are employed. In its properties the material is quite different from cut flakes but if it is fine enough to be added in small amounts to the flakes without problems. As Fig. 5.33 shows over-size particles are separated and reshredded. Some reduction is effected casually and without control in the pneumatic conveyors. In spite of the strong correlation between mill capacity and moisture content (cf. Fig. 5.36) the after-reduction usually is carried out prior to drying because very dry chips are liable to fractures across the grain and more dust results. Table 5.13 lists some data for mills used for after-reduction.

Table 5.13. Data for After-Reduction Mills

Type	Diameter		Revolutions	Power consumption		Remarks
	mm	in.	per minute	PS	kW	
Condux-Werk, Wolfgang near Hanau, W.-Germany						
CSK 650	600	25.6	1,850	61	45	Sieve perfo-
65/36	650	25.6	1,850	75··· 82	55···60	ration for face
900	900	35.4	1,450	109···122	80···90	layers, 6 mm
						cross slits
Alpine AG, Augsburg, W.-Germany						
40/32 K	400	15.7	2,800	6.8···13.6	5···10	Core layers
63/50	630	24.8	1,800	16.3···34	12···25	30···40 mm
80/63	800	31.5	1,400	27.2···47.6	20···35	square holes
100/100 L	1,000	39.4	800···1,000	40.8···85.6	30···63	

An interesting machine, in principle related to the circular saw chipper (No. 1c in the foregoing Section 5.2.1.4.4), is the defiberizer (Allis-Chalmers) designed to reduce wood into fine, wooly fiber bundles. Its two hoppers reciprocate over two high speed combined heads which contain a succession of steel needles so spinned that centrifugal force causes them to swing outward radially. These card the surface of the lowest wood bolt in the hopper, reducing it to fiber — at a power consumption of about 60 kW/ton. Fig. 5.45 shows a scheme of this defiberizer.

Chipping consequently is not the only possible method to produce useful particles for the manufacture of particleboards. *Fahrni* (1940/43) recognized the favorable influence of cut flakes in the face layers on board quality, especially surface smoothness and bending strength. It was his intention to produce thin decklayer chips, but it is not possible to cut thinner than about 0.2 mm($\sim$5/64 in.).

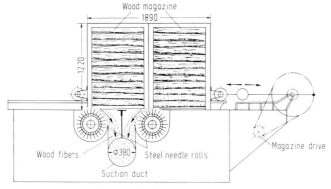

Fig. 5.45. Scheme of the Allis-Chalmers defiberizer

Contrarily the thickness of fibers and fiber bundles for the production of fiberboard is in the range from 0.01 mm to 0.1 m ($\sim$0.0004 to 0.004 in.).

The theoretical consideration in utilizing fibers, at least for the deck layers of particleboard, is very difficult because in practice:

a) Dry ligno-cellulose fibers tend very much to felting and clogging;
b) Addition of binder with the usual machines for chips is nearly impossible;
c) Mat laying is much more difficult.

The development of new or additional machines would have raised the costs of fiber preparation and thereby the costs of the end products. Recently these difficulties were considerably reduced because it became evident that shorter and less finer fiber material is sufficient for a remarkable improvement of the surface quality of particleboard. Normal machine units can be used after some modifications.

For the production of such short fiber material sawdust is very appropriate. The term "chipfiber material" (Spanfaserstoff) has been introduced into the German literature (*Himmelheber* and *Kull*, 1969). The production of the chipfiber material is similar to that of fiber material but thermal treatment is not necessary. An advantage of the cold defibration is the production of short fiber bundles.

Disk-type mills are mainly used for the defibration, they were earlier described and illustrated (Fig. 5.35). A view of a "refiner" for the manufacture of chipfiber material is shown in Fig. 5.46, of the grinding disks of another defibrator in Fig. 5.47. Finest particles or chips may be produced without much dust and with relatively low power consumption by means of special disk mills or sieve mills (Fig. 5.35 and 5.34). Most important is the proper choice of the sieve. An example is shown in Fig. 5.48. The main reduction results in the first milling zone whereas in the adjacent sieving zone some after-reduction and sifting occur. The "ratio of slenderness" (length to thickness) of the individual particles decreases because their length is reduced but their thickness is kept almost unchanged. The moisture content of the chips prior to the after-reduction should range between 40 and 60% (based on oven dry weight); if the moisture content is higher the power consumption remarkably increases and clogging of the sieves may happen (Fig. 5.49). For too dry chips the output increases but the particle quality worsens. Deck-layers consisting of very fine particles have a somewhat higher density than those consisting of conventional shavings or flakes.

Recently the sanding dust from chipboard finishing, formerly considered as a troublesome waste, became a useful material in special processes for face layers (*Moralt*, BP 1,097,656, 16. 3. 1959, and BP 1,453,397, 30. 3. 1961) or blended (10 to 20% in weight) with chipfiber material for the manufacture of entire chipboards (*Himmelheber*, BP 1,198,539, 23. 4. 1958. BP 1,183,240, 29. 12. 1962). Perhaps the cheapest way to improve board quality is proper size separation by screening or sifting (cf. Section 5.2.3).

The development of particleboard production with respect to their structure led from unilayer and three-layer board to multilayer board and to graded (graduated) density board. Fig. 5.50 illustrates the variation of the density within the thickness of graded board. The production of graded density board is a problem of mat-laying and will be dealt with in Section 5.2.8. Fig. 5.50 clearly shows that graded density board may have a relatively low density in the middle of the core and a rather high one near the surfaces and in the outer zones. Such boards may have the same average density, but the board manufacturer can adjust the density distribution according to the wishes of the consumers. If the faces have a higher

Fig. 5.46. Asplund defibrator, type LVP 36, fiberizing unit, for the production of fiber material for board surface

Fig. 5.47. Grinding disks of a defiberizer, Voith-Kunz-Hombak

density the corresponding average bending strength is higher. Some other advantages are: The close, nearly even surface of the board is suitable for veneering and painting. The resistance against absorption and swelling is higher, as is the resistance against spread of flame. *Lynam* (1968) concludes: "There are of course practical limits dictated by certain standards of shear strength, delamination strength and screw holding power."

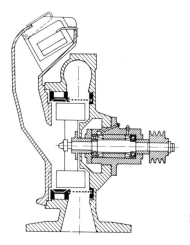

Fig. 5.48. Reduction zone and sieve zone in a special sieve mill, Ultraplex-cross stream mill, Pallmann, Zweibrücken

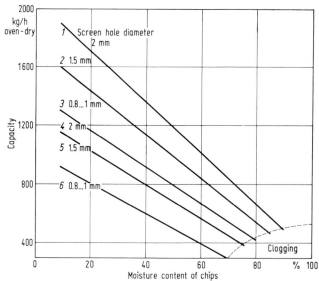

Fig. 5.49. Capacity versus moisture content for various types of chips and screen holes of an Ultraplex chipper, Alpine, Augsburg

5.2.1.4.6 Summary. Mitlin (1968) divides the chippers into four broad classes as follows:

a) Chipping machines for production of surface or core layer chips suitable for operation with round logs and high grade wood residues such as thinnings, edgings, trimmings, and peeler cores; chip thickness and length can be controlled by setting the cutters:

b) Coarse chippers for use with low class waste for conversion into core chips;

c) Desintegrators for the production of chips of different types and shapes from low class waste;

d) Mills for grinding residues of a planer mill or other shavings and for reduction of chips from class a) machines.

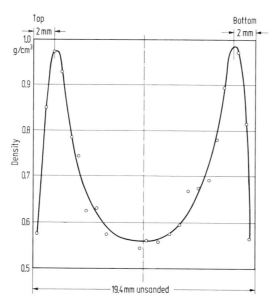

Fig. 5.50. Density versus board thickness of graded particleboard. From *Plath* (1973)

Though cutters are set for a desired chip thickness, in practice considerable variations around an average are obtained, generally with a normal binomial distribution as shown in Fig. 5.51 (*Mitlin*, 1968).

The development of machines for preparing particles has been slow, but there is a need for high capacity machines of the cutter type capable of handling small

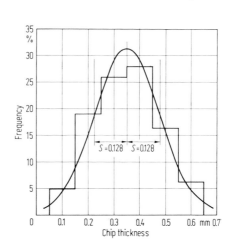

Fig. 5.51. Chip thickness distribution (Step diagram according to *Mitlin*, 1968) bell-shaped curve. From *Kollmann* (1972)

pieces of wood (FAO, 1958/59, p. 69). The output of any machine is influenced by the nature of the raw material and the type of particles desired. Machines in use at present range in capacity from about 400 to 2,000 kg/h (880 to 4,400 lb./h) based on dry weight.

5.2.2 Drying of Chips

5.2.2.0 General Considerations. Drying is a very important stage in the manufacture of particleboard (*Kollmann*, 1952a, 1952b; *Mitlin*, 1968, p. 82—87; *Klamroth* and *Hackel*, 1971). The continuously increasing quality requirements of particleboard are one reason for the trend to smaller tolerances of final moisture contents. Furthermore, the economy and security of chip dryers has been increased. The industry developed in the course of the years many types of chip dryers. As mentioned in Section 5.1.1.1 a great variety of wood sizes and wood species, raw wood and industrial residues are reduced to chips.

Particles fed into the dryers have different moisture contents, occasionally only 16%, normally between about 35 and 120%, based on oven dry weight. The moisture content can frequently vary within a short space of time. The thicknesses of suitable particles for board manufacture may be as a rule between 0.04 mm and 1 mm (0.0016 and 0.04 in.) and more. Classification of the chips is necessary. A modern dryer with its most important data, e.g. temperature and drying time of chips in the machine, has to be quickly adapted to the initial moisture content. Only in such a way high waste can be avoided.

The moisture content after drying usually should be between 3 to 6%, occasionally it varies between 5 and 12%. It depends mainly upon type and amount of the binder and upon the degree of wetting of the surface layers prior to pressing. Too wet chips cause steam blisters in the board core during hot pressing. The result is an adverse influence on the development of adequate strength. Especially the strength perpendicular to the board surface is extremely low. Too wet chips also need longer times of pressing and reduce the production capacity. If the chips are too dry, the following disadvantages exist:

a) Risk of fire in the dryer;

b) Dangerous electrostatic charging of the tubes if the chips are pneumatically conveyed;

c) Disturbing development of dust in the whole plant;

d) The edges of the board prior to sizing have the tendency to crumble;

e) The very light, over-dried chips are blown away from the surfaces when the hot press is closed. Chips that are either too wet or too dry also influence the behavior of the additive and lead to longer press times for different reasons.

The differences in the quality and behavior of chips are less than before because highly efficient chipping machines are available; nevertheless, there frequently appear splinters, flakes, fines and dust.

The drying process depends on wood species and chip thicknesses for the same drying conditions. Fig. 5.52 shows the effect of wood species and initial moisture content on drying time for convection drying (*Klamroth* and *Hackel*, 1971). For usual final moisture content, chip thickness and normal drying conditions softwood chips need a drying time of approximately 100 sec, hardwood chips about 200 sec. Generally the drying conditions are determined by the supply of heat and the heat transmission to the chips. The heat can be transferred to the chips by direct contact with the heated plates, tubes etc., to a smaller degree by convection with the aid of the gaseous drying medium, or by radiation, and finally by a combination of all these transmission types. Contact drying needs the longest

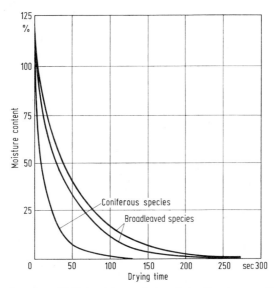

Fig. 5.52. Effect of wood species and initial moisture content on drying time for convection drying. From *Klamroth* and *Hackel* (1971)

drying time, convection drying has remarkably shorter drying times. The effect of chip thickness on drying time for two final moisture contents by convection drying is illustrated in Fig. 5.53. Fig. 5.54 shows a typical drying rate curve, obtained, when air of constant properties passed uniformly over a drying solid. The line OA respresents the period of the drying where the rate is falling in a nearly proportional manner. The line AB shows approximately the still decreasing falling rate, but at

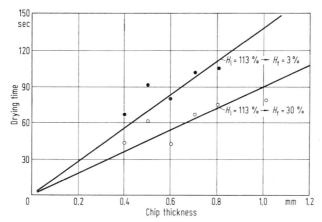

Fig. 5.53. Drying time versus chip thickness for the same initial moisture content ($H_i = 113\%$) and two final moisture contents ($H_f = 3$ and 30% respectively). From *Klamroth* and *Hackel* (1971)

point B the critical moisture content is reached, and the horizontal line BC represents the part of the drying process with constant drying rate. The moisture is now diffusing from the inside to the surfaces of the wood at a rate not less than the rate of evaporation from the surfaces. Similar curves were published by *Kollmann* and *Schneider* (1960) and *Kamei*, taken from *Krischer* and *Kröll* (1963) (Fig. 5.55).

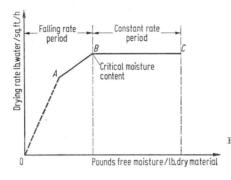

Fig. 5.54. Typical drying rate curve, schematically. (After *Mitlin*, 1968)

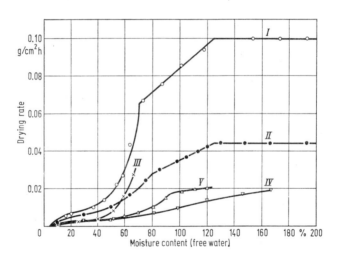

	Curve	Rel. humidity %	Air velocity m/sec.	Air temperature °C
Pine	*I*	40	6.14	40
Pine	*II*	40	0.63	40
Cypress	*III*	20	1.42	30
Cypress	*IV*	60	1.42	30
Cypress	*V*	60	1.42	40

Fig. 5.55. Drying rate curves as affected by the content of free moisture. From *Kamei* after *Krischer* and *Kröll* (1963, p. 267).

Such drying rate curves show how the drying is affected by the moisture content. For very high moisture contents the drying rate is not always constant. For example *Kamei* pointed out that cypress wood may show quite different relationships. Such curves may be observed as typical in a normal sense for normal drying conditions. Not only wood species, but also size of specimens, air velocity and air temperature have an influence on the drying rate which at present cannot be calculated in a reliable method. The moisture content expressed as the weight of free water, based on the weight of oven dry substance, reaches high values. In laboratory experiments the water content of the samples to be dried and investigated is increased by soaking in a high vacuum. Nevertheless, a simple calculation is possible using for the content of free water H_K the term:

$$H_K = \frac{1.5 - \varrho_0}{1.5 \cdot \varrho_0} \tag{5.1}$$

where ϱ = density of oven dry wood (*Kollmann*, 1936, p. 63/64).

If, for example, we take $\varrho_0 = 0.49$ g/cm³ for the average oven-dry density of European pinewood, we can calculate $H_K \approx 1.38$ or 138%. The corresponding calculation for European red beech wood with an oven-dry density of $\varrho_0 = 0.68$ g/cm³ gives the maximum content of free water $H_K = 0.80$ or 80%. The densities of wood species, of course, vary.

The drying process is influenced by the gradient of temperature from the drying agent to the chips, by the gradient of water vapor pressure between chips and drying agent, by the time in which the chips are exposed to the drying conditions, and finally by the possibilities of chemical reactions between wood and drying agent (with respect to fire risk). The shorter the drying times, the higher the drying temperatures can be chosen. A logical consequence is that for high oxygen content and low water vapor content of the drying agent the drying temperatures must be kept lower; consequently, the drying times will increase.

5.2.2.1 Dryer Types. Chip drying is easier than drying of solid wood (planks, boards and other large-sized pieces), it is easier than drying of fiberboard. Particles for the manufacture of particleboard are small (walnut-sized, even smaller), and disintegrated. Therefore, the drying times are relatively short.

There exists a great variety of wood chip dryers but the following classification is adequate:

a) Rotary drum dryers, Tube-bundle dryers;
b) Multiple band dryers;
c) Contact dryers;
d) Turbo-dryers;
e) Burned waste gas stream dryers;
f) Suspension-type dryers.

The following explanations are brief. More details are published by *Kröll* (1950), *Kollmann* (1952a, 1952b, 1955, p. 524—529, 1966, p. 165—199), *Johnson* (1956, p. 145—153), *Mitlin* (1968, p. 82—87), *Klamroth* and *Hackel* (1971).

a) Rotary drum dryers, Tube-bundle dryers, Rotary jet dryers. At the beginning of the particleboard industry mainly rotary drum dryers, best known from other industries, were used for chip drying. There are many kinds of rotary dryers, but most widely used were those of the direct type (Fig. 5.56).

In this type of dryer, the drum contains vanes or flights attached to the inner periphery (Fig. 5.57). As the drum rotates these flights lift the chips and sift them down through the stream of hot gases that pass axially through the drum (*Johnson*, 1956). The construction is such that the chips fall from one vane to a lower one and they move in a lengthwise direction toward the discharge end. Usually an automatic oil burner is the medium for producing the hot gases for drying. How-

ever, coal, gas or a mixture of sawdust from the particleboard manufacture with
oil or gas can be used as fuel. The drying time is relatively long (about 30 min).
Hence it follows that the drum contains a rather high amount of combustible
material. Another disadvantage of the one way-drum dryer is that the drum due
to the vanes cannot be properly inspected and that its cleaning, especially after
extinguishing procedures, is not easy.

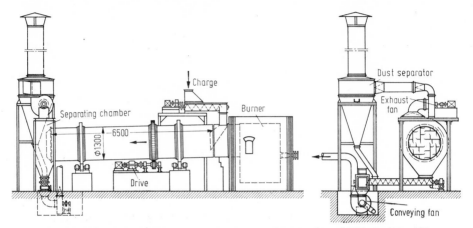

Fig. 5.56. Schematic side and front view of a rotary drum dryer for chips. Schilde, Bad Hersfeld

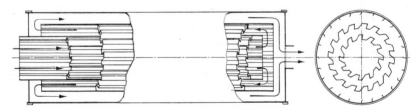

Fig. 5.57. Various vane or flight arrangements inside a rotary chip dryer. From *Johnson* (1956) after American
Society of Mechanical Engineers

A variance of the drum dryer is the three-way drum dryer (Fig. 5.58), well
known from the drying of foodstuffs and appropriate for chip drying. The principle of parallel movement of drying agent and drying good is maintained, but
the pneumatic component of the good transport is increased by aerodynamic
improvements. So the drying time can be reduced to about 8 to 20 min. The three
way-drum dryer (circuit of three ways in series, shortening the total length of the

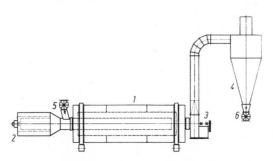

Fig. 5.58. Scheme of a three-way drum dryer
for chips

dryer) is especially suited for chips of wood residues with a low moisture content. It is widely used in the American particleboard industry. The tube-bundle dryers are simple, efficient and reliable. The specific heat consumption is relatively low.

Ponndorf constructed a horizontal tube-bundle dryer used in very many particleboard plants. The working method is illustrated in Fig. 5.59. The principal features are:

1. Steel sheet case with the trough a in which the bundle b of jointless tubes revolves. The tube-bundle forms with the tube-plates c and the headstocks d a recently welded unity. To this unit also the hollow pivots e belong, which rotate in the bearings f.

2. Heating medium is mostly hot water but also steam or waste gases from an oil burner are used (heat entry left side of the sketch);

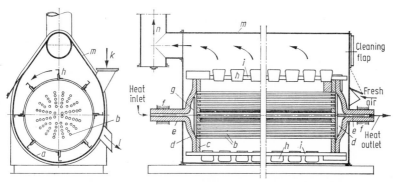

Fig. 5.59. Schematic cross and longitudinal section through a tube-bundle dryer. Ponndorf, Kassel

a Drying chamber, b Tube bundle, c Tube plate, d Head stock, e Hollow pivots, f Bearings, g Heat exchanger, h Radial blades (wings), i Blade carrier, k Chip inlet, l Chip outlet, m Hood, n Exhaust chimney

3. Exchanger g which distributes the heat equally to all tubes. Low specific heat consumption, on the average 750 kcal/kg water evaporated (1,350 BTU/lb.). Exit of the left heating medium through the hollow pivot e on the right side of the sketch.

4. On the periphery of the tube-bundle b rows of radial blades (wings) h are fixed with a slight twist on a cylindrical support fastened on both headstocks d. The effect of these blades is not only a stirring but also a whirling movement which slowly conveys the chips from the inlet k on the front of the dryer to the outlet l behind the dryer. Because the chips are moved pneumatically, rotary valves in the in-feed ducts prevent the admittance of cold air into the dryer.

5. The trough a is closed by a hood m which is well insulated against heat losses by means of glass wool and fiber-hardboard. On the top of the hood the exhaust chimney n is fastened.

6. Specialities: There are three windows behind flaps for observation and regulation of the chip flow; preheating of the fresh air up to 80 to 90 °C (176 to 194 °F) if the entering chips are very wet; generally the tube bundles are driven by an electromotor connected with a gear or a chain wheel; possible stepless regulation of rpm in the range of $\pm50\%$ in order to adapt the dryer output to the initial moisture content of the chips; low power consumption 5.5 kW ($\sim$7.4 HP) of for a dryer with about 7 m ($\sim$23 ft.) length; delivery of the dryer ready

for service set-up without foundation on a frame of profile irons; practically fire-proof because there is no contact between drying medium and chips to be dried, nevertheless automatic sprinklers controlled by a thermostat may be built into the hood.

7. In double-dryers two parallel-directed tube bundles are in the case. Uniform charging of both tube bundles is secured by means of a "pants-like" device with an electrically steered distributing flap.

8. The desired output (capacity) is a more or less hyperbolic function of the initial moisture content of the chips. Ponndorf offers a great variety of such dryers.

Similar types of tube-bundle dryers are constructed by Schilde (Fig. 5.60) and (even as twin types) by Tromag (Fig. 5.61).

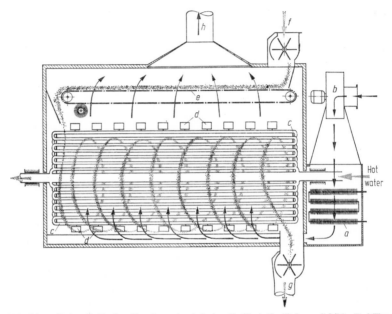

Fig. 5.60. Schematic longitudinal section through a tube-bundle ("rotation") dryer. Schilde, Bad Hersfeld

Recently rotary jet tube dryers came into fashion. The working scheme is illustrated in Fig. 5.62. The chips are moved spirally along the axis of the tube, adjustable baffles permit the control of the residence time and built-in rakes avoid aggregation.

b) Multiple band dryers. In the early years of the chipboard industry multiple band dryers (for instance equipped with 3 or 5 bands and in principle known from veneer drying) were used. These dryer types were well suited for smaller plants — power consumption between about 10 and 34 kW (~13.6 and 46.2 HP), specific heat consumption per 1 kg (2.2 lb.) evaporated water 820 to 960 kcal, but three disadvantages existed:

1. The mesh belt conveyors tended to trouble: tearing, entangling of the sieve netting;

2. The dryers contain relatively large amounts of chips, and therefore in the case of fire this can become uncontrollable;

3. The capacity of multiple band dryers is limited because the dimensions of bands and temperatures applied (length about 10,000 mm ∼33 ft., width 2,760 mm ∼9 ft., height 3,300 mm ∼11 ft., drying temperature 120 °C ∼248 °F) cannot be increased.

These are the reasons why other more efficient dryers for chips are used.

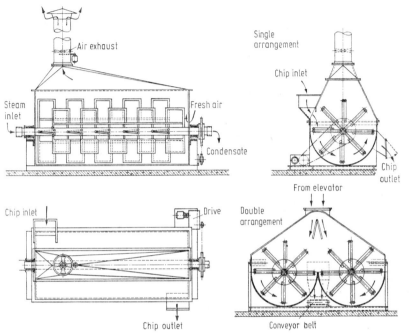

Fig. 5.61. Schematic longitudinal and cross section through a tube-bundle dryer with shovels; above: single row, below: double row. Tromag, Kassel

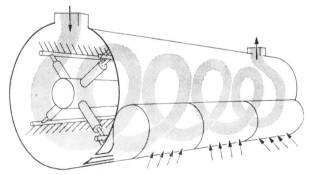

Fig. 5.62. Scheme of chip flow in a jet-tube dryer. From *Klamroth* and *Hackel* (1971)

c) *Contact dryers*. Fig. 5.63 shows a contact dryer (Hildebrand). The material to be dried enters the upper side of the dryer trough, a cone-shaped inlet, and falls on the top vibrating plate. The chips are then moved forward by the vibration of the plate. The speed of vibration can be regulated. At the end of the first plate, the chips fall onto an underlying plate and a reversal of direction is effected. In this manner, the chips move back and forth through the dryer while the flow of the hot air over moving layers of chips is both concurrent and countercurrent.

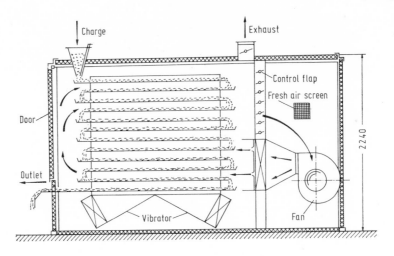

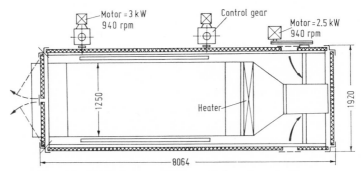

Fig. 5.63. Scheme of a chip contact dryer. Hildebrand, Oberboihingen

Advantages listed by the manufacturer:

1. Simplicity in close design and use of good heat insulation;

2. Heat requirement 950 kcal/kg (1,710 BTU/lb.) water evaporated;

3. Slight power requirements (only one ventilating fan, small motion force necessary for the vibration;

4. Easy access to the dryer by three doors, visual checking of the drying operation through two windows;

5. Complete and uniform drying as a result of continuous tumbling of the particles.

d) *Turbo-dryers.* Fig. 5.64 shows two sections and the segments of a turbo-dryer (Büttner). This dryer is cylindrical in shape, and consists of heating elements (sources of heat are steam, hot water or exhaust gases), turbines to produce the flow, circular shelves consisting of tiltable plates and driving motors for the turbines and the shelves (0.1 to 1.0 rpm). "When the shelf has completed almost one full revolution, the individual plates are tilted, one at a time. This allows the chips on the plate to fall to the shelf immediately below. The procedure is repeated, from shelf to shelf, until the chips have been revolved on the lowest shelf in the dryer. They are then discharged from the machine. The dumping of the chips by the tilting plates serves to thoroughly mix and expose them to the hot air current which circulates constantly through the unit" (*Johnson,* 1956, after *Kollmann* 1955, p. 526).

Economic data: Average temperature (steam heating) inside the dryer is approximately 100 °C (212 °F), exit temperature about 90 °C (194 °F). For hot water heating the temperature just before the entry of the chips is about 180 °C (356 °F), near the discharge outlet 160 °C (320 °F). Heat requirement when using an especially efficient system of hot water is about 900 kcal/kg (1,620 BTU/lb.) of moisture removed from the chips. When steam is used, the requirement is 1.8 kg (lb.) of steam per kg (lb.) of moisture evaporated . The depth of chips on each shelf ranges from 40 to 50 mm (∼1.6 to 2 in.), the drying time is between 15 and 45 min, depending on species of wood, type of chips, moisture content of the wood, and temperature applied.

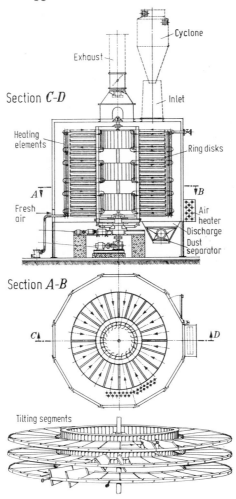

Fig. 5.64. Schematic view of a turbo-dryer. Büttner, Krefeld

e) *Waste-gas stream dryers.* Waste-gas stream dryers (Fig. 5.65) are designed for higher outputs (Schilde). The figure shows the method of operation: Feeding of material to be dried *m* — hot gases produced in combustion chamber *a* — transport of the chips in conducting pipe for gas *d* — partial separation of chips, a

portion of fines is conveyed through pipes by the waste gas g — cyclone separators
for fines f — heavier chips (minus the waste gas and fines) are conveyed through
pipes h — at the end of three cycles the chips are dried and free of fines. It
is evident that this dryer — burnt and no more established — had the following
characteristics and advantages:

1. Efficient drying as well as removal of fines;
2. High production rate;
3. Heat requirement about 900 kcal/kg (1,620 BTU/lb.) of evaporated water;
4. Power supply about 60 kW (82 HP), oil consumption approximately 180 kg/h
(397 lb./h);
5. Drying time about 20 to 25 sec to $H \approx 9\%$.

Recently another German firm (Overhoff & Altmayer) constructed a multiple-
stage waste-gas stream dryer (*Kollmann*, 1966, p. 185).

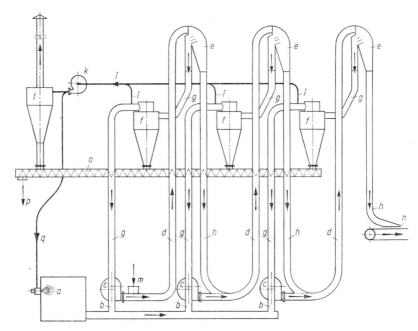

Fig. 5.65. Working scheme of a waste-gas stream dryer.

a Combustion chamber, *b* Hot gas ducts, *c* Fan, *d* Uptake, *e* Deflection hoods, *f* Cyclone, *g* Gas return, *h* Down
pipe, *i* Waste-gas duct, *h* Waste-gas fan, *l* Cyclone, *m* Charge, *n* Discharge, *o* Dust screw conveyor, *p* Dust discharge, *q* Waste gas supply. Schilde, Bad Hersfeld

Advantages of waste-gas stream dryers are:

1. Simplicity of construction;
2. Few revolving parts;
3. Simple maintenance;
4. Small charge of chips;
5. Extremely short residence times of the chips in the dryer;
6. Low prime costs;
7. Little space occupied;
8. Automatic control equipment;
9. High efficiency, especially due to gas return.

As a rule stream dryers are used as pre-**dryers** in which chips with very high initial moisture content are dried to about 60% m. c. It is advantageous to combine stream dryers with other dryer types which finish the drying process (Fig. 5.66). A special control gear permits by-passing the stream dryer during the summer season when wood with low moisture content is available.

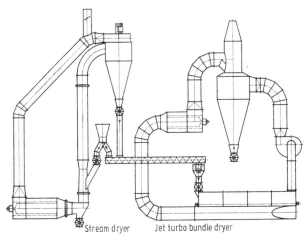

Stream dryer   Jet turbo bundle dryer

Fig. 5.66. Two-stage drying plant consisting of a stream dryer as a predryer and a jet-tube dryer for final drying (Büttner-Schilde-Haas). Scheme of *Klamroth* and *Hackel* (1971)

f) *Suspension-type dryers.* It is not easy to define correctly the construction and methods of operation of suspension-type dryers. In physics suspension means the state in which particles of a solid (wood particles) are mixed with a fluid (in this case a mixture of hot air and vapor) but are undissolved.

The details of the Bronswerk dryer, which belongs to the class of suspension dryers, and its method of operation are shown in Fig. 5.67 and its legend. The Bronswerk dryer is essentially a pneumatic dryer. The hot gases produced by the burner m and its associated ventilator l are mixed with re-circulated air from the exhaust blower k in the combustion chamber a where any remaining dust is burnt. The hot gases enter the drying chamber b and the chips are fed through the rotary valve c. Radial blades d_1 fitted to the rotor cage d, give the air-chip mixture a radial component. By the interaction of this radial component and the adjustable axial flow component a spiral air flow results in the interior of the cylindrical drying chamber. The first part of the dryer is used for drying only, and the second part for drying and classifying the chips. Two rotor cages f_1, f_3 serve for this purpose. The result is that the two chip fractions are discharged separately through rotary valves (g for fines, h for normal chips). The bulk of the air (nearly saturated at its exit temperature) is mostly re-circulated by the blower k to the drying chamber, but some of it must be discharged through the exhaust gas exit i.

A typical suspension-type dryer is manufactured by the German firm Keller. Fig. 5.68 shows a schematic cross section through such a dryer, which is classified as a whirling dryer. The wet material, the particles to be dried, is introduced into the top of the dryer where it is met by the rising hot-gas stream. This gas stream produces a whirling effect and at the same time holds the particles of a certain specific weight in suspension. It follows, then, that heavier particles will not be

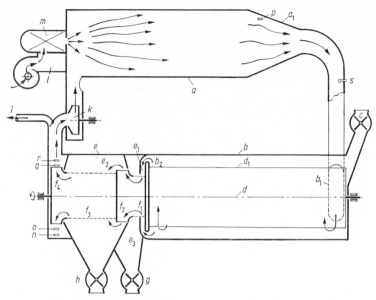

Fig. 5.67. Schematic section through a Bronswerk-dryer (Roodor).

a Combustion chamber, *a₁* End of combustion chamber, *b* Drying chamber, *b₁* End of the drying chamber, *b₂* Baffle *c*, *g*, *h* Rotary valves, *d* Rotor, *d₁* Rotor sheet blades, *e* Final drying with classification, *e₁* First classification stage *e₂* Second classification stage, *e₃* Smoothing part, *f₁* Rotor rod cage, *f₂ f₄* Pressure stages, *f₃* Second rotor rod cage *i* Exhaust gas exit, *k* Rotary blower, *l* High-pressure ventilator, *m* Burner, *n* Regulator, *o* Maximum thermostat, *p*, *q* Thermostats, *r* Overload-thermostat, *s* Water sprinklers

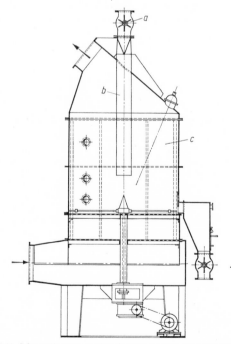

Fig. 5.68. Schematic cross section through a single-step suspension dryer.
a Rotary valve, *b* Down pipe, *c* Suspension chamber

held in suspension but will fall to the bottom of the machine where they are caught on a screen. A mechanical stirring apparatus moves slowly through the chips on the screen, conducting them to the outer edge where they are discharged. The coarse chips can then be used as the middle layer of a sandwich-type board or processed in the reduction units. The chips of proper size, which were held in suspension originally, will gradually become lighter due to the decrease in moisture content, and will then rise with the upward flowing gas stream to be conveyed to a cyclone.

Further development of the single-stage dryer has led to the design of a two-stage model (Fig. 5.69). Another suspension-type dryer (Büttner-Schilde-Haas)

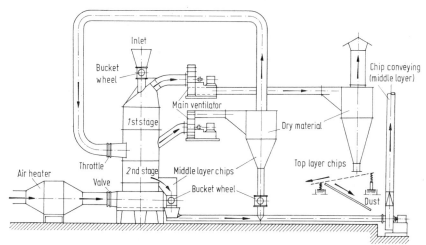

Fig. 5.69. Two-stage suspension type dryer, type Keller-Peukert, Leverkusen

is schematically drawn in Fig. 5.70. The moist chips enter the dryer at the inlet *a* as shown in the figure and are introduced via the puffer-bin *b* and the hopper below by means of a bucket wheel or rotary valve *c* into the rising hot air column *d*; all arrows in the sketch show the movement of chips and air. The heavier and larger particles sink down, countercurrent to the column of air, passing a screen *e*, and are discharged by the screw conveyors *i* for coarse chips. The correctly sized particles, on the other hand, are carried up by the column of air. At the top of the dryer, the duct is curved so that a complete reversal of direction is effected. The particles in the down duct *f* in their downward path have to flow around curving baffles *g*. A powerful whirling motion is imparted to the chip-air mixture by these baffles. The useful chips, thus dried, are finally deposited on a rapidly running endless mesh-belt conveyor which forwards them to the chip outlet. A part of the chips may fall down through the open spaces of the mesh network, still useful chips are carried by another endless belt conveyor to the chip outlet. Air circulation is through a closed loop and the reheated air is moved by the ventilator *k* into the air heater *l* which is located in the lower part of the dryer. Exhaust gases containing some dust are removed by the exhaust gas pipe or chimney duct *m* (*Kollmann*, 1955, 1966, *Johnson*, 1956, p. 152/153). This type of suspension dryer ("Flugband-Trockner") has a maximum evaporation capacity of 970 kg (~2,140 lb.) water per hour. The construction height is relatively great. This dryer type is scarcely used in new plants.

Summarizing it may be stated that rotary nozzle dryers with fractioning (screening) effect are more modern. Therefore Fig. 5.71 shows a schematic section through such a dryer (Büttner-Schilde-Haas). In the interior of a cylindrical fixed drying chamber a with a very well heat-insulated case rotates the conical nozzle tube b. Material feed and discharge are through rotary valves c and g. Adequate adjustment of the angles of the nozzle apertures effect a spiral chip flow from the

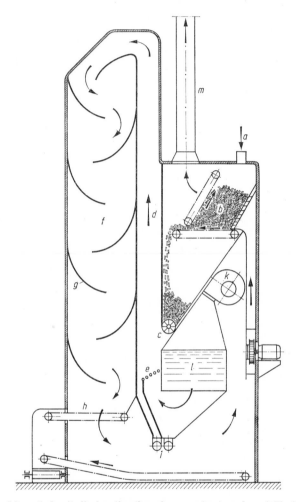

Fig. 5.70. Schematic longitudinal section through suspension-type dryer ("Flugbandtrockner"). Büttner-Schilde-Haas, Krefeld.

a Material inlet, b Puffer-bin, c Rotary valve (bucket wheel), d Hot air duct, e Screen, f Down duct, g Buffles, h Endless belt for chip removal, i Screw conveyor for coarse chip discharge, k Ventilator for air circulation, l Air heater, m Exhaust chimney

charge valve c to the exit side d. Around and along the periphery of the rotary nozzle tube, vanes e are fitted which cause lifting and purling of the chips to be dried. The result is a high thermal efficiency and thereby very short drying times. Coarse chips leave the dryer at the end of the rotor via the outlet h. At the exit side d, thin and light chips are first led by the air stream to the cyclone f followed by the thicker more slowly drying chips. The rotary valve g discharges the separated material. The air circulation is illustrated in Fig. 5.71. The hot gases — poor of oxygen — are produced in the combustion chamber i in which usually gas, oil or a mixture of oil and dust is burnt. The hot gases are mixed with moist return air k by means of the fan l. A part of the return air is removed by the exhaust duct m, the other part of the return air serves for classifying n. The path of the inert hot gases is shown by the letter o.

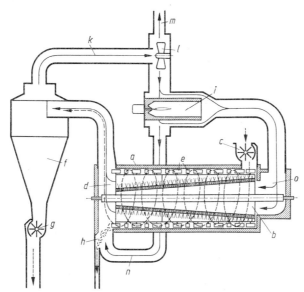

Fig. 5.71. Schematic section through a rotary nozzle dryer with fractioning effect.
a Cylindrical drying chamber, b Rotating conical nozzle tube, c and g Rotary valves for charge and discharge respectively, d Exit side of the dryer chamber, e Vanes, f Cyclone, h Outlet for coarse material, i Combustion chamber, k Moist air return, l Fan, m Exhaust duct, n Air for classifying, o Hot gas inlet

There are various sizes of rotary nozzle dryers with power consumptions from about 17 to 66 kW, evaporation capacity between 500 and 4,000 kg/h (∼1,100 and 8,800 lb./h). The specific heat consumption per unit of evaporated water, ranges between 850 and 900 kcal/kg (1,530 BTU/lb. and 1,620 BTU/lb.).

Many dryers have facilities for size separation, but there are often reasons to carry out drying operations and size separation independently.

5.2.3 Size Separation (Classifying) by Screening or Sifting

Generally classifying is carried out independently of the drying operation (*Kull*, in *Kollmann*, 1966, p. 199—216). As an example the Keller-dryer should be quoted, actually being used in some plants as a pneumatic classifier only.

Apart from pneumatic separation, screens are, of course, used in the particle board industry. Screens separate according to surface area and not to thickness. Therefore they should only be used to separate "small shavings from large shavings or alternatively for the extraction of fines" (*Mitlin*, 1968, p. 87). Screens are used to a large extent as flat screens (Fig. 5.72) or drums which are commonly slightly inclined and moved by oscillation or rotation. The oscillation may be done by shaking. The holes in a drum screen vary in their size from small to larger diameters (Fig. 5.73). Screening is not only combined with drying but

also with milling and chipping. The screens should be installed in a box in order
to avoid disturbance by escaping dust.

It is usual to classify wood particles by means of air. Fig. 5.74 shows the possible modifications:

a) Suspension sifting;
b) Throw sifting;
c) Wind sifting.

There is always a relative velocity between particles and air (*Rackwitz* and
Obermaier, 1962). The separation of the particles is effected by their surface weight.
Shape and size are of minor importance. A relatively high scattering occurs, and
therefore the sifting is not sharp. In a horizontal air-flow used for wind-layered or

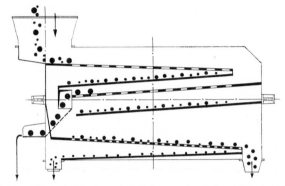

Fig. 5.72. Scheme of the material passage in a flat screen classifying system. MIAG, Braunschweig

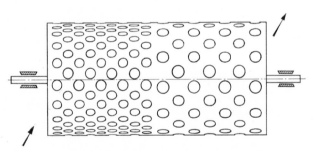

Fig. 5.73. Scheme of a drum screen. From *Kull*, in *Kollmann* (1966, p. 208)

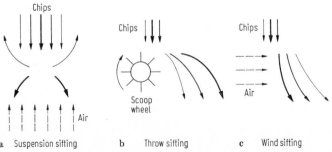

a   Suspension sifting   b   Throw sifting   c   Wind sifting

Fig. 5.74a—c. Methods for air circulation of wood particles. a) Suspension sifting; b) Throw sifting; c) Wind
sifting

throw-layered particleboard, the defects are rather rare. "By several series of connected sifting-procedures, however, a relatively exact separation can be achieved without substantial loss of thin particles" (*Rackwitz* and *Obermaier*, 1962).

The simplest case of wind sifting is the free fall (Fig. 5.74c). Thin chips or flakes have a lower sinking speed due to their higher air resistance than thicker, heavier chips. This principle is without practical importance for the particleboard industry because continuous feeding is not possible. In the suspension sifters the path of falling chips and a strong air stream are countercurrent. The sinking speed of the light thin chips is lower than that of the coarser, thicker particles.

A suspension sifter with two sifting-stages is shown in Fig. 5.75 (Type Keller). There are various possibilities of constructing wind sifters. The rising wind sifter, developed by a particleboard plant in West Germany (Moralt) and used in many

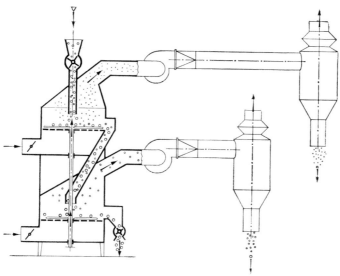

Fig. 5.75. Schematic section through a suspension sifter with two stages controlled by a flap in the air inlet ducts left side of the figure. Keller-Peukert, Leverkusen

experiments over several years, is especially space-saving (Fig. 5.76). The wind sifters in which the air stream is conducted across the falling direction of the particles drives them more or less sideways according to the air resistance; the lighter the particles the more they are driven sideways whereas the thick particles fall down with very little side deflection (Schilde, Himmelheber). It is possible to create closed air circulation, thus avoiding troubles by dust in the factory (Fig. 5.77). A recent type of a gravity wind sifter is the zig-zag classifier (Alpine), (*Kollmann*, 1966, p. 212, 213). Fig. 5.78 shows such a classifier with nine ducts directed in parallel.

The throw sifters are characterized by a rotating roll with radial pins as is shown schematically in Fig. 5.74a. In the throw sifters the particles, obtained by the activity of the pin rolls, have a horizontal initial velocity which decreases more or less as a function of the air resistance. Particles fall down in parabolic paths, the heaviest with the longest distance from the axis of the roll, the lightest with the smallest distance. Another possibility is to use centrifugal forces in wind sifters (Fig. 5.74 b). These centrifugal forces can be effected by a drift current of the air entering the sifting chamber or primarily by the rotation of a sling disk.

Wind classifying is especially suitable for the manufacture of particleboard, since for high-quality board, mainly for the deck layers, only thin chips should

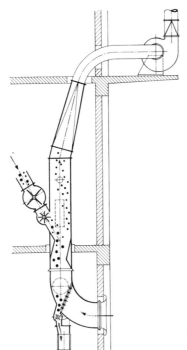

Fig. 5.76. Schematic section through a rising wind sifter. Moralt, Bad Tölz

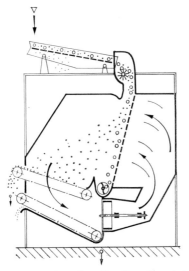

Fig. 5.77. Schematic section through a wind sifter with completely closed air circulation. Type Himmelheber, Büttner-Schilde-Haas, Krefeld

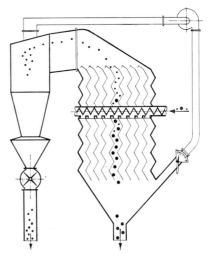

Fig. 5.78. Schematic section through a gravity wind sifter, zig-zag wind sifter with multiple air ways arranged in parallel. Alpine, Augsburg

be used, separated from thick particles (*Kull*, 1955). The necessity of proper sifting in the particleboard industry has long been underestimated (*Steiner*, 1954). The development of special wind sifters for wood chips was not simple because the aerodynamic conditions and relationships had to be investigated. Another problem came from the fact that some dryers were also wind sifters. Wind sifting and throw sifting, furthermore, were connected with mat forming. Wind sifters work efficiently only if a high capacity can be achieved which is valid also for pneumatic conveyors (*Knorpp*, 1963).

5.2.4 Conveying of Chips

Various types of conveyors are used in particleboard plants at the different stages of manufacture. Their investment costs, their working costs and their costs for repair and spare parts determine the production costs to a remarkable extent. From this point of view much care is necessary in the proper choice of conveyors. It is more or less a task of mechanical engineering which is not directly linked up with the principles of wood science and technology.

Conveying of the chips is a serious problem (*Steiner*, in *Kollmann*, 1966, p. 239/280), because not only the above mentioned economical calculations must be made, but also particle destruction of the chips should be excluded.

Principally there are pneumatic and mechanical conveyors available in many types. It should be kept in mind that e.g. after the chipping machines the chips are obtained discontinuously.

Mechanical conveyors of the following types are used in the particleboard industry:

a) Belt conveyors, Fig. 5.79 shows three types;

b) Screw conveyors (Fig. 5.80);

c) Scraper conveyors;

d) Trough chain conveyors, mainly used for vertical transportation of the chips. Fig. 5.81 illustrates position and bearing of the chains in the trough;

e) Swing conveyors.

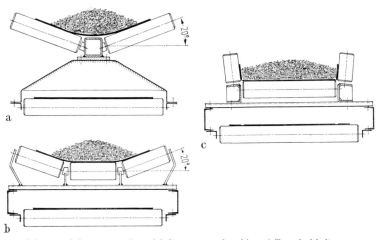

Fig. 5.79 a – c. Schemes of the cross section of belt vonveyors for chips. a) Troughed belt conveyor, supported by two inclined rolls (V-shaped, angle of roll axis to the horizontal plane 20°; b) Troughed belt vonveyor supported by 3 rolls; c) Troughed belt conveyor supported by 3 rolls, one series of horizontal bottom rolls and two series of steeply inclined side rolls. Type Schenck, Darmstadt. From *Haas*, in *Kollmann* (1966, p. 245)

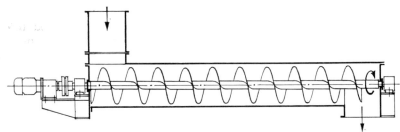

Fig. 5.80. Schematic section through screw (spiral) conveyor for chips.
Type Schenck, Darmstadt. From *Haas*, in *Kollmann* (1966, p. 245)

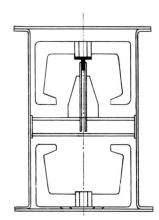

Fig. 5.81. Schematic cross section through a trough chain conveyor
showing position and bearing of the chains in the trough. From
Haas, in *Kollmann* (1966, p. 245)

Pneumatic conveyors are usable for all transports of chips within the particle-board plant. Pneumatic conveyors offer the following advantages:

a) Simple mechanical device;

b) Low investment costs;

c) Low requirements of space since the largest part of the device, the cyclone can be installed outside of the building, for instance on the roof;

d) Possibility to convey the chips over a distance vertically as well as horizontally or inclined, climbing or falling, and even in bows.

The pneumatic conveyors have the following disadvantages:

a) Necessary movement of large amount of air with high velocity and therefore high energy consumption;

b) Chips with a high moisture content and simultaneously with large area sizes may cause blockage in narrow bows as well as in the fan. Freshly blended chips cannot be conveyed pneumatically.

One has to distinguish between pressure, suction and mixed pressure and suction devices. The pressure is low up to a water column of about 250 mm (about 10 in.)

5.2.5 Storage of Chips (Wet and Dry Bins or Silos)

5.2.5.0 General Considerations. In all particleboard plants several temporary 'in-process-storage-units' are required. These units serve the following purposes:

a) Control of the flow of material from one operation to another. "The surge bin is usually equipped to discharge the material in metered quantities compatible with the capacity of the unit which receives it. Surge bins normally store small amounts of material, since capacities of the various components in any one phase of the process are coordinated" (*Johnson*, 1956, p. 135);

b) Buffer between two phases of production, to make them independent of each other. A breakdown in any machine in the particle preparation will not disturb the following production, unless the break-down persists until the buffer supply of material is exhausted" (*Johnson*, 1956, p. 135);

c) Labor cost reduction because hogs, dryers, and screens posses two to three times the capacity of the remaining portion of the system;

d) The aim in both bin design and discharge mechanisms is to overcome the arching of the chip mass; constricting areas should be eliminated.

A thorough knowledge of the properties of the chips is essential for bin selection. "Bins which work very satisfactorily on dry planer mill shavings may be quite unsuitable for wet prepared shavings, and alternatively some bins suitable for wet shavings may produce dust with dry chips" (*Mitlin*, 1968, p. 90).

Bins can control the flow of material either by volume or by gravity. Fig. 5.38 shows schematically in which ways various bins are coordinated with the whole manufacturing process.

Johnson (1956) writes: "In virtually all particleboard production systems, several temporary in-process-storage-units are required. Collectively, these units serve the following purposes":

a) Control (regulation) of the flow of materials from one operation to another. In this function, for example, surge bins may receive a batch output from one operation and dispense it to the next operation in a continuous flow. Conversely, a continuous flow input may be changed into a batch output. These bins are located between the two operations involved, they are elevated to reduce space requirements and to take advantage of gravity flow;

b) Equipment of the bin to discharge the material in metered quantities. Surge bins normally store small amounts of material. Ready stockpile of chips for one production shift or even only for several hours causes difficulties. In the production plants incorporated bins with a volumetric capacity of 40 m^3 ($\sim$1,413 cu. ft.) are already large units. With a bulk density of the chips of about 60 to 80 kg/m³ (3.75 to 5.0 lb./cu. ft.) the stored chip mass weighs 2.4 to 3.2 metric tons. This is not a real supply because medium sized and large chipboard plants consume for the production of the core layers 4 to 7 metric tons/per hour and of the face layers 3 to 5 metric tons per hour (*Steiner*, in *Kollmann*, 1966, p. 217);

c) Storage bins equalize the flow of raw material;

d) Labor cost reduction. If the preparation equipment, from hogs through dryers and screens, possesses two or three times the capacity of the remaining portion of the system, labor can be utilized at its potential, to produce in one shift sufficient particles to supply the process for two or three shifts (*Johnson*, 1956, p. 135).

e) The filling density should be equal in full as well as in nearly empty bins. Different, especially excessive compression of the chip materials would disturb a uniform and controllable discharge. This important demand cannot or only in a limited manner be accomplished in vertical bins;

f) First-in-first-out-material-flow should be provided. For blended chips this is a necessity;

g) The employment of bins is absolutely recommended for blended chips in order to make to some extent gluing and mat-laying independent of each other;

h) Bin batteries of two to six units permit forming and observing of a chosen checkable chip mixture from chips of different kinds and properties;

i) Very accurate chip flow is achieved when a small buffer bin is introduced, the discharge of which is directly controlled by a belt weigher. "Their function is to even out the flow, and according to chip characteristics they may be of the rotary table, twin worm, moving bottom, or fluted roll type" (*Mitlin*, 1968, p. 91).

Principally, horizontal, vertical and rotary type bins are available.

5.2.5.1 Horizontal Storage Bins.

Horizontal bins, developed in West Germany and especially suitable for prepared shavings, are by far the most common. These bins resemble a cubic room, the sidewalls are slightly beveled to the top, and possess windows for observing the movement of chips inside. The bottom of the bin is a wide belt. Particles are fed into the upper end of the bin, building up a predetermined height. These bins overcome the problems of bridging or arching of the chip mass due to the high chip slenderness (ratio of length to thickness up to 50/1), and they serve "to remove material from the bin in controlled quantities" (*Johnson* 1956, p. 136). In the horizontal bins there is always a cooperation of feeding after the entry of chips by means of belts, distributing by scrapers, controlling by flaps or brush rolls; the latter also clean the belt conveyor on the bottom. The bevel flap stops the outlet conveyor when the discharge head is full. Various possibilities of construction are illustrated in Fig. 5.82:

a) Bins which are not only storing but controlling (Fig. 5.82a) (Himmelheber, Schenck).

b) Silo-machines (Schnitzler-Baltex, Würtex, Fig. 5.82b) are similar to the foregoing types, with a combination of belt conveyors on the bottom of the box, running slightly horizontally and then very much inclined.

c) Mixing or blending bins (Himmelheber, Moralt). The aim is to mix the inlet material prior to the outlet. This may be necessary if into one bin chips of different kind are fed. Fig. 5.82c shows that the bottom belts do not only loosen the infed chips but also transport them to the outlet.

d) Fig. 5.82d shows that mixing bins can work in series.

5.2.5.2 Vertical Storage Bins.

Vertical storage bins may be constructed as slightly conical cylinders or as rectangular boxes in cross section. The discharge is secured by bottom rolls (*Blache*, 1936), screw conveyors, or belt conveyors which give a uniform output because of their low storage capacity.

A vibrator type discharger for vertical bins (Simon Handling Engineers Ltd., U. K.) tends to overcome bridging by using an ample outlet and "imparting vibrations directly into the material lying directly above the gyrating conical baffle of the activator. As the discharge flows over the perimeter of the conical baffle, short-circuiting down the centre of the bin is eliminated" (*Mitlin*, 1968, p. 91).

A cylindrical-type dry-chip storage silo has been developed in the U.S.A. by The Chipcraft Co., Morristown/Tenn. A completely different construction has been adopted by Ponndorf Maschinenfabrik KG, Kassel. This early type (Fig. 5.83) is characterized by the following basic features:

a) Absence of a conical bin bottom;

b) Chip discharge by a screw conveyor. One end of this conveyor is located under a protection hood on the vertical centre line of the bin, whilst the drive end is supported on a carriage rotating around the bin;

c) Control of the discharge by variation of the screw and carriage speeds;

d) Usable for wet and dry chips.

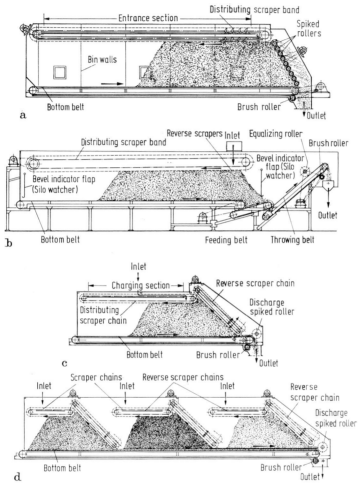

Fig. 5.82 a – d. Schemes of horizontal storage bins. From *Steiner*, in *Kollmann* (1966)

5.2.5.3 Rotating Storage Bins. Occasionally rotating bins are used. According to *Mitlin* (1968, p. 91) "this type of bin has a horizontal cylindrical shell fitted with longitudinal flights, seals are provided between the shell and stationary end plates. Inlet chutes are fitted to the top of the end plates, and screw conveyors which are open at the top run from the centre of the drum through the end plates from which they are supported. Chips cascade into these conveyors, the speed of which is far in excess of feed rate requirements, and are discharged through the ends. Changes in feed rate are obtained by varying the rate of rotation of the bin." (*Mitlin*, 1968, p. 91).

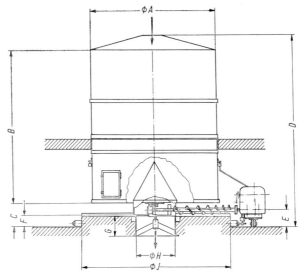

Fig. 5.83. Scheme of a vertical storage bin. Ponndorf, Kassel

5.2.6 Dosing (Weighing and Volumetric Dosing)

A continuous-type chip glue mixer should guarantee constant chip infeed. Continuous-type mixers require such a constant chip infeed; otherwise adding of proper amounts of resin would be incalculable. Constant flow of chips can be adjusted by means of a liquid metering pump. The gravimetric weigh-system has a high degree of control and accuracy. An American system is illustrated

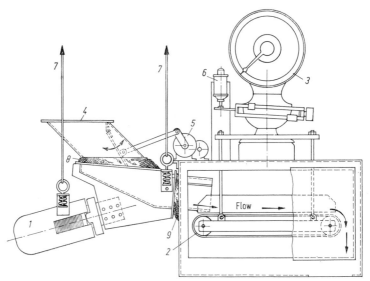

Fig. 5.84. Schematic view of a typical gravimetric chip-feeder. Syntron Co., Homer City, Pa., U.S.A. From
Johnson (1956)

1 Vibratory feeder, *2* Scale-suspended, constant-speed belt conveyor, *3* Weighing scale, *4* Supply chute and gate, *5* Flush control, *6* Motion transmitter, *7* Feeder suspension cables, *8* Dust seal, between supply chute and feeder, *9* Dust seal between feeder and cabinet

in Fig. 5.84. The sketch together with legend illustrate how a typical gravimetric chip feeder works (Anonymous, without data):

Supply chute *4* for chips — electro-magnetic vibrator *1* — constant speed belt conveyor *2*, a weighing scale *3* indicating, controlling and metering the desired rate of chip flow — flush control *5* — motion transmitter *6* — feeder suspension cables *7* — dust seals *8* and *9*.

If mat-forming is combined with dosing, there are the following possibilities:

a) *Volumetric dosing.* Variations in the weight of the scattered chips are difficult to be controlled. Probably the first spreading device with pure gravimetric control was built by Himmelheber (*Himmelheber, Steiner* and *Moralt*, 1957, 1960).

Fig. 5.85 explains the spreading device with the gravimetric dosing of the chips controlled by radioactive rays. The legend shows the path of the chips

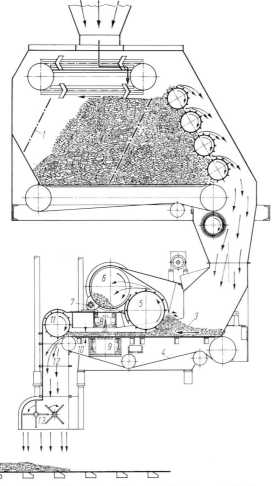

Fig. 5.85. Schematic view of a spreading device with gravimetric dosing controlled by radioactive rays. From *Himmelheber, Steiner* and *Moralt* (1957, 1960)

1 and *2* Boundary limits of bin, *3* Predosed chip-carpet, *4* Dosing band, *5* Backward roll, *6* Conveyor screw, *7* Adjusting axis for *5* and *6*, *8* Irradiation chamber, *9* Chamber for measuring ionization, *10* Wide-dosed chip carpet, *11* Stripping wheel, *12* Chip duct, *13* Pair of loosening wheels

through a small bin and the various endless belts conveying the chips, the picked rollers at the right side of the bin, the screw conveyor distributing the chips in the width direction, the ray emitter and the measurement chamber for the ionization effect; finally the chips are loosened and spread by suitable rollers.

The rays emitted by the radioactive isotopes pass through the mat; this mat consists of wood, glue and moisture. Together with the weight per area of the dosing band and index number the weight of unit area is continously obtained. By this sophisticated construction it is possible to adapt the roller (5 in Fig. 5.85) which shows that the correct amount of material is retained to the feeding speed of the dosing band and of the forming belt on which the transport cauls are electrically coupled.

5.2.7 Chip Blending

5.2.7.0 General Considerations. In the manufacture of particleboard accurate blending of chips with liquid (occasionally powder) glues is very essential. In order to prepare the glue ready for spraying and to transport it to the mixing station several accessory installations are necessary[1]. The glue must be mixed completely and homogeneously with a hardener or other additional agents within a short time (maximum 10 min). The prepared glue mixture is transported from the glue mixer to one or two storage pressure vessels (Fig. 5.86). By pressing

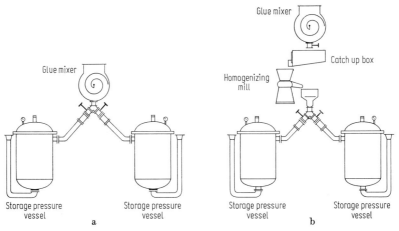

Fig. 5.86a, b. Glue preparation plant. a) For liquid glue; b) For powdery glue. Draiswerke, Mannheim-Waldhof

air into the upper part of the vessel the glue is fed to the spray nozzles. The preparation of glues in powder form in mixing machines may lead to agglomerations which can be avoided only by using high-speed mixing machines. These may cause undesired foam formation and therefore to avoid deposits in the feed lines (impairing an accurate dosing), the glue mixture, after leaving the mixing machine, enters a homogenizing mill where all agglomerates are completely dissolved.

Liquid glue has to be spread in a relatively small amount on a great mass of chips, i.e. on a large surface. The addition of glues contributes re-

---

[1] The autor is indebted to *Drais-Werke*, Mannheim-Waldhof, for information.

markably to the self-costs of particleboard (35 to 60%) the strength properties of which depend upon uniform gluing.

Blending of the chips is a mixing of the liquid glue with the chips. Very essential is the ratio of the liquid glue amount to the chip surface. Commonly 8 to 12 g liquid glue must be distributed per 1 m² chip surface. The distribution should be equal with respect to the strength properties of the board.

The amount of chips should be determined volumetrically. Dosing of glue is continuously secured by a belt weigher. It is possible to tune automatically the glue dosing and the amount of chips.

Blending should be performed in a technical process. The following methods are usual (*Schnitzler*, 1971):

a) Jet method: The liquid glue is distributed by means of centrifugal or turbulent jets; it is dispersed into very fine rays. The fine drops of the glue meet the chips kept in a tumbling movement;

b) Centrifugal spraying method;

c) Roller glue laying method.

It may be repeated that efficient chipboard manufacture to a high extent depends on effective blending of the chips. Glue consumption is a vital cost factor. The principal modes of action are cited in the paragraphs a) to c). It must be added that there are:

d) Batch operations;

e) Continuous operations.

Complete glue preparation plants are the necessary complements to chip and glue mixing machines. Modern chip and glue mixers provide uniform and economic blending of wood chips, flax shavings, straw, coconut fibers, rice shells, bagasse and particles of other organic substances (*Johnson*, 1956, p. 155—162; *Engels*, in *Kollmann*, 1966, p. 280—303; *Mitlin*, 1968, p. 92—93; *Schnitzler*, 1971).

5.2.7.1 Batch Operation. Chip and glue mixers for batch operation are simple to operate and easy to maintain. They are primarily vertical or horizontal vessels which are equipped with an apparatus in their interior for tumbling or agitating the chips while the liquid glue is sprayed onto the chip surfaces from above. Shape and rotational speed of the mixing element are determined by the type of chips to be handled. Discharge of the blended chips is through a bottom flap, manually or pneumatically operated. There are one to four two-component turbulent flow nozzles which can be adjusted from outside. The pressure vessel and gear pump for metering the adhesive are enclosed in the casing at the end of the machine frame. If required the chips can be volumetrically batched using the mixer trough itself for this purpose. The complete gluing cycle (evacuation, loading and spraying) is centrally controlled. The basic batch-type mixers have changed little over many years; however, some refinements have taken place:

a) Inrease of the rotational speed of the agitator shaft;

b) The tools for agitating chips an dglue are improved by a combination of spiral with prickle mixers (Fig. 5.87) or by application of ploughshare mixers helically displaced and rotating so quickly that the content is hurled (Fig. 5.88):

c) Spraying is accomplished by a series of nozzles or atomizers of various types.

The results of refinements are elimination of "dead" spots in the mixture, less sticking of particles on inner surface of the vessel and complete disappearance of the glue stain which was prevailing in earlier models.

d) Pneumatical discharge of the chip-glue mixture;

e) Equipment of the mixer with a large, removable lid to facilitate cleaning after each working period.

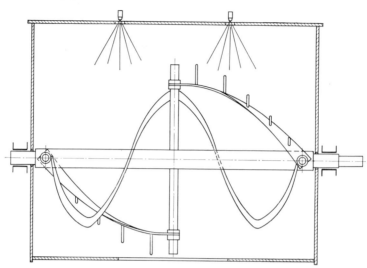

Fig. 5.87. Schematic longitudinal section through a gluing machine with combined spiral and prickle (pin) mixers and spraying by two nozzles. Drais-Werke, Mannheim-Waldhof

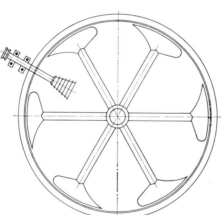

Fig. 5.88. Schematic cross section through a gluing machine with 6 rotating plough-shape mixers and multi-disk centrifugal sprayer. Lödige, Paderborn

5.2.7.2 Continuous Operation. There are various types of chip and glue mixers for continuous operation. It is not possible to describe them in detail in this book. Fig. 5.89 shows an early machine still used at present in Novopan-plants. This machine mainly consists of cylinders with relatively large diameters covered by means of several "doctor rolls" with a thin glue film. The chips are passed by steps with the aid of loosening and whirling rolls over the upper side of the cylinders. The glue film is transferred on the surfaces of the chips by rubbing and wiping. Machines of this type have two disadvantages:

1. If the mixture of chips is heterogeneous the coarser chips absorb more glue;

2. The dosing of the glue, expressed by the thickness of the film, depends on several factors such as viscosity and temperature of the glue, or the fouling of the apparatus which cannot be controlled.

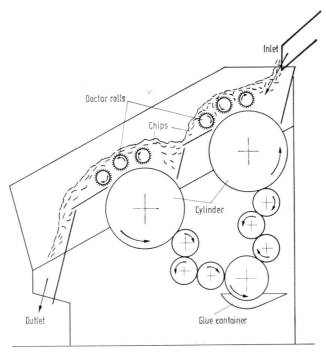

Fig. 5.89. Scheme of a cylinder or roller gluing machine. System Fahrni

The centrifugal spraying method converts the liquid glue into finest drops. The disks effecting the centrifugal force run at high speed. The result is an interaction between the high speed glue rays and the moving chips. Fig. 5.90 shows an example of such a glue and chip mixer (Type Fahrni-Institut AG, Zürich). The dried chips volumetrically dosed from a bin are continuously controlled by a belt weigher. A distributing plate and an accelerating rotor produce a chip paraboloid inside which a central high pressure turbulent nozzle causes a glue cone. Advantages of this machine:

1. Use of only one nozzle with a relatively great diameter, therefore no danger of impediment; easy changeability of nozzles without interrupting of the machine;

2. Short residence time of the chips in the mixer;

3. Very low amount of chips in the gluing zone;

4. Relatively low space requirements for the machine;

5. Easy survey;

6. Fine accessibility of the various aggregates of the machine.

Another glue mixer consists of vertical cylindrical vessels; in their interior are stirrers of high speed, the liquid glue is sprayed in the center of the chips which helically rotate by means of a high pressure pump (80 to 120 kp/cm^2 $\sim$1,140 to 1,700 lb./sq. in.). The mixers are intensively cooled. Machine parts in touch with the resinated chips are protected against corrosion. Incrustations of the material are avoided. It is possible to use one, two or three aggregates (Teuto-burger Maschinenfabrik, Detmold-Pivetsheide, W.-Germany, cf. *Schnitzler*, 1971).

A further type of chip and glue mixer consists of a series of horizontal gluing chambers (diameter 470 mm ~19.5 in.) with axial high speed stirrers. Fig. 5.91 shows the simplified scheme of the chip movement through the machine (Gebr.

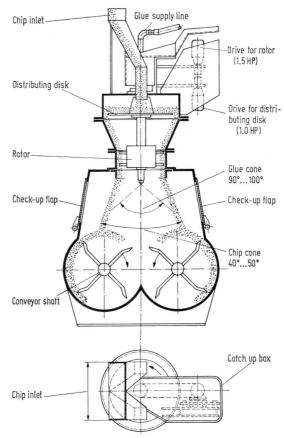

Fig. 5.90. Centrifugal glue sprayer. Type Fahrni-Institut, Zürich

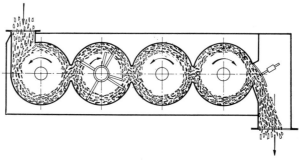

Fig. 5.91. Schematic longitudinal section through a gluing machine with a series of 4 horizontal gluing chambers and axial high speed stirrers and glue supply by tube sprayers rotating with high speed. Lödige, Paderborn

Lödige, Paderborn, Westfalia). The glue is not supplied by nozzles but by fine tubes mounted on the shaft and centrifuging the glue at their ends in very small droplets (60 to 120 μm). The transit time through the whole mixer requires only seconds.

A recently developed mixer for resinating chips and fines is the model Turbo-plan (Draiswerke, Mannheim-Waldhof, W.-Germany). Fig. 5.92 shows constructional details and mode of operation. The horizontally arranged, cylindrical mixer trough is divided into an upper an a lower half, locked together by toggle-lever fasteners. The hinged upper part is counterbalanced by a weight. Each of the semicylindrical halves is jacketed for water cooling. Correct dimensioning and the incorporation of baffles ensure intense turbulence of the flow and thus effective cooling.

An angular flanged joint proves a dustproof connection between the inlet duct of the mixer trough and the chip feed. The liner of the mixer trough consists of high-grade alloy steel which is both corrosion- and acid-resistant. The generously dimensioned hollow shaft is carried in bearings at both ends and is dynamically balanced. With the exception of the paddles, welded in position in the inlet zone, the mixing arms for moving, turning and mixing the chips are bolted to the shaft and secured by lock nuts. The mixing arms are individually rotatable and adjustable in height, to suit the precise nature of the feedstock, throughput rate and the like. In the secondary mixing zone in the second half of the mixer the shaft and mixing arms are water-cooled. The resin syrup flows into the hollow shaft through an inserted distributor tube. By virtue of the centrifugal effect the binder is expelled from the hollow shaft through the tubular nozzles and dispersed into a fine mist.

Control equipment for the flow of chips and resin in the main consists of the following units:

a) Chips batch weigher with dial indicator can be installed on the inlet or outlet side of the mixer. The dial indicator permits precise visual control of the feed rate and contents of the hopper. A belt weigher with dial indicator can be installed instead of the batch weigher.

b) The Drais-patented binder preparing unit with continuous mixing tank has proved its performance in many years of practical application. The ratio of the individual binder constituents can be varied by adjusting the length of stroke of the four-piston metering pump. Delivery rates are controlled by a variable speed gear unit. The flow of resin syrup or of the resin and hardener mixture is measured by an oval wheel type meter with converter for co-ordination with the feed rate of chips.

c) The metering, regulating and other electrical units for controlling and monitoring the system are accomodated in a control console.

Characteristics are: Best possible control of resin distribution — adhesive consumption remarkably reduced by graduated layer resination — atomizing air and positive pressure obviated by centrifugal application of resin syrup — low power consumption — minimum floor space requirement — uncomplicated design — no cleaning problems.

For the chip-glue mixing processes it is advisable to check the glue composition at reasonable intervals, and this can be done by measuring the gelation time (*Mitlin*, 1968, p. 104). A detailed discussion of the various advantages of the new gluing processes for chips has been published by *Schnitzler* (1971):

1. No supply of spraying air for dispersion of the glue in the gluing chamber; the problems of exhaust thus do not arise,

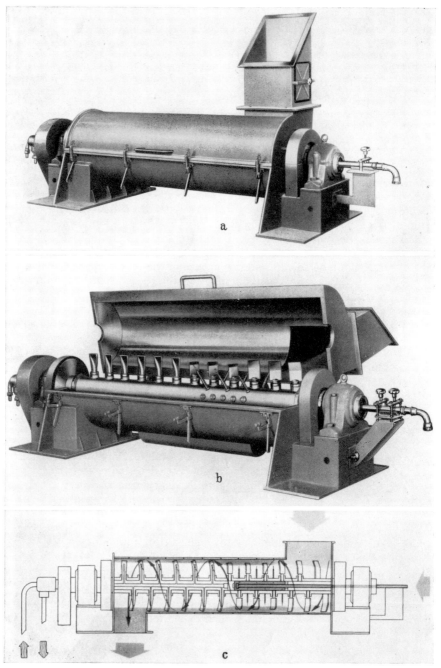

Fig. 5.92a—c. Continuous chip and glue mixer, model Turboplan. Drais-Werke, Mannheim-Waldhof.
a) View of the machine with closed trough, ready for use; b) View of the machine, half of the trough
with the inlet duct is hinged up. The interior and mixing mechanism are readily accessible from all sides; c) Schematic longitudinal section through the mixer showing its method of operation

Table 5.14. Data for Various Types of Drais-Turboplan Mixers

Types		K-TT 80	K-TT 175	K-TT 350	K-TT 700	K-TT 1400
Overall dimensions: Length	mm/in.	1,250/89	2,900/114	3,400/134	3,900/153	4,600/181
Width	mm/in.	2,080/42.5	1,405/55.2	1,635/64.3	1,785/70.2	2,075/81.6
(hood open) Height	mm/in.	550/21.6	655/25.8	775/30.4	920/36.1	1,045/41.2
Feed opening: Length	mm/in.	300/11.8	350/13.8	400/15.7	450/17.7	500/19.7
Width	mm/in.	200/7.9	250/9.8	300/11.8	350/13.8	400/15.7
Discharge opening: Length	mm/in.	100/5.9	200/7.9	250/9.8	300/11.8	350/13.8
Width	mm/in.	340/17.7	500/19.7	560/22.0	640/25.2	710/27.9
Throughput capacity	kg/h	250···2,000	500—4,000	1,000··· 8,000	2,000···16,000	4,000···32,000
	lb./h	550···4,400	1,100···8,800	2,200···17,600	4,400···35,200	8,800···70,400
Motor rating	kW	5 ···10	7.5···15	11 ···22	18.5···45	37 ···75
	HP	3.6··· 7.3	5.6···11.0	8.1···16.4	13.8···33.6	27.6···56

2. No excess pressure for the glue dispersion by use of centrifugal tubes;

3. Less power consumption because the amount of chips in the machine is small, incrustations do not occur, low heating of chips in the chamber due to friction;

4. The small volume of the machines permits intensive cooling of through and hollow shafts;

5. Forced rotation of the chips and formation of a closed annular chip mat moving around the inside of the trough. The binder is dispersed into a fine mist and the wiping effect between the chips produce closed glue films on the surfaces of the chips. The conditions of glue coating are very clear;

6. No evaporation of water;

7. Simple arrangement of the centrifugal tubes, the variations in glue dosing are minimum. This permits working with the theoretically calculated minimum glue amount;

8. The dosing of glue and chips can be controlled by a belt weigher;

9. Not only incrustations but also agglomerations of glue and chips which cause glue spots on the surfaces of the pressed boards are excluded;

10. The machine construction is remarkably smaller than that of older models; the newer types are cheaper and require less space in the production line.

5.2.8 Mat-Laying (Chip-Spreading)

5.2.8.0 General Considerations. The purpose of mat-laying is to prepare a consistently uniform mat or carpet of resinated chips, ready for pressing. This process is most important for the manufacture of high-quality particleboards. The trend to erect larger plants induced the suppliers of machines for this industry to continuously increase the efficiency of the chip spreaders. The use of several chip spreaders is not desirable with respect to more complicated conveyors, higher requirement for space, and especially the complicated control of the whole establishment (*Trutter* and *Himmelheber*, 1970).

The manufacturers of particleboard demanded highest possible uniformity of weight per unit area of spread chips. Some difficulties arose because the deck layers consisted of more and more finer chips up to fibrous material. Another difficulty was that due to economic reasons various wood species and grades were used. This variability influenced the size of chips and therefore the area weight of the spread particles. Although the plants tried to reach a uniform mixture of the different wood assortments, variations of the spread weight of $\pm 15\%$ occurred. The chip spreaders immediately record such variations and compensate them.

5.2.8.1 Distribution of the Particles, Manually or by Auxiliary Devices. In this connection it is worth-while to give a short historical retrospective view which is based on the introduction to U.S. Patent No 2,737,997 (application December 1, 1953, granted Mar. 13, 1956, inventors *Himmelheber, Steiner, Kull*) (p. 1, column 1, line 20 to 53):

"In the manufacture of boards from wood particles it is known to loosely deposit a heap of particles, preferably coated with a binding agent, onto a surface bordered by a frame of the size of the board to be produced, and to then distribute the particles manually or by auxiliary devices over the frame-bordered area before subjecting the resulting mat to subsequent fabricating operations which by pressure, or pressure and heat, reduce the mat thickness to that of the desired board and simultaneously cure the binding agent to convert the product into a solid body.

Considerable difficulties have been encountered with this method to produce articles of uniform density and uniform strength. These difficulties are due to the fact that, even with a seemingly uniform distribution of the wood particles on the mat-forming area, the resulting mat has localities of a rather loose texture as well as localities of denser packing or lumpy texture. Such irregularities are inevitable with particle materials of a pronounced interlacing or felting tendency. A uniform distribution as regards density, height and texture over the entire length and width of a mat is particularly difficult to obtain with wood particles intentionally cut to the shape of elongated shavings of good interlacing properties. While such shavings are favorable for high-quality boards, it is infeasible to prevent such shavings from intertwining, interweaving or lumping prior to being deposited on the mat-forming area. The storage silos, bins and conveying devices always necessary for large-scale manufacture tend to promote such a premature interlacing with the result that the known particle-depositing and mat-forming devices are inadequate to produce uniformly textured mats."

5.2.8.2 Terms, Failures, Volumetric and Gravimetric Dosing of the Material. Deck layers consisting of very fine particles should have a uniform structure. Modern screening or sifting equipment has markedly improved the uniformity in deck layers.

Spreading of chips by mechanical means, by throwing or blowing produced always a wedge-shaped beginning of the mat.

Fig. 5.93 shows schematically what happens: the wedge is characterized by three geometric parameters: spread-length, the spreading-angle and mat thickness. If the spreading angle is too steep, the spread-length becomes too short, and board properties are affected in the spread direction; the shorter the spread-

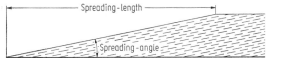

Fig. 5.93. Illustration of the terms spread-length and spreading-angle. From *Trutter* and *Himmelheber* (1970)

length the less is the parallel arrangement of the particles in the whole carpet to its middle-plane. Too steep spreading-angles cause horizontal shear forces which may result in the rise of destructive forces within the press. Proper dosing of the material in mat-laying is decisive. In recent times many, mostly patented chip-spreading devices have been developed that in this book only a short survey can be given. The history has been explained by *Johnson* (1956) and the modern development by *Steiner*, in *Kollmann* (1966, p. 303—323) and *Mitlin* (1968, p. 93 to 96).

Steiner (in *Kollmann*, 1966, p. 305—306) especially discussed possible failures in chip-spreading. This is a principal question which is illustrated in Fig. 5.94. Part a) shows how probably wrong spreading angles in the feed direction of the chips can cause too high edges on the one side and too low ones on the other side. This means too high density in the figure at the left side, too low density at the right side and a heterogeneous board after pressing. Part b) shows that between two transport cauls in the case of continuous spreading a distance between two following cauls may cause failures as illustrated. Part c) shows how failures according to a) and b) can be eliminated by edging prior to pressing.

26*

The material of the cut off edges is led back to the bin or the upper part of the spreading apparatus. Part d) shows the above-mentioned disadvantage of a too steep spreading angle. Part e) shows the correct manner of spreading; the chips should as far as possible lie parallel to the plane of the transport cauls.

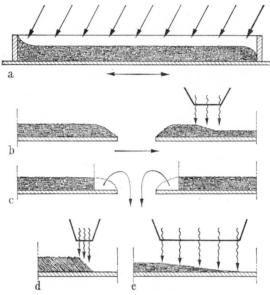

Fig. 5.94a—e. Failures in chip-sprading. From *Steiner*, in *Kollmann* (1966, p. 305). a) Too high border (left side of the frame) and sloped border (right side of the frame) effect too high and too low density respectively of the pressed board after frame-forming in one feed direction and/or sloping incidence of chips; b) Sloped narrow borders, using continuous forming, chips falling down through the distance of two following transport cauls (plates); c) Removal of the according to a) and b) insufficiently spread parts of the carpet by edging or trimming (not hatched parts) prior to press loading; d) Steep, like roofing tiles laid down chips due to too narrow chip exits of the spreading machine in feed direction or too large spreading amount of the machine; e) Favorable position of spread chips as parallel as possible to the transport caul

For the dosing of the resinated chips the following systems are applied:

1. Purely volumetric dosing with the aid of a dosing bin. The accuracy of spread is controlled by a weigher after the mat-laying device (Fig. 5.95). This system which is acceptable in plants of small to medium capacity, has the following advantages:

a) Control of the output of the mat-laying system;

b) Control of the weight of total spread chip amount and thus insight into the relationship between output of mat and speed of cauls;

c) Low investment cost and foolproveness.

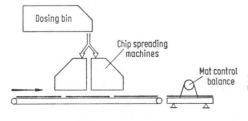

Fig. 5.95. Scheme of spreading with volumetric dosing and mat control weigher. From *Trutter* and *Himmelheber* (1970)

2. Control of the weight of chips. If a belt weigher is applied the exit device of the dosing bunker can be automatically regulated. By use of correct device (Fig. 5.96) the spreading apparatus works volumetrically. If the spread weight varies within determined quantity limits, then the chip-spreaders could run almost empty or be fed above the desired control value. If the bulk weight varies, the chip-spreaders could be fed below or above the desired control value.

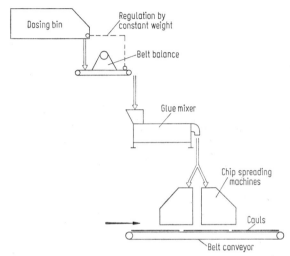

Fig. 5.96. Scheme of spreading with predestined weight and volumetric output of the spreaders. From *Trutter* and *Himmelheber* (1970)

5.2.8.3 Spreading Machines and Spreader Heads. Fig. 5.97 illustrates a chip-spreading machine of high capacity. The chips are uniformly distributed over the whole width of the apparatus by means of a worm feed. The details may clearly be seen in Fig. 5.97:

The steering flap and the various picker rolls in the pass-out head guarantee, in cooperation with the bottom belt, a uniform delivery of chips and an equal thickness of the mat. Variations in the design are possible.

There are various designs of spreader heads. In connection with the chip-spreading machine, type Schenck, Darmstadt, W.-Germany, spreading heads according to Fig. 5.98 (*Trutter* and *Himmelheber*, 1970) may be used:

a) The trowing roll is often applied for spreading normal face layers. There exists an effect of chip separation — variable by changes of the revolutions per minute and the spread-length — in this sense that the finer particles go into the lower layers and the coarser ones into the upper layers. Graded boards are produced, with fine outside layers, by a twin arrangement of two such spreader heads, throwing the chips in opposite directions.

b) The double roll dissolution head (each roll in the cross section star-shaped) causes expanded spreading without separation and with a small spreading-angle. The spreading is symmetrical and the apparatus can produce "cores" as a single machine.

c) The three roll spreader head, consists of star-like shaped rolls revolving in the same sense as indicated by arrows in Fig. 5.98 c. The spreading is still more expanded. This spreader head is suitable for high capacities, its separation effect is low. The boards produced are very homogeneous.

d) The spreader head for fibers has been developed for fiberlike and very fine ground particles. Combined rolls above the preformed mat serve to avoid clogging. These spreader heads are applied when very fine surfaces are required, especially in 5-layer boards.

e) Wind-spreading chambers effect a strong separation of the chip material in cases where the differences in chip size are great. The spreading effect can be regulated by amount and velocity of the air. For graded boards a twin arrangement is necessary.

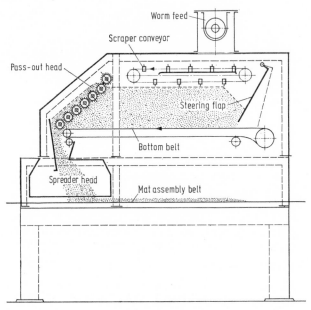

Fig. 5.97. Scheme of a chip spreading machine, manufactured by Schenck, Darmstadt.
From *Trutter* and *Himmelheber* (1970)

Vibrating mat-laying equipments of various design were in use in some older factories. For instance the Bartrev-process (*Fischbein*, 1950) used for carpet laying, a vibrating mesh tray, acting with a substantially vertical motion. Changes in thickness were obtained by altering the level of the tray above the feed belt, and the frequency was adjusted to suit each particular material. In some American installations vibrating screens have been used in conjunction with the belt and levelling roll type of equipment to further even out the material before deposition on the caul plates. Furthermore various types of vibratory feeders are in use for small surface chips in a number of plants (*Mitlin*, 1968).

Spreading machines can also work with a combination of weight and volume dosing. Wind spreading has been applied successfully by Bison-Werke, Bähre & Greten, Springe, W.-Germany. The scheme of such a spreader is shown in Fig. 5.99. After dosing the total chip material is fed centrically into the spreading machine. In this machine are two blower systems working in opposite directions. The air streams carry the fine particles farther than the heavy, coarse ones, which fall down nearly vertically. The result is that deck layers consist nearly exclusively of finest particles and the rather porous cores of coarse material.

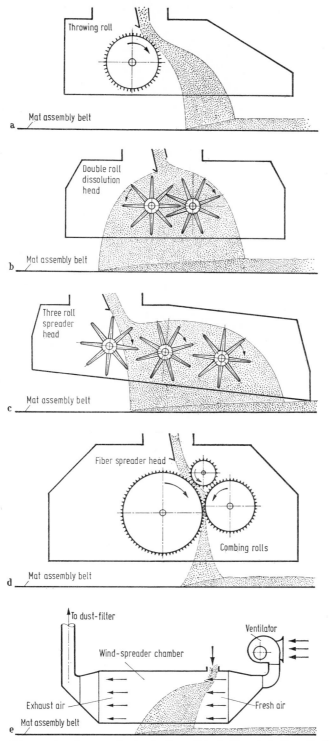

Fig. 5.98. a—e. Scheme of various spreader heads. Types Schenk, Darmstadt. From *Trutter* and *Himmelheber* (1970)

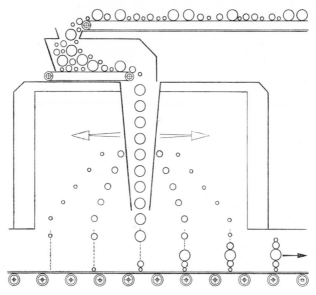

Fig. 5.99. Scheme of a chip spreading machine using wind (horizontal air flow). Type Bähre & Greten, Springe. From *Trutter* and *Himmelheber* (1970)

The combination of four spreaders is shown in Fig. 5.100. The wind spreading device consists of two symmetrical halves. By this arrangement it is possible to spread different chips, glued in distinct manner and with different moisture content separately and controlled for face layers and cores.

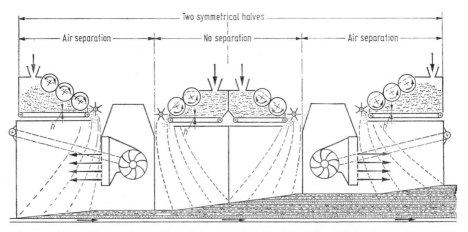

Fig. 5.100. Scheme of a spreading machine with four spreader heads. Type Bähre & Greten, Springe. From *Trutter* and *Himmelheber* (1970)

5.2.8.4 Resultant Mat. The resultant mat is very loose and thick. It is poured 3 to 20 times as thick as the final thickness of the boards, depending on the type of chips and the wood species used (*Kollmann*, 1955). The relationship between the initial and final thickness largely depends on the dimensions of the chips and the density of the wood or wood-mixture. Mats must be prepressed and pressed.

5.2.9 Prepressing, Wetting, Flat Pressing (Presses and Associated Equipment), Methods of Operation

5.2.9.0 General Considerations. As has been already explained the resin-coated particles, poured on the cauls, form a very loose and thick mat. The mat is poured up to 20 times as thick as the final thickness of the boards. During the pouring operation, and with any vibration of the loose mats, there is a marked tendency for the finer particles to sift through and settle nearer the bottom (*Johnson*, 1956, p. 163). The result is that the chip-glue mixture is lacking in homogeneity which changes most of the physical properties (especially the mechanical ones) for the worse; the appearance of the finished boards is also changed.

Particleboard plants are usually referred to as flat pressed or extrusion systems. *Mitlin* (1968, p. 96) writes: "This is a definition which leaves to be desired, and it may be preferable to talk in terms of boards in which the longitudinal axis of the particle is essentially parallel to the surface or perpendicular to the surface."

The present Section 5.2.9 is exclusively concerned with flat-pressed systems consisting of static, single or multi-daylight presses. More recently continous presses came into use; although they produce flat-pressed board, they belong principally to the category of extrusion presses. The predominant extrusion processes are dealt with in Section 5.2.12. So it is evident that to some extent the terms are intermingled. It may be added that at the moment continuous presses still demand considerable knowledge about theories on compression rates, heat transfer and the resin curing process (*Mitlin*, 1968, p. 96). It is assumed that suitable glued chips are fed on cauls, belts or sieves to a prepress or immediately to a hot press. It is further assumed that the mats have at least the desired width and that a gravimetric weighing-system achieved a high degree of control and accuracy.

As a rule the chip cake on its way from the prepress to the hot press is wetted on both surfaces to improve surface quality and bending strength of the board and to increase the rate of heat transmission into the core. The wetting methods are:

a) Spraying water on its upper surfaces of mat;
b) Spraying water on the transport cauls.

In the hot press, pressure and elevated temperatures convert the loose chip cake into a rigid strong board with desired properties. The process appears to be rather simple but in reality the total effect of all substantial factors resulting during hot pressing of particleboard is neither theoretically nor experimentally clear. The problem is difficult because too many factors are simultaneously combined.

K. van Hüllen (in *Kollmann*, 1966, p. 332—372) lists them as follows:

1. Structure of chips with respect to kind of materials used (wood, bagasse, flax shives etc.);
2. Type and dimensions of chips, flakes, milled chips, ratio of slenderness (= chip length to chip thickness), chip surface area;
3. pH value of the raw material;
4. Structure of the chip cake (uni- or multi-layers, spread height);
5. Pretreatment of chips;
6. Glue coating of chips (proportion of resin, type of adhesive, distribution and dispersion of binders, additives such as water-repellents, fire retardants, insecticides etc.);

7. Position of the chips in the cake;

8. Distribution of chips over the board area (homogeneity, exactness of spreading);

9. Prepressing (pressure, duration, temperature etc.);

10. Moisture content and its distribution over thickness and area of the board;

11. Manner of heating and temperature in the hot press;

12. Pressure;

13. Pressing cycle (pressing rate or speed);

14. Program of pressure application and press closing respectively;

15. Construction of press (stiffness, accuracy);

16. Effects of the chip cake during the transport especially in the automatic press loader.

The survey given above permits the conclusion that most causes of defective production occur in front of the hot press. The hot press is not able to equalize or compensate completely all defects originated in advance. Nevertheless it seems to be possible to improve the manufacture of particleboard by research and further development of machines in such a way that high-quality boards leave the press, which do not need finishing such as sanding, conditioning etc.

5.2.9.1 Prepressing. The following methods of prepressing are, in general, applied:

a) Compression of the mat between the platens of a single-opening hydraulic cold press. This press may be stationary or movable on rails back and forth; the path of the press is equal to the desired length of the ready particleboard after hot pressing. The system is characterized by continuous spreading of the chips on an endless belt. The movement of the prepress in action on its way forth is synchronized with the belt speed. At the end of the parallel movement of belt and prepress the latter opens and returns with increased speed to its inlet position. The scheme of operation is illustrated in Fig. 5.101 (*Trutter* and *Himmelheber*, 1970), showing the system Pagnoni, Monza, Italy.

b) Utilization of a system of loaded rolls or rollers through which the caul with its mat is passed. Another variation is as follows (Fig. 5.102): Continuous spreading of chips on plastic foils in series — movement on an endless belt — sawing to length — weighing — prepressing — hot pressing in a multi-daylight press, system Bähre & Greten (*Trutter* and *Himmelheber*, 1970).

The compression roller systems have, of course, both advantages and disadvantages (*Johnson*, 1956, p. 164). Advantages are:

1. Low initial cost of construction;

2. Simple, inexpensive, and continuous operation.

Disadvantages are:

1. The pressure is applied to the mat only through "point contact" of the rollers with the mat;

2. The actual pressure applied may be somewhat indeterminate because of the nature of the mat;

3. Particles tend to flow or squeeze out from under the roller as pressure is exerted;

4. Flow is evidenced along the edges of the mat, "where particles tend to spill over the caul rim. The density of the board is reduced along the edges, and these defective edges must then be trimmed off."

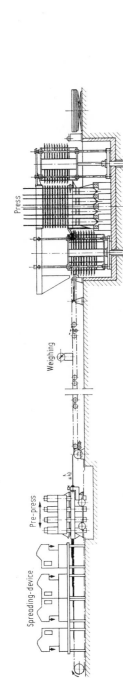

Fig. 5.101. Scheme of the chip mat and cake transport on an endless belt with movable prepress. System Pagnoni, Monza, Italy. From *Trutter* and *Himmelheber* (1970)

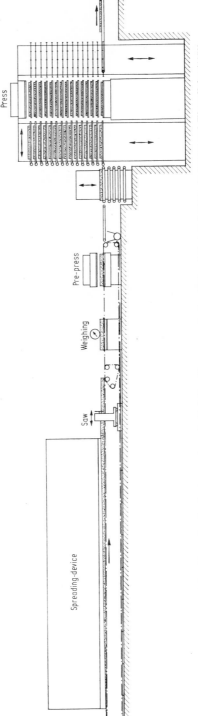

Fig. 5.102. Scheme of the chip mat and cake transport on plastic foils with prepress. System Bähre & Greten, Springe, in *Trutter* and *Himmelheber* (1970)

c) Application of chip mat and cake transport on an endless belt with a continuously operating prepress consisting of a relatively short pair of endless belt conveyors after which a saw follows (Fig. 5.103), system Siempelkamp, Krefeld, W.-Germany, shown by *Trutter* and *Himmelheber* (1970).

Some particleboard production lines with tray-belt loading operate without prepress and without caul plates in the hot press (Section 5.2.9.3).

5.2.9.2 Single and Multi-Daylight Presses[1]

5.2.9.2.0 General Considerations. The total process in a hot platen press mainly can be subdivided into four particular proceedings (*K. van Hüllen*, in *Kollmann*, 1966, p. 355/356):

1. Compression of the ship cake to raw board with small deviations from the desired dimensions, especially in thickness;

2. Simultaneous production of the necessary gluing pressure between the individual chips;

3. Heating of chips to the desired gluing temperature. Evaporation of water with the aim to reach an equilibrium moisture content;

4. Gluing of the individual loose chips to a rigid board.

The complex theory and practice of hot pressing of particleboard has been investigated over years by many researchers; only a condensed chronological list of names with respect to their publications referring to press conditions is possible: *Kull* (1954), *Rackwitz* (1954), *Klauditz* (1955), *Strickler* (1959), *Buschbeck* and *Kehr* (1960, with a very comprehensive bibliography), *Polovtseff* (1961), *Kunesh* (1961), *Neusser* (1962), *Grasser* (1962), and *K. van Hüllen*, in *Kollmann* (1966, p. 355 to 358).

5.2.9.2.1 Relative Investment Costs. There are important differences between the use of single daylight presses and of mult-daylight presses in particleboard manufacture. The reasons for the differences, however, are not technical but economical connected with the relative investment costs for the two press systems. The price of a single daylight press increases with length slightly less than proportionally, and with increasing width considerably more than proportionally. Doubling the length of a single daylight press means approximately doubling the price. Doubling the width, however, means considerably more than doubling the price. The relationship between pressing area of a single daylight press and its capacity is more or less linear, doubling of the capacity with a single opening press leads under some circumstances to more than doubling the investment costs.

In contrast to single daylight presses for multi-daylight presses the investment costs per square meter of pressing area decrease with an increasing number of daylights, as shown in Table 5.15.

Due to comparatively high investment for single daylight presses it is necessary to take advantage of all technological possibilities to obtain the highest possible capacity with the press. The "fast technology" applied to single daylight presses is not the result of efforts towards an exceptionally perfect technical development, but the forced result of economic circumstances.

---

[1] The following explanations are based on a publication by *R. Hesch*, Board Manufacture, Pressmedia Ltd., Ivy Hatch, Sevenoaks, Kent, U. K., Vol. XII (1969) p. 59—73, p. 79—82. The author is indebted to Dr. *Hesch* for his courtesy.

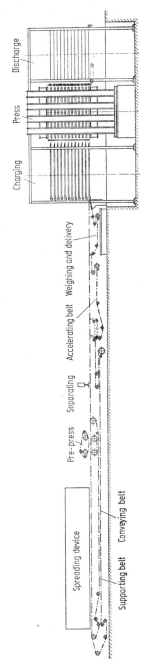

Fig. 5.103. Scheme of the chip mat and cake transport on an endless belt with continuously working movable prepress. System Siempelkamp, Krefeld.
From *Trutter* and *Himmelheber* (1970)

Table 5.15. Investment Costs for Hydraulic Presses per Square Meter of Usable Hot Plate in
Relation to the Number of Daylights
(Siempelkamp & Co., Krefeld, 1969)

Press No	No. of daylights	Length of press		Total pressing area		Specific pressure		Index of investment per sq. m.
		mm	ft.	sq. m.	sq. ft.	kp/cm²	lb./sq. in.	pressing area of hot plate %
1	1	3,050	10	6	64	23	327	100
2	2	5,450	19.1	20	214	23	327	67
3	2	7,315	24	27	288	25	355	64
4	3	3,660	12	20	214	25	355	61
5	3	7,320	24	40	428	30	426	56
6	4	3,660	12	27	288	25	355	52
7	4	4,880	16	36	385	25	355	48
8	6	5,500	18	60	642	30	426	41

Notes:

a) The comparison comprises complete press installations including forming line, charging and discharging equipment, hydraulic system and electrical equipment;

b) All presses have a finished board width of 1,830 mm for comparison;

c) Presses No. 1, 2 and 3 are without, presses No. 4 to No. 8 with simultaneous closing device;

d) Calculations were not available with the same specific pressure for all presses. In view of the fact that the specific pressure of press No. 1 is 23 kp/cm², i.e. 327 lb./sq. in., in contrast to 30 kp/cm², i.e. 426 lb./sq. in., for press No. 8 the decrease in investment costs is even more marked than shown in the table.

5.2.9.2.2 Technological Aspects. With single daylight presses extremely short pressing times can be achieved (Table 5.16) by:

1. Extremely high press temperatures;
2. Low moisture content of chips;
3. Utilization of urea-formaldehyde resins with a high content of formaldehyde and a higher proportion of hardener, permitting short curing times.

Table 5.16. Comparison of Technological Conditions Applied on Single Daylight Presses and
Multi-Daylight Presses
(*Deppe* and *Ernst*, 1965, *Hesch*, 1969)

		Single daylight presses	Multi-daylight presses
Temperatures of hot plates	°C	180···210	135···170
	°F	356···410	275···338
Moisture of chips after gluing %	Surface layer		14··· 18
	Core layer	9··· 11	10··· 14
Curing time in min per mm of raw board thickness		0.13···0.2	0.25···0.4

Particleboard plants using single daylight presses appear less complicated compared with multi-daylight presses. From the mechanical, hydraulic and electrical point of view they are less susceptible to trouble.

The functional advantages of single daylight presses are accompanied by disadvantages which appear after some years as symptoms of age. Single daylight presses are always longer than multi-daylight presses, and normally also wider. Thermal stresses and nonuniform mechanical strains lead to stressing and distortion of the press table, press ram and with ageing to increased thickness variation of the raw boards, i.e. higher sanding losses.

A further problem of single daylight presses is the steel belt which is a substantial part of the press. Steel belts with widths of 1,800 mm and more, as mainly required today, can at present only be produced by welding. Due to the high thermal and mechanical stress to which the steel belts are subjected, difficulties are frequently caused by weld fractures. Special skill is required to repair the weld, and this operation has to be frequently repeated.

As a result of the high stresses and of repeated welding the steel belts warp with advancing years, especially near the weld. Together with the tolerances of the press itself, caused by age, this finally results in considerable sanding losses.

Single daylight presses are also hardly suitable when it is intended to manufacture very thick low density boards of 40 to 45 mm in densities ranging from 300 to 400 kg/m³ ($\sim$18.7 to 25 lb./cu. ft.). For wood such low densities cannot normally be reached as there are only a few suitable species for manufacture of boards with densities lower than 450 kg/m³ (28 lb./cu. ft.). But with bagasse, jute chips, flax and hemp shives the manufacture of boards with densities of about 300 kg/m³ (18.7 lb./cu. ft.) and satisfactory properties is possible and attractive.

Summarizing, *Hesch* (1969) concludes that the advantages of single daylight presses due to less complicated design and operation are opposed by the disadvantages of aging.

5.2.9.2.3 Effect of Chip Moisture Content. Chip moisture influences the following properties of particleboard:

1. Strength;
2. Surface quality;
3. Separation of formaldehyde;
4. Required specific pressure.

A detailed investigation on the influence of moisture on the different quality characteristics of chipboard was carried out by *Kollmann* (1957). He found that water, added to the surface layer of particleboard, is the cheapest means of improving strength and surface quality of particleboard. *Fahrni* published similar findings in 1942 in a patent application (DBP 967,328).

In his investigations on the influence of moisture variations in chip mats before pressing on the properties of particleboard *Kollmann* (1957) comes to the conclusion that bending strength reaches its maximum at a surface layer chip moisture content of 18 to 20%. Depending on species of wood, shape of chips, type of resin and other conditions, the optimum moisture content may change within a small range. The order of values found by *Kollmann* corresponds to industrial experience and may be considered as a guide line.

Kollmann started his investigations with a uniform moisture content of 17.5% in the core and surface layers. Whilst he maintained the core moisture content of 17.5% constant for further comparison, he raised the humidity of the surface layers in his first series I of experiments by spraying water onto the chip mat. In contrast to this the increase in moisture content in experimental series II was obtained by air conditioning.

The results of *Kollmann*'s investigations are summarized in Fig. 5.104. It is shown to what extent the bending strength can be improved by moistening the chips. Curve II demonstrates that the bending strength increases more rapidly if the chip moisture content is raised by conditioning rather than by spraying.

In principle the results of *Kollmann* (1957) are confirmed by a very exact investigation carried out by *Plath* (1967). One of the points proved is that under maintenance of all other technological conditions an increase in moisture content of the surface layer from 12% to 20% results in an improvement of bending strength in the range of 218 to 227 kp/cm², i.e. 3,100 to 3,220 lb./sq. in., to 225 to 263 kp/cm, i.e., 3,620 to 3,730 lb./sq. in.

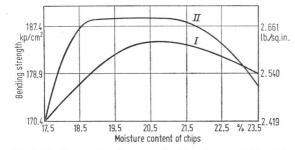

Fig. 5.104. Change of bending strength of three layer particleboard in relation to change of moisture content of surface layer chips. *I* = Increase of surface layer humidity by spraying water onto the surface, *II* = Increase of surface layer humidity by conditioning of chips. From *Kollmann* (1957)

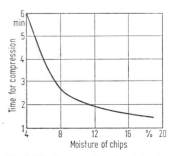

Fig. 5.105. Pressing time in relation to moisture content of chips when hot pressing single layer particleboard. From *Kehr* and *Schoelzel* (1967)

Similar results were obtained for tensile strength in the above-mentioned investigations. For boards with a thickness of 16 mm (i.e. 5/8 in.) a density of 600 kg/m³ (37 lb./cu. ft.), a pressing time of 6 minutes and a moisture content of 12% for surface layer chips and 8% for the core layer chips, *Plath* (1967) found a tensile strength of 5.68 kp/cm² (80.7 lb./sq. in.). By increasing the moisture content of the surface layer to 20% and the core layer to 10% the tensile strength was increased to 6.21 kp/cm² (i.e. 88.3 lb./sq. in.), that is about 10%. Furthermore it was found that the differences in tensile strength caused by moisture content variations are the higher the shorter the pressing time. The influence of chip moisture content on board strength becomes the more important the thicker the chips and the higher the density of the wood species used. The reason for the these improvements of strength of boards may be explained by the plasticising effect of moisture. An extremely dry chip is brittle and results in a predominantly "punctiform" gluing with the surrounding chips. In contrast to the inflexibility of too dry chips, particles with 14 to 20% humidity show a substantial increase in their plasticity. Glued chips lying across and next to each other are converted under pressure into a compact mass. The sufficiently moist and plastic chips give a glue line with greater surface, that means a higher degree of efficiency of the glue.

A sufficiently high moisture content and concomitant good plasticity of surface layer chips are one of the most important prerequisites to obtain good and closed board surfaces. A gradient of moisture content of surface and core layers is desirable in the manufacture of high-quality particleboard. Small differences in moisture content between surface and core layers effect only a slight graduation of density over the cross-section of board. The surface of such boards will be of average quality, the tensile strength perpendicular to the surface will be above average. In contrast, when choosing a high moisture content of surface layers of about 18 to 20% and a moisture content of 10 to 12% for the core layer, the result will be a much greater density variation in the cross-section. The surface layers will become much more compressed due to the plasticity of the moist chips whereas the dry and brittle core layer is less compressed and remains more porous. The result will be a board with an above average surface quality, increased bending strength, but reduced tensile strength due to the porous core layer.

A further increase in surface quality can be achieved by spraying water on the surface of the chip mat. The influence of the described actions on the surface quality (smoothness or roughness) are shown by *Kollmann* (1957).

Excessive *separation of formaldehyde* from cured boards is regarded as annoying because of the irritations provoked on eyes and mucous membranes. According to *Plath* (1967) the splitting-off of formaldehyde is higher when using chips with high moisture content. On the other hand the subsequent separation of formaldehyde decreases with increasing pressing time, however without completely compensating the influence of moisture. This disadvantage can be moderated by adding more hardener especially to the core layer and by the addition of urea to the binder.

When employing multi-daylight presses with moderate temperatures and sufficiently long pressing times normally urea-formaldehyde resins with a comparatively low proportion of formaldehyde can be used.

Finally the influence of chip moisture content on the *required pressure* shall be dealt with. Investigations in this respect were carried out by *Kehr* and *Schoelzel* (1967). The results show a nearly hyperbolic relation between time of compression and moisture content of chips. At constant specific pressure the time required for compressing the mat increases with decreasing chip moisture content. (Fig. 5.105).

There is a direct relation between the pressing time and the specific pressure of the press. The conclusion is that for the same rate of compression a higher specific pressure is required when using dry chips. For the achievement of identical raw board density, assuming the same pressing time, single daylight presses have to be designed for a higher specific pressure.

5.2.9.2.4 Effect of Press Platen Temperature. An *increase in temperature in the hot press* is effecting the "fast technology", i.e. an acceleration of resin cure and accelerated evaporation of the moisture contained in the mat.

Damage to the wood or urea-formaldehyde resin due to the high temperature does not normally occur. Although the thermal decomposition of wood already starts at about 170 °C (338 °F) the necessary time of exposure at this temperature is much longer than the chipboard pressing time. Only in the case of breakdowns, when the board in the press cannot be discharged in time, temperatures above 170 °C may cause thermal decomposition which causes a considerable decrease in strength. The effect of the thermal decomposition of the urea-formaldehyde resin is of greater importance than that of wood.

5.2.9.2.5 Effect of Curing Time on Physical Properties of Particleboard. The attempt to reduce the pressing time by elevated temperatures is faced with a second difficulty which has so far only been partially solved. Wood is a highly insulating material. In order to achieve satisfactory curing in the core of particleboard sufficient heat has to be transferred to it. Because of the low thermal conductivity of wood the cake of particles itself is not suitable as a heat carrier. Therefore the heat transfer depends primarily on the moisture of the surface layer chips or on additional water sprayed onto the surface. This is vaporized by the heat provided by the platens and diffuses as steam into the board. Within the board the steam condenses on the cooler particles and heats them up by the heat of condensation. During the whole pressing operation a temperature gradient remains between the core and surface of board (Fig. 5.106). Instructive is Fig. 5.107 showing the re-

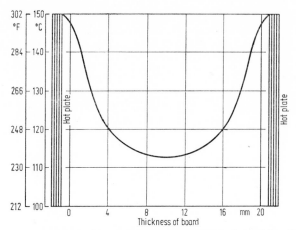

Fig. 5.106. Temperature gradient along the cross-section of a 20 mm (0.787 in.) particleboard after a pressing time of 9 minutes. From *Grasser* (1962)

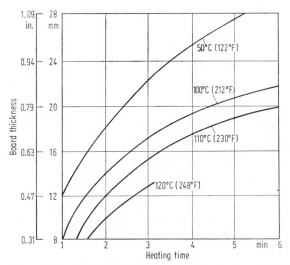

Fig. 5.107. Heating time for particleboard of different thicknesses when using a platen temperature of 150 °C (302 °F). From *Grasser* (1962)

quired heating time for particleboard of different *thicknesses*. The temperature gradient between surface and core of particleboard limits the possibilities of accelerating the curing of the core (*Grasser*, 1962). Time is lacking with single day-light presses. As a consequence of using the "fast technology" lower figures for tensile strength perpendicular to the surface are achieved, compared to boards produced in multi-daylight presses with sufficient pressing time.

Accordingly the reduced tensile strength is the consequence of the comparatively low plasticity of the chips due to low moisture content and the insufficient curing of the synthetic resin in the core. These conclusions correspond with the observations of industrial practice and with the investigations of *Kollmann* (1957), *Plath* (1967) and *Stegmann* and *v. Bismarck* (1967).

Stegmann and *v. Bismarck* (1967) in the course of their investigations also looked into the related problem of whether, and if so to what extent, an after-cure of the boards takes place in a hot stack to improve the tensile strength perpendicular to the surface when the initial pressing time of the boards was insufficient. Their findings are that a subsequent cure during storage of raw boards does not take place.

Instructive findings with respect to the influence of curing time on physical properties of particleboard were obtained by *Rayner* (1968). They are summarized in Table 5.17.

Table 5.17. *Properties of 3/4″ Chipboard Cured at 145°C (293°F) Spacer Thickness = 0.753*

Pressing time min		4	4 1/4	4 1/2	4 3/4
Thickness after 48 h storage	mm in.	21.1 0.831	19.4 0.783	19.2 0.777	18.9 0.764
Springback (as percentage of spacer thickness)	%	10.4	4.0	3.2	1.5
Swelling (24 h immersion in water)	%	15.6	14.9	14.3	13.2
Transverse tensile strength	kp/cm² lb./sq. in.	1.69 24	3.30 47	5.83 83	7.05 98
Density	g/cm³ lb./cu. ft.	0.628 39.2	0.642 40.0	0.648 40.5	0.652 40.7

Almost all properties of particleboard are improved by longer curing time. The springback of particleboard after 48 h storage is reduced from 10.4% to 1.5% as pressing time increases from 4 to 4 3/4 min. Low tolerances allow lower sanding losses. Swelling is also reduced by extending the curing time. The most important relation however is shown between curing time and transverse tensile strength. The prolongation of curing time of 3/4 minutes results in an improvement from 1.6 kp/cm² ($\sim$23 lb./sq. in.) at 4 minutes to 6.9 kp/cm² (98 lb./sq. in.) at 4 3/4 minutes.

5.2.9.2.6 Influence of Quantity of Hardener and Proportion of Formaldehyde in the Urea Resin on Quality of Particleboard. Besides the already discussed increase in chip moisture and increase in temperature, the chemical composition of the urea-formaldehyde resin[1] and the choice of type and quantity of hardener are important means of reducing curing time.

The molar ratio of neutral formaldehyde to urea in UF-resins varies from 1:1.5 to 1:2.0 according to *Mörath* (in: *Kollmann*, 1966, p. 59).

Free formaldehyde irritates the mucous membranes of nose and eyes and even the skin. Excessive subsequent separation of formaldehyde from particleboard therefore reduces board quality and must be avoided.

---

[1] In accordance with the technical literature the following abbreviations are used: for urea-formaldehyde UF, also uf, for phenol-formaldehyde PF, also pf.

The greater the excess of formaldehyde in UF-resins, the quicker the curing. Because of this, it is necessary to use UF-resins with a considerable excess of formaldehyde when employing single daylight presses, in order to obtain as short curing times as possible. On the other hand the subsequent separation of formaldehyde is the higher, the shorter the pressing time (*Plath*, 1967).

For the sake of completeness, it has to be mentioned that an increase in temperature advances the curing rate of resins and therefore counteracts the splitting-off of formaldehyde. The high temperatures as used with single daylight presses are mainly effective in the surface layers but not in the core. The boards are often discharged from the press with an under-cured core and emit considerable quantities of unreacted formaldehyde for some time. (*Plath*, 1966, 1967, 1968).

Another important auxiliary of particleboard technology is the hardener. Type and quantity are of great influence on curing time and quality of particleboard. These relations were investigated thoroughly by *Plath* (1967). Only the main results shall be summarized here:

a) A comparatively high quantity of hardener is required in order to get good curing and good physical properties when employing single daylight presses with short curing times;

b) With regard to subsequent separation of formaldehyde an improvement can be clearly observed when increasing the quantity of hardener. However, for the surface layer an extension of the curing time shows a greater effect on formaldehyde separation than an increase in the quantity of hardener. In contrast to this for the core layer lower formaldehyde separation is obtained by increasing the quantity of hardener than by an extended curing time due to the temperature gradient.

To summarize the different parameters, it has to be stated in agreement with *Deppe* and *Ernst* (1964, 1965), *Hauser* (1965), *Berger* (1964), and *Plath* (1966, 1967, 1968) that an extreme reduction of curing time by all the methods described will always result in particleboard with considerably higher subsequent separation of formaldehyde compared to boards produced by using a moderate technology.

It can be concluded that multi-daylight presses are to be preferred if particleboard with low subsequent separation of formaldehyde and high physical properties is to be produced under economic conditions.

5.2.9.2.7 Other Technological Aspects. Besides urea-formaldehyde, phenolic resins have recently gained greater significance because of their higher resistance to moisture. The increasing application of particleboard for housing purposes will probably result in an increasing demand for particleboard made with phenolic resins.

Phenolic resins normally require a much longer curing time than urea resins and temperatures above 170 °C (338 °F). The curing time for phenolic resins is generally still a little over 0.5 min per mm of raw board thickness. Recently new types of phenolic resins have become available with curing times of about 0.35 min per mm.

Compared to urea-formaldehyde resins, for single daylight presses the trend to phenolic bonded particleboard means a decrease in capacity of 50% or more due to the long curing time of phenolic resins. In contrast to this multi-daylight presses are only slightly affected in capacity when using phenolic resins as they are planned for longer pressing times from the very beginning. An increase in temperature to the level required for phenolic resins is no problem for modern multi-daylight presses.

In several less industrialized countries, which have to import phenolic resins, efforts are made to replace phenolic resins by domestic tannin resins which show a moisture resistance comparable to phenolic resins.

With a view to the possible application of phenolic and tannin resins, for economic reasons, single daylight presses can hardly be considered for particle-board manufacture any more.

The control of single and multi-daylight presses has been continuously improved during the last years. The available time for loading and unloading is short, and therefore the platens should close rapidly on the loose mats until shortly before the contact with the press platens. The last millimeters prior to the contact the movement of the platens must be slow to avoid blowing out of

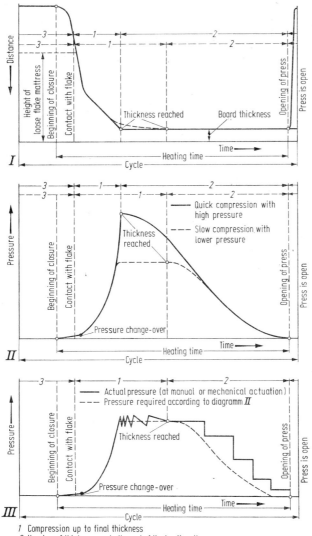

1 Compression up to final thickness
2 Keeping of thickness up to the end of the heating time
3 Additional operations

Fig. 5.108. Diagrams of pressure and temperature versus pressing time

chips; subsequently the closed position of the hot press should be reached as quickly as possible (Fig. 5.108). The same principle is valid for the reduction of the heating time and especially the ventilating of the press in connection with the opening procedure.

The press platens are heated preponderantly by hot water or recently more and more by oil as heat carrier. Steam heating is less suitable with respect to the desired high constancy of temperature over the platen area. With oil as heat carrier the press can be heated without difficulties up to 300 °C (572 °F) using a pressure-less working method.

The number of daylights or openings of hot presses has been increased though the heating times were reduced and the platen sizes became larger and larger. In connection with the high temperatures, to obtain short heating times, it is necessary to close the presses rapidly after loading. Such a procedure is especially important for caulless systems because in this case the cakes of chips are directly brought up to the hot platens. Therefore simultaneously closing devices were developed which permit the closing of all openings at the same time.

Moreover, this equipment produces more uniform conditions of temperature and pressure for the platens of all daylights. This is also valid for the ventilating, that is, the removal of the pressure. A critical point for simultaneously closing devices is the compensation of variations of chip cake thicknesses from one opening to another. Such variations may cause remarkable stresses in the carrying devices for the single hot platens. By means of a pressure oil unit (Fig. 5.109) the stresses

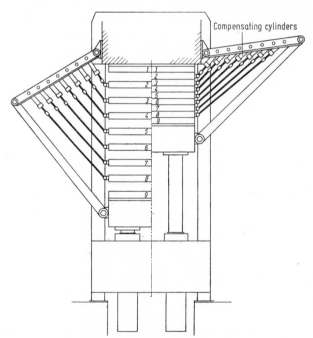

Fig. 5.109. Scheme of an oil-hydraulically operated simultaneously closing device. Type Becker & van Hüllen, Krefeld

are equalized. The ability to compensate is so great, that, even when one cake between two platens is absent, the press will not be damaged. Other systems use springs as elements for compensation. The oil-hydraulically operated simultaneously closing devices also permit pressing without stops. The following advantages are reached:

a) No costs for stops (distance bars);
b) Abolishment of time for the exchange of stops for various board thicknesses;
c) No influence of badly calibrated or dirty stops;
d) Possibility to produce any desired board thickness at any time.

Finally a few essentials of the heated platens are given. "The heating medium may be circulated through holes which are bored in solid steel platens or the platens may be constructed of fabricated steel and contain large steam chambers of labyrinthic design. The heating medium is introduced into the vertically moving platens by means of 'slip' pipes or, in some cases, flexible hose. The platens should be thicker and more rigid than those used for plywood manufacture to prevent warping or springing which would lead to nonuniform thickness of the pressed board and possible damage to the platens themselves. Sufficient support should be provided to prevent lateral movement of the platens." (*Johnson*, 1956, p. 167—170).

5.2.9.3 Caulless Pressing Systems[1]

5.2.9.3.0 General Considerations. Caulless pressing systems for the manufacture of particleboard are proven over a period of more than ten years in many plants up to a maximum capacity of 150 metric tons per day. Nevertheless at present the majority of board plants are still being built to use rigid or flexible transport caul plates to receive the board in the forming line and remain with the unpressed material through the pressing operation. The most frequently stated arguments in opposition to caulless systems are greater mechanical complexity, increased precure problems, lesser tolerance for short-fibered residual wood materials and greater resin content.

There are three types of caulless systems:

1. The tray-belt moulding system;
2. The tablet system;
3. The continuous belt system.

5.2.9.3.1 Tray-Belt Moulding System. There is a choice of:

a) Tray-belt plants with single daylight press for maximum board sizes and for surface-treated particleboard;
b) Tray-belt plants with loading device and multi-daylight press with simultaneously closing device.

These two types of plants work with mat formation in the reversing system, tray-belts travelling singly or in pairs or jointly in endless sequence for the mat formation.

For higher capacities, tray-belt plants for continuous mat formation are available in which the empty tray belts, ready to receive the mats, are returned in the lower part of the forming train.

---

[1] The author is indebted to Mr. *E. F. Steck* and Messrs. Siempelkamp & Co., Krefeld, W.-Germany; the paper, partly verbally quoted, partly condensed by the author or enlarged, was presented at the Fourth Washington State Symposium on Particle Board, March 1970, Pullman, Wash.

The essential element of the tray-belt moulding system is a retaining box, or mould, similar in principle to a deckel box with rigid side retaining walls but with three special features (Fig. 5.110):

a) A bottom belt which serves as a floor surface covering the full area inside the frame,

b) A front gate with movable pivot levers near the front edge of the frame,

c) A movable rear wall located inside of the rigid frame.

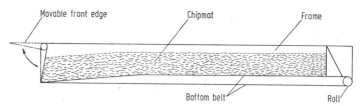

Fig. 5.110. Principle of the tray-belt moulding frame. From *Steck* (1970)

Fig. 5.111 shows some examples of tray-belt moulding systems. In the case of the single-opening press line, the forming station is normally in front of the press. Thus, the line is compact and mechanically simple. During the pressing cycle, the tray-belt mould passes underneath the spreading machines in a forward and reverse direction. In making a 3-layer board, the bottom of the conveyor belt is first spread with face layer material followed by the first layer of core material. The central diagram of Fig. 5.111 illustrates a multi-opening press-line. In a special version of this system for shorter pressing cycles, the frames can emerge from the loading cage in tandem with the other and then back into the various decks of the loading cage.

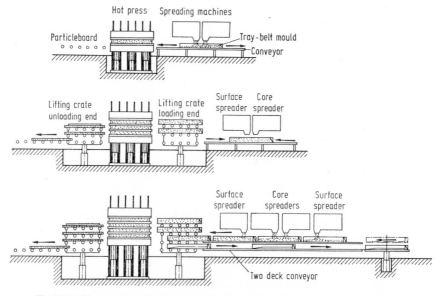

Fig. 5.111. Schemes of possible arrangements of tray-belt molding systems. From *Steck* (1970)

Another variation is shown in the bottom diagram (Fig. 5.111) in which a two level conveyor is used to transport the tray-belt moulds through a forming line and back to the press loader.

Board sizes as large as 2,240 mm by 12,800 mm ($\sim$7 ft. by 42 ft.) are produced with a version of this system. In Italy 4 mm-thick particleboard is made in four-opening presses using the tray-belt moulding system at a rate of 27 press cycles per hour.

By changing the molds, board size can be varied in both length and width within the limits of the hot press design. Since no prepress is used and a minimum of mechanical support equipment is required to complete the line, a captive plant with as little as 30 metric tons of daily capacity is economically practical.

The tray-belt mould has its own limitations. The transporting frame assembly is a complex mechanism. Time separation for press loading, loader guiding principles and forming sequence functions must be compensated by speed controls accelerating and decelerating in frame transport. There is no opportunity to reject the mat before pressing. Hence, the normal maximum capacity of this system would be about 160 to 170 metric tons per day.

5.2.9.3.2 Tablet System. In some ways the tablet caulless loading system is similar to the tray-belt moulding system because a rigid frame encircles the mat. There is also a flow of the frame below the mat spreading stations. The primary difference occurs in the floor for the frame which is a rigid plate or tablet made of reinforced plastic instead of an integral conveyor belt. The similarity ends because the tablet system enters in a stationary prepress: Cycle time and pressure are long enough and high enough to produce a firm mat. In its cold, compacted state, the mat can be handled without degradation by sliding it over suitably smooth surfaces to its place in the multi-opening hot press. Fig. 5.112 illustrates a self-reversing tablet line. One frame on its supporting plate, or tablet, moves to the forming line, while a second tablet and frame assembly is simultaneously moving into the elevating transfer station in front of the prepress. The third level in this station compensates the time the frame requires to make the forward and reverse path under the chip spreaders.

To be effective in producing a cake which can be safely transported, experience indicates that the prepress pressure should approximate the pressure of the hot press therefore, at least in the range of 25 to 28 kp/cm² $\approx$ 350 to 400 lb./sq. in. A distinct holding time between 15 to 18 sec for the total prepress cycle time is required. The prepress platen fits inside of the frame within close tolerances. Before full pressure is exerted on the mat, the frame is stripped off the tablet surface in an upward direction so that lateral expansion of the mat is not restrained. Normally, the prepress platen is exactly the same size as the pressed board after trimming.

A conveying mechanism inside the prepress slides the cake off the original conveying tablet to transfer to accelerating stations behind the prepress (cf. Fig. 5.112). The means should be examined by which the cake is deposited on the hot platen from its resting place on the loader plates. Fig. 5.113 shows on the top (a) the last prepressed cake sliding or being pushed at its trailing edge by the preloader onto the transport plate of the press loader. In the center sketch (b) one can see the loader plates conveying the cakes into the hot press and pushing out the pressed boards. In the bottom sketch (c) it is illustrated how the loader plates are withdrawn from underneath the mats. Cross bars restrain the mats from following.

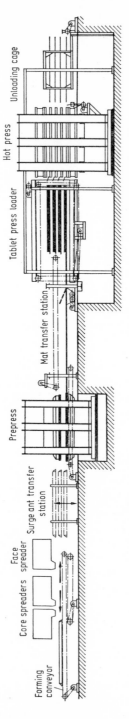

Fig. 5.112. Tablet loading particleboard system with self-reversing line. From *Steck* (1970)

The frictional forces between the mat and the carrier plate determine the pressure exerted by the trailing edge of the mat on the restraining cross bar, and it is predominantly the length of the mat which determines the specific pressure between cross bar and mat. If the forces are too high the mat will crumble. Then, two possibilities exist:

1. Improvement of the mat structure by increasing the resin content, or,
2. Decreasing the mat length.

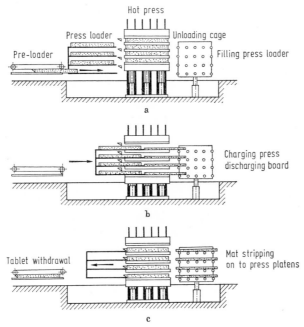

Fig. 5.113 a—c. Operation-stages in loading and unloading the hot press using the tablet system. From *Steck* (1970)

In practice the mat length is normally restricted to between 3,660 mm and 4,270 mm ($\approx$ 12 and 14 ft.) in the tablet system. The resin content is not higher than in any caul type or caulless system. The tablet system is adaptable to a broad range of plant capacities. For instance a self-reversing forming line with 1,520 mm by 3,660 mm (5 by 12 ft.) board size, made in a 4-opening press, has a capacity of something in excess of 60 metric tons per day with the same machines and in the same line length. The line width and press can be further widened to 3,660 mm $\approx$ 12 ft. thereby increasing the capacity to 150 metric tons.

The normal modification of the line for a capacity higher than 100 metric tons includes the addition of further openings in the hot press and an additional parallel transport system for the reassembly of the frame and transport plates. Modifications are also made to the number and location of the mat formers. When the frame is separated from the cake and the tablet in the prepress, both units exit from the prepress toward the hot press but at separate elevations. The frame is at a higher level than is the plate with the cake resting on it. At the first station behind the prepress, the frame is cross-transferred to a position above the parallel return conveyor.

The tablet and cake are meanwhile carried toward the hot press loader. At a separate station, the cake is slid from the tablet to the receiving plate of the press loader, after which the tablet is like wise cross-transferred to the parallel return conveyor. When the tablet arrives at the frame cross transfer position, on the return conveyor a conventional frame and tablet are reunited. The transport plates never enter the hot press; the transport line is considerably shorter than the full additional length of the press loader + hot press + unloader + caul cooling + cleaning station; the return parallel conveyor line is a part of the forming line. Forming on the return conveyor is done to limit the internal shear stresses in the board which occur when the mat is layed in only one direction. In this merry-go-round system, the first forming machine for a 3-layer board is a tace layer former. Next come core layer forming units, depending on the capacity of the plant. At the end of the parallel return conveyor line, a cross transfer carries the partially filled frame assembly to a position in line with the remaining forming heads. Closest to the prepress is a face layer former in the case of 3-layer board. In this fashion, the lower half of the mat is layed with particles oriented in one direction and the upper half of the mat with chips oriented in the opposite direction. In the self-reversing line, the same structure of mat build-up occurs. Even in homogeneous board, reverse orientation of particles in each mat is desirable. Since multiple forming heads are used with this system, it is the practice to treat the surface layers and core layers separately for moisture and resin curing characteristics for better performance in the hot press.

The schematic arrangement in Fig. 5.114 shows a comparison of the same line at floor level versus an elevated level. Whether at floor level or elevated, the tablet system offers some advantages. Most prominent is the minimal trim loss in finishing the board.

There is no caulless system that exerts less bending stresses on the mat after forming than does the tablet system. Vertical drop from loader plate surface to platen surface is only 3.2 mm ($\sim$1/8 in.). Due to the proximity of the top surface former to the prepress, there is less possibility for the sifting of fine surface particles

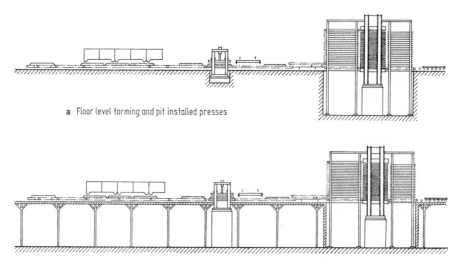

a Floor level forming and pit installed presses

b Elevated forming and press line

Fig. 5.114. a) Floor level and b) elevated lines for the tablet loading merry-go-round system.
From *Steck* (1970)

into the underlying core layers and no sifting occurs after mat consolidation in the prepress.

In both the tablet system and the tray-belt moulding lines, the procedure for filling the frames is one of overspreading before and behind the frame ends.

In any system, the tablet type plant has limitations. One is that of board length. Another is minimum board thickness. The normal minimum considered to be safe is 9.5 mm ($\approx 3/8$ in.) with dry refined wood as raw material. Under certain selective conditions, that minimum might be as low as 6.4 mm ($\approx 1/4$ in.), but a careful analysis of both raw material and the properties required in the finished product is necessary. The prepress is a machine of major magnitude and represents a significant portion of the total cost of the line; it is the limiting factor in the productive capacity of the system.

5.2.9.3.3 Continuous Belt System. For plants of larger capacity, there is a new caulless process, a continuous forming belt system.

In recent years, there has been a trend toward larger board sizes and plants of higher capacities. Larger boards can better serve certain specialized applications. Edge trim and cut-to-size losses decrease in percentage of raw board produced as the panel size becomes larger.

Two obstacles had to be overcome to make plants of very high capacity feasible. The first was industry's acceptance of making board wider than four or five feet. The forming machine makers, the glue producers and the process technologists had to convince themselves as well. The second major problem was one arising from the discontinuous pressing process. It was obvious that when the press was closed on a load of board, a number of mechanical devices were required to stock, transport, elevate, etc. the mats being formed during the hot press cycle. Even though some acceleration-deceleration systems have been employed, the end result was still one in which either the forming process was discontinuous or approaching discontinuity by taking large swaths of the mat out of the product line after forming. Obviously, continuous forming without furnish recirculation losses has advantages in both speed and quality control of the mat.

A system which is not necessarily predicated on the manufacture of wider than normal board, but which does provide continuous forming with minimum mat loss is schematically depicted in Fig. 5.115. Typical components are:

1. Forming heads or mat spreading machines;
2. Belt conveyor on which the mat is formed continuously;
3. Heavy supporting conveyor for the forming conveyor;
4. Permanent magnet;
5. Metal detector;
6. Continuous prepress;
7. Flying cut-off-saw (mat-separating device);
8. Mat accelerating belt section;
9. Mat preloader and weight control station;
10. Mat reject station;
11. Press loader;
12. Heat curtain doors;
13. Board ejectors;
14. Hot press;
15. Discharging rolls;
16. Unloading cage;
17. Board unloading device;
18. Board discharge nip rolls.

The mat remains at the same elevation from the beginning of the forming line until it is in the press loader.

In diagram 1 the leading edge of the mat on the left has arrived at the leading edge of the forming conveyor and is separated from the mat on the separating conveyor by only the 2 in. (≈ 51 mm) which disconnect the mats in this system, regardless of board length. The figure shows three conveyors in tandem:

The forming conveyor is at left. For a specified board thickness and density its speed is constant. The conveyor section in the middle is the acceleration station. It must run at two speeds:

a) The speed of the forming conveyor during the period of time a single mat is in process of transfer from the forming conveyor until that mat is totally on the accelerating conveyor, and

b) at a higher speed to create the time for the separation between mats required to accomodate the capacity of the total press line.

At the right of the diagram is the preloading conveyor which receives the mats from the acceleration conveyor and delivers them to the press loader. The preloading conveyor must similarly run at two speeds:

a) The high speed of the acceleration conveyor during the period of mat transition, and

b) the speed required to load the press loader in balance with the capacity of the total press line.

From top to bottom, the diagrams of Fig. 5.115 illustrate the progressive movement of succeeding mats from left to right.

In diagram 1 at the top one mat is on the forming conveyor, one on the accelerating conveyor, and one on the preload conveyor.

In diagram 2 the higher velocity of the separating conveyor has created an ampler separation between the mats. Meanwhile the leading edge of the mat on the forming conveyor has advanced, still at constant forming speed. To prevent a gap between the forming and mat separating conveyor, the front nose of the forming conveyor has advanced at the same velocity as the forming belt conveyor.

In diagram 3 the nose of the forming belt conveyor has advanced even further, but at this point, the previous mat has traveled beyond the mat separating belt section. Between the second and third diagram, the nose of the forming conveyor was advanced, the trailing edge of the mat separating conveyor was correspondingly advanced. With the mat off of the separating belt, one can now bring the leading and trailing edges of these two conveyors back toward the original starting position.

That movement occurs in diagram 4 and establishes the starting position once again. Transition from the separating station to the press loader prefeeding station is performed at the same speed as the prefeeder. Transition from the forming belt to the separating station occurs at the speed of the forming line. It is only the speed of the separating station which changes to either the speed of the preceding or the following conveyor. Synchronization is automatic.

The diagrams indicate some new items in mat handling. The continuous forming belt is an important element; it is a specially woven, prestretched material, but is somewhat similar to the loading belt for the tray-belt molding system. The conditions of full support around tight radii are the same, but in the case of the continuous belt system, those radii occur at the leading edges of the forming

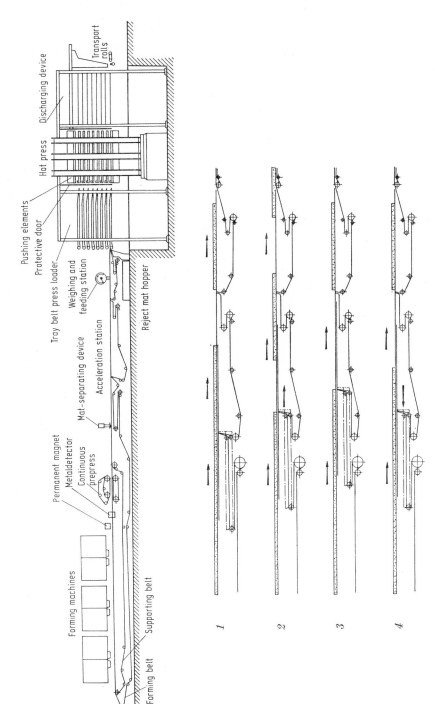

Fig. 5.115. Continuous forming belt system. From *Steck* (1970)

conveyor structure, the separating conveyor structure, and the preloader structure. The supporting surfaces are of polished metal, and the life expectancy of the transport belts is therefore longer.

Transport of the mat in the continuous belt forming system is very smooth and deceptively slow. The lack of vibration is the result of full support of the mat over its entire length and width during transport.

Indigenous to the performance of the belt forming system is the performance of the press loading mechanism because it could easily destroy the meticulous details of wood preparation, the excellent work of well executed and well operating forming machines, and the care given the mat in its transport through the preparatory system to the point of transition of the mats to the platen surfaces. Again there is a similarity to the previously described tray-belt moulding system and the continuous belt forming system. In the continuous belt line, there is no restraining frame around the periphery of the tray-belt. Otherwise, the same design criteria apply: minimal vertical drop through a long gradual slope to a relatively sharp nose, full support for the belt, etc. In the continuous belt system there is no front door at the press loader to push the board into the receiving cage as fresh mats are introduced into the press. Normally, a separate pushing mechanism is used to eject the pressed boards out of the hot press far enough to be picked up by nip rolls in the receiving cage.

Since the first plant of the continuous belt type has been started in the fall of 1968 in Northern Germany, more than twenty additional plants of this type have been ordered. The trend toward larger boards in the world market is evident.

The number of press openings ranges from a minimum of four to a maximum of twenty. Daily capacities range from a low of 130 metric tons to a high of 1,000 metric tons.

Arguments have been advanced that board thickness variation increases with press width due to structural limitations of the press. This has not been found to be the case. With proper equipment, properly operated, thickness uniformity is practically independent of width and within acceptable limits.

5.2.9.3.4 Calender Particleboard Process; System Mende-Bison. Smoothly spread, the particles blended with resin are moved to the dosing bin of the forming station. The chip bulk to be spread is controlled and regulated by a γ-ray-control-system. The whole manufacturing process is schematically shown in Fig. 5.116. Spreading is effected by the wind-spreading-system; thus a continuous structure of the board cross section is obtained. The chip mat rests upon an endless stell belt, the velocity of which is regulated according to board thickness (1.6 to 6.00 mm $\approx$ 1/16 to 15/64 in.). Immediately after the wind section, the steel belt with the chip mat is revolved around a heating drum (temperature about 240 °C $\approx$ 464 °F). Heated cylinders produce the pressure. Final board thickness is adjusted by calibrating the position of the drums (allowance ± 0.15 mm $\approx$ 1/160 in.). Feed rate is up to 25 m/min ($\approx$ 82 ft./min). For cooling, the cured hardened endless board band is moved from the unit, consisting of forming station and roll press, to the trim saw where the required board sizes are cut continuously. Boards, fabricated by this method in widths from 1,300 mm to 2,100 mm ($\sim$4 ft. 3 in. to 6 ft. 11 in.) need no surface sanding and are used as plywood cores or, if sheathed, as wall and roof sheathings, furniture backs, furniture boards, and drawer bottoms.

Output of a calender particleboard plant (without regard to gauge) is between 60 and 130 m³ ($\sim$2,120 and 4,600 cu. ft.) daily, depending on width of board.

The manufacturer of the plant indicates the following advantages:

a) Low investment costs;
b) Greatest possible economy;
c) Low space requirement.

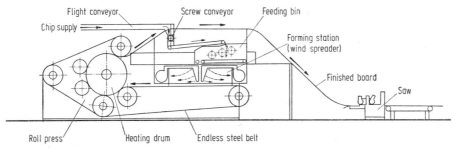

Fig. 5.116. Scheme of the manufacture process of particleboard according to *Mende*. Bisonwerke Bähre & Greten, Springe

5.2.10 Board Finishing (Sizing and Sanding or Thicknessing)

5.2.10.0 General Considerations. After leaving the press, and its associated equipment, boards are generally accumulated on a pallet, "on a hydraulic table, and then stacked for periods varying from a few days to a fortnight" (*Mitlin*, 1968, p. 99) in order to mainly equalize the moisture content. In reality the phenomena occurring are rather complex. During this time physical and chemical changes take place in the boards such as:

a) Approach of the temperature to an equilibrium;

b) Equalization of the average moisture content and elimination of moisture content gradients as much as possible;

c) The moisture movement from the board to the atmosphere is accompanied by heat transfer. A mathematical treatment of the problem is of dubious value because too many interior and exterior variables are involved such as initial temperature and moisture content of the board after pressing, under certain circumstances wetting of board surface after pressing, changes of conditions in the surrounding atmosphere, release of heat from the board due to conduction, convection, and radiation.

If the board leaves the press with about 8 to 9% m. c. — based on oven-dry weight — it will loose moisture whilst it cools. This moisture then has to be regained to create an equilibrium with the atmosphere. If the board emerges from the press at a slightly higher moisture content, no moisture has to be regained after cooling. In this case conditioning will be speeded up considerably.

d) After-cure (which will take a considerable time).

Before turning to the mechanical operations necessary in board finishing quality control and board standards have to be verified. The demand for high board quality is continuously increasing. Therefore a modern finishing line ought to include provision for uninterrupted quality control of properties (foremost being weight and thickness) which can be checked by non-destructive tests (Symposium, Spokane, Wash., U.S.A.). Quality control is not strictly a function of particleboard manufacture but of its operation research. It can therefore be omitted in this chapter, but a few references to publications may be given: *Grant* (1964), *Keylwerth* (1955, 1959), *Noack* (1966), *Plath* (168, 1971), and *Sachs* (1969).

5.2.10.1 Sizing. The problem of sizes in planning a particleboard factory is a first priority consideration since the capacities of all other machines and installations must be coordinated to the sizing plant. At first a decision is necessary as to whether sizing should be done immediately after pressing or after conditioning. Too early edging is unfavorable if the particleboards are still in a hot condition. Efficiency must be considered, for example with respect to idle times which should be in the most favorable ratio to the real sawing times.

The object of sizing is to produce a chipboard with determined width, length and rectangular edges. There are many systems of sizing, and though it is a rather simple operation, the variety of constructions is great. Usually, circular saws are arranged rectangularly. The saws themselves are well known, and it is not necessary to describe them. Fig. 5.117 shows a ground plan for saws in a rectangular arrange-

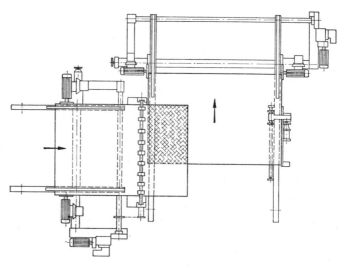

Fig. 5.117. Scheme of the horizontal projection of panel board sizing unit. From *Wehner*, in *Kollmann* (1966, p. 379)

ment. The first and the second of these saws are edging in the length direction and after having changed the direction with a 90 degree turn cutting across the middle axis of the board is accomplished. The feed speed in such plants goes up to 40 m/min ($\sim$130 ft./min). A special and occasionally difficult problem is the production of so called fixed dimensions.

5.2.10.2 Sanding. Particleboard out of the press in most cases is not fully usuable by some consumers with particular requirements respecting surface structure and thickness. Especially for the manufacture of furniture, chipboards are desired which allow very thin veneer to be glued directly to the surface.

After many experiments sanding machines with 2 to 4 cylinders are generally used. In West Germany the tolerances in thickness should not exceed 0.3 mm ($\sim$0.012 in.). Fig. 5.118 shows schematically the longitudinal cross section through a sanding machine with 4 cylinders. The necessary explanation is given in the legend.

The urgent requirement for high-efficiency sanding machines for the large scale production of particleboard, blockboard and similar panels led to the development of wide-belt sanding machines[1] (Fig. 5.119). Particular features are the high operating rate and excellence of surface finish which are achieved at feed rates which can be controlled infinitely between 0 and 40 m/min (0 and $\sim$130 ft./min), thickness tolerance: ± 0.1 mm (0.004 in.).

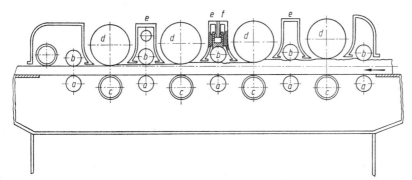

Fig. 5.118. Schematic longitudinal cross section through a sanding machine with 4 cylinders. *a* Lower feed roll, *b* Upper elastic feed roll, *c* Counterpressure rolls, *d* Sanding cylinders, *e* Upper bed connection with guide plates, *f* Air cushion device. From *Wehner*, in *Kollmann* (1966, p. 383)

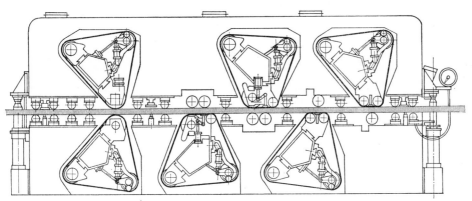

Fig. 5.119. Schematic longitudinal cross section through a wide-belt sanding machine. Bison-Werke Bähre & Greten, Springe

The panels are calibrated and subsequently fine-ground on both sides in one pass, whereby surface removal can be equal or varied on each face, even on panels with very dissimilar surfaces. The sanded areas are freed from dust by rotating brushes and powerful air suction inside the machines.

The machines are of heavy steel construction which ensures absolutely steady operation without vibration. The upper part of the machine can be adjusted vertically by means of a push-button controlled hydraulic lifting mechanism, and swinging spacers facilitate rapid conversion to any panel thickness. At maximum lift, the opening is about 300 mm ($\sim$12 in.) so that the inner components are adequately accessible for maintenance and attention. The sanding units in the top and bottom sections are supported on heavy-duty brackets and sanding belts can be changed quickly without having to dismantle any parts

---

[1] The author is grateful for information by Bisonwerke Bähre & Greten, Springe, W.-Germany.

of the machine. The belts are tensioned by tension rolls, the amount of tension being adjustable and indicated on a gauge. For the purpose of controlling the belts laterally the tension rolls are deviated axially at brief intervals by means of an oscillating cylinder so that the sanding belt vibrates continuously in a horizontal plane.

Coarse calibration sanding is carried out by contact rolls which ensure high output coupled with precise calibration. The fine surface finish is achieved with easily interchanged sanding shoes fitted with a graphited belt. Opposite the sanding heads there are pneumatically controlled pressure rolls which can be adjusted by means of handwheel, so that the opening of the sanding machine can be set to the closest thickness tolerances without having to stop the machine. Working widths are 1,300 mm to 2,600 mm (∼51 to 102 in.).

Long sanding belt life is granted. The drive motors are located at the back of the machine where they are easily accessible: They are directly-coupled polyphase induction motors fitted with a countertorque brake, allowing them to be stopped quickly in the event of breakdown or stoppage. Overload relays are incorporated as an additional safety measure.

The particleboard panels are moved through the sander by feed rolls covered with low-abrasion plastic material, with floating pressure rolls on the opposite side. These rolls are powered by a DC motor, the feed rate being infinitely adjustable with a push-button control unit, whilst an electronically controlled feed adjustment is for automatic operating, the feed rate varying in accordance with the load on the sanding belt motors. The range of regulation can be preset on contact ammeters between minimum and maximum marks, and the relevant feed rate is easily read from a dial gauge.

5.2.11 Particleboard — Plastic Combinations, Printing on Particleboard

5.2.11.0 General Considerations. Wood-plastic and/or wood-polymer combinations are recently developed materials for a wide range of applications. These combine the specific physical, mechanical and chemical properties of wood and wood based materials with those of plastics. The processes are described by *Adams*, and coworkers (1970), *Autio* and *Miettinen* (1970), *Heyne* (1970), *Kenaga* (1970), *Boiciuc* and *Petrican* (1970), *Meyer* and *Loos* (1969), *Langwig*, *Meyer* and *Davidson* (1968, 1969), *Proksch* (1969), *Mäkinen* (1968), and *Burmester* (1967).

The application of solid wood-plastic combinations of this type is limited. It is not yet clear whether this or a similar technique is suitable to produce special particleboard. For greater resistance against weathering, coating under certain circumstances is necessary.

The improvement of the surfaces of particleboard with laminates is widely used. *L. Plath* (1971) discussed the requirements which depend on the type of laminates and the processing methods applied for that purpose. Of considerable importance are the following conditions:

1. Properties of the chips;
2. Distribution of density in the boards;
3. Moisture content;
4. Quality of gluing;
5. Variations of thickness; tolerances ≤ 0.3 mm (0.012 in.);
6. Smoothness and uniformity of the board surfaces;
7. Resistance to the pressure applied in the lamination process.

The main purposes are:

1. Improvement of the physical and mechanical properties of the board surfaces;

2. Increase in the modulus or rupture (bending strength) of the composite (sandwich) board;

3. High resistance to chemical actions (corrosion) on the deck laminates;

4. Decorative effect;

5. Resistance to aging.

Laminated particleboards are mainly used in the manufacture of furniture or parts of prefabricated houses

5.2.11.1 Types of Laminates for Particleboard. The following types of laminates are used (*L. Plath*, 1971):

1. Laminates for particleboard

a) Melamine impregnated (alpha-cellulose-) papers;

b) Diallyl phthalate impregnated papers;

c) UV-hardened polyester primings with subsequent lacquering;

d) Pigmented polyester lacquers;

e) Rolled on primings with imprinted figuring or patterns of fabrics of many woods (such as walnut and Douglas fir). This type of photomechanic reproduction of figuring (texture) of valuable woods was originally patented as Masa- and Tarso-processes (German patent DRP 386005; *Kollmann*, 1936, p. 668—670) and applied first in Germany and Finland.

2. Laminates or foils glued on

f) High pressure laminated sheets;

g) Papers with precondensed aminoplasts;

h) Priming foils impregnated with aminoplasts and lacquering;

i) Vulcanized fiber coated with aminoplasts;

j) Thermoplastic foils (PVC-foils).

The great variety of available types of laminates with thicknesses in the range of (0.4) to 1.6 to 3.2 mm (0.016) to (1/16 to 1/8 in.) requires a proper choice. This is not always easy because standardized specifications or even prescriptions are still lacking. Requirements and applications of laminated particleboard are manifold.

There are numerous reciprocal effects or interactions between the laminates and the particleboard. Any defects observed in processing, or complaints of consumers (giving useful hints of further development) may be cleared up most advantageously by means of microscopic investigations combined with special staining techniques. In addition, topochemical spot analysis and the usual chemical, physical and technological tests adjusted to the problem in question should be carried out. The results of very many detailed and careful examinations have led to the conclusion that for all types of laminates mainly the surface quality and the firmness of the deck layers of the particleboard is decisive. Uniform and tight surfaces are obtained if decklayers consist of very fine chip material (length of particular chips below 2.5 mm (about 3/32 in.) and even dust. Fine particles pass a screen with mesh apertures of 1.0 mm ($\sim$0.04 in.), dust a screen with mesh apertures of 0.3 mm ($\sim$0.012 in.) (*Neusser* et al., 1969, *Kehr* and *Jensen*, 1970).

5.2.11.2 Structure, Manufacture, and Testing of Laminated Particleboard. Particleboard overlaid with synthetic resin impregnated papers must be laminated on both sides. One-sided lamination makes the board sooner or later concave on the laminated side. Symmetrical lamination (number of films or paper carriers, technological data, equal proportion of resin coating and volatile constituents) is necessary. The method is expensive but safe. Fig. 5.120 shows the possibilities of the structure of laminated particleboard. Structure I is the cheapest method but only practicable if the surface quality is high and the color of the rough particle-

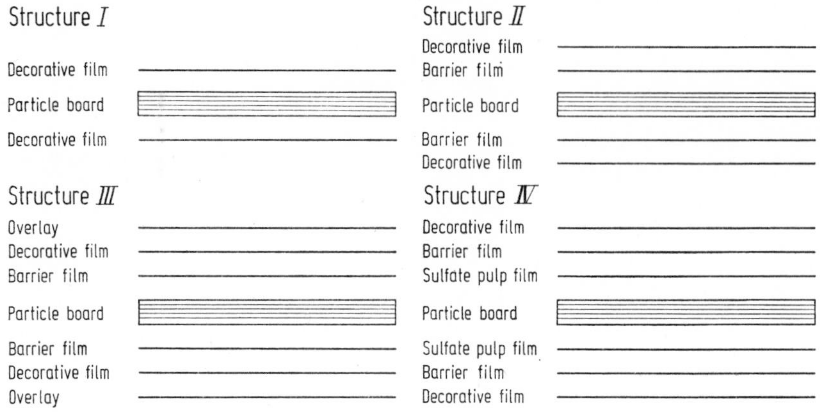

Structure *I*

Decorative film	————————————
Particle board	▦▦▦▦▦▦▦
Decorative film	————————————

Structure *II*

Decorative film	————————————
Barrier film	————————————
Particle board	▦▦▦▦▦▦▦
Barrier film	————————————
Decorative film	————————————

Structure *III*

Overlay	————————————
Decorative film	————————————
Barrier film	————————————
Particle board	▦▦▦▦▦▦▦
Barrier film	————————————
Decorative film	————————————
Overlay	————————————

Structure *IV*

Decorative film	————————————
Barrier film	————————————
Sulfate pulp film	————————————
Particle board	▦▦▦▦▦▦▦
Sulfate pulp film	————————————
Barrier film	————————————
Decorative film	————————————

Fig. 5.120. Possibilities of assembling laminates on particleboard. From *Enzensberger*, in *Kollmann* (1966, p. 414)

board is bright. Structure II is normal but only vertical application of the laminated board is possible because the resistance to wear is low. Structure II is excellent due to high resistance to all mechanical stresses. Structure IV shows additional embedding of a phenolic resin impregnated sulfate kraft paper applicable to particleboard of lower surface quality.

Adhesives for gluing *decorative plastic laminates* on particleboard are urea-formaldehyde resins, PVA-glues, or contact adhesives on the basis of neoprene (*Rayner* in *Houwink* and *Salomon*, 1965, p. 330—333). Gluing pressures range according to the glue types between 1 and 6 kp/cm² ($\sim$14 and 85 lb./sq. in.). If PVC-hard foils are used the pressures range between 2 and 8 kp/cm² ($\sim$28 and 113 lb./sq. in.). *PVC-laminated* particleboards have besides their decorative effect the following properties:

1. Impermeability for water and moisture;
2. High wear-resistance;
3. High resistance to chemical influences (corrosion by many acids and alkalies, salts, gases, oils, solvents);
4. Scratch resistance;
5. Very high resistance to aging and to light;
6. Low resistance to heat (temperatures above 80 °C (176 °F)) and to glowing cigarettes.

Plastic veneer-imitations consisting of alpha-cellulose papers impregnated with polyester resins showing figures of wood and printed on by rotogravure:
The plastic veneer-imitations can either be self-gluing (pressure $\geqq 8$ kp/cm² $\sim$113 lb./sq. in.; temperatures between 125 °C and 180 °C (257 and 356 °F) at the maximum, pressing times for 125 °C temperature about 2 min, for 180 °C about 35 sec) or they are wet glued on the particleboard (pressure $\geqq 3$ kp/cm² ($\sim$43 lb./sq. in.), temperatures between 95 °C and 160 °C (203 °F and 320 °F),

pressing times between 35 sec for 160 °C and 4 to 6 min for temperatures in the range of 95 °C to 110 °C (203 °F to 230 °F). Pressing with the shortest pressing times possible is desirable because the chipboards themselves are scarcely heated, and therefore subsequent ventilation is not necessary.

The surfaces of the veneer-imitations must be lacquered in one or several working operations.

Lamination of particleboard with *impregnated papers* on the basis of *diallyl phthalate resins* (as an example the method of the American Food Machinery & Chemical Corp., Dapon Dept.) presumes uniform and extra fine surfaces, a moisture content of the raw board of about 5% (based on oven-dry weight), tolerances in thickness not exceeding ± 0.125 mm (~ 0.005 in.) and a density of raw board of approximately 0.64 g/cm³ (~ 40 lb./cu. ft.). Generally for this type of lamination flat pressed three-layer or graded particleboards yield more uniform surfaces than extruded board.

Lamination of particleboard with polyester films is similar to the above described process (Goodyear Tire & Rubber Co., Akron, Ohio, 1959). The films are thermoplastic: softening point about 91 °C (~ 196 °F), melting point about 176 °C (~ 349 °F). Printing, embossing, or pigmenting is possible. The films are relatively thick, and the particleboard overlaid with them are used in furniture manufacture and for interior structures.

5.2.11.3 Printing on Particleboard. Direct printing of wood figure or of fancy patterns on particlebaords before or after their surface treatment is a special kind of surface improvement. There are approved systems in Western Europe as well as in the U.S.A. These systems (already listed in Section 5.2.11.1 under number 1., e) are not substantially different. Printing on particleboard is possible up to a width of 2,500 mm (~ 8 ft. 2 in.) and up to a thickness of 60 mm ($\sim 2\ 3/8$ in).

A typical working operation is shown in the block diagram Fig. 5.121. The printing process is a combination of offset and rotogravure printing (*Ross*, 1960,

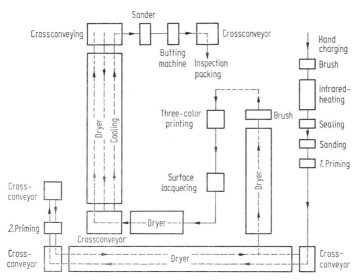

Fig. 5.121. Block diagram of the various stages of printing on particleboard. From *Enzensberger*, in *Kollmann* (1966, p. 422)

Schmutz, 1964). Printed particleboards are used mainly for vertical planes of furniture and railroad cars as well as for interior structures (wall linings, built-in furniture, houses, doors etc.).

5.2.12 Extruded Particleboard

5.2.12.0 General Considerations.. A vertical extrusion press unit for the production of a new type of particleboard was developed by *Otto Kreibaum* in Lauenstein (Hannover), West-Germany (Okal) in the years 1947 until 1949. In this period after the war a shortage of wood existed in Germany. The idea of *Kreibaum* was to produce particleboard in a progressive, continuous, and simple operation from hardly utilizable wood residues. Low production costs should especially guarantee profitableness of the manufacturing plant.

The following fact indicate the peculiar quality of the extrusion process, especially the Okal process:

1. Utilization of wood residues and vegetable waste substances, as far as they may be reduced to chips. Particularly may be mentioned saw mill residues, outside-planks, slabs, paring wastes, veneer wastes, cores from rotary-cutting of veneer, cudgels, knotty wood, logs and split bits of wood, refuse from circular saws and double squaring machines, etc. Assortments of larger diameter are split before they are chopped. Any species of wood, inclusive of all hardwood may be treated in the extrusion process;

2. The extrusion process (again mainly the Okal process) is one of the first methods in the world for processing ordinary sawdust, whereby boards of excellent compactness can be produced;

3. The percentage of binding agents required by the extrusion press board is fairly low. If urea-formaldehyde resin is used, the solid content of resin will be about:

5% if chips from larger wood residues are used,

8% if sawdust is raw material;

4. Purchase costs of the extruders are comparatively low with respect to the productivity of the system;

5. Boards of various shapes may be manufactured, for instance tubular (fluted) boards.

Extruded boards may be used for manufacturing of furniture, interior decoration and store fixtures, of table tops and wall boards, radio and TV cases and music-boxes, bill boards, doors and various building purposes, as well as in shipbuilding and vehicle construction. Tubular boards are particularly adapted for use in the manufacture of prefabricated houses.

5.2.12.1 Vertical Extruders. The course of operations from raw wood to completed particleboard according to the Okal-extrusion press method is shown schematically in Fig. 5.122. Feeding of the raw wood *1* is done by a belt conveyor *2* via a metal detector *3* in the chopping rotor *4* for coarse cutting. The wet chips produced are transported by the pneumatic conveyor plant *5* into the wet chip bin *6*. A vibrating gutter conveyor *7* carries the wet chips after repeated magnet separation of iron particles *8* to the chip reducing machine *9* which produces finer chips from the larger chips. The chip reducing machines *9* drop the wet chips straight into the dryer *10*. The moisture which is evaporated from the chips in the dryer is carried off by a ventilator, whilst a cyclone separates the sucked off dust.

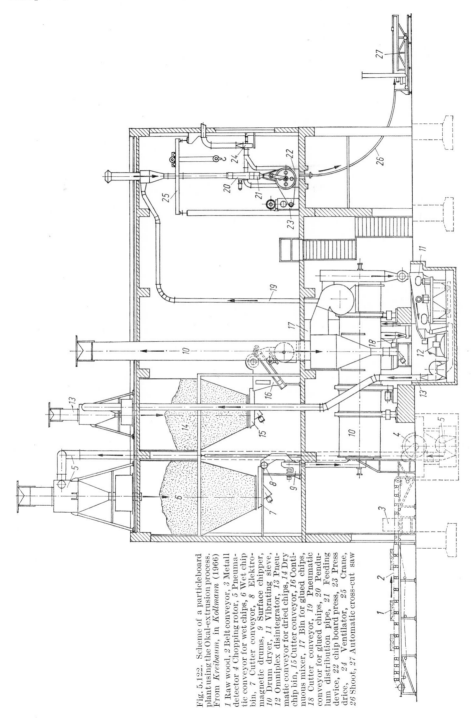

Fig. 5.122. Scheme of a particleboard plant using the Okal-extrusion process. From *Kreibaum*, in *Kollmann* (1966)

1 Raw wood, *2* Belt conveyor, *3* Metall detector *4* Chopping rotor, *5* Pneumatic conveyor for wet chips, *6* Wet chip bin, *7* Cutter conveyor, *8* Elektromagnetic drums, *9* Surface chipper, *10* Drum dryer, *11* Vibrating sieve, *12* Omniplex disintegrator, *13* Pneumatic conveyor for dried chips, *14* Dry chip bin, *15* Cutter conveyor, *16* Continuous mixer, *17* Bin for glued chips, *18* Cutter conveyor, *19* Pneumatic conveyor for glued chips, *20* Pendulum distribution pipe, *21* Feeding device, *22* chip board press, *23* Press drice, *24* Ventilator, *25* Crane, *26* Shoot, *27* Automatic cross-cut saw

The dried chips are screened *11*. Still too coarse chips pass a hammer mill *12*. The dust being collected is conducted to the boiler house for combustion. The chips which are suitable for manufacturing purposes are pneumatically carried *13* to the dry chips bin *14*.

A vibrating gutter conveyor *15* carries the dry chips via a dosing apparatus to the continuous mixer *16* in which the synthetic resin binding agent is mixed with the chips.

From the mixed chip bin *17* arranged underneath, into which the glued chips drop directly, the chips which are now ready to be pressed, are transported via a gutter conveyor *18* through a pneumatic conveyor plant *19*, over a distributing pipe *20*, to the press feeding device *21*.

In the chipboard special press *22* with its drive *23* the Okal chipboard is produced as an endless band *26*, being cut to the proper size by the automatic cross cut saw *27*.

Moving of the completed chipboards can either be done by trucks or on pallets by lift conveyors.

According to the desired volume of the plant to the number of presses and their size the following approximate values are required, viz:

Heat: Hot water circulating heating system 180 °C (356 °F) of
  12 atmospheres ($\sim$170 lb./sq. in.) with a boiler pressure of about 1,800,00 kcal/h ($\sim$7,100,000 BTU/h);
Current: Three phase 3×380 V, 50 Hz;
Installed: About 350 kW;
Compressed air: 8 atmospheres, about 10 l/min;
Space: manufacturing buildings about 500 m² ($\sim$600 sq. yds.);
  storage of boards about 1,500 m² ($\sim$1,800 sq. yds.);
  veneering plant about 1,500 m² ($\sim$1,800 sq. yds.);
Staff: 5 workmen (for the Okal plant) excepting lumber transport and moving of boards.
  Output of the Okal plant, fitted with two presses (board width 125 cm $\sim$ 49 in.) for a board thickness of 19 mm ($\sim$3/4 in.) and three-shift operation 7,500 metric tons per year.

In the U.S.A. *Crafton* (1956) developed in cooperation with Chipcraft Co., Morristown, Tenn. a vertical extrusion process. The material flow is shown in Fig. 5.123. The system is completely automatic, and the entire operation requires only three workers per shift. The chipcraft extruder will permit extrusion of either solid or fluted boards of 1.24 m (49 in.) in width and 3.66 m (12 ft.) in length.

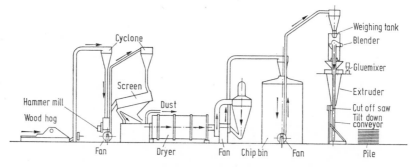

Fig. 5.123. Schematic diagram of a Chipcraft vertical extrusion process

Board thicknesses up to 41 mm (1 5/8 in.) are possible (*Johnson*, 1956, p. 34/35, 172—175). A conveyor-fed knife hog reduces the raw material initially. The large chips that are produced are pneumatically conveyed to a cyclone on the roof of the plant. Air is exhausted from the cyclone and subsequently the chips are screened. The coarser particles are then gravity-fed to a hammer mill. The finer wet chips are blown by a fan to a cyclone and from there conveyed to another screen which separates the usable chips from fines and dust which are removed. The following major components included in the processes are: dryer feed — drum dryer — pneumatic conveyor plant — storage cyclone — storage bin — pneumatic feed to a feed cyclone — metering section — chip glue mixer — feeding funnel — curing section of the extruder. The resinated chips are compacted by the pressure exerted by the extruder ram. Sufficient opposing force necessary for the compaction is caused by the frictional resistance between the board and platen surfaces. The holes, or flutes, in tubular boards are produced during the extrusion operation according to the (previous) Okal-system as well as according to the Chipcraft-system "by means of electrically heated rods which are fixed at the upper end of the extruder and extend downward for a length of twelve feet between the platens. The ram head contains a set of holes corresponding to the number and size of the tubes used. The heated tubes extend through the holes in the ram head and are rigidly fixed above it. The lower end of the tubes must remain free; however, the ram itself serves to restrain movement during operation. Formed board holds the extreme lower ends of the tubes in place. Extrusion of 30 mm (1 3/16 in.) board takes place at the rate of approximately 610 mm (24 in.) per minute (*Johnson*, 1956, *Anonymous*, 1956). Some types of tubular boards are illustrated in schematic cross-sections in Fig. 5.124. The cured

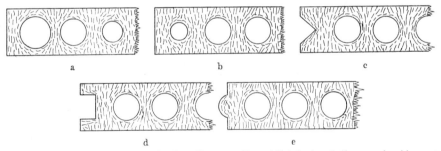

Fig. 5.124a—e. Types of tubular boards in schematic cross sections. a) Tubular boards (from regular chips or saw dust). Thicknesses: 25 to 50 mm (5/18 to 2 in.); b) Board with different diameters of tubes for parts with different loads, such as e.g. house bulding; c) Pressed on triangular profile (towards the inside); d) Pressed on square profile (towards the inside); e) Pressed on half round profile (towards the outside)

continuous board leaves the extruder guided vertically downward and is cut to length by an automatic cut-off saw which is located directly beneath the extruder. A tilting conveyor brings the cut piece through 90 degrees swinging to a horizontal position where it is released and falls onto pallets or is stacked. The mechanism is restored to its original vertical position ready to receive the next board. The entire receiving, cutting, and stacking operation is completely automatic.

5.2.12.2 Horizontal Extruders. *Mörath* (in *Kollmann*, 1966, p. 526—529) reports that already before World War II in the "Research Institute for Plywood and Other Wood Products" at Berlin *Dribbusch* developed a horizontal extruder for particleboard on a laboratory scale. In the U.S.A. the Lane Co., Inc., of

Altavista, Virginia, became interested, in 1951, in the extrusion process and set out to develop a horizontal type extruder, based on preliminary studies of the Englishman *Bibby*. The pilot pres began operation in about August, 1952, and the experience, gained with it enabled the company, in 1953, to design a press capable of producing some 3.3000 to 40.000 m³ (~1.2 to 1.4 million cu. ft.) of 19 mm (3/4 in.) board per 120 hours a week (*Johnson*, 1956, p. 176).

In Fig. 5.125 a schematic diagram of the Lanewood process is shown[1]. The system utilizes plant kiln-dried wood scraps which are transferred to a hammer-mill via a belt conveyor. The residues are reduced in the hog and then they are

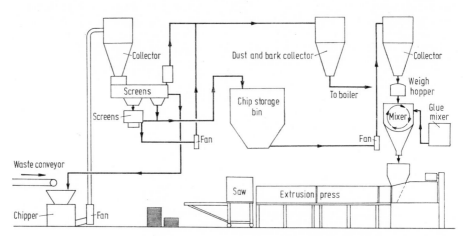

Fig. 5.125. Schematic layout of Lanewood particleboard process

pneumatically conveyed to a collector on the roof. From here they fall into a triple-deck screen and are classified. The oversize splinters (> 6.4 mm ~ 1/4 in.) are rejected by the screen and are pneumatically conveyed back to the mill for re-processing. Acceptable particles (usually 1.6 to 6.4 mm ~ 1/16 to 1/4 in.) are sent to the storage bin for use. Particles which pass through the screens (< 1.6 mm ~ 1/16 in.) fall into a fines separator. The dust is separated from the fines and a controlled amount of the fines is sent to the same storage as the normal particles. If an excess quantity of fines appears, a portion is passed to the boiler.

From the storage bins, usable particles are removed by screw conveyor on a belt to a pneumatic conveyor which transfers them to a collector. From this collector, the particles enter a weigh hopper. The particles are automatically weighed out in batches and then dumped into a tumbler-type mixer. Liquid adhesive, mixed in a glue mixer, is automatically sprayed into the mass of particles as it is tumbled. At the completion of the mixing cycle, the chip-glue mix is dumped into a hopper above a reciprocating ram at the infeed end of an extruder press. On each back-stroke of the ram, the resin-coated particles are gravity-fed with the assistance of agitators into the ram chamber and are pushed forward between heated platens in the extruder on the forward stroke of the ram. As this action es repeated, increment after increment of board is formed and pushed toward the outfeed end of the press, where it emerges as a continuous board. An automatic cut-off saw unit, located just beyond the orifice of the extruder, cuts the board to predetermined lengths. After the board is sawed, it falls onto a roller conveyor.

---

[1] The author is indebted to Lane Co. for information.

The next length of board, that is cut, falls atop the previous one, thus building up a stack which an operator ultimately moves down the conveyor from where they are taken to storage.

The power requirement for the press, including the saw, is listed as 25 HP, and the steam requirement is about 223 kg ($\sim$490 lb.) per hour at 182 °C ($\sim$360 °F) and 9.8 kp/cm² (140 lb./sq. in.). The output of the machine can be widely varied and is largely determined by the number of strokes per minute of the reciprocating ram. The length of the stroke is adjustable. Both of these features are desirable when making different thicknesses of stock. In addition to making various thicknesses of board, suitable means have been provided for producing different widths. Provisions have also been made to facilitate inspection of the platen surfaces. Both the upper and lower platens are made up of three sections. The section nearest the ram is only about 60 cm ($\sim$2 ft.) in length while each of the last two sections is about 210 to 240 cm ($\sim$7 to 8 ft.) in length.

5.2.13 Molded Particle Products

5.2.13.0 General Considerations. Molded products are formed in their final shapes in one operation without further working. The items are produced completely, no joints by gluing, dovetailing, dowelling, slots and feathers, screwing, nailing, fastening with bolts etc. are necessary. Curved boards and beams produced by sawing solid wood do not belong to this category nor do molded parts made of solid wood which are well known (*Forssman*, 1924, *Müller*, 1930, *Hoitz*, 1934, *Perry* 1944, 1497, *Taylor*, 1944, *Heebink*, 1946, *Stevens* and *Turner*, 1948, *Simonds*, *Weith* and *Bigelow*, 1949, *Kollmann*, 1951, 1955, *Wood*, 1963), manufactured for example by:

a) Bending;
b) Carving, turning, milling etc.;
c) Bonding thin veneers with synthetic resin adhesives into preformed shapes.

The bonding may be effected in one of three ways:

1. On a mechanical or hydraulic press, using specially shaped male and female forms or molds, and rigid pressure;
2. By the vacuum process in which one form only and rubber bag or sheet is required;
3. On specially constructed molding and shaping presses or pressure cylinders (autoclaves, whereby fluid pressure — by heated compressed air or steam — is exerted).

As far as solid wood is concerned working and molding is limited due to its natural structure (anisotropy and inhomogeneity), its physical properties and their changes under the influences of variations of the surrounding atmosphere. After development and ample application of particleboards, promoted by their excellent properties, the idea was obvious to manufacture not only sheetlike boards but also molded parts.

For the production of such molded particle items, essentially the same process as used for the manufacture of particleboard is suitable: pressing of resinated wood chips. With respect to the desired shape, tools or molds are necessary. The quality of molded parts depends to a large extent upon the tools and the precision with which the tools are manufactured (machined and polished) and used. The mold, applied in conjunction with suitable pressing and heating equipment, is the tool with which the raw material comes into contact when producing the piece. Molding of plastics has been known for more than a century (1950 rubber, 1870 celluloid, 1909 phenol-formaldehyde resins, 1919 casein, 1927 cellulose acetates, 1928 urea-formaldehydes, 1940 melamine formaldehydes; these approximate data show when commercial products were introduced in the U.S.A.) (*Simonds* et al. 1949, p. 6). The acrylic plastics stem back to the early work of *Röhm* in Germany in 1901.

In the U.S.A. (and probably in other industrialized countries) no less than about 15 various plastics-processes are used (*Simonds*, *Weith* and *Bigelow*, 1949, p. 12). The most important are casting, extrusion, injection molding, compression molding, and laminating. Most raw plastics are granulars or in sheet form passing gradually from a solid to a liquid condition as the temperature is raised. The terms "fusing point" and "softening points" have lately come into use but the nature of the transition when heat is applied to resins as the origin of cured plastics pieces depends on many variables.

Mixtures of wood particles and binders are different in their behavior during processing and in their properties after curing. Even saw dust and shavings need some preparation. The raw material must be uniform for industrial molding processes. Hence follows that either only one wood species (such as spruce or poplar or other indigenous suitable wood species) or mixtures of various wood species in always the same proportions are treated. Recommended thicknesses for the chips are between 0.1 and 0.3 mm ($\sim$0.004 and 0.012 in.) and lengths between 10 and 30 mm ($\sim$0,4 and 1.2 in.) (*Haas*, 1959, *Kollmann*, 1955). The specific volume of the resinated chips in uncompressed state is much more greater than that in the compressed state. Besides the size of the chips, their moisture content and its variation during pressing to molded products must be carefully observed. As binders only thermosetting resins or, in one process the lignin as the natural chemical constituent of the wood, are suitable. The amount of solid resin, based on the of the oven-dry weight, is depending on the process used in the range between 4 and 20%. The water added with the binder is converted into steam in the hot press-operation. This steam must be either condensed by subsequent re-cooling or conducted away. The molds must consist of special steel and their interior surfaces must be burnished if the product to be molded in the hot press receives a coating of resin-impregnated paper. The pressing time depends on the type of binder.

The costs of the tools and the production rate mainly determine the economy of processes for the manufacture of molded particle products. The following demands should be considered:

a) Large quantities of articles should be produced of one size and shape and sold in order to keep the proportion of the tool costs to the total production costs low;

b) The machines for the preparation of the raw material and for pressing should be completely utilized;

c) The process must be technically developed to such an extent that many and various products are available on the market;

d) Such standard articles should be chosen which would cause high waste of timber and high costs using conventional woodworking and/or combined structures. The following products have been proven in service (*Werzalit*, 1964, 1965, 1966):

Cases (cabinets) or base plates for phonographs, radio- and television sets, tape recorders,

Components of kitchen furniture, chair- and stool- seats and -backs, doors for cupboards,

Containers for transport and storage,

Instrument panel cover (of rare wood) in cars (e.g. *Porsche*),

Lighting (illuminating) ceiling (PAG Presswerk AG, Essen, W.-Germany),

Loudspeaker-halves,

Partition and wall elements,

Prefabricated building elements,

Profiled moldings,

Radiator covers,

Shelves,

Shoe-lasts,

Shutterings, for concrete,

Spacing rings for stucking barrels,

Supports for grind stones,

Tablets and trays,

Tops for camping tables, dining tables, dish-washers, kitchen counters, school desks,

Window sills.

Mainly three processes were successful (Fig. 5.126).

5.2.13.1 Thermodyn Process[1].

The Thermodyn process has been developed by *Runkel* and *Jost* (1948). It permits the manufacture of hard molded bodies on the basis of vegetable raw materials, which contain resin-forming or effective carbonyl-groups. Their latent ability for cendensation by application of pressure and heat is released. The process flows without or with only very small amount of added binders. *Wilke* (1951) investigated the thermal plasticizing (softening or melting) of wood. In the Thermodyn process wood chips, preferably with a moisture content (based on onven-dry weight) in the range between 10 and 17% are charged into molds, cold-prepressed with a pressure of about 180 kp/cm² ($\sim$2,550 lb./ sq. in.) and then put into a gas-tight closed mold manufactured of special steel. The press platens have channels for the heating medium (hot water) as well as for the cooling water. Fig. 5.127 shows a schematic cross section with the tightening grooves. The temperatures applied are between 160 and 290 °C (320 and 554 °F) and pressures between about 200 and 300 kp/cm² ($\sim$2,840 and 4,250 lb./ sq. in.). There are two reaction phases:

a) Hydrolysis by which the chemical constituents of the wood are broken down and condensable vapors and gases result. The acetic acid and the formic acid intensify the hydrolysis by which the lignin-carbohydrate linkage is split and aldehydes as well as furfural come into being. The lignin becomes active as a natural binder;

b) In the second phase the gases react upon the changed wood substance, condensing upon and producing a plastic mass. Some aldehydes may be further split off, for instance succinic dialdehyde and formic acid; the latter in combination with the water in the wood causes further hydrolysis of the carbohydrate components.

The kinetics of the alternate reactions are elucidated in Fig. 5.128. Reference to the general chemistry of wood, cellulose, and lignin may be made to *Brauns*,

---

[1] The Thermodyn-Process has been further developed and is applied by two West-German firms (Vereinigte Schulmöbelfabriken, Tauberbischofsheim, and by Gebr. Oldemeier, Falkenhagen). The principal method is based on the original patents, but the know-how is essentially improved and private.

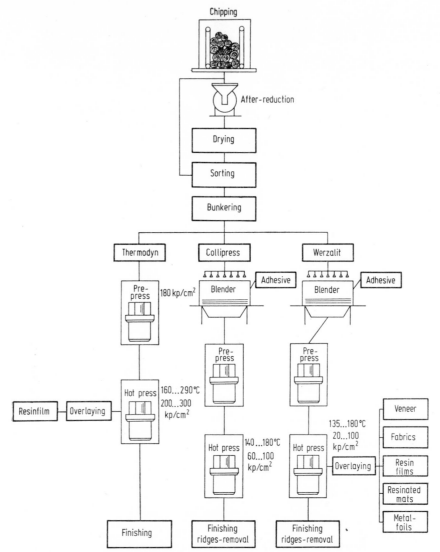

Fig. 5.126. Schematic survey on processes for the production of molded particle articles

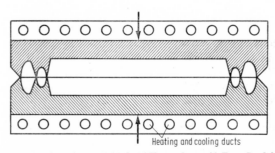

Fig. 5.127. Schematic cross section through a gastight closed Thermodyn-mold. From *Runkel* and *Jost,* in *Kollmann* (1966, p. 432)

1952, *Cote*, 1968, *Hägglund*, 1951, *Wise* and *Jahn*, 1952. Information about the Thermodyn process is summarized by *Kollmann*, 1955, p. 552—555, and *Haas*, in *Kollmann*, 1966, p. 432—434.

Influencing the properties and qualities of products are the following facts which have limited the application of the process in the industrial practice:

a) Completely gastight construction of the mold, this permits only the production of sheet-like parts with small profiles;

b) Utilization of the natural binding constituents of the wood;

c) High pressure and hence high density in the range between 1.1 and 1.3 g/cm³ (∼69 and 81 lb./cu. ft.);

d) Necessity of re-cooling after hot-pressing to temperatures below 100 °C (212 °F), practically to about 40 °C (104 °F). Re-cooling is uneconomical with respect to heat consumption and production rate;

e) The wood structure will be destroyed and some wood species which are easily affected by higher temperatures such as maple, oak etc. change to a less desired color.

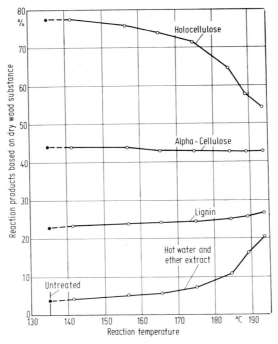

Fig. 5.128. Dependence of the yield of reaction products based on the oven-dry weight of wood, upon the reaction temperature. From *Wilke* (1951)

5.2.13.2 Collipress Process[1]. Cut, calibrated chips are mixed with a small amount of synthetic resin binder and additives to make the products suitable for the tropics, cold storage and water resistant etc. This material is pressed to the shape of boxes and cases with one open side. The hot press for this material possesses one vertical and several horizontal pistons which press the chip mixture

---

[1] The author is indebted to Mr. *K. van Hüllen* for information.

against a steel mold. The pressure applied is in the range between 60 and 100 kp/cm² (~850 and 1,420 lb./sq. in.). The temperature is between 140 and 180 °C (284 and 356 °F), the duration of pressing is only 2 to 5 minutes according to the thickness. The following characteristics of the products of the Collipress society, an affiliate of the reputable German firm Becker & van Hüllen, Krefeld, may be mentioned:

a) Tool and press are a unit;

b) Only packing cases are manufactured;

c) Improvement of the surfaces is not necessary;

d) The hollow bodies are removed from the press after curing of the binder without re-cooling;

e) The cross section of the containers may be rectangular or circular;

f) Areas which have to sustain higher stresses can be more compressed or reinforced by ribs;

g) The density of the container walls is between 0.8 and 1.1 g/cm³ (~50 and 68 lb./cu. ft.);

h) The woodparticle containers serve for transport and storage of mass-goods such as bottles, ammunition etc. Nowadays plastics are preferred for such purposes.

5.2.13.3 Werzalit Process[1].

The Werzalit Process[2] is by far the most important process for the industrial manufacture of molded particle products. The invention goes back to the year 1954 and was made by the firm. J. F. Werz Jr. KG in Oberstenfeld near Stuttgart, W.-Germany. Since that time many patents and patent applications have followed in W.-Germany and in many other countries. Works licensed by the Werzalit group are established all over the world. In some countries the Werzalit products have their own trade names, for instance "Formwood" in England or "Molpar" in Japan. Fig. 5.129 reproduces a schematic process flow chart in which by reason of simplification symbolic detail sketches are used and the pre-forming is left out, Fig. 5.120 shows a part of a modern plant.

The early stages of the Werzalit process follow closely the conventional methods for particleboard manufacture. Fresh felled, debarked logs or thinnings are cross cut to length of about 100 cm (~3 ft.) for the chipping machine. The cross-cut logs, after having passed a metal detector, are then fed into the chipping machine. From there they are conveyed pneumatically to a wing beater mill which reduces the chips or flakes into shreds. A pneumatic conveyor system takes the shreds to a steam-heated or oil-heated dryer in which the moisture content is brought down to 2 to 3% (based on oven-dry weight). The installation, consisting of a chipping machine, a wing beater mill, and a dryer, has a capacity of about 6 metric tons an hour. The shreds are then sorted into three sizes by means of a rotary sieve and stored in hoppers from which suitable blends can be made.

---

[1] The author is indebted to J. F. Werz jr. KG, Oberstenfeld, and Mr. *E. Munk* for information.

[2] Information is given in: Ciba Technical Notes 238: Mouldings from wood chips and Aerolite, Duxford Cambridge, Oct. 1962; D. *Alan*: Shaped components produced by moulding technique, Woodworking Industry, London, Jan. 1962; Formwood Ltd., Coleford, Glos., U. K. Circulars; Werzalit Information Apr. 1964, Ed. 1965, 1966, 1967, 1968, 1969, 1970, Werzalit-Pressholzwerk, Oberstenfeld, W.-Germany; *Haas*, in *Kollmann*, 1966, p. 437—440.

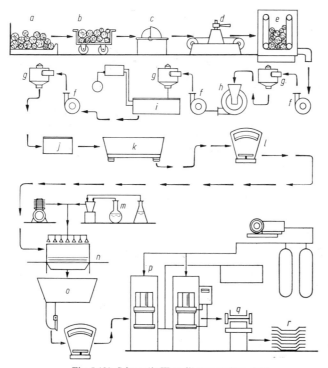

Fig. 5.129. Schematic Werzalit-process flow chart.

a Wood stock, *b* Conveying (e.g. by trucks), *c* Crosscut saw (e.g. wide bandsaw), *d* Metal detector, *e* Chipping machine, *f* Fans, *g* Cyclones, *h* Wing beater mill, *i* Steam-heated dryer, *j* Chip sorting derice, *k* Bunker, *l* Metering device (by weighing the chips), *m* Glue preparation station with glue pump, *n* Batch mixer, *o* Mobile container, *p* Hydraulic hot presses with molds, *q* Hand operated sanding machine, *r* Storage and dispatch of ready-for-use-articles

Fig. 5.130. View on a part of a modern Werzalit-plant

Mixing of the shreds with a urea-formaldehyde resin or according to the requirements urea-melamine or phenolic resin is carried out in batches in a mixer in
which the glue is sprayed on the shreds while they are tumbled. The size of the
shreds used and the proportion of resin added varies according to the type of
the manufactured article, the required strength, and the fineness of finish.
Since color does not matter in the ultimate product, a colored hardener is often
employed so that its addition can easily be controlled. Bins are provided for the
storage of the resin-covered shreds. The life of the mixture in storage is, for
example, 50 to 60 h at 21 °C ($\sim$70 °F).

The articles to be produced, including overlaid or complicatedly shaped items
are then formed in two stages:

1. A preform of the shape of the article is made by placing the particles into
a mold. The resinated particles are cold pressed to a moderate density so that
the "biscuit" is fairly solid and can be readily handled. The resin is not cured.

2. The final pressing operation is done in a hot press, fitted with chrome steel
male and female molds of great accuracy and high finish. When the articles receive
decorative finishes — either melamine resin based paper film or wood veneer —
these are applied during the final pressing operation. This has two advantages:

   a) No extra tooling is required for the preparation;
   b) Decorative finish and core become one integral resin-bonded structure.

The mass production of primary parts (pallets, shutter boards, etc.) is fully
automated. The bulk of the resinated chips is blown in the heated press tool. A screen
in front of the fill box of the tool separates air and chips, and meters exactly
and quickly the charging process. In this way, dependent on the thickness of the
wall of the pressed part, 25 to 50 press-cycles per hour may be attained.

Pressures vary according to the nature of the article; thinner components
generally are pressed to a higher density to obtain appropriate strength. After
removal from the press most articles need no finishing, some require only slight
sanding where the pressing process has produced jagged edges. As has been
listed in Section 5.2.13.0 the range of articles manufactured by the Werzalit

Fig. 5.131. TV-case molded as an integral part

process is very wide. Interesting is the TV-case shown in Fig. 5.131, molded in one piece, a commendably rigid and stable component, complete with all grooves, openings and perforations. Spacers for the use between the inner and outer shells of aluminium beer barrels can be manufactured with good stability and excellent dimensional accuracy. A few more recent articles may be mentioned: Concrete shuttering elements, especially shutter boards for sacrifice formwork of ribbed floors (system Kaiser Decken GmbH, Frankfurt/Main, W.-Germany), weather-resistant building elements, such as balcony, garage door and facade paneling, sidings, pallets, roof boarders, window-parapets, garden-fences, one-way-packages or containers, corrugated boards made of bagasse etc.

A few more remarks should be given about one-way-containers because here a future large field of application exists for Werzalit. *Munk* (in Werzalit-Information 1967) states that each year approximately 3 billion containers for fruit, vegetable and other agricultural products are required throughout the world.

The large distances between the individual producer countries and the consumer centers make it imperative that for storing and transporting such products one-way-containers are made available which have to meet the following demands:

1. Low price;
2. Minimum outside dimensions for maximum net contents, and light weight;
3. Stacking capacity;
4. Stiffness (rigidity) and high strength without crushing or displacing contents, even at high stacking heights;
5. Physiologically unobjectionable, odorless;
6. Quick cooling capacity of filled containers, even when stacked high and close;
7. No formation of condensate, even in changing climates, the container material must be hygroscopic;
8. Durability of the container, even under extreme climatic changes, for example at a cold storage climate of $+1\,°C$ ($\sim 34\,°F$) and 99% relative humidity;
9. Destructibility of container, that is, breaking and burning of the material must be possible with normal means;
10. Simple and quick filling and closing of the filled container must be guaranteed;
11. Simple, easy and non-damaging handling of the empty and the filled container must be asssured;
12. Adding effective advertising, printed or written, is required. The total appearance of the container should exert a favorable influence on the consumer.

It is very difficult to give leading values for the mechanical properties of molded particle products because too many factors such as wood species, dimensions of particles, type and proportion of the binding agent, pressure, finish, type and thickness of laminates or overlays, have an influence which can be adapted to a wide range of requirements according to the final use. Generally the strength values are superior to those of particleboard. Fig. 5.132 above shows, the bending strength versus density, below, the screw holding power versus density. Fig. 5.132 compares the strength of Werzalit-parts not overlaid with that of Werzalit-parts overlaid with melamine-impregnated-paper. In the whole range of densities, the bending strength σ_{bB} of parts not overlaid increases in a parabolic curve with increasing density whereas the bending strength of melamine impregnated paper-overlaid parts is higher for the same density but increases appro-

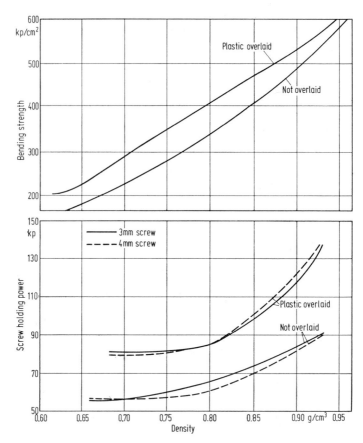

Fig. 5.132. Bending strength versus density and screw-holding power of Werzalit parts. From *Haas*, in *Kollmann* (1966)

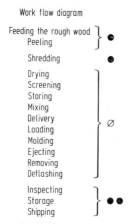

Fig. 5.133. Work flow diagram in an automatically operated Werzalit-plant. From *Haas*, in Werzalit-Information (67, p. 24)

ximately in a linear manner only. The screw-holding power is also higher for laminated than for non-overlaid parts but the influence of density may be neglected in the range between 0.65 and 0.80 g/cm³ (∼41 and 50 lb./cu. ft.).

Doubtlessly Werzalit-products enjoy international reputation. The possibilities of applying automation provide considerable advantages in industrial countries. The simplified work flow diagram in Fig. 5.133 shows that beginning with drying of shreds up to deflashing of the molded parts only supervising but no operating personnel is employed.

5.3 Some Additional Layouts

As has been described in systematic detail of principal aspects in the foregoing Section 5.2, flat-pressing and extrusion processes are applied in the particleboard manufacture. On an industrial scale they were invented and introduced by pioneers (*Fahrni, Himmelheber* and coworkers, *Greten, Kreibaum*), who were quick to develop new machines and specialized procedures. These were dealt with in the foregoing Section, which contains some layouts of complete works or essential parts of them. Reference may also be made to the schematic diagrams of particleboard manufacture in Fig. 5.38 and Fig. 5.39.

Under these circumstances the survey may be completed only by some additional layouts. The author knows that a fully exhaustive description and illustration of particleboard manufacture would need its own textbook. Therefore the readers should be aware that Chapter 5 is only one chapter of this book about wood science and technology applied to the various wood-based products. Added to the book "Holzspanwerkstoffe" (Wood particle materials), edited by *Kollmann*, are lists of manufacturers of machines and plants for the particleboard industry. Further lists contain data about locations, types of board, and capacities of particleboard manufaturers in W.-Germany, and in the whole world.

The rapid technical progress has to be observed which led not only to many new inventions but also to some alterations in the structure of the industries building such machines. It came on the one hand from close cooperation even to trust of reputable machine manufacturers on the other hand from some factories, which utilized particleboard in their main production program (for instance furniture) or had wood residues such as veneer and plywood — who developed machines or processes for particleboard production for their own use only. Some factories were dissolved, other firms stopped their activity in the field of particleboard equipment such as Hermal, München, W.-Germany, Miller Hofft, Richmond, Va., U.S.A., Glomera of Fisch-Sueffert AG, Basel, Switzerland. Books about particleboard are edited by Holz-Zentralblatt-Verlag (1954), *Johnson* (1956), *Scheibert* (1958), *Kollmann* (1955, 1966), *Deppe* and *Ernst* (1964), *Plath* (1963). Important articles dealing with particleboard production are published in the following periodicals: Holz als Roh- und Werkstoff, Springer-Verlag, Berlin-Heidelberg-New York, (first Vol. 1937/38), Forest Products Journal, Madison, Wisc., U.S.A. (Journal of the Forest Products Research Society), (first Vol. 1951). Holztechnologie, Dresden, Institut für Holztechnologie und Faserstoffe, (since 1960), Board Manufacture, Pressmedia Ltd., Ivy Hatch, Sevenoaks, Kent, England, (since 1957), Deutsche Gesellschaft für Holzforschung, Entwicklung und Herstellung von Holzspanplatten (Development and Manufacture of Particleboard) Berichte (Reports) No 2/1956, No 1/1957, No 1/1961, No 2/1965, Dosoudil, Formstabilität von Holzspanplatten, Ber. No 1/1965, Beuth-Vertrieb, Berlin 15; a card index covering all aspects of particleboard has been worked out by the Wilhelm Klauditz-Institut, Braunschweig, W.-Germany. The following additional layouts are self-explanatory (Figs. 5.134, 5.135 A, B, 5.136, 5.137, 5.138, 5.139).

5.4 Properties of Particleboard

5.4.0 General Considerations

In the introduction it was mentioned that many inventors ardently desired to create artificial boards on the basis of ligno-cellulose materials, preferably of wood wastes. Because suitable binders and machines for reduction, sorting, bunkering, drying, blending with binders, mat-spreading, and finishing, were not available for a long time, about a century was needed to achieve first promising results. Then the "explosion" of production and application of particleboard happened as mentioned previously. Doubtlessly the specific properties of particleboard opened the path to the markets, though the present consumers were reluctant at the beginning.

At first the specific gravity (density) of particleboard was of greatest interest and easily determined. The large possible sizes of particleboards were an asset, the manufacture by dry processes, the fine workability of particleboard, the ease of gluing, veneering and overlaying them. Later on special types of particleboard were made fire resistant. In summarizing, it can be stated that the modern manufacture of particleboard is really a technical evolution and very interesting in comparison with other sheet-like wood based material such as plywood, corestock, sandwich boards, and insulation and hard fiber (building) board. The competition in these fields is severe and occasionally undergoes unforeseen changes. But the trend of increasing capacity and consumption, still described by an approximately parabolic function, exists; the intersection-point on the S-shaped growth curve is not yet reached (cf. Fig. 5.5 and 5.7). New technologies (mainly in using phenolic resin glues, water proof overlays, sealing of board edges, development of standardized elements for prefabricated houses, for instance of the Okal-type) made particleboard, originally nearly exclusively used for interior work, nowadays attractive for exterior application. More details are given in the following Sections but a few economic aspects should not be omitted:

a) The particleboard industry consuming ligno-cellulose raw materials of low grade contributes favorably to the national economy and to the efficient management of cheap organic resources;

b) In manufacturing plants with high production capacities, up to 1,000 metric tons daily, the proportion of the costs of labor is relatively very low because most steps in production are fully mechanized or even automated. This may be desirable in countries with full-employment (e.g. in W.-Germany);

c) The prices of the various types of particleboard are well balanced with other competitive sheet-like construction and building materials;

d) Transport and shipping are easily done with appropriate freight rates;

e) Material wastes can be kept low if standardized boards are used and if the flow of materials at the point of utilization is well planned (programmed);

f) Many necessary operations in the application (for instance in furniture and cabinet making and in building, constructing etc.) such as gluing, lacquering and varnishing, sawing, drilling, jointing etc. are simple.

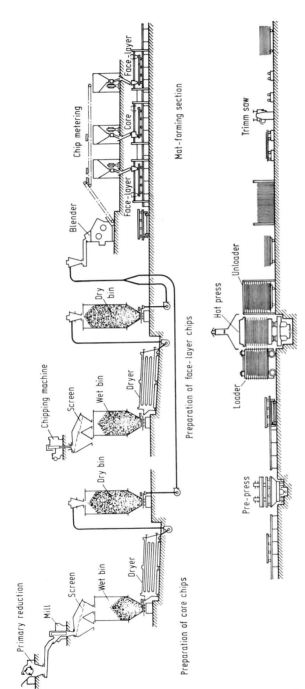

Fig. 5.134. Scheme of a work for manufacture of three-layer particleboard. System Fahrni-Novopan. From *Kollmann* (1952)

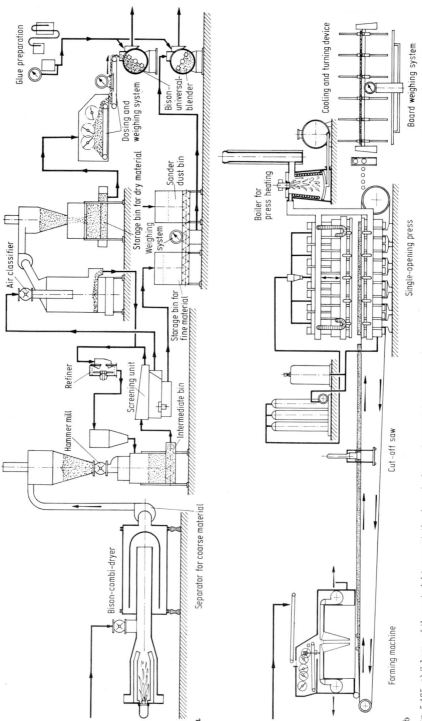

Fig. 5.135. a) Scheme of the material transportation in a single-opening particleboard plant; b) Scheme of single opening press operation. Bison-Werke Bähre u. Greten, Springe

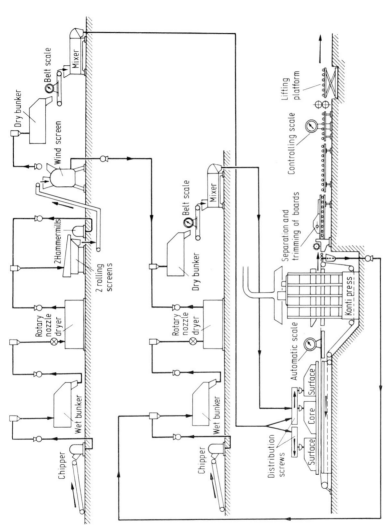

Fig. 5.136. Material flow sheet of a particleboard plant. System Kontipress, Becker & van Hüllen, Krefeld

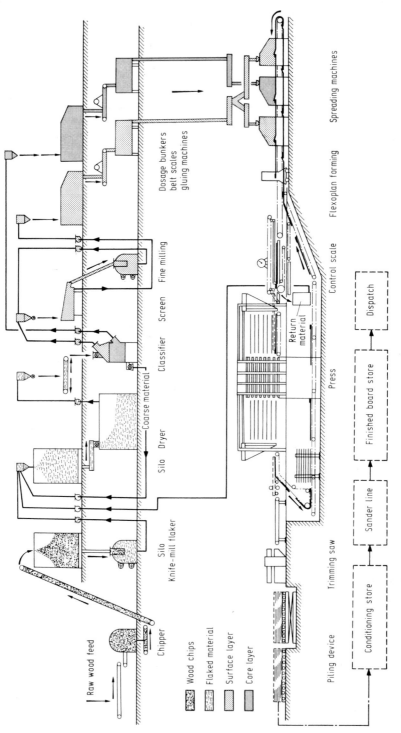

Raw wood feed

Chipper

Wood chips

Flaked material

Surface layer

Core layer

Silo   Knife-mill flaker

Silo   Dryer

Coarse material

Classifier   Screen   Fine milling

Dosage bunkers
belt scales
gluing machines

Spreading machines

Flexoplan forming

Control scale

Return material

Press

Finished board store

Dispatch

Sander line

Conditioning store

Trimming saw

Piling device

Fig. 5.137. Scheme of a particleboard plant for the production of three-layer boards. System Laboratorium Himmelheber, Baiersbronn

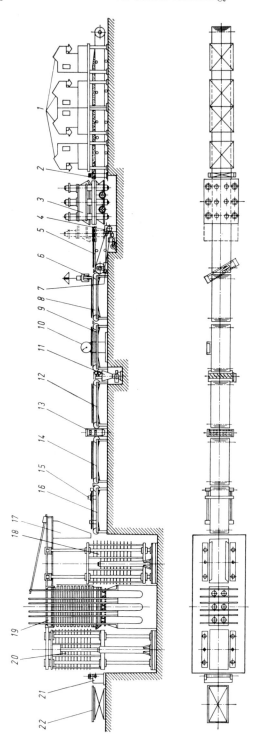

Fig. 5.138. Scheme of the production of particleboard. System Fratelli Pagnoni, Monza, Italy

1 Stationary spreaders, 2 Magnetic ejector of metallic parts, 3 Patented movable prepress with movable side-frames, 4 Pre-press motorization and transfer unit, 5 Conveyor belt made of special rubber, 6 Transversal saw unit and chips suction device, 7 Belt's brushing unit, 8 Belt conveyor for board transfer, 9 Belt conveyor, 10 Balance for boards weight control, 11 Trap for elimination and destruction of out-of-standard boards, 12 Belt conveyor, 13 Board spraying on both faces, 14 Belt conveyor, 15 Device for introduction of boards on the tablets in the openings of the loading lift, 16 Belt conveyor, 17 Loading arm, 18 Loading lift with tablets or belts, 19 Hydraulic press with heating platens, 20 Unloading lift, 21 Motor-driven rollers for board ejection, 22 Piling platform

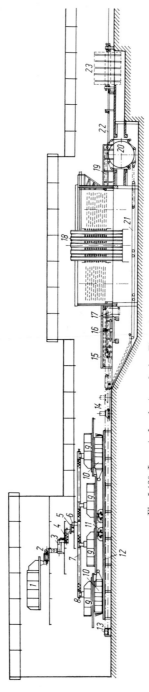

Fig. 5.139. Layout of a plant producing Flexoplan board. System Schenk-Flexoplan

1 Dosing belt bunker, *2* Belt weigher, *3* Gluing machine, *4* Screw conveyor, *5* Vibration feeder, *6* Distributor, *7* Screw conveyor, *8* Feeding screw conveyor, *9* Spreading bunker, *11* Spreader heads, *12* Spreading wall, *13* Forming belt, *14* Cross cut saw, *15* Forming belt, *16* Check weigher, *17* Recovery of miss-spread mats, *18* Heating press, *19* Press discharge and rawboard separation, *20* Storing drum, *21* Return run, *22* Rawboard transport, *23* Rawboard cooling

5.4.1 Factors Affecting the Properties of Particleboard

Thorough investigation of the various factors affecting the properties of particle-board led to a continuous improvement of their quality. The following factors are mainly to be considered:

a) Species of wood (coniferous species, broadleaved species or more com-mercially specified: softwoods and hardwoods; or, from another point of view: indigenous and imported woods; mixtures of two or more species), fiber structure, density, hardness, compressibility;

b) Form and size of raw wood (logs, slabs, edgings, scantlings, end trims, chips, planer shavings, excelsior (Brit.: wood-wool), sawdust, veneer wastes or scraps, peeler cores;

c) Wood with bark or debarked wood;

d) Non-wood ligno-cellulose materials (cf. Section 5.1.1.3 and 5.1.1.4) such as flax shives, bagasse and other vegetable materials;

e) Type and size of particles (cut, disintegrated, and ground particles, chips flakes, shreds, fines and dust, slenderness ratio (i.e. ratio of length to thick-ness) "I"-chips identified as long flat chips, strands, "S"-chips identified as shorter, curly (helix) chips, surface area of flakes;

f) Method of chip drying (temperature, drying time, minimum remaining chip mousture);

g) Chip screening and separating, chip size distribution expressed by frequency curves;

h) Type and amount of binding agents, for instance urea-formaldehyde or phenol-formaldehyde resins, amount expressed as ratio of solid resin weight to weight of oven-dry chips; catalysts, extenders, fillers, additives, such as water-repellents, fungicides, insecticides, fire retardants;

i) Method of chip spreading, structure of particleboard (uni-layer, multi-layer, graded board, particle orientation);

k) Moistening of chips prior to pressing (steam shock), final moisture content of board (average and moisture gradient), conditioning;

l) Curing conditions and cycles (temperature, pressure, time);

m) Thickness of board;

n) Sand content of particleboard (reducing their machinability by cutting tools);

o) Surface quality (smootheness, roughness, waviness, sanding properties);

p) Primings, coats of varnish or lacquer;

q) Laminating, veneering, overlaying.

This long list shows how many factors affect the properties of particleboard. Some may be freely chosen, such as species of wood, shape of raw wood, debarking or not debarking, non-wood ligno-cellulose materials, type of reduction and sub-sequent machines and methods in board manufacture including for example chip drying, separation and blending with binders. Type and amount of binding agents are determined by economic considerations and by the adaptation of the particle-board to their final use. Method of chip spreading is decisive for the structure of the boards. Curing conditions and cycles must be suitable for the particular purpose. Thickness of board and further treatment must be appropriate to special require-ments of the products. Because practically all the effects mentioned are inevitable and to some extent mixed, a great deal of science and experience are needed, together with advanced studies and experiments, to insure the optimum effect of any raw material and of any condition applied to a given process. "Therefore, optimum conditions in one case may be entirely different from those in another. Research must continue to evaluate each variable while keeping in mind that production techniques are of necessity far removed from laboratory experiments" (*Johnson*, 1956, p. 181).

5.4.2 Testing of Particleboard, Standards

The widespread acceptance of particleboard is known and is based on the product capabilities. Initially there was a certain wariness by consumers which at the beginning retarded the particleboard industry. Some disappointments have occured because the new products were applied for unreasonable uses. In order to allow fair judgements, it was and is necessary to develop and use methods of test and to establish a set of standard values for the properties of particleboard.

At present there are many standards in various countries available for the guidance of people who are concerned with chipboard production and use. Since Europe has been engaged for a longer period with particleboard development, earlier and more advances have been made there than in the U.S.A. Methods of test for particleboard have been published in Western Europe in England, France, Sweden, W.-Germany, in Eastern Europe in the ČSSR, East-Germany, Poland, USSR.

The International Organization for Standardization (ISO) dealt with the problems in its Technical Committees ISO TC 89, 92 and 151, and so did the American Society for Testing and Materials (ASTM). Many countries have industrial standards pertaining to the testing of particleboard, but their degree of acceptance is not known. In the U.S.A., a set of standards was prepared by representatives of the particleboard industry and members of the National Woodwork Manufacturers Associations. Reference is also made to explanations in Section 5.3.0.

"It must be stressed, here, that absolute values from various sources using different methods of test are almost worthless in the comparison of properties of two or more particleboard brands" (*Johnson*, 1956, p. 268).

In the testing itself three major considerations are evident:

a) *Sampling and preparation of specimen.* — The choice of sample must be made so that a representative specimen is obtained. The specimen must be conditioned, for instance in Germany in normal climate (DIN 50014, Dec. 1959) with $20\,°C \pm 2$ grd temperature and $65 \pm 3\%$ relative humidity of the air to weight constancy. Duration of conditioning as a rule 3 to 5 days. Prior conditioning, depending on the type of test, additional preparation of samples or specimen may be required, such as gluing, laminating or veneering, shaping etc.

b) *Apparatus.* — The machines used must be capable of performing the various tests under the specified conditions (size, shape, and spacing of supports and clamps, rate of loading). The materials to be used in special tests (nails, screws, abrasives etc.) must be of a specified type.

c) *Reporting.* — The report must be clear and accurate and contain the following statements (according to DIN 52360, April 1965):

1. Purpose of use: interior work, protected from weather influences, exterior work, exposed to weather;
2. Methods of pressing: flat-pressed board, extruded board (with or without tubular holes);
3. Type of chips (raw material cf. Section 5.4.1 a and d);
4. Size of chips (cf. Section 5.4.1 e);
5. Structure of the cross section (board construction, cf. Section 5.4.1 i);

6. Surface quality (cf. Section 5.4.1, 0);

7. Visual examination of surfaces: color, color defects, other defects;

8. Veneers (if present): wood species, veneer thickness, type of veneer (rotary-cut or sliced);

9. Lamination or overlays (structure, type, compressed etc.);

10. Specialities: for instance fungi or insect proof, fire resistant.

The following properties may be tested (*Kollmann*, 1949; Furniture Development Council, England, in: *Johnson*, 1956, p. 268/269):

a) *General properties*

*1. Dimensions (length, width, thickness) of particleboard ex factory, accuracy to size and tolerances (cf. German standard specification DIN 68760, Draft Dec. 1971, DIN 68761, Draft Sept. 1970);

*2. Color;

*3. Surface quality and surface treatment.

b) *Physical properties*

*4. Density, weight per unit area after conditioning;

*5. Moisture content, based on oven-dry weight, ex factory (or at time of delivery);

*6. Absorption and swelling in moist air or liquid water, movement due to moisture content change;

7. Permeability for water, other liquids, air and gases;

8. Thermal expansion;

9. Specific heat;

*10. Thermal conductivity, diffusivity;

11. Radiation of heat (emissivity);

12. Electrical properties (electrical resistance, electrical conductivity, dielectric properties);

*13. Acoustical properties (sound insulation and absorption).

c) *Mechanical (elastic and strength) properties*

*14. Modulus of elasticity and of rigidity;

*15. Tensile strength parallel to plane of board;

*16. Tensile strength perpendicular to plane of board, de-lamination strength ("Abhebefestigkeit"), internal bond;

*17. Compression parallel and perpendicular to plane of board, buckling;

*18. Bending strength parallel and perpendicular to plane of board;

*19. Deflection under sustained load, creeping;

*20. Shear strength parallel and perpendicular to the plane of board;

*21. Behavior in torsion;

*22. Behavior in impact bending or puncture tests;

*23. Hardness.

d) *Technological properties*

*24. Surface stability;

*25. Machining properties, workability;

26. Bending capacity;

*27. Nail holding (power);

*28. Screw holding (power);

29. Gluability;

30. Paintability.

e) *Resistance against destruction*

*31. Abrasion (wear) resistance,
*32. Resistance against decay (fungi);
*33. Resistance against insects including termites;
 34. Resistance against chemicals (corrosion);
*35. Reaction to fire.

There is still some disagreement concerning the properties which are relevant in particleboard and "there is sure to be much more contradiction concerning the relative importance of those properties" (*Johnson*, 1956, p. 269). In this chapter only such most important properties, marked in the foregoing catalogue by an asterisk (*), are discussed in a condensed form. The principles of physics, mechanics, and rheology of solid wood which are outlined in Volume I of this book (p. 160 to 419) give a reliable basis for conclusions, deductions and transformations from theories developed for solid wood to those necessary for the science of wood-based materials.

5.4.3 Density, Weight per Unit Area

The density influences most physical, mechanical, and technological properties of particleboard. The consumers are interested in low density with adequate strength. Density is the quotient between sample mass M and sample volume V. It is measured in g/cm^3 or kg/m^3 (1 kg/m^3 = 0.062,43 lb./cu. ft.). The symbol for density is ϱ with an index showing the moisture content in percentage, e.g. ϱ_{10}.

According to the German Standard Specification DIN 52361 (Apr. 1965) the density ϱ of a particleboard specimen after conditioning is calculated as follows:

$$\varrho = \frac{M}{V} = \frac{M}{A \cdot s} \ [\text{g/cm}^3], \tag{5.2}$$

where M = mass in g, V = Volume in cm^3, A = surface of test piece in cm^2, s = thickness of test piece in cm.

If more (n) specimens are taken from one particleboard then it is possible to estimate its average density $\bar{\varrho}$ as the arithmetic mean of the single values ϱ:

$$\bar{\varrho} = \frac{1^n \sum \varrho_1 \cdots n}{n} \ [\text{g/cm}^3]. \tag{5.3}$$

A dependent property is the weight per unit area which is calculated as the ratio of mass M to surface area A.

It is necessary that the board is bright-pressed or carefully sanded. *Lynam* (1959) and *Plath* (1963) showed by experiments that the density is highest on the deck layers and decreases to the core. *Walter* and *Wiechmann* (1961) pointed out that high density on the deck layers increases the mechanical properties. *Teichgräber* (1958) (Fig. 5.140) showed that the distribution of density in commercial, quality-supervised particleboard varies. The particleboards investigated were manufactured by four different firms and were sanded. The minimum values for the density always occurred in the center of the board. It is possible to vary the density within the thickness of graded board to a predetermined pattern with the same total amount of basic material and the same average density of the board (Fig. 5.141). In reality the curves are not completely smooth, there are small jumps due to irregularities in manufacture. The higher density in the surface layers effects correspondingly higher bending strength, closer and more even surfaces for veneering laminating, or painting, higher resistance to absorption and swelling, and higher resistance to ignition and spread of flame. There are, of course, practical limits dictated by certain standards of shear strength, delamination resistance, internal strength, face strength and nail- and screw-holding power.

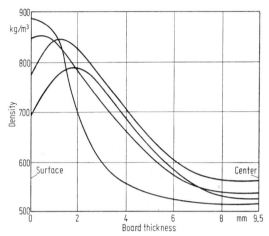

Fig. 5.140. Dependence of density on board thickness (19 mm) of 4 commercially supervised particleboard. From *Teichgräber*, in *Kollmann* (1966, p. 539)

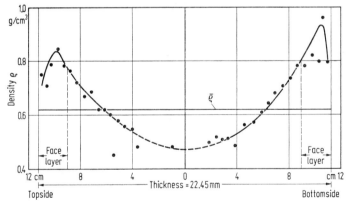

Fig. 5.141. Distribution of density versus thickness of a multilayered board. After *J. Horn* 1965, from *E. Schwab* 1969

Summarizing it may be stated that for particleboard, as for solid wood, a strong correlation exists between average density and absorption and swelling (cf. Fig. 5.15a), permeability, thermal conductivity, sound insulation, modulus of elasticity, bending strength (cf. Fig. 5.1, 5.2, 5.3, 5.15b) shear strength (cf. Fig. 5.16a), behavior in torsion, impact bending, hardness, machining properties, nail- and screw-holding power, abrasion resistance and reaction to fire. In many cases the mechanical properties increase in direct proportionality to the increase of density. Because many variables cooperate in strength properties of particleboard, often more complicated curves (slightly parabolic, hyperbolic, or even S-shaped) are observed. If any breaking strength is symbolized by σ_B, then the general law exists:

$$\sigma_B = C \cdot \varrho^n + A, \tag{5.4}$$

where C and A are constants $n \geqq 1$ is an exponent, to be determined by experiments. *Johnson* (1956, p. 182) discussing the curves in Fig. 5.3 (published by *Klauditz*, 1952) writes: "Although the hardwoods may be much heavier than the softwoods, the factors of hardness and compressibility are the most significant. Softwoods, because they are more pliable and compressible than hardwoods will usually form denser, more compact boards under a given pressure. By virtue of the resulting closer contact of the individual softwood particles, stronger boards are obtained through better bonding ... lighter boards of softwood possess strength equal to hardwood boards."

The pressure p exerted on the chip carpet during the hot pressing operation determines the compactness or density ϱ of the board, if the other factors such as particle size, moisture content, resin amount, etc. are considered equal, the formula holds

$$\varrho = C_1 \cdot \log p + C_2, \tag{5.5}$$

where C_1 and C_2 are constants to be determined experimentally. Relationships of this type were found by *Turner* and *Kern* (1950).

5.4.4 Moisture Content, Absorption and Swelling

The relative moisture content H (e.g. at the time of delivery) is the quotient of the difference of the mass M_w of the wet sample and the mass M_0 of the sample dried at a temperature of $103\,°C \pm 2\,°C$ ($117\,°F \pm 3.6\,°F$) to the mass M_0

$$H = \frac{M_w - M_0}{M_0} \cdot 100 \ [\%] \tag{5.6}$$

According to the German Standard Specification DIN 52361 (Apr. 1965) constant weight is reached if the sample mass compared with the foregoing weighing during a period of four hours does not change more than 0.1%. Another German Standard Spezification (DIN 50014, Dec. 1959) compiles the climates as shown in Table 5.18:

Table 5.18. Normal Climates

Symbol	Temperature		Relative humidity		Atmospheric pressure
	°C	°F	Nominal value	Usual deviation %	mbar
20/65	20 ± 2	$68 \ \pm 3.6$	65 }	± 3	800 to
23/50	23 ± 2	73.4 ± 3.6	50 }		1,000

The normal moisture content of particleboard generally is lower than that of solid wood. This has little influence on swelling and shrinking. The problem has been investigated by many scientists: *Elbert* (1961), *Fickler* (1962), *Klauditz* and coworkers (1960), *Lawniczak* and *Nowak* (1962), *Liiri* (1960), *Müller* (1962), *Runkel* (1951), *Teichgräber* (1958). *Müller* (1962) especially testing the dependence of thickness swelling of uni-layer board with and without increased amounts of water-repellents (Fig. 5.142).

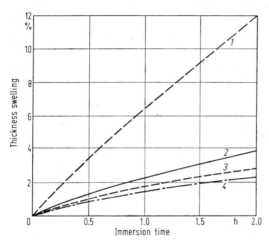

Fig. 5.142. Dependence of thickness swelling on immersion time and on the amount of water-repellent added in a unilayer particleboard. From *Müller* (1962)

1 Without addition of paraffin emulsion, *2* With 0.25% solid paraffin based on oven-dry wood, *3* With 0.50% solid paraffin, based on oven-dry wood, *4* With 0.75% solid paraffin, based on oven-dry wood

The determination of the normal moisture content needs controlled air-conditioned testing rooms. The measurements are not easy and are expensive, but in the normal climate (20/65, Table 5.18) the normal moisture content ranges between 8.5% and 11.0% and is in the average about 9.6%.

The hygroscopic behavior of particleboard was investigated early. Results, obtained by *Lawniczak* and *Nowak* (1962) are shown in Fig. 5.143. It should be realized that the moisture content of samples without water-repellents will increase by approximately 50% and the thickness swelling in about the same order. It is remarkable how relatively less re-drying reduces the thickness swelling. The moisture content for both normal board and those with addition of water-repellents after 48 h of drying (at $103 \pm 2\,°C$ or $221 \pm 42\,°F$) reaches nearly the zero value.

In West Germany the test methods for determining the thickness swelling are standardized (DIN 52364, Apr. 1965). The highest swelling after two hours' immersion, related to the original thickness, should not exceed the values given in Table 5.19.

Table 5.19. Swelling of Particleboard

Type of board	Density kg/m³ (lb./cu. ft.)	Thickness mm (in.)	Swelling in % of original thickness max.
Flat pressed boards	$750 \rightarrow 450$ ($47 \rightarrow 28$)	$6 \rightarrow 25$ ($\sim 1/4''\cdots 1''$)	$10\cdots 6$
Extruded boards without holes		$6 \rightarrow 25$ ($\sim 1/4''\cdots 1''$)	~ 3.1

The remarkable difference between the swelling behavior of solid wood and particle board is shown in Figs. 5.144 and 5.145.

The values in longitudinal direction were determined by *Teichgräber* in *Kollmann* (1966, p. 544). They are compared with earlier data found by *Mörath* for pine wood (in *Kollmann*, 1951). In the moisture content range investigated the

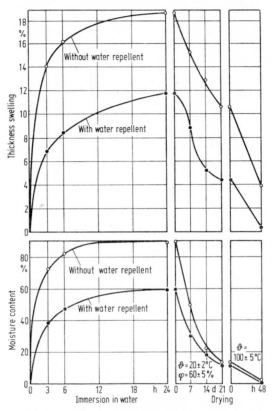

Fig. 5.143. Dependence of thickness swelling and water absorption in three-layer particleboard untreated and treated with water-repellents on the duration of immersion time and drying time respectively. From *Lawniczak* and *Nowak* (1962)

relationship for solid pine wood was linear whereas for flat-pressed particleboard (19 mm ∼ 3/4 in. thick) a concave curve against the abscissa exists. The average linear change of length for oven-dry particleboard up to the normal moisture content was determined to be 0.42%. The relatively higher thickness swelling of flat-pressed particleboard illustrated in Fig. 5.145 (*Teichgräber* in *Kollmann*, 1966, p. 545). It is remarkable that the measured points scatter widely especially for moisture contents above 12% (based on oven-dry weight). After conditioning in normal climate the average thickness swelling from the oven-dry condition was 4.15%.

Extruded particleboard, due to its structure, demonstrate less swelling of thickness (6 to 25 mm thick veneered board 3 to 4% after 24 h water storage). The expansion of extruded board width after 24 h immersion in water is about 26.5% for untreated board and 19.1% for board containing 0.5% solid paraffin (based on oven-dry wood) (*Müller*, 1962).

The absorption which causes the swelling in humid air is of importance for the utilization of all types of wood based material and should be reduced by suitable manufacturing techniques and/or by addition of water-repellents. Compared with solid wood the hygroscopic isotherms for particleboard are lower and this is practically essential in the range of 50 to 70% relative humidity (Fig. 5.146). The figure shows that an effective quality improvement is reached.

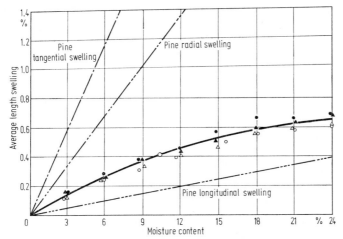

Fig. 144. Average change of length in board plane of 19 mm thick, quality-supervised particleboard versus moisture content, starting from conditioned state. From *Teichgräber*, in *Kollmann* (1966, p. 544). (For comparison: dotted lines for solid pine wood, in tangential direction, radial direction, and longitudinal direction, by *Mörath*, in *Kollmann*, 1951, are given.)

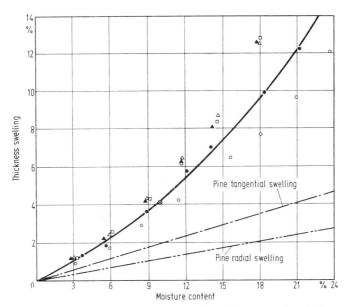

Fig. 5.145. Average thickness swelling of 5 different, 19 mm thick quality-supervised particleboard versus moisture content, starting from air-conditioned state. From *Teichgräber*, in *Kollmann* (1966, p. 543). (For comparison: dotted lines for solid pine wood, in tangential direction and radial direction, by *Mörath*, in *Kollmann*, 1951, are given.)

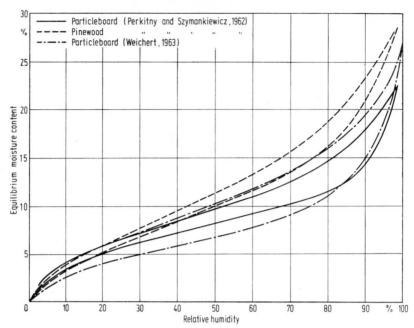

Fig. 5.146. Hygroscopic isotherms for desorption after adsorption of particleboard and for comparison solid pine wood. From *Perkitny* and *Szymankiewicz* (1962) and *Weichert* (1963)

5.4.5 Thermal Conductivity

Low thermal conductivity is important for wood based materials since low heat consumption in normal dwellings during the winter season and good protection against too high temperatures during the summer season can be obtained only by low thermal conductivity values. *Kollmann* and *Malmquist* (1956) carried out experiments. The results (explained in detail in Vol. I, p. 246—250) are summarized in Fig. 5.147. Particleboard with an average density of 0.6 g/cm³ (37.5 lb./cu. ft.) has a thermal conductivity nearly one third lower than solid wood across the grain. Fiberboard has a still lower thermal conductivity, but their density must be considered.

Fig. 5.148 shows the influence of moisture content on thermal conductivity at various temperatures. Light board produced in the flat process ($\varrho_{10} \leq 450$ kg/m³ or 28.1 lb./cu. ft.) have a thermal conductivity of about 0.06 kcal/m h °C (1 kcal/ m h °C = 0.672 BTU/ft. g °F), light extruded board have a higher thermal conductuvity of about 0.12 kcal/m h °C. In comparison with other building materials these values are good.

5.4.6 Acoustical Properties

The acoustical properties of solid wood (sound transmission in wood, sound wave resistance) as well as acoustics of buildings (sound transmission loss for various types of construction) are dealt with in a condensed manner in Vol. I of this book, p. 274—285. The acoustical properties of buildings and rooms do not belong to the principles of wood science and technology but are problems concerning the projecting, designing and performing of architects and engineers with the aim of guaranteeing adequate sound protection and satisfying audibility.

The fundamental principles of the acoustical properties of buildings were outlined by *Cremer* and *Zemke* (1964). Sound is an integral and vital part of any environment. Before 1900 acoustical research was engaged only with the pure physics of sound. Subsequently the direct application to architecture has been studied and adopted into the science of sound control.

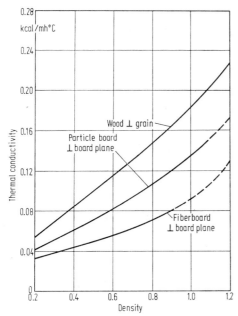

Fig. 5.147. Dependence of thermal conductivity of wood across the grain, particleboard and fiberboard on density. From *Kollmann* and *Malmquist* (1956)

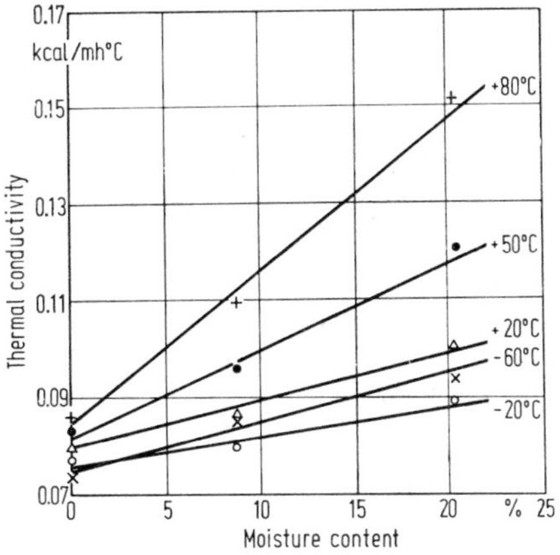

Fig. 5.148. Thermal conductivity of particleboard at various temperatures in dependence on the moisture content. From *Kühlmann* (1962)

Sound is a vibration in an elastic medium (*Hopper*, 1969). It is the result of alternating pressures and particle movements within this medium. The elastic media to be considered in architecture are fluid (air and liquids) or solid (building materials and the earth). When molecules throughout an elastic medium move back and forth in opposite positions, a single full circuit is called a *cycle*. The *amplitude* is the distance from the normal ("rest") position to an extreme. The number of cycles completed in one second is the frequency (Hz, formerly cps). Measured by time, the disturbance caused in the medium, is called a *wave*, and the distance between adjacent regions where such a molecule displacement occurs, is known as the *wave length*. The molecules within the medium only vibrate, but the effect, or sound, of the vibration moves rapidly and travels great distances.

Hearing is the subjective response to sound. The ear is simply a receiver. Pitch is the ear's response to frequency. Low frequencies are identified as low in pitch, high frequencies as high.

Loudness is the sensation produced in the ear in response to varying sound pressures and intensities. The ear is much more sensitive to low levels of acoustic energy, or low sound, than to high levels or loud sounds. This variation in sensitivity is, roughly, logarithmic. A sound of intensity I_0, usually a sound at the threshold of audability has a loudness of 0 decibels (db). The decibel is equal to one tenth of a bel — the unit named in honor of *Alexander Graham Bell* (1847 to 1922).

The formula giving the loudness l in decibels of a sound of intensity I is:

$$l = 10 \log \frac{I}{I_0} \tag{5.7}$$

On the decibel scale the approximate loudness of some common sounds are as follows: rustel of leaves 10 decibels, a quiet house (ordinary conversation) 40 decibels, loud radio music in room 80 decibels, riveting 95 decibels, a jet aircraft, starting, about 120 and more decibels (threshold of pain). Loudness of more than about 120 decibels begins to be felt as pain, rather than heard. Curves of equal loudness as a function of frequency according to ISO-Recommendation No. 352 are represented in Fig. 5.149.

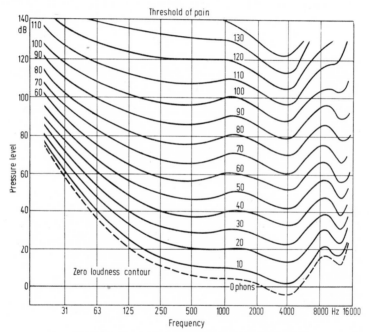

Fig. 5.149. Curves for equal loudness. ISO proposal No. 352

Architectural acoustics has the following practical aims:

a) *Wanted sound* should be heard distinctly in suitable loudness as close a copy of the original source as possible.

b) The loudness level anywhere in the space should be identical with the level of the source. With this in mind absorptive, reflective, and amplifying materials are necessary.

c) Wanted sound should be evenly distributed. The variation between "good" and "bad" seats in an auditorium should be eliminated.

d) *Reverberation* is the persistence of sound measured in seconds, after the sound source has ceased. A room with a short reverberation time is considered "dead". The right amount of reverberation is called "brilliance". No reflected echoes should be heard.

e) *Unwanted sound* is *noise*, produced by distorted sound.

f) Sharp "peaks" of sound should be eliminated.

The architect must know the following:

g) Sound travels move readily with less transmission loss through light, porous materials and constructions than through heavy, dense materials (Fig. 5.150).

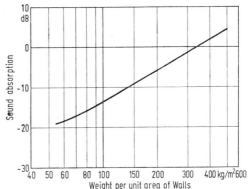

Fig. 5.150. Sound absorption in dependence on the weight of unit area of walls. DIN 4109, Bl. 5

h) Sound travels practically are undiminished through any openings.

i) Sound-absorptive materials are usually ineffective in preventing sound transmission through building elements (walls, ceilings, floors).

k) Every building material (including particleboard) and construction can be a useful tool toward good acoustics or sound control. Sound insulation can be improved by multi-shell instead of unilayer constructions. There are two fundamental systems of double-shell constructions:

1. Combination of a heavy thick shell with a relatively light, thin front-shell;
2. Combination of two light thin shells.

Principally there is an air-space between both shells, in which mats consisting of glass fiber, cork fillers, mineral fibers, etc. are suspended as sound absorbers (Fig. 5.151).

Summarizing and based on studies of many examples in practice, it can be stated that it is possible in almost any case to use particleboard for sound-insulating wall-elements, doors, sound-reflectors and sound-absorbing building parts. The fine machining properties and the adaptability to the aesthetic wishes of the designer are favorable.

5.4.7 Mechanical Properties

5.4.7.0 General Considerations. The manifold applications of particleboard in furniture manufacture, for building purposes and for other uses, for instance in shipbuilding, for containers, as parts of railway-carriages, automobiles, trucks, etc., require adequate elasticity, rigidity, strength, shock resisting ability and hardness. The mechanical properties also must be sufficient for handling and transporting, for high nail and screw holding power, and for proper machining. The mechanical properties have been continuously improved from the beginning about 25 years ago. At its introduction in handicraft and industry particleboard was not used for load-bearing or highly stressed structural elements. Particleboard was mainly used for coreboard in the manufacture of furniture, for panelling, ceilings, partitions and sub-flooring. These trends lasted over about a decade, but then increasing interest in prefabricated houses, at first built on a large scale in Sweden, made particleboard attractive for these new types of houses.

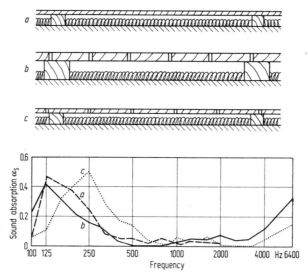

5.151 a – c. Sound absorption in dependence on frequency for various flooring constructions.
a) Particleboard, 8 mm thick, on lath grit 20 mm × 30 mm with lateral distance of 65.5 cm, between the laths 20 mm thick Silan mat;
b) Particleboard, 20 mm thick with holes (10 mm diameter, whole distance 125 mm), on lath grit 80 mm × 50 mm, with lateral distance of 65.5 cm, between the laths a 30 mm thick Silan mat;
c) Particleboard, 8 mm thick, with holes (10 mm diameter, whole distance 125 mm) on lath grit 20 mm × 30 mm, lateral distance 65.5 cm, between the laths a 20 mm thick Silan mat.
Institut für Technische Akustik, TU Berlin (1961)

In West Germany in addition to the Standard Specification DIN 1052 (Timber Structures, Oct. 1969) permissible stresses for wood-based materials were agreed upon for the first time. In compiling the necessary material it became evident that reliable data were scarce and scattering. Therefore in 1963 in the Institute for Wood Research and Technology of the University of Munich twelve types of commercial particleboard were tested systematically. The results of these tests and many data, obtained by the authors listed, are taken into consideration in Table 5.20.

It must be pointed out that multi-layer and graded particleboard are typical "combined board". The density and with it most mechanical properties vary remarkably from the core to the outer layers, as has been shown in Figs. 5.140 to 5.144.

Suitably made particleboard are not weakened by short exposure to water or vapor absorption, by swelling or shrinking, by attacks of fungi, insects and even fire. The various mechanical properties are correlated to the type and structure

Table 5.20. Normal Average Ranges of Mechanical Properties of Particleboard Glued with Urea-Formaldehyde Resin

		Flat-pressed	Remarks	Extruded
Thickness	mm	4 → 80	Most common 19 mm	Without holes 6 → 25 / 15/64 → 1 / With holes 23 → 120 / 29/32 → 4 3/4
	in.	5/32 → 3 5/32	3/4	
Density	g/cm³	High 0.8 → 1.2 ; Medium 0.4 → 0.8		Without holes 0.58 → 0.62 / 36 → 39 / With holes 0.30 → 0.45 / 19 → 28
	lb./cu. ft.	50 → 75 ; 25 → 50		
Modulus of elasticity	kp/cm²	Tension 38,000 ← 10,000 ; Bending 45,000 ← 12,000	Bending (*Keylwerth*, 1959) density 0.5 g/cm³ 18,000 ← 13,000 / 255,000 ← 185,000 density 0.9 g/cm³ 64,000 ← 42,000 / 900,000 ← 600,000	
$E = 18_{22}$	lb./sq. in.	540,000 ← 142,000 ; 640,000 ← 170,000		
Modulus of rigidity G_{44} G_{55} G_{66}	kp/cm² lb./sq. in.	2,300 / 32,500 / 2,100 / 30,000 / 11,800 / 167,000	Averages for cubes 64 mm × 64 mm × 64 mm *Albers* (1970)	
Tensile strength across the decklayers	kp/cm²	8 ← 3	20 ← 15	Without holes 27 ← 15
$\sigma_{tB}\perp$	lb./sq. in.	114 ← 43	280 ← 210 (*von Thielmann* and *Munz*, 1960)	380 ← 210
Tensile strength parallel to decklayers	kp/cm²	350 ← 35		
$\sigma_{tB}\|$	lb./sq. in.	5,000 ← 500		

Table 5.20 (Continued)

	Flat-pressed		Remarks	Extruded
Compressive strength parallel to decklayers $\sigma_{cB\|}$	kp/cm²	280 ← 100		Without holes
	lb./sq. in.	4,000 ← 1,420		100 ← 80
				1,420 ← 1,140
Bending strength (modulus of rupture) σ_{bB}	kp/cm²	310 ← 120	Increased by veneering (1 mm thick Okoumé) ∥ outface fiber up to 300% ⊥ outface fibers up to −25%	
	lb./sq. in.	4,400 ← 1,700		
Long duration bending strength 10⁴ to 10⁶ loading cycles	kp/cm²	100 ← 40		
	lb./sq. in.	1,420 ← 580		
Fatigue bending strength > 10⁶ loading cycles σ_{fB}	kp/cm²	50 ← 40		
	lb./sq. in.	710 ← 580		
Shear strength τ_s	kp/cm²	∥ 100 ← 75		
		⊥ 200 ← 165		
	lb./sq. in.	∥ 1,420 ← 1,065		
		⊥ 2,840 ← 2,340		
Resistance of decklayers to separation from the core	kp/cm²	18 ← 8		
	lb./sq. in.	256 ← 114		

Note: The following publications have been considered:
Albers (1971), ASTM D 1037 (1964) *Avale* (1964), *Brodeau* (1965), *Bryan* (1960, 1962), *Bryan and Schniewind* (1965) *Burmester* (1968), *Burrows* (1961), *Cizek* (1961), *Clouster* (1962), *Deppe* (1963), DIN 52362 (1965), DIN 52364 (1965), *Elmendorf and Vaughan* (1958), *Fischer and Merkle* (1963), *Flemming* (1957), *Gillwald and Luthardt* (1966), *Gratzl* (1956), *Grzeczynski and Bakowski* (1963), *Heebink* (1960), JIS 5908 (1957), *Jorgensen and Murphey* (1961), *Keylwerth* (1956, 1958, 1959), *Klauditz* (1955), *Klauditz and Stegmann* (1957), *Klauditz, Ulbricht and Kratz* (1958), *Kollmann* (1949), *Kollmann and Krech* (1961), *Kratz* (1969), *Kratz and May* (1963), *Kufner* (1966, 1968, 1969, 1970), *Liiri* (1961), *Meinike* (1960), *Möhler* and *Ehlbeck* (1968), *Nedbal* (1961), *Neusser* (1963), *Oertel* (1967), *Özen* (1971), *Perkitny T. and J.* (1966), *Post* (1958), *Shen and Carroll* (1969, 1970), *Shen and Wraugham* (1971), *Stofko* (1960), *Svoreman* (1968), *Treiber* (1965), *Walter and Deppe* (1960), *Walter and Rinkefeil* (1960), *Winter and Frenz* (1954), *Withington and Walters* (1969).

of particleboards, to their density, moisture content, and particularly to the nature of overlays glued to their surface.

There exist many types of particleboard with various thicknesses. Therefore the mechanical and related properties must be interpreted as wide ranges. Table 5.20 gives a survey of up-to-date data. The table cannot be complete and changes may be expected in the future. Nevertheless some determining and guiding values are given in the Table.

5.4.7.1 Theoretical Approach to Particleboard Mechanics. Particleboard and also high polymers display both perfect elastic solid (Hookean) and perfect liquid-like (Newtonian) behavior depending upon the time scale of measurement, the moisture content which they contain and the temperature to which they are subjected. The nature of visco-elastic behavior can be evidenced as a reaction to various time-dependent stresses or strains. The simplest method of describing the visco-elastic behavior of particleboard (and polymers) is with a creep experiment in which a shear stress τ_{xy} is suddenly applied at time zero and held constant during the test. The strain γ_{lx} is measured as a function of time. The strain increases monotonically with time as shown in Fig. 5.152 (*Sharma*, in: *Schmitz*, 1965, p. 148—199).

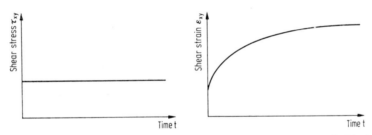

Fig. 5.152. Stress and strain as a function of time in a creep experiment. From *Sharma*, in *Schmitz* (1965)

Pure elastic solids display instantaneous deformation as a time-independent function of stress. Perfect liquids show flow that progresses linearly with time. Particleboard shows retarded elastic deformation due to the interplay of elastic and viscous effects. *Keylwerth* (1958) introduced the mechanics of multi-layer particleboard. He published stress-strain-diagrams for three-layer particleboard (Fig. 5.153) and showed the approximated distribution of stresses over the cross section of a bent three-layer board (Fig. 5.154). Corresponding diagrams and additional explanations are given in Section 5.4.7.5. *Keylwerth* (1958) assumed that "the most characteristic property of a good three-layer board is a relatively high bending and shearing stiffness".

Fig. 5.153 indicates that within a determined range of stresses (below the so-called proportional limit) particleboard as well as plywood may be considered as anisotropic-elastic systems. The literature on elasticity and visco-elasticity is extensive, and cannot be reviewed within the space available here. However, the reader is referred to the following authors: *Alfrey jr.* (1948), *Eirich* (1956, 1958), *Ferry* (1961), *Gross* (1953), *Hearmon* (1961), *Houwink* (1937), *Love* (1927), *Tobolsky* (1960). Mechanics and rheology of wood are treated in Vol. I of this book (1968, p. 292—321), and by *Keylwerth* (1958). The enlarged Hooke's law in a

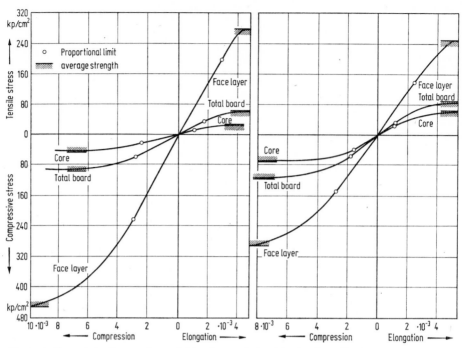

Fig. 5.153. Stress-strain diagrams for particleboards manufactured by two firms. The behavior in the various layers of board is considered. From *Keylwerth* (1958)

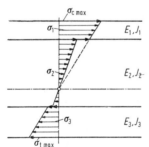

Fig. 5.154. Linear stress and strain distribution over the cross section of a bent three-layer particleboard. From *Keylwerth* (1958)

coordinate-system with the main axes x, y, z, usually determined as in Fig. 5.157, can be expressed by the matrix equation (5.8) (*Plath*, 1972):

$$
\begin{array}{c c c c c c c}
 & \sigma_x & \sigma_y & \sigma_z & \tau_{yz} & \tau_{zx} & \tau_{xy} \\
\varepsilon_x & s_{11} & s_{12} & s_{13} & 0 & 0 & 0 \\
\varepsilon_y & s_{21} & s_{22} & s_{23} & 0 & 0 & 0 \\
\varepsilon_z & s_{31} & s_{32} & s_{33} & 0 & 0 & 0 \\
\gamma_{yz} & 0 & 0 & 0 & s_{44} & 0 & 0 \\
\gamma_{zx} & 0 & 0 & 0 & 0 & s_{55} & 0 \\
\gamma_{xy} & 0 & 0 & 0 & 0 & 0 & s_{66}
\end{array}
\tag{5.8}
$$

The hatched diagonal line calls special attention to the principal strains s_{kk} for extension (contraction) and shear. The figures s_{hk} below this diagonal characterize the principal strains in the perpendicular directions (lateral strains). In engineering practice the reciprocal values, called moduli of elasticity E, and of rigidity G, are used according to the following definitions:

$$E_x = E_{11} = 1/s_{11}, \qquad E_y = E_{22} = 1/s_{22}, \qquad E_z = E_{33} = 1/s_{33} \qquad (5.10)$$

and

$$G_{yz} = G_{44} = 1/s_{44}, \; G_{zx} = G_{55} = 1/s_{55}, \qquad G_{xy} = G_{66} = 1/s_{66}. \qquad (5.11)$$

The matrix shows that the principal strains in the perpendicular directions are reflected with inverse indices with regard to the diagonal. Because a rhombic crystalline system is assumed (*Voigt*, 1928) the following numerical equality is valid:

$$s_{21} = s_{12}, \qquad s_{31} = s_{13}, \qquad s_{32} = s_{23}. \qquad (5.12)$$

The ratio of lateral strain to longitudinal strain is called Poisson's ratio. Some values for solid wood are given in Table 7.2, p. 298, in Vol. I of this book.

It may be mentioned that, in isotropic materials such as steel, there is a definite relation between the modulus of elasticity, Poisson's ratio and the modulus of rigidity. Such a relation does not exist for anisotropic materials such as wood, plywood, particleboard and fiberboard. For these materials the value of the modulus of rigidity is independent of the other elastic constants.

In wood based materials tensile and compressive strengths are in most cases different. Therefore the type of loading must be characterized by various indices: 11, 22, 33 for tension, -11, -22, -33 for compression. A distinction between the corresponding moduli for tension and compression is not necessary. For the calculation of strains and stresses due to tension, compression and shear 15 properties, listed in Table 5.21 are necessary:

Table 5.21. Moduli of Elasticity and Rigidity (Elasticity in Shear) and Corresponding Stresses

Tension and compression			Shear		
Direction	Modulus	Stresses	Plane	Modulus	Stress
x	E_{11}	$\sigma_{11}, \sigma_{-11}$	y,z	G_{44}	τ_{44}
y	E_{22}	$\sigma_{22}, \sigma_{-22}$	z,x	G_{55}	τ_{55}
z	E_{33}	$\sigma_{33}, \sigma_{-33}$	x,y	G_{66}	τ_{66}

Plath (1972) investigated the problems of moduli and stresses in bending simply supported, long plates. As a rule, besides external loads P, moments M are effective and can be represented as vectors acting vertically with respect to the planes yz, zx and xy. The moments $M_{yz} = M_{44}$, $M_{zx} = M_{55}$, and $M_{xy} = M_{66}$ are illustrated in Fig. 5.155. Schematically, plates (bars) loaded by forces P_z vertically to the plate plane are shown in Fig. 5.156. These loading cases correspond to the usual (standardized) tests for deflection, modulus of elasticity E_{44} or E_{55}, and bending stress σ_{44} or σ_{55}. The conditions for plates placed upright are illustrated in Fig. 5.157. In these cases the moduli of elasticity and maximum stresses for anisotropic plates depend upon the position of the axis of the plate. For instance, for plywood plates, with the grain direction of the deck veneer parallel to the plate axis the moduli and maximum stresses are higher than the corresponding values for plates with deck veneers whose fibers are oriented perpendicular to the plate axis.

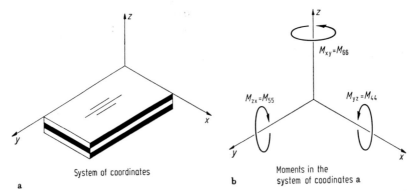

Fig. 5.155. a) System of coordinates; b) Moments in the system of coordinates a. From *Plath* (1972)

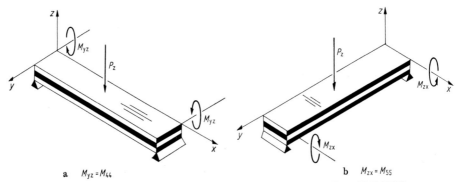

Fig. 5.156. Plates, loaded vertically to the plate plane by forces P_z. From *Plath* (1972)

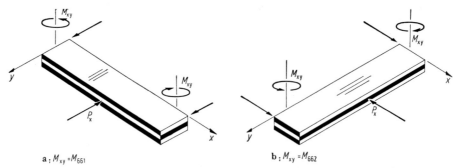

Fig. 5.157 a, b. Plates, loaded by forces P_y (upright placed plate). From *Plath* (1972)

Plath (1972) remarks that in these cases double indices are no more sufficient and recommends the addition of a third index-figure as shown in Table 5.22.

For particleboard, which is quasi-isotropic in the xy-plane, the differences in the elastic and strength properties characterized by the indices 661 and 662 can be neglected, but not for plywood.

Table 5.22. General View of Bending Properties
(*Plath*, 1972)

Position of plate	Moduli	Stresses
Flat	E_{44}	σ_{44}
	E_{55}	σ_{55}
Upright	E_{661}	σ_{661}
	E_{662}	σ_{662}

A comparison of Table 5.21 with Table 5.22 leads to the statement that for a complete description of the elasto-mechanical properties of anisotropic wood based materials the following properties are necessary:

> 7 moduli of elasticity,
> 3 moduli of rigidity,
> 13 maximum stresses (strengths),
> 3 Poisson's ratios.

From most of these 26 various material properties only the orders of magnitude, but not the statistical distributions are known. It should be mentioned that the results of bending tests depend strongly on the geometry of the test specimen and the type of loading especially for thin plates of boards ($s < 5$ mm ∞ 0.2 in.). Standard specifications for testing wood based materials are still incomplete, especially considering their growing application in engineering. For particleboard only a few elastic constants and strengths are of common interest and these can be varied in a wide range by laminating.

5.4.7.2 Elasticity and Rigidity. In practice the modulus of elasticity E_b is determined together with bending strength and deflection in the same test. The modulus E_b (as a rule identical with E_{11} in Table 5.21, and using three-point loading as shown in Fig. 5.159 A) is calculated as follows:

$$E_b = \frac{\Delta P \cdot L_s^{\,3}}{4 \cdot b \cdot s^2 \,\Delta f} \;[\text{kp/cm}^2],$$ (5.13)

where

ΔP = load difference in the proportional range of the stress-strain diagram in cm,
L_s = span between supports = 20 s in cm,
b = width of test piece in cm,
s = measured thickness of test piece in cm,
f = deflection in cm, corresponding to load difference ΔP.

In contrast to plywood the numerical values for E_{11} and E_b are not equal for particleboard (*Plath* 1971). The resulting modulus of elasticity E^* for a parabolic distribution of density across board thickness, for tension, and compression is:

$$E^* = \frac{E_1}{\varrho_1}\left[\varrho_0 + \frac{\varrho_1 - \varrho_0}{m+1}\right],$$ (5.14)

for bending

$$E^* = \frac{E_1}{\varrho_1}\left[\varrho_0 + \frac{3(\varrho_1 - \varrho_0)}{m+3}\right].$$ (5.15)

31*

Evaluating the results of bending tests, the influence of shear deformation is neglected. The true modulus of elasticity in bending is somewhat greater. *Plath* writes:

$$E_b = \frac{E_b \text{ eff (measured)}}{\eta_s}, \tag{5.16}$$

where the factor η_s is the shearing effect figure.

For η_s *Plath* deduced the following formula:

$$\eta_s = \frac{G/E^*}{G/E^* - \left(\frac{s}{L}\right)^2 \cdot 1,2}, \tag{5.17}$$

which is based on the formula for the deflection f in bending (cf. p. 497) (*Plath*, 1972).

The modulus of elasticity has been remarkably increased in the course of technical development. For flat-pressed particleboard, with an average density $\varrho \approx 650$ kg/m³ (~ 41 lb./cu. ft.) (FP/Y according to German standard specification DIN 68761, Bl. 1, Sept. 1970) E is about 30,000 kp/cm² (455,000 lb./sq. in.). Light particleboard with a density ϱ up to 450 kg/m³ (LFP DIN 68761, Bl. 2, February 1963) (28 lb./cu. ft.) have a modulus of elasticity E of only about 10,000 kp/cm² (142,000 lb./sq. in.).

If particleboards have to carry only their own weight, and are not loaded to produce buckling — which is commonly the case — the modulus of elasticity is of limited importance. *Keylwerth* (1959) showed for unilayer particleboard that the modulus of elasticity increases with density, however, with more scattering (Fig. 5.158). *Teichgräber* (in *Kollmann*, 1966, p. 562/563) found a strong correlation between modulus of elasticity E_b and bending strength σ_{bB} for four commercial flat-pressed particleboard (Fig. 5.159). He calculated the linear regression equation as follows:

$$E_b = 190 \cdot \sigma_{bB} - 7,018 \text{ [kp/cm²]}. \tag{5.18}$$

Extruded board is of interest for building elements only when veneered or otherwise overlaid. They have a modulus of elasticity between about 15,000 and 33,000 kp/cm² ($\sim 210,000$ and 470,000 lb./sq. in.).

The modulus of rigidity G for shear parallel or perpendicular to the board plane depends mainly upon the statistical orientation of the chips within the board. Flat-pressed board have a higher rigidity than extruded board. Close face layers of flat-pressed three-layer board increase the modulus of rigidity as does veneering of extruded board. *Keylwerth* (1958), in experiments carried out in cooperation with *Ranta* (Diploma-study, University of Hamburg, 1957), obtained values listed in Table 5.23.

The method of testing has a distinct influence. *Keylwerth* (1958) used the diagonal loading of quadratic plates (recommended by *Nadai* for isotropic plates, cf. *Bergsträsser*, 1927) shearing parallel to the board plane by means of a device, developed in the U.S. Forest Products Laboratory (1950). *Albers* (1971, 1972) published the results of direct measurements of the modulus of rigidity, using shear cubes. Some results are compiled in Table 5.24.

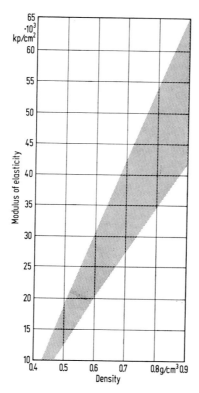

Fig. 5.158. Modulus of elasticity of unilayer particleboard in dependence on density. From *Keylwerth* (1959)

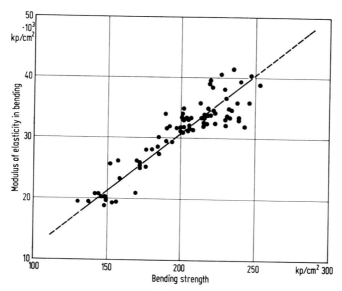

Fig. 5.159. Modulus of elasticity in bending of 4 flat-pressed particleboard versus bending strength.
From *Teichgräber*, in *Kollmann* (1966, p. 562)

Table 5.23. Moduli of Rigidity (*Keylwerth*, 1958)

Material	Modulus of rigidity		
	Symbol	kp/cm²	lb./sq. in.
Blockboard (German abbr. "ST")	G_{66}	3,400	49,000
Laminboard (German abbr. "STAE")	G_{66}	3,000	42,700
Three-layer flat-pressed particleboard (German abbr. FP/Y)	G_{66} G_{44}	5,300, 6,500 1,040⋯1,210	74,000, 92,000 14,800, 17,000
Extruded board (German abbr. SP)	G_{66} G_{44} G'_{55}	450 430⋯720 1,280⋯1,840	6,400 6,100⋯10,200 18,205⋯26,000
Extruded board, veneered with Limba, 0.7 mm thick	G_{66} G_{44} G_{55}	2,000 650⋯1,210 1,930⋯3,150	28,400 9,200⋯17,000 27,500⋯44,802

Table 5.24. Moduli of Rigidity (*Albers*, 1971)

Material	Modulus of rigidity		kp/cm² (lb./sq. in.)
	$G_{yz} = G_{44}$	$G_{zx} = G_{55}$	$G_{xz} = G_{66}$
Particleboard, flat-pressed (German abbr. FPY) 16 mm thick, PF-glued, $\varrho = 0.67$ g/cm³ (42 lb./cu. ft.)	2,300 (32,700)	2,100 (29,900)	11,800 (167,800)
Particleboard, extruded (German abbr. SPV) $\varrho = 0.62$ g/cm³ (39 lb./cu. ft.)	800 (11,400)	3,500 (49,800)	1,700 (24,200)
Likewise with tubular hole (German abbr. SPR) $\varrho = 0.33⋯0.40$ g/cm³ (21⋯25 lb./cu. ft.)	730⋯ 1,100 (10,400⋯15,600)	590 ⋯ 1,230 (8,400⋯17,500)	1,200 (17,000)
Beech-plywood, 17 mm thick PF-glued, $\varrho = 0.76$ g/cm³ (47 lb./cu. ft.) Structure: 2.5 − 1.1 − 2.5 − 1.1 − 2.5 − 1.1 − 2.5 − 1.1 − 2.5 mm	5,500 (78,200)	3,000 (42,700)	8,000 (113,800)

If the shear cubes are turned around the fixed z-axis the s'_{66}-values can be measured. Fig. 5.160 demonstrates that for particleboard $s'_{66} = s_{66} = $ const, whereas for plywood a strongly marked angle-dependence exists.

5.4.7.3 Tensile Strength

5.4.7.3.1 Tensile Strength, Parallel to Board Plane. Tensile strength $\sigma_{bB\parallel}$ parallel to particleboard plane is tested only occasionally. It is regarded as not important, and standardized tests do not exist. Nevertheless, this strength

property determines the bending strength to a large extent, is suitable to eluci-
dating the connection between structure and strength of particleboard, and follows characteristic general laws for the mechanical behavior of particleboard, for
instance, with respect to density and moisture content. *Keylwerth* (1958) published
the schematic arrangement of the tensile test on a three-layer board (Fig. 5.161).

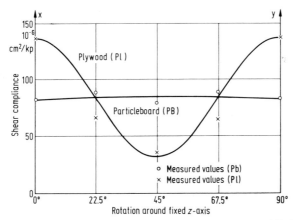

Fig. 5.160. Shear compliances for particleboard and plywood in rotation around the fixed *z*-axis.
From *Albers* (1971)

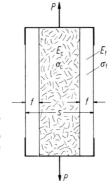

Fig. 5.161. Schematic arrangement of the tensile test on a three-layer board.
From *Keylwerth* (1958)

E_c Modulus of elasticity in the core, E_f Modulus of elasticity in the face
layer, σ_c Tensile strength in the core, σ_f Tensile strength in the face layer,
s Total board thickness, f Face layer thickness, P Tensile load

In the elastic range the following expression holds:

$$\frac{\sigma_c}{E_c} = \frac{\sigma_f}{E_f} \tag{5.19}$$

with symbols as used in Fig. 5.161.

With respect to the equilibrium of forces the resulting modulus of elasticity
E_{res} of the whole board is as follows:

$$E_{res} = \frac{\sigma_{res}}{\varepsilon_{res}} = \frac{\sigma_{re}}{\sigma_c/E_c} = \frac{\sigma_{res}}{\sigma_f/E_f} = E_c + \lambda(E_f - E_c) \tag{5.20}$$

If $E_f \gg E_c$, then we have in approximation

$$\sigma_{res} \approx \sigma_f \cdot \lambda \tag{5.21}$$

where λ is the proportion of the sum of the two face layers in thicknesses $(2f)$ to the thickness s of the whole board or the sheeting ratio (,,Beplankungsverhältnis") using the symbols of Fig. 5.161

$$\lambda = \frac{2f}{s}. \tag{5.22}$$

It should be noticed that E_{res} and ε_{res} are only fictitious values for calculations but not real material properties.

The tensile strength, $\sigma_{tB\parallel}$ parallel to the board plane, is proportional to board density (e.g. ϱ_{10}) (Fig. 5.162). To be on the safe side the following formula, computed by *Kollmann*, can be applied:

$$\sigma_{tB\parallel} = 333\varrho_{10} - 133 \text{ [kp/cm}^2\text{]}, \tag{5.23}$$

where $\varrho_{10} =$ density at 10% moisture content in g/cm³.

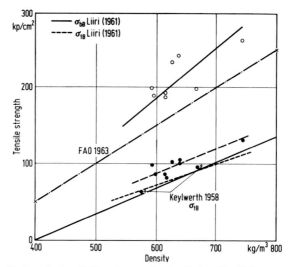

Fig. 5.162. Tensile strength of particleboard versus density. Designed by *Kollmann*, using data of authors as indicated

For high density particleboard ($\varrho_u = 1.1$ g/cm³ $\sim$69 lb./cu. ft.) manufactured in the laboratory *Klauditz* (quoted by *Winter* and *Frenz*, 1954) obtained $\sigma_{tB\parallel}$ = 460 kp/cm² ($\sim$6,5000 lb./sq. in.). *Winter* and *Frenz* (1954) explained on the basis of experiments that the influence of moisture content on the tensile strength, parallel to the board plane, is practically constant in the range between 0 and 15% moisture content; above 15% a clearly indicated decrease follows. Especially important is the orientation of the chips within the board. *Keylwerth* (1959) and *Klauditz* and coworkers (1960) investigated the production of particleboard with oriented strength. The result was, for example, for a board with beechwood flakes, oriented almost parallel to the board plane $\sigma_{tB\parallel}$ = 600 kp/cm² ($\sim$8,500 lb./ sq. in.) at a density of $\varrho = 0.8$ g/cm³ (50 lb./cu. ft.). The ratio of strength to density (breaking length, ,,Reißlänge") is, to be sure, only 8.8 km instead of 20 km in the average for solid wood.

5.4.7.3.2 Tensile Strength Perpendicular to Board Plane, Face Strength.
Testing of tensile strength perpendicular to the plane of the board ($\sigma_{tB}\perp$) has been
commonly done since the early days of particleboard manufacture. With this test,
the layer with the lowest coherence is determined. From this point of view, this
test contributes essentially to quality control in the plants. There are various
methods and devices to test the tensile strength perpendicular to the plane of board.
As an example the arrangement according to the German Standard Specification
DIN 52365 (Apr. 1965) is shown in Fig. 5.163a. The specimens are glued between two

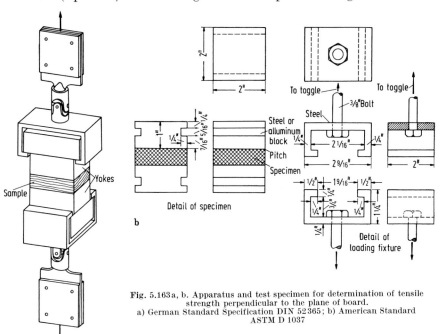

Fig. 5.163a, b. Apparatus and test specimen for determination of tensile
strength perpendicular to the plane of board.
a) German Standard Specification DIN 52365; b) American Standard
ASTM D 1037

prismatic yokes (blocks) made of hardwood, steel or aluminium alloy. Fig. 5.166b
shows specimen and loading fixtures according to American Standards (ASTM D
1037-64). In W.-Germany the test specimens shall be 50 mm × 50 mm × thickness
of the board. In the U.S.A. they shall be 2 in. (~51 mm) square by the thickness
of the finished board. In the U.K. (British Standard B. S. 1811, 1961/1967) test
pieces are prescribed which shall be 1.5 in. (~38 mm) × 1.5 in. (~38 mm) × the
thickness of the board. In England the blocks of suitable wood shall be glued
without the application of heat using a suitable adhesive. In the Institut für
Holzforschung und Holztechnik in Munich short-time high-frequency curing of
the glue is used to quickly set the joint. The gluing of the specimens with the two
blocks is time-wasting. Therefore direct measurements were attempted, but the
results are falsified by accompanying bending and shear stresses. Note-worthy
are proposals to use ring-shaped specimen (*Grzeczynski* and *Bakowski*, 1963,
Japanese Standard Specification JIS 5908 (1957)).

Failure should, as far as possible, occur in the test piece and not in the wood
block or in the glue line. Quite another problem is the testing of the face strength
or internal bond ("Abhebefestigkeit") which gives information on the resistance
of the face layers to separation from the core. It may happen, but rarely, that the
failure occurs in the face layer (Fig. 5.164). *Fahrni* (cf. *Teichgräber*, in *Kollmann*,

1966, p. 557) developed a method by which an annular small groove with a limited depth (e.g. 0.5 mm) is milled in the board; on the round surface encircled by the groove a steel-piston is cemented, and then tension is applied to this piston.

Fig. 5. 164. Fractures of tensile tests of particleboard perpendicular to the board plane. Left side: Fracture in the core, Right side: Fracture in the face layer. From *Kollmann*, in Triangel Spanplattenwerke (1964, p. 35)

This method was standardized in the German Standard Specification Draft DIN 52366 (Sept. 1972). The value for the tensile strength $\sigma_{tB\perp}$ is the quotient of the maximum load $P_{\max}$ [kp or lb.] and the nominal cross-sectional area A [cm² or sq. in.]:

$$\sigma_{tB\perp} = \frac{P_{\max}}{A} \text{ [kp/cm}^2 \text{ or lb./sq. in.]}. \tag{5.24}$$

The following facts may be mentioned:

a) The tensile strength $\sigma_{tB\perp}$ of flat-pressed particleboard increases proportionally to increasing density; thickness of board and more importantly the orientation of chips have an influence. This is shown in Table 5.25.

Table 5.25. Minimum Values of Tensile Strength $\sigma_{tB\perp}$ for Particleboard
(DIN 68761, Bl. 1, Draft Sept. 1970)

Type	Thickness mm (in.)	Tensile strength kp/cm² (lb./sq. in.)
Flat-pressed	6···13	4.0
	(∼1/4···12)	(56.9)
	>13···20	3.5
	(∼1/2···13/16)	(49.8)
	>20···25	3.0
	>25···32	2.4
	(∼13/16···1)	(42.7)
	(∼13/16···1/4)	(34.1)
	>32···50	2.0
	(1/4···2)	
Extruded without holes not standardized	6···25	15.0
	(∼1/4···1)	(213.3)

Plath (1963) remarks that the unequal distribution of density in the mostly used particleboard of medium density commonly does not permit an average to be used as measure for their technical properties. Here, reliable information could be gathered only from major publications which include extremely high and extremely low densities. In the case of medium density ($0.45 < \varrho < 0.75$ g/cm³ or $28 < \varrho < 47$ lb./cu. ft.) the effect of the density of the raw material, as compared to other effects, plays a minor role, but Fig. 5.165 (*Neusser*, 1960) shows the general trend mentioned earlier.

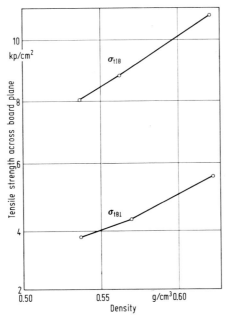

Fig. 5.165. Tensile strength perpendicular to the board plane $\sigma_{tB\perp}$ and internal bond σ_{tIB}. From *Neusser* (1960)

b) The face strength σ_f is as much as two to three times as high as $\sigma_{tB\perp}$ (Fig. 5.168 and the following data given by *Teichgräber* in *Kollmann*, 1966, p. 557; *Teichgräber* tabulates values for $\sigma_{tB\perp}$ from 3.8 to 5.2 kp/cm² ($\sim$54 to 74 lb./sq. in.) and for σ_f from 9.8 to 18.3 kp/cm² ($\sim$140 to 260 lb./sq. in.).

c) The function $\sigma_{tB\perp} = f(H)$ when $H =$ moisture content in %, based on oven-dry weight, apparently has a maximum at $H \approx 10\%$ (Fig. 5.169, *Perkitny*, 1962).

5.4.7.4 Compressive Strength. Particleboards are practically never loaded parallel to the board plane. Laboratory tests, nevertheless, have been carried out (*Winter* and *Frenz*, 1954, *Keylwerth*, 1958). Typical stress-strain diagrams showing the behavior (extension $-\varepsilon\%$ or compression $+\varepsilon\%$) under tensile or under compressive stresses are given in Fig. 5.153. Prismatic specimens were tested having a cross-section of s mm $\times s$ mm (where $= s$ board thickness in mm) and a height of $3 \times s$. Thin boards with $s < 19$ mm ($\sim$3/4 in.) should first be glued together but with low pressure to avoid densification. *Keylwerth* (1958) investigated separately the stress-strain behavior of face layers and cores of three-layer board.

On the average, the compressive strength parallel to the board plane is:

$$\sigma_{cB\parallel} = \frac{P_{max}}{A} \; [\text{kp/cm}^2] \tag{5.25}$$

where P_{max} = maximum (breaking) load [kp or lb.], A = effective cross section. It ranges between 80 and 200 kp/cm² ($\sim$1,140 and 2,840 lb./sq. in.). The dependence upon moisture content (m. c.) is similar to that of solid wood; in a range between 5 and 12% moisture content corresponding to a hygroscopic equilibrium at a

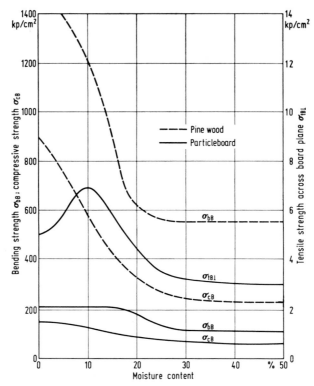

Fig. 5.166. Dependence of bending strength, compressive strength and tensile strength perpendicular to the board plane on moisture content. Solid curves for particleboard, dotted curves for pine wood. From *Perkitny* (1962)

relative humidity of the surrounding air of about 30% to 80% at about 20 °C (68 °F) temperature (cf. Fig. 5.146), one can assume that $\sigma_{cB\parallel}$ decreases about 25 to 30%, so that at 30% m. c. the compressive strength $\sigma_{cB\parallel}$ is at the minimum $0.7 \times \sigma_{cB\parallel_5}$. The numerical index notes the m. c. According to DIN 68760 (Oct. 1972) the moisture content H of flat-pressed particleboard for common uses, for instance for furniture, radio or TV cabinet or for container manufacture should be $H = 9 \pm 4\%$, based on oven-dry weight.

Likewise the compressive strength perpendicular to the board plane is of minor importance except when particleboards are applied as sub-floors, or if the pressure applied during veneering or laminating operations, causes deformation in thickness.

The problem is a complicated one because several effects, such as type and density of board, moisture content, temperature, kind of loading, for instance uniformly over the whole surface or only partially (Fig. 5.167), have to be considered. Compressive strength perpendicular to the board plane is closely related to hardness. Extra heavy particleboard ($\varrho \approx 1.1$ g/cm³ or 69 lb./cu. ft.) with chemically hardened surfaces can be produced, but this type of board is not widely used.

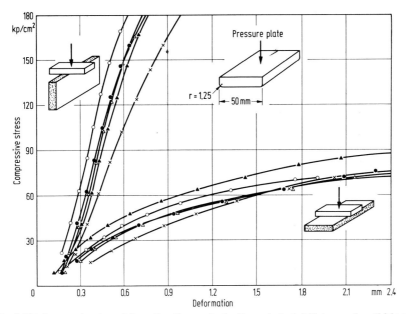

Fig. 5.167. Compressive stress-deformation diagram for loading as indicated (flat-pressed particleboard)
From *Teichgräber*, in *Kollmann* (1966, p. 559)

One large West-German manufacturer gives $\sigma_{cB\perp}$-values between 90 and 140 kp/cm² (~1,280 and 1,990 lb./sq. in.) at a temperature of 20 °C (68 °F) for flat-pressed board with normal densities between 630 and 700 kg/m³ (~39 and 44 lb./cu. ft.). If the temperature of the press platens is raised at the time of test to 100 °C (212 °F) then the $\sigma_{cB\perp}$-values decrease to 15 to 25 kp/cm² (~213 to 356 lb./sq. in.). For medium density board (say $\varrho = 520$ kg/m³ or 33 lb./cu. ft.) the perpendicular compressive strength is 50/5 kp/cm² (~711 to 71 lb./sq. in.) at a temperature of 20 °C (68 °F)/100 °C (212 °F) respectively. Short time loading (about 8 min at room temperature, 1 min pressing time at 100 °C (212 °F) of the press plates is assumed.

5.4.7.5 Bending Strength, Creep, Fatigue

5.4.7.5.1 Bending Strength, Modulus of Rupture. The bending strength or modulus of rupture is the most important mechanical property of particleboard with respect to their practical application as structural elements or in combination with them. It is known that neither in beams or plates of solid wood nor in comparable samples of wood based materials the stress distribution over the cross section is linear according to *Navier*.

The usual test application for bending strength and deflection is illustrated in Fig. 5.168 Diagrams showing the distribution of shear along the neutral plane and of compression and tension on the upper and lower surfaces, respectively, from end to end of a simple beam are given in Vol. I of this book, Fig. 7.108,

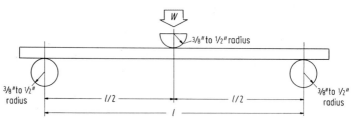

Fig. 5.168. Diagram showing application of test for bending strength and deflection. BS 1811 (1961)

p. 362. It is well known that, for static bending tests of particleboard, center loading is standardized. In this case according to elementary mechanics the tensile and compressive stresses σ developed at any point of the beam (or plate) may be calculated using the following formula[1]:

$$\sigma = \frac{M y}{I},\tag{5.26}$$

in which

σ = stress in tension or compression [kp/cm² or lb./sq. in.],
M = bending moment induced by the applied load [kp cm or lb. in.],
y = distance from neutral axis [cm or in.],
I = moment of inertia of the section [cm⁴ or in.⁴].

For a rectangular, normally loaded section

$$I = \frac{b \cdot s^3}{12},\tag{5.27}$$

in which

b = width [cm or in.],
s = depth (board thickness) [cm or in.].

The bending strength σ_{bB} for each test piece can be calculated as follows

$$\sigma_{bB} = \frac{3 P_{\max} \cdot L}{2 b \cdot s^2} \text{ [kp/cm² or lb./sq. in.]},\tag{5.28}$$

where

$P_{\max}$ = ultimate failure load [kp or lb.],
L = span between centres of supports [cm or in.],
b = width of test piece [cm or in.],
s = mean thickness of test piece [cm or in.].

The ratio of span L to board thickness or height s has a great influence on the values calculated, using formula (5.24). If the ratio is lower than 15 shear stresses occur besides bending stresses. The result of this superposition is a reduction

---

[1] The symbols in the formulae are different in various books and standards, for instance for strength and stress σ_{bB} in U.S.A. S (Wood Handbook, For. Prod. Laboratory, Washington, D. C., 1955, p. 264, *Wangaard*, 1950, p. 41, *Marin* in: *Schmitz*, 1965, p. 122), s in: *Schmidt* and *Marlies*, 1948, p. 382, f in British Standard BS 1811 (1952) and in the same sources for ultimate failing load $P_{\max}$ (or Vol. I: P) also P, W, for span l also L, for width b also w, for mean thickness s also d or h etc.

of the total breaking strength. Therefore most national standards prescribe a ratio L/s between 14 and 16. A smaller ratio (e.g. 10, but not less than 200 mm, DIN 52362) (April 1965) saves material in the bending tests, but leads to incorrect (too low) values.

The range of values for the bending strength of particleboard is listed in Table 5.20. Figs. 5.1, 5.2 and 5.3 show the dependence upon density ϱ; there is (at least in a limited range of moisture content) a relationship:

$$\sigma_{bB} = A \cdot \varrho^n + B, \qquad (5.28\,\mathrm{a})$$

where the exponent $n \approx 1$ or a little lower or higher than 1. An approximate regression line (for widely scattering test points measured by *Liiri*, 1961, Fig. 5.169) follows the Eq. (5.28 b), computed by *Kollmann*:

$$\sigma_{bB} = 0.626\varrho - 187. \qquad (5.28\,\mathrm{b})$$

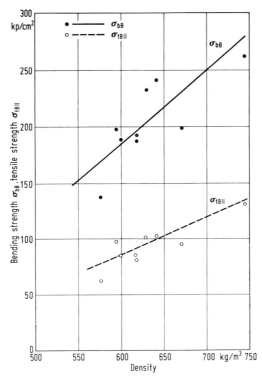

Fig. 5.169. Bending strength and tensile strength of particleboard versus density. From *Liiri* (1961)

The calculated bending strength decreases with increasing board thickness. For particleboard of high quality up to a thickness $s = 28$ mm (1 1/8 in.) (*Deppe* and *Ernst*, 1964, p. 247) a linear relationship between σ_{bB} and s exists (Fig. 5.170) computed by the author of this chapter:

$$4 < s\ 29\ \mathrm{mm}\ (\sim 0.16 < s < 1\ 1/8\ \mathrm{in.})$$
$$\sigma_{bB} = 5.625s + 325\ [\mathrm{kp/cm^2}], \qquad (5.29)$$

where the thickness s is measured in mm.

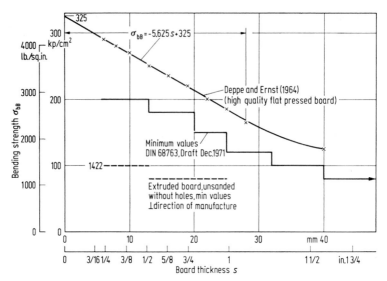

Fig. 5.170. Bending strength of particleboard in dependence on board thickness. From *Deppe* and *Ernst* (1964)

The difference between the bending strength σ_{bB} and the tensile strength parallel to the plane of board $\sigma_{tB\parallel}$ determines the characteristic behavior of particleboards in bending. The average ratio $\sigma_{bB}/\sigma_{tB\parallel}$ for flat-pressed particleboard in the air-dry condition can be calculated, based on *Liiri*'s data, to be 2.13.

This quotient is not a ratio of exact breaking stresses because, following the usual standardized tests, the so-called bending strength is computed using *Navier*'s theory which assumes that in bending the stresses are distributed linearly and symmetrically over the cross section of the bent beam. This assumption is

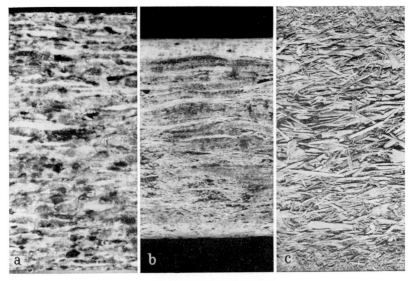

Fig. 5.171a—c. Cross sections through particleboard.
a) Uni-layer board; b) Three-layer board; c) Graded board (Bison). From *Greten*, in *Kollmann* (1966, p. 441)

justified for isotropic and homogeneous beams, but for anisotropic (or only quasi-isotropic) particleboard (5.171) this holds only for unilayer board up to the proportional limit (Fig. 5.172a). For three-layer and for graded particleboard (5.172b and c) the stress distribution is much more complicated. In reality sharp jumps, as shown at the boundaries between face layer and core, do not exist; for graded board, there exists a steady transition. This knowledge should also be applied to the scheme of stress distribution over the cross section in a pine wood beam loaded with a constant bending moment where jumps exist due to the variations of *Young*'s moduli in earlywood and latewood (*Ylinen*, 1943, cf. Vol. I, p. 371, Fig. 7.124).

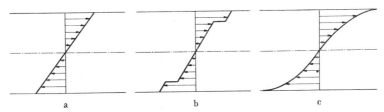

Fig. 5.172a–c. Distribution of stresses over the cross section of bent particleboard according to Fig. 5.171. a) Uni-layer board; b) Three-layer board; c) Graded board. From *Greten*, in *Kollmann* (1966, p. 442)

5.4.7.5.2 Deflection, Modulus of Elasticity. The stiffness of a solid body, used either as a beam or a long (or intermediate) column, is a measure of its ability to resist deformation or bending. It is expressed in terms of the modulus of elasticity (cf. Section 5.4.7.2) and applies only within the proportional limit. The total deflection f may be taken from stress-strain-diagrams (Fig. 5.153) and may be calculated according to *Bach* and *Baumann* (1942) (see also p. 480)

$$f = \frac{P \cdot L^3}{48 \, E \, I} + 0.3 \, \frac{P \cdot L}{b \cdot s \, G}. \tag{5.30}$$

The second term in Eq. (5.30) expresses the influence of shear forces and may be neglected if $L \geq 15s$. Then the modulus of elasticity of a simple beam loaded at the center and resting on supports at either end is

$$E = \frac{P_1 \cdot l^3}{48 f I}. \tag{5.31}$$

in which

E = modulus of elasticity [kp/cm² or lb./sq. in.],
L = length of span (distance between supports) [cm or in.],
f = maximum deflection at midspan [cm or in.], corresponding to the load P_1.
P_1 = any load at or below the proportional limit [kp or lb.],
I = moment of inertia of the section [cm⁴ or in.⁴].

For a beam or plate of rectangular cross section this equation becomes

$$E = \frac{P_1 \cdot L^3}{4f \cdot b \cdot s^3}, \tag{5.32}$$

in which b and s are the width and depth of the beam, respectively, in [cm or in.]

Modulus of elasticity values usually are derived from static bending tests, but they can also be deduced from endwise compression tests on short columns or from tensile tests:

$$E = \frac{P_1}{A\,\varepsilon} = \frac{\sigma_1}{\varepsilon}, \tag{5.33}$$

in which

P_1 = any load at or below the proportional limit [kp or lb.],
A = area of cross section [cm² or sq. in.],
ε = ratio of total compression or elongation to distance of measurement points prior to test (strain).

The values computed from such uniaxial tests are similar to those obtained from static bending tests when the bending specimens are long enough in relation to depth so that shear deformations are small; they are, however, somewhat higher than the values obtained from the standard static bending specimens.

Formulae and their derivation applicable to other types of loading are presented by *Forsaith* (1926) and by *Brown, Panshin* and *Forsaith* (1952), (see also *Timoshenko* 1945, *Timoshenko* and *MacCullough* 1949).

5.4.7.5.3 Creep, Relaxation. Particleboard as well as solid wood and high polymers have been known to display both perfect solid-like (elastic solid) and perfect liquid-like (Newtonian liquid) behavior, depending upon the time of loading and the temperature to which they are subjected, and, in the case of hygroscopic bodies, the moisture content which they possess up to the so-called fiber saturation point. The nature of the viscoelastic behavior can be clearly understood, when they are subjected to various time dependent stress or strain patterns.

In rheology, i.e. the theory of deformation and flow of matter, elastic behavior is compared to a spring and plastic behavior to a dashpot (*Schmidt* and *Marlies*, 1948, *Kollmann*, 1955, p. 13—19, Vol. I of this book, p. 315—321 *Kufner*, 1970). A usual model consists of elastic and viscous elements acting in series and in parallel. Models and calculations are given in the quoted publications.

Typical for the deformation in visco-elastic bodies are time-strain curves such as shown in Fig. 5.173a. If a stress is applied at zero-time t_0 there is an instantaneous elastic deformation OA. This is followed by a retarded deformation (creep) AB ($= \varepsilon_{el}$) under constant stress. On removal of the stress at time t_1 an instantaneous elastic recovery BC_1 ($= OA = \varepsilon_{ea}$) takes place, followed by a retarded partial creep recovery C_1D at the time t_2. The absolute strain value is $\overline{C_1C_2}$. After the time t_2, further recovery is so insignificant that it can be neglected. Thus $\overline{DE}$ represents the permanent plastic deformation ε_{pl} at the left end of the loading-unloading cycle. The large differences between individual wood species and wood based materials in their structure and resultant mechanical properties are reflected in very different forms of creep and creep recovery curves. A typical example of time-deflection curve for a 5-layer particleboard is shown in Fig. 5.173b.

In studying the problem and neglecting the fact that, as a rule, the strain-time curves indicate irregular jumps probably due to instantaneous dislocation and even partial fractures in the microstructure (Figs. 5.174 and 5.175), three types of curves may be distinguished (Fig. 5.176):

1. *Transition-creep*, characterized by decreasing strain velocity $d\varepsilon/dt$ which finally becomes zero, meaning $\varepsilon = $ const;

2. *Stationary creep.* After a starting period $d\varepsilon/dt$ it remains constant in time;

3. *Tertiary creep*, characterized by continuously growing strain velocity $d\varepsilon/dt$ until fracture occurs.

Further special theoretical and experimental investigations were carried out by *Kollmann* (1952, 1957, 1961, 1962), *Ivanov* (1958), *Wellisch, Marker* and *Sweeting* (1961), *Sauer* and *Haygreen* (1963), *Müller* (1966), *Oertel* (1967), and *Kufner* (1970). General introductions are given by *Houwink* (1957), *Reiner* (1958, 1968), and *Hult* (1972).

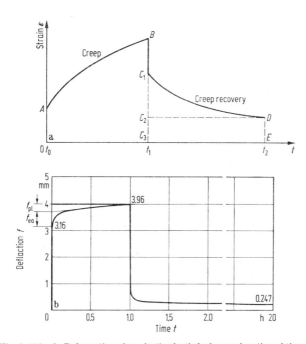

Fig. 5. 173a, b. Deformation of an elastic-plastic body as a function of time.

a) Theoretical curve; phenomena and symbols as explained in the text. From *Kollmann* (1962); b) Deflection-time diagram in rheological short-time experiment on graded particleboard, 19 mm thick, density 0.74 g/cm³, moisture content 10%. From *Kufner* (1970)

The remarkable influence of moisture content on creep of particleboard has been reported by *Kufner* (1970) and *Kollmann* (1972). In bending tests the spontaneous deflection f_0 immediately after loading and the deflection f_t accompanying the fracture (or the deflection f_t at the end of the experiment) were measured. Best fitting regression equations are for boards with 10% moisture content (based on ovendry weight):

$$\frac{f_t}{f_0} = 1.291 + 0.115 \log t + 0.034 \, (\log t)^2 \tag{5.34}$$

for board with 20% moisture content:

$$\frac{f_t}{f_0} = 1.678 + 0.159 \log t + 0.148 \, (\log t)^2. \tag{5.35}$$

These functions were extrapolated up to 10^4 (10,000) h (Fig. 5.177).

32*

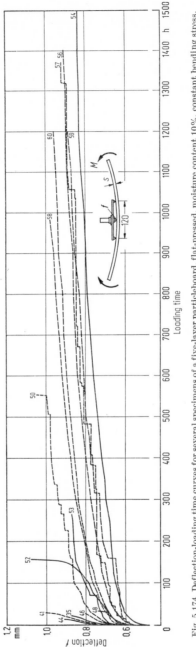

Fig. 5.174. Deflection-loading time curves for several specimens of a five-layer particleboard, flat-pressed, moisture content 10%, constant bending stress.
From *Kufner* (1970)

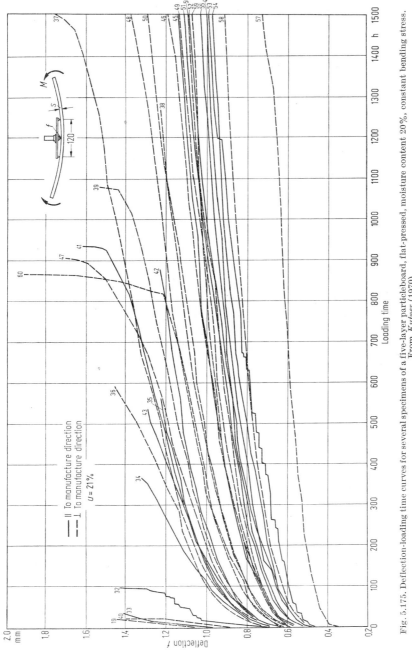

Fig. 5.175. Deflection-loading time curves for several specimens of a five-layer particleboard, flat-pressed, moisture content 20%, constant bending stress. From *Kufner* (1970)

5.4.7.5.4 Fatigue, Endurance. The increasing application of particleboard for
building elements especially in prefabricated houses, for concrete-forms, in
shipbuilding and for bodies of light trucks, is concomitant with improved board
quality and higher manufacturing standards. This means also that particleboard
should have adequate durability (endurance) under long-term loading and under
oscillating stresses.

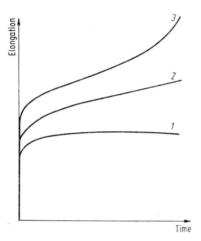

Fig. 5.176. Types of deformation
(elongation)-time curves

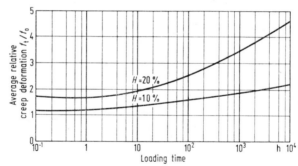

Fig. 5.177. Average relative creep deformation of 19 mm thick particleboard in dependence on loading time.
From *Kufner* (1970)

Static long-term loading, in its influence on failures, depends upon the interaction of elasticity and plasticity. Time and creep within wood and wood based
materials as well as in metals lead to areas of slip and spreading cracks until
fracture occurs (*Timoshenko* and *MacCollough*, 1949, p. 364). The relationship of
long-term stress at the proportional limit and of bending strength as compared
with corresponding values obtained in standard bending tests is shown by *Markwardt* (1930) for Sitka spruce (Vol. I of this book, p. 375, Fig. 7.130). It is to be
seen that the static fatigue bending strength after a duration of loading of more
than 6 months is about 70% of the static bending strength obtained in tests
within 0.5 and 2 minutes (BS 1811:1961). *Roth* (1935) found a ratio between
65 and 70%. To be on the safe side for particleboard 50 to 60% may be considered
adequate.

The relations between structure, particle size, internal glue-joints etc.,
together influencing elasticity and plasticity of polymers and particleboard, are

complex and complicated. Fig. 5.176 shows the three main types of strain-time curves for wood and wood based materials. Curves of the type 2 and 3 were observed for balsawood beams by *Draffin* and *Mühlenbruch* (1937). The highest ultimate load was observed for a curve of the type 3 (high creep) with 66%, the lowest for a curve of the type 2 (limited creep) with 58%.

Of even more practical importance than long-term static stresses are dynamic stresses. The most common types of dynamic stress (Fig. 5.178) are the following (*Schmidt* and *Marlies*, 1948, p. 409/410):

1. *Alternating stress* which fluctuates, usually sinusoidally, between two limits. Vibrations usually induce such stresses (for instance in walls, partitions, roofs, ceilings etc.). An alternating stress is characterized by both a mean stress σ_m and an alternating component σ_a.

2. *Reversed stress* which reverses in sign during part of the cycle. It usually refers to the case of alternating stress in which the mean stress is zero, i.e. the stress alternates between two limits equal in magnitude and opposite in sign.

3. *Impact stress* which is characterized by a high rate of change with time, for wood only a few milliseconds (*Kollmann*, 1940, Vol. I of this book, 1968, p. 392/393) and not reversible. On impact, the kinetic energy of colliding bodies is absorbed by deformations and/or cracks in the material.

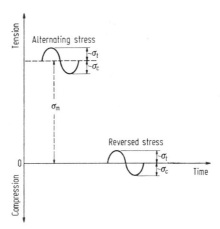

Fig. 5.178. Types of dynamic stresses.
From *Schmidt* and *Marlies* (1948)

Kollmann and *Krech* (1961) carried out experiments on particleboard specimens, applying cyclic tensile or bending stresses. Reversed bending is not only the simplest way in fatigue testing but the results are of high interest for application in the practice. It is usual to take several specimens and to test them at various loads. A plot is made of the applied alternating stress against the number of cycles, required to cause the failure. The plot is usually semilogarithmic for convenience (Fig. 5.179). There is a fatigue range AB in which, the lower the value of the repeated stress, the greater is the number of cycles required to produce failure. Finally a point B is reached beyond which any further reduction in the stress does not cause failure even if the cycles are repeated infinitely. These diagrams are known as stress-cycle or *s-n* (also *S-N*) diagrams. Another term is Woehler diagram in honor of *A. Woehler*, a railway engineer who was the first (1851) to investigate the phenomenon of fatigue (see account of his work in "Engineering", Vol. XI, 1871). An arrow near the terminus of an S-N curve indicates that a specimen has not ruptured during test. The stress corresponding to the transition point B is known both as the fatigue limit (fatigue strength) and as the endurance limit.

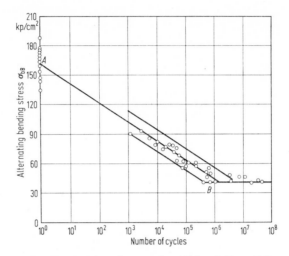

Fig. 5.179. Wöhler-curves (S-N-diagrams) for a flat-pressed particleboard. From *Kollmann* and *Krech* (1961)

The fatigue strength of particleboard apparently is reached after 10^6 stress cycles. In reversed bending the value ranges between 22 and 25% of the static short time bending strength. For comparison it may be mentioned that for solid spruce wood the ratio varies between 24 and 29%, for pine wood between 22 and 38%. It must, of course, be considered that the absolute fatigue bending strength of particleboard is only between 40 and 47 kp/cm² ($\sim$570 and 670 lb./sq. in.) and therefore much lower than that of solid wood and densified wood laminates (for example Compreg). An unfavorable influence on the fatigue behavior of particleboard arises from thermosetting cured binders which are not uniformly distributed and which are more or less brittle. The situation is complicated by the fact that some materials do not possess a definite fatigue strength. Fatigue strengths can only be taken as "helpful semiquantitative guides for design purposes" (*Schmidt* and *Marlies*, 1948, p. 423) since many factors influence them, for example, size, shape, presence of notches, surface conditions, properties of the material, damping effect etc.

5.4.7.6 Impact Strength, Punctural Resistance. Impact strength, called also "shock resistance ability" has for a long time been regarded as a dynamic property giving an answer to the question of whether a material is tough or brittle. For solids of high internal cohesion, such as steel, brass, special aluminium alloys, special types of resin impregnated densified wood laminates, some wood species like ash, hickory and bongossi, some plastics, also reinforced with glass fibers or sheathed with fabrics, resin impregnated papers etc., two main test methods are used:

a) Pendulum hammers exerting one breaking blow, on solid or notched specimens (according to *Charpy* and *Izod*);

b) Hatt-Turner-tests with stepwise repeated drops of a given weight from increased heights.

For some rather brittle materials such as zinc and many plastics, Charpy- and Izod-tests are carried out on specimens without notches. Details about these tests and their results on solid wood are given in Vol. I of this book, p. 379—394.

These tests, modified and used on specimens without notches, also may be applied to particleboard. The specimens are plates, their dimensions are similar to those of static bending tests. The ratio of the span l [cm or in.] to board thickness s has a great influence on the results. It is usual to express the specific impact strength I_s as the quotient of the totally consumed dynamic energy E_t [kp cm or lb.-ft.] to the cross-section A [cm² or sq. in.]:

$$I_s = \frac{E_t}{A}. \tag{5.36}$$

Kollmann (1937) found that for solid wood and wood laminates minimum I-values are obtained with $L/s \approx 12$. *Winter* and *Frenz* (1954) came to approximately the same result (Fig. 5.180). They recommended a pendulum with an

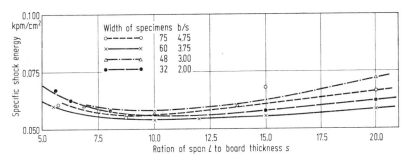

Fig. 5.180. Specific shock energy in dependence of the ratio span to board thickness, 16 mm thick flat-pressed particleboard, dimensions of specimen as indicated, density 0.66 g/cm³, moisture content 9.5%. From *Winter* and *Frenz* (1954)

original potential energy of 1.6 kp cm and specimens with a width b of 50 mm (≈ 2 in.) for board thicknesses $s < 13$ mm ($\approx 1/2$ in.) and a width b of 75 mm (≈ 3 in.) for board thicknesses $s > 13$ mm (1/2 in.). They measured for 25 mm (≈ 1 in.) thick particleboard average specific values $I_s \approx 6.8$ kp m/cm² (Fig. 5.181). These tests are not standardized and their results are of minor and doubtful importance for the practical application of particleboard.

Impact strength tests on quadratic specimens are of greater value (*Dosoudil*, 1950, *Lewis*, 1956, *Winter* and *Heyer*, 1957, BS 1811, 1961). *Winter* and *Heyer* described the testing apparatus, the experiments and the evaluation. The scattering of the results was large, but the impact strength expressed in kp cm or kp m was roughly proportional to the weight per unit area (kg/m²) in the range of relative values between 100 and 220%. The schematic arrangement of the test piece, supporting frame, and block of hemispherical mild steel is shown in Fig. 5.182.

According to BS 1811:1961, the weight of the block together with any associated falling parts shall be 10 lb. (~ 5 kg). The block shall be allowed to fall first from a height of 1 in. (25.4 mm) measured from the upper surface of the test

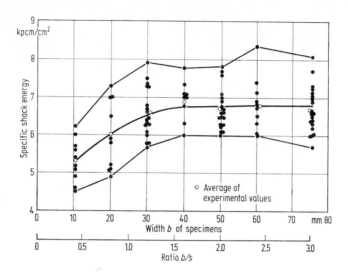

Fig. 5.181. Specific shock energy as affected by width of specimens and ratio of span to board thickness. From *Winter* and *Frenz* (1954)

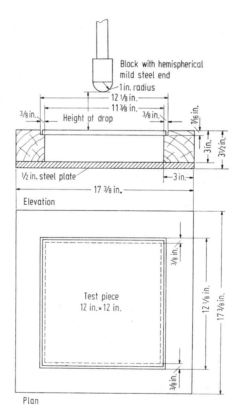

Fig. 5.182. Testing device for impact strength tests. BS 1811 (1961)

piece and then from successive heights in increments of 1 in. (25.4 mm) until the failure of the test piece occurs. The failure may take the form of complete fracture, puncture (punctural resistance) or deformation of the test piece. Deformations shall be deemed a failure when a straight edge placed across the convex surface of the test piece in any direction, in contact with one edge, is deflected by 1/4 in. (6.4 mm) or more from contact with the opposite edge of the board.

The height of fall in inches causing failure shall be recorded for each test piece.

5.4.7.7 Shear Strength

5.4.7.7.0 General Considerations. Shear strength and torsional strength of particleboard is seldom needed in design. This applies also to solid wood (Wood Handbook, U. S. Forest Products Laboratory, 1955, p. 81; Vol. I of this book, p. 394—402). The moduli of rigidity are listed in Table 5.20. The shear strength of flat-pressed particleboard perpendicular to the plane of the board is relatively high and amounts to 165 to 200 kp/cm² (~2,340 to 2,840 lb./sq. in.).

The shear strength parallel to the board plane depends on many factors, such as:

a) Structure and density of board;
b) Thickness of board;
c) Manufacture of board;
d) Type, amount, and distribution of binder;
e) Moisture content;
f) Method of test (compression-shear, tension-shear, torsion-shear);
g) Shape and size of specimen, etc.

Because of these many influential factors the results of shear tests are not reproducible and comparable, and they scatter widely. On the other hand there exist some simple and fast shear techniques, the results of which are closely related to the internal bond strength. This can be applied to predict some important strength properties of particleboard and is therefore generally suited for quality control and field work.

5.4.7.7.1 Compression-Shear Tests. A historical review shows how many shear tests have been developed (Vol. I of this book, p. 397—399). Cubes, prisms, some with incisions or notches to produce shear failures in one plane, or crosses and special shaped specimens to produce shear failures in two planes are used.

The resistance to shear in the plane of the board determined by a compressive shear test is standardized in the British Standard 1811:1961. The arrangement of the test piece, the shearing jig and the loading device are schematically shown in Fig. 5.183. Test pieces shall be prepared and conditioned as usual. Each test piece shall be 3/4 in. (~19 mm) × 3/4 in. (19 mm) × the thickness of the board. The time from the initial application of the load until failure of the test piece is 0.5 to 2 min. The shear strength τ_s (lb./sq. in. or kp/cm²) shall be calculated as follows:

$$\tau_s = \frac{P_f}{A} \qquad (5.37)$$

where

P_f = ultimate failing load in lb. or kp,
A = measured cross-sectional area in sq. in. or cm².

Notched compression-shear test specimens as recommended by *Bousquet* (1970) for solid wood (Fig. 5.184) can also be used for particleboard. A notched particleboard specimen for compression-shear parallel to plane is simple to fabricate and to test, without requiring an expensive shear tool and testing machine. The mean shear strength of solid wood (red pine, *Pinus resinosa* Ait.) increases with longer shear plane lengths l (for instance if for $l = 0.5$ in. $\tau_s = 100\%$, for $l = 1.5$ in. $\tau_s = 132\%$).

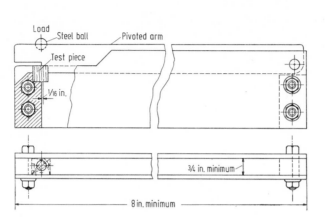

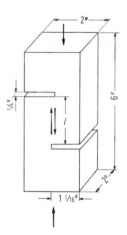

Fig. 5.183. Shear test jig according to BS 1811 (1961)

Fig. 5.184. Geometry of notched compression-shear specimen. From *Bousquet* (1970)

Shen and coworkers (1968) established a correlation between compression-shear strength and tensile strength perpendicular to the surface. They proposed a compression-shear test, parallel to the center plane of the board, on a simple square or rectangular specimen in lieu of the conventional internal-bond strength tests which need gluing of the test specimens to metal or hardwood blocks.

5.4.7.7.2 Tension-Shear Tests. Tension-shear tests are not standardized. Early experiments were carried out by *Winter* and *Frenz* (1954). They used specimens cut from boards with the following dimensions: length = 200 mm (∼8 in.), width = 40 mm (∼1 1/2 in.), thickness = board thickness. The specimens were notched on both opposite surfaces with a distance l between the two notches. The depth of the notches (depth of cut) was approximately half the thickness of the board, the width of the notches was equal to the actual width of kerf (2 mm ∼0.08 in.). Fig. 5.185 shows the influence of the notch distance on the test results. The curves for the absolute upper and lower limits obtained by *Winter* and *Frenz* (1954) were transformed into relative, dimensionless curves, mainly because the shear strength values were much too low at this time. Nevertheless, the quoted authors stated that the tension-shear strength is closely related to the tensile strength perpendicular to the plane of the board.

5.4.7.7.3 Torsion-Shear Tests. Torsion-shear techniques have been developed in the U.S.A. (*Shen* and *Carroll*, 1969, 1970; *Carroll* and *Wrangham*, 1970, *Shen*, 1970) and are proposed for standard tests and for quality control:

a) A 1-sq. in. maple block, 1/2 in. (~13 mm) thick, is glued to the center of the surface of a 3-sq. in. particleboard. The specimen is clamped horizontally in a bench wise and the maple block sheared off by hand with a 0 to 50 lb.-ft. (~0 to 7 kpm) torque wrench. It is necessary to glue backing pieces 1/2 in. (13 mm) thick onto specimens less than 3/8 in. (~10 mm) thick to avoid buckling;

b) Two 1-sq. in. particleboards are glued face-to-face and are placed with the glue line horizontally in a bench wise. The load is applied with a torque wrench;

c) A 1 1/2-in. (~38 mm) diameter metal cylinder is glued with a suitable cement (urea-formaldehyde resin catalyzed for a room-temperature cure) and then a torque wrench is attached. The torque required to produce failure is, however, too great for manual operation;

d) Twisting of particleboard plate specimens of various sizes using a torque wrench.

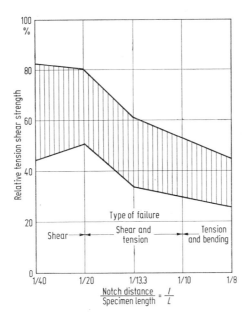

Fig. 5.185. Influence of relative notch distance on relative tension-shear strength. From *Winter* and *Frenz* (1954) values converted by *Kollmann* to dimensionless figures

The last mentioned technique is the simplest and has been investigated thoroughly by *Shen* (1970). He found the following summarized results:

Specimens with parallel particle orientation yielded torque values about 13% higher than the specimens with perpendicular particle orientation. *St. Venant's* equation for the shear strength τ_s:

$$\tau_s = \frac{12\,M}{b \cdot s^2}\left(3 + 1.8\,\frac{s}{b}\right) \quad [\text{lb./sq. in.}], \qquad (5.38)$$

where

M = maximum torque [lb.-ft.],
b = specimen width [in.],
s = specimen thickness [in.].

must be modified. For example for specimens of 2 in. width the shear strength
can be expressed by the equation

$$\tau_s = 12M \left(\frac{1.5}{s^2} + \frac{0.45}{s} \right). \qquad (5.39)$$

A very close correlation exists between torsion-shear strength and internal
strength bond (I. B.).

Torsion-shear tests on one-sq. in. specimens of different thicknesses show a
strong linear correlation between torsion-shear strength and layer density in
graded board. *Shen* and *Carroll* (1970) computed a correlation coefficient of 0.89.
Density and strength profiles of a 1 in. and 1/2 in. graded board, made by the
same manufacturer, are shown in Fig. 5.186. Also a general correlation between

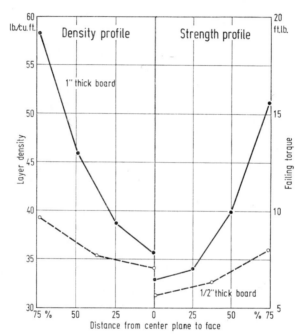

Fig. 5.186. Density and strenght profiles of a 1 in. and 1/2 in. graded board, made by the same manufacturer.
From *Shen* and *Carroll* (1970)

bending strength or modulus of rupture and failing torque (Fig. 5.187) was found
for flat-pressed particleboard made by three manufacturers. The general correlation
was relatively weak (0.7) as a result of the combined differences among manufacturers and thicknesses. When the comparison was restricted to a single board
thickness or to a single manufacturer the correlation coefficient was greatly improved to 0.95.

5.4.7.8 Hardness. Hardness is defined as the resistance of a solid body against
the penetration of another solid body by force (*Kollmann*, 1951, p. 909—926,
Vol. I of this book, p. 403—408). For metals the impression of a hardened steel
ball (10 mm diameter, 500 kg load for soft metals, 300 kg for hard metals) on
plane smooth surfaces, not too near the edges, delivers clear, reproducible figures,

as the quotient of force applied to area of indentation. The method is called Brinell-hardness test after the Swedish engineer *Brinell* (1849—1925). There is a strong correlation between Brinell-hardness H_B and tensile strength σ_{tB} ($\sigma_{tB} \approx 0.33$ to $0.36\ H_B$ kp/mm²).

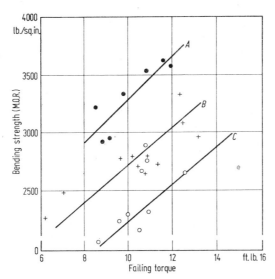

Fig. 5.187. Bending strength (modulus of rupture) versus failing torque of flat-pressed 3/4 in. particleboard, made by 3 manufacturers (*A, B, C*). From *Shen* and *Carroll* (1970)

There are many other proposals and procedures for hardness tests:

a) *Scratch-tests* mainly in mineralogy, according to *Mohs*. The Mohs-scale of hardness is based on 10 common minerals and was developed in 1822. The softest mineral with the lowest hardness is talcum designated as 1, the hardest is diamond with 10;

b) *Vickers-test*, using as indentator or penetrator a diamond cut and polished to the shape of a square-based pyramid with an angle of 136° between pairs of opposite faces. The load applied to the indentator can be varied from 1 to 120 kg and is applied continuously for about 10 sec;

c) *Knoop-test* similar to the Vickers-test;

d) *Rockwell-test*. The hardness is determined by the following procedure: A steel ball or spheroconical diamond is placed between the specimen and an apparatus that induces pressure. The machine exerts two distinct loads on the tool which are projected onto the specimen: the major load usually with 60, 100 or 150 kg (~132, 220 or 331 lb.), the minor load with 10 kg (≈22 lb.). The arithmetical difference between the depth of penetration caused by the major load and that caused by the minor load is subtracted from an arbitrary constant, and the resultant figure expresses the hardness of the specimen.

For anisotropic or quasi-isotropic, inhomogeneous and hygroscopic materials such as wood and particleboard the hardness values are rather uncertain. *Nördlinger* (1881) has already pointed out that the hardness of wood depends on the indenting tool employed. He came to the conclusion that no "absolute", but only a relative hardness value is reasonable. *Büsgen* (1904) tested the hardness of wood by impressing a steel needle to a depth of 2 mm (~0.08 in.) (Vol. I of this book, p. 403).

The procedure never found general application, but the Janka-hardness test (1906, 1908, 1915) is used in the United States of America as well as in the United Kingdom. According to BS 1811:1961, indentation is measured after the Janka-test. Each test piece shall be approximately 2 in. × 2 in. × at least 1 in. thick and shall consist of specimens built up to the required thickness if necessary. Each test piece shall be held in a suitable chuck. A steel indentator with a hemispherical end 0.144 ± 0.002 in. diameter shall be forced into the center of the test piece to the depth of 0.022 in. The load required shall be recorded in lb. and is a value of the Janka-hardness. For particleboard the test is without much theoretical and/or practical value. The load applied is closely related to the crushing strength perpendicular to the grain σ_{cB}. *Janka*, in evaluating the results of hardness tests of 280 wood species, found the following empirical relationship between hardness H_J and crushing strength σ_{cB}:

$$H_J = 2 \times \sigma_{cB} - 500 \; [\text{kp/cm}^2]. \tag{5.40}$$

Friction, shearing and cleavage greatly influence the results of the Janka-test. There is, of course, a nearly linear relationship between end hardness and side hardness of woods and their density (*Mörath*, 1932, *Trendelenburg*, 1939, *Ylinen*, 1943). In Vol. I of this book, p. 403, the problem is explained in detail. In France the hardness test according to *Chalais-Meudon* (AFNOR B 5-33, 1942) is standardized for wood using the indentation of a steel cylinder of 30 mm in diameter. For particleboard hardness is without any clear physical sense, not only because of the several reasons mentioned above, but also because of their special structure, since the highest density and strength and therefore also the highest hardness exist in the face or deck layers. If a ball according to *Janka* is forced deeply into the specimen to be tested, an integral value will be reached which does not clarify the hardness as a surface property.

5.4.8 Technological Properties

5.4.8.0 General Considerations. Technological properties are very important in the practical application of particleboard. They influence opinion and experience of particleboard-consumers partly immediately at the time of delivery, partly during storage and transport, partly when the board is used for products such as furniture, packaging and in construction as elements of cars, trucks, carriages, buildings, especially prefabricated houses etc. The technological properties are many and different. They depend not only on the principal type (platen-pressed and extruded particleboard), but also on their structure (uni-layer, multi-layer, graded board), on the raw material used (wood or other ligno-cellulosic materials, mainly flax shives and bagasse), on the types, sizes and orientation of the particles within the board, on wetting prior to final pressing, on the types and amounts of binders and additives etc. The technological properties are related to nearly all physical and mechanical properties. The moisture content and its variations have a remarkable effect. Aging must be considered. Some technological properties are of higher importance than others. Standardized methods for testing and comparing technological properties do not exist (or only as suggestions or drafts, usually in a simplified manner). Doubtlessly the technological properties are not problems of fundamental but of applied research. Instead of "principal" only "arbitrary" clarification is possible. The publications on technological properties are numerous. They are written by scientists of forest products laboratories as well as by engineers of particleboard factories. Some manufacturers have developed their own methods and use them in connection with continuous quality control and for improving their products. Therefore, experts interested in any technological property must study the literature concerned, or even carry out experiments themselves. In the following Sections important technological properties are dealt with in a condensed manner.

5.4.8.1 Surface Quality. The surface quality of particleboard is of unquestionable importance for all finishing operations like painting, lacquering, veneering, coating with foils or laminates, printing, etc. In the remarks to the German Standard Specification DIN 68761 (Draft, Sept. 1970) the following is included: "For some properties of utilization there are at present no reliable test methods giving uniform (reproducible) values. To these problems belong: density of surface (tightness, porosity), local surface strength, homogeneity of surface (smoothness, roughness), local strength of cut edges, behavior in overlaying (veneers, films, lacquers). For the choice of the most suitable particleboard for a determined purpose, the worker also has to use his own practical experience." Testing, characterizing and qualifying the surfaces of particleboard is a more practical than a scientific task. *Elmendorf* and *Vaughan* (1958) gave a survey on apparatuses for measuring the surface smoothness of wood. Other authors dealing with methods of measuring the smoothness of wood and particleboard surfaces are *Rinkefeil* (1956), *Ehlers* (1956/58), *Flemming* (1957), *Aleksandrov* (1958), *Walter* and *Deppe* (1960), *Heebink* (1960), *Neusser* (1963), *Fischer* and *Merkle* (1963), *Deppe* and *Ernst* (1963).

Principally there are mechanical and optical measuring apparatuses or combinations of both types. A simple method is photographing wood and particleboard surfaces. The Glossimeter consists of a glass plate upon which thin strips of metal foils have been mounted in the form of a grid. The test specimens are placed before the tilted grid. Light from a lamp in the back of the grid is allowed to pass through the glass, thereby throwing shadows on the panel below. The surface shows a difference in light reflectivity or gloss which is influenced by the surface smoothness. Differences in surface contours can be discovered by use of a thin-flat ribbon of light reflected from the specimen and observed through a microscope. In highlighting tests a source of light is placed at an acute angle to the panel surface, so that the light streams across the grain, and then the surface is photographed from above with a sharp focus. Another method utilizes microphotographs of wood or (more difficult to be obtained) particleboard surfaces. The waviness of thick panels may be measured by means of a micrometer (*Marian* and *Suchsland*, 1956, *Hann*, 1957), the touching surface of which contains a steel ball, 1/4 in. in diameter. An amplifier recorder was used to magnify the strains in the pick-up head. The differences in surface contours were recorded on a moving paper chart. The size of the steel ball was far too great to record minute scratches produced by sandpaper.

The German Forster-apparatus and the English Talysurf-apparatus measure surface irregularities by means of a stylus. Types and dimensions of the stylus-tips are important. For example, profiles of the surfaces can be obtained in which the height of irregularities is magnified 1,000 times and the width 100 times. Typical profile curves of particleboard surfaces are shown in Fig. 5.188, obtained by using the apparatus developed by *Forster* and built by Leitz (*Kollmann*, 1957). Another apparatus has been designed by *Fischer* and *Merkle* (1963). Here, the roughness can be recorded also in a vertical magnification by means of a stylus. Fig. 5.189 shows the effect of wetting over various times as well as the influence of different board types. Surface quality can be roughly estimated also by observing or photographing

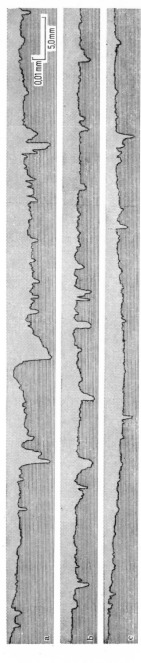

Fig. 5.188a—c. Profile curves of the surfaces of flat-pressed particleboards, manufactured with various pressing techniques. Curves picked up with Forster apparatus, built by Leitz.

a) Moisture content of chips in face layers and core equal; b) Moisture content of face layer chips, increased by spraying with water. From *Kollmann* (1957)

the gently blackened surfaces, the extension of coloured drops, or pneumatically. All methods are casual and not reproducible.

5.4.8.2 Accuracy of Dimensions, Surface Texture and Stability. Tests of surface quality of particleboard are related to the following additional tests:

a) *Accuracy of board dimensions.* Straightness, squareness, width, and thickness of boards may be measured according to British Standard B. S. 1811: 1961. The so-called "dubbing value" for each corner can be determined as the difference of the board thicknesses measured by centering a micrometer at two points, A and B, as indicated in Fig. 5.190. "Dubbing" is a thickness deficiency

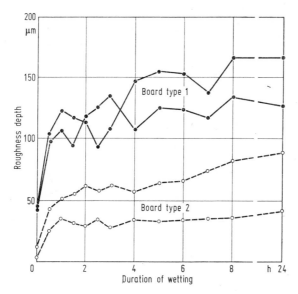

Fig. 5.189. Roughness depth of two types of flat-pressed particleboard in dependence upon duration of wetting. From *Walter* and *Deppe* (1960)

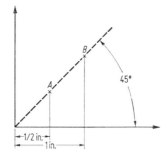

Fig. 5.190. Pattern for the measurement of board edge thickness. BS 1811 (1961)

along the edge of a board, caused by faulty sanding. The required test pieces shall be conditioned as specified (exposure for 24 hours immediately before testing to the air of a well ventilated room or to an atmosphere maintained at a relative

33*

humidity of $65 \pm 5\%$ at a temperature of $25 \pm 2\,^{\circ}\mathrm{C}$ until they are substantially constant in weight).

b) *Surface texture* can be roughly determined on a selected area of board surface of $10\,\mathrm{cm} \times 20\,\mathrm{cm}$ ($\sim$4 in. $\times$ 8 in.), by gently rubbing a soft black crayon flat over the test area which will show up the most obvious surface irregularities by blackening the spots; or by measuring the maximum vertical distances between peaks and valleys over 10 path lengths with the aid of a simple apparatus as shown in Fig. 5.191 (B. S. 1811, Amendement No 3, 31st March, 1967). The dial micrometer may be replaced by a sensing probe with a 3 mm tip, coupled to an automatic recording device.

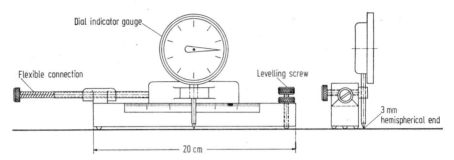

Fig. 5.191. Apparatus for surface texture test. BS 1811 (1961), Amendment 3, 1967

c) *Dishing* (dished deformation) is important for the use of particleboard as building elements in prefabricated houses and other building constructions mainly under differential humidity conditions. A typical three-dimensional diagram is re-

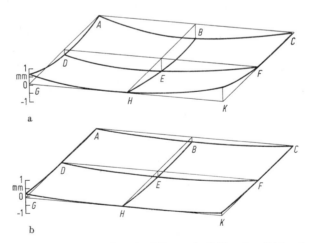

Fig. 5.192 a) Typical three dimensional diagram for dishing of particleboard;
b) Three-layer particleboard, cut chips, sanded for further processing.
From *Walter* and *Rinkefeil* (1960)

produced in Fig. 5.192. There are many contributions and rules for appropriate tests (DIN 52367, Draft, Apr. 1954, Proposals by *Keylwerth*, 1956, *Gratzl*, 1956, *Walter* and *Rinkefeil*, 1960, *Dosoudil*, 1954, 1965). The following general conclusions may be drawn:

1. For platen-pressed particleboard the dishing decreases under similar environmental influences with increasing number of layers or if they are graded;

2. The radii of curvatures are greater perpendicular to the direction of manufacture than parallel to this direction;

3. Accuracy of dimensions is the better, the higher board density is;

4. Extruded particleboard shows a very low and uniform thickness swelling and therefore a fairly good form stability.

5.4.8.3 Machining Properties. All types of particleboard can be worked with standard hand and machine tools used in woodworking trades and industries. Consequently no special treatment or technique is required. It must, of course, be considered that particleboard consists on the average of 90% of wood, 8% of artificial resins as binding agents, and about 2% of additives of different kinds for various purposes (mainly water repellents and preservatives). Particleboard is available in a wide range of densities (0.4 to 1.2 g/cm³ or 25 to 75 lb./cu. ft.), the raw material may be softwoods, hardwoods or a mixture of them; geometry of particles and structure of board (unilayer, multilayer, graded board) vary as do normal sizes and thicknesses, the latter between 4 and 80 mm (0.16 and 3.15 in.). The boards may be veneered with hardwood or overlaid with plastic sheets.

In spite of this diversity, particleboards have working properties and uses similar to those of seasoned timber or plywood. In addition particleboards have certain advantages over natural wood: their large area (for instance 244 cm × 122 cm or 8 ft. × 4 ft.), the absence of knots and faults, and uniformity from board to board. Special precautions are taken in manufacture (cf. Section 5.2.1.1) to exclude metal, stone and other impurities which might damage tools in manufacture.

Particleboard may be hand or machine sawn, milled, moulded, planed, sanded, mortised, chiselled, drilled, rebated, jointed. There are, however, some considerations which must be made regarding ease of machinability (*Johnson*, 1956, p. 264—266):

a) The *resin content* of a chipboard has a dulling effect on tools. Urea resin is more brittle and abrasive than phenolic resin. *Bridges* (1971) found that the rate of dulling is directly proportional to *board density* (see also *Akers*, 1966) and to the *silica* (*grit*) *content*. Increasing percentage of resin solids content from 5 to 8% had no appreciable effect on tool dulling (abrasiveness) but at the 9 to 11% level the abrasiveness was almost double that observed at the 5 to 8% level. For machining large quantities *tungsten carbide tipped saws* are recommended as much longer runs are then possible between successive re-sharpening.

b) Particleboard can be cut equally well on straight line edging and cross-cut *circular saws*. Some types of teeth for circular saw blades for cutting particleboard are illustrated in Fig. 5.193 (*Golbs* and *Fentzahn*, 1956). Ordinary high speed steel tools lose their initial sharpness rapidly in contact with the synthetic resins cured in the boards. Tungsten carbide possesses higher hardness and therefore higher abrasion resistance and higher resistance to heat than high-speed steel. The clearance angles of carbide tipped saw teeth are much smaller (10 to 15°) than on high-speed steel (∼40°) thus giving greater support to the cutting edge. Approximate intervals between correct re-sharpening of saws may be taken to be:

Carbide tipped saws   4,560 m ∼ 15,000 ft.
Steel plate saws   915 m ∼ 3,000 ft.

Carbide tipped saws normally require no setting as this is provided by the shape of the cross section of the teeth. Requirements for circular saws machining particleboard are listed in Table 5.26.

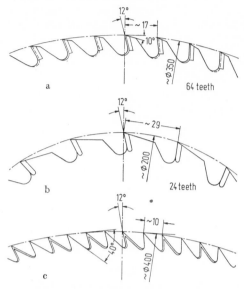

Fig. 5.193a — c. Types of teeth for circular saw blades for cutting particleboard.
a) and b) Hard metal tipped; c) High speed steel.
From *Golbs* and *Fentzahn* (1956)

Table 5.26. Machining Requirements for Particleboard
(Airscrew Co. & Jicwood, in *Johnson*, 1956, *Lindner*, 1961)

Spindle speed	rpm	3,000···6,000
Diameter	mm	400, 350, 300, 250
	in.	16 14 12 10
Teeth number	Edging	90···40
	Cross-cut	30
Peripheral	m/sec	{50···95
(tooth) velocity		{50···60
	ft./min	10,000···19,000
		10,000···12,000
Clearance angle		10°···12°30′···15°

Tipped saws usually have a thicker blade which is stiff and less liable to vibration. This is a factor improving the quality of cutting. It is re-commended to use a power feed on the saw bench, to ensure a constant pressure at the cutting point and a uniform *feed speed* up to 15 m/min (∼50 ft./min) (*Smith*, in *Mitlin*, 1968, p. 123).

Where power feeding is employed it is possible to use "climb" sawing (*Lubkin*, 1957, *Koch*, 1964, 1972, p. 869, Vol. I of this book, 490, 491, 503) where the cut is made in the same direction as the feed.

For sizing particle board, saw benches are fitted with extension tables to support the large sheets. Highly efficient panel sizing machines are characterized by conveyors and transfer devices at right angles to each other to trim both dimensions, and by punch card control.

The depth of the saw, protruding out of the saw bench, should be 13 mm (1/2 in.) greater than the thickness of the board being cut. A tendency of the board to ride over the saw indicates that the saw blade needs re-sharpening.

The cutting force P_s exerted by a saw blade is directly proportional to the amount of material (volume of kerf) V_k removed by each tooth. The following formula may be used:

$$V_k = \frac{f \cdot k \cdot c}{i} \tag{5.41}$$

where

f = feed speed,	c = cutting (or peripheral) velocity,
k = width of kerf,	i = number of teeth

c) *Band-sawing* of particleboard follows usual timber practice; almost any type will provide successful cutting.

d) *Milling (moulding), profiling, grooving,* and *routering* is similar to normal woodworking. Satisfactory results can be obtained with high-speed steel cutters, but it is advisable to use tungsten carbide tipped tools if long production runs are desired. Suitable speed for spindle 6,000 and 7,200 rpm, for routering 24,000 rpm.

Pahlitzsch and *Jostmeier* (1964a, 1964b) after having studied in earlier experiments the dulling effect in milling particleboard (*Grube* and *Alekseev,* 1961, *Prokes,* 1961, *Bier* and *Hanicke,* 1963) investigated the blunting behavior of knives made of high-speed steel or hard alloys during milling of particleboard. They observed that pine particleboard blunted knives more rapidly (about 100%) than did poplar particleboard, though at an equal state of knife bluntness, poplar required more cutting power than pine. Knives blunted by poplar showed fairly regular wear along the cutting edges whereas pine caused irregular wear. Tool wear depends primarily on the density of the various layers in the board. Also *Shen* (1971) reports that the highest cutting force, in cutting with a circular saw, is developed by the high-density, high-strength surface layers, and the minimum cutting force is usually found in the center plane of the board. Idealized patterns of cutting force versus depth of cut for homogeneous and sandwich board with surface layers stronger than the core are shown in Fig. 5.194. *Shen* (1971) found in his study that a strong positive linear correlation exists between cutting force as a dependent variable and modulus of rupture, surface strength and hardness. *Pahlitzsch* and *Jostmeier* (1964b) explained the deviations from normal abrasiveness relationships mainly by the presence and amount of silicates, which accounted for the faster and more irregular wear observed for the pine particleboard. The tool-wearing effect increases with higher cutting velocities on the one hand, with longer feed-paths on the other hand. Fig. 5.195 shows the maximum cutting force as a function of the cutting speed which for sharp as well as for blunt knives reaches a minimum at about $c \approx 40$ m/sec ($\sim$7,840 ft./min). The most economical cutting speed may be lower, and must be calculated from case to case as the last mentioned authors demonstrated. Generally, reference may be made to machining tests with common hardwoods and softwoods, in which the following machining factors were varied: feed rate, cutterhead speed, number of knife cuts per inch, cutting angles, knife material (*Davis* and *Nelson,* 1954).

The results are of value for particleboard molding.

Shaped profiles can be machined on a router or spindle-molder with tipped cutters, using templets or jigs. Suitable speed for profiling is 6,000 rpm, for routering 24,000 rpm.

e) *Face planing* of particleboard is in most cases neither necessary nor recommended. *Edge planing* with normal overhand planers is satisfactory. "It is sometimes advantageous to allow a newly produced board to 'set' for at least a day before planing. In this period, the resin hardens further, lessening the chance of chip pull-out" (*Johnson*, 1956, p. 265).

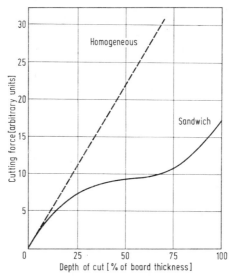

Fig. 5.194. Relation between cutting force and depth of cut for an ideal panel of homogeneous construction and an ideal panel of sandwich construction with surface layers stronger than the core. From *Shen* (1971)

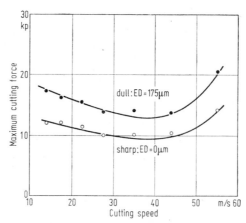

Fig. 5.195. Dependence of maximum cutting force on cutting speed for cutting with dull and sharp tool edges. From *Pahlitzsch* and *Jostmeier* (1964)

f) Particleboards are *sanded* to flatten and smooth their surfaces and, in some cases, for thicknessing. A detailed discussion of machining with coated abrasives is available (*Kollmann*, 1955, p. 740—757, 1968, Vol. I of this book, p. 528—533; *Koch*, 1964, Chapter 11).

Sanding as one of the final operations in particleboard production is dealt with in Section 5.2.10.2 of this Volume. Planers give a low quality finish, but permit to equalize substantial thickness variations between boards. In modern manufacture, raw boards show a fairly uniform thickness, finishing requirements are not exceptionally high, and for multilayer and graded board minimum amounts of material should be abraded from the surface.

The following types of particleboard sanders are usual:

1. *Drum sanders* with four and even six heads cutting *on the top* (Fig. 5. 118);

2. *Drum sanders* cutting *on the bottom*;

3. In the U. S. A. most sanders cut with half the heads on the top and half on the bottom;

4. *With wide belt sanders* (Fig. 5.119) a high class finish is obtained. *Koch* (1972, p. 895/899) reports the following details for a four-head particleboard sander (manufactured by Tidland Machine Co.): Abrasive belts travel counter to feed direction—first two heads opposed, next pair staggered—feed speeds up to 61 m/min. (~200 ft./min.)—top rolls driven independently of the bottom rolls— 25 HP (~19 kW) drive feed rolls—125 HP (~93 kW) each sander head;

5. *Combination of a drum and two vertical belt sanders* for high class surface finish at high feed speeds;

6.) *Combination of a planer and drum sander* when considerable thickness variations are expected (*Mitlin*, 1968, p. 102).

The development of complete mechanical *finishing lines,* in which dimensioning saws, sanders, eventually turning devices, automatic board feeders and stackers are combined, secure continuous high output with minimal labor requirement. A trend is to build up finishing lines from standardized work and feed units. Chippings, saw dust and sander dust are collected separately with the aim of keeping the machines and the air inside and outside of the workshop clean, and to return them as a production raw material thus improving process economics.

A few data concerning the technology of machining particleboard with coated abrasives may be given, based on the instructive summarized explanations by *Koch* (1972, p. 894—899) and the findings of the quoted authors:

1. The first top and the first bottom heads do the major cutting job. Cuts of 0.38 to 0.75 mm (0.015 to 0.03 in.) are common. Coarse grit (compared to fine) produce rougher surfaces, remove more wood particles per unit of time (*Pahlitzsch* and *Dziobek,* 1959, 1961) and cause a greater temperature rise at the surface (*Franz* and *Hinken* 1954). There is apparently a positive linear correlation between depth of cut and power required and between feed rate and power demand (cf. *Stewart,* 1970).

2. Belt speeds of 25 to 34 m/sec (~55,000 to 6,750 ft./min.) are common. Belt velocity is positively correlated with rate of wood removal (*Franz* and *Hinken,* 1954, *Hayashi* and *Hara,* 1964) and power consumed (*Nakamura,* 1966). High belt speeds yield smoother surfaces and consume less energy per unit volume of wood removed than low speeds (*Ward,* 1963). Temperature of the work-piece increases proportionally with belt velocity.

3. It is preferable to feed panels against the direction of belt travel (*Ward,* 1963, *Seto* and *Nozaki,* 1966).

4. Belts are expected to last 50 to 100 h. As the belt dulls with use, the rate of wood removal decreases and the energy consumed per unit volume removed increases.

5. It appears that the net belt power is linearly proportional to depth of cut (*Koch*, 1972, p. 899).

6. Increased pressure of the belt against the workpiece increases rate of wood removal and workpiece temperature.

g) Hand and *machine boring* (*drilling*) is a common operation with particleboard whenever screws, bolts, dowels or rungs are required in assembling particleboard and wood components. Experience allows appropriate selection of bit type, tool material (high-speed steel or hard alloys) and feed speed (chip thickness). Information about bit types, nomenclature and fundamental aspects, figures and data for southern pine wood are published by *Woodson* and *McMillin* (1971) and by *Koch* (1972, p. 918—936). The following facts are valid also for boring of particleboard:

1. The velocity v of the cutting edge varies with the spindle speed (expressed in rpm) and the distance r from the axis of rotation:

$$v = 2\pi r \cdot n. \tag{5.42}$$

A one-in. (25.4 mm) diameter bit, rotating at 3.600 rpm, has a cutting velocity of 942 ft./min (4.8 m/sec).

An 8 mm ($\sim$5/16 in.) diameter bit, consisting of high-speed steel, rotating at 18,000 rpm has a cutting velocity of 7.5 m/sec ($\sim$1,476 ft./min.). Small bits made of hard alloy are too brittle for use.

2. The thickness of the undeformed chip (t) is directly proportional to the feed speed and inversely proportional to the number of cutting lips and the spindle speed.

3. The net horsepower requirement at the spindle is a positive linear function of the torque and the rotational speed of the spindle. To this must be added the no-load idling losses of motor and spindle assembly, and the power (normally only a fraction of a HP) to overcome thrust when advancing the bit.

4. Least energy is consumed boring a hole if bits cut thick chips.

5. When boring perpendicular to plane of board, torque is primarily correlated with bit diameter, board density and chip thickness. When the thickness of chips is held constant, torque is constant for usual cutting velocities.

6. Chips formed in dry particleboard—and this is the normal case because moisture content varies in a small range of about 9% (based on oven-dry weight)—are broken to small particles.

7. Thrust (feed force) increases with increasing bit diameter. With flat-cut bits the differences of thrust with diameter are not significant.

8. Hole quality improves with decreasing chip thickness, spindle speed has no effect in wood boring.

h) Particleboard first was *lipped* (banded) by having the panel edges grooved and the lippings were made with a tongue on which glue was spread. The lipping (banding) was clamped to the panels until the glue had set. Disadvantages were:

1. Use of many clamps;
2. Considerable labor;
3. Waste of time (a few hours);
4. High percentage of rejects unless very accurate machining was maintained.

With radio frequency heating the glue can cure in a very short time (10 to 60 sec). Therefore, more complicated and effective clamping methods can be applied. "As the lippings (bandings) are held in their correct position by the press, there is no need for them to be tongued and the panels to be grooved" (*Pound*, in *Mitlin*, 1968, p. 126). This allows a saving in material and labor. Most of the radio frequency energy goes into the glue line. The lippings (bandings) may be applied to two opposite edges or to all four edges of the panel. Pressure is mostly obtained from air hoses housed in trays.

Chipboard *joints* must not be too delicate or thin, because particleboards are usually weaker than solid wood. Mortise and tenon or dovetail joints should be avoided, if possible (*Akers*, 1955). *Hunt* (in *Mitlin*, 1968, p. 129—137) gives information on the joint design (corner joints, edge joints, face joints, T-joints), on methods of static or impact loading and possible fracture lines, on the choice of adhesives and the maximum glue line thickness. In edge joints, the highest average joint strengths are obtained with urea-formaldehyde resins. Under long-term loading both polyvinyl acetate and contact adhesives tend to creep which can result in failure.

5.4.8.4 Nail-Holding and Screw-Holding Ability

5.4.8.4.1 Nails. In furniture manufacture nailing has been replaced nearly without exception by the more economical and attractive gluing methods, but nails kept or even extended their position as jointing elements in carpentry, for interior architecture, and for the construction of panellings and bearing parts in prefabricated houses. Whereever nailing is a success for solid wood or plywood nailing of particleboard also may be applied (*von Thielmann* and *Munz*, 1960).

One of the outstanding characteristics of wood and particleboard is the ease with which pieces can be nailed together. Nail joints (and fastenings), however, are the weakest parts of timber constructions.

In particleboard as well as in wood many factors affect the nail-holding power:

a) *Direction of driving* (in wood: angle to the grain, driving into side-grain or into end-grain lumber, in particleboard: angle to the board plane);

b) *Density of the wood* or particleboard;

c) *Moisture content* and changes in it. Important for wood, practically not for particleboard;

d) *Diameter of the nail* because the nail and its distance from the edge of the board should not be of a size to cause evident splitting;

e) *Depth of penetration;*

f) *Types of nails* (cylindrical, square, triangular, longitudinally grooved, annularly grooved, spirally grooved, barbed shanks (U.S. Dept. Agric. Wood Handbook, Washington, D.C., 1955), helically threaded nails (*Stern*, 1951);

g) *Types of nail points* (common, long and sharp, blunt, blunt and tapered);

h) *Material of nail* (steel, aluminium-alloy, bronze);

i) *Surface treatment of nail shanks* (cement coating, hot- or electro-galvanizing, zinc-coating, chemical etching);

k) *Prebored holes* slightly less in diameter than that of the nail;

l) *Nail spacing.*

This rather condensed compilation of nails and nailing criteria shows that there are great variations in design of nails, made by different manufacturers as well as in other factors greatly affecting nail performance.

The principal mechanical properties of nails of interest to the nail user are:

1. *Driving resistance;*
2. *Withdrawal resistance;*

3. *Lateral resistance* (lateral load-carrying capacity);
4. *Creep properties;*
5. *Splitting characteristics.*

Due to the manifold, diverse and complex relationships in nailing and in mechanical properties of nails the following data and curves do not represent absolute values, but show relative figures and general tendencies. The nail-holding power is expressed as the maximum force [kp or lb.] to overcome the withdrawal resistance either as the relative nail holding power, meaning the quotient of the maximum force to the total penetration [kp/cm or kp/mm or lb./in.] or as the specific nail-holding power, meaning the quotient of the maximum force to the area of reference, i.e. the nail surface enclosed by the particleboard [kp/cm² or lb./sq. in.].

Ad a) *Minimum* requirements for the *specific nail-holding power* in W.-Germany (particleboard with a density $\varrho_{11} \approx 10\%$, bright nail length 70 mm) *perpendicular to the plane of board* are 24 (to 36) kp/cm², parallel to the plane of board 12 (to 18) kp/cm². Evaluating publications by *Winter* and *Frenz* (1954), *Liiri* and *Haavisto* (1961), *Kratz* and *May* (1963), the values in Table 5.27 could be calculated:

Table 5.27. Quotients of Specific Nail-Holding Power Perpendicular to the Plane of Board $\sigma_{N\parallel}$ to Relevant Figures Parallel to the Plane of Board $\sigma_{N\perp}$

Material nailed	Type of nail		Reference
	Common wire nail	Nail with grooved shank	
Particleboard[1]	1.34···1.50	1.31···1.43	*Kratz* and *May*
,,	1.37	—	*Liiri* and *Haavisto*
,,	3.47		*Winter* and *Frenz*

[1] Density 600 kg/m³ ($\sim$38 lb./cu. ft.)

The specific nail-holding power in softwoods perpendicular to grain is about two times, parallel to grain 1.5 to 1.9 times greater, in hardwoods $\perp$ about 4 times, $\parallel$ about 3 times greater as compared with particleboard.

Ad b) The *relative and specific nail-holding power* increases linearly or slightly curvi-linearly with increasing *density* of the board. Perpendicular to the plane of the board, an increase in density ϱ from 550 to 750 kg/m³ (34 to 47 lb./cu. ft.) means roughly an increase of nail-holding power of about 200%. In wood, the ultimate load IP per linear inch of penetration is proportional to ϱ_0^n, where $n = 2.5$.

Ad c) *Moisture content* and its changes are important for nail-holding ability in wood but practically not in particleboard.

Ad d) For solid wood the withdrawal resistance is directly proportional to the *diameter of the nail*. Unfortunately the relationships for particleboard are much more complicated due to their lower cohesion (internal bond). Experimental results are scarce and not recent (*Winter* and *Frenz*, 1954). But a common theoretical consideration permits the conclusion that thicker nails probably will not be better embedded in the (as compared with wood) looser structure of particleboard. Therefore the specific nail-holding power will decrease (perhaps accompanied or caused by microsplitting phenomena) with increasing nail diameter. Apparently there exists an influence of board thickness in the sense that maxima may occur (Fig. 5.196). Parallel to the plane of board the specific nail-holding power decreases continuously with increasing nail diameter (Fig. 5.197). This fact supports the "micro-splitting" theory as do the force versus drive-in or pull-out path diagrams with their characteristic jumps or irregularities (Fig. 5.198).

Ad e) Increasing *depth of penetration* increases the specific nail-holding power, but along the plane of the board apparently maxima will be reached. If behind the nail point an intact part of the particleboard is left, the nail-holding power becomes higher because the internal splitting effect is lessened (Fig. 5.199).

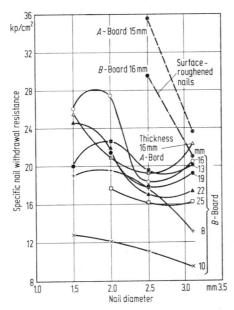

Fig. 5.196. Specific nail withdrawal resistance for nails with various diameters, two types of flat-pressed particleboard, density 0.59 g/cm³, moisture content 11%, thicknesses as indicated. From *Winter* and *Frenz* (1954)

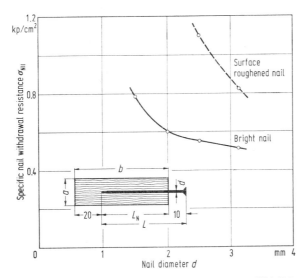

Fig. 5.197. Specific nail withdrawal resistance, parallel to board plane in 25 mm thick flat-pressed particleboard, density 0.59 g/cm³, moisture content 11%, bright nails and surface roughened nails. From *Winter* and *Frenz* (1954)

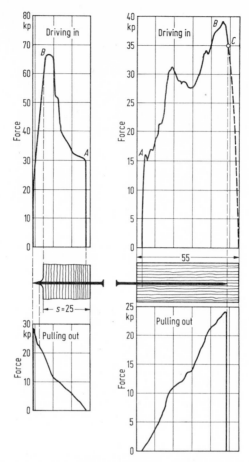

Fig. 5.198. Force-way diagrams of driving-in and pulling-out 2.5 mm thick wire nails, geometry as indicated. From *Winter* and *Frenz* (1954)

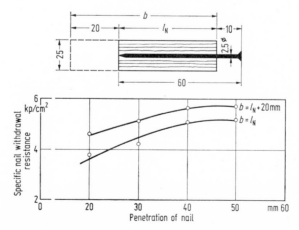

Fig. 5.199. Specific nail withdrawal resistance in dependence upon penetration of nail, 25 mm thick, flat-pressed particleboard, density 0.59 g/cm³, moisture content 11%, geometry as indicated. From *Winter* and *Frenz* (1954)

Ad f) Doubtlessly *nails with grooved or threated shanks* have a higher nail-holding power than common wire nails. Removing of such nails will damage the board structure.

Ad g) Common conical *nail points* are usual.

Ad h) In most cases *steel nails* are adequate.

Ad i) *Surface treatment of nail shanks* for particleboard is not a common practice. If nails are pulled out a long time after driving, the extraction force may be doubled as a result of corrosion.

Ad k) *Preboring of holes* for nails in particleboard is not necessary if the technique of nailing is unobjectionable.

Ad l) Perpendicular to the plane of the board, the minimum distance of nails with a diameter D from the edge can reach $(3 \times D)$ without splitting danger. Spacing of nails should be $(10$ to $20 \times D)$. If nails are driven into a particleboard under an angle of 45° or less to the plane of board or parallel to it, care and experience are necessary to avoid splitting. *Winter* and *Frenz* (1954) observed the boundary conditions illustrated in Fig. 5.200.

"Lateral resistance is a measure of the load required to cause failure when a joint is loaded so that adjacent faces of mating members define a shear plane" (*Koch*, 1972, p. 1224, 1254—1261). The lateral resistance provided by nail joints between particleboard and wood or other supporting materials is of minor im-

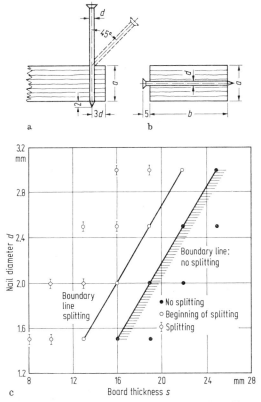

Fig. 5.200a — c. Splitting of particleboard as caused by nailing.
a) Direction of driving perpendicular to board plane and under an angle of 45 degrees; b) Nail driven in parallel to board plane; c) Boundary lines for the diameter of the nails in relationship to board thickness. Flat-pressed board, moisture content 10%. From *Winter* and *Frenz* (1954)

portance. In general, lateral resistance is less affected by board properties and nail variables than withdrawal resistance. Expanded application of particleboard, especially with higher densities and overlaid with laminates, may make lateral resistance of nail joints more important. In this case the theoretical design of a nailed or bolted joint under lateral load should be considered (*Kuenzi*, 1953.)

5.4.8.4.2 Screws. Reliable and reproducible data of particleboard screw-holding ability is difficult to obtain because:

a) No standards exist for testing this property;

b) Many factors as for nail-holding ability, and listed in this connection are of influence;

c) The variations of this property tested by the same method within one board are great;

d) The correlation between nail and screw holding power on the one hand and internal bond (tension strength perpendicular to the plane of board) on the other is fairly strong and leads to the question whether the results of such tests are necessary;

e) If data are given, exact definitions of type of screw and penetration depth are absolutely necessary;

f) A calculation of the specific screw-holding power is impossible; one can only distinguish between the absolute screw-holding power [kp or lb.] if the whole screw is used and the relative screw-holding power as the quotient of the maximum length withdrawal force to the total depth of penetration [kp/cm or lb./in.].

Nevertheless, some general statements are possible:

1. Absolute and relative screw-holding power perpendicular to the plane of the board is 100 to 125% greater than the corresponding values parallel to the plane of the board;

2. Relative particleboard screw-holding ability in the density range from 0.55 g/cm³ to 0.80 g/cm³ (34 lb./cu. ft. to 50 lb./cu. ft.) is usually 25% less than that of softwood ($\perp$ plane of board and grain direction respectively), and 55 to 60% less than that of hardwood;

3. Screw-holding ability perpendicular to the surface is, for the most part, directly proportional to overall board density;

4. Screw-holding power parallel to the surface is controlled by the closeness of contact of individual particles;

5. The optimum pilot-hole size should be approximately 75% the diameter of the threaded section of the screw (*Wissing*, 1959);

6. Screw-holding ability is greatly (100% and more) raised by the application of resin to the pilot-hole prior to insertion of the screw (*Johnson*, 1956, p. 264);

7. Greater strength is obtainable when longer and larger screws are screwed into the edge of the board, provided that not splitting occurs;

8. Another possibility is to insert and glue a wood plug in the board edge to receive the screw;

9. Extruded boards need crossbands and veneers or other facings which reinforce the structure for screw-holding ability.

5.4.9 Resistance to Destruction

5.4.9.1 Abrasion (Wear) Resistance. Abrasion resistance (cf. Vol. I of this book, p. 409—412) is a very important mechanical property for floorings, table tops etc. Abrasion is caused by various and irregularly changing factors: slow and quick walking, straight or circling, rubbing and friction, intensified by sand, dirt and other extraneous matter, sudden shocks, vibrations, effects of changes in moisture content and temperature, wetting and corrosion by chemicals. The phenomena of abrasion under conditions in practice are so complex and so different that a uniform standardized test is not possible.

Therefore tests for abrasion are of comparative value only (*Wangaard*, 1950). The wear is measured as loss of weight or loss of thickness of a specimen (for instance wood, particleboard, hardboard, plastic etc.) when scrubbed by abrasives such as fine quartz sand, propelled either by compressed air or by means of a superheated steam jet, sandpaper, grinding discs, hard metal scrapers, steel brushes or a combination of such tools. The conditions must be controlled. Some abrasion test machines simulate the wear produced on floors in service. Investigations on the wear resistance of wood, wood based material and other materials for flooring has been published by *Kollmann* (1961, 1963) with a list of references and Standard Specifications. The most important machines and procedures for abrasion tests are described and illustrated there, and own experiments are reported.

The test specimens may be rather small (1 in. × 1 in. × 1/4 in. thickness for the National Bureau of Standards-abrader (ASTM D 395-55), 2 cm square by 1 cm thickness for the Du Pont de Nemour & Co-abrader, 40 mm × 40 mm × thickness for the Austrian Wood Research Institute-abrader, 50 mm × 50 mm × 25 mm for the Kollmann-abrader, annular abrasion surface of 3,400 mm² for the Taber-abrader, 2 in. × 3 in. × thickness for the U. S. Navy-type-(later Olsen-) wearometer, 200 mm × 200 mm × thickness according to *Egner* and DIN 51945, Aug. 1937 etc.). Another method is to test whole panels (for instance 36 in. by 21 in.) for the abrasion machine developed in the British Forest Products Laboratory, Princes Risborough, or 1,500 mm by 1,800 mm used by *Thunell* and *Perem* (1948) in the Swedish Wood Research Institute, Stockholm. In the U. K. the resistance of particleboard as flooring material to point loads in static and impact tests is determined on specimens 8 ft. by 4 ft. (B. S. 1811: 1961, Amendment No. 2, July 1966). These properties are instructive, but not related to abrasion resistance.

The abrasion of wood and wood based materials can be remarkably reduced by the application of oil, lacquers and sealers on their surfaces. For rough wood samples of various species, *Chaplin* and *Armstrong* (1936) used a machine with a stamper and an abrader which moved slowly from side to side, whilst the panels to be tested slid backward and forward below the tools. They found that the wear-resistance A, expressed as the reciprocal value of the loss in thickness, (cf. *Kollmann*, 1951, p. 933/934) complies with the following linear function:

$$A = \beta \cdot \varrho_0 + \alpha \qquad (5.43)$$

for ϱ_0, in the range of about 0.6 to 1.1 g/cm³ (~38 to 69 lb./cu. ft.), is the density at zero % moisture content, β and α are constants (for A in [in.$^{-1}$] $\beta = 356$ and $\alpha = -186$).

Equation (5.43) is identical with that developed by *Ylinen* (1943) for the Brinell-hardness. Some exceptions were noted. They included Australian jarrah with a high content of brittle resinous substances, accelerating the wear in a similar manner to the pressure of grit during the test, and, in another way, Burma teak with a high content of natural oils acting as lubricants and retarding the wear.

Fig. 5.201 shows the average relative abrasion resistance of various flooring materials, untreated and sealed as obtained by *Kollmann* (1961, 1963), using a Taber-abrader.

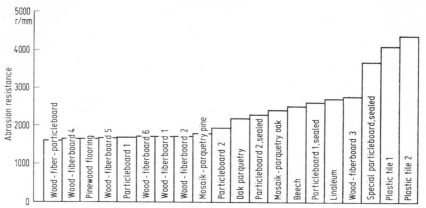

Fig. 5.201. Average abrasion resistance of various flooring materials, determined with the Taber-abrader. From *Kollmann* (1961/1963)

5.4.9.2 Resistance to Biological Attack

5.4.9.2.1 Fungi (Decay). In the literature and in practice it was often assumed that due to the presence of artificial resin binders particleboards are more resistant to attack by wood rotting fungi than solid wood, but experimental investigations and practical experience have proved the following:

a) The *binding agent* alone ensures no protection worth mentioning;

b) The *type of resin* has a pronounced influence on the rotting. Urea-formaldehyde resins may even promote rather than inhibit attack by wood rotting basidiomycetes, probably because the nitrogenous material present in the urea resins may have a stimulating effect. The whole matter is still not without contradiction. It has been mentioned earlier (Section 5.1.2.3) that *Clark* (1960) found that doubling the urea resin content of particleboard nearly doubled its decay resistance whereas a similar trend using *phenolic resins* failed to appear. Phenolic resins do not produce a stimulus, but neither do they confer any significant resistance to attack by fungi (*Evans*, in *Mitlin* 1958, p. 70).

c) The *fungi* react differently upon *urea or phenolic resins* as binders. *Klauditz* and *Stolley* (1954) found that particleboard glued with phenolic resins are less attacked by *Coniophora cerebella* than board glued with urea resins. The conditions are reversed for *Poria vaporaria* (Table 5.28).

Table 5.28. Attack of Wood Rotting Fungi on Normal Particleboard
(*Klauditz* and *Stolley*, 1954)

Type of binding agent	Test-fungus	Average loss of weight by the fungus attack %
Phenolic resin	*Coniophora cerebella*	18···22···26
	Poria vaporaria	7···11···19
Urea resin	*Coniophora cerebella*	52···54···56
	Poria vaporaria	2··· **5**···15
	Merulius lacrymans	**39**

Material tested: particleboard of spruce, 8 g of solid resin/100 g of ovendry chips, board density 0.8 g/cm³ (50 lb./cu. ft.). Test method: blocks of 5 cm × 2.5 cm × thickness (1—1.5 cm) in Kolle-flasks, according to DIN 52 176.

d) The strength of particleboard is reduced by the *loss of weight* in about the same proportion as with solid wood.

Finally it may be referred to Section 5.1.3.3 and to Tables 5.6 to 5.8 where the preservation of particleboard with fungicides and insecticides is dealt with. *Willeitner* (1969) reported on laboratory tests concerning the resistance of particleboard against wood-destroying fungi.

5.4.9.2.2 Insects, Termites. In moderate climate, under normal conditions, insects hardly attack particleboard in service in dwellings, expecially in furniture. Nevertheless, experiments showed that it is perhaps better to consider particleboard as "resistant" rather than "immune" to attacks by common insects. Furniture beetles (mainly *Anobium punctatum* de Geer) under certain circumstances "well bore into and survive within chipboard for many months, though obviously on a meagre or possibly unsuitable diet as reflected in a low or even negative rate of growth" (*Evans*, in *Mitlin*, 1968, p. 70). Certainly *Anobium punctatum* is able to damage decorative surface veneers. In England the *Lyctus*-powder-post beetles cause greater damage than all the other wood-invading insects together (*Broese van Groenou, Rischen, van den Berge*, 1952). The *Lyctus*-powder-post beetles exclusively attack some hardwood species in which the vessels are wide enough that they can lay in their eggs. This is not the case in beech and in softwoods. It is thinkable that the edge structure of the cores of particleboard is liable to attack. There are also some indications that the house long-horn beetle (*Hylotrupes bajulus* L.), which is by far the most serious pest for built in lumber in Germany and Denmark, may also attack particleboard. This is not yet clear, because the invasion is not noticeable some years after the first infection, on account of the very long period which the larvae need for their development. The behavior of particleboard, untreated and preserved, against organisms has been studied more recently by *Becker* (1969), *Becker* and *Deppe* (1969), *Becker* and *Griffioen* (1969).

Doubtlessly the increasing production of particleboard in tropical countries (Africa, Asia, South-America) made their termite-resistance very important. In tropical countries only termite-proof particleboard should be used (*Neumann*, 1970). Various types of particleboard were investigated in laboratories with regard to the attack of various groups of termites, including *Reticulitermes flavipes, Reticulitermes lucifugus, Heterotermes indicola* (*Schmidt*, 1953, 1954, 1955, 1960, *Schmidt* and *Nehm*, 1972). Some general conclusions are noticeable:

a) All normal (not protected) particleboard on the market are unsuitable for practical application in termite inhabited regions.

b) The edges are predominantly attacked because they have a looser structure, especially in the core, offering favorable areas for the attack. The tighter face-layers show much higher resistance against termites.

c) The boards exhibit different degrees of attacks.

d) The bonding agents (urea or phenolic resins) have no distinct influence on the termite resistance.

e) Larger particles seem to favor termite attack.

f) Among the wood species tested Makoré showed a poisonous effect. It is known that various species of wood have satisfactory or even very good termite-resisting properties. Teak wood (*Tectona grandis* L. f.) is generally known to be termite-proof. *Wolcott* (1946a, 1946b) has shown that the resistance of teak is due to its content of tectoquinone which substance is strongly termite-repellent. He also (1946a) published a list of wood species, classified in the order of their

resistance against West Indian dry-wood termite, *Cryptotermes brevis*. Further
works dealing with the resistance of a large number of wood species against insects,
termites included, have been published by *den Berger* (1923), Council Scientific
and Industrial Research, Australia (1936), *Cox* (1939), Forest Products Research
Laboratory (1941, 1943), and *Seifert* (1942).

g) Treatment of particleboard with chemical preservatives against termites is
advisable.

5.4.9.3 Reaction to Fire

5.4.9.3.0 General Considerations. The danger from fire in buildings are mostly
to life and to commercial values. There is an increasingly unfavorable gap between
fire damage and the premium amounts of fire insurance, especially for industrial
objects. Fire statistics are problematic. The building constructions change considerably and quickly. Material values and production resources are more and
more concentrated in spatial unities. Previously brickwork prevailed; it was
incombustible and it could be assumed that even after fire, important building
parts had not lost their stability and strength so that they were available for reconstruction. In the meantime, modern constructions, consisting of coated steel,
concrete and even building parts containing combustible materials, can fulfil
high requirements of fire resistance, but the buildings are frequently used in an
improper and even careless manner. Therefore, if combustible materials are
applied for building, furnishing, equipping or for storing, strict building codes
and fire regulations exist throughout the world with relatively severe legal provisions for instance in W.-Germany.

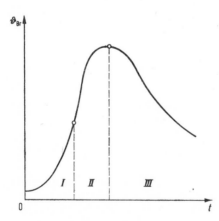

Fig. 5.202. Schematic curves of temperature during a
detrimental fire.
I Phase of ignition; *II* Fully developed fire; *III* Combustion end; From *Knublauch* and *Rudolphie* (1972).

Each detrimental fire has its origin in a spatially closed limited range. Ease
of ignition, rate of flame spread, and rate of heat release depend on the energy
supplied and released, which leads to raised temperatures. Thermal radiation and
convection of hot gases heat the air in the burning section, and the surfaces of
building parts bordering this section. Combustible materials yield combustible
gases as the result of pyrolytic decomposition. In case the temperature
in the room concerned reaches everywhere about 650 °C ($\sim$1,200 °F), the fire often
suddenly spreads within fractions of a second. This critical moment is called
"flash over". There-upon the not yet ignited combustible materials catch fire. The

course of temperature during a fire is schematically shown in Fig. 5.202 (*Knublauch* and *Rudolphie*, 1972).

The height of temperature in the fire room and the duration of the heat depend on many factors. The most important are load of combustible material per unit area, expressed in Mcal/m², size of windows as a measure of ventilation, volume of the room, type and size of walls, partitions, ceilings, floorings, type and heating-value of the combustible materials.

The basis for a critical examination of the behavior in fire are fire tests of:

a) Building materials, and
b) Building parts, as reported by *Thomas* (1972).

There are many test criteria, test conditions, and test methods to characterize the non-combustibility or combustibility of building materials on the one hand, and structural elements, like partitions and floors, on the other. Both areas cover the traditional fire testing in various countries. A third question is how materials, i.e. combustible wall linings, will contribute to the development of a fire.

Although there are considerable difficulties in agreeing on apparatuses to measure the combustibility of a material in the context of a building fire, there is little disagreement in general terms. Similarly, the furnaces for testing the fire resistance of structural elements may differ in their details (dimensions, sources of heat), but the basic concept is fairly uniform. Fire tests have been developed with particular hazards of fire situations in mind. Mentioned may be only the British Standard B. S. 476: Part 3 (Spread of flame) which also exists in a slightly modified form in the Netherlands and in New Zealand, with the aim to represent a fire in a corridor. Later research on the development of fire in rooms led to "enclosure" or "boxtests" (Sweden, the Netherlands, U.K.). ASTM E 162 measures the spread of flame. The heat released by the burning is measured by the temperature of the combustion gases, as also in the French "epirateur"-test. The U. S. Forest Products Laboratory, Madison, Wisc. (1953), has described the following fire-test methods used in research:

a) Spread of flame tests (fire tube tests, modified Schlyter test, horizontal furnace test, crib test, roof corner test, sidewall fire test, corner wall fire test, inclined panel test, SS-A-118a fire test;

b) Flame penetration test (vertical panel tests, 20 in. by 20 in. furnace, 10 ft. by 10 ft. furnace).

In W.-Germany a revised edition of DIN 4102 on "Behavior in Fire of Building Materials and Building Parts", was edited in February 1970. The designations by building authorities are given in Table 5.29.

Building parts are tested in their behavior to fire in furnaces in which the temperature is raised, following a normal curve, e.g. within 30 min to 821 °C (1,542 °F), within 60 min to 925 °C (1,697 °F).

Building materials are classified according to their behavior in fire as shown in Table 5.29. Wood based materials cannot range in class A of non-combustible materials. Even building boards manufactured of non-combustible materials, but-containing small amounts of artificial resins or other bonding agents hardly correspond to the test conditions. For wood based materials such as particleboard the most important aim is to be classified as "hardly inflammable". The necessary test installations are developed (*Wolgast*, 1963). Fig. 5.203 shows the test device used in W.-Germany (*Kollmann* and *Teichgräber*, 1961, DIN 4102, Feb. 1970).

Combustible solid materials can be classified as "hardly inflammable" (class B 1) if the following conditions are reached:

1. No sample should be completely burnt;

2. The average length of the not burnt surface should be 15% at the minimum;

3. The average temperature of the smoke gases should be below 250 °C or 482 °F;

4. The samples should not behave disfavorably, e.g. with respect to maximum height of flames, after-burning, after-glowing, time and duration of ignition, type of flame spread, condition of the samples after the test.

Table 5.29. Behavior of Building Materials and Building Parts in Fire

Building materials		
Building material class	Designation by building authorities	
A	Non-combustible building materials	
A 1		
A 2		
B	Combustible building materials	
B 1	Hardly inflammable materials	
B 2	Normally inflammable materials	
B 3	Easily inflammable materials	
Building parts		
Fire resistance class	Fire resistance duration, minutes	Designation by building authorities
F 30	$\geq$30	Fire retardant
F 60	$\geq$60	
F 90	$\geq$ 90	Fire resistant
F 120	$\geq$120	
F 180	$\geq$180	Highly fire resistant

The test should be completed eventually by determining the loss of weight of the samples, by observations of burning drops, of falling down of burning parts etc.

Throughout the world there have been carried out very many small-scale and large-scale experiments of measuring the "reaction to fire". The knowledge about the materials and the constructions involved has been furthered but the test results were often opposite and the variations between laboratories considerable. Tests should provide information to be used in some real situation, but there does not exist a large enough amount of data relating to real fires, and all tests contain certain arbitrary characteristics and can be "much more discriminating than any statistical data about the performance of materials" (*Thomas*, 1972). It is a merit of ISO/TC 92, W. G. 4, to work on international standardization of test procedures. It might be necessary to have more than one test.

5.4.9.3.1 Aspects of Tests. Aspects on which fire tests would be required are as follows:

a) *Ease of ignition* (ignitability). Several factors have to be chosen, for example, heat sources (convective or radiant), orientation of the heated specimen, thermal and atmospheric environment, use with or without pilot ignition, definition "what is ignition"?

b) *Spread of flame*. Attempted is some absolute measure of flame spread. For some classes of materials ignitability and flame spread are highly correlated. Of interest is the assessment of spread in various orientations. Preheating of the specimen can drastically affect the relative performance of a composite (thermal inertia of protecting surface). Rate of spread is a partial measure of rate of growth or extent of fire.

c) *Rate of heat release* is a basic characteristic of a material (in a defined atmospheric and thermal environment). The results are functions of the thermal inertia of the apparatus and so comparative only. An enclosure is advisable. Measurement as time dependent output for a chosen thermal line dependent input. Intention of assessing the contribution of the tested material to propagating the fire in other materials as well as in itself (feedback);

d) *Smoke obscuration*. Produced from the material, its thermal exposure and the ventilation conditions. The hazard from smoke is a function of the rate at which a material becomes involved in a fire.

5.4.9.3.2 Wood and Wood Based Panels in Fire. The cell walls of wood (wood based materials mainly consist of wood) are composed of three principal chemical materials, cellulose, hemicellulose and lignin (Vol. I of this book, p. 55—78).

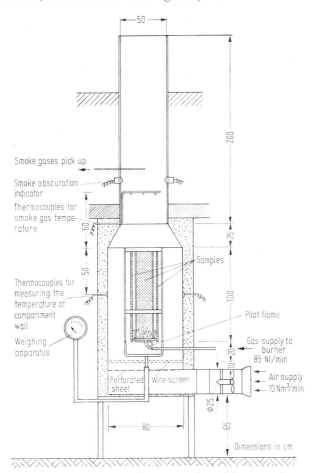

Fig. 5.203. Apparatus for testing combustible materials. From *Kollmann* and *Teichgräber* (1961)

Normal hardwoods and softwoods usually both contain $42 \pm 2\%$ of cellulose. The lignin content of hardwoods varies between 18 and 25%, while in softwoods the range is 25 to 35%. All principal chemical wood constituents are polymeric. Cellulose contains linear macromolecules of considerable length and has a partly crystalline structure. Hemicelluloses are relatively low-molecular weight polysaccharides. Lignin, the encrusting substance in wood, is a three dimensional polymer. The atoms of all principal wood constituents are carbon, hydrogen and oxygen. Therefore wood is combustible. It is only possible to reduce ignitability, spread of flame, rate of heat release and flame penetration by chemical preservation, but is impossible to make ligno-cellulose materials completely non-combustible.

The reaction to fire of building materials depends not only on their chemical composition but on many factors such as

a) Density (porosity, internal surface);

b) Thermal conductivity and diffusivity;

c) Specific heat;

d) Moisture content;

e) Special chemical compounds (extractives);

f) Shape and dimensions of the fire test sample (especially proportion of surface to volume);

g) Surface quality (emissivity, cf. Vol. I, p. 256/257);

h) Position and arrangement of the sample in the room;

i) Air supply and smoke gases deduction (ventilation);

k) Reflection of the environment.

Patzak (1972) analyzed the combustion processes in wood under physico-chemical aspects. He studied the effects of geometry and surface quality of specimen and of the surrounding atmosphere. Thermogravimetry, differential thermo-analysis, pressure analysis and measuring of temperature around the specimen were applied simultaneously and on the same specimen.

Kollmann and *Teichgräber* (1961) published the results of fire tests on spruce boards and sheetlike wood based materials (Fig. 5.204). One can see that thin plywood has the worst properties (highest raise of temperature of combustion gases); thicker spruce board are better. Particleboard of the same thickness did not behave well, if they were not impregnated with fire retardants. If they are impregnated—and this is valid also for thin plywood—the temperature of the combustion gases is much lower. The best behavior was observed with excelsior (wood wool) board because they contain a rather high amount of mineral substances. In Fig. 5.204 the drawn lines show the success of impregnation with fire retardants against the dotted lines for unimpregnated materials.

The tests have proved that the thickness of the samples has a great influence on the classification with respect to fire resistance (Fig. 5.205).

A proper classification of composite and sandwich construction is absolutely necessary because they are applied to an increasingly larger extent in modern buildings as linings, facings, panellings, sheathings, for fittings, furnishing, as elements in prefabricated houses etc. The following should be considered:

1. Hardly inflammable materials glued or cemented together are not inflammable;

2. Treated particleboard made to be hardly inflammable may lose this advantage by painting, lacquering, surface-coating, especially veneering.

Doubtlesly, model tests should create theoretical and practical approaches to the reduction of fire hazards.

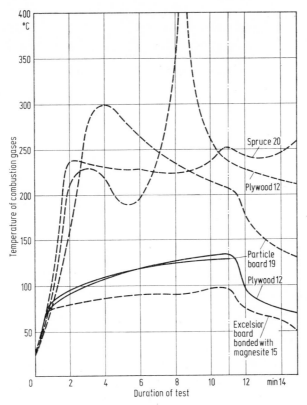

Fig. 5.204. Dependence of the temperature of combustion gases on duration of test for wood based materials as indicated. From *Kollmann* and *Teichgräber* (1961)

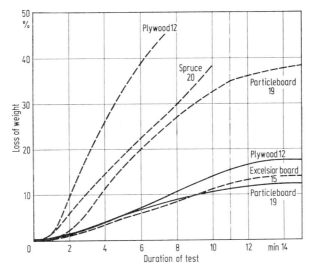

Fig. 5.205. Dependence of loss of weight on duration of test for wood and wood based materials, as indicated From *Kollmann* and *Teichgräber* (1961)

Literature Cited

Aaro, P., (1961) Über den Festgehalt der wichtigsten finnischen Schichtholzsortimente. Kom. ex Inst. Quaest. Forest Finnl., Ed. 14 (5).

Adams, D. G. and coworkers, (1970) Bending strength of radiation produced Southern pine wood-plastic combinations. For. Prod. J. **20**: 25—28.

Akers, L. E., (1955) Chipboard and its use in furniture. Furniture Development Council, Research No A-4.

—, (1966) Particleboard and hardboards, p. 51. Pergamon Press, London.

—, (1968) Particleboard in building. In: *Mitlin, L.*, Particleboard manufacture and application, p. 159—164. Pressmedia Ltd., Ivy Hatch, Sevenoaks, Kent.

Alan, D., (1962) Shaped components produced by moulding technique. Woodworking Industry, London.

Albers, K., (1971) Gleitzahlmessungen an Holzwerkstoffen. 1. Mitt.: Direktmessung an Schubwürfeln. Holz als Roh- und Werkstoff **29**: 178—183.

—, (1972) Gleitzahlmessungen an Holzwerkstoffen. 4. Mitt.: Biege-Schubversuche an Verbundwerkstoffen. Holz als Roh- u. Werkstoff **30**: 182—190.

Aleksandrov, A., (1958) Einfluss der Oberflächengüte von Spanplatten auf die Qualität beim Furnieren. Derev. Prom. **7**: 7—9.

Alfrey, T., Jr., (1948) Mechanical Behavior of polymers. Interscience, New York.

Anonymous, (1956) Extrudes 12,000 ft. of chipcore for 3,000 flush doors daily. Wood and Wood Products, Vance Publ. Corp. Chicago.

ASTM D 1037, (1964) Standard methods of evaluating the properties of wood-base fiber and particle panel materials.

Autio, T., Miettinen, J. K., (1970) Experiments in Finland on properties of wood polymer combinations. For. Prod. J. **20**: 36—42.

Avale, M., (1964) Kriechversuche an Bücherregalen. FESYP, Techn. Comm., 13th Session, Stockholm.

Becker, G., (1969) Versuche an Holzspanplatten mit Hausbockkäferlarven und Termiten. Material und Organismen, Beih. **2**: 27—41

—, *Deppe, H. J.*, (1969) Zum Verhalten von unbehandelten und chemisch geschützten Holzspanplatten gegen Organismen. Holzforsch. Holzverwert. **21**: 103—108.

—. *Griffioen, K.*, (1969) Verhalten und Schutz von Holzspanplatten und anderem Plattenmaterial gegen Organismen. IUFRO Symp., London; Material und Organismen, Beih. 2.

Bengtson, K., (1970) Aspects of the consumption of fiber building boards and other sheet materials for the building industry. Defibrator News 1, Stockholm.

den Berge, L. C., (1923) Grondslaag voor de classifatie der Nederlandsch-Indische timmerhoutsoorten. Tectona **16**: 602.

Berger, V., (1964) Harnstoffbindemittel mit herabgesetztem Gehalt an freiem Formaldehyd. Holztechnologie **5**: 79—82.

Bergsträsser, M., (1927) Bestimmung der beiden elastischen Konstanten von plattenförmigen Körpern. Z. techn. Phys. **8**: 355—359.

Bibby, R. D., (1958) A study of the characteristics, manufacture and use of particle board from wood and other vegetable fibres. Fiberboard and Particle Board. Vol. IV, Paper 5.43. FAO, Rome.

Bier, H., Hanicke, P., (1963) Die spezifische Schnittkraft als Funktion der Schneidenabstumpfung beim Fräsen. Holztechnologie **4**: 158—162.

Blache, H., (1936) Vorrichtung zum Zuführen von faserstoffhaltigen Kunstharzmassen zu der Dosiervorrichtung einer Presse. DRP 680372.

Blomquist, R. F., (1959) A progress report. Glues and gluing 1958. For. Prod. J. **9**: 59—67.

—, (1960) An international look at glues and gluing, 1959. For. Prod. J. **10**: 62—70.

—, (1961) Progress in 1960: glues and gluing processes. For. Prod. J. **11**: 77—85.

—, (1962) Progress in glues and gluing processes 1961. For. Prod. J. **12**: 49—58.

Boiciuc, M., Petrican, C., (1970) Dimensionsstabilisierung von Rotbuchenholz durch Anlagerung von Styrol. Holztechnologie **11**: 94—96.

Bornstein, L. F., (1956) Success story of resin-bonded wood waste. Wood Working Digest, July, p. 46.

Bott, R., (1955) Betrachtungen über die industrielle Holzentrindung. Holz als Roh- und Werkstoff **13**: 147—160.

Bousquet, D. W., (1970) Preliminary study of a notched compression shear test. For. Prod. J. **20**: 29—30.

Brauns, F. E., (1952) The chemistry of lignin. Academic Press Inc., New York

Bridges, R. R., (1971) A quantitative study of some factors affecting the abrasiveness of particleboard. For. Prod. J. **21**: 39—41.

Brodeau, A., (1965) Déformation des complexes formés par les panneaux de particules et les stratifies décoratifs. Centre Technique du Bois, Paris.

Broese van Groenou, H., Rischen, H. W. L., van den Berge, J., (1952) Wood preservation during the last 50 years. 2nd Ed., p. 17. A. W. Sijthoff's Uitgevermaatschappij N. V., Leiden, Neth.

Brown, H. P., Panshin, A. J., Forsaith, C. C., (1952) Textbook of wood technology. Vol. II, p. 229—410. McGraw-Hill Book Co. Inc., New York, Toronto, London.

Browning, B. L., (1963) The chemistry of wood. Interscience Publishers, New York.

Bryan, E. L., (1960) Particle board under long-term loading. For. Prod. J. **10**: 200—204.

—, (1962) Dimensional stability of particleboard. For. Prod. J. **12**: 572—576.

—, *Schniewind, A. P.*, (1965) Strength and rheological properties of particleboard. For. Prod. J. **15**: 143—148.

BS 1811, (1961) Methods of test for wood chipboards and other particle boards. Brit. Standards Institution, London.

Büsgen, M., (1904) Zur Bestimmung der Holzhärten. Z. Forst.-Jagdwes. **36**: 543—562.

Burmester, A., (1967) Zur Vergütung von Holz durch strahlen-polymerisierte Kunststoff-Monomere. Holz als Roh- und Werkstoff **25**: 11—25.

—, (1968) Untersuchungen über den Zusammenhang zwischen Schallgeschwindigkeit und Rohdichte, Querzug- sowie Biegefestigkeit von Holzspanplatten. Holz als Roh- und Werkstoff **26**: 113—117.

Burrows, C. H., (1961) Some factors affecting resin efficiency in flake board. For. Prod. J. **11**: 27—33.

Buschbeck, L., Kehr, E., (1960) Untersuchungen über die Eignung von Kunstharzbindemitteln verschiedener Rohstoffbasis zur Herstellung von Spanplatten. Holztechnologie **1**: 29—38.

—, —, (1960) Untersuchungen über die Verkürzung der Preßzeit beim Heißpressen von Spanplatten. Holztechnologie **1**: 112—123.

Carroll, M. N., Wrangham, B., (1970) The measurement of surface strength in particleboard. For. Prod. J. **20**: 28—33.

Carson, (1932) USP 2033411.

Chaplin, C. J., Armstrong, F. H., (1936) Abrasion of floors, testing relative resistance to wear of flooring timbers. Wood **1**: 576—581.

Ciba A. G., (1962) Mouldings from wood chips and Aerolite. Ciba Technical Note 238, Duxford, Cambridge.

Čížek, L., (1961) Dauerfestigkeit und rheologische Eigenschaften von Holz und Holzwerkstoffen. Holz als Roh- und Werkstoff **19**: 83—85.

Clark, I. W., (1960) Decay resistance of experimental and commercial particle boards. U.S. For. Prod. Lab., Rep. No 2196, Madison, Wisc.

Clouser, W. S., (1962) Determining the compressive strength parallel to surface of wood composition boards. Mat. Res. and Standards **2**: 996—999.

Côté, W. A., Jr., (1968) Chemical composition of wood. In: *Kollmann/Côté*, Principles of wood science and technology. Vol. I, p. 55—78. Springer-Verlag, Berlin, Heidelberg, New York.

Council Scientific and Industrial Research (1936) Termites (white ants). Council Sci. Ind. Res., Trade Circ. **36**. South Melbourne.

Cox, H. A., (1939) A handbook of empire timbers. London.

Crafton, J. M., (1956) Extruded chipcore board. For. Prod. J. **6**: 173—175.

Cremer, L., Zemke, H.-J., (1964) Bau- und Raumakustik. In: Bauen mit Spanplatten, p. 89 to 113. Triangel-Spanplattenwerke, Triangel, West Germany.

CSP 56350, (1936) F. Pfohl.

CSP 67763, (1940) Dyas.

Davis, E. M., Nelson, H., (1954) Machining test of wood with the molder. For. Prod. Res. Soc., Preprint No 561.

DBP 841055, (1948) Runkel and Jost.

DBP 967328, (1942) Fahrni.

DBP 975846, (1947/1962) Triangel.

DBP 1088697, (1957) Himmelheber and Steiner.

DBP 1097656, (1959) Moralt.

DBP 1127573, (1960) Himmelheber and Steiner.

DBP 1183240, (1962) Himmelheber.

DBP 1198539, (1958) Himmelheber.

DBP 1453397, (1961) Moralt.

Deppe, H.-J., Ernst, K., (1963) Zur Messung der Oberfläche von Spanplatten. Holz als Roh- und Werkstoff **21**: 129—132.

—, —, (1964) Technologie der Spanplatten. Holz-Zentralblatt-Verl., Stuttgart.

—, —, (1965) Probleme der Preßzeitverkürzung bei der Herstellung von Holzspanplatten. Holz als Roh- und Werkstoff **21**: 441—445.

DIN 52362, Bl. 1 (1965) Biegeversuch, Bestimmung der Biegefestigkeit.
DIN 52364, (1965) Bestimmung der Dickenquellung.
Dosoudil, A., (1950) Dynamische Festigkeit von Platten. Svensk Papperstidning **53**: 609—612.
—, (1954) Entwurf DIN 52367. Prüfung von Holzspanplatten, Bestimmung des Stehvermögens.
—, (1965) Untersuchungen über die Formstabilität von Holzspanplatten. Dt. Ges. f. Holzforschung, Ber. No 1/65. Beuth-Vertrieb, Berlin.
—, (1969) Studien über das hygroskopische Verhalten von Holzspanplatten mit besonderer Berücksichtigung der Dimensionsstabilität. Holz als Roh- und Werkstoff **27**: 172—179.
Draffin, J. O., Mühlenbruch, C. W., (1937) The mechanical properties of balsa wood. ASTM Rep. 1937-96.
DRP 407526, (1924) Forssmann Holzblech G.m.b.H.
DRP 465755, (1924) Forssmann Holzblech G.m.b.H.
DRP 680372, (1936) Blache.
DRP 692159, (1940) Pfohl.
Dyas, (1940) CSP 67763.
Ehlers, W., (1956/1958) Die Oberfläche des Holzes, ihr Wesen und die Bestimmung ihrer Güte unter besonderer Berücksichtigung des Verfahrens nach Forster-Leitz. Diss. Univ. Hamburg. Holz als Roh- und Werkstoff **16**: 49—60.
Eirich, F. R., (1956/1958) Rheology: Theory and application. Vol. I, 1956, Vol. II 1958. Academic Press, New York.
Eisner, K., Kolejak, M., (1960) Herabsetzung des Quellens bei Spanplatten. Dřevo **15**: 363 to 365.
El'bert, A. A., (1961) Povyšenie vodostojkosti stružečnyh plit. Derev. Prom. **10**: 21—22.
Elmendorf, A., Vaughan, T. W., (1958) A survey of methods of measuring smoothness of wood. For. Prod. J. **8**: 275—282.
Engels, K., (1966) Beleimung und Imprägnierung der Späne. In: *Kollmann, F.*, Holzspanwerkstoffe, p. 280—303. Springer-Verlag, Berlin, Heidelberg, New York.
Enzensberger, W., (1966) Beschichten und Bedrucken von Spanplatten. In: *Kollmann, F.*, Holzspanwerkstoffe, p. 392—424. Springer-Verlag, Berlin, Heidelberg, New York.
Evans, W. J., (1968) Protection of particleboard against fungal and insect attack. In: *Mitlin, L.*, Particleboard manufacture and application, p. 69—73. Pressmedia Ltd., Ivy Hatch, Sevenoaks, Kent.
Fahrni, F., (1942) DBP 967328.
—, (1942) Die Holzspanplatte. Holz als Roh- und Werkstoff **6**: 277—283.
Ferry, J. D., (1961) Viscoelastic properties of polymers. Wiley, New York.
Fickler, H. H., (1962) Hydrophobierung von Spanplatten mit kationaktiven Dispersionen. Holz als Roh- und Werkstoff **20**: 280.
Fischbein, W. J., (1950) Production of wood waste boards by a continuous process. Northeastern Wood Utilization Council., Bull. No 31, p. 131.
Fischer, A., Merkle, G., (1963) Oberflächenprüfung von Holzspanplatten. Holz als Roh- und Werkstoff **21**: 96—104.
Flatscher, J. H., (1929) Handbuch des Sägebetriebes. Parey, Berlin.
Flemming, H., (1957) Die Bestimmung der Oberflächengüte an schwierigem Material. Z. f. wirtschaftl. Fertigung **56**: 67.
Food and Agriculture Organisation, Rome (1957/1958) Fibreboard and particle board. Report of an international consultation on insulation board, hardboard and particle board (Geneva, 21. 1. to 4. 2. 1957). Food and Agriculture Organisation of the United Nations and Economic Commission for Europe.
Food Machinery and Chemical Corp., Dapon Dept. (1960) Data for Industry No 2.
Forest Products Research Laboratory, (1941) A handbook of home-grown timbers. London.
—, (1943) Recognition of decay and insect damage in timbers for aircraft and other purposes. London.
Forsaith, C. C., (1926) Technology of New York State Timbers. N. Y. State Coll. For., Techn. Publ. No 18, Syracuse.
Forssmann Holzblech GmbH, (1924) DRP 407526 and 465755.
FP 679708, (1929) de Samsonow.
Franz, N. C., Hinken, E. W., (1954) Machining wood with coated abrasives. For. Prod. J. **4**: 251—254.
Furniture Development Council, (1956) List of properties. In: *Johnson, E. S.*, Wood particle board handbook, p. 268—269.
Gillwald, W., Luthard., H., (1966) Beitrag zur Dauerstandfestigkeit von Vollholz und Holzspanplatten. Holztechnologie **7**: 25—29.
Golbs, H., Fentzahn, F., (1956) Die Verarbeitung der Holzspanplatte im Möbelbau. Holz als Roh- und Werkstoff **14**: 68—74.

Gomme, D. L., (1968) Particleboard in the furniture industry. In: *Mitlin, L.*, Particleboard manufacture and application, p. 156—159. Pressmedia Ltd., Ivy Hatch, Sevenoaks, Kent.

Goodyear Tire and Rubber Co., Akron/Ohio, (1959) Videne — a polyester laminating and surfacing film. Paper No 59-A-316.

Graham, F., (1958) Electronic detection of metal hidden in wood. Wood working **60**, Issue 3, 29—32.

Grant, E. L., (1964) Statistical quality control. 3rd Ed., p. 561. McGraw-Hill, New York.

Grasser, M., (1962) Temperaturverlauf in industriell gefertigten Spanplatten während des Pressvorgangs. Holz-Zbl. **88**: 137.

Gratzl, A., (1956) Zur Messmethodik bei der Prüfung des Stehvermögens plattenförmiger Körper. Internationaler Holzmarkt **54**: 13—17.

Greten, E., (1966) Bison-Verfahren. In: *Kollmann, F.*, Holzspanwerkstoffe, p. 441. Springer-Verlag, Berlin, Heidelberg, New York.

Gross, B., (1953) Mathematical structure of the theories of viscoelasticity. Hermann & Co., Paris.

Grube, E., Alekseev, A. V., (1961) Spezifische Schneidarbeit beim Fräsen von Spanplatten. Derev. Prom. **10**: 7—8.

Grzeczynski, T., Bakowski, S., (1963) Verfahren zur schnellen Bestimmung der Zugfestigkeit senkrecht zur Plattenebene. Holz als Roh- und Werkstoff **21**: 495—496.

Haas, H., (1959) Das Spanholzpressteil. Holz **13**: 170—172.

—, (1966) Erzeugung von Holzspanformteilen. In: *Kollmann, F.*, Holzspanwerkstoffe, p. 432 to 434. Springer-Verlag, Berlin, Heidelberg, New York.

Hann, R. A., (1957) A method of quantitative topographic analysis of wood surfaces. For. Prod. J. **7**: 448—452.

Hauser, H., (1965) Vergleichende Betrachtung der Herstellung von Spanplatten in Mehretagenpressen, Einetagenpressen und Endlosfertigung. Holztechnologie **6**: 60—63.

Hayashi, D., Hara, O., (1964) Studies of surface sanding of lauan plywood. Wood Ind. **19**: 9, 16—21.

Hearmon, R. F. S., (1961) An introduction to applied anisotropic elasticity. Oxford University Press, Oxford.

Heebink, B. G., (1946) Fluid-pressure molding of plywood. U. S. For. Prod. Laboratory Rep. No. R 1624. Madison, Wisc.

—, (1960) A new technique for evaluating showthrough of particleboard cores. For. Prod. J. **10**: 379—388.

Herdey, O., (1958) Synthetic resin and other additives used in the manufacture of particle boards. Fibreboard and Particle Board, Vol. II, Item 4.18, FAO, Rome.

Hesch, R., (1968a) Annual plants as raw material for the particleboard industry. Holz als Roh- und Werkstoff **16**: 129—140.

—, (1968b) Manufacture of particleboard from bagasse. In: *Mitlin, L.*, Particleboard, manufacture and application, p. 47—59. Pressmedia Ltd., Ivy Hatch, Sevenoaks, Kent.

—, (1969) Single and multi-daylight presses. Technology and economics of particleboard manufacture in relation to the employed press system. Board manufacture **12**:59—63, 79—82.

—, (1970) Bagasse as raw material for particleboard manufacture. Licht's Sugar Yearbook.

—, (1972) Trinidad sets up bagasse particleboard plant.

Heyne, D., (1970) Betrachtungen zu den Herstellungs- und Anwendungsmöglichkeiten von Holz-Plastkombinationen. Holztechnologie **11**: 53—56.

Himmelheber, M., (1948) Arbeiten der Holig-Homogenholz-Werke GmbH auf dem Gebiet der Holzwerkstoffe. Dt. Ges. f. Holzforschung, Mitt. Nr. 37, p. 111.

—, (1949) Entwicklung und Stand der Homogenholz-N- und -T-Verfahren. Dt. Ges. f. Holzforschung, Tagungsbericht Braunschweig, p. 40.

—, (1958) BP 1198539.

—, (1962) BP 1183240.

—, *Kull, W.*, (1969) Die moderne Spanplattenfertigung: Rohstoffeinsatz und Plattenqualität. Holz als Roh- und Werkstoff **27**: 397—406.

—, *Steiner, K.*, (1957) DBP 1088697. Vorrichtung zur Herstellung von Holzspanplatten oder dergl.

—, —, (1960) Zusatzpatent zu DBP 1088697. Vorrichtung zur Herstellung von Holzspanplatten oder dergl.

—, —, (1960) DBP 1127573.

—, —, *Kull, W.*, (1953) USP 2737997.

Höchli, O., (1956) Der Beitrag der schweizerischen Faser- und Spanplattenindustrie zur Erweiterung des Angebots an hochwertigen Holzwerkstoffen. Diss. Handelshochschule St. Gallen.

Hoitz, B., (1934) Über die Verwendung dünnsten Sperrholzes für Verpackungszwecke. Sperrholz **6**: 41—46.

Holz-Zentralblatt-Verlag (1954) Die Platte — ein Holzwerkstoff. Holzwirtschaftl. Jahrbuch No 4. Stuttgart.

Hopper, R. F., (1969) Acoustics. The american people's enzyclopedia, Vol. I. 93—96. Grolier Inc.

Horn, J., (1965) Untersuchungen über die Rohdichte-Profile bei Holzspanplatten. Diplomarbeit, Hamburg.

Houwink, R., (1937) Elasticity, plasticity and structure of matter. MacMillan, New York.

—, (1957) Elastizität, Plastizität und Struktur der Materie. Breitkopff, Leipzig, Dresden.

van Hüllen, K., (1966) Vorpressen, Befeuchten, Pressen. In: *Kollmann, F.*, Holzspanwerkstoffe, p. 332—372. Springer-Verlag, Berlin, Heidelberg, New York.

Hult, J., Ed. (1972) Creep in structures. IUTAM Symposium 1970. Springer-Verlag, Berlin, Heidelberg, New York.

Hunt, D. G., (1968) Wood chipboard joints. In: *Mitlin, L.*, Particleboard manufacture and application, p. 129—138. Pressmedia Ltd., Ivy Hatch, Sevenoaks, Kent.

Interwood AG, (1947) Swiss Patent 276790 Holzspäne enthaltender Werkstoff sowie Verfahren und Vorrichtung eines solchen Werkstoffes.

Iwanow, M., (1958) High elastic deformation of wood. Comp. Wood **5**: 51—56.

Jalava, M., (1929) The measuring on round, piled woodgoods. Kom. ex Inst. Quest. Forest. Finnl. Ed. 13, H. 8.

Janka, G., (1906) Die Härte des Holzes Cbl. ges. Forstwes. **9**: 193—241.

—, (1908) Über Holzhärteprüfung. Cbl. ges. Forstwes. **11**: 443—456.

—, (1915) Die Härte der Hölzer. Mitt. Forstl. Versuchswes. H. 39. Wien.

Jensen, U., Kehr, E., (1970) Untersuchungen zur Verarbeitung von Feingut bei der Spanplattenherstellung. Holztechnologie **11**: 97—100.

Johnson, E. S., (1956) Wood particle board handbook. North Carolinc State College, School of Engineering, Raleigh/N. C.

Jorgensen, R. N., Murphey, W. K., (1961) Particle geometry and resin spread. For. Prod. J. **11**: 582—585.

Kamei, S., (1937) Untersuchung über die Trocknung fester Stoffe (III). Memoirs of the College of Engineering, Kyoto Imperial University Vol. **X**: 65—11.

Kehr, E., Jensen, U., (1970) Herstellung und Eigenschaften5von Spanplatten mit Feinstpartikel-Deckschichten. Holz als Roh- und Werkstoff **28**: 385—391.

—, *Schölzel, S.*, (1967) Untersuchung über das Pressdiagramm zur Herstellung von Spanplatten. Holztechnologie **8**: 177—181.

Kenaga, D. L., (1970) The heat cure of high boiling styrene type monomers in wood. Wood Fiber **2**: 40—51.

Keylwerth, R., (1955) Studien über die Anwendung mathematisch-statistischer Methoden in Holzforschung und Holzwirtschaft. 5. Mitt.: Untersuchungen über die Gleichmässigkeit von Spanplatten und Faserplatten. 6. Mitt.: Statistische Qualitätskontrolle. Holz als Roh- und Werkstoff **13**: 215—221; 266—271.

—, (1956) Dimensionsstabilität und Gleichmässigkeit von Möbel- und Türplatten. Holz als Roh- und Werkstoff **14**: 353—360.

—, (1958) Zur Mechanik der mehrschichtigen Spanplatte. Holz als Roh- und Werkstoff **16**: 419—430.

—, (1959a) Statistische Methoden der Qualitätskontrolle im Holzindustriebetrieb. Dt. Gesellsch. f. Holzforschg. Mitt. Nr. 44.

—, (1959b) Erreichte und erreichbare Verminderung der Anisotropie in Holzwerkstoff-Platten. Holz als Roh- und Werkstoff **17**: 234—238.

Klamroth, K., Hackel, J., (1971) Trockungsanlagen für die Spanplattenindustrie. Holz als Roh- und Werkstoff **29**: 449—455.

Klauditz, W., (1947) Untersuchungen zur Kennzeichnung industrieller Holzabfälle. Morphologische Kennzeichnung von Sägespänen. Inst. f. Holzforschung, Braunschweig.

—, (1949) Untersuchungen über die Ergiebigkeit von Phenol- und Harnstoff-Formaldehyd-Kunstharz-Bindemitteln bei der Herstellung von Holzspanplatten. Institut für Holzforschung, Braunschweig, Ber. No 14/1949.

—, (1952) Untersuchungen über die Eignung verschiedener Holzarten, insbesondere von Rotbuchenholz, zur Herstellung von Holzspanplatten. Institut für Holzforschung, Braunschweig, Ber. No 25/1952.

—, (1955a) Entwicklung, Stand und holzwirtschaftliche Bedeutung der Holzspanplattenherstellung. Holz als Roh- und Werkstoff **13**: 405—421.

—, (1955b) Zur Kenntnis der Druckverleimung von Holzspänen zu Holzspanplatten in beheizten hydraulischen Etagenpressen und Beschleunigung der Beleimungsvorgänge durch

Einstellung zweckmässiger Feuchtigkeitsgehalte des beleimten Spangutes und durch Erhöhung der Temperatur bei der Verleimung. Institut für Holzforschung, Braunschweig, Ber. No 45/1955.

—, (1960a) Untersuchungen über die Verleimungsvorgänge bei der Holzspanplattenherstellung. Dt. Gesellsch. f. Holzforschg., Mitt. Nr. 47, p. 87—90.

—, (1960b) Untersuchungen über die Quellung und Herabsetzung der Quellung von Holzspanplatten sowie über die Herstellbarkeit von wetterfesten Holzspanplatten. Dt. Gesellsch. f. Holzforschg., Mitt. Nr. 47, p. 91—93.

—, (1966) Entwicklung der Herstellung, der Eigenschaften, der Produktion und des Verbrauches von Spanplatten. In: Kollmann, F., Holzspanwerkstoffe, p. 1—27. Springer-Verlag, Berlin, Heidelberg, New York.

—, and coworkers (1960) Herstellung und Eigenschaften von Holzspanwerkstoffen mit gerichteter Festigkeit. Holz als Roh- und Werkstoff 18: 377—385.

—, Stegmann, G., (1957) Über die Eignung von Pappelholz zur Herstellung von Holzspanplatten. Holzforschung 11: 174—179.

—, Stolley, I., (1951) Holzschutz bei Holzspanplatten. Verein f. techn. Holzfragen, Tagung Braunschweig 1951, p. 159, Braunschweig-Kralenriede.

—, —, (1954) Entwicklung und Herstellung von termitenfesten Holzspanplatten. Holz als Roh- und Werkstoff 12: 185—189.

—, Ulbricht, H. J., Kratz, W., (1958) Über die Herstellung und Eigenschaften leichter Holzspanplatten. Holz als Roh- und Werkstoff 16: 459—466.

Knorpp, W., (1963) Späneabsaugungsanlagen. Holz als Roh- und Werkstoff 21: 268—272.

Knublauch, E., Rudolphie, R., (1972) Das Risiko voraussagen. Die theoretische Ermittlung des Wagnisses einer Feuerversicherung. VDI-Nachr. 49: 9, 50: 10.

Koch, P., (1964) Wood machining processes. Ronald Press Co., New York.

—, (1972) Utilization of the Southern Pine. Vol. II: Processing, p. 755—948. Agric. Handbook No 420. U. S. Dept. Agric. For. Serv., Washington, D.C.

Kollmann, F., (1936) Technologie des Holzes. 1st Ed., Springer-Verlag, Berlin.

—, (1937) Über die Schlag- und Dauerfestigkeit der Hölzer. Mitt. Fachaussch. f. Holzfragen 17: 17—30. VDI-Verlag, Berlin.

—, (1940) Die mechanischen Eigenschaften verschieden feuchter Hölzer im Temperaturbereich von —200 bis +200 °C. Forsch. Ing. Wes. H. 403, VDI-Verlag, Berlin.

—, (1949) Eigenschaften, Prüfung und Klassifizierung von Holzfaser- und Holzspanplatten. Svensk Papperstidning 52: 251—260.

—, (1951) Technologie des Holzes und der Holzwerkstoffe. Vol. I., p. 416, 913. Springer-Verlag, Berlin, Göttingen, Heidelberg.

—, (1952a) Herstellung halbschwerer Holzspanplatten im Trockenverfahren. Holz als Roh- und Werkstoff 10: 121—134.

—, (1952b) Der Einfluss der Zeit auf die mechanischen Eigenschaften der Hölzer. Holz als Roh- und Werkstoff 10: 187—197.

—, (1952c) Herstellung und Eigenschaften von Holzspanplatten der Norddeutschen Homogenholz-Gesellschaft mbH, Triangel. Holz als Roh- und Werkstoff 10: 463—468.

—, (1954) Stand der Technik bei der Herstellung von Holzspanplatten. Holz als Roh- und Werkstoff 12: 117—134.

—, (1955a) Some aspects of wood technology. Lecture Notes. The Manager of Publications, Delhi.

—, (1955b) Technologie des Holzes und der Holzwerkstoffe. Vol. II. Springer-Verlag, Berlin, Göttingen, Heidelberg.

—, (1955c) Automatisierung bei der Herstellung von Holzspanplatten. Holz als Roh- und Werkstoff 13: 421—433.

—, (1957a) Über Unterschiede im rheologischen Verhalten von Holz und Holzwerkstoffen bei Querdruckbelastung. Forsch. Ing. Wes. 23: 49—54.

—, (1957b) Über den Einfluß von Feuchtigkeitsunterschieden im Spangut vor dem Verpressen auf die Eigenschaften von Holzspanplatten. Holz als Roh- und Werkstoff 15: 35—44.

—, (1958) Resin Application and Equipment for Blending Resin with Raw Material Particles. Fibreboard and Particle Board. Vol. IV. Item 5.30. Food and Agriculture Organisation, Rome.

—, (1961a) Untersuchungen über den Abnutzungswiderstand von Holz, Holzwerkstoffen und Fussbodenbelägen. Forsch. Ber. Nordrhein-Westf. Nr. 1043, Westdt. Verlag, Köln-Opladen.

—, (1961b) Rheologie und Strukturfestigkeit von Holz. Holz als Roh- und Werkstoff 19: 73—80.

—, (1962) Über das rheologische Verhalten von Buchenholz verschiedener Feuchtigkeit bei Druckbeanspruchung längs der Faser. Materialprüfung 4: 313—319.

Kollmann, F., (1963) Untersuchungen über den Abnutzungswiderstand von Holz, Holzwerkstoffen und Fussbodenbelägen. Holz als Roh- und Werkstoff **21**: 245—256.

—, (1964) Mechanische und technologische Eigenschaften von Spanplatten. In: Bauen mit Spanplatten, p. 25—49. Triangel Spanplattenwerke GmbH, Triangel.

—, (1966) Holzspanwerkstoffe. Springer. Verlag, Berlin, Heidelberg, New York.

—, (1972) Kriechen von Holz und Holzwerkstoffen. Holztechnologie **13**: 88—95.

—, *Côté, W. A.*, (1968) Principles of wood science and technology. Vol. I Solid wood, p. 528 to 533. Springer-Verlag, Berlin, Heidelberg, New York.

—, *Krech, H.*, (1961) Zeitfestigkeit und Dauerfestigkeit von Holzspanplatten. Holz als Roh- und Werkstoff **19**: 113—118.

—, *Malmquist, L.*, (1956) Über die Wärmeleitzahl von Holz und Holzwerkstoffen. Holz als Roh- und Werkstoff **14**: 201—204.

—, *Schneider, A.*, (1960) Der Einfluss der Belüftungsgeschwindigkeit auf die Trocknung von Schnittholz mit Heissluft-Dampfgemischen.

—, *Teichgräber, R.*, (1961) Beitrag zur Prüfung der Brandeigenschaften, insbesondere der Schwerentflammbarkeit, von Platten aus Holz aund Holzwerkstoffen. Holz als Roh- und Werkstoff, **19**: 173—186.

—, —, (1962) Abhängigkeit der Querzugfestigkeit der Spanplatten vom Anteil an Feingut. Holz als Roh- und Werkstoff **20**: 404—405.

Kratz, W., (1969) Untersuchungen über das Dauerbiegeverhalten von Holzspanplatten. Holz als Roh- und Werkstoff **27**: 380—387.

—, *May, H. A.*, (1963) Über das Schrauben- und Nagelhaltevermögen von Holzspanplatten. Möbelkultur **15**: 744—745.

Kreibaum, O., (1966) Okal-Verfahren. In: *Kollmann, F.*, Holzspanwerkstoffe, p. 519—526. Springer-Verlag, Berlin, Heidelberg, New York.

Krischer, O., *Kröll, K.*, (1963) Die wissenschaftlichen Grundlagen der Trocknungstechnik. 2nd Rev. Ed., p. 318. Springer-Verlag, Berlin, Göttingen, Heidelberg.

Kröll, K., (1950) Die Vorgänge in Trocknungs- und Erwärmungstrommeln für rieselfähige Güter. Springer-Verlag, Berlin, Göttingen, Heidelberg.

Kühlmann, G., (1962) Untersuchung der thermischen Eigenschaften von Holz und Spanplatten in Abhängigkeit von Feuchtigkeit und Temperatur im hygroskopischen Bereich. Holz als Roh- und Werkstoff **20**: 259—270.

Künzelmann, E., (1960) Über das Verhalten von Holzfaser- und Holzspanplatten gegenüber holzzerstörenden Pilzen und Feuer sowie die Möglichkeiten des Schutzes. T. I, II. Holztechnologie **1**: 44—51, 129—134.

Kuenzi, E. W., (1953) Theoretical design of a nailed or bolted joint under lateral load. U. S. Dept. Agric. For. Serv., For. Prod. Lab. Rep. No D 1951, Madison, Wisc.

Kufner, M., (1966) Entwicklung eines Verfahrens zur Prüfung des Formänderungsverhaltens von plattenförmigen Holzwerkstoffen. Holz als Roh- und Werkstoff **24**: 4—9.

—, (1968) Festigkeitswerte von Holzspanplatten und deren Schwankungen. Holz als Roh- und Werkstoff **26**: 253—260.

—, (1969) Die Prüfung des Saugvermögens von Spanplatten-Oberflächen. Holz als Roh- und Werkstoff **27**: 378—380.

—, (1970) Das Kriechen von Holzspanplatten bei langzeitiger Biegebeanspruchung. Holz als Roh- und Werkstoff **28**: 429—446.

Kull, W., (1954) Die Erwärmung von parallelflächigen Stoffen zwischen Heizplatten und die Bestimmung der Heizzeit bei der Holzverleimung, insbesondere bei der Spanplattenherstellung. Holz als Roh- und Werkstoff **12**: 413—418.

Kull, W., (1955) Der Weg zur Qualitätsspanplatte. Holz-Zbl. **81**: No 48, Messe-Sonderheft: 18—21.

—, (1966a) Zerkleinerung und Zerspanung des Holzes. In: *Kollmann, F.*, Holzspanwerkstoffe, p. 129—163. Springer-Verlag, Berlin, Heidelberg, New York.

—, (1966b) Sichtung und Sortierung. In: *Kollmann, F.*, Holzspanwerkstoffe, p. 199—209 Springer-Verlag, Berlin, Heidelberg, New York.

Kumar, V. B., (1968a) Manufacture of Chipboards from Tropical Hardwoods. In: *Mitlin, L.*, Particleboard manufacture and application, p. 41—47. Pressmedia Ltd., Ivy Hatch, Sevenoaks, Kent.

—, (1968b) Problems of chipboard manufacture in developing countries. In: *Mitlin, L.*, Particleboard manufacture and application, p. 4—6. Pressmedia Ltd., Ivy Hatch, Sevenoaks, Kent.

Kunesh, R. H., (1961) Das unelastische Verhalten von Holz. Ein neuer Weg zur verbesserten Plattenformung. For. Prod. J. **11**: 395—406.

Langwig, J. E., Meyer, J. A., Davidson, R. W., (1968) Influence of polymer impregnation on mechanical properties of basswood. For. Prod. J. **18**: 33—36.

—, —, —, (1969) New monomers used in making wood-plastics. For. Prod. J. **19**: 57—61.

Lawniczak, M., Nowak, K., (1962) Der Einfluß hydrophobierender Imprägniermittel auf feuchtigkeitsbedingte Formänderungen der Span- und Flachschäbenplatten. Holz als Roh- und Werkstoff **20**: 68—72.

Levon, M., (1931) The wood waste in the sawmill industry and its utilization. Komex Inst. Quaest. Forest. Finnl. Ed. 16, Helsinki.

Liiri, O., (1960) Investigations on the effect of moisture and wax upon the properties of wood particle board. Paperi ja Puu **42**: 43—56.

—, (1961) Investigations on properties of wood particle boards. Paperi ja Puu **43**: 3—18.

—, *Haavisto, S.*, (1961) Some questions associated with the use of wood particle board. Paperi ja Puu **43**: 219—230.

Lindner, J., (1961) Werkzeuge für die Bearbeitung von plattenförmigen Werkstoffen. Holz als Roh- und Werkstoff **19**: 150—155.

Love, A. E. H., (1927) A treatise on the mathematical theory of elasticity. 4th Ed. Cambridge University Press, New York.

Lubkin, J. L., (1957) A status report on research in the circular sawing of wood, Vol. I. Central Res. Lab., American Machine and Foundry, Greenwich, Conn.

Lynam, F. C., (1959) Factors influencing the properties of wood chipboard. J. Inst. of Wood Sci. **4**: 14—27.

—, (1968) Factors influencing the properties of wood chipboard, p. 28—41. In: *Mitlin, L.*, Particleboard manufacture and application. Pressmedia Ltd., Ivy Hatch, Sevenoaks, Kent.

Mäkinen, A., (1968) Oberflächenimprägnierung von Holz und Holzwerkstoffen für Bauzwecke. Paperi ja Puu **50**: 669—675.

Marian, J. E., (1958) Adhesive and adhesion problems in particle board production. For. Prod. J. **8**: 172—176.

—, *Suchsland, O.*, (1956) Waviness of lumber and chipcore board. Paper and Timber No 6—7.

Marin, J., (1965) Mechanical relationships in testing for mechanical properties of polymers. In: *Schmitz, J. V.*, Testing of polymers, p. 87—146. John Wiley & Sons, New York, London.

Markwardt, L. J., (1930) Aircraft woods: their properties, selection and characteristics. Nat. Advisory Comm. Aeron. Rep. No 354.

Marra, G. G., (1960) Binders for particle boards. Symposium on adhesives for the wood industry. U. S. For. Prod. Lab., Rep. No 2183.

Meinecke, E., (1960) Über die physikalischen und technischen Vorgänge bei der Beleimung und Verleimung und der Holzspäne bei der Holzspanplattenherstellung. Diss. T.H. Braunschweig.

Meyer, J. A., Loos, W. E., (1969) Processes of, and products from, treating southern pine wood for modification of properties. For. Prod. J. **19**: 32—36.

Miller, H. C. L., (1953) Chipcore — its characteristics and production. For. Prod. J. **3**: 149 to 152.

Mitlin, L., (1968) Particleboard manufacture and application. Pressmedia Ltd., Ivy Hatch, Sevenoaks, Kent.

Möhler, K., Ehlbeck, J., (1968) Versuche über das Dauerstandverhalten von Spanplatten und Furnierplatten bei Biegebeanspruchung. Holz als Roh- und Werkstoff **26**: 118—124.

Mörath, E., (1932) Studien über die hygroskopischen Eigenschaften und die Härte der Hölzer. Hannover.

—, (1966a) Bindemittel. In: *Kollmann, F.*, Holzspanwerkstoffe, p. 64—68. Springer-Verlag, Berlin, Heidelberg, New York.

—, (1966b) Sonstige Flachpressverfahren. In: *Kollmann, F.*, Holzspanwerkstoffe, p. 510 to 519. Springer-Verlag, Berlin, Heidelberg, New York.

Moralt, A., (1959) BP 1097656.

—, (1961) BP 1453397.

Müller, H., (1962) Erfahrungen mit Paraffin-Emulsionen als Quellschutzmittel in der Spanplattenindustrie. Holz als Roh- und Werkstoff **20**: 434—437.

Müller, O., (1930) Holzblech — seine spanlose Formung zu Hohlkörpern. Diss. T. H. Dresden. Herm. Prinz, Bückeburg, W.-Germany.

Müller, R. K., (1966) Das Verhalten einiger Kunststoffe bei periodischer Be- und Entlastung. VDI-Ber. No 103: 85—88. Düsseldorf.

Munk, E. E., (1967) The throw-away package or lost container. Werzalit-Information, p. 29 to 38. Oberstenfeld, West Germany.

Nakamura, G., (1966) Studies on wood sanding by belt sander. II. Industrial test of plywood sanding by wide belt sander. Tokyo Univ., Agric. and Technol. Exp. Stat., For. Bull. **5**: 17.

Nedbal, F., (1961) Beitrag zum Prüfen des Schrauben- und Nagelhaltevermögens von Spanplatten. Drevársky Výskum **6**: 19—24.

Neumann, C., (1970) Über die Möglichkeiten der Herstellung und Verwendung von Spanplatten in Entwicklungsländern. Diss. Univ. Hamburg.

Neusser, H., (1960) Spanplattenprüfung in Versuchsanstalt und Betriebslaboratorium. Materialprüfung **2**: 295—300.

—, (1962) Über die Veränderung des Leimes während des Produktionsvorganges in Spanplatten, bzw. über einige Einflussfaktoren auf die Verleimungsqualität. Holzforschung u. Holzverwertung **14**: 88—96.

—, (1963) Die Oberflächenqualität von Spanplatten, ihre Ursachen und ihre Prüfung. Holzforschung, Holzverwertung **15**: 83—88.

—, *Krames, U.*, *Zentner, M.*, (1969) Spanplatten mit verschiedenen Deckschichtausführungen als Trägermaterial für Beschichtungen verschiedener Art. Holzforschung und Holzverwertung **21**: 56—61.

Nicholls, W. H., (1968) Phenolic resin binders. In: *Mitlin, L.*, Particleboard manufacture and application, p. 25—27. Pressmedia Ltd., Ivy Hatch, Sevenoaks, Kent.

Noack, D., (1966) Statistische Qualitätskontrolle. In: *Kollmann, F.*, Holzspanwerkstoffe. Springer-Verlag, Berlin, Heidelberg, New York.

Nördlinger, H., (1881) Anatomische Merkmale in wichtigsten deutschen Wald- und Gartenholzarten. Stuttgart.

Oertel, J., (1967) Untersuchungen über Kriechverhalten, Spannungsrelaxation und Quellungsdruckspannung an Holzspanplatten. Holztechnologie **8**: 119—125.

Özen, R., (1971) Formstabilität einschichtiger Holzspanplatten, hergestellt mit verschiedenen Bindemitteln. Diss. Univ. München.

Pahlitzsch, G., *Dziobek, K.*, (1959) Untersuchungen über das Bandschleifen von Holz mit geradliniger Schnittbewegung. Holz als Roh- und Werkstoff **17**: 121—134.

—, —, (1961) Über das Wesen der Abstumpfung von Schleifbändern beim Bandschleifen von Holz. Holz als Roh- und Werkstoff **19**: 136—149.

—, *Jostmeier, H.*, (1964 a) Beobachtungen über das Abstumpfungsverhalten beim Fräsen von Spanplatten. Holz als Roh- und Werkstoff **22**: 139—146.

—, —, (1964 b) Weitere Beobachtungen über das Abstumpfungsverhalten und den Einfluss der Schnittgeschwindigkeit beim Fräsen von Spanplatten. Holz als Roh- und Werkstoff **22**: 424—429.

—, *Mehrdorf, J.*, (1962) Herstellen von Schneidspänen mit Flachscheiben-Spanern. 1. Mitt.: Einfluss von Spanungsdicke und Holzfeuchtigkeit auf die Erzeugung von Holzspänen. Holz als Roh- und Werkstoff **20**: 314—322.

Patzak, W., (1972) Zur Theorie des Brandgeschehens von Holz. VDI-Forschungsheft No 552. VDI-Verlag, Düsseldorf.

Perkitny, T., (1962) Beiträge zur Ermittlung der Qualität von Spanplatten. Holztechnologie **3**: 64—70.

—, *Perkitny, J.*, (1966) Vergleichende Untersuchungen über die Verformungen von Holz, Span- und Faserplatten bei langdauernder konstanter Biegebelastung. Holztechnologie **7**: 265—270.

—, *Szymankiewicz, H.*, (1962) Hygroscopic balance of chip boards. Przemysl Drzewny **13**: 6—8.

Perry, Th. D., (1944) Moulded plywood. Wood **9**: 107—113.

—, (1947) Modern plywood. Sir Isaac Pitman & Sons Ltd., London.

Pfohl, F., (1936) CSP 56350.

—, (1937) Swiss Pat. 193139. Platte für Möbel- und Bautischler und Verfahren zu ihrer Herstellung.

—, (1940) DRP 692159.

Plath, E., (1963 a) Einfluss der Rohdichte auf die Eigenschaften von Holzwerkstoffen. Holz als Roh- und Werkstoff **21**: 104—108.

—, (1963 b) Die Betriebskontrolle in der Spanplattenindustrie. Springer-Verlag, Berlin, Göttingen, Heidelberg.

—, (1968) Quality control in the German chipboard industry. In: *Mitlin, L.*, Particleboard manufacture and application, p. 111—114. Pressmedia Ltd., Ivy Hatch, Sevenoaks, Kent.

—, (1971 a) Beitrag zur Mechanik der Holzspanplatten. Holz als Roh- und Werkstoff **29**: 377—382.

—, (1971 b) Auswertung der Betriebskontrolle in Spanplattenwerken. Holz als Roh- und Werkstoff **29**: 393—396.

—, (1972) Kennzeichnung der elastomechanischen Eigenschaften von Holzwerkstoffen. Holz als Roh- und Werkstoff **30**: 91—94.

—, (1972) Holzwerkstoffe. In: Wendehorst, Baustoffkunde, Curt E. Vincentz, Hannover, p.150.

Plath, L., (1966) Bestimmung der Formaldehyd-Abspaltung aus Spanplatten nach der Mikro-diffusions-Methode. Holz als Roh- und Werkstoff **24**: 213—318.

—, (1967a) Einfluss von Presszeit und Presstemperatur auf die Formaldehyd-Abspaltung. Holz als Roh- und Werkstoff **25**: 63—67.

—, (1967b) Einfluß der Härter-Zusammensetzung auf die Formaldehyd-Abspaltung. Holz als Roh- und Werkstoff **25**: 169—173.

—, (1967c) Einfluß der Feuchtigkeit im Spanvlies auf die Formaldehyd-Abspaltung. Holz als Roh- und Werkstoff **25**: 231—238.

—, (1968a) Versuche über die Formaldehyd-Abspaltung bei Spanplatten. Holz-Zbl. **94**: 1487.

—, (1968b) Einfluß der Härtungsbeschleunigung und Reifezeit auf die Formaldehyd-Abspaltung. Holz als Roh- und Werkstoff **26**: 125—128.

—, (1971) Anforderungen an Spanplatten für die Beschichtung mit Kunststoffen. Holz als Roh- und Werkstoff **29**: 369—376.

Polovtseff, B., (1961) Practical aspects of wood chipboard densification patterns. Wood Chipboard. The Airscrew Co. and Jicwood Ltd., Weybridge.

Post, P. W., (1958) Effect of particle geometry and resin content on bending strength of oak flake board. For. Prod. J. **8**: 317—322.

—, (1961) Relation of flake size and resin content to mechanical and dimensional properties of flakeboard. For. Prod. J. **11**: 34—37.

Pound, J., (1968) Particleboard lipping. In: *Mitlin, L.*, Particleboard manufacture and application, p. 125—129. Pressmedia Ltd., Ivy Hatch, Sevenoaks, Kent.

Prokes, S., (1961) Einfluß der Ausgangsqualität von Schneiden aus Werkzeugstahl und aus gesintertem Hartmetall auf die Abstumpfung. Holztechnologie **2**: 234—238.

Proksch, E., (1969) Die Herstellung von Polymerholz aus Buchenholz. Holzforschung **23**: 93—98.

Pungs, L., Lambertz, K., (1957) The application of high frequency heating in the particle board industry. FAO/ECE Board Consultation, Paper 5.36.

Rackwitz, G., (1954) Ein Beitrag zur Kenntnis der Vorgänge bei der Verleimung von Holzspänen zu Holzspanplatten in beheizten hydraulischen Pressen. Diss. TH Braunschweig.

—, (1955) Zur Kenntnis der Vorgänge bei der Verleimung von Spanplatten in beheizten Pressen. Deutsche Ges. f. Holzforsch., Ber. 2/1955: 10—12.

—, *Obermaier, M.*, (1962) Die Sichtung von Holzspänen. 1. Mitt.: Grundlagen der Sichtung und die Sichtung im waagerechten Luftstrom. Holz als Roh- und Werkstoff **20**: 27—38.

Rayner, C. A. A., (1965) Synthetic organic adhesives. In: *Houwink, R., Salomon, G.*, Adhesion and adhesives, 2nd Rev. Ed., Vol. I: Adhesives, p. 186—353. Elsevier Publ. Co., Amsterdam, London, New York.

—, (1966) Resin adhesives in wood chipboard — some comparative features. Wood **31**: 41.

—, (1968) Amine formaldehyde resins. In: *Mitlin, L.*, Particle board manufacture and application, p. 11—25, Pressmedia Ltd., Ivy Hatch, Sevenoaks, Kent.

Reiner, M., (1958) Rheologie. In: Handbuch d. Physik, VI. Elastizität und Plastizität Springer-Verlag, Berlin, Göttingen, Heidelberg.

—, (1968) Rheologie in elementarer Darstellung. Carl Hanser-Verlag, München.

Rinkefeil, R., (1956) Oberflächenprüfung in der Holzindustrie. Holzindustrie **9**: 62—67.

Ross Engineering, I. O., (1960) New panel finishing line. Wood and Wood Products **65**: 32—35.

Runkel, R. O. H., (1951) Zur Kenntnis des thermoplastischen Verhaltens von Holz. 1. Mitt. Holz als Roh- und Werkstoff **9**: 41—53.

—, *Jost, J.*, (1948) DBP 841055. Thermodyn-Verfahren.

Sachs, L., (1969) Statistische Auswertungsmethoden. Springer-Verlag, Berlin, Heidelberg, New York.

de Samsonow, M. A., (1929) FP 679708.

Sarkanen, K. V., Ludwig, C. H., (1971) Lignins. Wiley Interscience, New York.

Satow, T., (1930) USP 2007585.

Sauer, D. J., Haygreen, J. G., (1963) Effects of sorption on the flexural creep behavior of hardboard. For. Prod. J. **18**: 57—63.

Schaeffer, R. E., (1970) Cure rate of resorcinol and phenol-resorcinol adhesives in joints of ammonium salt-treated southern pine. U.S.D.A. For. Serv. Research Paper 121, For. Prod. Laboratory, Madison, Wisc.

Schaffrath, H., (1957) Elektrisches Metallsuchgerät für die Spanplattenindustrie. Holz als Roh- und Werkstoff **15**: 146—147.

Scheibert, W., (1958) Spanplatten. Herstellung, Verarbeitung, Anwendung. VEB Fachbuchverlag, Leipzig.

Schmidt, A. X., Marlies, Ch. A., (1948) Principles of high-polymer theory and practice. McGraw Hill Book Company Inc., New York, Toronto, London.

Schmidt, H., (1953) Studien an Holzwerkstoffen in der „Termitenprüfung". 1. Mitt.: Eigenschaften und Bewertung der Versuchstermiten (*Reticulitermes*). Holz als Roh- und Werkstoff **11**: 385—388.

—, (1954) Studien an Holzwerkstoffen in der „Termitenprüfung". 2. Mitt.: Formen des Termitenangriffs an Furnier-, Span- und Faserplatten. Holz als Roh- und Werkstoff **12**: 44—46.

—, (1955) Bemerkungen zur Methodik von Termitenversuchen an Holzerzeugnissen. Z. Weltforstwirtsch. **18**: 222—224.

—, (1960) Ein Termitentest an Sägespänen verschiedener Holzarten. Holz als Roh- und Werkstoff. **18**: 325—328.

—, *Nehm, A. B.*, (1972) Holzspanplatten im Termitentest. Holz als Roh- und Werkstoff **30**: 174—177.

Schmitz, J. V., (1965) Testing of polymers. Interscience Publishers, New York, London, Sydney.

Schmutz, Mfg. Co., (1964) Description of the printing process, System Schmutz (combination of offset and rotogravure printing). Schmutz Mfg. Co. Inc., Louisville.

Schnitzler, E., (1971) Neue Techniken der Spänebeleimung. Holz als Roh- und Werkstoff **29**: 382—389.

Schwab, E., (1969) Flammenausbreitung auf Oberflächen von Holzwerkstoffen. Mittlg. d. Deutsch. Ges. f. Holzforsch. No. 56, p. 18—22; Fig. 4.

Seifert, L., (1942) Untersuchungen über die Termitenfestigkeit tropischer Nutzhölzer. Kolonialforstl. Mitt. **5**: 438—439.

Seto, K., *Nozaki, K.*, (1966) On the machining of plywood surface with coated abrasives; the performance of drum sander und drum type wide belt sander. Hokkaido For. Prod. Res. Inst. Rep. 49.

Sharma, M. G., (1965) Theories of phenomenological viscoelasticity underlying mechanical testing. In: *Schmitz, J. V.*, Testing of polymers, p. 147—199. Interscience Publ., New York, London, Sydney.

Shen, K. C., (1970a) Correlation between internal bond and the shear strength measured by twisting thin plates of particleboard. For. Prod. J. **20**: 16—20.

—, (1970b) Correlation between torsion-shear strength and modulus of rupture of particleboard. For. Prod. J. **20**: 32—36.

—, (1971) Evaluating particleboard properties by measuring saw-cutting force. For. Prod. J. **21**: 47—55.

—, *Carroll, M. N.*, (1969) A new method for evaluating the internal strength of particleboard. For. Prof. J. **19**: 17—22.

—, —, (1970) Measurement of layer strength distribution in particleboard. For. Prod. J. **20**: 53—55.

—, —, *Wrangham, B.*, (1968) Study on compression shear strength and relationship to internal bond properties of particleboard. Can. Dept. Fish. Forest, For. Prod. Lab., Ottawa. Inf. Report.

—, *Wrangham, B.*, (1971) A rapid accelerated aging test procedure for phenolic particleboard. For. Prod. J. **21**: 30—33.

Simonds, H. R., *Weith, A. J.*, *Bigelow, M. H.*, (1949) Handbook of plastics. 2nd Ed., p. 637 to 645, 926—930. Van Nostrand Co. Inc., Toronto, New York, London.

Smith, A., (1968) Machining of board materials. In: *Mitlin*, Particleboard manufacture and application, p. 121—124. Pressmedia Ltd., Ivy Hatch, Sevenoaks, Kent.

Smith, I. W., (1948) Mechanical methods of bark removal. For. Prod. Res. Soc., Proc. Nat. Ann. Meet., 1948, p. 119—129.

Sorg, K. A., (1959) Der Einsatz von Metallsuchgeräten in der Holzindustrie. Holz als Roh- und Werkstoff **17**: 397—402.

Steck, E. F., (1970) Caulless pressing systems for the manufacture of particleboard. Fourth Washington State University Symposium on Particleboard, Pullman, Wash.

Stegmann, G., (1958) Entwicklung und Herstellung von Holzspanplatten. Holz als Roh- und Werkstoff **16**: 360—362.

—, *v. Bismarck, C.*, (1967) Zur Preßzeitverkürzung bei der Herstellung harnstoffharzgebundener Spanplatten. 1. Mitt.: Einfluß spezieller Preßbedingungen als Voraussetzung für die Anwendung kurzer Preßzeiten. Holzforschung u. Holzverwertung **19**: 53—60.

—, *Storck, W.*, (1963) Die Verwertung von schwachem Waldholz und Industrierestholz für die Herstellung von Span- und Faserplatten sowie Zellstoff und Holzschliff. Holz-Zbl. **89**: 2547—2548, 2550—2553.

Steiner, K., (1954a) Über Verfahrenstechnik und Maschinen zur Herstellung hochwertiger Holzspanplatten. Holz als Roh- und Werkstoff **12**: 343—348.

—, (1954b) Zerspanungsmaschinen für die Holzspanplattenfabrikation. Holzwirtschaftl. Jahrbuch Nr. 6/7, Holz-Zentralblatt-Verlags GmbH, Stuttgart.

Steiner, K., (1966a) Naß- und Trockensilierung. In: *Kollmann, F.,* Holzspanwerkstoffe, p. 217—239. Springer-Verlag, Berlin, Heidelberg, New York.

—, (1966b) Förderung der Späne. In: *Kollmann, F.,* Holzspanwerkstoffe, p. 239—280. Springer-Verlag, Berlin, Heidelberg, New York.

—, (1966c) Einstreuen und Formen. In: *Kollmann, F.,* Holzspanwerkstoffe, p. 303—332. Springer-Verlag, Berlin, Heidelberg, New York.

Stern, E. G., (1951) Nails and screws in wood assembly and construction. Virgina Polytechnic Institute, Wood Res. Laboratory, Blacksburg, Va.

—, (1967) Nails — definitions and sizes, a handbook for nail users. Virginia Polytechn. Inst., Wood Res. Laboratory, Bull. No 61. Blacksburg, Va.

Stevens, W. C., Turner, N., (1948) Solid and laminated wood bending. H. M. Stationery Office, London.

Stewart, H. A., (1970) Abrasive vs. knife planing. For. Prod. J. **20**: 43—47.

Stofko, J., (1960) Abhängigkeit der mechanischen Eigenschaften beim Spanholz von den geometrischen Spanformen. Drevársky Výskum **5**: 241.

Strickler, M. D., (1959) Einfluß der Preßbedingungen und der Holzfeuchtigkeit auf die Eigenschaften von Spanplatten aus Douglasienholz. For. Prod. J. **9**: 203—215.

Svarcman, G. M., (1968) Vergleich der Festigkeit von Holzspanplatten mit verschiedenem Aufbau. Derev. Prom. **17**: 5—8.

Swiderski, J., (1960) Zur Technologie der Flachsspanplatten-Erzeugung. Holz als Roh- und Werkstoff **18**: 242—250.

Swiss Patent 193139, (1937) Pfohl.

Taylor, J. P., (1944) Fabricating wood aircraft "skins". Electronics, Apr. 1944: 102—108.

Teichgräber, R., (1958) Eigenschaften von Holzspanplatten. Rund um die Spanplatte. Sdh.: Eigenschaften, Verarbeitung. Holz-Verlag E. Kittel, Mering.

—, (1966) Eigenschaften und Eigenschaftsprüfung. In: *Kollmann,* Holzspanwerkstoffe, p. 530—579. Springer-Verlag, Berlin, Heidelberg, New York.

v. Thielmann, C. A., Munz, W., (1960) Handbuch der Spanplattenverarbeitung. Holz-Verlag, E. Kittel, Mering.

Thomas, P. H., (1972) The development of tests for measuring "reaction to fire". ISO/TC 92, Working Group 4. ISO International Organization for Standardization.

Thunell, B., Perem, E., (1948) Undersökning av avnötningen hos olika trägolv. Svenska Träforskningsinstitutet, Trätekniska Avd., Medd. 19, Stockholm.

Timoshenko, S., (1945) Strength of materials. Vol. II. Van Nostrand Co. Inc., Toronto, New York, London.

—, *MacCullough, G. H.,* (1949) Elements of strength of materials. 3rd Ed. Van Nostrand Co. Inc., Toronto, New York, London.

Tobolsky, A. V., (1960) Properties and structure of polymers. Wiley, New York.

Treiber, H., (1965) Die räumliche Anpassung von hochpolymeren Stoffen, insbesondere von Holzspangemischen. Holz als Roh- und Werkstoff **23**: 319—331.

Trendelenburg, R., (1939) Das Holz als Rohstoff. J. F. Lehmann, München.

Triangel Holzwerkstoff GmbH, Triangel (1947/1962) DBP 975846. Holzspanbauteil und Vorrichtung zur Erzeugung seiner Späne.

Trutter, G., Himmelheber, M., (1970) Die moderne Spanplattenfertigung. Stand der Technik bei den Fertigungsverfahren. Holz als Roh- und Werkstoff **28**: 85—101.

Turner, H. D., Kern, J. D., (1950) Relation of several formation variables to properties of phenolic-resin bonded wood waste hardboard. U. S. For. Prod. Laboratory, Rep. No R 1786. Madison, Wisc.

U. S. For. Prod. Laboratory, (1955) Wood handbook. U. S. Dept. Agric. Handbook No 72. Washington, D. C.

USP 796545, (1901) Watson.

USP 2007558, (1930) Satow.

USP 2033411, (1932) Carson.

USP 2737997, (1953) Himmelheber, Steiner and Kull.

Voigt, W., (1928) Lehrbuch der Kristallphysik. Teubner, Leipzig, Berlin.

Vorreiter, L., (1943) Handbuch der Holzabfallwirtschaft. 2nd Ed., Neudamm.

Walter, F., (1960) Untersuchungen zur Dickenquellung bei Spanplatten. Holztechnologie **1**: 67—72.

—, *Deppe, H.-J.,* (1960) Die Formbeständigkeit der Oberflächen von Spanplatten. Holztechnologie **1**: 79—82.

—, *Rinkefeil, R.,* (1960) Ein Beitrag zur Bestimmung der Formbeständigkeit von Holzwerkstoffplatten. Holztechnolgie **1**: 159—163.

—, *Wiechmann, H.,* (1961) Dichteuntersuchungen an Faser- und Spanplatten. Holztechnologie **2**: 172—178.

Wangaard, F. F., (1950) The mechanical properties of wood. John Wiley & Sons Inc., New York; Chapman & Hall Ltd., London.

Ward, D., (1963) Abrasive planing challenges your knife cutting techniques. Hitchcock's Wood Working Dig. **65**: 29—32.

Watson, H. F., (1901) USP 796545.

Wehner, E., (1966) Fertigbearbeitung. In: *Kollmann, F.*, Holzspanwerkstoffe, p. 372—388. Springer-Verlag, Berlin, Heidelberg, New York.

Wellisch, E., Marker, L., Sweeting, O. J., (1961) Viscoelastic properties of regenerated cellulose sheet. J. Appl. Polymer Sci. **5**: 647—657.

Werz, K. G. Jr., (1964, 1965, 1966) Werzalit-Information 1964, 1965, 1966. J. F. Werz KG, Oberstenfeld, W.-Germany.

Wilke, K.-D., (1951) Untersuchungen über die thermische Plastifizierung von Holz. Diss. Univ. Hamburg.

Willeitner, H., (1969) Über die Laboratoriumsprüfung von Holzspanplatten gegen Pilzbefall. Material und Organismen, Beih. **2**: 109—122.

Winter, H., (1949) Eigenschaften, Prüfung und Verwendung von plattenförmigen Holz-halbzeugen unter besonderer Berücksichtigung der Holzfaser- und Holzspanplatten. Mitt. Deutsche Ges. f. Holzforschg., H. 37, p. 133, Stuttgart.

—, *Frenz, W.*, (1954) Ein Beitrag zu den Prüfverfahren für die Kennzeichnung der Eigenschaften von Holzspanplatten. Holz als Roh- und Werkstoff **12**: 348—357.

—, *Heyer, S.*, (1957) Untersuchungen zur Entwicklung von Prüfverfahren für die Kennzeichnung der Eigenschaften von Holzspanplatten: Schlagversuche an Platten. Holz als Roh- und Werkstoff **15**: 51—58.

Wissing, A., (1955) Spanskivor, board, plywood och lamelträ — fyra foradlade trämaterial. Svenska Träforsknings-Institutet, Träteknik, Stockholm.

—, (1959) Erfarenheter från provning av spånskivor. Svenska Träforsknings Institutet, Trätekn. Medd. 110 B: 85.

Wittington, J. A., Walters, C. S., (1969) Withdrawal loads for screws in soft maple and particleboard. For. Prod. J. **19**: 39—42.

Wittmann, O., (1971) Zur Hydrophobierung von Spanplatten. Holz als Roh- und Werkstoff **29**: 259—264.

Wolcott, G. N., (1946a) What to do about *polilla*? Univ. Puerto Rico, Agric. Exp. Station, Bull. 68.

—, (1946b) Factors in the natural resistance of woods to termite attack. Caribb. Forester **7**: 121—124.

Wolgast, W., (1963) Schwerentflammbare Baustoffe. Z. d. Vereinig. z. Förderung d. Dt. Brandschutzes e. V. H. **2**: 1—11.

Wood, A. D., (1963) Plywoods of the world, their development, manufacture and application. W. A. K. Johnston and G. W. Bacon Ltd., Edinburgh, London.

Woodson, G. E., McMillin, C. W., (1971) Machine boring of southern pine wood. III. Effect of six variables on torque, thrust and hole quality. U.S.D.A. Forest Service, Southern For. Exp. Station, Final Report. Alexandria, La.

Wyss, O., (1957) Einige Wirtschaftlichkeitsfragen der Spanplattenindustrie. Holz als Roh- und Werkstoff **15**: 58—61.

Ylinen, A., (1943) Über den Einfluß der Rohwichte und des Spätholzanteils auf die Brinellhärte des Holzes. Holz als Roh- und Werkstoff **6**: 125—127.

Zieger, E., (1960) Technologie der Holzentrindung. VEB Fachbuchverlag, Leipzig.

6. FIBERBOARD

6.0 Types of Fiberboard

There are various types of fiberboard[1], and unfortunately the terminology is not yet internationally standardized. Neverthelesss, one fundamental definition generally is recognized (FAO, 1958/1959, p. 4): "Fiberboard is a board generic term encompassing sheet materials of widely varying densities manufactured from refined or partially refined wood fibres or other vegetable fibres. Bonding agents and other materials may be incorporated in the manufacture of the board to increase strength, resistance to moisture, fire or decay or to improve some other property." In the technical sense the ISO-definition is more precise:

"Sheet material generally exceeding 1.5 mm in thickness, manufactured from ligno-cellulosic fibers with the primary bond from the felting of the fibers and their inherent adhesive properties. Bonding materials and/or additives may be added".

FAO, OEEC and ISO used formally various definitions but the differences between them are without practical importance.

The classification of fiberboard into types is based on:

a) Type of *raw material* and method of fiber production;
b) Method of *sheet formation*;
c) *Density* of product (kg/m³ or g/cm³ or lb./cu. ft.);
d) Kind and place of *application*.

Perhaps, the best factor for classifying fiberboard is the density. This is internationally recognized. There is a rather simple difference between pressed and not pressed sheets but the range of qualities is wide and there is an overlapping. Another point which should be taken into consideration is the fact that wood fibers are blended occasionally with mineral fibers (such as asbestos), plastics and with other chemical additives. Some half-hard fiberboard contain no less than 20% of thermoplastic bonding agents.

The general term "fiberboard" is not only adopted for use in the publications of FAO and ECE, but is generally understood in the literature and in the industry. FAO (1958/1959) gives a survey on the nomenclature.

Fiberboard is manufactured from separated fibers or bundles of them. This is a fundamental difference between fiberboard and particleboard. Fiberboard is produced by interfelting of fibers in such a way as to produce a mat or sheet. The properties of the mat depend on the characteristic natural bonding of fibers between themselves. Chemicals may be added during manufacture with the aim to improve the cohesion and the water resistance. Typical of fiberboard is the wide range in densities between 0.02 and about 1.45 g/cm³ ($\sim$1.25 and 90 lb/cu. ft.).

---

[1] Literature: *Segring* (1947), Voith Maschinenfabrik, Heidenheim, W.-Germany (1948), Defibrator AB, Stockholm (1952), *Stamm* and *Harris* (1953), *Kollmann* (1955), *Neusser* (1957), *Kaila* (1958).

Fiberboard are used mainly as panels for insulation and as covering materials in buildings and constructions, where flat sheets are necessary and a moderate strength is needed. Fiberboard are also used as parts in doors, cabinets, cupboards, furniture, etc.

Fiberboards may be classified according to their density ranges into five types as shown in Table 6.1. The limits to reach are approximate. Classifications may vary from one country to another. These types and their ranges are, however, suitable for general understanding and discussion of the subject.

Table 6.1. Classification of Fiberboard According to Density

Fiberboard	Density	
	g/cm³	lb./cu. ft.
Non-compressed		
Semi-rigid insulation board	0.02···0.15	1.25··· 9.5
Rigid insulation board	0.15···0.40	9.5 ···25
Compressed		
Intermediate or medium density fiberboard (half-hard)	0.40···0.80	25···50
Hardboard	0.80···1.20	50···75
Special densified hardboard	1.20···1.45	75···90

In the ISO-recommendation 818 (Fiber building board, Sept. 1968; definition and classification, 1973) the following figures are agreed upon:

Type of board	soft	medium		hard
Density g/cm³	≤0.35	>0.35	≤0.80	>0.80

6.1 History and Development

The origin of wood fiberboard goes back to Japan where as early as in the 6th century B. C. heavy papers were used for the construction of walls for small houses. In Europe a patent was granted to the British inventor *Clay* in 1772 for the application of "papier maché" (*Neusser*, 1957) not only for use in dwellings, furniture, doors, but also for carriages. The idea of using the new material for big stiff building elements was evident. Since the middle of the 19th century, the proposals for use of fiberboards have greatly increased. More than 200 patents were issued in this field between 1858 and 1928 (*Rossmann*, 1928), more than 600 patents until 1957 (*Neusser*, 1957).

In spite of these early developments the actual fiberboard industry started near the beginning of the 20th century in England and in the U.S.A. The development up to 1926 was rather sporadic and without a remarkable increase in capacity. The historical development of fiberboard industry is shown condensed in Tables 6.2 and 6.3. There may be some uncertainties or even some gaps in the summaries, but in general they show the sequence of the development.

Table 6.2. Summary of the Historical Development of the Fiberboard Industry.
Kaila (1968), *Kollmann* (1934, 1955), *Lampert* (1967), *Neusser* (1957) and authors separately
quoted

1898	First plant established for the production of halfhard fiberboard using a paperboard machine with four cylinders. "The Patent Imperable Millboard Co.", built by *Sutherland* in Sundbury-on-Thames, England. Still operating according to the same method and with similar machines producing laminated half-hard board ("Sundeal "using waste-paper (*Kaila*, 1968, p. 821);
1901	Production of insulation board for building purposes in Minnesota, U.S.A. (*Lampert*, 1967, p. 18);
1908	Erection of a corresponding plant to that in Sundbury-on-Thames by *Sutherland j.* in Trenton, N.Y., U.S.A. (*Kaila*, 1968, p. 821);
1909	Production of Tentest insulation board in Thorold, Ont., Canada (*Kaila*, 1968, p. 821);
1914	Erection a of pilot plant for the production of insulation board on the basis of groundwood wastes (Minnesota & Ontario Paper Co.) in International Falls, U.S.A. (*Kaila*, 1968, p. 822);
1914	The Leikam-Josephstadt AG in Vienna, Austria, applied for a patent covering the air-transport of fibers and in 1918 the mat-forming of dry fibers using a kind of litter-strewing machine is described.
1916	Installation of Insulite Co. plant, equipped with Bauer-mills, in Minnesota, U. S. A., for production of thick insulation board from coarse ground fibers (*Kaila*, 1968, p. 822);
1921	Installation of Celotex plant, using bagasse as a raw material for insulation board, in Marrero, Louis., U.S.A. (proposal by *B. Dahlberg*, managing director of Minnesota & Ontario Paper Co.) (*Kaila*, 1968, p. 822);
1926	Founding of the Masonite Co. with a plant in Laurel, Miss., U.S.A. *Mason* (a collaborator of *Edison*) improved the invention by *Lyman* (1858) to separate wood fibers by the expansion of hot water, steam or compressed air. Wood wastes were used as raw material. Steam explosion converted wood chips into fibers, which were hot-pressed to hardboard of high quality without addition of artificial binding agents;
1929	Establishment of Masonite AB, hardboard plant, by Nordmalings Ångsåg AB in Nordmaling, Sweden;
1930	Started Midnäs AB, hardboard plant in Sweden using ground wood according to engineer *Johnson*;
1931	Invention of *Asplund*, Sweden, to continuously defibrate wood chips under pressure and steam (170 to 175 °C, corresponding to 338 to 347 °F); first mill opened in 1932 at Johannedal by Svenska Cellulosa AB;
1932	Installation of the first fiberboard-plants in Germany: Großsärchen and Gutenberg;
1938	Simultaneous experiments conducted in Europe and in the U. S. A. to convert conventionally produced wet process insulation board by pressure into hardboard;
1943	Development by Weyerhaeuser Timber Co. and 1945 Plywood Research Foundation on an industrial scale of transport of fibers and mat-forming by air. The ideas for dry and semi-dry methods in the U.S.A. came from *Heritage, Evans* and *Neiler* (*Kaila*, 1954, 1958, *Sandermann* and *Künnemeyer*, 1956, 1957, *Nuolivaara*, 1961, *Swiderski*, 1963);
1951	Establishment of Anacores Veneer Inc. plant, based on method of Plywood Research Foundation, for production of hardboard on a semi-dry process; Gebrüder Künnemeyer in Horn-Lippe, W.-Germany, established a plant for semi-dry process, original capacity 90 metric tons per day (*Sandermann* and *Künnemeyer*, 1956, 1957);
1952	Weyerhaeuser Timber Co. completed pilot plant for hardboard production in dry process;
1959	Pilot plant for the semi-dry manufacture of fiberboard (annual capacity 5,000 metric tons) in Pravenec, ČSSR (*Swiderski*, 1965);
1960	Plant of Isorel in St. Dizier, France, projected capacity 300 metric tons per day (Siempelkamp & Co., 1961);
1961	Hardboard factory, 10,000 metric tons per year, started in Japan (*Lehotsky* and *Nagy*, 1963).

Table 6.4. Comparison of Wet, Semi-Dry and Dry Hardboard Processes (*Kaila*, 1968, p. 1049)

Consumption per metric ton of product	Wet process Scandinavia	Semi-dry process U.S.A.	Dry process U.S.A., Japan, France
Raw water (cu. meters)	10··· 25 hardboard 10··· 20 insulation board	10	7
Steam (metric tons)	2.8···3.8 hard board 4.0···4.5 insulation board	2.5···3.0 } hard board	2.6···3.5 } hardboard
Electric power (kWh)	380···450 hardboard 550···650 insulation board	500···600	500···600
Yield (%)	65··· 82 hardboard 90··· 95 insulation board	85	85
Waste water and pollution	Greatest	Considerably less	Considerably less

Table 6.3 indicates the beginning of board industries in various countries.

The trend to install in the future more mills producing hardboard by the semi-dry and dry process is evident. *Lehotsky* and *Nagy* (1963) reported that until the end of the year 1962 about 17 plants operating on a semi-dry or a dry method were established, approximately half of them in the U.S.A.

Table 6.3. Beginning of Fiberboard Industries in Various Countries and the Percentage Consumption of Different Raw Materials in the Year 1955 (OEEC 1956, The Fiber Building Board Industry, 1956)

Start year	Country	Raw material %					
		Solid wood		Residues from saw-mills, small wood	Wastes from paper-fabrication	Straw	Bagasse
		Coniferous wood	Broad-leaved wood				
1898	England	8	30	48	5		9
1908	U.S.A.		10	78	10		2
1909	Canada			90	10		
1928	France			90	10		
1929	Sweden	25		75			
1931	Finland			100			
1932	Switzerland	90		10			
1933	Norway	97	3				
1936	Italy			100			
1937	Austria	5		95			
1937	Denmark	100					
1938	Germany		15	84	1		
1939	Belgium			100			
1948	Holland			10···30		70···90	

Main characteristics for the conventional wet and for the more recent semi-dry and dry processes for the production of fiberboard are given in Table 6.4. Production costs are not included as they are not within the scope of this book and because they vary from country to country and from time to time. Observations, however, suggest that hardboard produced, using semi-dry or dry methods, may be about 10% more expensive than similar boards made by a wet method.

Historical and statistical data from various sources do not harmonize, and there are some gaps but on the whole the picture is clear that:

1. Rapid *development* of the *capacity* of the fiberboard industry in industrialized countries has occurred since 1922 (Fig. 6.1).

2. Relatively high percentage of *residues* from sawmills and wood in small dimensions is used *as raw material*.

3. Remarkable differences exist in the *consumption of fiberboard* and net national income per capita in various industrialized countries (Fig. 6.2). The consumption depends primarily on the local conditions, climate, level of income, technical production, competing products, customers service, market conditions and research efficiency. The Scandinavian countries lead in per capita production. In Sweden, in 1962 (*Kaila*, 1968, p. 832) the building industry consumed 85% of the fiberboard production (35% for repairing, 15% for new buildings, 35% for joineries, prefabricated constructions, and acoustic board). In Finland the use of

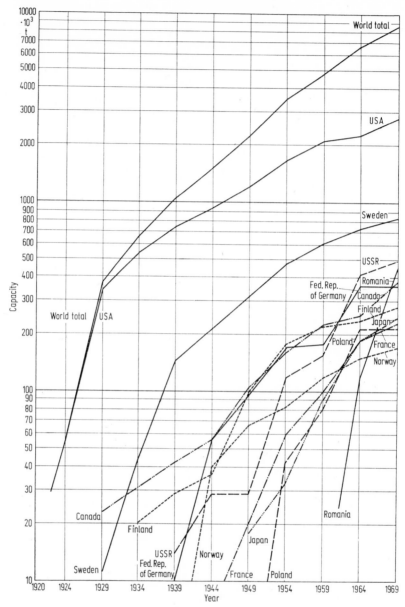

Fig. 6.1. Development of the capacity of the fiberboard industry in industrialized countries and in the world between 1922 and 1969. Defibrator AB, Stockholm (1969)

fiberboard in 1957 was estimated as follows:

	Hardboard	Insulation board
Private homes	60%	60%
Repairing of old building	30%	40%
Furniture industry	10%	

In Japan the following figures are given for fiberboard consumption in 1961:

	Hardboard	Insulation board
Buildings	50%	80%
Electric industry	14%	—
Packing cases	—	13%

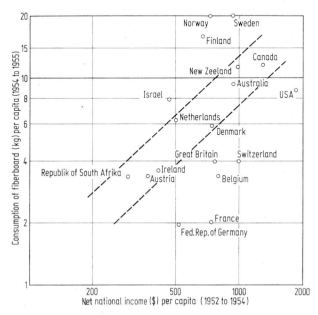

Fig. 6.2. Relationsships between consumption of fiberboard per capita and net national income per capita in various industrialized countries for the average of the years 1954/1955. FAO, Rome (1959)

Doubtlessly in the U.S.A., in West-Germany, in France, in Belgium and in Great Britain there are possibilities for a considerable increase in the consumption of fiberboards.

4. The largest *amount of exported fiberboard* is from the Scandinavian countries to Western countries. In the year 1960 about 800,000 metric tons were exported in the whole world, 90% from Europe, Sweden contributing 50%, Finland 15%. The next important exporters were Norway, South Africa, Poland, France and the Federal Republic of Germany. The U.S.A. and Canada together exported only 5%. In the year 1962 imports of hardboard to the U.S.A. were 15%, based on their own production. A high export of fiberboard is closely linked with the fact that the freight is comparatively low. The great proportion of export is due to relatively low production cost of standard and improved board (including such special properties as are smoothness, paintability).

In some cases security against attacks by insects and against easy ignition or against water absorption is wanted and can be provided.

5. The *price* of fiberboard has tended to decrease since about 1952, especially in comparison with sawn lumber (Fig. 6.3) and other sheetlike forest products (plywood, particleboard, sandwich board).

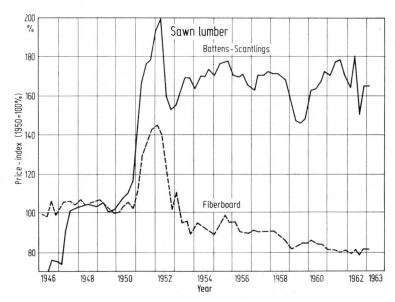

Fig. 6.3. Development of Swedish export values (expressed as price-indices) of fiberboard and sawn lumber (battens, scantlings) 1946 to 1963. Defibrator AB, Stockholm (1963)

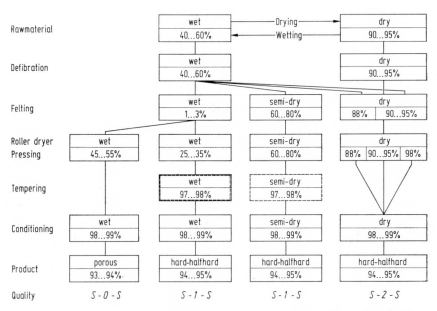

Fig. 6.4. Flow diagram for the manufacture of various types of fiberboard. From *Kaila* (1968)

6.2 General Outline of Processes

The methods of manufacture may be roughly classified as wet, semi-dry and dry processes. There exists a great variety of processes that aim to simplify manufacture or improve the products and enlarge its marketability. Figs. 6.4 to 6.6 show the general steps of wet, semi-dry and dry processes in the flow-diagram form. The sources for these slightly revised flow-diagrams extend back to publications of a period of about two decades (*Mörath*, 1949, *Lampert*, 1967, *Kaila*, 1968).

The diagrams show the flow of material, mainly fibers, water, and chemicals, and they show the sequences and connections of operations, but they do not show the particulars about the series and arrangements of machinery. Equipment information is given in Figs. 6.7 to 6.11, which show the widespread possibilities of fiberboard manufacture, without going into details or variations.

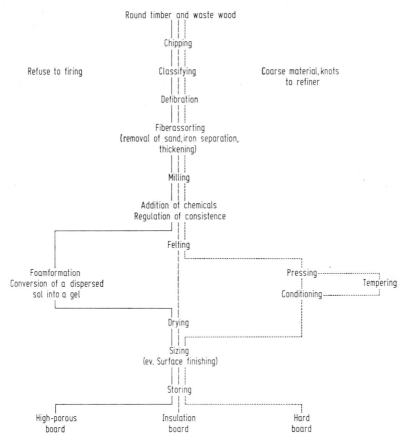

Fig. 6.5. Flow diagram for the manufacture of high porous insulation board by the wet processes.
From *Mörath* (1949)

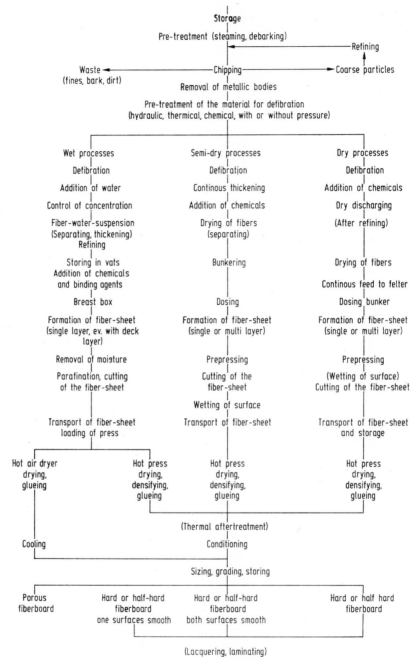

Fig. 6.6. Flow diagram for the production of various types of fiberboard using wet, semi-dry and dry processes
From *Lampert* (1967)

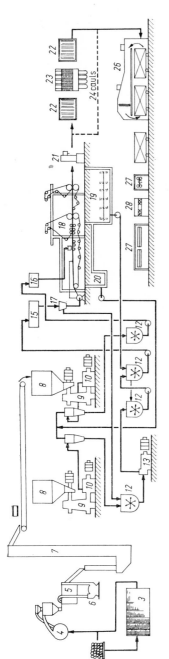

Wet-method: hard-board S-1-S

Fig. 6.7. Wet-method hardboard, S-1-S (smooth on one surface, manufactured using a screen in pressing). *From Kaila* (1968)

1 Raw wood with bark; *2* Debarked raw wood; *3* Debarking; *4* Hog; *5* Screen; *6* After-reduction; *7* Elevator; *8* Bin; *9* Preheater; *10* Defibrator; *11* Grinder; *12* Stock chest; *13* Refiner; *14* Knot catcher; *15* Additives; *16* Fiber dosing; *17* Classifier; *18* Dewatering machine; *19* Edge fiber chest; *20* Waste water chest; *21* Trim saw; *22* Press loader and unloader; *23* Hot press; *24* Cauls; *25* Roller dryer; *26* Heat treatment; *27* Saws; *28* Conditioning

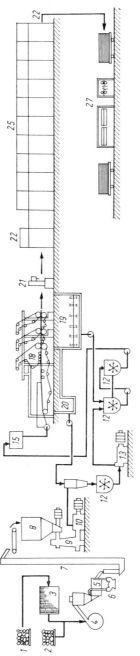

Wet-method: insulation-board S-0-S

Fig. 6.8. Wet-method insulation-board, S-0-S. From *Kaila* (1968)

1 Raw wood with bark; *2* Debarked raw wood; *3* Debarking; *4* Hog; *5* Screen; *6* After-reduction; *7* Elevator; *8* Bin; *9* Preheater; *10* Defibrator; *11* Grinder; *12* Stock chest; *13* Refiner; *14* Knot catcher; *15* Additives; *16* Fiber dosing; *17* Classifier; *18* Dewatering machine; *19* Edge fiber chest; *20* Waste water chest; *21* Trim saw; *22* Press loader and unloader; *23* Hot press; *24* Cauls; *25* Roller drier; *26* Heat treatment; *27* Saws

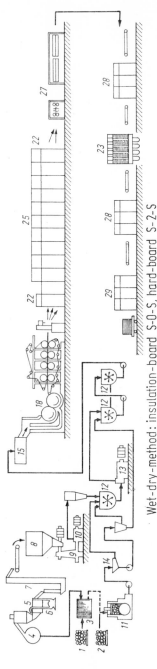

Wet-dry-method: insulation-board S-0-S, hard-board S-2-S

Fig. 6.9. Semi-dry method: insulation board S-0-S, hardboard S-2-S (smooth on two surfaces, manufactured without screen). From *Kaila* (1968)

1 Raw wood with bark; *2* Debarked raw wood; *3* Debarking; *4* Hog; *5* Screen; *6* After-reduction; *7* Elevator; *8* Bin; *9* Preheater; *10* Defibrator; *11* Grinder; *12* Stock chest; *13* Refiner; *14* Knot catcher; *15* Additives; *16* Fiber dosing; *17* Classifier; *18* Dewatering machine; *19* Edge fiber chest; *20* Waste water chest; *21* Trim saw; *22* Pressloader and unloader; *23* Hot press; *24* Cauls; *25* Roller dryer; *26* Heat treatment; *27* Saws; *28* Conditioning; *29* Board tempering

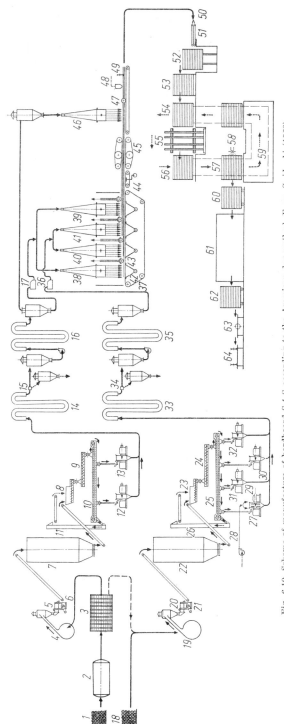

Fig. 6.10. Scheme of manufacture of hardboard S-1-S according to the American dry method. From *Swiderski* (1963)

1 Raw wood for deck layers; *2* Steaming vat; *3* Debarking drum; *4* and *19* Hogs; *5* and *20* Screens; *6* and *21* After-reduction; *7* and *22* Chip bin; *8* and *23* Dosing bin;
9 and *24* Digester; *10* and *25* Distributing conveyor; *11* and *26* Elevator for waste fibers; *12, 13, 27, 30, 31, 32* Bauer double disc-refiner; *14* and *33* Hot air pre-dryer;
15 and *34* Fiber classifier; *16* and *35* Hot air main dryer; *17* and *36* Equalizing vat; *18* Raw wood for core layer; *28* Air supply; *29* Glue supply; *37* Traveling belt screen;
38 and *39* Felter for deck layers; *40* and *41* Felter for core layer; *42* Suction device; *43* Pin roll; *44* Automatic band weigher; *45* Pre-press; *46* Felter for finest particle layer;
47 Edging saw; *48* Automatic thickness and density control; *49* Cross-cut saw; *50* Tipple; *51* Two level board pick-up; *52* Twenty daylight elevating platform; *53* Caul supply;
54 Press loader; *55* Hot platen press; *56* Unloader; *57* Position of unloader to render board to conditioning trucks; *58* Compressed air nozzle; *59* Cooling channel for cauls;
60 and *62* Conditioning trucks; *61* Conditioning channel; *63* Edging saw; *64* Cross-cut saw

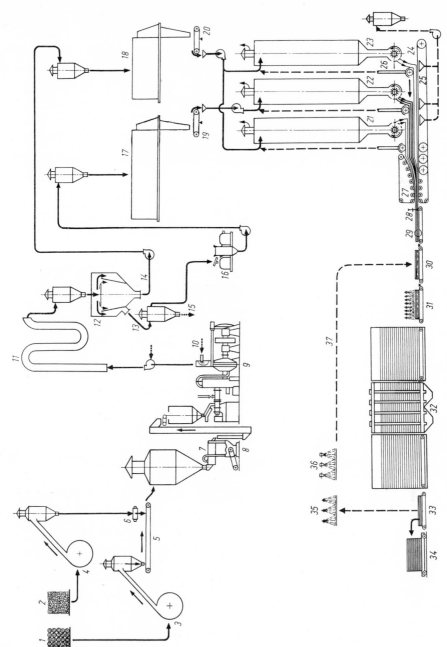

Fig. 6.11. Scheme of manufacture of hardboard in the semi-dry method. From *Swiderski* (1963)

1 Rounds logs; 2 Residues from sawmill; 3 and 4 Hog; 5 and 6 Conveyor for chips; 7 Screen; 8 After-reduction; 9 Defibrator; 10 Additive supply;
11 Hot air dryer; 12 Centrifugal classifier for fibers; 13 Coarse fibers; 14 Fine fibers; 15 Classifier for fiber bundles; 16 Bauer-refiner; 17 and
18 Equalizing bin; 19 and 20 Band weigher; 21, 22, 23 Felter; 24 Traveling belt screen; 25 Suction device; 26 Nylon brush roll; 27 Prepress;
28 Cross-cut saw; 29 Edging saw; 30 Loading of caul with prepressed board; 31 Wetting of mat; 32 Hot platen press; 33 Separation of board and caul;
34 Piling of board; 35 Washing of cauls and screens; 36 Drying of cauls and screens; 37 Caul and screen return

6.3 Raw Materials

6.3.1 Wood

The fiberboard industry started with the aim to convert more or less worthless or very cheap wood residues (e.g. slabs and edgings, not saw dust) into more valuable and applicable products. The prerequisite was the acquisition of a proper technology, mainly taken over from the paper industry. The fiberboard industry is rather young but had a rapid development within the last three decades with various changes in the procedures and more in the needs for raw materials. A survey may be given:

a) *Pulpwood* and *fuelwood*;

b) *Wood of inferior quality* or even damaged wood, not wanted by sawmills, veneer plants or pulp mills, is suitable for the fiberboard industry;

c) *Thinnings* or small logs up to 50 mm ($\sim$2 in.) in diameter are a suitable raw material;

d) *Residues* from sawmills and veneer plants are useful;

e) *Waste from agriculture* (straw, flax shives, hemp shells);

f) Tropical *grasses, bamboos.*

g) *Wastepaper*;

h) *Bark.*

According to a report of FAO, "International Consultation on Plywood and Other Wood-Based Panel Products", published in 1963, the fiberboard industry used 50% wood of small dimensions and 44% residues from wood-working industries in the world. In West-Germany after World War II inferior forest raw materials (e.g. stumps, branches) were used by the board industry, but had to be thoroughly washed since they were dirty.

The fiberboard industry would prefer softwoods but hardwoods usually are cheaper and easier to obtain. For these reasons more and more blending is applied. In the Scandinavian countries the proportion of hardwoods may vary between 10% and 100%. In subtropical and tropical countries indigenous broadleaved species are utilized for the manufacture of hardboard. Nearly all of the species available (for instance 36 in Mexico, 10 in India) are applicable. In Finland in six plants trials on the utilization of wood in small dimensions were carried out during 1956/57 (*Siimes* and *Liiri*, 1959). A summary of the results follows:

a) All examined wood species (pine, birch, poplar, and alder) are suitable as a raw material for the production of hardboard, even when mixed;

b) Wood in small dimensions (thinnings, tops, and forest waste) is a good raw material for the hardboard industry, even if mixed in a wide range of proportions;

c) Bark is not detrimental in manufacturing hardboard (with the exception that birch-bark occasionally causes difficulties);

d) The moisture content of softwood, if the conventional processes are used, should not be below 50%;

e) The properties of hardboard scarcely depend on the kind of raw wood;

f) Some wood species require variations in processing to give products with optimum properties. The situation regarding availability of raw materials should control the manufacturing processes;

g) Insulation board should be produced from small-dimensioned softwoods;

h) Some kinds of small-dimensioned wood (thinnings, tops, limbwood, waste veneer and veneer cores, slabs, edgings, cut-offs) are converted in the fiberboard plant by chippers (mainly disc chippers) into chips of appropriate size.

Sawdust, shavings and sanding dust abraded from wood surfaces in sanding operations originally were regarded as "fillers", but they recently gained more importance for use in board manufacture. Sawdust particles from gang saws are coarse and can be pulped in the usual manner (*Renteln*, 1951, *Helge*, 1963). Between 10 and 30% of sawdust can be added in half-hard board and hardboard production without changing the board properties noticeably. A mesh-analysis of typical sawdust particles from cutting coniferous wood in a gang saw is given in Fig. 6.12 (*Klauditz* and *Buro*, 1962). The main fractions R_3 and R_4 together make up 69% of the weight. Their specific surface amounts to about 1.5 and 3.25 m²/100 g, whereas flakes, especially from spruce wood for particleboard production, have an average specific surface of about 1.5 m²/100 g, consuming less amount

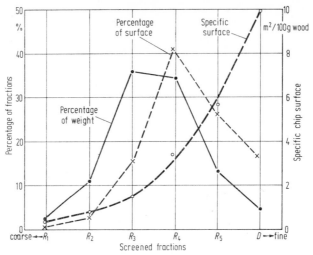

Fig. 6.12. Mesh analysis of typical sawdust particles from cutting coniferous wood on a gang saw. From *Klauditz* and *Buro* (1962)

of glue. Sawdust with short fibers (fiber particles) and sanding dust from veneer plants are suitable materials for forming outer layers. Shavings may be used technically in the same manner as sawdust, but the economy is limited by the transportation costs.

Sawdust has become more and more important as a raw material also in the fiberboard industry as a consequence of the increasing scarcity of fiber raw materials (Defibrator-News, Stockholm, Sweden, 1:1973). Earlier, however, it was not possible to utilize sawdust as a high-grade raw material. Thanks to developments in machinery, processes and material handling this is possible today and there are mills using up to 100% of sawdust as a raw material for their products.

The sawdust should be screened before being fed into the Defibrators. Oversized material may be desintegrated and returned to the screen, or fed into a chip based Defibrator line.

Silos, conveyors, chutes, funnels, etc. have to be designed to suit the special characteristics of sawdust so that bridging and other feeding troubles are eliminated.

Sawdust screw feeders for Defibrator preheaters must have suitable compression ratios. The new types of screw feeders (S 24 and ADI 12) can be used for sawdust and ordinary chips as well. Older types of screw feeders have to be equipped with special screws and screw pipes to be able to handle sawdust.

By increased peripheral speed and increased number of bars in the disc pattern, the possibility for a sawdust particle to pass untreated through the discs is practically eliminated. The speed of the Defibrator, when treating sawdust, is normally 1,500 rpm (1,800 rpm at 60 Hz of electric current). The disc clearance has to be considerably smaller than for defibering chips, and thus the precision and stability requirements of the Defibrator are much greater.

Pulp from sawdust is very suitable as to layer for hard, half-hard as well as insulation board. In this case, the pulp has to be refined in two or three stages after the Defibrator, depending on the requirements of the board surface.

Today, there are mills producing insulation board and half-hard fiberboard of high quality from 100% of sawdust and hardboard with up to 40% of pulp from sawdust.

For pulping of spruce sawdust, typical operational data for a Defibrator are:

Capacity:	100 metric tons per 24 h,
Steam pressure:	6.5 at,
Steam consumption:	550···650 kg/metric ton pulp,
Power consumption:	130 kWh/ton.

The structure of wood and of the wood cell walls have been dealt with in detail in Volume I of this book, p. 1—54. Softwoods and hardwoods differ greatly in structures. The fibers of hardwood are shorter, and the thickness of their cell walls is on the average greater than in coniferous species. A problem is that structure and dimensions of fibers vary remarkably in the same stem. Coniferous species have approximately 0.6 to 0.8 millions fibers per cm³, broadleaved species have between 2 and 3 millions fibers per cm³. Table 6.5 gives some values:

Table 6.5. Cell Dimensions of Softwood and Hardwood (*Kaila*, 1968, p. 857)

Kind of cells	Length mm	Diameter μm	Wall thickness μm
Tracheids			
Pine	1.8···3.8	26···34	1.5···3.4
Spruce	2.2···3.8	30···39	2.5···7.5
Birch	0.8···1.6	14···40	—
Alder	0.7···1.6	—	—
Vessel segments	0.4···1.2	100···400	1.6···3.0
Parenchyma cells	0.01···0.2	10···100	2.0···4.5

The length of the fiber is not a distinguishing characteristic of the properties of a board, but it has an influence on the felting properties. Fig. 6.13 gives a survey on the fiber length of wood species suitable for the manufacture of fiberboard.

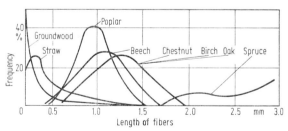

Fig. 6.13. Frequency of length of fibers in various wood species. From *Steenberg* (1962)

There is no doubt that some properties of boards produced depend on the fiber length as well as on the wall thickness. The stiffness of fibers is based on the wall thickness x on the one hand and on the diameters of fiber lumina and their shape in cross section on the other (Fig. 6.14). Fibers of broadleaved species behave quite

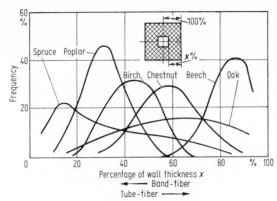

Fig. 6.14. Fibers frequency and relative wall thickness in various wood species. From *Steenberg* (1962)

different from those of coniferous species. It is much easier to form a sheet of fibers consisting only of softwood tracheids.

Wood contains the following chemical constituents: cellulose, wood polyoses (pentosans and hexosans) and lignin. Extraneous constituents are chiefly ash, resins, protein. Table 6.6 gives a summary of the composition of pine, spruce, birch and poplar. All values are expressed as the percentage of moisture-free unextracted wood, except where otherwise stated.

The main constituent of fibers is cellulose which is crystalline ("Microfibrillar orientation of the cell wall" in Vol. I of this book, p. 21—26). Monoclinic units of cellulose (*Meyer* and *Misch*, 1937) form cellulose chains. A single cellulose chain may run through several crystalline regions (called crystallites or micelles) as well as through amorphous zones in between ("fringed micellar structure", *Mark*, 1967). Fig. 6.15 shows the structural differences between a hardwood (birch) and a softwood (spruce) (*Meier*, 1955, *Wardrop*, 1957, 1958; *Emerton* and *Goldsmith*, 1956. *Liese*, 1960; *Frey-Wyssling* and *Mühlethaler*, 1965; *Harada*, 1965). The relationship between cellulose and water is of importance. Higher temperatures

Table 6.6. Chemical Composition of Four Wood Species
(*Clermont* and *Schwarz*, 1951)

	White spruce %	Jack pine %	White birch %	Aspen poplar %
Extractive content				
Cold water	1.36	2.18	2.02	1.48
Hot water	2.22	3.69	2.17	2.75
Ethyl ether	2.12	4.30	2.36	1.89
1% caustic soda	12.5	16.3	20.1	19.3
Acetyl	1.08	1.08	4.94	3.41
Methoxyl	5.07	4.97	5.93	5.47
Pentosans	8.00	10.13	22.0	17.2
Ash	0.22	0.19	0.29	0.38
Lignin[1]	26.96	27.38	18.48	18.12
Holocellulose[2]	72.0	68.0	79.3	80.3
Alpha-cellulose[3]	50.24	47.52	40.97	49.43
Polyoses[1]	16.39	16.18	27.25	21.18
Uronic anhydride	4.48	3.67	4.86	4.97

[1] Corrected for ash.

[2] Corrected for ash, lignin and extractives.

[3] Corrected for ash and lignin.

break down the chains of the cellulose molecules with an accompanying loss of
water of constitution. Cellulose adsorbs water. At elevated temperatures, alkalies
and water cause a more rapid swelling of cellulose, but to a lesser degree. The

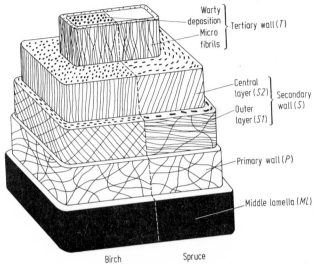

Fig. 6.15. Structure of hardwood fibers (left side: birch) and soft wood fibers (right side: spruce).
From *H. Meier* (1955)

question whether cellulose or lignin is the principal source of fiber adhesion in
fiberboards has not yet been completely clarified.

The fiber pulp discharged, for instance from a defibrator, is refined in a disc refiner or in a continuous beater. The degree of refining depends on the sort of
board desired. Pulp freeness or drainage time can easily be adjusted to the different types of board. According to Defibrator AB (1952) for the manufacture
of hardboard the pulp is refined to a Defibrator drainage time of 20 to 30 sec,
half-hard board possibly requires 25 to 30 sec, pulp for insulation board a drainage
time of 60 to 80 "Defibrator-seconds".

6.3.2 Bark

Bark is a residue from solid wood industries and to a limited extent from
fiberboard and particleboard industries. Bark has been used mostly as a fuel, but
in this case its moisture content, based on the wet weight, should be less than
60% (*Virtanen*, 1963). Bark from softwoods is cheap since it is not profitably
utilized by sawmills. Originally debarked wood was preferred to wood with bark
for the manufacture of fiberboard and particleboard, but there are exceptions
which need explanation (FAO, 1958, 1959, p. 38/39, *Kaila*, 1968, p. 842—848):

Permissible amount of bark as a raw material depends upon the following
details:

a) Density of board desired (hardboards permit more bark than insulation
boards);

b) Effect of bark on the manufacture processes. Bark fractions may have
adverse effects on pulping, may cause foaming in wet processes and difficulties
in pH control and introduce higher contents of dirt and extractives;

c) Effect of bark on the appearance, physical and strength properties of the
finished product;

d) Influence of bark of different species on added binders;

e) Influence of bark on heat-tempering of fiberboard.

According to FAO (1958/1959, p. 38/39) "most insulation board and the surface layers of flat-pressed particle board are manufactured from bark-free wood; for hardboard and for extruded particleboard as much as 15 percent of bark and in some cases even more is used. In wet process fiberboard the final product may be darker because of the tannin present in the bark components or the bark may appear as dark flecks in the final board." *Kaila* (1968, p. 843) reports that the bark content within the fiberboard industry may vary between 0 and 45%. *Anderson* (1956) manufactured experimental hardboards without binders from 8 wood species containing 50% of bark; three species (Douglas fir, ponderosa pine and western hemlock gave good results for water resistance and bending strength. One plant used 45% Douglas fir bark in the production of standard hardboard and found that even for insulation board up to 25% bark can be added.

In the Scandinavian countries hardboard may contain 10 to 15% bark, depending on the density of finished product. In mills manufacturing insulation board no bark is utilized, except in rigid insulation board treated with asphalt or other bitumens. Sapwood residues from sawmills and veneer plants, that contain between 15 and 30% of bark and are mixed with wood chips, give boards with adequate properties. The utilization of bark is an economic problem because the yield will be reduced from about 89.5% for debarked wood to 82% or even less than 75% for spruce wood with 50% bark. The reduction in yield due to bark content in the pulp mainly depends on wood species and amount of bark on the wood (maximum about 50%).

Anderson (1956, 1957) and *Anderson* and *Helge* (1957) carried out a series of investigations on the influence of some wood spezies and barks on the properties of hardboard. He used a laboratory Asplund-defibrator with a steam-pressure of 10 kp/cm² (142 lb./sq. in.) and a total treating time of 4 min. A summary of his results as quoted by *Kaila* (1968, p. 843—845) is given in Table 6.7.

Table 6.7. Properties of Hardboard Manufactured from Separately Pulped Wood and Bark (*Anderson*, 1957)

Raw materials		Untreated		Heat-treated 3 h at 165 °C (329 °F)	
		Water absorption %/24 h	Bending strength kp/cm²	Water absorption %/24 h	Bending strength kp/cm²
Without water repellent					
Spruce	100% wood	53.0	560	27.8	611
	50% wood, 50% bark	51.6	347	46.5	425
Pine	100% wood	56.8	503	26.7	378
	50% wood, 50% bark	20.1	295	16.9	270
Birch	100% wood	66.3	546	62.3	542
	50% wood, 50% bark	62.8	251	56.6	265
With 1/2% paraffine + 1% alum.					
Spruce	100% wood	39.2	523	17.4	607
	50% wood, 50% bark	41.2	387	27.8	418
Pine	100% wood	29.1	321	19.5	378
	50% wood, 50% bark	15.6	270	14.3	267
Birch	100% wood	52.3	560	43.1	551
	50% wood, 50% bark	45.7	268	—	298

The following conclusions are possible:

a) Hardboard consisting of 100% of wood without water repellent has in the non heat-treated condition, a relatively high water absorption. If 50% of wood and 50% bark are blended the water absorption for non heat-treated spruce and birch hardboard remains practically the same. For pine, 50% bark reduces the water absorption remarkably.

b) Heat-treatment reduces the water absorption of hardboard without water repellents, for 100% spruce or pine as raw materials, roughly to the half but not for birch. An addition of 50% of bark is unfavorable for spruce board, it is almost negligible for birch board, but is definitely favorable for pine board, probably due to the action of some chemical constituents of the pine bark under the influence of heat and pressure as binders.

c) Water repellents reduce the water absorption, especially if a heat-treatment follows. Again the bark content of pine hardboard lowers the water absorption;

d) The bending strength of hardboard consisting of 100% of spruce or pine wood without water repellents and heat-treatment is about 60% respectively 70% higher than that of hardboard manufactured from a mixture of 50% wood and 50% bark. The corresponding figure for birch wood is about 210%. Heat-treatment has less effect than expected. Differences of bending strength between hardboard made only from wood and that made from 50% wood and 50% bark become smaller as a result of heat-treatment.

Summarizing it may be said, that the interrelationships are complicated, that bark can be used in the manufacture of hardboard, that the effects should be carefully investigated, especially in new plants, and that a thorough control of raw material by the factory is necessary.

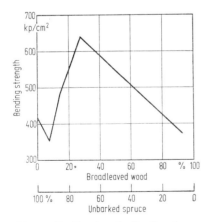

Fig. 6.16. Bending strength of hardboard as a function of the ratio of unbarked spruce wood to broadleaved wood in one mixture; no binders used. From *Carlsson* and *Elfving* (1957)

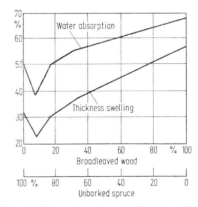

Fig. 6.17. Water absorption and thickness swelling of hardboard as a function of the ratio of unbarked spruce wood to broadleaved wood in one mixture; no binders used. From *Carlsson* and *Elfving* (1957)

A mixture of broadleaved wood and unbarked spruce wood in the ratio of about 30%:70% may lead to a maximum in bending strength (Fig. 6.16) whereas the lowest values for water absorption and thickness swelling—that means highest water resistance—will be obtained for the ratio of 10%:90% (Fig. 6.17).

It may be mentioned that a considerable amount of spruce bark in hardboard prevents sticking to the polished upper cauls in the multi-platen press if temperatures above 200 °C (392 °F) are applied. Separately pulped spruce bark may be added to the wood fiber pulp up to an amount of 30 to 40% without causing difficulties. The effects on important properties of hardboard are shown in Fig. 6.18 and Fig. 6.19.

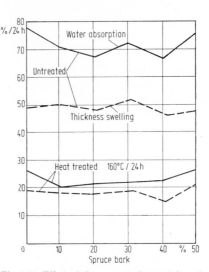

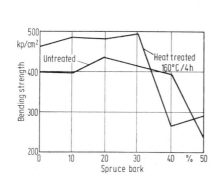

Fig. 6.18. Effect of the amount of separately and mildly pulped spruce bark on the bending strength of hardboard; no binders used. From *Helge* (1963)

Fig. 6.19. Effect of the amount of separately and mildly pulped spruce bark on the water absorption and thickness swelling of untreated and heat-treated hardboard; no binders used. From *Carlsson* and *Elfving* (1957)

Producing hardboard by the semi-dry or dry processes, generally of 30 to 50% debarked wood is used as a raw material. Referring to the foregoing remarks the question of sizing once more may be discussed. Sizing of hardboard should increase strength as well as reduce water absorption. The chemical additives (e.g. paraffin, wax, tall oil-derivates, artificial resins) increase the production costs to a varying degree. Improved processing techniques and/or final heat treatment may be cheaper and therefore more attractive. Bark of pine and especially of Douglas fir contain waxes and phenolic substances acting as water-repellents and strength-increasing agents during the hot pressing. Bark of spruce and birch does not develop "natural sizing" as does bark of pine.

6.3.3 Modified Wood Raw Materials

Extracted wood raw materials are available in some areas of the world, (FAO, 1958/1959, p. 39) e.g. licorice root (from which the licorice is extracted), spent turpentine chips from the production of naval stores (in Southern States of the U.S.A.), chips from tannin-extracted chestnut, extracted bark. Though the properties of the lignocellulosic material are somewhat changed by distillation or extraction due to the high temperatures, causing some hydrolysis, a satisfactory production of boards from these materials is possible in most cases.

Waste paper is an important raw material for insulation board and, to a lesser extent, for hardboard. Waste paper may be used for board as:

a) Main fibrous ingredient;

b) Blended with freshly pulped wood fibers;

c) Repulped material, added in high proportion to bagasse, with the aim to improve stiffness, wet strength and appearance of insulation board (Colonial Sugar Refining Co., Building Materials Division, 1957) (FAO, 1958/1959).

6.3.4 Non-Wood Fibrous Raw Materials

The non-wood fibrous materials suitable for fiberboard —, and also particle-board — manufacture are agricultural residues, chiefly from the fibrous part of annual plants. Some of these materials are already widely used in various areas of the world. Others are at present used only in limited amounts, but have growing possibilities for the future in regions with insufficient wood supply. The following points have to be taken into consideration:

a) Continuous variations of agricultural conditions (crop quantities, methods of harvesting, considerable price fluctuations); all these factors determine availability, quality and costs of these fibers;

b) Low density of piles, difficult storage, handling and transportation;

c) Impurities such as husks, weeds, leaves, debris, dirt, and for some residues a rather high amount of silicates, causing difficulties in gluing.

The principal non-wood fibrous raw materials are (FAO, 1958/1959, p. 42—46):

1. Bagasse, available in such parts of the world where sugar cane is an important crop. Bagasse is not a cheap raw material since it is normally used as a fuel in the sugar mills. Further, it must be collected over a relatively short harvest period (2.5 to 3.5 months), baled, protected against self-ignition, mould and fungi and stored in quantities sufficient for one year's board production. Bagasse has an elastic tough fiber and is a good raw material for insulation and half-hard board. The increasing importance of bagasse for particleboard production is explained in Section 5.1.1.4.

2. Flax shives are an important raw material in several parts of the world for particle board (cf. Section 5.1.1.3). It is possible to use them, under some circumstances in a combination with wood or mixed with wood pulp, for the production of insulation board. The storage problems are similar to those for bagasse. Another essential point is the question of continous and sufficient supply.

3. Cereal straw from wheat, rice, or other grains is suitable, partly in combination with wood fiber, for insulation board or hardboard. Varying quantities, problems in storing, rapid deterioration of wet straw bales induced some factories, which started with straw, later to change to wood.

4. Cotton stalks, particularly in areas of Africa and Asia where other cellulosic raw materials are rather scarce, may be of growing importance for fiberboard and particleboard manufacture.

5. Corn stalks were used for many years in the U.S.A. for the production of insulation board. "However, because of the high cost of collection, baling and handling, this operation has now changed to wood as its primary raw material" (FAO, 1958/1959).

6. Other fibers, representing potential raw materials for fiberboards, are bamboo, papyrus, coconut fiber residues, palmleaf ribs etc.

6.4 Storage of Raw Material

A fiberboard plant needs raw material for one year's production. To make one metric ton of fiberboard between 1,020 and 1,230 kg of raw wood are necessary (*Kollmann*, 1955, p. 559). If calculated in volume, 8 to 9 m³ (283 to 318 cu. ft.) are used for the production of 1 metric ton of board. The specific gravity of loose ovendry softwood chips is between 120 and 145 kg/m³ (7.5 and 9.0 lb./cu. ft.), that of hardwood (beech) chips about 200 kg/m³ (12.5 lb./cu. ft.). With a moisture content up to 120% the corresponding figures are 250 and 305 kg/m³ (15.6 and 19.0 lb./cu. ft.) or 400 kg/m³ (25.0 lb./cu. ft.). Loose sawdust (spruce) has a specific gravity (ovendry) of 140 kg/m³ (8.7 lb./cu. ft.), (green) of 280 kg/m³ (17.5 lb./cu.ft.), loose shavings (ovendry) 65 kg/m³ (4.1 lb./cu. ft.), (green) 95 kg/m³ (6.0 lb./cu. ft.).

The proportion of bark, based on the volume of the stem with bark, is listed in Table 6.8.

Table 6.8. Proportion of Bark, Based on the Volume of the Total Wood in %

Wood species	Reference		
	Müller (1923)	*Flury* (1928)	*Trendelenburg* (1939)
Coniferous woods			
Spruce	10	6.4···14.9	7···10···18
Fir	10···12	8.3···12.3	8···11···19
Pine	10···16	10.1···16.8 ∅ <20 cm	20···30 >30
		16···18···20	15 9···11···14
Larch	16···22	17.0···21.9	20
Broadleaved woods			
European beech	6··· 8	5.4··· 9.9	6··· 7··· 9
Oak	10···20	*Quercus pedunculata*:	
		8···18	
		Qu. sessiliflora:	
		15···20	
Ash	12···14		11
Birch	13···17		10
Alder			8···11···14

The number of round logs per 1 rm (35.3 cu. ft.) depends on the log diameter and on the deviations from the ideal cylindrical form due to bows, knots, and taper (Fig. 6.20). Logs with small diameters are unfavorable since the measurement-loss is relatively high and the costs for handling and peeling increase.

The size of the stock pile is influenced by the conditions of the particular locality, by the form of raw material, by the moisture content of the wood and the process of air drying. As a rule slabs, edgings, cut-offs, veneer cuttings and other industrial wood residues require smaller storage facilities than round logs. FAO (1958/1959, p. 40) has pointed out that wood storage for board manufacture is very important and that the wood yard or equivalent facilities should be engineered as any other part of the plant.

Particular attention should be paid to:

a) Appropriate size;
b) Protection from contamination;
c) Maintaining a moisture content that will give optimum manufacturing conditions;

d) Assuring the turn over of stock in a systematic economical manner;

e) Protection from deterioration by decay, fungi or other attack;

f) Permitting, handling and transfer of raw material at minimum cost;

g) Providing adequate protection from spontaneous ignition, adequate means of fire control and isolation from the rest of the plant by fire walls or distance.

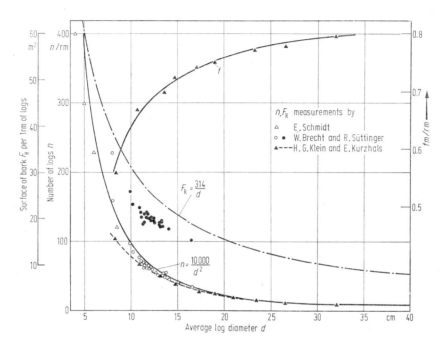

Fig. 6.20. Number of round logs and proportion of bark per 1 rm (35.3 cu. ft.). From *Kollmann* (1951)

6.5 Raw Material Preparation

6.5.1 Debarking of Wood[1]

For manufacturing hardboard usually unbarked wood is used with the exception of the material for special light faces. For these faces also coarse, in general barkfree sawdust from gangsaws is used. Cores from veneer plants are a fine raw material for outer faces. For the production of insulation board wood with bark is not appreciated since a bright lustre is esteemed on the market.

The wood may reach the mill in log form or as chips. Only occasionally the bark may have been removed in the forest after felling. In most cases the wood arrives with bark, and the first, therefore, is to remove as much bark as possible. With wood chips, bark removal presents a special problem. However, under some circumstances separation is successfully carried out.

For debarking logs (cf. Section 5.2.1.3) a variety of processes and machines are available which may be classified into the following groups:

---

[1] Bark removal is called barking or debarking.

1. Mechanical debarking machines,
 1.1 Machines with knives,
 1.2 Friction or abrasion type machines,
  1.2.1 Abrasion stem by stem (mutual friction between the logs and against the debarking irons inside the drum),
  1.2.2 Abrasion of stem by chains or tools.
2. Hydraulic type debarkers,
3. Steam explosion (as practised in the Masonite-process for defibration, but not used yet for debarking.

According to *Smith* (1948) debarking machines generally demand the following:

1. Barking should be *continuous*;

2. Barking should be done *without rotation* of the log to avoid difficulties with crookedness;

3. Mechanical handling should *reduce labor costs*;

4. Bark should be removed completely *without loss of good wood*, regardless of season or length of soaking;

5. *Power* should be efficiently used. Observations on existing machines—excepted the streambarker, which for the same barking speed needs three or four times more power installed than all other usual barker types—show that 120 HP for 100 ft./min(30 m/min) of barking speed is average.

6. The *machine* should be as *simple* as possible to ensure a low first cost and freedom from mechanical troubles.

Various machines are available for bark removal from logs and slabs:

a) *Knife debarkers* with cutter heads are used in wood preserving industries where the relatively high wood loss (8 to 20%) is not too serious. There are rotating disks in which the knives run radially outward (3,000 rpm). (Fig. 6.21 shows the Efurd-peeler with two rotating cutter heads, mounted in tandem for roughing and finishing respectively).

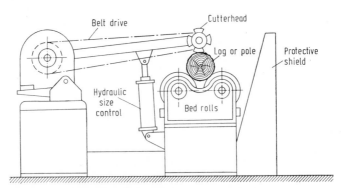

Fig. 6.21. Multiple cutterhead peeler of Efurd Machine and Welding Co., Bossier City, La. From *I. W. Smith* (1948)

b) *Friction debarkers.* Many types of machines and methods have been developed for *cambium separation*. Three methods of breaking the cambium are employed: tension (low power consumption, use of hand-tools, application immediately after trees are felled), crushing (by means of hammers, chains and rolls) and shear, exerting tangential forces against the bark; the tools resemble blunt-edged scrapers (Fig. 6.22) and are pressed against the bark, with a pressure of 7 kp/cm² (∼10 lb./sq. in.).

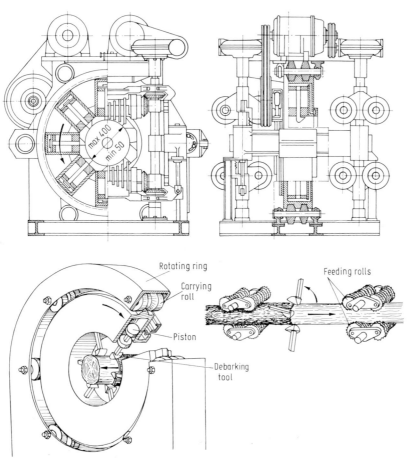

Fig. 6.22. Scheme of debarking machine, type Cambio. From *Kollmann* (1955)

Drum debarkers are rotating cylinders in which the logs are rolled or tumbled against each other at low speed; water is sprayed on them to wash the bark away. A great disadvantage of the drum barker is "brooming" or splintering at the ends of the logs. Drums vary from 1.5 to 6.0 m (∼4.9 to 20 ft.) in diameter and from 2.8 to 22.0 m (9.2 to 72 ft.) in length.

Some debarkers are *multiple pocket peelers* (Fig. 6.23). The Thorne debarker, one of the older friction debarkers for pulpwood consists of three sheet-metal pockets wide enough to accomodate the pulpwood length, i.e. 4 ft. 8 ft., 12 ft., etc. (1.2 m, 2.4 m, 3.6 m etc.). There are openings in the bottom and shafts below drive cams swinging up and down as the shaft rotates. The cams agitate the bolts in the pockets and push them into the next pocket. Water is sprayed on the logs as they rub against each other and they gradually work from pocket to pocket, finally emerging at the discharge end. For small capacities *single hopper log peelers* are available; a set of three or four endless chains with projections roll the logs (cf. Fig. 5.24) against each other in the hopper. Water jets spray the logs to wash off the loosened bark.

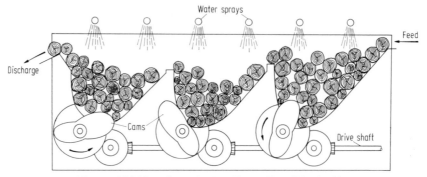

Fig. 6.23. Multiple-pocket peeler, type Thorne. From *Fobes* (1957)

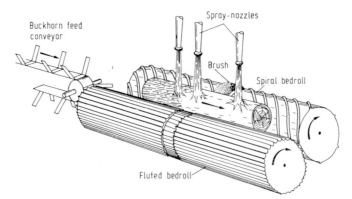

Fig. 6.24. Hydraulic debarking machine, system of Allis-Chalmers Manufacturing Co., Milwaukee, Wisc.
From *Kollmann* (1955)

c) *Hydraulic peelers* have proved successfull (*Holekamp*, 1956). Conveyors carry the logs (or slabs face down) under either fixed (Fig. 6.24) or oscillating nozzles. The water pressure is between 85 and 90 kp/cm² (1,210 and 1,280 lb./ sq. in.). Although the log rotates during de barking, some crookedness and irregularity can be tolerated. Softening of the bark by water storage is not necessary. The water consumption is high, amounting to 1,900 to 4,200 l/min (500 to 1,100 gallons/min), but the water can be reused. Also the power consumption (440 to 1,200 kW) is high. Logs varying in diameter between 100 and 1,800 mm (~40 and 700 in.) and with lengths from 2.5 to 6 m (8 to 20 ft.) can be peeled. Smaller types exist with an energy consumption between 350 and 600 kW. Cost of machines, power consumption and water requirements, however, limit their use to mills where large amounts of logs are to be peeled.

A survey on data for debarking methods and machines is given in Table 5.9.

6.5.2 Chipping, Magnetic Separating, Screening

Raw material available in the form of solid wood before being converted into fibers or fiber bundles must be chipped into pieces approximately of the same size as the chips needed in pulp mills. Wood in large dimensions must be chipped in log chippers which are usually constructed for round wood up to 350 mm ($\sim$14 in.) in diameter (cf. Section 5.2.1.4.2).

Smaller dimensioned logs, for example thinnings and slabs from sawmills, are chipped in slab chippers with adapted feeding devices.

It is not possible to obtain chips of uniform size in one operation. Therefore, the material must be screened; too large chips have to go to rechipping for further treatment and they are screened again.

The screened chips are stored in a bin (cf. Section 5.2.5). It is necessary to store sufficient chips for 18 to 24 hours operation.

Disturbances in the operation by pieces of a scrap iron must be avoided. Before entering the defibrator or a similar apparatus the chips should be led over a magnetic separator which removes iron particles (cf. Section 5.2.1.1). This separator can be installed either before or after the chip bins according to local conditions. Screening is done either in vibratory or in drum-type screens. Self cleaning properties of some types of screens are desirable.

6.5.3 Storing

The chips produced in a fiberboard plant are hurled upwards through a sloping sheet metal duct (cross section about 35 cm $\times$ 40 cm to 50 cm ($\sim$14 in. $\times$ 16 in. to 20 in.) to a cyclone where the dust is removed from the stream of air and chips

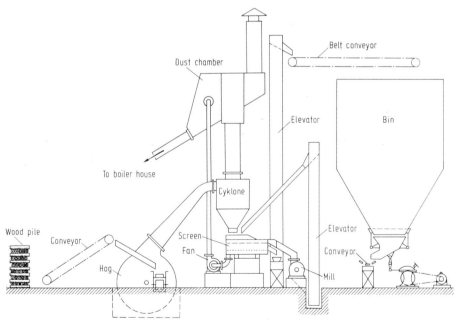

Fig. 6.25. A schematic survey of machines and procedures for the preparation of chips. From *J. M. Voith* (1948)

by centrifugal forces. The average size of chips for the production of fiberboard is between about 25 mm ($\sim$1 in.) in length and 3 mm to 6 mm ($\sim$1/8 to 1/4 in.) in thickness. From the outlet of the cyclone the dust-free chips fall onto a vibratory screen, essentially consisting of a box in which a frame is moved horizontally back and forth by means of a cam. In this frame two sives are built in, one above the other (area about 1.9 m × 2.0 m, corresponding to 75 in. × 79 in.) to separate coarse and fine particles. Knots and coarse chips rejected from the production are fed to a hammer-mill (cf. 5.2.1.4.3) for rechipping, dust and splinters are blown to the dust box from where they are transported together with the wood dust from the cyclone to the boiler house.

The screened chips for example come via an elevator and a conveyor or pneumatic system, to the concrete silo, the capacity of which must suffice at least for the consumption of 18, better 24 h. In its lower part the silo consists of a sheet iron cone ending in a tilted mouth piece which is automatically shaken in order to avoid bridge building and to ensure continuous discharge. Fig. 6.25 gives a schematic survey of machines and procedures for the preparation of chips.

6.5.4 Wetting

Green or wet wood can be defibrated more easily and with lower power consumption than dry wood. Not only for mechanical pulping but also for chemical-mechanical and for thermal-mechanical pulping a rather high moisture content of 40 to 60% (based on oven-dry weight) of the wood chips should be maintained or restored if it is less than 35% m. c. In this respect there are three possibilities (which may be combined):

a) Keeping the round logs wet by storing them in ponds or by water-spraying the piles on the wood yard;

b) Adapting the moisture content of the wood chips to the above values during the transport to or in the silos by means of spraying;

c) Pumping a suitable amount of water into the defibrators.

6.6 Pulping (Wet Processes)

6.6.1 Mechanical Pulping

There are two different ways for the production of mechanical pulp. One is the grinding of debarked logs, the other one is the refining of wood chips. In both methods a mechanical treatment leads to the defibration of the wood. Though recently only about 7% of the world production (1968) of mechanical pulp derives from the refiner defibration, this proportion will increase in the future as hardwoods and wastewoods can also be converted in refiners (*Giese*, 1968).

Logs of softwoods and low density hardwoods such as poplar can be defibrated only in grinders. In 1843 *F. G. Keller*, a weaver of Saxony, claimed for a German patent for a wood-grinding machine. Blocks of wood were pressed against a revolving wet grindstone. Also in recent machines the logs undergo the wet abrasive action of the circular surface of a grindstone. The stones vary from 1.4 to 2.3 m (55 in. to 90 in.) in diameter. They consist of a core of reinforced concrete on which a layer of concrete or ceramic bound stone is mounted. In the case of ceramic bound stone this layer is assembled of replaceable hexagonal elements (Fig. 6.26). The grinding material consists of quartz, silicon carbide or aluminium oxide.

The wood logs are pressed against the rotating grindstone in different grinder types by different mechanisms. In the discontinuously working magazine grinders and pocket grinders the logs are pressed against the stones by hydraulically moved pistons. In continuous grinders the mechanism consists of endless chains which press a vertical log stock against the grindstone. The less used ring grinder works with a rotating ring which is arranged excentrically to the grindstone. Fig. 6.27 shows a schematic representation of the different grinder types. The average pulp production per day amounts to 20 to 50 metric tons (22 to 55 short tons) in

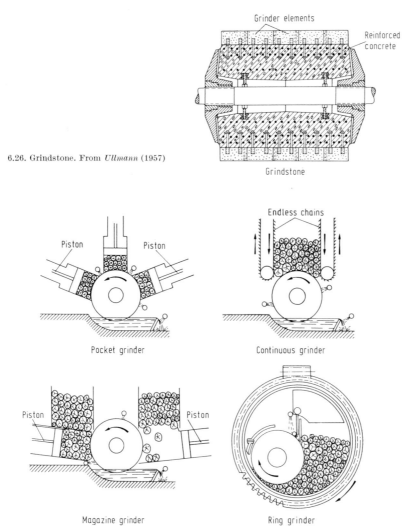

6.26. Grindstone. From *Ullmann* (1957)

Fig. 6.27. Different grinder types. From *Ullmann* (1957)

24 h. In some mills in the United States grinders were recently installed driven by motors of 8,000 to 10,000 HP which produce pulp under high load and by high speed grinding up to 120 metric tons (130 short tons) in 24 h (*Perry*, 1967).

The pressure with which the logs are pressed against the grindstone, the rotatory velocity of the grindstone, and the surface of the grindstone are important factors in determining the quality of the pulp. The pressure and the rotatory velocity cause a rise of temperature in the grinding zone. The desirability of the higher temperature for the mechanical separation of the fibers was recognized by *Klemm* (1957). The higher temperature softens the lignin and the polyoses in the middle lamella (intercellular substance) between the wood fibers and thus favors the separation of the fibers. Studies of the grinding process and of the thermal behavior of the cell wall components showed that temperatures of 170 °C to 190 °C (338 °F to 374 °F) as formerly assumed are not reached in the grinding zone and are also not necessary for the thermal softening. *Steenberg* and *Nordstrand* (1962) found that the maximum temperature in the log under grinding does not exceed 100 °C (212 °F). They calculated that 15 to 35% of the heat developed in the grinding zone is transported out of it by the heated surface layer of the grindstone.

In basic studies *Goring* (1963) found that a thermal softening of dry lignin and polyoses takes place at temperatures ranging from 130 to 190 °C (266 °F to 374 °F). Water uptake lowers the softening temperature to near or even below 100 °C (212 °F). That means that the temperature in the grinding zone is sufficient for at least partial thermal softening of the middle lamella of the wood fibers. In general, grinding is done at higher temperatures in the U. S. A. than in Europe (*Stamm* and *Harris*, 1953, p. 306).

The temperature in the grinding zone of refiners can be varied by the temperature of the shower water. Experimental refining resulted in considerably better defibration, lower content of sheaves, and better strength properties of the pulp with water at 85 °C (185 °F) than with cold water (*Birkeland*, 1969).

Klemm (1957) subdivides the grinding process into two basic steps. The primary process is the "initial grinding period", the moment during which one point of the stone surface passes the grinding zone (50 to 60 milliseconds). In this stages the pressure of the log against the grindstone has a fundamental influence on the defibration. If the pressure is too low, single fibers are split and the pulp gets too soft. If the pressure is too high fiber bundles and sheaves are produced and the resulting pulp is too coarse. The so-called "initial grinding factor" should be optimal, that means the grinding pressure has to be so that the pulp contains a high percentage of long undamaged fibers.

The secondary step on the grinding stone involves subsequent beating of the separated fibers. This beating of the fibers is highly influenced by the contour pattern of the stone surface. The energy consumed in grinding varies with the species, stone texture, temperature, and other factors. Coarser pulps used in board manufacture require about 50 HP days per ton. It has been estimated that 90 to 99.9% of the work done in grinding is converted into sensible heat (*Stamm* and *Harris*, 1953, p. 306).

6.6.2 Chemical-Mechanical Pulping

Occasionally it is useful to treat the raw materials, especially grasses, with chemicals prior to pulping in disc refiners or attrition mills. As chemicals, neutral sulphite cook, sodium hydroxide cook or a lime cook are used. The concentration (FAO, 1958/1959) of chemicals is about 1/10 or 1/5 that of a full cook, and yields are in the range between 70 and 85%, with grasses perhaps somewhat lower.

The chips are carried from the silo in conventional stationary or rotary ball type digesters in which they are steamed and cooked. Fig. 6.28 shows the scheme of pulp preparation for fiberboard manufacture using the steam cooking process (Voith, Heidenheim, W.-Germany). For the subsequent reduction to pulp there are different types of mills, e.g. Condux-mills, Biffar-mills (*Biffar, Belani,* 1942) (Fig. 6.29), Bauer-defibrators and Boija-Jung-defibrators (Boija-Jung, Svensk Patent 106602).

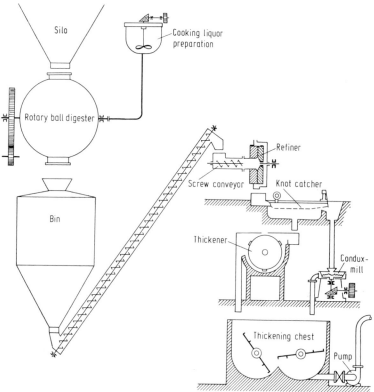

Fig. 6.28. Scheme of pulp preparation for fiberboard manufacture using the steam cooking process.
J. M. Voith (1948)

The corrosive action of the chemical agents must be taken into consideration. Frequently during the steaming operation small amounts of alkaline agents are added to the digesters, in order to neutralize some of the organic acidic components produced during the steaming, primarily by splitting the hemicelluloses. In this way the corrosion of equipment may be reduced.

6.6.3 Thermal-Mechanical Pulping

6.6.3.1 General Considerations. Thermal mechanical pulping involves a preliminary treatment of the raw material with heat prior to mechanical action. The aim of the pre-treatment is to reduce subsequent power consumption and to improve pulp qualities. Water soaking, steaming or steaming in conjunction with a mild chemical cooking action may be employed (FAO, 1958/1959, p. 53).

Water soaking and steaming soften the wood and the pulp produced has the following properties:

a) Fewer broken fibers and coarse fiber bundles;
b) More flexible fibers;
c) Better felting properties, thus producing a stronger board.

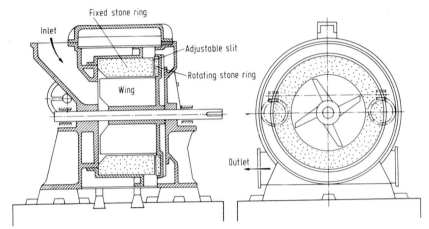

Fig. 6.29. Sections through a Biffar-mill. From *Kollmann* (1955)

The moisture content of the chips is of great importance in the defibration. Fresh wood and floated wood usually contain 40 to 60% moisture (based on oven-dry weight) which is just suitable for the defibration. If the moisture content in the wood is less than 35%, water has to be added either to the raw material or by pumping an adequate amount of water into the defibration process. Preheating is done either in fixed or in rotating vats. The fixed preheaters have the form of vertical or horizontal tubes. Feeding is effected by pistons or conical screws.

Thermo-mechanical processes are based on the softening effect produced in wood or other lignocellulose fibrous materials of appropriate moisture content when heated to temperatures of 150 °C to 180 °C (302 °F to 356 °F). The individual fibers in wood are bound together by the so-called "middle lamella" or inter-cellular layer (cf. Fig. 6.15 and Vol. I of this book, p. 21, Fig. 1.36, p. 32) containing a large amount of lignin. At about 150 °C (302 °F) the middle lamella is noticeably softened, and at temperatures reaching between 160 °C to 188 °C (320 °F to 370 °F) it is actually melting, loses its cementing power (Fig. 6.30) and exhibits plastic flow characteristics so that the fibers can be rubbed apart more easily than at normal temperatures. Steaming is more effective with broadleaved species then it is with coniferous in reducing the power consumption for defibering. During the heat treatment the lignin content does not change nor are the wood polyoses hydrolized appreciably. The pulp has been subjected to only limited chemical action and therefore its chemical composition is almost identical to that of the original wood. Therefore the original fiber structure and the chemical composition are preserved and the yield, based on dry weight of the raw material, amounts to as much as 90 to 93% or even 95% when debarked spruce or pine wood are pulped.

6.6.3.2 Asplund Process (Defibrator Method). The Asplund process, invented
by the Swede *Arne Asplund*, is based on the utilization of thermoplastic properties of lignocellulose raw materials and can be characterized as a continuous
operation. Various kinds of wood and other raw materials such as bagasse, straw

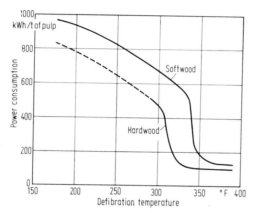

Fig. 6.30. Dependence of power consumption on defibration temperature. Defibrator AB, Stockholm (1949)

and bamboo lend themselves to this pulping method. Defibrator pulp has found
use in a great number of fiber products, such as insulation and hardboards, roofing
and flooring felts, bulk fiber and certain paper products.

The pulp produced by this method mainly consists of fibers with the original
fiber length preserved (Fig. 6.31 and 6.32). The mechanical separation takes

Fig. 6.31. Birch fibers produced by the Asplund-Defibrator. Defibrator AB, Stockholm

place at elevated temperatures varying for different types of fibrous raw materials,
but usually around 177 °C to 188 °C (350 °F to 370 °F). Higher steam pressure, and
with it higher temperatures, are unsuitable since the hydrolysis of the wood hemicelluloses increases rapidly with an increase in temperature (Fig. 6.33).

This process offers the following advantages:

a) Continuous operation (Fig. 6.34 and 6.35);

b) Low conversion cost on account of low power consumption averaging 125 to 175 kWh per metric ton of oven-dry pulp (Fig. 6.36), low steam consumption averaging 640 to 770 kg/metric ton (~1,400 to 1,700 lb./metric ton) and fully automatic operation requiring the supervision of only one man for every two machines. A theoretically calculated heat balance in the form of a Sankey-diagram is reproduced in Fig. 6.37;

c) High pulping yield (90 to 95%);

d) Uniform pulp of high freeness (Fig. 6.38).

e) Practically all kinds of ligno-cellulose raw materials can be processed.

Fig. 6.32. Pine fibers produced by the Asplund-process. Defibrator AB, Stockholm

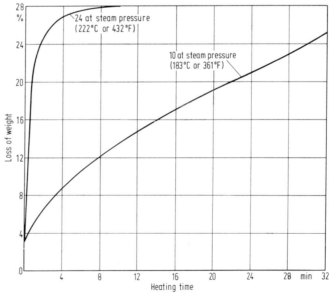

Fig. 6.33. Dependence of the loss of weight for pine wood chips on heating time and steam pressure.
Defibrator AB, Stockholm

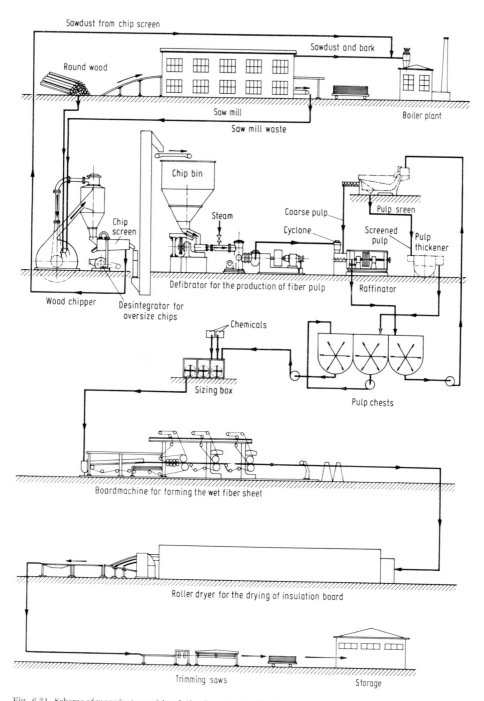

Fig. 6.34. Scheme of manufacture of insulation board using the Defibrator-process. Defibrator AB, Stockholm (1952)

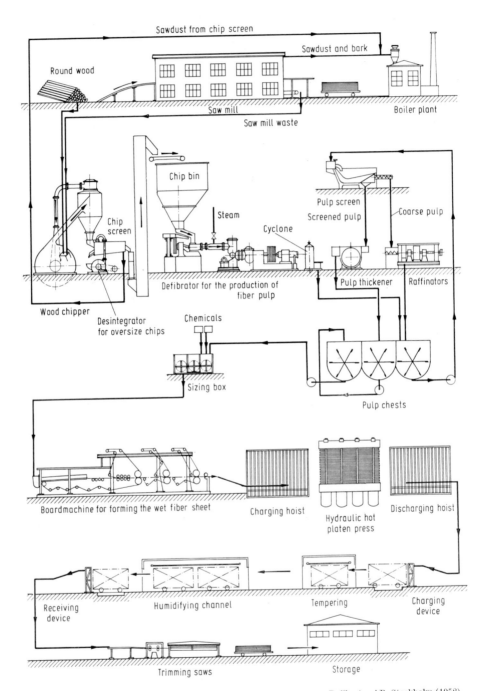

Fig. 6.35. Scheme of manufacture of hardboard using Defibrator process. Defibrator AB, Stockholm (1952)

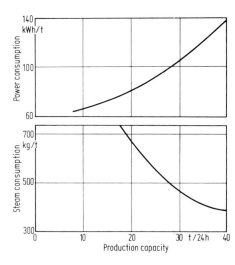

F g. 6.36 Consumption of energy and steam per metric ton in dependence of daily production capacity, using Asplund-defibrators and hardwood as raw material. From *Ögland* (1960)

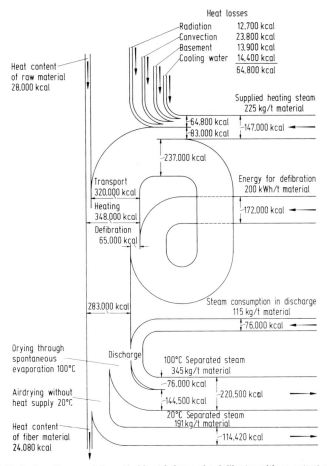

Fig. 6.37. Sankey diagram of theoretical heat balance of a defibrator with an output of 7 metric tons per 24 h. From *Kollmann* (1955)

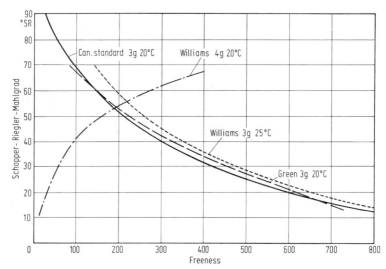

6.38. Dependence of Schopper-Riegler-Mahlgrad upon freeness values, measured by various test methods.
From *Kollmann* (1955)

6.6.4 Explosion-Process

The explosion-process is known as the Masonite-Process. Masonite is the trade name for various hydrolized wood fiberboard products. The raw material is usually coniferous wood, under circumstances blended with the wood from broad-leaved species up to 75%. *Kaila* (1968) reports that at present about 600,000 metric tons per year of explosion fibers are produced in seven plants, the oldest of which is the Masonite plant in Laurel/Miss., U.S.A., the newest is in Ukah, Cal., U. S. A., using chiefly redwood. The inventor *Mason* went back to a patent granted in the year 1858 to the American *Lyman* for his means of defiberizing. He developed his process starting in 1924 and various forms of the Masonite process were covered by about 75 patents, many of which have expired (a list of these patents is given by *Stamm* and *Harris*, 1953, p. 390, Quot. 32 and 33).

In the U.S.A. Masonite is chiefly made of longleaf pine (*Pinus palustris*), southern gums (*Nyssa* sp.) and more recently redwood (*Sequoia sempervirens*), but it has been made very successfully from spruce as well as from some hardwoods (*Boehm*, 1930). It may also be made from some vegetable fibers, although boards made from these fibers are generally inferior to those made from wood. Wood is received at Masonite plants in 3 classes:

a) Roundwood or small logs with the bark;
b) Trimmings and edgings from sawmills;
c) Other sawmill residues like bark, splinters, shavings, sawdust etc.

Wood, except the wood residues, is chipped to chips approximately 20 mm (3/4 in.) long. These chips together with the wood residues pass through chip screens which separate them to fine, medium and coarse fractions. The fines, practically all sawdust, are blown into the boilers house to be used a fuel. The medium-sized chips, 10 to 25 mm (3/8 to 1 in.) in length are sent to the chip bins. Longer chips run through hogs or chip crushers and are returned to the screens. Barks or knots are not separated because they do not interfere with the explosions process.

The unique feature of the Masonite-process is that the cellulose-lignin bond is broken and wood fibers are partially rapidly hydrolized under high steam pressure. Natural acids of the wood are developed. The explosion process takes place in high-pressure cylinders, about 1,500 to 1,800 mm (about 5 to 6 ft.) high and 500 to 600 mm (about 20 to 25 in.) in diameter. These cylinders are known as "guns". A longitudinal section through a gun is shown in Fig. 6.39.

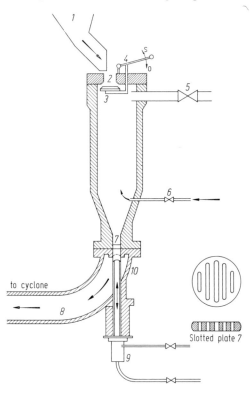

Fig. 6.39. A longitudinal section through a Masonite-fiber gun.
1 Wood chips from storage bin, *2* Inlet for chips, *3* Top valve, *4* Top valve manoeuvering equipment, *5* High pressure steam valve, *6* Steaming valve (*12* steaming inlets inside the gun), *7* Slotted plate; *8* Pipeline for shot fiber to cyclons, *9* Hydraulic manoeuvering equipment for bottom valve, *10* Bottom valve. Masonite AB, Rund-viksverken (1971)

The top valve seals against the inside of the top plate, the bottom valve seals against the discharge pot and is held tight by appropriate hydraulic pressure. The gun is charged with about 120 kg (260 lb.) of wood chips. It is then sealed and steam is admitted to raise the pressure to 40 kp/cm² (about ~570 lb./sq. in.) in about 30 sec (Fig. 6.40 (*a*)). After a steaming period under this pressure, of another 30 sec (*b*), the steam pressure is quickly raised to 70 to 80 kp/cm² (996 to 1,138 lb./sq. in.) (*c*), raising the temperature to 284 to 294 °C (543 to 562 °F) and held for only a few (about 5) seconds at this level before suddenly releasing the pressure (*d*). This is done by opening the hydraulic bottom valve, and accompanied by a loud "blast". "The gun is blown". The chips pass through the slits in the bottom valve, where they explode at once due to the high internal pressure. The structure of wood chips is completely broken and a mass of brown fluffy long fiber bundles is produced. The holding time at the high pressure is very critical and depends on the properties of the chips (wood species, size, moisture content).

Variations in the holding time of only half a second can vastly change the character of the product (*Boehm*, 1944). The exploded fibers pass into a cyclone where
water is added (Fig. 6.41) .The steam is separated and the fibers fall into a storage
chest. The stock at this point is known as "gun fiber". Clear hot water is led into
the stock for washing out the sugars, formed as a result of hydrolysis. At this stage

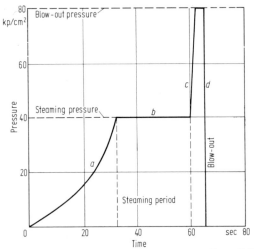

Fig. 6.40. Pressure-time diagram for the Masonite-process. From *Kaila* (1968)

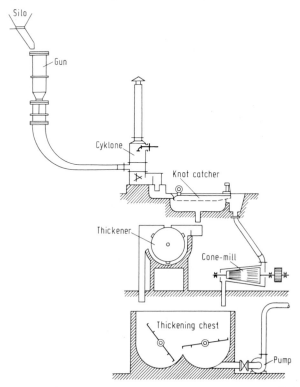

Fig. 6.41. Sequence of operations in the Masonite-process. From *J. M. Voith* (1948)

of the process and throughout the refining the entire mass is maintained at approximately 70°C (158°F). This elevated temperature serves a double purpose:

a) Facilitating the refining of the fiber;
b) Assisting in the penetration of sizing emulsions.

The thermo-chemical phenomena occuring during the Masonité-process are the following:

The acetic and formic acid, formed during the first steps of the procedure, effect the hydrolysis at the peak temperature finally attained. The heat softens the lignin and allows separation of the fibers by the sudden steam expansion. Very little breaking of fibers occurs. Hydrolysis and fibrillation are closely connected during the explosion procedure. The hemicelluloses, hydrolized to sugars, are easily removed by washing. The lignin released from its major bond with the carbohydrates is "activated" so that it is to some extent soluble in methylalcohol. In this chemical condition it can serve as a binding agent for the cellulose fibers when high pressure and heat are applied. The gun fiber is refined by passing through cone mills. The consistency of the stock is maintained by a regulator. The stock is screened to remove fibers not sufficiently refined for the next operation. The produced stock passes through a battery of refiners (*Wiener, Sprout-Waldron* or *Asplund*) adapted to the properties of the raw material. Finally the fibers enter the head box of a modified Fourdrinier-board machine, where water is added to bring the stock to a consistency of 1 to 2%. The pH value of the stock leaving the guns is about 4.5 due to the formation of organic acids. In order to secure proper refining as well as to lessen corrosion of the machinery it is desirable to adjust the pH to about 6.5. Small amounts of water-repellent agents, such as paraffine are frequently added to the mass during the refining. A wet lap, varying in thickness from 19 to 51 mm (3/4 to 2 in.) depending on the subsequent densification and the thickness desired, is formed on the endless wire screen of the Fourdrinier-machine. The wet lap passes over suction boxes and through press rolls. It is still hot and contains about 28 to 35% solids. The dewatered wet lap is cut into about 5.0 m (16 ft.) lengths. The travel speed of the screen in the Fourdrinier-board machine varies between 4 to 41 m/min (13 to 135 ft./min) depending on the product. The cut lengths are discharged into large movable racks each of which holds 20 panels and from the racks they are fed into the presses where they are pressed at 175° to 200°C (347 to 392°F) temperature and under pressure of 6 to 35 kp/cm² (85 to 497 lb./sq. in.). When making insulation board the presses are equipped with stops which hold the boards at the desired thickness. The boards are dried in the press requiring 15 to 30 min. The press is opened, the boards are piled again in the movable racks, and after inspection and grading they run through a humidifier, to raise the moisture content of the board to about 5%. Tempering by treating with various drying oils and heating with high temperatures after pressing improve the water resistance, strength properties and abrasion resistance of special products.

Masonite boards contain about 38% lignin in contrast to 18 to 22% for other hardboards and about 26% for softwoods (*Stamm* and *Harris*, 1953, p. 362). The explosion method is discontinuous, and one worker serves 6 to 7 guns. For the explosion process no electrical energy, but 1.6 to 1.8 metric tons of steam per board is required. A plant with a capacity of 50,000 metric tons of board per year needs 4 to 5 guns. One gun permits 22 to 35 shots per hour and gives 1.5 to 2.0 metric tons of dry fibers per hour. The hardboard produced by this method with the

Table 6.9. Average Physical Properties of Masonite-Boards[1]

Product	Thickness range mm	in.	Density g/cm³	lb./cu. ft.	Water absorption 24 h immerion[2] in% Thinnest stock	Thickest stock	Tensile strength kp/cm²	lb./sq. in.	Modulus of rupture kp/cm²	lb./sq. in.
Structural insulation	13	1/2	0.33	21		37.0	14	200	28	400
Standard presdwood	3.2···8	1/8···5/16	1.01	63	16.0	9.0	218	3,100	420	6,000
Tempered presdwood	3.2···8	1/8···5/16	1.98	67	13.5	4.5	330	4,700	690	9,800
Tempered hardboard	3.2	1/8	1.18	74	6.1		485	6,900	795	11,300
Presdwood tempertile	3.2	1/8	1.23	77	6.0		460	6,500	800	11,400
Lofting board	6.4···10.2	1/4···4	1.28	80		5.6	485	6,900	780	11,200
Die stock			1.38	86	1.9	0.2	540	7,700	880	12,500

[1] Data taken from Masonite-Corporation advertisements, compiled by *Stamm* and *Harris*, 1953, p. 363.
[2] Samples 6 by 8 in. (150 × 200 mm²) by thickness with 1 in. (25 mm) immersion in water at 20°C (68°F).

quality S-1-S (Smooth-one-surface, pressed using a screen) or S-2-S (Smooth-two-surfaces, pressed without screen) may have densities between 0.85 to 1.43 g/cm³.

Different figures are given in the literature for the yield up to 83% (*Kollmann*, 1955, p. 567). *Kaila* (1968, p. 897) lists the losses of raw material in % of oven-dry weight during the explosion-process as follows:

Screening losses (fine fibers)   10 to 15%,
Blow-out losses   8%,
Washing losses   5 to 8%.

6.7 Manufacture of Insulation Board

6.7.1 Sizing

Sizing of insulation board stock is performed with the aim to increase the resistance to water and/or to increase mechanical strength. The required strength can be obtained without additives so that sizing primarily is used to improve water resistance.

The principal types of water repellent sizing agents are rosin, paraffin and cumarone resin (a coal tar product). When the final product is used for exterior sheathing or other severe exposures, asphalt and asphalt emulsions are applied, in the most cases, however, rosin or paraffin emulsions or mixtures of them are used. Residues from tall oil distillation or by-products of the synthetic gasoline process, in general, have not been satisfactory. Commonly about 1% occasionally larger quantities of sizing agent, based on the board stock is added. Too much sizing may not necessarily improve resistance to water absorption (FAO, 1958/1959, p. 55). In certain woods the relatively high natural resin content contributes towards a reduction of water absorption. "To obtain the utmost advantage from the natural resin present, it is essential to use aluminium or ferric sulphate for precipitation and the pH should be as low as 4" (FAO, 1958/1959, p. 55).

The sizing agent is added continuously as an emulsion and well mixed with the pulp, often in separate mixing chests, immediately before the forming machine. In general, the procedures in use for sizing in fiberboard production are similar to those in paper manufacture.

6.7.2 Mat or Sheet Formation

The felting characteristics of the individual fibers have a profound influence on the mat formation. Felting capacity chiefly governs the strength properties of insulation board.

There are two basic forming processes, recognized as wet-felting and air-felting (cf. Fig. 6.7 to 6.11). For all insulation board and most hardboard the sheet or mat is formed from a pulp suspension in water with low consistency (around 1%) over a long period of development.

A principal problem in forming fiberboard is that of maintaining uniformity. Variations in consistency and quality of the pulp influence the thickness of the mat being formed and both pulp quality and mat thickness affect the drying time.

There are three basic methods by which insulation and hardboard mats may be formed:

a) *Deckle box-method*. The deckle box consists of a bottomless frame which can be raised or lowered onto a screen. A measured amount of stock, sufficient for manufacture of one sheet, is pumped into the deckle box and vacuum is applied from the bottom. After drainage of the bulk of the water, a pressure is applied from the top to further dewater and reduce the thickness of the sheet. Subsequently the deckle frame is raised and the sheet is conveyed to the dryer. (Fig. 6.42) shows the method of operation of a small construction-unit.

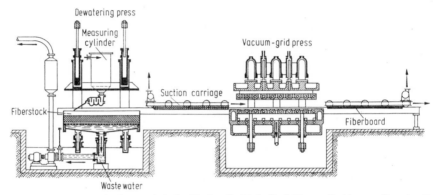

Fig. 6.42. Sequence of operation in manufacturing fiberboard with the Deckel box-method in a small construction-unit. From *J. M. Voith* (1948)

The method goes back to the old technique of papermaking where fibers mixed with water are lifted with a sievelike screen (a ,,mold") from the water in the form of a thin stratum, the water draining through the small openings of the screen. The first patent to make fiberboard in such a way by means of a machine was granted in 1913 to *Sutherland* (U.S. Pat. 1272566). The method has been highly developed in Germany by *Basler* (cf. *Kollmann*, 1955, p. 578, where many patents are quoted).

b) *Fourdrinier-method*. The Fourdrinier-machine (*Henry Fourdrinier*, 1766 to 1854) used in wet-felting and forming sheets for fiberboard, is essentially similar to machines used in papermaking. The wire, however, moves more slowly: 1.5 to 15 m/min (5 to 45 ft./min.) depending on the nature of the fiber and the coarseness of the mesh employed. Suction boxes or a rubber suction belt under the wire assist in dewatering the sheet. A typical Fourdrinier-machine is shown in Fig. 6.43.

c) *Cylinder-method*. The first successful attempt to make paper by a cylinder-type machine was in the U.S.A. in 1817 in the mill directed by *Gilpin* on Brandyvine Creek near Wilmington, Del. At present various-cylinder-machines are in use for fiberboard production but the most common is the single cylinder vacuum-type, consisting of a large drum filter, 2.4 to 4.3 m (8 to 14 ft.) in diameter (Dorr-Olivier Inc., 1957). The drum is covered with a coarse wire and rotates in a vat containing the pulp stock, delivered from the head box at a consistency between 0.75 to 1.5%. The sections which are immersed in the pulp suspension are subjected to vacuum. In the section leaving the pulp suspension, the vacuum is broken and a slight positive pressure assists the sheet in leaving the wire. The sheet is passed through a roll press which further removes water and controls the thickness. Then the sheet is cut to the desired length and fed to the dryer.

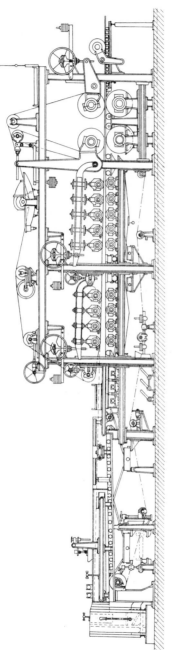

Fig. 6.43. Scheme of a Fourdrinier-machine for the production of fiberboard. From *J. M. Voith* (1948)

6.7.3 Drying

Using Fourdrinier-type machines the continous wet board is trimmed in width and cut into suitable lengths slightly greater than that required after drying. A variety of saws may be used for these purposes. For cutting to length, however, in general circular knife blades are used. They run across the top of the conveyor carrying the wet board, and they are "either moved longitudinally at the same speed as the board, or traversed on the bias at a suitable speed, so that the resulting cut is square with the center line of the board. Trimmings are returned to the system" (FAO, 1958/1959, p. 57).

The moisture content of boards leaving the forming machine may vary between 50 to 80%, based on total wet weight (*Kollmann*, 1955, p. 579, FAO, 1959, p. 57, *Kaila*, 1968, p. 951). The amount of water removal, to attain a moisture content of 1 to 3% (not exceeding 10%, based on oven-dry weight) should be evaporated as quickly as possible, otherwise warping and or delamination at the time of final trimming may occur. The consumption of steam for evaporation of 1 kg water amounts to 1.6 to 1.8 kg (*Segring*, 1947). The heat balance for a fiberboard dryer, in the form of a Sankey-diagram, is shown in Fig. 6.44 (*Segring*, 1947). The consumption of steam for the production of one metric ton of fiberboard may vary between 2.8 and 5.6 metric tons (*Kollmann*, 1955, p. 582, *Kaila*, 1968, p. 953).

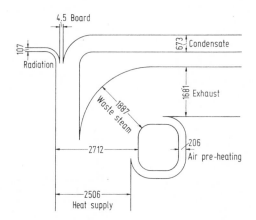

Fig. 6.44. Sankey diagram for the heat balance of a fiberboard dryer. From *Segring* (1947)

In some cases, but only occasionally, it may be efficient to use a predryer. An example is the predryer developed by *Lippke* (1959). Fig. 6.45 shows how this pre-dryer works. The wet fibermat is transported with the aid of an endless felt belt through a system consisting of an endless wire belt and cylinders or rolls. The wire belt is heated electrically (*Joule*'s heat) so that the fibermat reaches a temperature up to 100°C (212°F). Under the influence of heat and low mechanical pressure, the fibermat is brought to a moisture content of 40 to 50%, based on wet weight. The energy consumption amounts to about 70 kWh/metric ton.

Drying may be carried out by any of the following methods:

a) *Tunnel-kilns* using racks or carts to support the material (not effective for high production);

b) *Steam- or hot water-platen dryers* (usually for hardboard);

c) Continuous *multi-deck roller-type dryers*, with circulation of the drying agent (air and steam) according to Fig. 6.46 (*Öholm*, 1959). These roller-type dryers are still most extensively used. They have dimensions and number of decks as follows:

1. *Length* 36 to 330 m (~120 to 1,80 ft.), common length ranges from 45 to 90 m (~150 to 300 ft.). Distance of rolls 200 mm (~8 in.);
2. *Width*, commonly 3.66 m (12 ft.);
3. *Number of decks:* 3 to 20, commonly 8.

d) Continuous (normally) *one-deck roller-dryer with nozzle-circulation*.

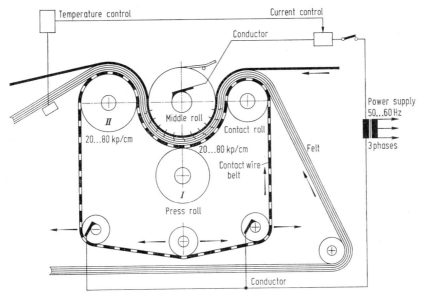

Fig. 6.45. Predryer for wet fiberboard mats, System. From *Lippke* (1959)

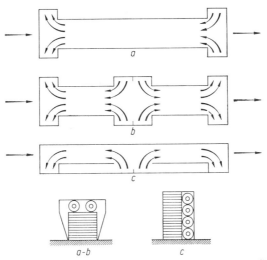

Fig. 6.46. Scheme of three air circulation methods in fiberboard dryers. From *Öholm* (1959)

This system gave a nearly revolutionary impetus to the drying of fiberboard and veneer. The drying agent is led perpendicular to the faces of veneer or fiberboard mats through a series of jet nozzles. The following advantages are attained:

a) Remarkably shorter drying periods;
b) Simpler loading due to a smaller number of decks;
c) Lower consumption of steam due to compact construction, uniform heat distribution and also better heat insulation;
d) Uniform drying over the whole working width;
e) Reduction of the danger of staining.

Fiberboard as well as veneer nozzle-dryers are constructed as already shown in Fig. 3.53. There is a definite sequence of nozzle jets and counteracting rolls. *Öholm* (1959) has discussed the drying of fiberboard. In this connection he published Fig. 6.47 showing typical curves for the reduction of moisture content for various dryers using different changes of temperature along the dryers. The

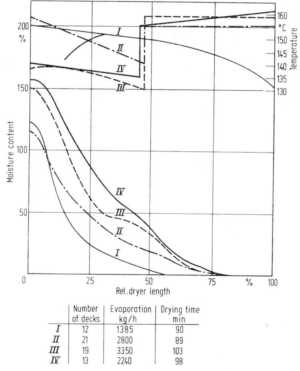

	Number of decks	Evaporation kg/h	Drying time min
I	12	1385	90
II	21	2800	89
III	19	3350	103
IV	13	2240	98

Fig. 6.47. Reduction of moisture content in four dryer systems, using different temperature curves. From *Öholm* (1959)

influence of the number of decks, the length of the dryer and the temperature can be estimated from the curves. Courses of temperatures and moisture content of mat as functions of drying time in a roller dryer are shown in Fig. 6.48 (*Holmgren*, 1948).

Heating is done either by steam with a pressure of 12 to 16 kp/cm² (170 to 228 lb./sq. in.) or by hot water with a temperature of about 200 °C (392 °F). The drying agent consists mainly of air retained for circulation. The heat is led to the fiberboard partly by convection and partly by radiation.

Loading of fiberboard dryers is done automatically; an example is given in Fig. 6.49.

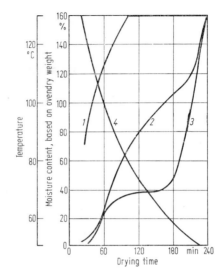

Fig. 6.48. Temperature and moisture content (based on oven-dry weight) versus drying time in a roller-type dryer.
1 Channel temperature, *2* Surface temperature of board, *3* Center temperature in board, *4* Moisture content. From *Holmgren* (1948)

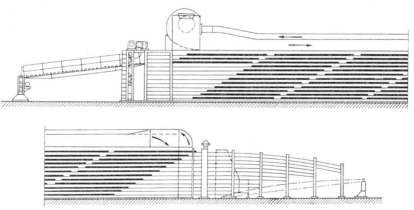

Fig. 6.49. Scheme for loading and unloading a continuous fiberboard dryer. From *Öholm* (1959)

Fiberboards are dried in various zones of the tunnel dryers giving the possibility of a proper control of air velocities, humidities and temperatures. Normally the temperatures used range from 120°C (248 °F) to about 190 °C (374 °F). In some dryers temperatures are above 190 °C (374 °F). On the average the total drying time for 13 mm thick insulation board in conventional roller dryers may range between 1.25 to 4 h and the moisture content of the board leaving the dryer is, as has been mentioned before, between 1 to 3%, based on oven-dry weight. In many cases it may be desirable to introduce air at room temperature into the last section of the dryer to lower the temperatures of the board.

Table 6.10 gives some data for an American roller dryer (length 180 m ∼590 ft., 8 decks, each with 8 sections as follows: 1 charge section, 5 dryer sections, 1 cooling section, 1 discharge section, heating by oil-burners, temperatures from inlet to outlet 230 — 230 — 220— 190 — 177 °C (446 — 446 — 428 — 374 — 351 °F).

Table 6.10. Date for Drying Insulation Board (*Kaila*, 1968, p. 961)

Thickness		Feed speed		Drying time	Board density	
mm	in.	m/min.	ft./min.	h	g/cm³	lb./cu. ft.
3.2	3/8	18.0	59	1.5	0.268	16.7
12.7	1/2	6.5	21.3	4	0.260	16.2
25.4	1	5.0	16.4	5	0.245	15.3

6.7.4 Additional Processing Operations

After the drying process the fiberboards are trimmed to size and either used as such for general purposes or further treated for special applications. Some of the board may be laminated, usually with the aid of asphalt to produce roof insulation board. For exterior sheathing, fiberboards are also coated or impregnated with asphalt.

Various types of acoustical tiles are produced with special equipment. Especially in the U.S.A. fiber insulation boards, are treated with a sealer coat to facilitate painting or finishing. For coating insulation board, various lacquers and paints are recommended.

Special treatments are sometimes given to insulation board. The most important is preservative treatment against insects and fungi with such chemicals as pentachlorophenol, pentachloro-naphthenates and several arsenic salts.

As in the case of wood, surface treatments by dipping or spraying provide only a low protection. The most effective treatment, however, is the distribution of the preservative throughout the board obtained by pressure impregnation.

6.8 Manufacture of Hardboard

6.8.1 Sizing

Sizing of stock for hardboards depends on the type of process: wet-pressing or dry-pressing. For wet-pressed boards the following additives may be used:

a) Paraffin wax emulsions (up to 1% of dry stock);

b) Tall oil derivatives in about the same proportion;

c) Phenol-formaldehyde resins, which not only increase strength but also to some extent water resistance.

The additives are fixed or precipitated on the fibers by aluminium or acids. The final pH ranges between 4.5 and 4.8.

For dry-pressed boards drying oils are usually added in emulsified form to the fiber suspension. They are precipitated on the fibers by ferric sulfate or a mixture of ferrous sulfate and sulfuric acid. The usual pH is between 4.5 and 6.5.

The addition of sizing materials depends upon the need for increased strength and water resistance versus the cost. When hardboard is subjected to a final heat-treatment, additives are not necessary. In the case of low grade raw materials additives may play an important role.

6.8.2 Mat or Sheet Formation

6.8.2.1 Wet Felting. Wet felting is chiefly used for forming the sheets or mats for hardboard. Techniques are similar to those employed in the wet manufacture processes of insulation board (Section 6.7.2). The mats are usually formed on a Fourdrinier-machine (Fig. 6.43).

Occasionally precompressing of the wet mat occurs prior to the hot pressing. It does not present difficulties as the sheet is quite "free" and loses water easily (FAO, 1958/1959, p. 60). In the continuous system of formation, pressing is accomplished by means of a series of roll presses. The wet sheets have an average moisture content between 65 and 70%, based on wet weight. They are then trimmed, cut to length and transferred to the hot presses there using loading and unloading devices.

6.8.2.2 Air Felting. In the air felting processes an air suspension of fibers is used instead of a water suspension. The literature distinguishes between "semi-dry" and "dry" processes (cf. Fig. 6.4, 6.6, 6.9 to 6.11).

Air felting processes were developed about in 1950. The products are still inferior in quality to that obtained by wet felting, but the methods offer advantages when fresh water is limited.

Raw material for use in air felting processes is normally pulped in the same way as for use in wet processes. Dry or nearly dry material is defibrated in disk mills. The chips are steamed prior to refining. The balance of moisture content is complicated since the moisture added by steaming is partly removed by the friction heat in the refiner (Fig. 6.50). Moisture control is necessary. Occasionally

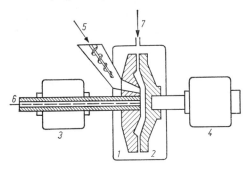

Fig. 6.50. Scheme of operation (refining, gluing and impregnation of fibers) in a Bauer double disk refiner. *1, 2* Grinding disks, *3, 4* Electromotors, *5* Chip feeding screw, *6* Paraffin or wax emulsion supply, *7* Glue supply (under 2 to 3 kp/cm² $\sim$ 28 to 43 lb./sq. in. pressure). From *Swiderski* (1963)

the pulped fibers are dried and usually classified by screens with the aim to separate usable fibers and fines. Additives such as wax and phenolic resins are used in air felting processes (see Fig. 6.9, 6.10 and 6.11). The wax is usually fed with the chips into the digester or may be sprayed on the fiber in emulsified form, after having passed the disk refiners. The amount of wax added is usually about 2.5%, based on the dry weight of the fibers (*Evans*, 1957). The air felting unit consists of a metering device which feeds the fibers to the felting unit. A diagram of a felter is given in Fig. 6.51. The felter may be equipped with a suction box. Other devices such as trimmers and the saw for cutting mats to length are similar to those on wet felters. *Swiderski* (1963) compared the wet, semi-dry- and dry processes for the manufacture of hardboard. He gave some figures for the consumption of raw material and energy (Table 6.11).

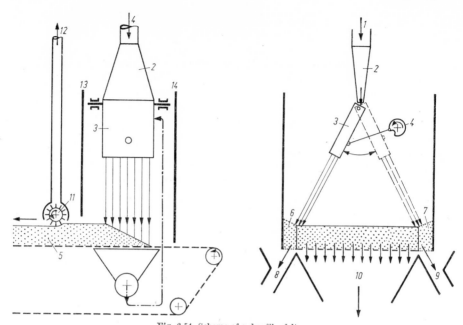

Fig. 6.51. Scheme of a dry fiberfelter.

1 Fiber supply, *2* Feeding funnel, *3* Swing nozzle, *4* Eccentric drive, *5* Traveling belt screen, *6, 7* Lateral fiber excess, *8,9* Suction duct, *10* Main suction box, *11* Toothed rolls, *12* Excess fiber return, *13, 14* Wall of spreading device. From *Swiderski* (1963)

Table 6.11. Consumption of Raw Material and Energy for the Production of Hardboard (3.2 mm Thickness)

	Consumption of				
	Raw wood m³/metric ton	Phenolic resin kg/metric ton	Paraffin kg/metric ton	Tall oil kg/metric ton	Electric energy kWh/metric ton
American procedure	2.1	15···25	20	240	560···610
Czecho-slovakian procedure	2.2	—	11···17	—	360

6.8.3 Hot Pressing

The moist sheets coming from the forming machine are cut by means of circular blades (without teeth) into the lengths corresponding to the dimensions of the platens of the hot press. Conveyors transport the wet sheets to the hot press where they are loaded one at a time into a charging hoist. Generally the hot press has twenty openings and therefore the hoist pushes simultaneously 20 wet sheets into the press. Similar hot presses are used in the particleboard and in the plywood industry. Some principles of pressing are outlined in Section 5.2.9.2.

The press is heated by hot water or steam, in a very few installations by electricity. Hot water is most efficient. The consumption of energy is approximately 8 to 10% lower than for steam heating. There are problems of condensation. Typical diagrams for press cycles are shown in Fig. 6.52. Pressures and temperatures for semi-dry and dry processes are higher than for wet processes. For dry processes the highest pressure amounts to 70 kp/cm² (996 lb./sq. in.) and the highest temperature to 260°C (500 °F). For the semi-dry processes the values are between those for dry and wet processes. The U.S. dry process allows the shortest possible press cycle that is a total of 2.5 min. for boards of 3.2 mm (1/8″). As is shown in Fig. 6.52 the semi-dry process requires 5 min and the wet process about 8 min to attain the same result. The "glueless" process is complicated in its press diagram. *Swiderski* (1963) expressed the opinion that in the future in dry processes a remarkable shortening of press cycles (260 °C = 500 °F, pressing time 75 to 90 sec) may be reached. If the mats are wet, a supporting wire screen is necessary so that the steam can escape. The screen makes an impression on the underside of the hardboard. It reduces not only the homogeneity, but also the trength of hardboard to some extent. After proper pressing the sheets are simultaneously unloaded onto a discharging hoist. From this hoist the hardboards, still with a temperature of approximately 125 to 130°C (257 to 266 °F) are moved to heat treatment chambers.

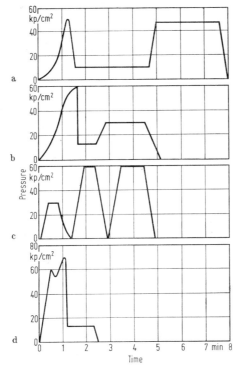

Fig. 6.52 a—d. Press diagrams for fiber hardboard, 3.2 mm (1/8″) thick.
a) Wet process 200°C (392°F) 50 kp/cm² (711 lb./sq. in.); b) Semi-dry process, 170°C (338°F) 60 kp/cm² (853 lb./ sq. in.); c) Glueless dry process 200°C (392°F), 60 kp/cm² (853 lb./sq.); d) American dry process 220°C (428°F), 70 kp/cm² (996 lb./sq. in.).

Hardboard is pressed in two ways:

a) With a screen back thus obtaining only one smooth side (S-1-S). For uses where thicknesses must be within close tolerances (e.g. in laminating or in flush door manufacture), the screen-back is often sanded or planed to caliper;

b) With both faces smooth (S-2-S) when a board is pressed from a dry mat.

6.8.4 Heat Treatment and Oil Tempering

The *heat treatment* is now quite universal in the hardboard industry. The heating of the hardboards at appropriate high temperatures increases some mechanical strength properties, especially the modulus of rupture, but the shock resistance (impact strength) is lowered. According to Swedish investigations the modulus of rupture, after heat treatment with hot air of 160 °C (320 °F) for 8 h, increases by approximately 25%. At lower temperatures a longer heat treatment gives approximately the same increase. There is a maximum temperature-time combination beyond which the strength properties begin to decrease. In practice the 1/8″ (~3.2 mm) thick hardboards are heated for about 5 h in the range of temperatures between 150 to 165 °C (302 to 329 °F), occasionally only between 120 to 140 °C (248 to 284 °F). The treatment is in most cases fully automated.

A relatively small amount of hardboard, not previously subjected to heat treatment, is treated by *oil tempering*. The aim is to improve the strength, water resistance, weathering properties and resistance against abrasion. Oil tempering is occasionally done by mixing an oil emulsion with the pulp. This method is not economical as the drainage rate is reduced, the speed of the forming machine lowered and plate lubricants must be used.

The conventional method of oil tempering is to impregnate the hardboard after hot pressing. The chemicals used can be drying oils such as linseed oil, tung oil, perila oil, soybean oil, tall oil or some alkyd resins. These additives are used in the range of 4 to 8% based on oven-dry weight of board. Petroleum products and phenolic resins, latex and lignin compounds are recommended, but used to a limited extent in practice.

Usually the hot pressed board is passed through a bath, containing the oil at a temperature just above the boiling point. Then the boards are exposed in a kiln to a temperature of 160 °C to 170 °C (320 °F to 338 °F) for 6 to 9 h. Under these conditions not only hardening of the oil, but also a reaction with the fiber takes place.

6.8.5 Humidification and Conditioning

Hot pressed hardboard and heat- or oil tempered hardboards have a moisture content considerably below the normal equilibrium moisture content. Therefore it is desirable to humidify the boards to a moisture content of about 5 to 8%. Warping should be avoided.

Humidification may be carried out in chambers or tunnels with various systems of loading and conveying. Normally the relative humidity is 80 to 85%, the temperature 38 to 50°C (~100 °F to 122 °F). The time required for humidification is normally between 5 to 6 hours. More recently a saturated atmosphere at 60 °C (140 °F) is recommended.

A special case is the conditioning of S-2-S boards by spraying with water. Thus the boards are preliminarily cooled and then immersed in water for approximately one hour. A moisture content between 10 and 12% is reached by this dipping method when the boards are stacked for approximately 30 h.

6.8.6 Trimming and Additional Processing

After conditioning boards are trimmed to the required size. When boards are oil tempered, trimming takes place before the tempering operation (FAO, 1959, p. 63). The waste caused by trimming roughly amounts 6 to 7% in volume.

Formerly it was burnt, but now it is frequently recovered by chipping, shredding, steaming and mild cooking. Repulped trim is generally of lower quality than the original fibers.

Additional manufacturing operations are used for various purpose. Improvement of the surfaces of the board should be mentioned: both plain and tile patterns are manufactured. Grooves are formed by special tools. Leather surfaces can be simulated. Paper overlays as well as perforated board are made.

6.8.7 Losses and Yield in Manufacture, Production Costs

Harstad (1956) measured the losses of raw material in the manufacture of hardboard made from: a) barkfree spruce chips, and b) chips from slabs with bark. Figs. 6.53 and 6.54 show his results. Screens for chips with bark must have greater openings (8 mm or 5/16 in.) than screens for chips of debarked wood (5 mm or 3/16″). The milling loss in defiberation amounted to 11.8% for barkfree chips and was increased to 31.5% for chips with bark. Based on the weight of bark only

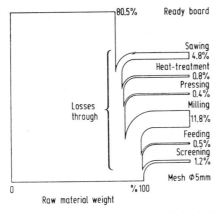

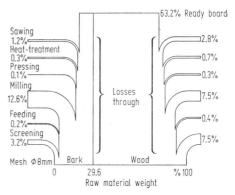

Fig. 6.53. Raw material losses in per cent of weight in the manufacture of hardboard made from bark-free spruce chips. Mesh 5 mm diameter. From *Harstad* (1956)

Fig. 6.54. Raw material losses in percent of weight in the manufacture of hardboard made from chips from slabs with bark. Mesh 8 mm diameter. From *Harstad* (1956)

the loss by defiberation was $(12.6/29.6) \times 100 = 42.5\%$. The bark content of 29.6%, based on the total weight of raw material reduced the overall yield, from 80.5% to 63.2%. Prior to trimming the yield in manufacturing insulation board (from chips with bark, losses of chips not included) is 88 to 96%, in manufacuturing S-1-S hardboard by conventional methods 88 to 94%, and by the Masonite-process 80%. Trimming the boards to the required size entails additional losses between 2 and 7%. The losses depend on a series of various factors as follows (*Kaila*, 1968, p. 1038):

a) Quality of the raw material;
b) Amount of bark;
c) Time, pressure and temperature of preheating;
d) Method of defiberation;
e) Cutting to length;
f) Secondary products and rejects;
g) Wastes in sawing to fixed dimensions;
h) Moisture content;
i) Fiber recovery.

Preheating must be carefully done to avoid excessive hydrolysis. Control is possible by measuring pH value (>3.5).

Cutting hardboard to normal sizes causes losses between 2 and 6%. Milling or any other reductions of these residues consumes much energy, the fibers are of poor quality (too short and brittle), and the color of the regained pulp is too dark and may cause black flecks in the boards. Hardboard wastes are very ignitable.

Residues from sawing insulation boards return to a refiner and are mixed with the primary pulp. This is the reason why the total yield in insulation board manufacture normally is 3 to 4% higher than in hardboard manufacture (*Kaila*, 1968, p. 1040). The losses mentioned above originate in a productive plant. The total economic efficiency of a fiberboard plant depends on further factors such as storage, handling, factory management, operating time, marketing, competition, and terms of sale.

There are interesting relationships between production costs per ton and per year as a function of daily mill capacity (Fig. 6.55, FAO 1958/1959).

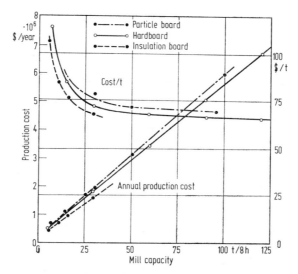

Fig. 6.55. Production costs as function of mill capacity. FAO (1958/59, p. 88) (The figures are out of date.)

Total annual production costs Z for fiberboards as well as for particleboards is an inverse function of mill capacity x:

$$Z = \frac{C}{x} + K, \tag{6.1}$$

where C and K are constants.

The first derivative of this hyperbolic function shows that the saving in production cost per ton for a given unit increase in capacity reflects the incentive to establish larger production units:

$$\frac{dZ}{dx} = \frac{-C}{x^2}. \tag{6.2}$$

The equation demonstrates that a high C value means a greater incentive to erect larger units than does a low value. The C values for the production costs corresponding to different types of fiberboard are relatively close together for fiberboard and particleboard.

6.8.8 Special Fiberboards

It is beyond the scope of this book to describe manufacture, properties and application of special fiberboards. Therefore they may only be listed alphabetically using nomenclature and definitions used by FAO (1958/1959):

a) *Acoustical board*, panels specially fabricated with numerous holes, grooves or other sound traps for interior use in ceilings and walls where reduction of sound reflection is desired (cf. Section 5.4.6). Usually they are backed with an insulation material;

b) *Bituminous board*, rigid insulation board treated with asphalt or other bitumen to make the board more water resistant. The addition may be done as emulsion, as powder or as melt. Boards for packaging contain 1 to 2% asphalt, boards for framing with plaster applied directly to the face 8 to 10% asphalt, boards for exterior walls 20 to 30% asphalt. Asphalt emulsion may be combined with resin glues and possibly wax emulsion. 1/2″ (13 mm) thick bituminous board with a basis weight of about 3.5 kg/m² (0.72 lb./sq. ft.) and a density of 0.26 g/cm³ ($\sim$16 lb./cu. ft.) have a tensile strength of about 14 kp/cm² ($\sim$200 lb./sq. in.) parallel and across machine direction, and a tensile strength perpendicular to surface of about 7 kp/cm² ($\sim$100 lb./sq. in.). Water absorption after a 2 h immersion is about 4.4 Vol.-% (after *Kaila*, 1968, p. 990). Bleeding of the asphalt additives should be prevented. Insulation board of sheathing grade is used for insulating the exterior of buildings. Boards treated with a high-melting-point-asphalt can be covered with granules of coloured stone, surfaces are similar to brick;

c) *Composite board*, any combination of different types of board, either with another type of board or with another sheet material. The composite board may be laminated in a separate operation or at the time of pressing. Exemples are hardboard-faced insulation board, and metal-faced hardboard.

d) *Decorative laminated fiber (building) board*, one or both faces overlaid with paper or other carriers impregnated with thermosetting resins. The overlays are hot pressed to the board. Quality requirements according to German Standard DIN 68751 (January 1973) are minimum bending strength (in the average in both machine directions) 500 kp/cm² ($\sim$7,000 lb./sq. in.), cigarette proof (permitting only staining and reduction of brightness), no permanent changes under the influence of hot cooking ware, water, steam, light or chemicals as listed in German Standard DIN 53799, resistance against indentation of a falling steel ball (test according to DIN 53799); no formation of cracks, and dimensional stability in changing climates;

e) *Embossed hardboard*, pressed with patterned cauls so that the surface receives a characteristic appearance, e.g. simulating some other material such as ceramic tile, textile fabrics or leather;

f) *Interior finish boards*, rigid insulation board—may be in the form of plank, board, panels or tile—for interior use with a factory-applied paint-finish.

g) *Perforated hardboard* with factory-punched or drilled holes, closely spaced, used for decorative displays, with special hangers for storage racks and walls, and for sound absorption;

h) *Prefinished panels* are compressed fiberboards with a factory applied, baked on enamel or similar finish (lacquer), e.g. blackboard, etc.;

i) *Prime-coated board*, fiberboard with an initial coating of a sealer or paint, to satisfy a particular use requirement or to enhance its use in a certain way;

k) *Roof insulation board*, rigid insulation board, manufactured for use as roof deck insulation.

6.9 Properties of Fiberboard

6.9.0 General Considerations, Tests, Uses

The properties of insulation board as well as half-hard board and hardboard vary considerably. There are differences not only due to the raw material used and the basic manufacturing process used but also resulting from subsequent processing such as gluing, impregnation, heat-treatment, and tempering. Furthermore the methods of testing and their evaluation vary. The specifications in various countries are different. Therefore it is very difficult to compare figures for physical, mechanical, or technological properties, obtained from various sources. Nevertheless, general surveys and trend analyses are valuable.

Markwardt in 1957 wrote in the introduction to a survey of testing methods for fiberboards: "The increasing importance of fiberboards as reflected by the continuing increase in production, the diversity of properties, uses and applications, and the significant contribution of this form of product to improved forest utilization, has been properly recognized by ... FAO". This organization has instigated the exchange of technical informations and research findings, an endeavor especially important and promising for the board industry.

Markwardt pointed out that insulation board, hardboard and half-hard board, and particleboard have overlapping uses and applications and that it is desirable to establish test methods for evaluating physical and mechanical properties, "that are applicable to all fiberboard regardless of type." There are no principal technical difficulties in this approach since test methods and equipment which have long been used for wood and other wood-based material may be applied or adapted.

From the standpoint of purpose of tests three distinct fields of application must be considered:

a) Methods of test for the general evaluation of physical and mechanical properties.

b) Quality control procedures, for use by manufacturers, in order to maintain the desired quality.

c) Acceptance tests, as a mean of establishing product quality in relation to specification requirements for general or for specific uses.

In the scope of this book only methods of test according to point a) will be briefly discussed under reference to Section 5.4.

Size and appearance of boards should be thoroughly described in connection with any test. If only small samples are available the size (length, width, thickness) of the original board must be declared by the manufacturer. It is not yet possible to determine exactly the color of a fiberboard so that only general remarks such as "white-yellow", "medium-brown" or "dark-brown" are used. The description of the surfaces is simpler: screen-back board, is ucommonly called "S-1-S" board, or two smooth surfaces, known as "S-2-S" hardboard. Heat treatment or oil tempering, respectively, can be needed and must be certified by the manufacturer. Subsequent tests of strength and water resistance make heat treatment or oil tempering evident.

Interior type insulation board is often furnished with a factory-applied paint finish. Boards for more severe service are usually darker in color, due to asphalt additives (or coatings) used. In a more recent type of plywood, face veneers are replaced by hardboard.

In Section 6.0 present definitions of various types of fiberboard are given and it is shown that limits and ranges of density are characteristics for soft or insulating board, half-hard board, and hardboard. For wood, plywood and particle board the density mainly influences the mechanical properties such as elasticity, strength, hardness, abrasion resistance, nail holding power etc. Other physical properties such as sorption and swelling, behavior during drying, thermal properties, electrical and acoustical properties also depend upon density. Physical and mechanical properties of fiberboard are tested in various countries according to national standards, for instance ASTM 1037, AFNOR, DIN. Recently ISO Technical Committee 89/SC 1 unanimously came to the following Standards or Recommendations: ISO 766, 767, 768, 769, 818, 819.

In comparison to wood and plywood mechanical tests of fiberboard are rarely made. Insulation board are used with respect to their low thermal conductivity and sound insulation purposes, half-hard board as underlayment for flooring, and hardboard for panelling, ceilings, and partitions. Factory finished hardboard in ceramic-tile or other patterns is used in increasing quantities as interior wall covering in kitchens and bathrooms and as an exterior covering for houses and other buildings, as a maintenance or modernizing material in house repair and improvement, in furniture and cabinet applications as flat sheets and curved to shape for door construction, perforated for backs of radio and television cabinets, for panelling in automobiles, truck bodies, railway passenger cars and freight cars, as a lining for concrete forms (shuttering), for novelties and special items such as cut-out letters for advertising, fixtures for stores, toys, games, work bench tops, trunks etc. The U.S. War Production Board listed more than 300 uses for hardboard (FAO, 1958/1959, p. 116—119).

Summaries of properties for insulation board and hardboard are presented in Table 6.12 and 6.13.

The properties of insulation board listed in Table 6.12. with FAO (1958/59) mentioned as reference, are based on a density range from 0.25 to 0.40 g/cm³ (15.6 to 25 lb./cu. ft.). The greatest quantity of insulation board is world-wide produced in this range. The lower figures according to *Kollmann* (1951) show that in West-Germany special attention has been paid to high heat insulation. Values of strength and other physical properties may be influenced to some extent by the test procedure used and data mentioned can be considered indicative only. Insulation board in the density range of 0.02 to 0.15 g/cm³ (1.25 to 9.4 lb./cu. ft.) has been produced (mainly by air-felting with a starch or similar binder) on a limited basis for special heat insulation, low cost padding, and cushioning.

In the U.S.A., in view of the large field of end-purposes for which the board is used, the quality of hardboard varies between wider limits, than in Europe, where the tendency on the market is towards more uniform qualities. In both areas, however, two basic types dominate: standard hardboard and oil tempered hardboard (also called superhardboard). In Scandinavia all standard hardboard is heat treated.

Table 6.12. Some Physical and Mechanical Properties of Insulation Board

Property	Values in metric units	Values in English units	Reference
Density g/cm³ or lb./cu. ft.	0.17···0.28 0.25···0.40	10.6···17.5 15.6···25	Kollmann (1951) FAO (1958/59)
Modulus of rupture (bending strength) kp/cm² or lb./sq. in.	10···27 15···55	142···384 213···782	Kollmann (1951) FAO (1958/59)
Modulus of elasticity in bending kp/cm² or lb/sq. in.	800···4,000 1,780···8,800	11,400··· 56,900 25,000···125,000	Kollmann (1951) FAO (1958/59)
Tensile strength parallel to surface kp/cm² or lb./sq. in.	5···16 15···35	71···228 213···500	Kollmann (1951) FAO (1958/59)
Tensile strength perpendicular to surface kp/cm² or lb./sq. in.	0.7···1.70	10··· 26	FAO (1958/59)
Water absorption (24 h immersion at 20°C) % volume % weight	5··· 15 30···100 15··· 60	5··· 15 30···100 15··· 60	FAO (1958/59) Kollmann (1951) FAO (1958/59)
Maximum thickness swelling Maximum linear expansion[1] % %	12··· 20 0.50	12··· 20 0.50	Kollmann (1951) FAO (1958/59)
Thermal conductivity kcal/h m °C BTU in./h sq. ft. °F	0.028···0.048 0.035···0.056	0.22···0.33 0.27···0.45	Kollmann (1951) FAO (1958/59)
Sound absorption of acoustical board[2] %	50···85	50···85	FAO (1958/59)

[1] Change in linear dimension when a board with equilibrium moisture content in an atmosphere of 50% R. H. is brought to equilibrium moisture content at 97% R. H. at 20°C (68°F).
[2] At the frequency of 522 Hz.

6.9.1 Density, Weight per Unit Area

Density influences most properties of fiberboard. The ISO-Recommendation R 818 (Sept. 1968) for classification of fiber building boards according to their density is mentioned in Section 6.0. Figures obtained in practice are compiled in Tables 6.1, 6.12 and 6.13.

Density and weight per unit area can be computed using Eqs. (5.2) or (5.3), p. 466. Details of measurement are explained in Section 5.4.3. For solid wood, plywood and particleboard a strong correlation exists between average density, water absorption and swelling, thermal and acoustical properties, elasticity and strength properties, technological properties, and reaction to fire. For fiberboard methods of manufacture, especially freeness of pulp, pH value, type and amount of glue and additives, temperature, pressure and pressing time of hardboard, heat treatment and oil tempering determine properties and behavior of the board in an altogether complex and complicated manner.

6.9.2 Moisture Content, Absorption and Swelling

Determination of moisture content should be carried out as described in Section 5.4.4. The test pieces may have nearly any shape and dimensions, but should have an area between 60 and 300 cm² and should preferably be 10 cm × 10 cm (ISO R 767, June 1968).

The hygroscopic behavior of fiberboard has been investigated many times (*Kollmann* and *Mörath*, 1934; *Vorreiter*, 1941; *Lundgren*, 1958, 1969, *Kaila*, 1968). Hysteresis loops for moisture content, length change and thickness change in complete relative humidity cycles 0 to 85 to 0% at 20°C temperature are reproduced in Fig. 6.56 for two types of fiberboard. Extrapolated to fiber saturation point, which is a rather dubious concept because the moisture content of

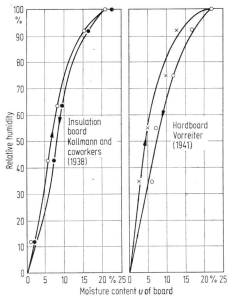

Fig. 6.56. Hysteresis loops for adsorption and desorption of insulation board and hardboard. From *Kollmann* (1938) and *Vorreiter* (1941).

Table 6.13. Physical and Mechanical Properties of Hardboard

Property		Values in metric units	Values in English units	Reference
Density	g/cm³ or lb./sq. ft.	Standard 0.90···1.05	55··65	FAO (1958/59)
		0.88···1.05	54··65	Kollmann (1951)
		Masonite 1.03···1.11	63··68	Kollmann (1951)
		Oil tempered 1.02···1.06	63··66	FAO (1958/59)
Modulus of rupture (bending strength)	kp/cm² or lb./sq. in.	Standard 300···550	4,300··· 7,800	FAO (1958/1959)
		280···550	3,900··· 7,800	Kollmann (1951)
		Masonite 620···780	8,800···11,100	Kollmann (1951)
		Oil tempered 450···700	6,400···10,000	FAO (1958/59)
Modulus of elasticity in bending	kp/cm² or lb./sq. in.	Standard 28,000···56,000	400,000··· 780,000	FAO (1958/59)
		25,000···40,000	355,000··· 570,000	Kollmann (1951)
		Masonite 67,000	950,000	Kollmann (1951)
		Oil tempered 56,000···70,000	800,000···1,000,000	FAO (1958/59)
Tensile strength parallel to surface		Standard 210···400	3,000···5,700	FAO (1958/59)
		140···330	2,000···4,700	Kollmann (1951)
		Masonite 290	4,100	Kollmann (1951)
		Oil tempered 450···550	6,400···7,800	FAO (1958/59)

Table 6.13 (Continued)

Property		Values in metric units	Reference
Water absorption 24 h-immersion at 20 °C	% weight or volume	Standard 14…25 10…30 Oil tempered 8…20	Kollmann (1951) } FAO (1958/59)
Maximum linear expansion[1]	%	Standard 0.60 Oil tempered 0.40	} FAO (1958/59)
Coefficient of thermal conductivity	kcal/m h °C BTU, in./h sq. ft. °F	Standard 0.13 1.10 0.095 Oil tempered 0.15 1.20	} FAO (1958/59) Kollmann Vol. I p. 249, extrapolated } FAO (1958/59)

[1] Change in linear dimension when a board with an equilibrium moisture content of 50% R. H. is brought to equilibrium moisture content at 97% RH at 20 °C (68 °F).

all wood-based and other hygroscopic materials increases rapidly when the humidity of air exceeds 80% (cf. Vol. I of this book, p. 192—195), the following order may be estimated:

	Moisture Content at FSP 20 °C %	Reference
Solid beechwood, untreated	28	*Weichert*, 1963
Solid beechwood, heat treated at 180°C	20	*Kollmann* and *Schneider*, 1963
Insulation board	~25 ⎫	
Hardboard	~18 ⎬	*Lundgren*, 1958
Oil tempered hardboard	~14 ⎭	

The remarkable differences in water absorption and swelling of 1/8 in. thick standard hardboard and oil tempered hardboard samples immersed in water at room temperature are reflected in Fig. 6.57. The reduction of water absorption of 1/8 in. thick oil tempered hardboard in 48 hours immersion tests as compared with standard hardboard in 24 hours immersion tests was shown by *Kumar* (1961) in frequency distribution graphs (Fig. 6.58).

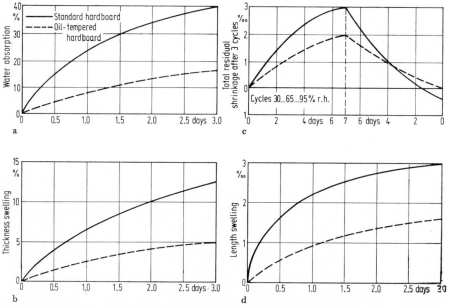

Fig. 6.57a—d. Changes of properties of 1/8 in. thick standard hardboard and oil tempered hardboard samples immersed in water at room temperature. From *Kaila* (1968, p. 1004)

Carlsson (1958) investigated the effects of manufacturing factors on the quality of hardboard. Some of his laboratory results may be quoted:

1. *Freeness* is a measure of the intensity of defibration. High freeness (e.g. in Defibrator-seconds) means partial destruction of the lignin lamella and thus better access to the hygroscopic hemicelluloses. The pulp becomes wet-beaten. Hence follows the production of hardboard with higher water absorption and thickness swelling (Fig. 6.59). It will be stated later that relatively high freeness improves the strength properties to some extent.

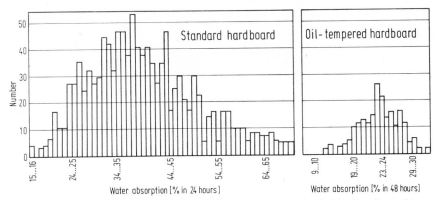

Fig. 6.58. Frequency distribution graphs of water absorption of 1/8 in. thick standard hardboard and oil tempered hardboard in water immersion tests (duration 24 respectively 48 h). From *Kumar* (1961)

2. *Sizing* is done by introducing the sizing agent into the pulp. The chief sizing agents are rosin, artificial resins of the phenolic type, paraffin, starch, vegetable protein (soybean glue), and albumin (*Höglund* 1957 in *Kaila*, 1968, p. 1104, Ref. 97). Sizing does not reduce the hygroscopicity of the fiber, as do heat treatment and oil tempering. It merely reduces the rate at which water and other polar liquids penetrate the fibers and the fibrous mat (*Stamm* and *Harris*, 1953, p. 331).

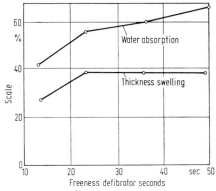

Fig. 6.59. Effect of milling intensity expressed as freeness in Defibrator-seconds on water absorption and thickness swelling of hardboard. Sprucewood with bark, no glue, no tempering. From *Carlsson* (1958)

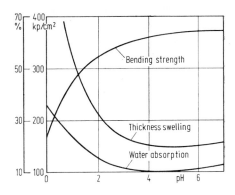

Fig. 6.60. Effect of pH value on properties of hardboard glued with 3% phenolic resin. From *Kaila* (1968, p. 998)

Phenolic resin is used in aqueous solution. The phenolic resin used in the wet process has 41% solids content, a specific weight of 1.190, pH = 10.5, a viscosity between 150 and 170 cP at 25 °C (= 77 °F) and a pot life of about 3 months at 25 °C (*Kaila*, 1968, p. 990). The sizing agents, or mixtures of them, are added to the pulp in the mixing box at a pH of 4.4 to 5.5. The effect of pH on water absorption, thickness swelling, and bending strength of hardboard is shown in Fig. 6.60. For constant pH = 4.5 up to 10% resin content the *bending strength* of hardboard increases approximately parabolically whereas *absorption of water* and *thickness swelling* decrease nearly hyperbolically with increasing resin amount (Fig. 6.61). Good results are obtained with 2 to 3% resin, but economic reasons usually limit the addition to 0.5 to 2%.

It is necessary to add an aluminum-fixing agent $(Al_2(SO_4)_3$ to precipitate the size. Fixing may also be effected by addition of sulfuric acid (H_2SO_4). Water absorption is less for addition of aluminum than of sulfuric acid (Fig. 6.62). The influence on bending strength is much less marked. The amount of aluminum needed for good sizing is "one to one and one-half times the weight of

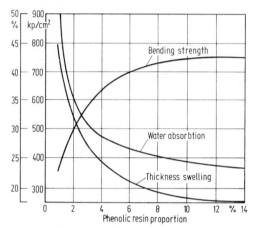

Fig. 6.61. Effect of phenolic resin proportion (pH = 4.5) on properties of hardboard. From *Kaila* (1968, p. 998)

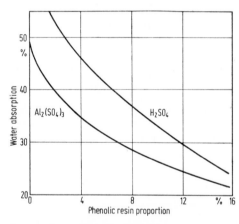

Fig. 6.62. Effect of phenolic resin proportion on properties of hardboard using $Al_2(SO_4)_3$ or H_2SO_4 as fixing agents. From *Kaila* (1968, p. 998)

rosin, which is considerably more that than required to just precipitate the rosin" (*Stamm* and *Harris*, 1953, p. 332). It is not yet completely clear to which extent the aluminum precipitates insoluble aluminum soaps and to which extent the electropotential charge on the cellulose is reversed, thus aiding precipitation. The relationships between aluminumsulfate addition to the pulp and water absorption, thickness swelling and strength properties of not tempered hardboard are illustrated in Fig. 6.63 (*Carlsson*, 1958). "Care should be taken not to use too large an excess of aluminum, as its presence in the final product, because of its acidic properties, causes a hydrolytic deterioration" (*Stamm* and *Harris*, 1953, p. 332) of the board.

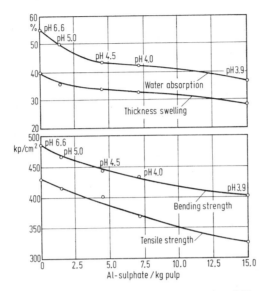

Fig. 6.63. Effect of aluminumsulfate addition to pulp on water absorption, thickness swelling, bending strength and tensile strength of not tempered hardboard. From *Carlsson* (1958)

3. In the wet process the *moisture content,* based on wet weight *of the fiber mats* prior to hot pressing, is in the range between 65 and 72%. There is a strong correlation between both water absorption and bending strength, and content of moisture or dry substance. *Carlsson* (1958) found that the water absorption of hardboard increases whereas the bending strength decreases with increasing proportion of solids (dry matter content) in the mat before it will be pressed (Figs. 6.64 and 6.65). In practice a compromise is necessary between heat economy and strength requirements.

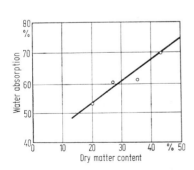

Fig. 6.64. Effect of dry matter content in the wet mat before hot pressing on the water absorption of finished hardboard, Spruce wood with bark, no glue no tempering. From *Carlsson* (1958)

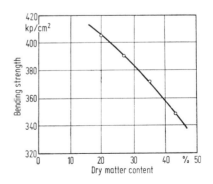

Fig. 6.65. Effect of dry matter content in the wet mat before hot pressing, on the bending strength of finished hardboard. Spruce wood with bark, no glue, no tempering. From *Carlsson* (1958)

4. Hardboards are pressed and dried in multiple-platen hydraulic presses at *pressing temperatures* of 170 °C to 240 °C (338 °F to 464 °F). The conditions of pressing will vary with the nature of the stock, the moisture content of the mats entering the press, and the final density. The effect of the pressing temperature on water absorption and thickness swelling is shown in Fig. 6.66 and a corresponding diagram reflecting bending strength versus pressing temperature is reproduced in Fig. 6.67 (*Carlsson*, 1958). Hardboard with a density of 0.95 (60 lb./

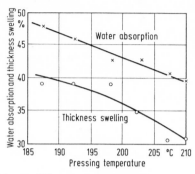

Fig. 6.66. Effect of pressing temperature on water absorption and thickness swelling of finished hardboard. Spruce wood with bark, no glue, no tempering. From *Carlsson* (1958)

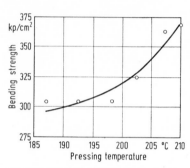

Fig. 6.67. Effect of pressing temperature on bending strenght of finished hardboard. Spruce wood with bark, no glue, no tempering. From *Carlsson* (1958)

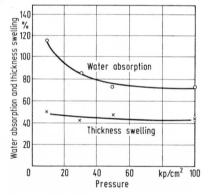

Fig. 6.68. Effect of pressure in hot pressing on water absorption and thickness swelling of finished hardboard. Spruce wood with bark, no glue, no tempering. From *Carlsson* (1958)

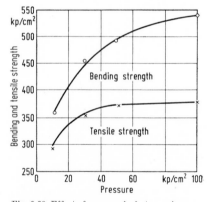

Fig. 6.69. Effect of pressure in hot pressing on bending strength and tensile strength of finished hardboard. Spruce with bark, no glue, no tempering. From *Carlsson* (1958)

cu. ft.) can, in general, be made at *pressures* less than 17.5 kp/cm² (∼ 250 lb./sq. in.). When densities much above 1.0 are desired, special high-pressure presses are needed. The relevant effects of pressure on water absorption, thickness swelling, and strength properties are shown in Figs. 6.68 and 6.69 (*Carlsson*, 1958). *Hechler* (1959) published results of laboratory experiments (Tables 6.14 and 6.15).

In order to avoid trapping of steam between the polished press platens and the boards which might cause an explosion of the board, a screen is inserted into the press with the board. Steam then can escape through the screen wires. An imprint of the screen is left on the back-surface of the board (quality S-1-S).

Table 6.14. Influence of Pressure, Pressing Time and Temperature on the Properties of Hardboards (*Hechler*, 1959)

Pressing temperature °C (°F)	Pressing schedule	Total press. time min	Pressure kp/cm²	Appearance	Density kg/m³	Thickness mm	Bending strength kp/cm²	Water absorption %	Thickness swelling %
160 (319)	1···4···1	6	28···14···28	Normal	948	2.88	391	56.2	26.2
	1···3···1	5	56···28···56	,,	1,015	2.65	488	60.2	29.2
180 (356)	1···4···1	6	28···14···28	,,	960	2.81	455	49.8	24.1
	1···3···1	5	56···28···56	,,	1,020	2.61	521	57.0	28.1
200 (392)	1···3···1	5	28···14···28	,,	950	2.80	488	46.9	20.1
	1···2···1	4	56···28···56	,,	1,031	2.55	550	54.0	25.9
220 (427)	1···2···1	4	28···12···28	Spots	968	2.78	520	45.6	21.8
	1···1···1	3	56···28···56	,,	997	2.49	572	40.0	23.5
	1···1···1	3	56···28···56	,,	—	—	—	—	—
240 (464)	1···2···1	4	28···14···28	Overdone	1,061	2.14	307	25.2	11.6

Table 6.15. Influence of Pressing time on the Properties of Hardboard (*Hechler*, 1959)

Pressing temperature °C (°F)	Pressing schedule	Total press. time min	Pressure kp/cm²	Appearance	Density kg/m³	Thickness mm	Bending strength kp/cm²	Water absorption %	Thickness swelling %
180 (356)	1··· 1···1	3	58···28···58	Normal	952	3.30	291	107.2	71.5
	1··· 2···1	4	58···28···58	,,	965	2.78	407	75.0	51.0
	1··· 3···1	5	58···28···58	,,	964	2.52	483	73.4	47.1
	1··· 4···1	6	58···28···58	,,	983	2.40	504	68.4	51.2
	1··· 6···1	8	58···28···58	,,	982	2.30	456	56.7	32.5
	1··· 8···1	10	58···28···58	,,	005	3.05	404	58.5	38.0
	1···10···1	12	58···28···58	,,	022	3.10	516	53.8	32.9

When the smooth surfaces are wanted (quality S-2-S), the mats must be partially dried before pressing between two smooth cauls. The higher the density of the final board, the greater the proportion of solid fibers originally present thus the shorter the time for pressing. 1/8 in. (3.2 mm) thick hardboard can, in general, be press-dried in a 10 to 15 min cycle.

5. *Heat treatment* causes chemical changes in fiberboards as follows:

a) Under adequate temperature-time conditions a loss of water of constitution occurs;

b) The initial thermal degradation of wood constituents results in furfural polymers of breakdown sugars that are less hygroscopic than the hemicelluloses from which they are formed (*Stamm*, 1964, p. 317);

c) A cross-linking reaction in which water is eliminated between hydroxyl groups on two adjacent cellulose chains, with the formation of ether linkages (*Stamm* and *Hansen*, 1935) is now well disproved;

d) The square of the reduction in swelling is proportional to the weight loss (*Stamm*, 1959);

e) Wood and fiberboards acquire considerable decay resistance by heating (*Stamm* and *Baechler*, 1960).

Back and *Klinga* (1962) investigated the various reactions which take place during dimensional stabilization of paper and fiber building boards by heat-treatment. Fig. 6.70 shows that a treatment with 170 °C (338 °F) is more effective

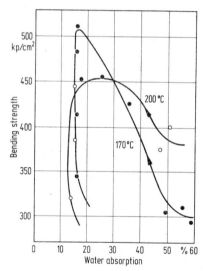

Fig. 6.70. Effect of pressing temperature on water absorption and bending strength of 1/8 in. thick hardboard. From *Back* and *Klinga* (1962)

than with 200 °C (392 °F). A relative minimum of water absorption corresponds to a maximum of bending strength. Static strength properties such as bending strength and tensile strength of 1/8 in. (3.2 mm) hardboard treated with hot air at 190 °C (374 °F) temperature reached maximum values after about two hours hardening, whereas the shock resistance (impact work) continuously decreased with increasing heating times (Fig. 6.71). Dimensional stabilization, which is most important from a practical point of view, of 1/8 in. (3.2 mm) hardboard (tested by a sequence of three cycles 100 — 65 — 100 — 65 — 100 — 65% relative humidity) was continuously improved by increased temperatures and prolonged heating periods (Fig. 6.72).

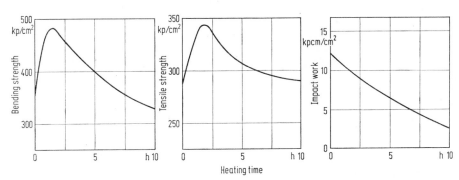

Fig. 6.71. Effect of heating time during hot pressing on mechanical properties of hardboard. From *Back* and *Klinga* (1963)

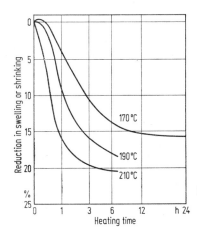

Fig. 6.72. Effect of heating time and temperature on dimensional stabilization (reduction in swelling and shrinking of dimensions in board plane) of 1/8 in. thick hardboard after three cycles 100 − 65% relative humidity. From *Back* and *Klinga* (1962, 1963)

6.9.3 Thermal Conductivity

Insulation board is widely used as a heat insulator and provides the following advantages:

1. Efficiency because of its low density caused by the many minute air spaces (pores) in the board itself;

2. Barrier against heat losses, effecting fuel savings which may justify the cost of the insulation;

3. Uniform surface temperature that closely approximates the air temperature. Cool spots on wall or ceilings collect dirt. Dirt patterns give rise to the need for frequent redecorating of the surface;

4. Protection from outside heat in summer. Insulation reduces the power required for air-conditioning;

5. Insulation board may be fabricated with vapor barriers and with sealants for the joints between the boards, or the vapor barrier may be a separate membrane;

6. Heat insulation is advantageous when insulation board is used with respect to its structural properties such as for interior walls, ceiling finish, sheathing for wooden frame construction, and roof insulation;

7. Sandwich panels in houses and other buildings often require heat insulation properties which are obtained when insulation board is used for the cores.

The dependence of thermal conductivity on density is discussed in Vol. I, p. 246—250 and is also shown in Fig. 5.150. *Teesdale* (1955) reported in a very instructive publication on "Thermal Insulation Made of Wood-base Materials, Its Application and Use in Houses". In the preface is mentioned: "The selection of the wrong kind, the improper thickness, or ... faulty installation methods is uneconomical of the home owners' income as of the Nation's resources." After an introduction to thermal insulation and heat transfer problems such as inside surface temperature, the report discusses fuel savings, dirt patterns, condensation within walls and roofs and on interior wall surfaces, ventilation in attics and roofs, effect of humidity on comfort and health and finally fire hazard and its control. Tables of coefficients of transmission (English symbol U) of frame walls, of wood frame ceiling below attic, of pitched roofs, and of solid doors are given. Sketches of various kinds of insulation of sandwich construction, of methods of insulating ceilings, and of ventilating attics and roof spaces are illustrated.

6.9.4 Acoustical Properties

The amount of insulation board being used for acoustical application is great. Insulation board in itself is a fair sound absorber but the coefficient of absorption depends on board thickness, and in a slightly parabolic manner on frequency (cf. Fig. 6.149, p. 284 in Vol. I of this book). When the surface of the board is broken by holes, slots or other traps for sound waves, insulation board becomes one of the most effective means of controlling sound levels in buildings (Fig. 6.73 and Table

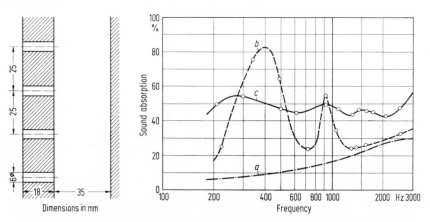

Fig. 6.73. Sound absorption of a 18 mm (3/4 in.) thick fiberboard in dependence on the frequency. *a* Board alone, *b* Board with holes and in distance from the solid wall as indicated, *c* construction as in *b*, but the hollow space between acoustic board an solid wall filled with mineral wool. From *Wintergerst* and *Klupp* (1933)

6.12). Attractive patterns for all types of holes have been developed, which provide very pleasant architectural designs. Acoustical tiles made from insulation board very often are provided with tongued and grooved edges which allow concealed nailing. Careless painting after installations may penetrate the sound traps and destroy much of the effectiveness of the treatment (FAO, 1958/59, p. 113).

A discussion of the acoustical properties of fiberboards needs to distinguish between sound transmission and sound reflection (cf. Section 5.4.6.). Insulation board, hardboard, plywood, and particleboard may be used as components in special constructions for reducing sound transmission. The mass and the continuity of construction, and the use of double-shell constructions affect the over-all transmission.

Perforated hardboard is becoming very popular; its use for acoustical purposes is mentioned in Section 6.8.8. under a) and g).

6.9.5 Mechanical Properties

6.9.5.0 General Considerations. Moderate price, a wide range in desired density, elasticity and strength make fiber building boards interesting as construction materials. There is a multiplicity of uses for hardboard more diversified than for insulation board. The choice of board for a particular application is a matter of balancing costs against the requirements for use (FAO, 1958/59, p. 116). Because of local factors usage varies from country to country and from one locality to another. New applications are being developed continuously. It is considered good practice to condition fiberboards before fixing them in place.

Advantages are the availability in large flat sheets (for hardboards the possibility to bend or post-form to curved shapes), the possibility to fasten them easily by nailing, screwing, cementing, or with special systems or clips. Where construction is adequate, the boards have sufficient rigidity as long as loading time is limited and irreversible dimensional changes due to wetting-drying cycles are not to be expected.

Another advantage is the possibility of making insulation fiberboard more water-sesistent by impregnating with asphalt or by coating it. Standard quality hardboard is sufficiently durable for nearly all interior or exterior purposes, particularly if protected with a paint or other film. In Scandinavia all standard hardboard is heat treated Tempering hardboards by treating with a drying oil and then subjecting to heat slightly increases the density but imparts superior strength and water resistance.

Insulation boards are often used for dual purpose (structural properties and heat insulation), for instance for interior walls, ceiling finish and sheathing for wooden frame construction. The very smooth and hard surface of at least one surface (quality S-1-S) makes hardboard ideal as form-lining for concrete work, when a smooth face for architectural concrete is required. Form system with a hardboard face can be used many times. Hardboard, especially tempered hardboard is important as an exterior cover and siding material for houses and becomes more and more important for prefabricated houses. Hardboard is used as panelling, shelving and door material. Large quantities are used in furniture and allied manufacture. Panelling in automobiles, truck bodies, railway passenger cars and freight cars has been mentioned in Section 6.9.0. A growing use for half-hard fiberboard is as underlayment for flooring. Hardboard itself is used as flooring. Good maintenance with a film of wax is necessary to avoid excessive wear. Prefabrication systems include acoustical board, bituminous board, composite board, embossed hardboard, multi-layer board, prefinished and lacquered board, etc.

6.9.5.1 Elasticity. The modulus of elasticity of fiberboards may be determined according to methods used for testing particleboard as described in Section 5.4.7.2. Average ranges of the moduli of elasticity of insulation board and both standard and oil tempered hardboard are given in Tables 6.12 and 6.13.

Various makes of fiberboard differ greatly in their elastic and strength properties. In Australia (Div. of For. Prod., C. S. I. R. O., Melbourne, letter to the author and *Kloot*, 1954) this problem has been thoroughly investigated. A typical load-deflection curve for static bending of Masonite is shown in Fig. 6.74. Frequency polygons for the modulus of elasticity parallel to sheet length and perpendicular to sheet length respectively are reproduced in Fig. 6.75.

The modulus of elasticity varies not only between various makes of fiberboard but within boards of the same make due to thickness variations. Moduli of elasticity must be calculated from the results of short-term tests.

Fig. 6.74. Typical load-deflection curve in static bending of Masonite hardboard

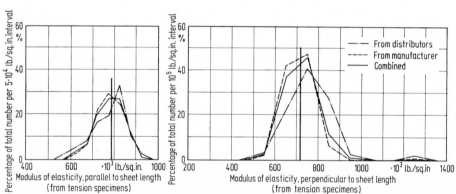

Fig. 6.75. Frequency distribution polygons for the modulus of elasticity parallel to sheet length and perpendicular to sheet length of hardboard manufactured from Eucalypt-wood. Australia, Div. For. Prod., C. S. I. R. O, letter communication

In this case the modulus of elasticity E_b is directly proportional to the density ϱ_0 in the ovendry condition (Fig. 6.76). In hardboards the influence of shear stresses on the behavior in bending is considerable. The ratio of span to height (thickness) should be about 35 to minimize shear stresses which reduce modulus of elasticity as well as modulus of rupture (Fig. 6.77). The introduction of fixed-E-values in engineering calculations is common, but, in reality, hardboards have no modulus of elasticity but only a modulus of deformation resistance

$$E = \frac{d\sigma}{d\varepsilon}, \tag{6.3}$$

where $d\sigma$ is the stress at a point P of the stress-strain curve and $d\varepsilon$ the actual deformation at this point (cf. Vol. I of this book, p. 358/359 and *Lundgren*, 1957).

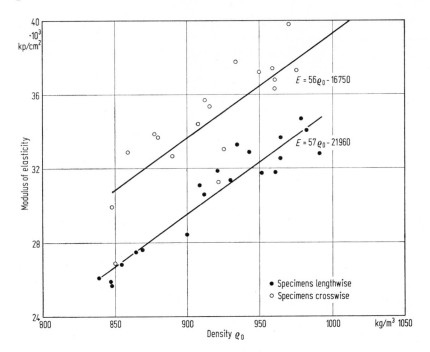

Fig. 6.76. Modulus of elasticity in bending versus density for hardboard. From *Kollmann* and *Dosoudil* (1943)

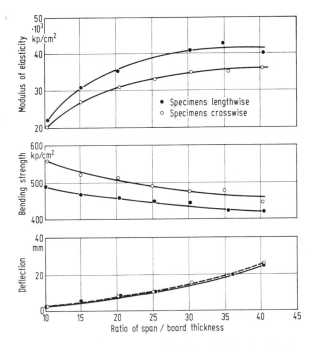

Fig. 6.77. Deflection, bending strength and modulus of elasticity of hardboard as functions of the ratio span to board thickness. From *Kollmann* and *Dosoudil* (1949)

Due to creep, $d\sigma/d\varepsilon$ decreases with increasing loading time. Oil tempered hardboard of very high quality has an initial modulus of elasticity of 60,000 kp/cm² (~860,000 lb./sq. in.). After one-year-loading the E-value is reduced to about 16,000 kp/cm² (~230,000 lb./sq. in.). Corresponding figures for normal standard-hardboard are 42,000 kp/cm² and 9,300 kp/cm² (~600,000 and 143,000 lb./sq. in.) (*Lundgren*, 1957).

Buckling calculations can be carried out using the following equation:

$$A \cdot \varepsilon = \frac{4\pi^2 I}{l^2}, \tag{6.4}$$

where A = Area of cross section in cm², ε = strain in cm/cm, I = least moment of inertia in cm⁴, l = buckling length in cm.

Lundgren (1957), assuming that an increase in relative humidity from 45 to 80% causes a swelling of the board of about 0.15%, calculated the following buckling lengths l for rigidly clamped fiber hardboards:

Thickness a	in.	1/8	1/4	1/2
	mm	3.2	6.3	12.7
Buckling length l	in.	5.9	11.8	23.6
	cm	15	30	60

6.9.5.2 Bending Strength, Creep, Fatigue

6.9.5.2.1 Bending Strength or Modulus of Rupture. Bending strength or modulus of rupture is by far the most commonly measured mechanical property of fiber building boards. International Standard ISO 768 (Sept. 1972) was drawn up by Technical Committee ISO/TC 89. The test piece shall be rectangular of the following dimensions:

Width b: approximately 75 mm,
Length L: 25 times the nominal thickness, plus approximately 50 mm.

The test pieces shall be conditioned to constant mass (reached when the results of two successive weighing operations, carried out at an interval of 24 h, do not differ by more than 0.1% of the mass of the test piece) in an atmosphere of a relative humidity of 65 ± 5% and a temperature of 20 ± 2 °C.

The width and thickness of each test piece shall be measured in accordance with ISO 766 as follows:

a) The thickness at three points along the transverse axis, one in the middle, the other two 15 mm from the edges, as shown in Fig. 6.78;
b) The width on the same axis as shown by the arrows in Fig. 6.78.

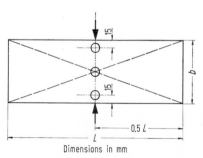

Fig. 6.78. Width and points for measurement of thickness of fiberboard pieces for bending tests. ISO 766 (1972)

Dimensions in mm

It has been shown in the foregoing Section 6.9.5.1 by Fig. 6.77 that for $L/a = 25$ the modulous of elasticity and the bending strength are still somewhat too low due to the effect of shear stresses. This effect is eliminated only for $L/a = 35$, but test pieces with such dimensions are too large and not economical and furthermore the deflection under the load may become too great. The test piece is placed on two supports and the load is applied in its centre until failure.

The testing apparatus (Fig. 6.79 a) essentially has two parallel cyclindrical supports, adjustable in the horizontal plane, having a length exceeding 75 mm and a diameter D of

15 ± 0.5 mm, if the thickness of the test piece is $\leqq 7$ mm,
30 ± 0.5 mm, if the thickness of the test piece is > 7 mm or $\leqq 20$ mm,
50 ± 0.5 mm, if the thickness of the test piece is > 20 mm.

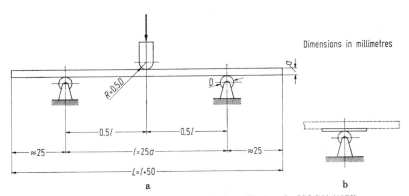

Fig. 6.79 a, b. Arrangement for bending tests on fiberboards. ISO 766 (1972)

The apparatus, furthermore, has a loading head, placed parallel to the supports and equidistant to them, adjustable in the vertical plane, and has the same length and diameter as the supports.

For testing soft boards it is recommended to place a steel plate with a thickness < 1.0 mm on each cylindrical support, as shown in Fig. 6.79 b.

The bending strength σ_B, expressed in Newtons per square millimeter, of each test piece is given by the equation:

$$\sigma_B = \frac{3PL}{2ba^2} \tag{6.5}$$

where

P = maximum load, in Newtons, to the nearest $1N$;
L = distance between the supports, in millimeters, to the nearest 1 mm;
b = width of the test piece, in millimeters, to the nearest 0.1 mm for test pieces of medium and hard boards, to the nearest 0.5 mm for test pieces of soft boards;
a = thickness of the test piece in millimeters:
   for specimens of hard and medium boards:
   to the nearest 0.01 mm for test pieces of $\leqq 7$ mm thickness, ·
   to the nearest 0.05 mm for test pieces of > 7 mm thickness;
   for test pieces of soft boards:
   to the nearest 0.1 mm for test pieces of all thicknesses.

The bending strength of each test piece shall be expressed to

the nearest 0.5 N/mm² for medium and hardboards,
the nearest 0.1 N/mm² for softboards.

The bending strength of each group of test pieces of a board (see 6.5) is the arithmetic mean of the bending strength values of the relevant test pieces, rounded to

0.5 N/mm² for medium and hardboards,
0.1 N/mm² for softboards.

In the U.K. often the breaking load is reported because it provides a direct measurement of the flexural strength of the sheet. *Lundgren* (1969, p. 100) remarks: "Would it not be far better to give the breaking moment instead of the breaking load?" The present method involving different span widths for different thicknesses gives breaking loads which do not allow comparison of the flexural strength in boards of different thickness. 3.2 mm hardboard has a breaking load of about 4 kp/cm while 12.5 mm porous board has only 1 kp/cm because of the fourfold greater span width in testing. Despite the difference in the breaking load, the two categories have the same actual flexural strength, that is, a breaking moment $N = PL/4b$ kpcm/cm width (Nmm/mm width).

If breaking moment were to replace breaking load, the problem of exact specifications of span widths for different types of boards would also be avoided.

Note:
$$\sigma_b = \frac{3Pl}{2ba^2} = \frac{Pl}{4W}.$$

where W = section modulus (in American publications the symbol Z is used),

a, b, l have the usual meanings, $W = \dfrac{A \cdot a}{6}$ in cm³,

where $A = a \cdot b$ = cross section in cm².

Lundgren (1969, p. 100) published mean values of the strength of insulation board and hardboard, obtained when various brands and qualities were procured and tested in Sweden at intervals over a period of one year (Table 6.16).

Table 6.16. Mean Values of the Strength of Insulation Board and Hardboard
(*Lundgren*, 1969)

Type and thickness mm or in.	Quality	Breaking load kp/cm width	Breaking moment kpcm/cm width	Modulus of rupture kp/cm²
Insulation (porous)	Strong	1.4	10	36
12.5 ~1/2	Weak	0.7	5	18
Hard	Strong	4	6	450
3.2 ~1/8	Weak	3	8	320

The bending strength of all types and qualities of fiber building boards depends on factors such as wood morphology, conditions of manufacture and external physical conditions. A condensed survey follows:

a) *McMillin* (1968a, b, 1969) investigated the effects of *gross wood characteristics* of Loblolly pine (*Pinus taeda* L.) refiner groundwood on density and mechanical properties of wet formed hardboard. He found that most properties were improved by using fiber, refined from wood having short, slender *tracheids* with thin walls. Bending strength (modulus of rupture, common abbreviation in the paper "MOR") is a complex function of lumen diameter, tracheid diameter, cell wall thickness, and tracheid length. It increases with decreasing tracheid length for all levels of cell wall thickness and tracheid diameter (Fig. 6.80). The level of the relationship of bending strength to tracheid length increased with decreasing tracheid length for all levels of cell wall thickness. The slope remained essentially unchanged. Similar relationships exist between modulus of elasticity and tracheid length.

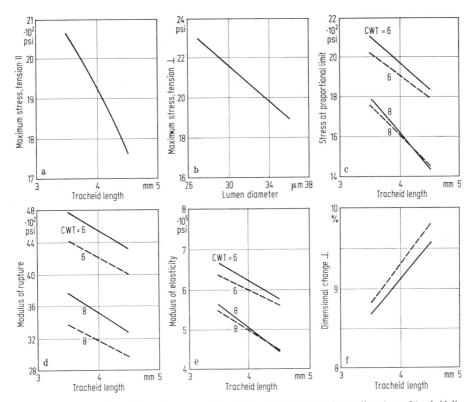

Fig. 6.80 a—f. Effect of tracheid length, average cell wall thickness (CWT in μm), lumen diameter, and tracheid diameter on the strength and dimensional stability of fiberboard. Dotted lines plot tracheids having outside diameter of 49 μm. Solid lines plot tracheids having outside diameter of 42 μm. From *McMillin* (1969)

b) The intensity of defibration expressed as *freeness* in Defibrator-seconds has a remarkable influence on bending strength as well as on tensile strength. Fig. 6.81 shows an increase of strength of wet formed hardboard made of sprucewood with bark but without addition of binding agents and not heat treated. An optimum value of freeness was about 35 Defibrator-seconds. Higher freeness slightly reduced again the bending strength and kept the tensile strength approximately constant (*Carlsson*, 1958).

c) Higher *pressure* on the originally wet mat causes more squeezing out of water and more compression. *Pressure* and *temperature* in the hot platen presses are probably the most important characteristics in board manufacture (cf. Figs. 6.66 to 6.69). The *density* of all finished fiberboards depends on the pressure exerted in the pressing process. Between density ϱ of fiberboard (as well of solid wood and particleboard cf. Vol. I of this book, p. 307, 308, 327, 343, 345—347, 367,

386 and Fig. 6.78) and modulus of elasticity E and strength σ exists a linear relationship. Generally the following equations holds:

$$E = a(\varrho - b),\tag{6.6}$$
$$\sigma = a\varrho + b,$$

where E = modulus of elasticity of σ = static short-term strength, and a and b are constants which must be determined and depend on the units used.

This general empirical law is valid for insulating fiberboard as well as for hardboard. There is, of course, a rather wider scattering of points (Fig. 6.82 and

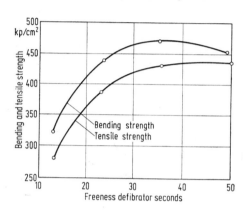

Fig. 6.81. Effect of milling intensity, expressed as freeness in Defibrator-seconds on bending strength and tensile strength of hardboard. Spruce wood with bark, no glue, no tempering. From *Carlsson* (198)

Fig. 6.83) (*Luxford*, 1955; *Kollmann* and *Dosoudil*, 1949 and 1956). Straight lines are obtained for insulation board tested dry or after soaking 6 and 48 hours respectively. The influence of the orientation of the test pieces to the working direction of the Fourdrinier-machine, lengthwise or crosswise, is shown in Fig. 6.84. There is a clear average difference between bending strength of longitudinal and transverse specimens apparently due to a predominant longitudinal fiber orientation in the machine direction. The slope of the regression lines for the transverse specimens is steeper, and therefore the differences are smaller the higher the density is.

Flexure testing of fiberboard contains many *sources of error* (*Lundgren*, 1959):

a) The calculation of the bending strength is based on *Navier's equation*, the conditions of which (linear stress distribution over the cross section, equal tensile and compressive breaking stress) is in no sense fulfilled (cf. for solid wood Vol. I of this book, p. 361/362). It has been calculated that the deviation from triangular stress distribution in the section gives breaking stress values which are two high by up to 15%.

b) The *E-moduli* for tension and compression are not equal;

c) Large *deflections* in breaking cause inwardly directed, moment-increasing reactions, which are not taken into account. This entails an undervaluation particularly of hardboards with high toughness;

d) When the *deformation rate* is constant the loading time up to failure is longer for tough board which reduces the apparent modulus of rupture;

e) *Compressed fiber building board* has a greater density and strength in the *outer layers*, which increase strength more in bending than in tension;

f) When the smooth side (quality S-1-S) is bent concave, a higher strength is attained than when it is bent convex (*Lundgren*, 1969, p. 103/104). The difference

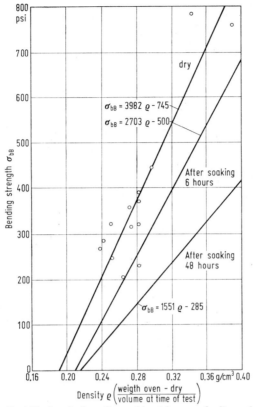

Fig. 6.82. Bending strength of fiberboard, dry, after soaking 6 hours and after soaking 48 h versus density. From *Luxford* (1955); regression lines are calculated by *Kollmann*

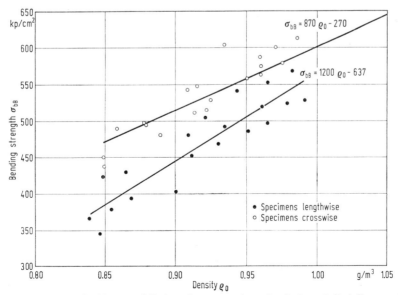

Fig. 6.83. Bending strength of hardboard (test specimens crosswise or lengthwise as indicated) versus density. From *Kollmann* and *Dosoudil* (1949)

is about 10% on average for 3.2 mm standard hardboard, but can be 20% for sheets with a particularly dense and strong face on the smooth side. If this assymmetry is absent, the strength is the same in both directions despite of the mesh pattern on the reverse side. On the other hand the pattern has another effect worth mentioning. If a large number of types of fiber building board are tested and the strengts along and across the machine direction is compared, no constant difference is found even if certain types or individual specimens do show an anisotropy of about 10% (Fig. 6.83). If the reversible moisture movements are determined for the same types, however, a clear average difference is apparent between longitudinal and transverse specimens (Fig. 6.84). The latter have on the average about 8% greater movements which appears to be due to a predominantly *longitudinal fiber orientation*. This should, however, be reflected in correspondingly higher strength in this direction, something which the particular flexure tests did not confirm. Hence the question of the *mesh pattern* arises. Coarse continuous mesh pattern provides an effective failure indication in all directions.

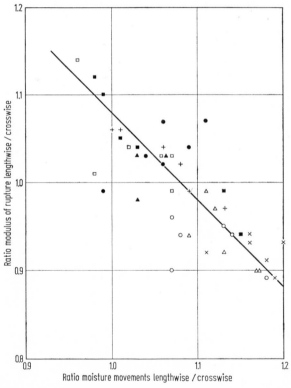

Fig. 6.84. Ratio modulus of rupture lengthwise/crosswise versus ratio moisture movements lengthwise/crosswise
From *Lundgren* (1969)

A method was chosen, that of gluing together test pieces two by two with the mesh sides laid against each other, so that these surface came into the neutral axis in bending and thereby received practically no influence in comparative tests lengthwise and crosswise. By this method of testing the ratio between bending strength crosswise and lengthwise dropped by about 6% compared with the results from single test pieces. Almost complete equivalence was obtained between mechanical and hygroscopic anisotropy. Both confirm that in general a predominant fiber orientation exists in the longitudinal direction of the panel (machine direction).

Modulus of rupture (bending strength) of wood-based sheet materials depends on the *moisture content* qualitatively in the same sense as solid wood (cf. Vol. I of this book, p. 367). Fig. 6.85 shows curves obtained in laboratory tests. In the range between about 6 and 16% moisture content a linear relationship is justified. In this case the following equation may be used, where σ_1 and σ_2 are breaking stresses within this moisture content range at moisture contents H_1 and H_2, respectively:

$$\frac{\sigma_1}{\sigma_2} = \frac{b - H_2}{b - H_1}. \tag{6.7}$$

The following numerical values for the constant b may be used:

Standard hardboard	Constant b
Bending strength	29
Tensile strength	36

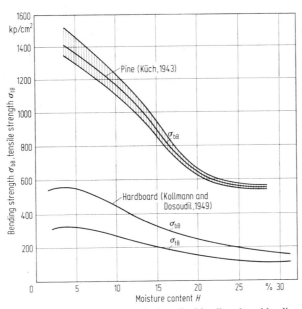

Fig. 6.85. Dependence of bending strength and tensile strength of hardboard, and bending strength of pine wood on moisture content.

With increasing temperature ($-20° < 0° < +80°C$) the bending strength decreases linearly. Roughly estimated, a decrease of -4% σ_{bB} for $10°C$ increase in temperature may be expected.

Lundgren (1969, p. 105/106) tested plywood (7 mm ≈ 5/16 in. thick), particleboard as well as oil tempered hardboard (6.4 mm ~ 1/4 in. thick) by considerably amplifying the traditional bending tests. Testing has been carried out not only with specimens brought to a moisture content equilibrium at 32, 65, 90 and 95% relative humidity but also with specimens immersed in water for 24 hours and 12 days (288 h), and tested immediately in the wet condition, and with specimens immersed for 24 h and subsequently dried 7 days in normal climate (20°C, 65% r. h.). Tests strips were taken out parallel to the face grain of the veneers in the

case of plywood and, as a rule, transversely to the machine direction for other
types of board. The results are reported graphically in Fig. 6.86. The large effect
of moisture on strength can be seen. Most striking is the steep decline of the curves
at over 80% rel. humidity. Strength remained pretty well unchanged for boards
2 to *5* within the interval 30 to 80% which covers most normal atmospheric conditions. The strength of plywood is drastically reduced at humidities of over 90%.

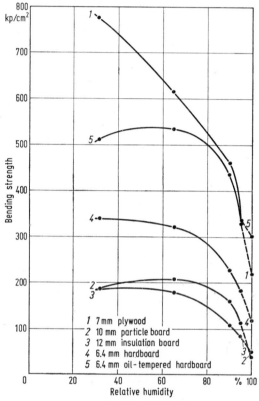

Fig. 6.86. Dependence of bending strength of wood-based sheet-like materials, as indicated, on relative humidity.
From *Lundgren* (1969)

Failure patterns in the sheet materials are as follows:

a) Plywood	dry	tension failure,
	moist	primary crushing of the compressed surface veneer,
b) Particleboard		compressive fractures in bending where the compressive strength is noticeably lower than the tensile strength,
c) Hardboard		
moderately tempered		inaudible compression failures, without clear indication when the breaking load is exceeded (impression of being "tough"),
effectively heat treated		normal tension failure, produced instantaneously and audibly especially if it occurs on the dense smooth side.

6.9.5.2.2 Creep and Relaxation. Strength in fiber building board is greatly dependent upon time. According to *Lundgren* (1957) tensile tests are more suitable than any other tests to explain rheological phenomena. Typical creep curves for various board types are shown in Fig. 6.87. It is evident that hardboard is a visco-elastic material. The elastic deformation is characterized by one elastic constant whereas the viscous deformation continuously increases under the load even if the stress is low. Prolonged loads may eventually cause failure (*Reiner,* 1955).

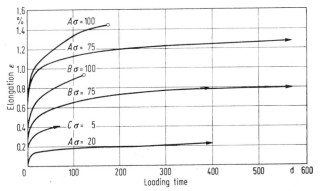

Fig. 6.87. Elongation of fiberboard (*A* and *B* hardboard from two manufacturers, *C* insulation board) as a function of loading time in long-term tensile tests. From *Lundgren* (1957)

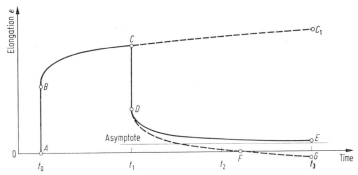

Fig. 6.88. Typical creep curve with one cycle of loading, unloading and recovery. From *Lundgren* (1969)

Loading taking place at time t_0 (Fig. 6.88) causes the instantaneous elastic deformation *AB* (*Lundgren*, 1969, p. 187). Under constant load the deformation increases to *C* at time $(t_1 - t_0)$ and would have continued to increase to point C_1, if the load had not been removed at *C*. Unloading brings about instantaneous recovery *CD*. As a first approximation most authors assume that *AB* is equal to *CD*. This would be correct if ideal elasticity governed by Hooke's law, up to the Point *B* existed (*Schmidt* and *Marlies*, 1948, *Kollmann*, 1962). In reality deformations, for instantaneous elongation ε, of most solid bodies are never completely and instantaneously recovered after unloading. There remain small residual deformations (*Baumann*, 1922) at zero-stress. The assumption of ideal (Hookean) elastic deformation is useful for the evaluation of material tests and in practical engineering applications, but such ideal elasticity is not a reality in a strict scientific sense. Thus *Lundgren* (1969, 0.188) states that the recovery *CD* is somewhat less than AB. The elastic after-effect subsequently follows the curve DE with the asymptote which is the quasi-permanent deformation caused by plasticity (flow) inevitable for every loading-unloading cycle.

Experiments have shown that the residual deformations are reduced if the test pieces are subjected to varying humidities; their practical significance therefore may be considered to be slight. In order to obtain more reliable relationships between creep and recovery *Lundgren* (1969, p. 189—194) carried out special experiments with alternate loading and unloading at different loading levels. Fig. 6.89 shows two such experiments in which test pieces A and B (3.2 mm $\approx$ 1.8 in. hardboard) were alternately loaded to $\sigma_t = 50$ and 110 kp/cm^2 (≈ 711 and 1,565 lb./sq. in.). The deformation after 4 cycles (1.649 h) was calculated and compared with the observed deformation. Both test pieces were left unloaded for 12,000 h (1.4 years) after the load cycles and then the "quasi-permanent" deformation (0.040% for A and 0.042% for B) was determined. The calculations showed a good agreement with observations for both specimens A and B. The manifold problems of creep and relaxation are discussed in detail in Section 5.4.7.5.3.

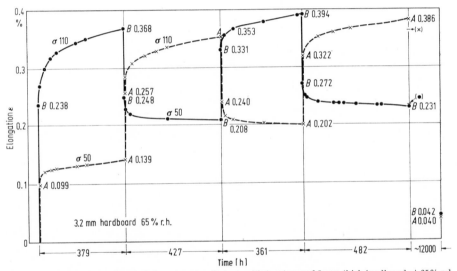

Fig. 6.89. Rheological experiments in long-term tensile tests with two types of 3 mm thick hardboard at 65% rel. humidity up to about 12,000 h. From *Lundgren* (1969)

The insight into the nature of deformation is better if a logarithmic scale is chosen for the loading time. *Lundgren* (1957) published such instructive diagrams for tension tests with standard hardboards (Fig. 6.90) and with oil tempered hardboard (Fig. 6.91). Fig. 6.92 shows curves extrapolated to breaking tensile stresses of four kinds of 3.2 mm hardboard after very long loading times. The values refer to hardboards conditioned at 73% relative humidity. If the relative humidity becomes higher than 83%, catastrophic failure of load bearing hardboard building elements could be observed. However, such high values of the relative humidity are rare and usually of short duration so that the hardboard will not reach the moisture equilibrium. *Lundgren* (1957) concluded: In practice only 15% of the strength shown by short time tests can be used, which reduce the competitive ability against plywood.

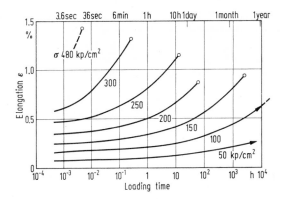

Fig. 6.90. Elongation versus loading time in long-term tensile tests of standard hardboard.
From *Lundgren* (1957)

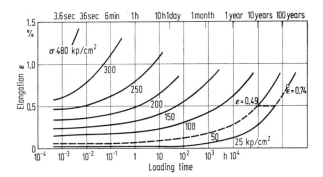

Fig. 6.91. Elongation versus loading time in long-term tensile tests of oil tempered hardboard.
From *Lundgren* (1957)

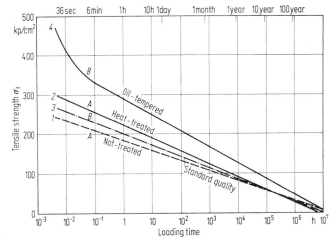

Fig. 6.92. Fatigue tensile strength of various qualities of hardboard. From *Lundgren* (1957)

6.9.5.2.3 Fatigue, Endurance. Pulsating loads, vibrations etc. which cause fatigue of the material are extreme examples of change of load. *Kollmann* and *Dosoudil* (1949/1956) studied the problem in detail and found the fatigue strength of hardboard varying bending stress to be about 22% to 24% of the static bending strength under short-term load (Fig. 6.93). The corresponding figure for wood is given as 30%. Comparative testing of fiber building board with 6 to 20% size content gave similar results. Because of internal friction (hysteresis) the temperature of the board rose with the result that its moisture content fell 2% which was thought to indicate some damping capacity. *McNatt* (1970) investigated the effect of rate, duration, and repetition of loading and obtained qualitatively similar results.

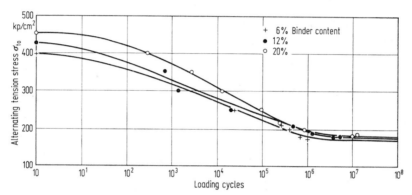

Fig. 6.93. Wöhler (S-N) diagrams for alternating tension stress tests of hardboard with various binder contents. From *Kollmann* and *Dosoudil* (1949/1956)

Design stresses for hardboard need to be developed to permit it, to be used confidently in construction as a load bearing element. For efficient structural design, the effects of rate of loading and duration of load must be known and appropriate factors applied to basic strength values.

The behavior of hardboard under different loading conditions was found to be very similar to the behavior of solid wood and particleboard. Hardboard specimens evaluated at different rates of loading decreased in strength approximately 8% for each tenfold increase of the time to maximum load. For hardboard under constant stress, a decrease of 8% in the level of stress increased the time fo failure by a magnitude of 10.

The possibility of more than 10 million repetitions of stress at or near design level during the life of a structure is extremely rare. Therefore, for practical purposes the fatigue strength of wood-based materials for 10 million cycles of stress can be used in the same way that endurance limit is used for steel.

Lewis (1951) reported that the modulus of rupture of insulation fiberboard decreased about 6 to 10% for each tenfold increase in the duration of test. *Youngquist* and *Munthe* (1956) included one hardboard in their study. A graph of the hardboard data indicates that the bending strength decreased about 9% for each tenfold increase in time to failure, although the data were limited and there was wide variation in the individual value. *Čižek* (1961) observed deflection under constant load for various I- and Box-beams with hardboard and particleboard shear webs. He concluded that the same long-time loading factor used for wood was valid for these materials.

Haygreen and *Sauer* (1969) obtained a linear relationship between constant bending stress and log of time to failure for untempered hardboard at 42 and 72% relative humidity. The slope of the line was slightly greater at the higher humidity, suggesting that creep rate is increased as the equilibrium moisture content is increased.

The effect of duration of load on the tensile strength parallel to the surface of 1/4 in. (~6 mm) tempered hardboard was evaluated by *McNatt* (1970) for material furnished by three manufacturers. Two of the boards were wet felted and wet pressed—one from Douglas fir and one from mixed southern hardwoods. The third board, of western pine and fir, was air felted and wet pressed.

The stress-number of cycles to failure (S—N) curves were determined in tension parallel to surface and interlaminar shear for 1/4 in.-thick tempered hardboard for the same materials used in the duration of load study. 12 in. (305 mm) squares were randomly selected from the material furnished by three different manufacturers. One pair of interlaminar shear specimens and one pair of tension parallel to surface specimens were cut from each square. Specimen lengths were perpendicular to the machine direction.

One specimen from each pair was designated as the control to be tested statically, and the other specimen was subjected to repeated loading. All specimens were tested at 75 °F (21 °C) and 64% relative humidity.

The tension-two-plate shear procedure was used to evaluate the interlaminar shear properties of both control and fatigue specimens. Details of specimen preparation are described in ASTM C 273-61. Epoxy resin was used to bond the specimen between steel loading plates. The tension parallel to surface specimens were prepared and tested according to ASTM Standard D 1037-64, except for the method used to grip the specimens during loading.

The loads were applied at a rate of 15 Hz until failure or to a maximum of 30 million cycles without failure.

The mean modulus of rupture values for the 1/4 in. (~6 mm) tempered hardboards, expressed as a percent of the static control values, are plotted as a function of the logarithm of time to maximum load in Fig. 6.94 (*McNatt*, 1970). The straight-line graph through the points was calculated by least squares regression with

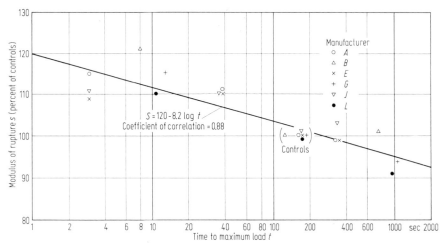

Fig. 6.94. Effect of rate of loading on the modulus of rupture (bending strength) of 1/4 in. thick tempered hardboard. From *McNatt* (1970)

time as the independent variable and modulus of rupture as the dependent variable. The equation of this line is:

$$\sigma_{Bt} = 120 - 8.2 \log t, \tag{6.8}$$

where t is mean time to maximum load in seconds. Coefficient of correlation was 0.88.

For comparison, the same type of equation for four species of wood, as reported by *Liška* (1955) is:

$$\sigma_{Bt} = 121 - 8.5 \log t. \tag{6.9}$$

Coefficient of correlation was 0.82.

Regression equation for the combined data for all strength properties of 1/4 in. ($\sim$6 mm) tempered hardboard is

$$\sigma_{Bgeneral} = 119 - 8.1 \log t. \tag{6.10}$$

The time to failure for three different 1/4 in. ($\sim$6 mm) tempered hardboards loaded at various constant stress levels is presented in Fig. 6.95. The logarithm of time to failure for the individual specimens is plotted as a function of the stress level. The straight line showing the trend of the data, passes as nearly as possible through the average of all points at each stress level. Included for comparison is the line showing the trend of long-time loading data obtained from clear, straight-grained Douglas fir bending specimens (*McNatt*, 1970).

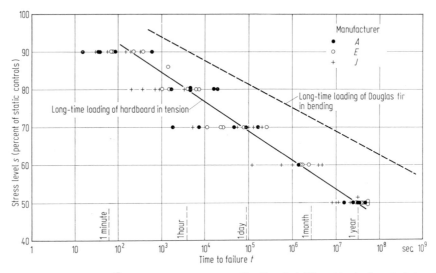

Fig. 6.95. Duration of load to failure for 1/4 in. thick tempered hardboard at different levels of constant stress in tension parallel to surface. From *McNatt* (1970)

Results of the fatigue tests are presented in Fig. 6.96. The S—N curve for these data indicates that the fatigue strength of 1/4 in. ($\sim$6 mm) tempered hardboard in interlaminar shear is approximately 40% of the static strength for 30 million cycles of stress. The S—N curves for hardboard interlaminar shear and wood glue-shear specimens evaluated by *Lewis* (1951) were similar. In both studies, the maximum repeated stress was a predetermined percent of the control strength and the minimum repeated stress was 10% of the maximum.

Working stresses for wood are based on the assumption that the strength of wood under continuous long-time loading is nine-sixteenths of the strength obtained from the standard laboratory tests. The nine-sixteenths factor is the result of studies on the rate of loading and long-time loading properties of solid wood. Obviously, data from these two methods of loading are not directly comparable, since in the first case the load is gradually increased at a constant rate until failure occurs and in the second case a load is applied and held constant until failure occurs. The trends of the two types of loading data for solid wood were similar and were combined to obtain working stresses for structural design (*Wood*, 1960).

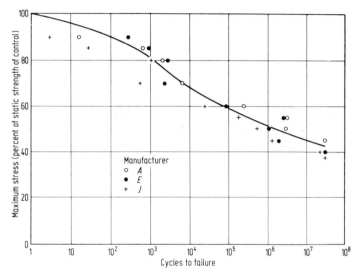

Fig. 6.96. Plot of individual tests resulting S−N curves for 1/4 in. thick tempered hardboard in tension parallel to surface fatigue. From *McNatt* (1970)

The dotted lines in Fig. 6.97 show how the rate of loading and long-time loading data for Douglas fir in bending were combined to obtain the hyperbolic curve representing the relationship between working stresses and duration of load. The equation for this curve, calculated by *Wood*, is:

$$S = \frac{108.4}{t^{0.04635}} + 18.3, \tag{6.11}$$

where S is the stress level expressed as a percent of the static strength and t is the duration of stress to failure in seconds.

The lines representing the two types of data for 1/4 in. (≈ 6 mm) tempered hardboard in tension parallel to surface are shown as solid lines in Fig. 6.97. In order to obtain an equation for hardboard similar to Eq. (6.11), information is needed on the strength for very rapid loading (impact strength). No such data are available for tensile loading; however, 10 hardboard bending specimens loaded to failure in approximately 0.06 sec (600 in. per min.) averaged 38% stronger than matched specimens loaded at the standard rate of stress. The equation for the hyperbola which passes as closely as possible through this point, through the

41*

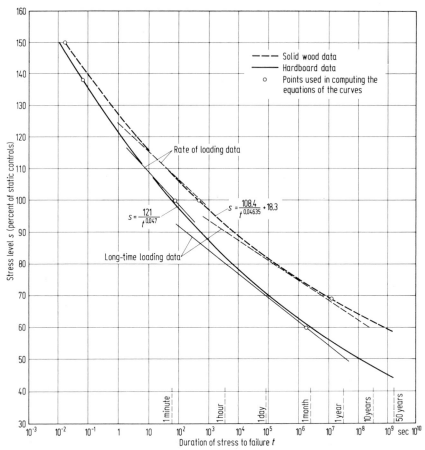

Fig. 6.97. Hyperbolic curves for the relationship between stress level and duration of stress to failure for solid wood and 1/4 in. thick tempered hardboard. From *McNatt* (1970); curves for wood by *Wood* (1960)

point indicating standard rate of loading, and through the point indicating the limit of the available long-time loading data is:

$$S = \frac{121}{t^{0.047}}, \tag{6.12}$$

where S and t are the same as defined for Eq. (6.11).

6.9.5.3 Impact Strength, Puncture Resistance. The principles of the meaning of impact strength or "shock resistance" and the various methods of testing are discussed in Section 5.4.7.6. *Dosoudil* (1950) carried out puncture tests on panels 25 cm × 50 cm (about 10 in. × 20 in.) area. In the practical range of densities of fiber building boards, the total energy to break through the panels increased approximately directly linear with increasing thickness of board. Insulation board with their low density and thicknesses from 8 mm to 14 mm (∼5/16 in. to 9/16 in.) consumed an energy between about 60 kpcm and 180 kpcm (about 4 lb. ft. and 13 lb. ft.). The corresponding values for hardboard with an average density of 0.8 and thicknesses from 2,8 to 4.8 mm (about 7/64 in. to 3/16 in.) were about 350 kpcm to 400 kpcm (25 lb. ft. to 29 lb. ft.). The puncture resistance of plywood was about 2 to 3.5 times higher than that of hardboard.

Testing strips of fiberboard by means of a pendulum hammer according to German standards DIN 52350 and DIN 7701 gave the results shown in Fig. 6.98. The upper curve in the diagram reflects the ratio of totally absorbed dynamic energy U_t to the cross section A of the test sample. It can be seen that there is a nearly positive linear dependence of U_t/A on the "slenderness" ratio l/a, but this is not meaningful. *Ylinen* (1944) proved theoretically that the impact work U_t should be related to the volume V of the test specimen which is stressed. He found:

$$U_t = \frac{1}{18} \, b \cdot a \cdot l \cdot \frac{\sigma_{bB}}{C}, \qquad (6.13)$$

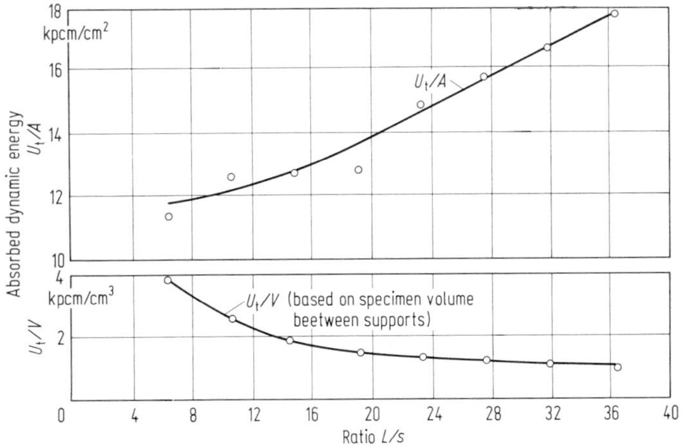

Fig. 6.98. Absorbed dynamic energy, reduced to area of cross section (upper curve) and to volume between supports versus ratio span to board thickness. From *Kollmann* and *Dosoudil* (1949/1956)

where the volume $V = b \cdot a \cdot l$ [cm³], the bending strength σ_{bB} [kp/cm²] and the constant C [kp/cm²] depends on various material properties.

It is to be seen from Fig. 6.98 that U_t/V for the hardboards tested decreases nearly hyperbolically with increasing values of l/a. The typical course of the curve for U_t/V is easily understood if one realizes that the bending strength σ_{bB} (computed with the formula of *Navier* which is rather doubtful for fiberboard) decreases also hyperbolically with increasing values of l/a (Fig. 6.99).

Shock resistance or toughness of fiberboard is of minor importance in practice. *Kollmann* (1951, p. 858) quotes the following values (obtained in cooperation with *Dosoudil*):

Table 6.17. Impact Work of Fiberboards

Type of board	Density kg/m³	Impact work kpcm/cm² lengthwise	crosswise
Light, with 32% binding agent	200	0.8	
Insulation, normal	190···305	0.8···**1.7**···2.8	0.8···**1.8**···3.0
"Pek-Pressholz" (hardboard, made of sawdust with phenolic resin under high pressure, no more for sale)	1,000···1,170	3.6···**3.8**···4.8	
Hard, standard	920···1,055	9.5···**11.9**···15.5	18.8···**13.0**···17.4

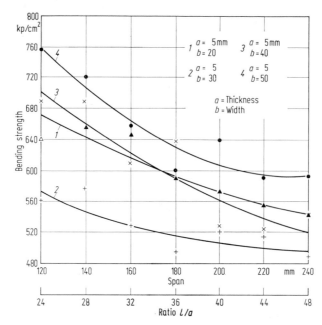

Fig. 6.99. Bending strength of hardboard, test specimens with various dimensions as indicated

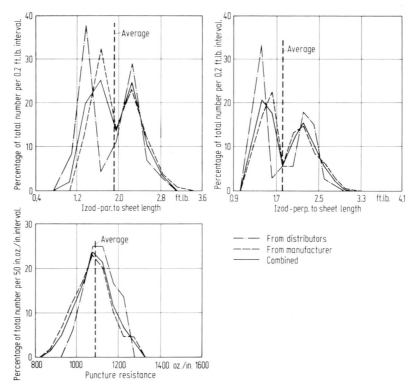

Fig. 6.100. Frequency polygons for Izod-values and puncture resistance of hardboard. Australia, Div. For. Prod.,
C. S. I. R. O., letter communication

Fig. 6.100 shows how uncertain the measured values of impact tests are. The frequency distribution for Izod-values do not follow normal (Gaussian) frequency distributions, they are irregular with random peaks.

6.9.5.4 Tensile Strength. In the U.S.A. all types of fiberboards are subjected to tensile tests whereas in Germany and other European countries as a rule only hardboards are tested in tension. No standards exist in Sweden. Because the boards to be tested have quite different tensile strengths (cf. Table 6.12, 6.13 and 6.18) and because the strength values vary and are scattered over wide ranges, comparisons are difficult or even misleading. The effect of size and shape of specimen on the tensile strength of fiberboard is remarkable (For. Prod. Lab., Madison, Wisc., 1948). The resulting recommendations were taken into consideration in drafting ASTM Standard 1037-49 T (64).

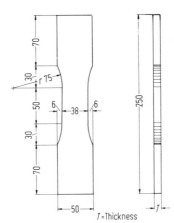

Fig. 6.101. Test specimen for tensile tests, dimensions in mm. ASTM 1037-65 T

According to this U.S. standard, specimens are tested dry and after 24 h water immersion. Half of the specimens are taken across and the other half parallel to the machine direction. Shape and dimensions of the specimens for testing *parallel to surface* are illustrated in Fig. 6.101. Loading is done continuously at a rate of 0.15 in./min. ($\sim$4 mm/min.). Procedures and type of failure are recorded and moisture content after testing is determined.

The tensile strength parallel to surface $\sigma_{tB\parallel}$ is calculated using the following equation:

$$\sigma_{tB\parallel} = \frac{P_{max}}{A}, \tag{6.14}$$

where P_{max} = maximum load (failure load) in kp (or lb.), $A = a \cdot b_1$ = cross section of the specimen in cm² (or sq. in.), a = thickness of the test piece in cm, b_1 = width of the test piece at the minimum cross-sectional area in mm. In the future the unit for σ_{bB} is N/mm². Tests of the tensile strength *perpendicular to surface* are carried out on specimens with the dimensions 2 in. × 2 in. × thickness ($\sim$50 mm × 50 mm × thickness). These specimens are glued with molten pitch to 1 in. ($\sim$25 mm) thick plates made of aluminium or steel of the same area as such of the specimens. The whole arrangement including the holding devices is shown

Table 6.18. Tensile Strength, Compressive Strength, Shear Strength of Fiber Building Boards

Property and units		Structural insulation board	Medium hardboard[1]	High density hardboard	Tempered hardboard	Special densified hardboard	Reference
Density	lb./cu. ft.	10...30	33...50	50...80	60...80	85...90	
	g/cm³	0.16...0.48	0.53...80	80...1.28	0.96...1.28	1.36...1.44	
Tensile strength parallel to surface	lb./sq. in.	200...500	—	3,000...6,000	3,600...7,800	7,800	
	kp/cm²	14...35	—	211...422	253...548	548	
Tensile strength perpendicular to surface	lb./sq. in.	10...25	—	—	160...450	500	
	kp/cm²	0.7...1.76	—	—	11...32	35	*Wayne C.*
Compressive strength parallel to surface	lb./sq. in.	—	—	1,800...6,000	3,700...6,000	26,500	*Lewis*
	kp/cm²	—	—	127...420	260...422	1,860	(1967)
Shear strength in plane of board	lb./sq. in.	—	—	—	430...350	—	
	kp/cm²	—	—	—	30...60	—	
Shear strength across plane of board	lb./sq. in.	—	—	—	2,800...3,400	—	
	kp/cm²	—	—	—	197...239	—	
Density	lb./sq. in.	16...18	—	—	—	—	
	g/cm³	0.26...0.29	—	—	—	—	
Tensile strength parallel to surface	lb./sq. in.	87...161	—	—	—	—	
	kp/cm	6...11	—	—	—	—	
Tensile strength perpendicular to surface	lb./sq. in.	137...155	—	—	—	—	
	kp/cm²	9.6...10.9	—	—	—	—	
Tensile strength parallel to surface	kp/cm²			80			Permissible stresses for wooden houses
Compressive strength parallel to surface	kp/cm²			40			
Compressive strength across plane of board	kp/cm²			30			DIN 1052, Table 2,
Shear strength in plane of board	kp/cm²			3			DIN 68750
Shear strength across plane of board	kp/cm²			15			and 68751

[1] For siding use high moisture resistance is required

in Fig. 6.102. The feed rate of the machine is 0.035 in./min (~0.9 mm/min). Failure must occur within the board.

An additional more recent survey on mechanical properties of five fiberboard is given in Table 6.18.

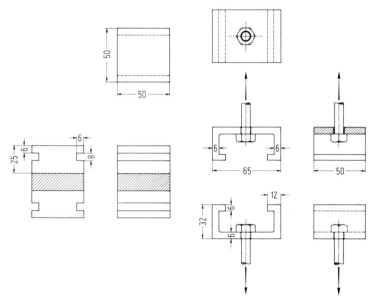

Fig. 6.102. Specimen in cribs for tensile tests across board plane; dimensions in mm. ASTM 1037-64

6.9.5.5 Compressive Strength. For the determination of compressive strength may be referred to ASTM D 1037-64. The problem is to prevent buckling of the relatively thin hardboard sheets in loading parallel to their plane. Testing can be made easy by gluing several sheets together by means of high viscosity glue which is only slightly absorbed and remains plastic. *Lundgren* (1969) used test pieces of 50 mm × 50 mm × board thickness glued together into blocks about 50 mm thick. By this method local variations of properties are balanced. The blocks were conditioned for a long time at 65% and 90% relative humidity, respectively, before testing. Table 6.19 contains some values for the compressive strength of hardboard.

The effect of relative humidity on compression strength is very much greater than on tensile strength and greater than on bending strength (*Lundgren*, 1969). Half-hard fiberboards show failure patterns in the form of splitting into thinner layers and the cracking of these layers (Fig. 6.103). Standard and oil tempered hardboard tend to exhibit shear failures with definite oblique cleavage planes which cause segmentation of the block (Fig. 6.104 and 6.105).

6.9.5.6 Shear Strength (Delamination). The shear strength of particleboard and fiberboard in different directions is difficult to determine with certainty. Even if the test method is properly used, the values obtained are not generally applicable in practice. This is due to effects of many factors involved in shear tests (cf. Section 5.4.7.7 and for solid wood *Kollmann*, 1951, p. 879—909, and Vol. I of this book, p. 394—402). *Lundgren* (1969, p. 126) distinguishes between:

a) *Layer-shear* as the effect of possible failure between layers in the plane of the panel. In construction layer-shear often leads to failure and expensive repairwork;

Table 6.19. Results of Compression Tests

Type of board Thickness mm/in.	Density g/cm³ lb./cu. ft	Moisture content	Compression in the plane of the sheet	normal to the sheet face[4]	Reference
Masonite standard 1.7 / 0.187	1.00 52.4	5.1···6.4···9.0	$\sigma_{cB}l$[1] 300 kp/cm² — 4,267 lb./sq. in. $\sigma_{cB}cr$[2] 295 kp/cm² — 4,169 lb./sq. in.	210/kp/cm² — 2,987 lb./sq. in.	Austral. C. S. I. R. O. letter communication
Insulation board 13 / 1/2	0.25 15.8	9.7	σ_{cB} 11.3 kp/cm² — 161 lb./sq. in.	1.4 kp/cm² — 19.9 lb./sq. in.	
Hardboard standard 4 / 0.157	0.98 60.5	8.2	$\sigma_{cB}l$ 220 kp/cm² — 3,129 lb./sq. in. $\sigma_{cB}cr$ 219 kp/cm² — 3,115 lb./sq. in.	130 kp/cm² — 1,849 lb./sq. in.	Kloot (1954)
Hardboard super 4.8 / 3/16	1.04 67.5	5.9	$\sigma_{cB}l$ 400 kp/cm² — 5,689 lb./sq. in. $\sigma_{cB}cr$ 390 kp/cm² — 5,547 lb./sq. in.	280 kp/cm² — 3,983 lb./sq. in.	
Hardboard half-hard 12 / 15/32	0.63 39.3	65% r. h. cond.[3] 90% r. h. cond.	$\sigma_{cB}cr$ 76 kp/cm² — 1.081 lb./sq. in. $\sigma_{cB}cr$ 35 kp/cm² — 498 lb./sq. in.	—	
Hardboard standard 6.4 / 1/4	0.97 60.6	65% r. h. cond. 90% r. h. cond.	$\sigma_{cB}cr$ 180 kp/cm² — 2,560 lb./sq. in. $\sigma_{cB}cr$ 102 kp/cm² — 1,451 lb./sq. in.	—	Lundgren (1969)
Hardboard oil tempered 6.4 / 1/4	1.02 64.2	65% r. h. cond. 90% r. h. cond.	$\sigma_{cB}cr$ 266 kp/cm² — 3,783 lb./sq. in. $\sigma_{cB}cr$ 213 kp/cm² — 3,030 lb./sq. in.	—	

[1] $\sigma_{cB}l$ = maximum crushing strength parallel to length of sheet.
[2] $\sigma_{cB}cr$ = maximum crushing strength perpendicular to length of sheet.
[3] r. h. cond. = conditioned at relative humidity in % as indicated.
[4] stress at 0.1 in. compression.

b) *Panel-shear* occurs if the panel is stressed in shear as it is in the web of a beam;

c) Shear stresses can cause cutting or *punching* perpendicular to the plane of the board.

Layer-shear is related to bending or tensile strength. It is about three times greater than the transverse tensile strength. Experiments carried out by *Kollmann* and *Krech* (1959) on hardboards of different kinds gave an average ratio of $1:2.76$ and a correlation coefficient of 0.94. It should therefore not be necessary to test both properties.

6.103. Compressive fractures in insulation board. Phot. *Lundgren* (1969)

Fig. 6.104. Compressive fractures in hardboard. Phot. *Lundgren* (1969)

Fig. 6.105. Compressive fractures in oil tempered hardboard. Phot. *Lundgren* (1969)

Shear tests for hardboards (*Lundgren*, 1969, p. 126) at favorable moisture levels and short loading times gave values of about 30 kp/cm² ($\sim$427 lb./sq. in.) which is many times more than can be utilized in practice.

Both layer-shear and panel-shear in fiber building board have been studied by *Granum* (1954). He obtained for 6.4 mm ($\sim$1/4 in.) standard hardboard a layer-shear strength of about 27 ± 3.5 kp/cm^2 ($\sim$384 $\pm$ 50 lb./sq. in.) under short term load. Similar tests carried out by *Lundgren* (1969) at the same time, (1954), gave somewhat lower values for standard hardboard, but about 45 kp/cm^2 ($\sim$640 lb./sq. in.) for oil tempered hardboard. Both *Granum's* and *Lundgren's* experiments have shown that tensile strength perpendicular to board plane and layer-shear strength are very sensitive to higher levels of relative humidity. The shear strength at 65% r. h. has only been reduced by 10 to 15% at atmospheric humidities around 90% and after redrying the board has recovered its strength. Panel-shear strength in hardboard is at least 63% of the tensile strength, and for oil tempered hardboard at least 60%.

Fiber building board and particleboard have about the same panel-shear strength in all directions. Therefore it does not matter how beam webs are cut from complete sheets.

6.9.5.7 Hardness. The many problems of defining and testing hardness are outlined in Vol. I of this book, p. 403—408, and with respect to particleboard in Section 5.4.7.8. Hardness is of practical interest for hardboard and super hardboard only. It is seldom discussed and tested. In most countries standards for hardness tests do not exist. In the U.K. the British standard (BS 1811) prescribes the Janka-test. This method has been used in Australia by *Kloot* (1954). He used specimens 2 in. by 2 in. by 2 in. (laminated) and measured in the usual manner the load which is necessary to force a hemispherical steel indentator of 0.444 in. (11.284 mm) diameter into the center of the test piece (normal to sheet face) to a depth of 0.222 in.

Kloot (1954) found values listed in Table 6.20.

Table 6.20. Janka-hardness of Fiber Board (*Kloot*, 1954, p. 25)

Type of board[1]	Janka-hardness lb.
Insulation	50.8···**65.2** ···80.1
Hardboard, standard	2,080···**2,740**···3,690
Hardboard, super	**3,120**

[1] Densities and moisture content as indicated in Table 6.19

6.9.6 Technological Properties

6.9.6.0 General Considerations. Technological properties of fiberboards depend upon their type, mainly upon their density, their binding agents and additives, their special treatments and their uses mentioned in Section 6.9.0. Reference should be made also to Section 5.4.8.0 because some remarks concerning particleboard are valid for fiberboard. Fiberboards play an important role as building materials, as sheathing, and are extensively used in U.S. house constructions (*Luxford*, 1955), for furniture manufacture, panelling, packing materials, as decorative materials, etc. Fiberboards are appreciated by "Do it yourself" hobby.

Desirable characteristics of all board types are (*Kaila*, 1968):

1. Equal qualities and equal properties in all directions within the plane of the sheet;

2. Standardized and large sizes adapted to the requirements of the building industries;

3. Easily handled, worked and fixed in usual applications;

4. Light and in various combinations suitable as construction material;

5. Qualification for architectural designs;

6. Flexible bent surfaces;

7. Decorative effect of painted or overlaid fiberboards;

8. Reduction of weights and costs of constructions.

Insulation fiberboards are extensively used for the following reasons:

1. The size of heating equipment can be reduced;

2. Space can be heated more quickly;

3. "Breathing" reduces condensation of moisture;

4. There is good resistance to wind pressure, especially for densities ≥ 0.45 g/cm² ($\sim$28.1 lb./cu. ft.);

5. Emissivity of heat radiation is reduced;

6. Stresses caused by temperature gradients are compensated;

7. Improved sound insulation (cf. Section 5.4.6) and removal of "sound-bridges";

8. Improvement of living conditions in metropolitan areas.

Half-hard fiberboards with densities between 0.50 and 0.65 g/cm³ ($\sim$31.2 and 42.2 lb./cu. ft.) are especially useful for partitions, as subfloring and as materials used in joineries in competition with particleboard and blockboard with densities around 0.65 g/cm³. They are not sensitive to variations of humidity. Fiberboards, like wood, have some weaknesses which limit their use. In such cases, special treatments and special products are available and technical development continues in cooperation of the fiberboard and the chemical industries.

6.9.6.1 Surface Quality. Surface quality or surface smoothness is of high importance for all wood-based materials. The problems and the various measuring devices are discussed in connection with particleboard (Section 5.4.8.1). For fiberboards profilographs of various types are available. *Lundgren* (1969, p. 206 to 209) describes their principles. They consist of a gauge rod which, mounted on a guide, moves laterally over the surface under examination and indicates its irregularities. After magnification, the movements of the gauge rod are recorded (Fig. 6.106). There are some problems:

a) Recording profilographs are expensive and need calibration and skilled handling;

b) The rate of measurement is low;

c) Evaluation of the graphs obtained calls for special calculating machines, etc., to be objective.

For the measurement of surface irregularities in paper and cardboard the Danish Bendtsen-method has long been in general use.

Experiments carried out by *Lundgren* (1969) in 1957 showed that the Bendtsen-apparatus could not be used without modification on fiber building board. In measuring fiber building board the smooth surface must be held absolutely flat because potential bowing would affect the air stream between the plane surface

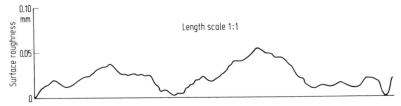

Fig. 6.106. Surface profile of a 3.2 mm thick hardboard of standard quality. From *Lundgren* (1969)

of the nozzle and the not entirely plane surface of the board. A special fixture was designed for this purpose (*Lundgren*, 1969).

To increase the speed of measurement, a new improved apparatus has been constructed. Features include semi-automatic dust extraction and an arrangement for measuring simultaneously with 10 parallel-coupled nozzles.

Some typical values are:

Board type	Surface smoothness (air volume) ml/min
3.2 mm hardboard standard quality	15···70
3.2 mm hardboard surface densified for enamelling	10···20
3.2 mm hardboard surface sanded	0···10
Particleboard of standard quality	0···10

6.9.6.2 Accuracy of Dimensions, Surface Texture and Stability. Accuracy of dimensions in panel size at the time of delivery is important. National standards exist which are a matter of agreements between manufacturer and board consumer. Discussion of such standards is beyond the scope of this book. *Lundgren* (1969) investigated the shape stability of fiber building boards. He could establish a striking relationship between buckling of installed fiber building board and bowing of the sheets in a free state in varying air humidity.

"The cause of bowing is the unequal expansion or contraction of the outer layer. This, in its turn, may be due to many factors:

1. Asymmetrical make-up of the sheet causes unequal movements during moisture changes;

2. If an unsuitable drying technique is used porous board will warp even during manufacture;

3. Because of hysteresis or different atmospheres on either side of the sheet both outer layers may acquire different moisture contents;

4. Unequal moisture changes cause asymmetrical residual shrinkage of the outer layers.

Bowing usually occurs in the two main directions of the sheet, so that the surface becomes doubly curved like the surface of a sphere or an ellipsoid" (*Lundgren*, 1969, p. 86).

The bowing in 3.2 mm hard board often varies rapidly with the atmsopheric humidity. When the latter increases above 90% r. h. the smooth side swells sufficiently to become concave.

"Porous board usually exhibits a contrary pattern of behaviour. The smooth side (the upper side of the sheet in manufacture) is covered with a more finely ground fibre layer with larger moisture movements which in high humidity makes the surface convex. However, owing to the greater thickness of porous board a given difference in moisture movements results in less bowing than in 3.2 mm (hard) board" (*Lundgren*, 1969, p. 88).

If fiber building board are subjected to abrupt moisture changes, differential swelling may be caused by differences in the absorption rate between the surfaces. If the smooth side of the board is painted, the bowing is more pronounced. The painting reduces absorption and swelling with the result that the smooth side first becomes concave. A similar but reverse process occurs when the mesh side is painted.

Wall-papering causes rapid one-sided wetting of fiber board. Commonly living rooms are well dried in advance of painting or wall-papering. The moisture from the wall-paper glue wets the fiberboard. Plasticised wall-papers retain moisture so that the board swells and buckles outwards. During drying the wall-papers shrinks and turns the board in opposite direction. The final equilibrium depends upon many circumstances including moisture movements at the board joints.

Another example of atmospheric influence is exposition to sunlight. A special, in practice important form of unequal wetting and resultant bowing takes place when board is moistened with water on the mesh side prior to being nailed to walls or other surfaces.

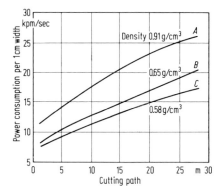

Fig. 6.107. Relative power consumption per 1 cm width versus cutting path in milling fiberboard of various densities. From *Back* and *Klinga* (1962)

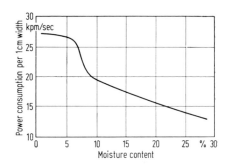

Fig. 6.108. Power consumption per 1 cm width in milling fiber hardboard as affected by the moisture content. From *Back* and *Klinga* (1962)

6.9.6.3 Working, Machining, Moldability of Hardboard.

Insulation board and hardboard are used in housing and for other typical uses in the same way as plywood and other sheet material. Fiberboards are attached with nails, screws or mechanical fasteners, cements or glue. Insulation board can be sawn, drilled and machined in the same way as any other rigid insulation sheet material.

Hardboard also works well with the usual woodworking tools. Because of the predominant lack of grain, one direction is worked as well as another (FAO, 1958/59, p. 118). Oil- or heat-treated (tempered) hardbord is harder and more brittle than the standard or utility grades. With respect to the quality of edges when correct scraper or planer knives are used the following order of merits holds:

	Tempered	Standard or utility	Lower density board
Edge condition	very smooth	clean	less clean

In manufacturing operations carbide-tipped saws and cutting knives are preferred. *Back* and *Klinga* (1964) investigated variables of fiber building board which influence the wear of cutting tools.

Fig. 6.107 shows how the relative power consumption in kpm/sec per width in milling hardboard increases with board density (roughly proportional) and with cutting path as linear length in meters ("footage" can be obtained by multiplying with 3.281). The influence of moisture content H (based on ovendry weight) is interesting (Fig. 6.108). There is a steep fall in power consumption (and tool edge wear) between $H \approx 5$ to 10%. Between about 10 and 30% moisture content the power consumption decreases approximately in a linear manner. Some hardboard behave abnormally due to a higher content of impurities (cf. Section 5.4.8.3), for example sand. Usually fiberboards contain only 0.1 to 0.2% mineral substances.

Heat treatment and reduction of moisture content improves the quality of the cut. The edges become smoother. Insulation board can be cut with much less power consumption than hardboard but the edges are rougher. Half-hard or medium density hardboard ($\varrho \approx 0.65 \text{ g/cm}^3$) is frequently utilized in joinery. Hardboard can be planed satisfactorily. *Koch* (1964, p. 295—311) published a picture (Fig. 6.109) of a specialized type of single surfacer designed to plane thin, flexible material such as tempered hardboard and explains: "The sheet material to be planed is introduced to and removed from the machine on belts, running over the infeed and outfeed tables. In the cutting zone, the work-

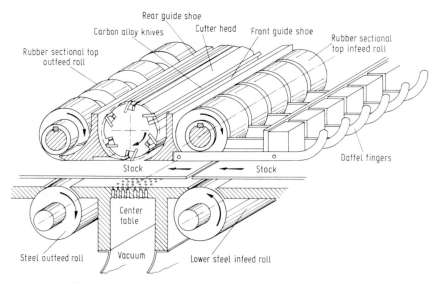

Fig. 6.109. Perspective drawing of a 50 in. single surfacer for very thin and flexible materials. Buss Machine Works Inc., in *Koch* (1964)

piece is driven by two pairs of powered rolls. The lower rolls are chrome plated and smooth. The upper rolls are sectional and faced with rubber. The cutterhead is equipped with as many as six carbide knives and rotates at 4,500 rpm with the feed (down-milling). Guide shoes are located on both sides of the cutterhead but are raised so that they are not in contact with the workpiece. The thin and flexible workpiece is held down during the sizing operation by means of a perforated vacuum table immediately under the cutterhead. If the carbide knives are jointed, a feed speed of 150 ft./min (~46 m/min) would produce 15 knife marks per inch; if unjointed, the feed rate would drop to 25 ft./min (7.6 m/min) for the same number of knife marks per inch."

Better planed surfaces can be produced on dense hardboards than on less dense boards. The finer the board components produced by defibration are, the easier it is to achieve a good quality planed surface.

Mass production of hardboard items largely stresses the sawblades and lower machine beds. Since about 1950 hardalloy-tipped sawblades are used for cutting hardboard. The much better behavior of such tools in comparison with those made of carbon steel, high-speed steel, and stellite is evident from curves plotted in

Fig. 6.110. After a relatively short cut of 25 m (about 82 ft.) in hard board the wear of the steel and stellite cutting edges increases rapidly whereas the wear of hard metals remains constant up to 500 m (1,650 ft.) tool paths (cutting lengths) (*Danielsson*, 1958, quoted in *Kaila*, 1968, p. 1097). Sawing 1/8 in. (3.2 mm) hardboard with hardalloy-tipped sawblades a cutting (running) time of 10,000 h required 10 to 15 knife sharpenings. With chromium-plated carbon-steel knives had to be sharpened 1,500 times over 10,000 h of cutting.

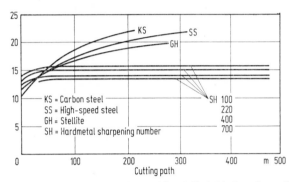

Fig. 6.110. Wear of milling tools, made of various metals as indicated in dependence of cutting path. From *Danielsson* (1958)

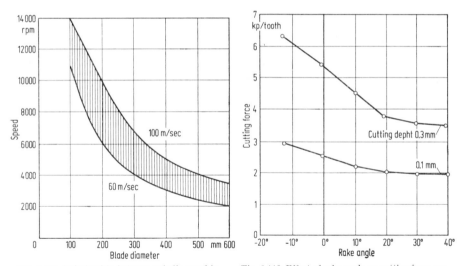

Fig. 6.111. Relationship between spindle speed in rpm and blade diameter (teeth tipped with hard metal) for two cutting speeds. From *Danielsson* (1958)

Fig. 6.112. Effect of rake angle on cutting force per tooth for two cutting depths. From *Helge* (1957)

Usual velocities of cutting for woodworking machines are 45 to 100 m/sec (about 8,860 to 19,700 ft./min). The relationships between revolutions per minute (rpm) and blade diameter are shown in Fig. 6.111 for two cutting velocities. Increased cutting speed increases output, improves quality of cut edges and, as a rule, increases running time.

The tooth geometry (mainly the rake angle and the clearance angle), the cutting depth, and the direction of sawing with respect to the direction of feeding (countersawing or climbsawing) affect the cutting force (cf. Vol. I of this book,

p. 500—504). Fig. 9.52 in Vol I (p. 503) shows (*Skoglund* and *Hvamb*, 1953) the following results:

a) For climbsawing, which is the most common form, the cutting force in kp per tooth decreases linearly, with increasing rake angle in the range between about −20 and +20 deg. For rake angles between about +20 and +40 deg. the cutting force remains nearly constant (Fig. 6.112). Absolute values were obtained in experiments with pine;

b) The clearance angle in the range between 20 and 50 deg. is negligible;

c) Power consumption by countersawing is somewhat higher than by climbsawing;

d) Cutting force increases with increasing cutting depth.

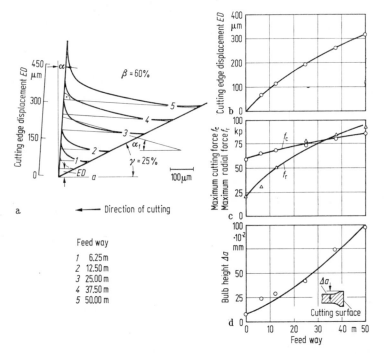

Fig. 6.113a−d. Effect of tool wear. Change of cutting edge profile, cutting edge displacement (ED), maximum cutting force and radial force, bulb height versus feed way. From *Pahlitzsch* and *Sandvoss* (1970)

The phenomena of wear of the cutting wedge during moulding of hardboards were recently investigated by *Pahlitzsch* and *Sandvoss* (1970, 1972). The following summarized results were obtained (details should be studied in the original publications):

1. Cutting force and radial force (the latter perpendicular to the former) acting at the edge of the cutting tool, increase degressively with increasing feed way; increase in radial force is markedly greater than in cutting force (Fig. 6.113c);

2. Blunting changes the shape of the cutting wedge (cutting edge displacement (Fig. 6.113a and b), radius of approximately 50 μm at the cutting edge, clearance surface chamfer of increasing length), increases the cutting forces and hence the power consumption of the moulder, and reduces the quality of the cutting surface;

3. The quality of the cutting surface is characterized by the bulb height (Fig. 6.113 d);

4. Smallest possible wedge angle β (cf. Vol. I of this book, p. 478), together with large clearance angle γ yields the largest feed ways within the service life of the tool;

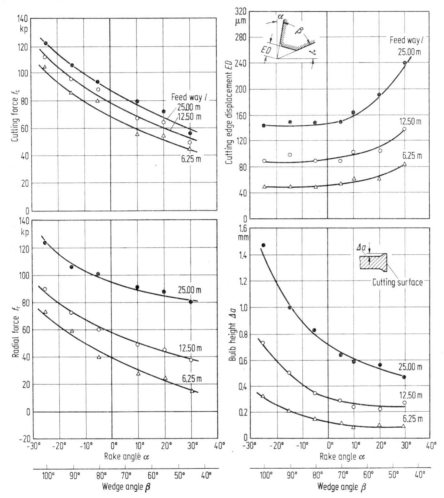

Fig. 6.114. Cutting force, radial force, cutting edge displacement, and bulb height versus rake and wedge angle. From *Pahlitzsch* and *Sandvoss* (1970)

5. For varying clearance and wedge angles, cutting edge displacement is unsuitable as a criterion for cutting quality and service life of the tool;

6. The depth of cut should be as small as possible because the cutting edge displacement increases degressively with increasing depth of cut;

7. The effect of the geometry (rake angle α, wedge angle β, and clearance angle γ) of the cutting wedge on cutting force, radial force, cutting edge displacement, and bulb height for various feed ways on special experiments (cutting velocity 38.4 m/sec, feed per tooth 1.5 mm, depth of cut 5 mm, diameter of the tool 122 mm, one cutting edge, hard metal) are shown in Fig. 6.114 and 6.115. Cutting force, radial force, and bulb height decrease degressively with increasing

rake angle, whereas cutting edge displacement increase progressively. Cutting force only to a small extent depends for higher feed ways on clearance angle; radial force f_r follows an equation of the type

$$f_r = -a \cdot \gamma + b \text{ [kp]}, \qquad (6.15)$$

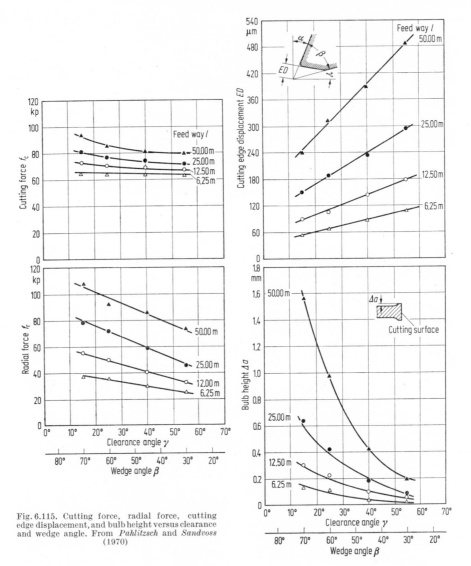

Fig. 6.115. Cutting force, radial force, cutting edge displacement, and bulb height versus clearance and wedge angle. From *Pahlitzsch* and *Sandvoss* (1970)

where in Fig. 6.114 for a feed way of 50 m, $a = \dfrac{6}{7}$, $\gamma = $ clearance angle (degr.), $b = 120$ [kp].

The formula for cutting edge displacement ED is as follows:

$$ED = c \cdot \gamma + d \text{ (mm)}, \qquad (6.16)$$

where $c \approx 6$ and $d = 152$;

8. Radial force, cutting force, cutting edge displacement and bulb height increase with cutting velocity as shown in Fig. 6.116. The effect of cutting velocity on radial force is more pronounced than that on cutting force;

9. Increased feed per tooth reduces hyperbolically radial force, cutting edge displacement, and bulb height, whereas cutting force is nearly proportionally increased (Fig. 6.117);

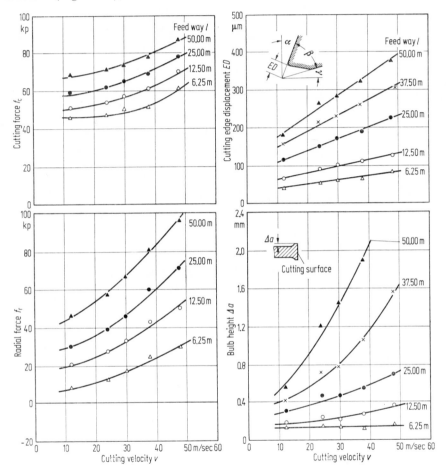

Fig. 6.116. Cutting force, radial force, cutting edge displacement, and bulb height versus cutting velocity. From *Pahlitzsch* and *Sandvoss* (1972)

10. Depth of cut increases degressively cutting force, radial force, and cutting edge displacement, progressively bulb height (Fig. 6.118).

Chipless forming (moldability) of hardboard has been studied at the Swedish Forest Products Laboratory, Stockholm, by *Back* and coworkers over a period of years (1963, 1966, 1967a, b, 1968, 1969, 1971). Some patents have been granted.

Molding of hardboard was studied. Strips and sheets of untreated and heat treated hardboard of various thickness were heated for time intervals of 5 to 150 sec between steel platens at temperatures between 300 and 400 °C (572 °F and 752 °F). The hot strips were then molded in forms to curves of 2, 4 and 8 cm, radii, while the sheets were formed in a tray mold with a smallest radius of curvature of 0.5 cm. The molding results were classified by visual inspection. Good moldability was obtained with the wire side convex after preheating in a press in the temperature range of 300° to 350 °C (572 to 662 °F) for 3 to 80 sec. The optimum preheating time was dependent on the temperature and on the bending direction as well as on the thickness and on the previous heat treatment

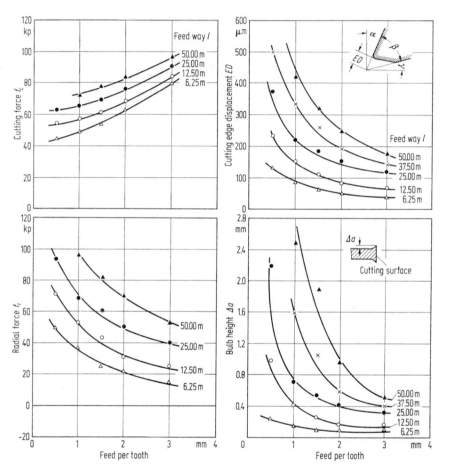

Fig. 6.117. Cutting force radial force, cutting edge displacement, and bulb height versus feed per tooth. From *Pahlitzsch* and *Sandvoss* (1972)

of the board. Since laminating is one way of improving the surface of molded trays, the effect of lamination with melamine-impregnated paper was tested with best results obtained with separate laminating after molding. Fig. 6.119 shows the preheating time required for optimum moldability of untreated and oil tempered 2.5 mm-Masonite strips and Fig. 6.120 shows the preheating time of Masonite strips versus heating temperature. Solid symbols represent bending with the wire side convex and open symbols bending with the smooth side convex.

The moldability M is presented as a percentage of the possible defects avoided and is defined by the expression

$$M = 100 \left[-1 \frac{\text{defect points}}{\text{max. defect points}} \right]. \qquad (6.16)$$

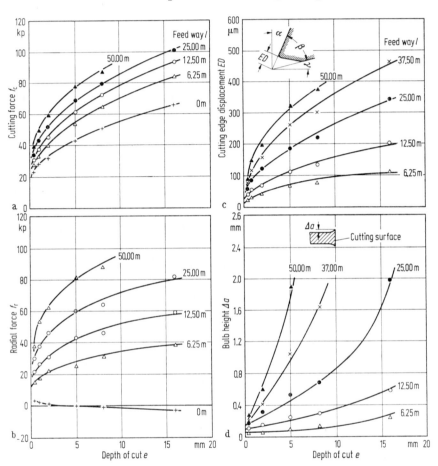

Fig. 6.118 a—d. Cutting force, radial force, cutting edge displacement, and bulb height versus depth of cut. From *Pahlitzsch* and *Sandvoss* (1972)

A minor discoloration of the material can occur during the preheating but this was not included in the evaluation.

The preheating time is critical for optimum moldability and the usual time interval is shorter when higher preheating temperatures are chosen. The rate at which the bending is carried out is normally not critical as long as rates between 5 and 15 cm/sec are used. It is evident that for most of the hardboard types examined, good moldability is obtained in the preheating temperature interval of 300 to 350 °C (572 to 662 °F) at a 4 cm (1.6 in.) bending radius, and in a narrower interval of 300 to 325 °C (572 to 617 °F) at a bending radius of 2 cm (0.8 in.).

The moldability is better when the wire side is convex. It is possible that the hardboard top layer, made from highly refined pulp, cracks more easily under

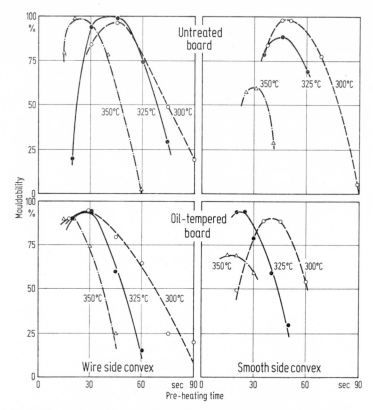

Fig. 6.119. Moldability of strips of 4 cm bending radius versus preheating at different temperatures. Above, untreated and below oil tempered 2.5 mm Masonite board. To the left the wire side is convex and to the right the smooth side is convex. From *Back, Didriksson, Johanson* and *Norberg* (1971)

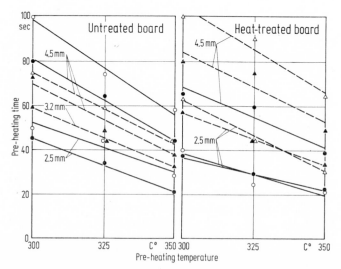

Fig. 6.120. Preheating time required for optimum moldability of strips versus preheating temperature. To the left untreated and to the right heat treated, the boards of Masonite pulp being oil tempered. Solid lines represent boards of Masonite pulp and dotted lines boards of Asplund pulp. Filled Symbols represent bends with the wire side convex and open symbols the smooth side convex. From *Back, Didriksson, Johanson* and *Norberg* (1971)

tensile than under compressive stresses. The moldability is better when the machine direction is parallel to the axis of curvature.

Normally, non-heat-treated hardboard has a better moldability than heat treated board. Satisfactory results have, however, been obtained with commercially heat treated board. The oil tempered boards of Masonite pulp had a better moldability than two types of heat treated board of Asplund pulp.

Conductive preheating of the boards was used in this investigation. Microwave techniques, however, give a faster and possibly a more homogeneous heating, so that there is less time for decomposition and perhaps "crosslinking" reactions to occur. No matter which preheating method is used, the time between preheating and molding should be as short as possible, and the heating and molding stations should therefore be close together.

Hardboards of Masonite pulp, expecially the non heat treated type, exhibited the best moldability. Generally, it is an advantage to use board with only slight heat treatment, probably with a higher pH, i.e. almost neutral, and if necessary, to heat treat the products after molding.

It might be mentioned that the thermal softening of lignocellulose products such as paper and hardboards can be employed for dry, hot smoothing (*Johanson* and *Back*, 1967) as well as for heat sealing of such products (*Johanson* and *Back*, 1968).

Lamination of trays with melamine-resin-impregnated paper was tried, and it was found that wrinkles and cracks were then hidden and an attractive surface obtained. If the laminating was carried out in the same step as the molding of the hot board, satisfactory adhesion was not obtained between the hardboard and the laminating paper, perhaps because of gas development from the hardboard.

Satisfactory results were, however, obtained when the laminating was carried out separately on the trays. It is in fact preferable from a production standpoint to laminate in a separate step, since the curing of the melamine resin requires 5 to 8 min at 140° to 150 °C (284 to 302 °F) while molding of the hot hardboard requires not more than 20 sec.

6.9.6.4 Nail- and Screw-Holding Ability. Nailed and screwed joints in competition whit gluing in building construction, especially in prefabricated houses, for interior architecture as for instance panelling, and in furniture and cabinet manufacture are discussed extensively in Section 5.4.8.4 with respect to particleboard. Because the factors affecting the nail- and screw-holding power in fiberboards are similar to those in particleboard it may be refereed to that section.

Nails and screws in particleboard and fiberboard do not give the same results as in solid wood and plywood. The reasons for the good behavior of particleboard and fiberboard in this respect are the following (*Lundgren*, 1969, p. 133—135):

a) Porous structure and tough fibers in insulation and half-hard fiberboards enable the material to recoil locally without disintegrating while the screw is driven in. Heat treated hardboard does not behave so favorably;

b) Splitting is prevented by the felting of fibers (varying fiber directions);

c) Relatively high transverse tensile strength counteracts delamination.

Particleboard and fiberboard provide a good hold perpendicular to the surface, especially if suitable pilot-holes are predrilled. Both panels are decidedly inferior to timber in insertion in the edges. The taper of ordinary wood screws produces a wedge effect. Long slender screws with a sharp thread give the best hold, but they are expensive and cause additional work. A few characteristic data for screw-holding power are given in Table 6.21.

Table 6.21. Screw-Holding Power at 65% r. h., Standard Values in Insertion
into Surface and Edge (*Lundgren*, 1969, p. 135)

Type	Sheathing screw 19 mm × 10	Wood screw 25 mm × 7	
	Withdrawal from surface kp	Withdrawal from edge kp	predrilling
Pine, timber 10 mm	115	125	⌀ 2.0 mm
Plywood, 7 mm	85		
Particleboard, 10 mm	55	75	⌀ 2.6 mm
Insulation board, 12 mm	85	40	⌀ 2.8 mm
Hardboard, 6.4 mm	50		
Oil tempered hardboard, 6.4 mm	80		

On the base of outdoor tests of several hundred square meters of different
types of board using different fastening techniques *Lundgren* (1969, p. 238) concluded:

a) Nailing should be done with 35 mm clout-head-nails, rust resistant, either
galvanized or of aluminum;

b) The spacing at the edges should not exceed about 79 to 100 mm.

6.9.7 Resistance to Deterioration

6.9.7.1 Abrasion (Wear) Resistance. Abrasion resistance is dealt with in
Vol. I of this book, p. 409—412 and for particleboard in Section 5.4.9.1. The
problem is of importance only for hardboard and super hardboard used for
floorings, table tops, in coaches and trucks, etc. Tempering and surface treatment
has a great effect on the endurance. Fig. 5.201 shows the average relative abrasion
resistance of various untreated and sealed flooring materials, including hardboard
(*Kollmann*, 1961, 1963). The behavior of particleboard and fiberboard with overlays of various artificial resins (melamine, PVC, polyesters) under the effects of
humidity, heat, cigarettes, chemicals, and sunlight has been tested and compared
by *Kollmann* (1960).

6.9.7.2. Resistance to Biological Attacks. Deterioration of wood and wood
based materials by microorganisms (bacteria, fungi, insects and marine borers)
is of great concern. The economic importance is high.

Some attempts have been made to estimate losses by decay for certain countries. *Lundgren* (1953) estimated the monetary losses due to decay of wood to
approximately $ 300,000,000 per year in the U.S.A.

Plywood, particleboard and fiberboard can be attacked by fungi. The lower
the density of the material and the lower the content of glue, the higher is the
danger and rate of deterioration. Wood is reduced by wood-rotting fungi to
simpler organic chemicals There are two major types of wood decay: brown-rots
and white-rots. White-rot fungi bleach the wood removing the lignin and holocellulose, reduce the static strength only to a moderate extent, but reduce the
toughness rapidly in the early stage. Preferred host-woods are hardwoods. Brown-
rot fungi produce a reddish brown color of the wood since chiefly the holocellulose
is removed. Shrinkage is abnormally high, static strength is greatly reduced and
to it is toughness, even immediately after the infection. In the main the hosts are
softwoods.

6.9.7.3 Behavior in Fire. Fiberboard as well as particleboard and solid wood consist of organic compunds which are mainly composed of carbon and hydrogen. Therefore, all these materials are combustible and it is impossible to make them incombustible, but incombustibility of building constructions is required only in special cases. Most building components are suitable if spread and penetration of flame are low (cf. Vol. I of this book, p. 149—152). *Metz* as early as 1936 showed that rate of combustion decreases nearly hyperbolically with increasing density. Retardation of ignition in seconds increases nearly as a parabolic function with density. From this point of view insulation board with low density will be rapidly ignited and rate of combustion is relatively high. On the other hand hardboard with an average density of 1.0 has a much lower rate of combustion and the ignition is retarded about three or even more times as compared to insulation board. *Metz* (1942) also found, that the maximum temperature developed increases to some extent with density.

Wood is a combustible material and its calorific value is theoretically and practically of interest. This calorific value averages for ovendry wood without variations amongst various species, 4,500 kcal/kg (8,100 BTU/lb.).

Fiberboard contains water and, together with the combustion gases, water vapor is produced. The heat or calorific value is then diminished by the heat of evaporation. The dependence of the so-called lower calorific value on the moisture content of wood, related to ovendry and moist weight, is shown in Vol. I, Section 5.4.3. The curves of temperature and the chemical phenomena in combustion of wood are described. Briefly discussed is also the problem of the lowest temperature which may lead to self-ignition of wood and wood based materials.

Kollmann as early as 1934 tested the behavior of fiberboard exposed to an open gas flame. It was remarkable that insulation board was easily ignited and that the time for the ignition of hardboard was longer. Oil tempered hardboard has a tendency of rapid delamination and nearly explosive spread of flame.

FAO (1957/58, p. 114) remarks in its report about "Fiberboard and Particleboard" that resistance to fire penetration and spread of flame are often an important consideration when insulation board is used in building "as manufactured it is more combustible than wood or other materials commonly used for interior finish; when used in buildings large enough to be rated for insurance purposes, it may require premiums higher than for other materials. It is common practice to use a fire resistant paint to protect the board. These paints may be of the intumescent type, so called because the foam to form an insulating layer of air bubbles and char when exposed to temperatures that could ignite the board. Insulation board so finished is much more satisfactory from the standpoint of fire exposure ... In the U.S.A. the industry producing paint-finished board employs only paints of the calss F (flame-resistant) type. Special boards impregnated with fire-resistant salts are also available."

Reference is necessary to Section 5.9.4.3 and to the report "Fire Hazard of Internal Linings" (*Hird* and *Fischl*, 1955). It is evident that the fire hazard of building boards depends, apart from the nature of the boards themselves, on the conditions under which they are used and, in particular, on the nature and amount of other combustible material with which they are associated.

At present the only method generally available for judging the relative merits of different types of board is the surface spread of flame test B. S. 476. The experiments described show that this test gives sufficient information to enable an assessment to be made of the suitability of any type of board for many situations. However, the highest classification of the surface spread of flame test includes boards with a wide range of performance, and where they are to be used as linings for both walls and ceilings of compartments, containing an appreciable amount of combustible material the surface spread of flame classification is not a sufficient indication of the fire hiazard.

The problems are under investigation by ISO/TC 92 "Fire Test on Building Materials and Structures". Working Group 4 (Australia) is concerned with spread of flame tests. A recent paper (*Grubits*, 1969) describes a new apparatus with a nichrome spiral heating element. The specimens tested, 10 cm × 10 cm, were

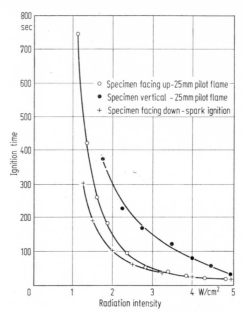

6.121. Ignition curve for hardboard. From *Grubits* (1969)

conditioned in an atmosphere of $55 \pm 2\%$ r. h. and $20 \pm 1\,°C$. Different methods of ignition with a gas flame were used. Four materials were tested. The results obtained for a fiber hardboard, 12 mm thick, are shown in Fig. 6.121. The effects of specimen position, of the radiation intensity and of the flame or sparkignition is evident. It could be claimed that the radiation intensity and the time to ignition is approximately a rectangular hyperbolic relationship.

Ignition, heat release, and noncombustibility of materials have been treated at an Symposium held at the National Academy of Sciences, Washington, D. C., 6th Oct 1971 (*Robertson*, 1972).

Literature Cited

Anderson, B., (1956) Anvendelse av bark i wallboardsindustrin. Norsk Skogindustri, No 2.
—, (1957) The influence of some Scandinavian barks on hardboard properties. Norsk Skogindustri No 5.
—, *Helge, K.*, (1957) The utilization of unbarked residues in hardboard. Norsk Skogindustri No 9.
Back, E. L., Didriksson, E. I. E., (1969) Four secondary and the glass transition temperatures of cellulose, evaluated by sonic pulse technique. Svensk Papperstidning **72**: 687—694.
—, —, *Johanson, F., Norberg, K. G.*, (1971) The dry, hot mouldability of hardboard. For. Prod. J. **21**: 96—100.

Back, E. L., Klinga, L. O., (1962) Reactions on dimensional stabilization of paper and fiber building board by heat treatment. Medd. Wallboardind. Centrallaboratorium No 14.

—, —, (1963) Reactions in dimensional stabilization of paper and fiber building board by heat treatment. Svensk Papperstidning **66**: 745—753.

Baumann, R., (1922) Die bisherigen Ergebnisse der Holzprüfungen in der Materialprüfungsanstalt an der Technischen Hochschule Stuttgart. Forsch.-Arb. Ing.-Wes., H. 231.

Belani, E., (1942) Über das Biffar-Verfahren in der Leichtbauplattenerzeugung. Suomen Paperi-ja Puutavaralehti, p. 440.

Biffar, A., DRP 286874 (cf. *Belani*).

Birkeland, S., (1969) Neues aus der Herstellung und Prüfung mechanischer Holzstoffe. Das Papier **23**: 705—711.

Boehm, R. M., (1930) The masonite process. Ind. Engng. Chem. **22**: 493—497.

—, (1944) Paper Trade Journal, **118**: 35

Carlsson, K. E., (1958) Inverkan av några tillverkningstekniska factorer på hårda träfiberskivors egenskaper. Svensk Papperstidning **61**: 128—139.

—, *Elfving, L.*, (1957) Tillsats av bark och lövved vid tillverkning av hårda träfiberskivor. Lecture on a Meeting, Oct. 10, 1957.

Čiček, L., (1961) Dauerfestigkeit und rheologische Eigenschaften von Holz und Holzwerkstoffen. Holz als Roh- und Werkstoff **19**: 83—85.

Danielsson, B., (1958) Formatsågning av träfiberskivor faner och spånplattor med hårdmetallklingor. Industria 24. 9. 58 (quoted in: *Kaila*, 1968).

Defibrator AB, (1952) Manufacture of wallboard. Defibrator AB., Stockholm, Brochure No 309—41 E: 10. 52.

—, (1963) Fibreboard industry and trade 1963. Defibrator AB, Stockholm, Brochure No 306—8, 1963.

—, (1969) Fiberboard industry and trade. Some statistical data. Defibrator AB, Stockholm.

Emerton, H. W., Goldsmith, V., (1956) The structure and the outer secondary wall of pine tracheids from kraft pulps. Holzforschung **10**: 108—115.

Evans, H., (1957) Air-felting formation for the manufacture of hardboard. FAO/ECE/Board consultation, Paper 5.8.

Flury, Ph., (1928) In: *Hitschmann's* Vademecum für Forst- und Holzwirtschaft, Wien.

Fobes, E. W., (1957) Bark-peeling machines and methods. U. S. For. Prod. Laboratory, Rept. No 1730. Madison, Wisc.

Food and Agriculture Organisation, Rome (1958/59) Fibreboard and Particleboard. Report of an International Consultation on Insulation Board, Hardboard and Particle Board. FAO and ECE, Geneva, 21. 1. to 4. 2. 1957. Food and Agriculture Organisation, FAO, Rome.

Frey-Wyssling, A., Mühlethaler, K., (1965) Ultrastructural plant cytology. Elsevier Publ., New York.

Goring, C.A.I., (1963) Thermal softening of lignin, hemicellulose and cellulose. Pulp Pap. Mag. Can. **64**: T-517-T-527.

Granum, H., (1954) Spikerlimte trekonstruksjoner. Teknisk Ukeblad No 33, Oslo.

Grubits, S., (1969) Results of ignitability tests. ISO TC 92/WG 4, Doc. 66 E, Oct. Australian Comm. Experim. Building Station.

Harada, H., (1965) Ultrastructure and organization of gymnosperm cell walls. In: *Coté, W. A.*, Cellular ultrastructure of woody plants. Syracuse University Press.

Harstad, L., (1956) Wallboardindustriens anvendelse og utnyttlese av råstoffet og forbrukernas kvalitetskrav till trefiberplater. Norsk Skogindustri No 6.

Haygreen, J., Sauer, D., (1969) Predictions of flexural creep and stress rupture of hardboard by use of a time-temperature relationships. Wood Science **1**: 241—249.

Hechler, E., (1959) Einige Betrachtungen über Holzfaserplatten. Wochenblatt f. Paperfabrikation No 20/1958, No 2/1959.

Helge, K., (1957) Undersökelser av forskjellige typer sirkelsagblad og forholdene under skjaering av harde trefiberplater ved justersagene. Norsk Skogindustri **2**: 46—54.

—, (1963) The use of bark and sawdust in hardboard manufacture. FAO/PPP/Cons., Paper 2.4, June 2.

Hird, D., Fischl, F. C., (1955) Fire hazard of internal linings. Dept. Sci. Ind. Res., Spec. Rep. No 22, London.

Höglund, E., (without data) Inverkan av 0-fiber och vissa stärkelse- och proteinhaltige ämnen på hårdboardegenskaperna. Examensarbete THF (på finska).

Holekamp, I. A., (1956) Comparison of barkers and cost. Pulp and Paper **30**: 90—94.

Holmgren, B., (1948) Självantändning i porösa träfiberplattor. Svensk Papperstidning **10**.

Johanson, F., Back, E. L., (1967) "Torr varmforming", en ny metod för formning av hårda träfiberskivor. Wallboardindustriens Centrallaboratorium, Medd. No 42 B, Stockholm.

Johanson, F., Back, E. L., (1968) Termisk jämning av träfiberskivors ytor genom kortvarig behandling mellan heta plattor. Wallboardindustriens Centrallaboratorium, Medd. No 44 B, Stockholm.

Kaila, A., (1954) Observationer beträffande avfallsfrågan och den nyaste industrin för hård board i U.S.A. Papperoch trä, No 3.

—, (1958) Den torra och halvtorra metoden för hård board. Småvirkeskommitténs meddelande, No 61.

—, (1968) Boardindustrin. In: Träindustriell Handbok, Vol II. p. 951. AB Svensk Trävarutidning, Stockholm.

Klauditz, W., Buro, A., (1962) Die Eignung von Sägespänen zur Herstellung von Holzspanplatten. Holz als Roh- und Werkstoff **20**: 19—26.

Klemm, K. H., (1957) Neuzeitliche Holzschlifferzeugung. Dr. Sändig-Verl., Wiesbaden.

Klinga, L. O., Back, E. L., (1964) Fiber building board variables influencing the wear of cutting tools. Svensk Papperstidning No 8.

Kloot, N. H., (1954) A survey of the mechanical properties of some fibre building boards. Australian Journal Appl. Science **5**: 18—35.

Koch, P., (1964) Wood machining processes. Ronald Press Comp., New York.

Kollmann, F., (1951) Technologie des Holzes und der Holzwerkstoffe. 2nd Ed., Vol. I/II. Springer-Verlag, Berlin, Göttingen, Heidelberg.

—, (1960) Untersuchungen der wichtigeren Gebrauchseigenschaften von kunstharzbeschichteten Holzfaser- und Holzspanplatten. Forsch. Ber. Nordrhein-Westfalen Nr. 905. Westdt. Verl. Köln-Opladen.

—, (1962) Furniere, Lagenhölzer und Tischlerplatten. Springer-Verlag, Berlin, Göttingen, Heidelberg.

—, (1967) Rheologisches Verhalten und Bruchgeschehen bei Holz und Holzwerkstoffen. Paperi ja Puu **49**: 173—188.

—, *Dosoudil, A.,* (1943) Vergleichende Prüfung von Homogenholz und Holzfaserplatten mit steigendem Bindemittelgehalt. Reichsanstalt f. Holzforschung, Eberswalde, Ber. No. 123.

—, —, (1949/1956) Holzfaserplatten. Ihre Eigenschaften und Prüfung, mit besonderer Berücksichtigung der Dauerfestigkeit. VDI-Forschungsheft No. 426. VDI-Verlag, Düsseldorf.

—, *Krech, H.,* (1959) Querzugfestigkeit und Scherfestigkeit von Holzfaser-Hartplatten. Holz als Roh- und Werkstoff **17**: 326—327.

—, *Mörath, E.,* (1934) Holzhaltige Leichtbauplatten. Mitt. d. Fachaussch. f. Holzfragen, H. 7. VDI-Verl., Berlin.

—, *Schneider, A.,* (1963) Über das Sorptionsverhalten wärmebehandelter Hölzer. Holz als Roh- und Werkstoff **21**: 77—85.

Küch, W., (1943) Der Einfluß des Feuchtigkeitsgehalts auf die Festigkeit von Voll- und Schichtholz. Holz als Roh- und Werkstoff **6**: 157—161.

Kumar, V. B., (1961) Neuere Untersuchungen an ölgehärteten Faserplatten. Holz als Roh- und Werkstoff **19**: 15—21.

Lampert, H., (1967) Faserplatten. VEB Fachbuchverlag, Leipzig.

Lehotsky, C., Nagy, V., (1963) Herstellung von Holzfaserplatten im Trockenverfahren. Dřevo **18**: 43—51.

Lewis, W. C., (1951) Fatigue of wood and glued joints used in laminated construction. Proceedings of For. Prod. Res. Soc. **5**: 221—229.

—, (1967) Insulation board, hardboard and particleboard. U.S. For. Prod. Lab., Madison, Wisc.

Liese, W., (1960) Die Struktur der Tertiärwand in Tracheiden und Holzfasern. Holz als Roh- und Werkstoff **18**: 296—303.

Lindgren, R. M., (1953) An overall look at wood deterioration. U. S. For. Prod. Laboratory, Rep. No. 1966. Madison, Wisc.

Lippke, P., (1959) Complete web dehydration. Pulp and Paper International, Feb. 1959.

Liska, J. A., (1955) Effect of rapid loading on the compressive and flexural strength of wood. U.S. For. Prod. Lab., Rep. No 1767. Madison, Wisc.

Lundgren, S. A., (1957) Hardboard as construction material — a viscoelastic body. Holz als Roh- und Werkstoff **15**: 19—23.

—, (1958) Die hygroskopischen Eigenschaften von Holzfaserplatten. Holz als Roh- und Werkstoff **16**: 122—127.

—, (1969) Wood-based sheet as a structural material, Part I. Swedish Wallboard Manufacturers' Association, Swedish Fiberboard Information 2.11. Stockholm.

Luxford, R. F., (1955) Properties of insulating fiberboard sheathing. U.S. For. Prod. Laboratory, Rep. No 2032, Madison, Wisc.

Mark, R. E., (1967) Cell wall mechanics of tracheids. Yale University Press, New Haven and London.

Markwardt, L. J., (1957) Survey of testing methods for fiberboard. Food and Agriculture Organisation FAO, Intern. Consultation, January 1957.

Metz, L., (1943) Holzschutz gegen Feuer. 2nd Ed., VDI-Verl., Berlin.

Meyer, K. H., Misch, L., (1937) Position des atomes dan le nouveau modèle spatial de la cellulose. Helv. Chim. Acta **20**: 232—244.

McMillin, C. W., (1968a) Morphological characteristics of Loblolly pine wood as related to specific gravity, growth rate and distance from pith. Wood Sci. and Technol. **2**: 166—176.

—, (1968b) Fiberboard from Loblolly pine refiner groundwood: Effects of gross wood characteristics and board density. For. Prod. J. **18**: 51—59.

—, (1969) Fiberboards from Loblolly pine refiner groundwood: Aspects of fiber morphology. For. Prod. J. **19**: 56—61.

McNatt, J. D., (1970) Design stresses for hardboard, effect of rate, duration, and repeated loading. For. Prod. J. **20**: 53—60.

Mörath, E., (1949). Die Holzfaserplatte. Holzforschung **4**: 14—25.

Müller, U., (1923) Lehrbuch der Holzmeßkunde. Berlin.

Neusser, H., (1957) Entwicklung und Stand der Faserplattenerzeugung. Holz-Zbl. **83**: 79—82.

Nuolivaara, I., (1961) Tillverkning av hård board enligt den våta, halvtorra och torra metoden. Småvirkeskommitténs meddelande No 115.

Ögland, N. J., (1960) Defibrering av lövved vid tillverkningen av hårda träfiberskivor i Unalitfabriken, Belgien. Svensk Papperstidning **63**: 333—334.

Öholm, G., (1959) Torkning av Fiberskivor. Svensk Papperstidning **62**: 411—415.

Pahlitzsch, G., Sandvoß, E., (1970) Verschleißuntersuchungen beim Fräsen von Faserhartplatten. Holz als Roh- und Werkstoff **28**: 245—254.

—, —, (1972) Einfluß der Schnittbedingungen auf die Beanspruchung und den Verschleiß der Schneide. Holz als Roh- und Werkstoff **30**: 133—143.

Perry, J. H., (1967) Pulp grinding at peak levels. Paperi ja Puu **49**: 753—756.

Reiner, M., (1955) Building materials, their elasticity and inelasticity. Vol. VI. Interscience Publishers, New York, London.

Renteln, H., (1951) Sågspån ett användbart råmaterial för wallboard och kartong. Svensk Papperstidning No 54.

Robertson, A. F., (1972) Ignition, heat release, and noncombustibility of materials. American Society for Testing and Materials, Philadelphia, Pa.

Rossmann, J., (1928) Wallboard patent history. Paper Trade J., **86**: 50.

Sandermann, W., Künnemeyer, O., (1956) Stand der neuen Verfahren zur Herstellung von Faserplatten nach dem Trocken- und Halbtrocken-Prozeß und Versuche über die wasserlöslichen Anteile. Das Papier **10**: 287—294.

—, —, (1957) Über Trocken- und Halbtrocken- (Dry- and Semidry-) Faserplatten. Holz als Roh- und Werkstoff **15**: 12—18.

Schmidt, A. X., Marlies, Ch. A., (1948) Principles of high-polymer theory and practice. McGraw Hill Book Company Inc., New York, Toronto, London.

Segring, S. B., (1947) Fiberplattor-framställning och egenskaper. Tekn. Tidskr. **6**: 121—132.

Siempelkamp & Co., (1961) Die Herstellung von Hartfaserplatten nach dem Trockenprozeß. G. Siempelkamp & Co., Krefeld, W.-Germany.

Siimes, F., Liiri, O., (1959) Klenved som råvara för board. Helsinki.

Skoglund, C., Hvamb, C., (1953) Tannvinklenes innvirkning på kraftverbruket ved sagning med og mot fibrene. Skogbrukets og Skogindustriens Forskningsförening, Medd. No 29, Oslo-Blindern.

Smith, I. W., (1948) Mechanical methods of bark removal. For. Prod. Res. Society, Preprint No 15.

Stamm, A. J., (1959) Dimensional stabilization of wood by thermal reactions and formaldehyde cross-linking. Tappi **42**: 39—44.

—, (1964) Wood and cellulose science. The Ronald Press Comp., New York.

—, *Baechler, R. H.*, (1960) Decay resistance and dimensional stability of five modified woods. For. Prod. J. **10**: 22—26.

—, *Hansen, L. A.*, (1935) Minimizing wood shrinkage and swelling. Ind. Engng. Chem. **27**: 1480—1484.

—, *Harris, E. E.*, (1953) Chemical processing of wood. Chem. Publ. Comp. Inc., New York.

Steenberg, B., (1962) Användning av lövvedsfiber i mälden. Suomen Puutalous No 2.

—, *Nordstrand, A.*, (1962) Production and dissipation of frictional heat in the mechanical wood grinding Process. Tappi **35**: 333—336.

Swiderski, J., (1963) Vergleich der Verfahren zur Herstellung von Hartplatten: Naß-, Halbtrocken- und Trockenverfahren. Holz als Roh- und Werkstoff **21**: 217—225.

—, (1965) Trockenverfahren zur Herstellung von Holzfaserplatten ohne zusätzliche Klebkraft. Przemysl Drzewny **14**: 7.

Teesdale, L. B., (1955) Thermal insulation made of wood-base materials. Its application and use in houses. U.S. For Prod. Laboratory, Rep, No. 1740. Madison, Wisc.

Teesdale, L. B., (1958) Report on progress in development of testing methods for fiberboards. U.S. For. Prod. Laboratory, Rep. No 2105, Madison, Wisc.

Meier, H., (1955) Über den Zellwandabbau durch Holzvermorschungspilze und die submikroskopische Struktur von Fichtentracheiden und Birkenholzfasern. Holz als Roh- und Werkstoff **13**: 323—338.

Trendelenburg, R., (1939) Das Holz als Rohstoff. München.

Ullmann's Enzyklopädie der techn. Chemie (1957). 3rd Ed. Urban & Schwarzenberg, München, Berlin.

Virtanen, P., (1963) Fuel properties of barking refuse from finnish tree species. Paperi ja Puu, Papper och Trä **45**: 313—330.

Voith, J. M., (1948) Die Defibrator-Methode. Maschinenfabrik J. M. Voith, Heidenheim/Brenz, W.-Germany, Printing Matter No 418.

Vorreiter, L., (1941) Untersuchungen über Masonite- und Kapag-Hartplatten. Holz als Roh- und Werkstoff **4**: 178—187.

Wardrop, A. B., (1957) The organization and properties of the outer layer of the secondary wall conifer in tracheids. Holzforschung **11**: 102—110.

Weichert, L., (1963) Untersuchungen über das Sorptions- und Quellungsverhalten von Fichte, Buche und Buchenpreß-Vollholz bei Temperaturen zwischen 20° und 180°C. Holz als Roh- und Werkstoff **21**: 290—300.

Wintergerst, E., Klupp, H., (1933) Grundlegende Untersuchungen über Schallabsorption. Z. VDI **77**: 91.

Wood, L. W., (1960) Relation of strength of wood to duration of load. U. S. For. Prod. Lab., Rep. No 1916. Madison, Wisc.

Ylinen, A., (1944) Begründung der Abänderungsvorschläge der Prüfnormen für Holz. Silvae Orbis **15**: 99.

Youngquist, W. G., Munthe, B. P., (1956) The effect of a change in testing speed and span on the flexural strength of insulating and structural fiberboards and a proposed new method of test. U.S. For. Prod. Lab. No 1717, Madison, Wisc.

CONVERSION FACTORS

In this book conventional metric units and English units are used. Normally equivalents of both units are given to facilitate reading of the book and comparing of its content with the available literature and own knowledge and experience. Necessary conversion is straining and may lead occasionally to errors. In the future the internationally standardized SI-units will remove these difficulties and troubles. Tables will be at disposal for requirements in theory and practice. The present survey is condensed.

1. Prefixes

Because some basic units may be of inconvenient sizes for some kinds of measurements and applications, names and symbols for them have been assigned to multiples and submultiples of units on a decimal system by attaching prefixes to the names of units as follows:

10^{12}	Tera	T
10^9	Giga	G
10^6	Mega	M
10^3	Kilo	k
10^2	Hecto	h
10	Deca	da
10^{-1}	Deci	d
10^{-2}	Centi	c
10^{-3}	Milli	m
10^{-6}	Micro	μ
10^{-9}	Nano	n
10^{-12}	Pico	p
10^{-15}	Femto	f
10^{-18}	Atto	a

One kilometer (km) is 1000 meters (m), one millimeter (mm) is 0.001 meter (m). The prefixes are appropriately applied to all kinds of units and retain their significance as in kilowatts (kW), megahertz (formerly also megacycles) (MHz), microbars (μ bar) etc.

2. Length

2.1 Conversion Factors for Metric System to English System

From	To	Multiply by
millimeter	inches	0.0394
centimeters	inches	0.3937
decimeters	inches	3.9370
meters	inches	39.3700
meters	feet	3.2808
kilometers	miles, statute	0.6214
kilometers	miles, nautical	0.5396

2.2 Conversion Factors for English System to Metric System

From	To	Multiply by
mils (0.001 of an inch)	millimeters	0.0254
inches	millimeters	25.4001
inches	centimeters	2.5400
feet (1 foot = 12 inches)	meters	0.3048
yards (1 yard = 3 feet)	meters	0.9144
miles, statute	kilometers	1.6093
miles, nautical	kilometers	1.8520

2.3 Conversion Table: Fractions of an Inch to Millimeters

Inch	Millimeters	Inch	Millimeters	Inch	Millimeters
1/32	0.794	3/8	9.525	23/32	18.256
1/16	1.588	13/32	10.319	3/4	19.050
3/32	2.381	7/16	11.112	25/32	19.844
1/8	3.175	15/32	11.906	13/16	20.638
5/32	3.969	1/2	12.700	27/32	21.431
3/16	4.762	17/32	13.494	7/8	22.225
7/32	5.556	9/16	14.288	29/32	23.019
1/4	6.350	19/32	15.081	15/16	23.812
9/32	7.144	5/8	15.875	31/32	24.606
5/16	7.938	21/32	16.669	1	25.400
11/32	8.731	11/16	17.462		

2.4 Conversion Table: Inches to Centimeters

Inch	Centimeters	Inch	Centimeters	Inch	Centimeters
1	2.54	13	33.02	25	63.52
2	5.08	14	35.56	26	66.04
3	7.62	15	38.10	27	68.58
4	10.16	16	40.64	28	71.12
5	12.70	17	43.18	29	73.66
6	15.24	18	45.72	30	76.20
7	17.78	19	48.26	31	78.74
8	20.32	20	50.80	32	81.28
9	22.86	21	53.34	33	83.82
10	25.40	22	55.88	34	86.36
11	27.94	23	58.42	35	88.90
12	30.48	24	60.96	36	91.44

2.5 Conversion Table: Meters to Feet — Feet to Meters

Meters	to	Feet	Feet	to	Meters
1		3.2808	1		0.3048
2		6.5617	2		0.6096
3		9.8425	3		0.9144
4		13.123	4		1.2192
5		16.404	5		1.5240
6		19.685	6		1.8288
7		22.966	7		2.1336
8		26.247	8		2.4384
9		29.528	9		2.7432
10		32.808	10		3.0480
15		49.213	15		4.5720
20		65.617	20		6.0960
25		82.021	25		7.6200
30		98.425	30		9.1440
35		114.83	35		10.668
40		131.23	40		12.192
45		147.64	45		13.176
50		164.04	50		15.240
60		196.85	60		18.288
70		229.66	70		21.366
80		262.47	80		24.384
90		295.28	90		27.432
100		328.08	100		30.480

2.6 Note

The antiquated use of Greek symbols for the designation of very small units is in contradiction to the SI-System, therefore it is recommended to write:

10^{-6} m = 1 μm = 1 micrometer, and *not* = 1 μ = 1 micron
10^{-9} m = 1 nm = 1 nanometer, and *not* = 1 mμ = 1 millimicron

1 Ångstrom (Å) = 10^{-10} m = 10^{-8} cm (in W.-Germany admissible until 31. 12. 1977)

3. Area

3.1 Conversion Factors for Metric System to English System

From	To	Multiply by
square millimeters	square inches	0.0016
square centimeters	square inches	0.1550
square decimeters	square inches	15.5000
square meters	square feet	10.7639
square meters	square yards	1.9560
square decameters[1]	acres	0.0247
square hectometers[2]	acres	2.4710
square kilometers	acres	247.1044

[1] Also called ares.
[2] Also called hectares.

3.2 Conversion Factors for English System to Metric System

From	To	Multiply by
square inches	square centimeters	6.4516
square feet	square centimeters	929.0341
square feet	square meters	0.0929
square yards	square meters	0.8361
square miles statutes	square hectometers[1]	258.9998
square miles statutes	square kilometers	2.5900

[1] Also called hectares.

3.3 Conversion Table Square Meters (Sq. M.) to Square Feet (Sq. Ft.) Square Feet (Sq. Ft) to Square Meters (Sq. M.)

Sq. M.	to	Sq. Ft.	Sq. Ft.	to	Sq. M.
1		10.764	1		0.0929
2		21.528	2		0.1858
3		32.292	3		0.2787
4		43.056	4		0.3716
5		53.820	5		0.4645
6		64.584	6		0.5574
7		75.348	7		0.6503
8		86.112	8		0.7432
9		96.876	9		0.8361
10		107.64	10		0.9290

4. Volume

4.1 Conversion Factors for Metric System to English System

4.1.1 Solid Measure

From	To	Multiply by
cubic millimeters	cubic inches	0.0006
cubic centimeters	cubic inches	0.0610
cubic decimeters	cubic inches	61.0240
cubic meters	cubic feet	35.3149
cubic meters	cubic yards	1.3080

4.1.2 Liquid Measure

From	To	Multiply by
milliliters	fluid ounces	0.0338
centiliters	fluid ounces	0.3382
deciliters	fluid ounces	3.3815
liters	gallons (Imperial)	0.2200
liters	quarts (U.S.)	1.0567
liters	gallons (U.S.)	0.2642
hectoliters	gallons (U.S.)	26.4178

4.2 *Conversion Factors for English System to Metric System*

4.2.1 Solid Measure

From	To	Multiply by
cubic inches	cubic centimeters	16.3870
cubic feet	cubic meters	0.0283
cubic yards	cubic meters	0.7646
cords	cubic meters	3.6246

4.2.2 Liquid Measure

From	To	Multiply by
fluid ounces	milliliters	29.5734
fluid ounces	liters	0.0296
pints (British)	liters	0.5683
pints (U.S.)	liters	0.4732
quarts (British)	liters	1.1365
quarts (U.S.)	liters	0.9463
gallons (Imperial)	liters	4.5460
gallons (U.S.)	liters	3.7853

4.3 *Conversion Table Cubic Meters (Cu. M.) to Cubic Feet (Cu. Ft.)*
Cubic Feet (Cu. Ft.) to Cubic Meters (Cu. M.)

Cu. M.	to	Cu. Ft.	Cu. Ft.	to	Cu. M.
1		35.315	1		0.028
2		70.630	2		0.057
3		105.945	3		0.085
4		141.259	4		0.113
5		176.574	5		0.142
6		211.889	6		0.170
7		247.204	7		0.198
8		282.519	8		0.227
9		317.834	9		0.255
10		353.149	10		0.283

4.4 *Note*

Solid measure

1 cubic foot = 1728 cubic inches
1 cubic yard = 27 cubic feet

Liquid measure, also called capacity

1 gallon   = 4 quarts
1 quart   = 2 pints
1 Imperial gallon = 1.20096 U.S. gallons

5. Mass

5.1 Conversion Factors for Metric System to English System

From	To	Multiply by
milligrams	grains	0.0154
grams	ounces, avoirdupois	0.0353
kilograms	pounds, avoirdupois	2.2046
metric tons	long tons	0.9842
metric tons	short tons	1.1023

5.2 Conversion Factors for English System to Metric System

From	To	Multiply by
grains	milligrams	64.7989
grains	grams	0.0648
ounces, avoirdupois	grams	28.3495
pounds, avoirdupois	kilograms	0.4536
tons, long	metric tons	1.0161
tons, short	metric tons	0.9072

5.3 Note

16 ounces (oz.)	= 1 pound avoirdupois (lb.)
U.S. 100 pounds	= 1 hundredweight (cwt.) = 45.36 kg
Brit. 112 pounds	= 1 hundredweight (cwt.) = 50.80 kg
U.S. 2000 pounds	= 1 short ton
1 long ton	= 2240 pounds (commonly Brit.)

6. Density

6.1 Conversion Factors for Metric System to English System

From	To	Multiply by
grams per cubic centimeter	troy ounces per cubic inch	0.5269
grams per cubic centimeter	ounces per cubic inch	0.5780
kilograms per cubicmeter	pounds per cubic feet	0.0624
tons per cubic meter	long tons per cubic feet	0.0271

6.2 Conversion Factors for English System to Metric System

From	To	Multiply by
troy ounces per cubic inch	grams per cubic centimeter	1.8980
troy ounces per cubic feet	grams per cubic meter	1.0984
troy pounds per cubic feet	kilograms per cubic meter	13.1810
long tons per cubic feet	tons per cubic meter	35.8816

6.3 Conversion Table Grams per Cubic Centimeter (g/cm³) and Kilograms per Cubic Meter (kg/m) in Pounds per Cubic Feet (lb./cu. ft.) and vice versa

g/cm³	kg/m³	lb./cu. ft.	lb./cu. ft.	kg/m³	g/cm³
0.200	200	12.49	12	192	0.192
0.220	220	13.73	14	224	0.224
0.240	240	14.98	16	256	0.256
0.260	260	16.23	18	288	0.288
0.280	280	17.48	20	320	0.320
0.300	300	18.73			
			22	352	0.352
0.320	320	19.98	24	384	0.384
0.340	340	21.23	26	417	0.417
0.360	360	22.47	28	449	0.449
0.380	380	23.72	30	481	0.481
0.400	400	24.97			
			32	513	0.513
0.420	420	26.22	34	545	0.545
0.440	440	27.47	36	577	0.577
0.460	460	28.72	38	609	0.609
0.480	480	29.97	40	641	0.641
0.500	500	31.21			
			42	673	0.673
0.520	520	32.46	44	705	0.705
0.540	540	33.71	46	737	0.737
0.560	560	34.96	48	769	0.769
0.580	580	36.21	50	801	0.801
0.600	600	37.46			
			52	833	0.833
0.620	620	38.71	54	865	0.865
0.640	640	39.95	56	897	0.897
0.660	660	41.20	58	929	0.929
0.680	680	42.45	60	961	0.961
0.700	700	43.70			
			62	993	0.993
0.720	720	44.95	64	1025	1.025
0.740	740	46.20	66	1057	1.057
0.760	760	47.45	68	1089	1.089
0.780	780	48.69	70	1121	1.121
0.800	800	49.94			
			72	1153	1.153
0.820	820	51.19	74	1185	1.185
0.840	840	52.44	76	1217	1.217
0.860	860	53.69	78	1250	1.250
0.880	880	54.94	80	1282	1.282
0.900	900	56.19			
			82	1314	1.314
0.920	920	57.43	84	1346	1.346
0.940	940	58.68	86	1378	1.378
0.960	960	59.93	88	1410	1.410
0.980	980	61.18	90	1442	1.442
1.000	1000	62.43			
			92	1474	1.474
1.050	1050	65.55	94	1506	1.506
1.100	1100	68.67			
1.150	1150	71.79			
1.200	1200	74.92			
1.250	1250	78.04			
1.300	1300	81.16			
1.350	1350	84.28			
1.400	1400	87.40			
1.450	1450	90.52			
1.500	1500	93.64			

7. Pressure

7.1 Comparison of Pressure Units still Used but Outlawed by SI-Units
(from „Hütte", Des Ingenieurs Taschenbuch, 27. Ed. Vol. I, Berlin 1941, p. 1200, Table 61)

Unit	kp/cm²	mm Hg	Physic. atm	bar	kcal/m³	inch Hg	lb./sq. in.
1 kp/cm²	1	735.559	0.9678	0.9807	23.42	28.591	14.223
1000 mm Hg	1.3595	1000	1.3158	1.3332	31.84	39.370	19.337
1 physic. atm	1.0332	760.000	1	1.0133	24.20	29.921	14.696
1 bar	1.0197	750.062	9.0869	1	23.88	29.530	14.504
1 kcal/m³	0.0427	31.408	0.0413	0.0419	1	1.237	0.607
10 inch Hg	0.3453	254.000	0.3342	0.3386	8.087	10	4.912
10 lb./sq. in.	0.7031	517.151	0.6805	0.6895	16.47	20.360	10

7.4 Conversion Table in SI-Pressure Units for Gases, Vapors, Liquids[1]

$$1 \text{ Pa} = 1 \text{ N/m}^2 \approx \frac{1}{9,81} \text{ kp/m}^2 = 0{,}102 \text{ kp/m}^2$$

	Pa	bar	kp/m²	at	atm	Torr
1 Pa = (1 N/m²)	1	10^{-5}	0.102	$0.102 \cdot 10^{-4}$	$0.987 \cdot 10^{-5}$	0.0075
1 bar = (= 0.1 MPa)	100000 = 10^5	1 (= 1000 mbar)	10200	1.02	0.987	750
1 kp/m² =	9,81	$9.81 \cdot 10^{-5}$	1	10^{-4}	$0.968 \cdot 10^{-4}$	0.0736
1 at = (= 1 kp/cm²)	98100	0,981	10000	1	0.968	736
1 atm = (= 760 Torr)	101325	1.013 (= 1013 mbar)	10330	1.033	1	760
1 Torr = $\left(= \dfrac{1}{760} \text{ atm}\right)$	133	0.00133	13.6	0.00136	0.00132	1

[1] Tables 7.4, 7.5, 7.6 and 10.3 are reprinted with special permission of Deutscher Normenausschuß (DNA), Berlin, from the book: W. Haeder and E. Gärtner, Die gesetzlichen Einheiten in der Technik, 2nd. Ed., Beuth-Vertrieb GmbH, Berlin 30, Köln, Frankfurt (Main). p. 42, 43, 49.

7.5 Conversion Table for Heights of Liquid Columns to SI-Pressure Units[1]

$$1 \text{ kp/m}^2 \triangleq 1 \text{ mm WS} \approx 1 \text{ daN/m}^2; \; 1 \text{ Torr} \triangleq 1 \text{ mm Hg};$$
$$1 \text{ Pa} = 1 \text{ N/m}^2 \approx (1/9.81) \text{ kp/m}^2 \approx 0.102 \text{ kp/m}^2$$

	µbar	mbar	bar	Pa (= N/m²)
1 mm WS ≐ △ 1 kp/m² ≈ 1 daN/m²	100	0.1	0.0001	10
1 m WS ≐ △ 0.1 at △ 0.1 kp/cm² ≈ 0.1 daN/cm²	100000	100	0.1	10000
10 m WS ≐ △ 1 at △ 1 kp/cm² ≈ 1 daN/cm²	1000000	1000	1	100000
1 mm Hg (mm QS) ≐ △ 1 Torr	1330	1.33	0.00133	133

[1] cf. Footnote to 7.4.

7.2 Conversion Table Kiloponds per Square Centimeter (kp/cm²) in Pounds per Square Inch (lb./sq. in.)

1 kp/cm² = 14.223293 lbs. per sq. inch or 1 sq. inch per lb. = 14.223293 cm²/kp

	0	1	2	3	4	5	6	7	8	9
0	0	14.2233	28.4466	42.6699	56.8932	71.1165	85.3398	99.5631	113.786	128.010
10	142.233	156.456	170.680	184.903	199.126	213.349	227.573	241.796	256.019	270.243
20	284.466	298.689	312.912	327.136	341.359	355.582	369.806	384.029	398.252	412.475
30	426.699	440.922	455.145	469.369	483.592	497.815	512.039	526.262	540.485	554.708
40	568.932	583.155	597.378	611.602	625.825	640.048	654.271	668.495	682.718	696.941
50	711.165	725.388	739.611	753.835	768.058	782.281	796.504	810.728	824.951	839.174
60	853.398	867.621	881.844	896.067	910.291	924.514	938.737	952.961	967.184	981.407
70	995.631	1009.85	1024.08	1038.30	1052.52	1066.75	1080.97	1095.19	1109.42	1123.64
80	1137.86	1152.09	1166.31	1180.53	1194.76	1208.98	1223.20	1237.43	1251.65	1265.87
90	1280.10	1294.32	1308.54	1322.77	1336.99	1351.21	1365.44	1379.66	1393.88	1408.11
100	1422.33	1436.55	1450.78	1465.00	1479.22	1493.45	1507.67	1521.89	1536.12	1550.34

7.3 Conversion Table Pounds per Sqare Inch (lb./sq. in.) in Kiloponds per Square Centimeter (kp/cm²)

1 lb. per sq. inch = 0.070307208 kp/cm² or 1 cm²/kp = 0.070307208 sq. inch per lb.

	0	10	20	30	40	50	60	70	80	90
0	0	0.70307	1.40614	2.10922	2.81229	3.51536	4.21843	4.92150	5.62458	6.32765
100	7.03072	7.73379	8.43686	9.13994	9.84301	10.5461	11.2492	11.9522	12.6553	13.3584
200	14.0614	14.7645	15.4676	16.1707	16.8737	17.5768	18.2799	18.9829	19.6860	20.3891
300	21.0922	21.7952	22.4983	23.2014	23.9044	24.6075	25.3106	26.0137	26.7167	27.4198
400	28.1229	28.8260	29.5290	30.2321	30.9352	31.6382	32.3413	33.0444	33.7475	34.4505
500	35.1536	35.8567	36.5597	37.2628	37.9659	38.6690	39.3720	40.0751	40.7782	41.4812
600	42.1843	42.8874	43.5905	44.2935	44.9966	45.6997	46.4028	47.1058	47.8089	48.5120
700	49.2150	49.9181	50.6212	51.3243	52.0273	52.7304	53.4335	54.1365	54.8396	55.5427
800	56.2458	56.9488	57.6519	58.3550	59.0580	59.7611	60.4642	61.1673	61.8703	62.5734
900	63.2765	63.9796	64.6826	65.3857	66.0888	66.7918	67.4949	68.1980	68.9011	69.6041
1000	70.3072	71.0103	71.7133	72.4164	73.1195	73.8226	74.5256	75.2287	75.9318	76.6348

7.6 Conversion Table for Mechanical Stresses (Strengths) and Moduli of Elasticity or Rigidity[1]

$$1 \text{ Pa} = 1 \text{ N/m}^2 \approx \frac{1}{9.81} \text{ kp/m}^2 = 0.102 \text{ kp/m}^2$$

	Pa	N/mm²	daN/cm²	daN/mm²	kp/cm²	kp/mm²
1 Pa = (= 1 N/m²)	1	10^{-6}	10^{-5}	10^{-7}	$0.102 \cdot 10^{-4}$	$0.102 \cdot 10^{-6}$
1 N/mm² = (= 1 MPa)	1 000 000	1	10	0.1	10.2	0.102
1 daN/cm² = (= 1 bar)	100 000	0.1	1	0.01	1.02	0.0102
1 daN/mm² = (= 1hbar)	10 000 000	10	100	1	102	1.02
1 kp/cm² = (= 1 at)	98 100	0.0981	0.981	0.00981	1	0.01
1 kp/mm² =	9 810 000	9.81	98.1	0.981	100	1

[1] cf. Footnote to 7.4.

8. Temperature

8.1 Conversion Factors Celsius (Centigrades, C) in Fahrenheit (F) and vice versa

$$C = \frac{5}{9} (F - 32);$$

$$F = 1.8 \, C + 32.$$

8.2 Conversion Table for Centigrades (C) in Fahrenheit (F)

C	F	C	F	C	F	C	F	C	F	C	F
−20	−4.0	+30	+86.0	+80	+176.0	+130	+266.0	+180	+356.0	+500	+932
−19	−2.2	31	87.8	81	177.8	131	267.8	181	357.8	550	1022
−18	−0.4	32	89.6	82	179.6	132	269.6	182	359.6	600	1112
−17	+1.4	33	91.4	83	181.4	133	271.4	183	361.4	650	1202
−16	3.2	34	93.2	84	183.2	134	273.2	184	363.2	700	1292
−15	5.0	35	95.0	85	185.0	135	275.0	185	365.0	750	1382
−14	6.8	36	96.8	86	186.8	136	276.8	186	366.8	800	1472
−13	8.6	37	98.6	87	188.6	137	278.6	187	368.6	850	1562
−12	10.4	38	100.4	88	190.4	138	280.4	188	370.4	900	1652
−11	12.2	39	102.2	89	192.2	139	282.2	189	372.2	950	1742
−10	14.0	40	104.0	90	194.0	140	284.0	190	374.0	1000	1832
− 9	15.8	41	105.8	91	195.8	141	285.8	191	375.8		
− 8	17.6	42	107.6	92	197.6	142	287.6	192	377.6		
− 7	19.4	43	109.4	93	199.4	143	289.4	193	379.4		
− 6	21.2	44	111.2	94	201.2	144	291.2	194	381.2		
− 5	23.0	45	113.0	95	203.0	145	293.0	195	383.0		
− 4	24.8	46	114.8	96	204.8	146	294.8	196	384.8		
− 3	26.6	47	116.6	97	206.6	147	296.6	197	386.6		
− 2	28.4	48	118.4	98	208.4	148	298.4	198	388.4		
− 1	30.2	49	120.2	99	210.2	149	300.2	199	390.2		

Table 8.2 (Continued)

C	F	C	F	C	F	C	F	C	F	C	F
0	32.0	50	122.0	100	212.0	150	302.0	200	392		
+ 1	33.8	51	123.8	101	213.8	151	303.8	210	410		
2	35.6	52	125.6	102	215.6	152	305.6	220	428		
3	37.4	53	127.4	103	217.4	153	307.4	230	446		
4	39.2	54	129.2	104	219.2	154	309.2	240	464		
5	41.0	55	131.0	105	221.0	155	311.0	250	482		
6	42.8	56	132.8	106	222.8	156	312.8	260	500		
7	44.6	57	134.6	107	224.6	157	314.6	270	518		
8	46.4	58	136.4	108	226.4	158	316.4	280	536		
9	48.2	59	138.2	109	228.2	159	318.2	290	554		
10	50.0	60	140.0	110	230.0	160	320.0	300	572		
11	51.8	61	141.8	111	231.8	161	321.8	320	608		
12	53.6	62	143.6	112	233.6	162	323.6	340	644		
13	55.4	63	145.4	113	235.4	163	325.4	360	680		
14	57.2	64	147.2	114	237.2	164	327.2	380	716		
15	59.0	65	149.0	115	239.0	165	329.0	400	752		
16	60.8	66	150.8	116	240.8	166	330.8	450	842		
17	62.6	67	152.6	117	242.6	167	332.6				
18	64.4	68	154.4	118	244.4	168	334.4				
19	66.2	69	156.2	119	246.2	169	336.2				
20	68.0	70	158.0	120	248.0	170	338.0				
21	69.8	71	159.8	121	249.8	171	339.8				
22	71.6	72	161.6	122	251.6	172	341.6				
23	73.4	73	163.4	123	253.4	173	343.4				
24	75.2	74	165.2	124	255.2	174	345.2				
25	77.0	75	167.0	125	257.0	175	347.0				
26	78.8	76	168.8	126	258.8	176	348.8				
27	80.6	77	170.6	127	260.6	177	350.6				
28	82.4	78	172.4	128	262.4	178	352.4				
29	84.2	79	174.2	129	264.2	179	354.2				

8.3 Thermodynamic or Kelvin-Temperature

According to SI the basic unit of temperature is Kelvin (K). The thermo dynamic temperature (T), called also Kelvin Temperature (formerly "absolute temperature") is a physical quantity based on the laws of thermodynamics.

The International Practical Temperature-Scale from 1968 starts at the absolute zero point of thermodynamics ($-273.15\,°C$). Any temperature ϑ in Celsius is based on the difference between a thermodynamic temperature T and the value $T_0 = 273,15\ K$:

$$\vartheta = T - T_0 = T - 273.15\ K.$$

Note: $1\ K = 1\,°C$ as scale-unit, but as temperature $1\ K = -272.15\,°C$.

For differences or intervals of temperature both scales of Kelvin or Celsius are equal,

for instance: $\Delta T = 20\ K = 20\ \text{grd} = 20\,°C,$

$\Delta T = (20.0 \pm 2)\ K$ or $T = 20.0\ K \pm 2\ K,$

$\Delta\vartheta = (20.0 \pm 2)\ \text{grd}$ or $\vartheta = 20.0\ \text{grd} \pm 2\ \text{grd}$ or

$\Delta\vartheta = (20.0 \pm 2)\,°C$ or $\vartheta = 20.0\,°C \pm 2\,°C.$

9. Thermal Properties

9.1 Thermal Expansion

The coefficient of linear thermal expansion α is defined as the relative change in length per degree change in temperature.

SI-unit for α:

$$\frac{m}{m\,K} = m/mK = 1/K = K^{-1},$$

further units:

for instance $\mu m/mK$, cm/mK, mm/mK.

Units for α used in earlier publications are:

$$1/grd = grd^{-1},\ \text{also}\ {}^\circ C^{-1}\ \text{or}\ {}^\circ F^{-1}\ (= 1.8\ C^{-1})$$

cf. Vol. I of this book, p. 240—245.

9.2 Specific Heat Capacity

The specific heat c is defined as the quantity of heat in calories needed to raise the temperature of one gram of substance $1\,^\circ C$ (cf. Vol. I of this book, p. 245/246) or the quantity of heat in kilocalories required to raise the temperature of one kilogram $1\,^\circ C$.

In the English system the unit quantity of heat (British thermal unit, Btu) is defined as that amount of heat which raises the temperature of one pound of water $1\,^\circ F$. It follows that 1 Btu = 252 calories.

SI-Unit for c:

J/kg K (Conversion Factors are listed in Table 10.2).

9.3 Thermal Conductivity

The therma lconductivity λ (in English-speaking countries symbol K) is the thermal energy per unit time which flows through a thickness of a substance with a given surface area under a steady-state temperature difference between opposite faces (cf. Vol. I of this book, p. 246—250).

To convert c.g.s.-units (cal/cm s $^\circ$C) into English units (Btu in./sq. ft. second $^\circ$F) multiply by 0.80525.

To convert English units (Btu in./sq. ft. hour $^\circ$F) into technical metric units (kcal/m h $^\circ$C) multiply by 0.12404.

In SI-Units a flow of heat is measured in Watt (1 W = 1 J/s); therefore the following Conversion Table according to DIN 1341 (Nov. 1971) may be applied:

	$\dfrac{W}{mK}$	$\dfrac{W}{cm\,K}$
$1\ \dfrac{W}{mK} =$	1	0.01
$1\ \dfrac{W}{cm\,K} =$	100	1
$1\ \dfrac{kcal}{m\,h\,K} =$	1.163	0.01163
$1\ \dfrac{cal}{cm\,s\,K} =$	418.68	4.1868

9.4 Heat Transfer

1 kcal/m² h K = 0.2048 Btu/sq. ft. h °F
1 Btu/sq. ft. h °F = 4.882 kcal/m² h K,
in SI-units:
1 kcal/m² h K = 1.163 W/Km²,
1 cal/cm² s K = 4.1868 W/K cm².

9.5 Diffusivity

The diffusivity factor a (in U.S. papers h^2) is defined as follows (cf. Vol. I of this book, p. 250—256):

$$a = \frac{\lambda}{c \cdot \varrho} \ (m^2/h),$$

where λ = thermal conductivity in kcal/m h °C,
c = specific heat in kcal/kg °C, and ϱ = density in kg/m³.
In English units
1 sq. in./s = 2.32 m²/h,
SI-unit for a (where λ in W/mK, c in J/kgK (Note: 1 J = 1 Ws) and ϱ in kg/m³) is m²/s.

10. Energy, Work, Heat Amount

10.1 Conversion Factors from Metric to British System

	Kilogram calorie kcal	British Thermal Unit Btu	Kilowatt hour kWh	Horsepower-Hour HPh
1 kcal =	1	3.968320	0.0011628	0.0015598
1 Btu =	0.251996	1	0.0002930	0.0003931
1 kWh =	860.01	3413	1	1.34142
1 HPh =	641.1	2544.1	0.745476	1

Note: 1 HPh = 1.0139 PSh; according to DIN 1309 is 1 kcal$_{14.5/15.5}$ ≅ 426.99 kpm ≅ 4186 Joule.

10.2 SI-Units

Energy, work and heat amount have the same SI-Unit:

J(1 J = 1 Nm = 1 Ws). 1 kcal = 4186.8 J ≈ 4200 J.

$$1 \ Nm = \frac{1}{9.81} \ kp \ m = 0.102 \ kp \ m.$$

10.3 Conversion Table for Units of Energy, Work and Thermal Energy[1]

$$1 \ Nm = \frac{1}{9.81} \ kp \ m = 0.102 \ kp \ m$$

	J	kJ	kWh	kcal	PSh	kp m
1 J = (= 1 Nm = 1 Ws)	1	0.001	$2.78 \cdot 10^{-7}$	$2.39 \cdot 10^{-4}$	$3.77 \cdot 10^{-7}$	0.102
1 kJ =	1000	1	$2.78 \cdot 10^{-4}$	0.239	$3.77 \cdot 10^{-4}$	102
1 kWh =	3600000	3600	1	860	1.36	367000
1 kcal =	4200	4.2	0.00116	1	0.00158	427
1 PSh =	2650000	2650	0.736	632	1	270000
1 kp m =	9.81	0.00981	$2.72 \cdot 10^{-6}$	0.00234	$3.7 \cdot 10^{-6}$	1

[1] cf. Footnote to 7.4.

AUTHOR INDEX

SUBJECT INDEX